High-Throughput Analysis

A Tool for Combinatorial Materials Science

High-Throughput Analysis

A Tool for Combinatorial Materials Science

Edited by

Radislav A. Potyrailo

GE Global Research
Schenectady, New York

and

Eric J. Amis

National Institute of Standards and Technology
Gaithersburg, Maryland

Kluwer Academic / Plenum Publishers

New York, Boston, Dordrecht, London, Moscow

Library of Congress Cataloging-in-Publication Data

High-throughput analysis: a tool for combinatorial materials science/edited by Radislav A. Potyrailo and Eric J. Amis.

p. cm.

Includes bibliographical references and index.

ISBN 0-306-47758-0

1. Materials science. 2. Combinatorial analysis. I. Potyrailo, Radislav A. II. Amis, Eric J.

TA401.3.H54 2004.

620.1′1—dc21

2003050646

ISBN 0-306-47758-0

233 Spring Street, New York, New York 10013

http://www.wkap.nl/

10 9 8 7 6 5 4 3 2 1

A C.I.P. record for this book is available from the Library of Congress

Printed in the United States of America

ACKNOWLEDGMENTS

Nothing begets good science as much as the development of a good instrument
Sir Humphrey Davy (1778–1829)

Science is infinite, problems are not
Tomas Hirschfeld (1939–1986)

We thank Ken Howell at Kluwer Academic/Plenum Publishers for inviting us to produce such a book and for encouraging us during the various stages of this project. We would like to express our appreciation to the authors for their efforts in preparing their chapters and to the referees worldwide for reviewing them. Each of us wishes to thank our colleagues and management for technical vision, inspiration, and support. R.A.P. thanks the leadership team at GE Global Research for supporting the whole GE Combinatorial Chemistry effort. E.J.A. thanks NIST management, in particular the Materials Science and Engineering Laboratory, for the inspiration to build the NIST Combinatorial Methods Program. We both also thank our "combi" colleagues for inspiring and creating the new ideas and innovations represented in this approach to science. Last, but not least, we would like to thank our families, who have allowed this book to become part of them for quite a while.

FOREWORD

This book, edited by Potyrailo and Amis, addresses a new paradigm-shifting approach in the search for new materials—*Combinatorial Materials Science*. One way to consider such an approach is to imagine an adventurous chef who decides to look for new entrees by cooking food ingredients in many pots using different combinations in every pot, and boiling, steaming, or frying them in various ways. Although most of the pots will not have the tastiest food ever devised, some recipes will taste intriguing, and some eventually will lead to a discovery of a new fascinating cuisine. Of course, having a skilled chef design the combinatorial formulation will certainly be helpful in ensuring a successful outcome. Similar to food, each engineering material is a complex product of its chemical composition, structure, and processing. Generally, each of these components matters—change one and you get another material. Most of these "new" materials will be less good than ones we use now since existing materials have been refined with the extensive work of scientists and engineers. At the same time if one prepares diverse materials like our adventurous chef, changing material composition, processing conditions and time, etc., some of these materials will be superior to existing ones and a few might represent breakout technology.

The task of *Combinatorial Materials Science* is to develop the tools necessary for synthesizing and screening new materials or formulations and then to use these tools in a miniaturized and automated discovery work flow. What is required in order to implement these integrated discovery (or optimization or formulation) work flows is the following: (1) design software to delineate all the elements of the library of materials to be synthesized; (2) automation software to control the robotic synthesis of the library; (3) instrument control software for the subsequent processing of the library, as necessary; (4) control software for the automated screening of the library for a wide range of physical or chemical properties depending on the intended application of the material; (5) data analysis software to provide objective metrics of the ranking of the new materials for their intended use; and (6) data mining software to help guide the synthesis of the next library generation. Furthermore, this methodology must be constructed *à priori* such that there are no bottlenecks, the mantra among the professionals being "screen in a day what you synthesize in that day *and* analyze in a day what you screen in that day."

Philosophically, this approach is in the spirit of combinatorial chemistry as applied to drug discovery in the pharmaceutical and biotechnology industries. Although less than 20 years old, this approach is now ubiquitous in these industries. Practically, however, the implementation of this philosophy to the materials sector is much less straightforward. Rather than a single technology platform or work flow that is required in drug discovery, in materials discovery there are a myriad of work flows which must be developed depending on the end use application of the material. In this case, however, the cliché that great challenges present great opportunities is absolutely apropos!

Symyx Technologies was the pioneer of this entirely new field of science and technology, and it is extremely gratifying to us to see the almost universal skepticism of just 7 years ago replaced by the preponderance of acceptance today that business as usual will never be the same. The corollary to this acceptance is the need for a primary reference source to introduce the scientific community to this new frontier of technology. This book, edited by Potyrailo and Amis, is well recommended on this basis.

Henry Weinberg (Symyx Technologies)

PREFACE

Over the last decade, combinatorial chemistry has become a well-established approach for the identification of properties of molecules for therapeutic applications. It provides a set of automated synthetic and screening tools for rapid optimization of molecular properties that are difficult to predict and design using existing knowledge. The high screening rate of drug candidates is the key aspect of the wide acceptance of combinatorial chemistry in pharmaceutical research. Indeed, from about 10^{18} pharmaceutically attractive molecules that could theoretically be made (which is at least 10^{10} times larger than the CAS database) [1], only an insignificant fraction of this amount can be characterized using conventional approaches. A good chemist can make and characterize only about 50 compounds per year [2]. Therefore, in order to analyze the myriad of potential candidates for a specific therapeutic target in a timely and cost-effective manner, analytical techniques and instruments have been developed [2]. With these tools, a typical analysis rate in a conventional combinatorial drug-discovery factory is millions of samples per year [3].

Since 1995, the successes of combinatorial chemistry in the pharmaceutical industry have been chased by an exponentially growing number of discoveries of new important materials in chemistry and materials science [4]. At present, these materials range from catalysts, polymers, and zeolites to luminescent and magnetoresistive compounds, to agricultural materials and high-temperature superconductors, dielectric, ferroelectric, and structural materials, and to many others. The number of useful materials in these new areas is predicted to dramatically exceed the number of pharmaceutically attractive candidates. For example, in organic chemistry, 10^{63} potentially stable structures are possible with up to 30 atoms of H, C, O, N, and S elements [5]. In materials science, about 3×10^3 binaries, 7×10^4 ternaries, 1×10^6 quaternaries, and 6×10^{12} decanaries are possible from 75 useful and stable elements in the periodic table, in addition to stoichiometric and structural diversity, and different processing conditions [6]. Similar to the situation in the characterization of drug candidates, traditional analytical approaches will never be able to deal with this extensive materials and process parameters space. The number of traditional analytical analyses performed *worldwide* does not exceed 10^{10} per year [7].

While some of the combinatorial concepts have been adapted from the pharmaceutical to materials research [8], new high-throughput analytical tools are urgently required to address a large number of materials-specific analysis needs. Currently, we observe that the field of high-throughput analysis tools in materials science is converging into an analytical subdiscipline, being driven by the demands of combinatorial practitioners and high-throughput experimentalists [9,10]. In their broad viewpoint, an ideal high-throughput materials analysis system should analyze small-size samples with high precision and accuracy, should require minimal sample preparation, and should provide rapid turnaround by utilizing either serial or parallel analysis of library elements during or after the reaction and testing. Selective examples can include analysis under extreme environmental conditions (temperature, pressure, electromagnetic radiation, etc.), characterization of nanometer-sized domains, determination of trace quantities of materials in the presence of closely related interferents, and so on. For these and many other reasons, analytical screening tools are being developed that often utilize non-traditional and conceptually new detection methods appropriate for a particular application. As a result, the need became apparent for a book that addresses and summarizes such high-throughput characterization tools beyond the pharmaceutical area.

When Ken Howell, Editor of Kluwer Academic/Plenum Publishers, approached one of us at Pittcon '00 and '01 to produce a book on instrumentation for combinatorial methods, it was clear that such a book should bear in mind the extreme diversity of disciplines involved in this field. Thus the first goal of this book is to equip the multidisciplinary practitioners and beginners with a treatment of the principles, methodology, and characteristics of instrumentation for high-throughput analysis in order to stimulate further research in this rapidly expanding area. The second goal is to provide the opportunity for these specialists from different disciplines to work more closely in the design and operation of more efficient high-throughput screening systems. Although a variety of excellent recent books are available on combinatorial chemistry, these primarily focus on the high-throughput and combinatorial synthesis aspects [11–14]. Different types of analytical instrumentation were described only for analysis of drug candidates [2,11].

Throughout the book, the attempt was made to present principles and concepts in a uniform style that can be applied by the reader who is determined to solve a specific analysis problem. In addition, enough practical detail is provided to aid the interested practitioner in the proper design of her/his own high-throughput analysis system for new materials targets. The book is also intended to demonstrate that the analytical tools for accelerated materials discovery do not always require an enormous capital investment, as has been the case of screening for pharmaceutically active molecules. Furthermore, the book illustrates that the advances in components miniaturization, electronics, computer, transducer technologies, and data analysis open new opportunities in the development of new concepts in high-throughput characterization.

Thus the book is intended to be a primary source on both fundamental and practical information of where high-throughput analysis technologies are now and where they are headed in the future. This book is addressed to the rapidly growing number of active practitioners and those who are interested in starting research in this direction, directors of industrial and government research centers, laboratory supervisors and managers, students, and lecturers. The book is structured to offer a basis for a high-throughput instrumentation course at advanced undergraduate or graduate level.

Radislav A. Potyrailo
Eric J. Amis
December 2002

REFERENCES

1. Hann, M., Hudson, B., Lewell, X., Lifely, R., Miller, L., Ramsden, N. Strategic pooling of compounds for high-throughput screening. *J. Chem. Inf. Comput. Sci.* **1999**, *39*, 897–902.
2. Swartz, M. E. (Ed.) *Analytical Techniques in Combinatorial Chemistry* **2000**, New York: Marcel Dekker.

3. Czarnik, A. W., DeWitt, S. H. (Eds.), *A Practical Guide to Combinatorial Chemistry* **1997**, Washington, DC: American Chemical Society.
4. Jandeleit, B., Schaefer, D. J., Powers, T. S., Turner, H. W., Weinberg, W. H. Combinatorial materials science and catalysis. *Angew. Chem. Int. Ed.* **1999**, *38*, 2494–2532.
5. Bohacek, R. S., McMartin, C., Colin, G. *Med. Res. Rev.* **1996**, *16*, 3–50.
6. Maier, W. F. Combinatorial chemistry—challenge and chance for the development of new catalysts and materials. *Angew. Chem. Int. Ed.* **1999**, *38*, 1216–1218.
7. Kellner, R., Mermet, J.-M., Otto, M., Widmer, H. M. (Eds.) *Analytical Chemistry.* Weinheim, Germany: VCH.
8. Jandeleit, B., Schaefer, D. J., Powers, T. S., Turner, H. W., Weinberg, W. H. Combinatorial materials science and catalysis. *Angew. Chem. Int. Ed.* **1999**, *38*, 2494–2532.
9. *Invited Symposium "Analytical Challenges and Opportunities in Combinatorial Chemistry,"* Pittsburgh Conference on Analytical Chemistry and Applied Spectroscopy, New Orleans, LA, March **2000**; *Invited Symposium "Analytical Tools For High Throughput Chemical Analysis And Combinatorial Materials Science,"* Pittsburgh Conference on Analytical Chemistry and Applied Spectroscopy, New Orleans, LA, March **2001**.
10. Vision 2020. Technology Roadmap for Combinatorial Methods, **2001**; Borman, S. *Chem. Eng. News* **2002**, *80 (May 6)*, 49–52; Dagani, R. *Chem. Eng. News* **2002**, *80 (Nov. 11)*, 58–60; http://polymers.msel.nist.gov/combi/index.html.
11. Bunin, B. A. *The Combinatorial Index*; 2nd ed., **1998**, San Diego, CA: Academic Press.
12. Miertus, S., Fassina, G. (Eds.) *Combinatorial Chemistry and Technology. Principles, Methods, and Applications*, **1999**, New York: Marcel Dekker.
13. Bannwarth, W., Felder, E. (Eds.) *Combinatorial Chemistry: A Practical Approach*, **2000**, New York: John Wiley.
14. I. Sucholeiki (Ed.) *High-Throughput Synthesis. Principles and Applications*, **2001**, New York: Marcel Dekker.

CD INFORMATION

As a special feature, a CD-ROM is included with the volume that contains supplementary information that is intended to compliment four of the chapters in the book. Nine separate video files demonstrate library preparations techniques, measurement methods, and analysis schemes. Video files are often used to illustrate scientific principles in undergraduate textbooks and by including them with this reference volume we have the same goal: to illustrate principles with simple examples. Just as we have set the goal of lowering the barrier to acceptance for combinatorial methods, we hope that the materials in this CD will serve to illustrate the simplicity and accessibility of many of the techniques discussed in this book. In addition, two detailed reports are provided in PDF format that discuss some particularly challenging analysis schemes.

CONTRIBUTORS

Jay Akhave Avery Dennison Corporation, Avery Research Center, 2900 Bradley Street, Pasadena, CA 91107, USA

Eric J. Amis Polymers Division, National Institute of Standards and Technology, Gaithersburg, MD 20899, USA

Hermann Bach Bayer Corporation, Coatings & Colorants Division, 100 Bayer Road, Pittsburgh, PA 15205, USA

Catherine A. Srebalus Barnes Eli Lilly and Company, Indianapolis, IN 46285, USA

L. Baumes Equipe de Recherche en Ingénierie des Connaissances,
Université Lumière Lyon 2, Bâtiment L, 5 avenue P. Mendès-France, F-69676 Bron, France

L. N. Brewer General Electric Company, GE Global Research Center, P.O. Box 8, Schenectady, NY 12301, USA

Maureen L. Bricker UOP LLC, 25 East Algonquin Road, Des Plaines, IL 60017, USA

Han-Ting Chang Symyx Technologies Inc., 3100 Central Expressway, Santa Clara, CA 95051, USA

Aaron Chavira Avery Dennison Corporation, Avery Research Center, 2900 Bradley Street, Pasadena, CA 91107, USA

Bret J. Chisholm General Electric Company, GE Global Research Center, P.O. Box 8, Schenectady, NY 12301, USA

Deryn Chu U.S. Army Research Laboratory, Adelphi, MD 20783-1197, USA

David E. Clemmer Indiana University, Department of Chemistry, Bloomington, IN 47405, USA

Alfred J. Crosby Department of Polymer Science and Engineering, University of Massachusetts, Amherst, MA 01003, USA

R. Davis National Institute of Standards and Technology, 100 Bureau Drive, Gaithersburg, MD 20899, USA

W. Demory National Institute of Standards and Technology, 100 Bureau Drive, Gaithersburg, MD 20899, USA

Olaf Deutschmann Universitat Heidelberg, Interdisziplinares Zentrum für Wissenschaftliches Rechnen, Im Neuenheimer Feld 368, 69120 Heidelberg, Germany

D. FarrusSeng Institut de Recherches sur la Catalyse – CNRS, 2 Avenue Albert Einstein, F-69626 Villeurbanne, France

Mark-Florian Fellmann School of Chemical Engineering, Purdue University, West Lafayette, IN 47907, USA

Jean M. J. Fréchet University of California, Department of Chemistry, Berkeley, CA 94720, USA

Yu Fu Department of Chemical Engineering, University of Delaware, Newark, DE 19716, USA

Charles A. Gambino Bayer Corporation, Coatings & Colorants Division, 100 Bayer Road, Pittsburgh, PA 15205, USA

Günter Gauglitz Institute of Physical and Theoretical Chemistry Auf der Morgenstelle 8, 72076 Tübingen, Germany

Peter Geissinger Department of Chemistry, University of Wisconsin-Milwaukee, P.O. Box 413, Milwaukee, WI 53201-0413, USA

Ralph D. Gillespie UOP LLC, 25 East Algonquin Road, Des Plaines, IL 60017, USA

Jeffrey W. Gilman Fire Research Division, National Institute of Standards and Technology, Polymer Building-224, mail stop 8665, 100 Bureau Drive, Gaithersburg, MD 20899-8665, USA

Michael E. Grove MicroFab Technologies Inc., 1104 Summit Avenue, Suite 110, Plano, TX 75074, USA

Jaime Grunlan Avery Dennison Corporation, Avery Research Center, 2900 Bradley Street, Pasadena, CA 91107, USA

Reed J. Hendershot Department of Chemical Engineering, University of Delaware, Newark, DE 19716, USA

John D. Hewes Honeywell International, Inc. Washington D.C. 20004, USA

Jennifer S. Holmgren UOP LLC, 25 East Algonquin Road, Des Plaines, IL 60017, USA

Ghassan E. Jabbour Optical Sciences Center, University of Arizona, Tucson, AZ 85721, USA

M. R. Jackson General Electric Company, GE Global Research Center, P.O. Box 8, Schenectady, NY 12301, USA

Rongzhong Jiang U.S. Army Research Laboratory, Adelphi, MD 20783-1197, USA

Thorsten Johann Max-Planck-Institut für Kohlenforschung, Kaiser-Wilhelm-Platz 1, 45470 Mülheim an der Ruhr, Germany

Christopher W. Johnson Department of Chemistry, University of Wisconsin-Milwaukee, 3210 N. Cramer Street, Milwaukee, WI 53211, USA

Alamgir Karim Polymers Division, National Institute of Standards and Technology, Gaithersburg, MD 20899, USA

T. Kashiwagi National Institute of Standards and Technology, 100 Bureau Drive, Gaithersburg, MD 20899, USA

Stefan Kleditzsch Universität Heidelberg, Interdisziplinäres Zentrum für Wissenschaftliches Rechnen, Im Neuenheimer Feld 368, 69120 Heidelberg, Germany

Jens Klein hte Aktiengesellschaft, Kurpfalzring 104, 69123 Heidelberg, Germany

Tetsuhiko Kobayashi Special Division for Green Life Technology, National Institute of Advanced Industrial Science and Technology (AIST), 1-8-31 Midorigaoka, Ikeda, Osaka 563-8577, Japan

Kerstin Kröger Institute of Physical and Theoretical Chemistry, Auf der Morgenstelle 8, 72076 Tübingen, Germany

Valentin Kulikov University of Regensburg, Institute of Analytical Chemistry, Chemo- and Biosensors, 93040 Regensburg, Germany

Steven S. Lasko School of Chemical Engineering, Purdue University, West Lafayette, IN 47907, USA

Jochen Lauterbach Department of Chemical Engineering, University of Delaware, Newark, DE 19716, USA

Mark Licon Avery Dennison Corporation, Avery Research Center, 2900 Bradley Street, Pasadena, CA 91107, USA

Phillip D. Lunney Bayer Corporation, Coatings & Colorants Division, 100 Bayer Road, Pittsburgh, PA 15205, USA

Wilhelm F. Maier Lehrstuhl für Technische Chemie, Universität des Saarlandes,

Im Stadtwald, D-66123 Saarbrücken, Germany

Ali R. Mehrabi Avery Dennison Corporation, Avery Research Center, 2900 Bradley Street, Pasadena, CA 91107, USA

J. Carson Meredith School of Chemical Engineering, 778 Atlantic Drive, Georgia Institute of Technology, Atlanta, GA 30332, USA

C. Mirodatos Institut de Recherches sur la Catalyse – CNRS, 2 Avenue Albert Einstein, F-69626 Villeurbanne, France

Vladimir M. Mirsky University of Regensburg, Institute of Analytical Chemistry, Chemo- and Biosensors, 93040 Regensburg, Germany

William G. Morris General Electric Company, GE Global Research Center, P.O. Box 8, Schenectady, NY 12301, USA

Marcel Murga Avery Dennison Corporation, Avery Research Center, 2900 Bradley Street, Pasadena, CA 91107, USA

Son Hoai Nguyen Symyx Technologies Inc., 3100 Central Expressway, Santa Clara, CA 95051, USA

M. Nyden National Institute of Standards and Technology, 100 Bureau Drive, Gaithersburg, MD 20899, USA

Gudbjorg Oskarsdottir School of Chemical Engineering, Purdue University, West Lafayette, IN 47907, USA

L. A. Peluso General Electric Company, GE Global Research Center, P.O. Box 8, Schenectady, NY 12301, USA

Miroslav Petro Symyx Technologies Inc., 3100 Central Expressway, Santa Clara, CA 95051, USA

Radislav A. Potyrailo General Electric Company, GE Global Research Center, P.O. Box 8, Schenectady, NY 12301, USA

Hariklia Reitz General Electric Company, GE Silicones, Waterford, NY 12188, USA

Jens W. Saalfrank Lehrstuhl für Technische Chemie, Universität des Saarlandes, Im Stadtwald, D-66123 Saarbrücken, Germany

J. W. Adriaan Sachtler UOP LLC, 25 East Algonquin Road, Des Plaines, IL 60017, USA

Adam L. Safir Symyx Technologies Inc., 3100 Central Expressway, Santa Clara, CA 95051, USA

Dennis Saunders Avery Dennison Corporation, Avery Research Center, 2900 Bradley Street, Pasadena, CA 91107, USA

Jens Scheidtmann Lehrstuhl für Technische Chemie, Universität des Saarlandes, Im Stadtwald, D-66123 Saarbrücken, Germany

Hans-Werner Schmidt Max-Planck-Institut für Kohlenforschung, Kaiser-Wilhelm-Platz 1, D-45470 Mülheim an der Ruhr, Germany

Stephan A. Schunk hte Aktiengesellschaft, Kurpfalzring 104, 69123 Heidelberg, Germany

Ferdi Schüth Max-Planck-Institut fürzKohlenforschung, Kaiser-Wilhelm-Platz 1, 45470 Mülheim an der Ruhr, Germany

Alan W. Schwabacher Department of Chemistry, University of Wisconsin-Milwaukee, 3210 N. Cramer Street, Milwaukee, WI 53211-3029, USA

Amit Sehgal Polymers Division, National Institute of Standards and Technology, Gaithersburg, MD 20899, USA

Michael Seidel Institute of Physical and Theoretical Chemistry, Auf der Morgenstelle 8, 72076 Tübingen, Germany

J. Shields National Institute of Standards and Technology, 100 Bureau Drive, Gaithersburg, MD 20899, USA

Christopher M. Snively Department of Chemical Engineering, University of Delaware, Newark, DE 19716, USA

Hui Su Ames Laboratory-USDOE and Department of Chemistry, Iowa State University, Ames, IA 50011, USA

Frantisek Svec University of California, Department of Chemistry, Berkeley, CA 94720-1460, USA

David Sýkora Department of Analytical Chemistry, Institute of Chemical Technology, Technická 5, 166 28 Prague, Czech Republic

David B. Wallace MicroFab Technologies, Inc., 1104 Summit Avenue, Suite 110, Plano, TX 75074, USA

Qi Wang National Renewable Energy Laboratory, Golden, CO 80401, USA

Pierre-Alain W. Weiss Lehrstuhl fuer Technische Chemie, Universität des Saarlandes, Im Stadtwald, D-66123 Saarbruecken, Germany

Douglas A Wicks University of Southern Mississippi, Box 10076, Hattiesburg, MS 39406, USA

Richard R. Willis UOP LLC, 25 East Algonquin Road, Des Plaines, IL 60017, USA

Ronald J. Wroczynski General Electric Company, GE Global Research Center, P.O. Box 8, Schenectady, NY 12301, USA

Yusuke Yamada Solid State Combinatorial Research Group, Special Division for Green Life Technology, National Institute of Advanced Industrial Science and Technology (AIST), 1-8-31 Midorigaoka, Ikeda, Osaka 563-8577, Japan

Edward S. Yeung Ames Laboratory-USDOE and Department of Chemistry, Iowa State University, Ames, IA 50011, USA

Yuka Yoshioka Optical Sciences Center, The University of Arizona, Tucson, AZ 85721

Torsten Zech hte Aktiengesellschaft, Kurpfalzring 104, 69123 Heidelberg, Germany

J.-C. Zhao General Electric Company, GE Global Research Center, P.O. Box 8, Schenectady, NY 12301, USA

CONTENTS

SENSORS AND SENSOR ARRAYS

ANALYSIS OF MECHANICAL, ELECTRICAL, AND THERMAL PROPERTIES

ELECTROCHEMICAL ANALYSIS

COMBINATORIAL LIBRARIES, DATA MANAGEMENT, AND SCALE-UP OF DISCOVERED MATERIALS

Chapter 1

Elements of High-Throughput Analysis in Combinatorial Materials Science

Radislav A. Potyrailo and Eric J. Amis

INTRODUCTION

Combinatorial materials science combines a small number of starting chemical reagents in all combinations defined by a given reaction scheme with the combinations of process conditions to yield a large number of well-defined products. Combinatorial methods have opened up the exploration of multidimensional chemical composition and process parameter space at a previously unavailable level of detail with significant time savings, much lower cost per "lead", and shorter time to market [1]. For these and many other reasons, combinatorial and high-throughput methods have been already successfully implemented for discovery of a wide variety of important materials ranging from catalysts, polymers, and zeolites to luminescent and magnetoresistive compounds, agricultural materials and high-temperature superconductors, dielectric, ferroelectric, and structural materials, to polymeric coatings and thin-films, and so on [2–8].

As indicated at several recent symposia on analytical instrumentation at Pittcon and ACS Meetings [9–11], the success of combinatorial and high-throughput methodologies strongly relies on the development of analytical tools with new and improved capabilities. This book outlines the role of high-throughput analytical instrumentation in the combinatorial discovery cycle, advantages and limitations of different sampling and measurement modes of materials libraries, and the prospects of mathematical tools for processing large amounts of data. For the first time, several examples from our and other laboratories illustrate the power of combinatorial and high-throughput methods in the discovery of new materials and process conditions by presenting the results of the scale-up experiments.

Radislav A. Potyrailo • General Electric Global Research, Schenectady, NY 12301, USA.
Eric J. Amis • National Institute of Standards and Technology, Gaithersburg, MD 20899, USA.
Contact authors: Radislav A. Potyrailo, email: Potyrailo@crd.ge.com, Eric J. Amis, email: eric.amis@nist.gov

High-Throughput Analysis, edited by Potyrailo and Amis
Kluwer Academic Publishers, New York, 2003.

THE RIGHT TIME AND RIGHT SPACE

One of the earliest examples of a comprehensive screening of materials space is the work of T. Edison and co-workers, who between 1878 and 1880 screened about 6×10^3 material candidates for the filament of incandescent light bulb. Two materials "hits" were discovered that included a fine platinum wire and a carbonized cotton thread (Figure 1.1).

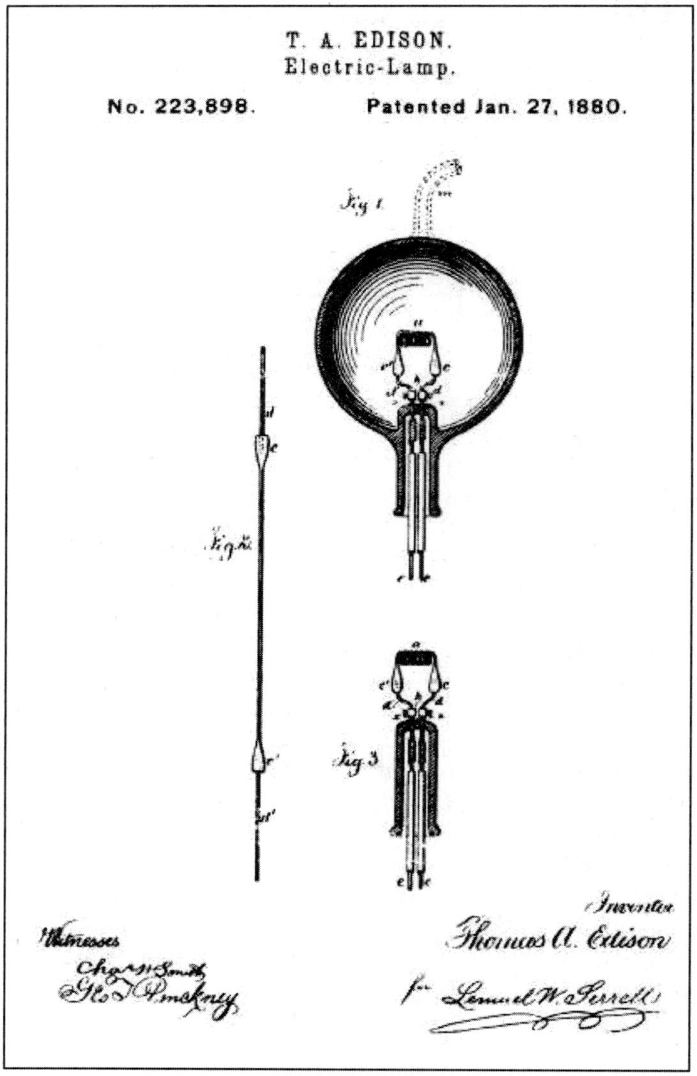

FIGURE 1.1. Early screening of diverse materials for properties of interest. Between 1878 and 1880 Edison and co-workers screened 6000 material candidates for the filament of incandescent light bulb. Two materials "hits" were discovered that included a fine platinum wire and a carbonized cotton thread.

Another classical example is the work of Mittasch, who between 1900 and 1909 performed 6.5×10^3 screening experiments on 2.5×10^3 catalyst candidates to find a catalyst for industrial ammonia synthesis [12]. Pioneering efforts in parallel materials synthesis followed by screening of properties of interest were reported in the mid-1960s. These efforts were focused on generation of libraries of materials with a gradient composition. Kennedy et al. [13] demonstrated a ternary alloy phase library that was fabricated using electron beam coevaporation and screened with electron diffraction. Two years later, Miller and Shirn [14] used a cosputtering technique to produce a film with a composition gradient of Au and SiO_2. Hanak [15] was the first to highlight the power of the parallel synthesis and screening methodology. His 1970 paper outlines the advantages of "synthesis of entire multicomponent system in one experiment" followed by "simple rapid non-destructive all-inclusive chemical analysis." While this "multiple-sample concept" in materials research was introduced more than 30 years ago, it has only recently became practical. The combinatorial methodology relies heavily on numerous innovations from different disciplines. At present, these advancements make possible computerized automated preparation of libraries and handling of picoliter-sized samples, rapid testing of materials against a variety of factors of interest, noncontact and nondestructive screening of resulting compounds, and efficient automated analysis, interpretation, and management of large amounts of data. Overall, in the early 1970s the levels of computerized support, automation in synthesis, and capabilities in screening were not adequate to make them effective tools for combinatorial materials studies [7].

The work by Xiang et al. [16] on superconductors, published in 1995, initiated a recent avalanche of advancements in numerous areas of combinatorial materials science and high-throughput screening (HTS) (Figure 1.2). Xiang and co-workers demonstrated discrete combinatorial libraries of materials made using masking methodology adapted from the biotechnology area [17]. In materials science, this masking method enables the production of a large number of different compositions by generating many permutations of precursors. At present, arrays of materials compositions are widely used in combinatorial materials research. These compositions are typically arrayed as two-dimensional gradient or discrete libraries with variations in materials composition, amount, processing,

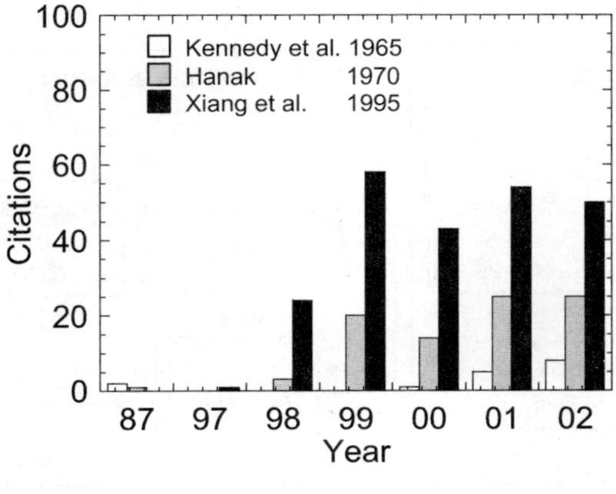

FIGURE 1.2. Increasing number of citations of pioneering works by Kennedy et al. [13], Hanak [15], and Xiang et al. [16]. Sources: SciFinder and ISI, December 2002.

and testing conditions as described in some representative references [3,16,18–25] and in chapters of this book. Recently, one-dimensional combinatorial libraries have been reported [26–29] where the materials composition is changing on a single support along one dimension. The screening is performed in sequence or in parallel from all regions of the one-dimensional library.

INSTRUMENTATION

The diversity of materials and process parameters space in combinatorial materials science is addressed with parallel synthesis, testing against single or multiple performance parameters, and screening of a large number of components. A schematic of this process is illustrated in Figure 1.3. High-throughput screening is the process for rapid automated assessment of single or multiple properties of a large number of samples, most often fabricated as a combinatorial library. A representative list of such materials properties can include chemical, morphological, mechanical, optical, electrical, biological, thermal, and many others. Thus, to address a large number of materials-specific analysis needs, a variety of high-throughput analytical tools are required. The materials discovery cycle may also include a performance-testing step (Figure 1.3). The testing process includes exposure of the library to an environment that imitates the end-use application and alters materials properties in a detectable manner [18,29]. Upon testing, the materials undergo changes that are impossible to predict quantitatively using existing knowledge. Testing examples include solar radiation, heat, mechanical stress, aggressive fluids, and others. An example of the materials discovery cycle is presented in Figure 1.4 where new properties of polymeric coatings are explored [21].

Ideally, a useful HTS system for combinatorial materials discovery should be capable of handling small samples with minimal sample preparation requirements and should provide rapid turnaround by using either serial or parallel analysis of library elements during or after the reaction and testing. Also, as with other types of analytical instruments, it should have high precision and accuracy of measurements, while analyzing much smaller sample size. Further, the data acquisition and instrument operation and communication

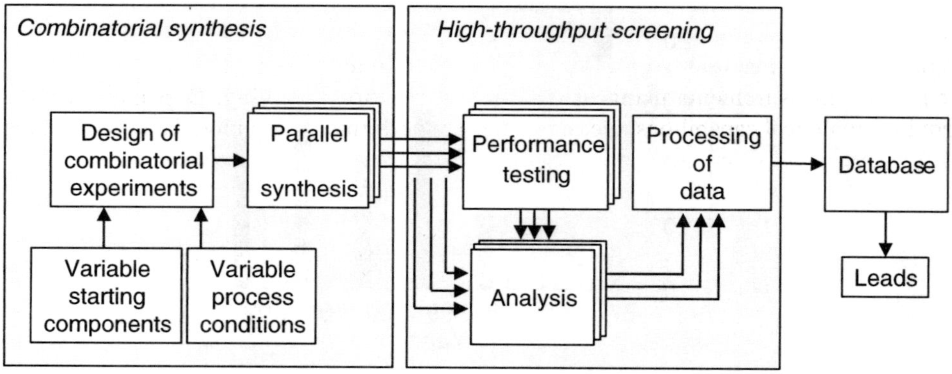

FIGURE 1.3. Combinatorial materials science and HTS discovery cycle. (Adapted from ref. 44. Copyright 2003 Elsevier.)

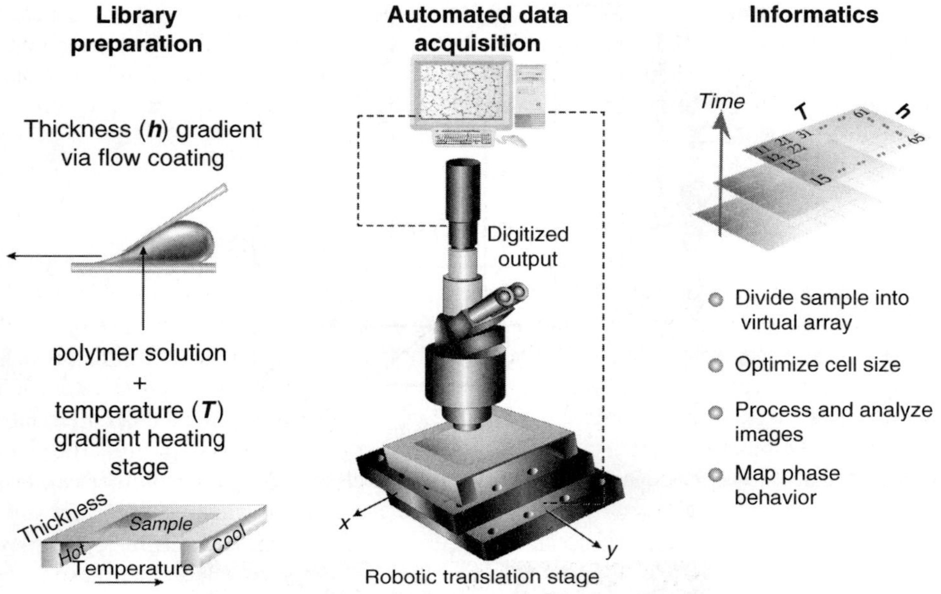

FIGURE 1.4. Schematic of combinatorial polymer materials science and coatings technology.

with other instruments of the discovery cycle should be automated. The data analysis should be efficient to handle large amounts of data utilizing either univariate or multivariate algorithms and should be capable of providing simple transfer functions for the property of interest. Finally, data management should be well ordered and organized. While numerous analytical techniques were implemented for the HTS of libraries of materials, no single HTS tool meets all these requirements. Thus the need exists to develop innovative screening tools that utilize conceptually new detection methods to meet the need for a particular measurement problem.

A research program with multiple diverse projects typically requires different analysis tools. For example, for efficient operation of our HTS systems across a variety of materials science projects, we developed an infrastructure schematically illustrated in Figure 1.5. Combinatorial samples are analyzed for different properties of interest by serial or parallel measurements using in situ or post-reaction sampling. Depending on the required analytical capabilities, measurements are performed using modular spectroscopic systems, sensors, and sensor arrays built in-house [29–34] as illustrated in Figure 1.6. A unifying theme across the components of this infrastructure is a common platform for data acquisition, communication between instruments, and data management.

Serial analysis of materials libraries is used when it is prohibitively complicated to multiplex multiple analyzers in a parallel system or the relatively long total analysis time is not an issue. Often, serial measurements provide more detailed information about materials property over parallel methods. For example, optical spectroscopic tools, such as electronic and vibrational spectroscopies, allow direct spectral analysis of materials [35]. A parallel analysis such as hyperspectral imaging can be potentially implemented

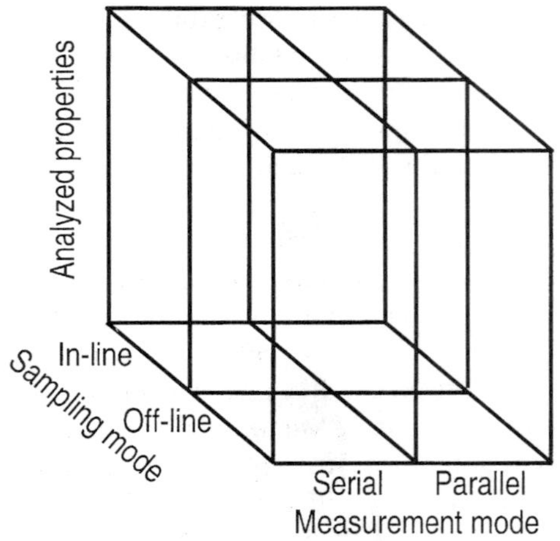

FIGURE 1.5. Sampling and measurement strategies for high-throughput analysis of combinatorial libraries.

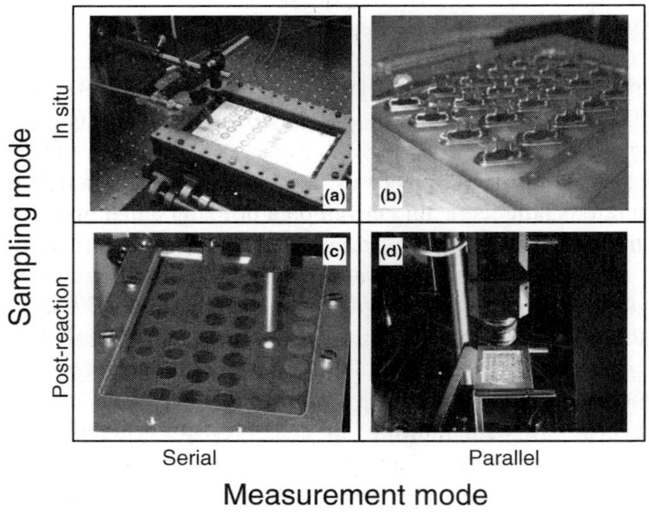

FIGURE 1.6. Illustrative examples of measurement and sampling approaches. (a) Screening of colorimetric and fluorescent chemical sensor materials. (b) In situ monitoring of mass and viscoelastic properties of thin films as a function of polymer composition using a 24-channel array of acoustic-wave transducers. (c) Scanning scattered light analysis for evaluation of abrasion resistance of libraries of organic coatings. (d) Fluorescence and reflected light imaging of adhesion and weathering of libraries of organic coatings.

instead [36], but at present may require extensive computational requirements if both high spectral and spatial resolution are required.

In general, parallel analysis has several attractive features in comparison with serial HTS tools. An obvious advantage is reduced acquisition time per sample, especially if the measurement time of a serial method is much longer than library fabrication time or any other step in the combinatorial cycle. Also, in situations when handling of samples to convert a two-dimensional combinatorial array into a one-dimensional sequence of samples for serial analysis is costly, generates a lot of waste, and induces errors at each sample-handling step, parallel analysis is very attractive. Another advantage of parallel HTS tools is in the monitoring of dynamic processes in the library when the screening rate of the serial method is much larger than time constants of measured dynamic processes.

Similar to any other field of analytical chemistry, the selection of an appropriate sampling approach is critical. In situ sampling provides several attractive options for HTS. Real-time observation of the reaction progress in combinatorial reactors can tremendously speed up the materials discovery process by providing previously unavailable information about the starting reaction components evolving into the reaction and the dynamics of reaction progress at each reaction phase. Monitoring of reaction components can provide valuable feedback information to control and rapidly optimize reaction parameters. Overall, in situ monitoring of combinatorial reactions includes all the attractive features of in-line detection methods, such as no sample preparation steps and thus a reduced number of contamination sources. Infrared, Raman, UV-visible, and fluorescence spectroscopies and imaging are evolving as the premier analytical methodologies for direct in situ monitoring of combinatorial reactions [35,37–39].

In situ sampling coupled with the adequate measurement mode can be further adapted for applications and reaction scales beyond HTS, as illustrated in Figure 1.7. For example, in situ measurements can be done on a 10^{-12}–10^{-6} L scale for combinatorial materials science screening, while a reaction volume as large as 10^4–10^5 L should be monitored at a manufacturing plant. While the diversity of goals for these in situ measurements is governed by the reaction scale, spectroscopic monitoring techniques can remain the same. Thus a measurement technique initially developed for HTS can be of use for process monitoring where measurements are performed in the pilot and manufacturing plants. Examples of different scales of reactors ranging from combinatorial [40] to scale-up [41] and laboratory [42] reactors are presented in Figure 1.8.

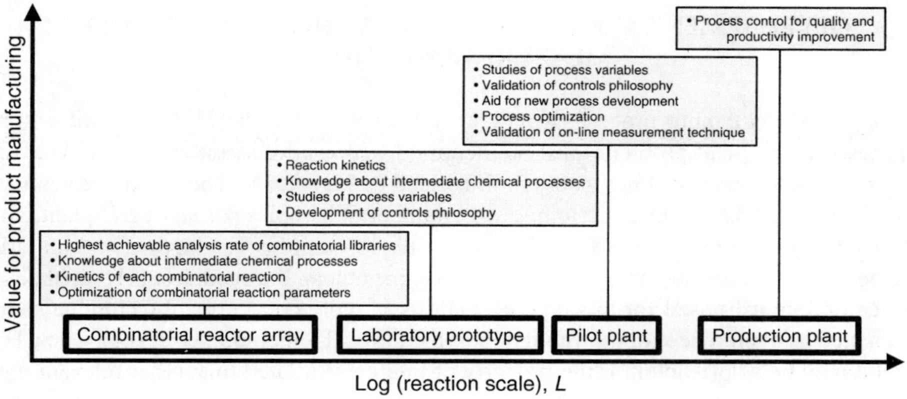

FIGURE 1.7. Needs for in situ monitoring of chemical reactions on different reactor scales.

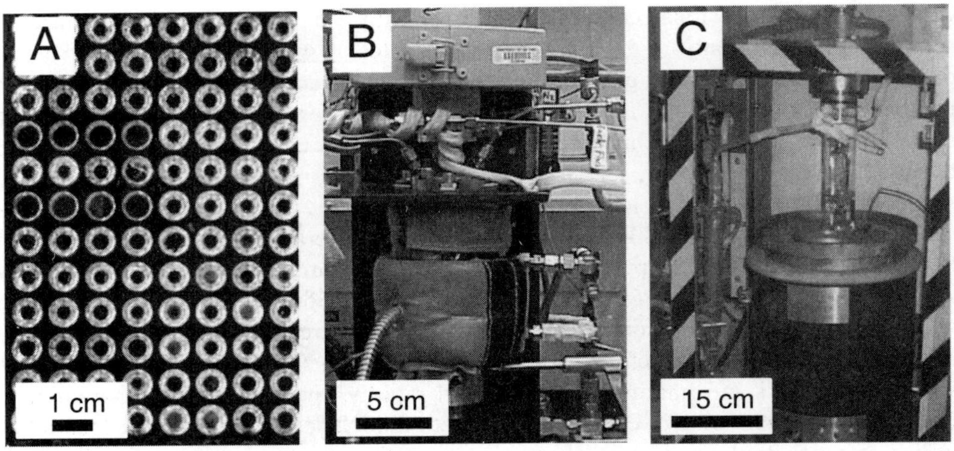

FIGURE 1.8. Chemical reactors of different scale: (A) combinatorial 96-microreactor array; (B) scale-up reactor; (C) laboratory reactor. (Adapted from ref. 44. Copyright 2003 Elsevier.)

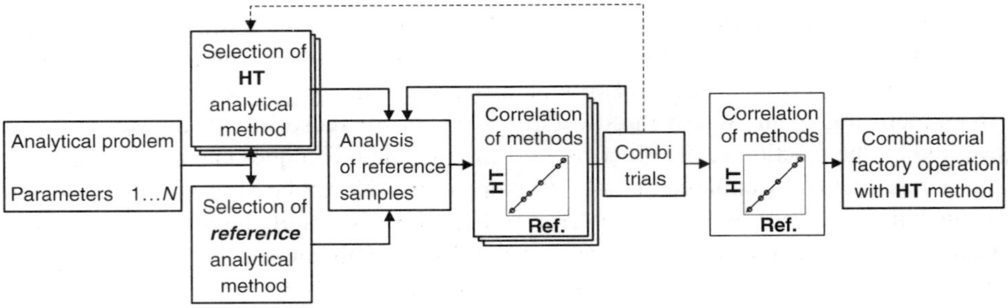

FIGURE 1.9. Strategies for development of new HTS tools.

DEVELOPMENT STRATEGIES OF NEW HIGH-THROUGHPUT MEASUREMENT TOOLS

In the development process of new analytical tools for the HTS of combinatorial libraries, an analytical problem must be identified with well-understood requirements for analysis. A schematic of this process is illustrated in Figure 1.9. These requirements can include the analysis turnaround, sample size, number of samples per analysis, quantitation precision and accuracy, need for the availability of samples for further manipulation (i.e., destructive or nondestructive analysis), and some others. If a candidate HTS method has not been previously used for this type of analysis, a more conventional technique should be selected as a reference for the validation of the new HTS method. Also, such a new HTS method may be a "predictor" if the property of interest is related to another relevant property, which is measured instead. Examples may include predicting the concentration of hydrogen gas using near-IR reflection sensors [43], and predicting molecular weight of polycarbonate and the concentration of the branched polycarbonate side product from measured fluorescence spectra [44].

A set of materials is further presented to the new analytical tools for the measurements, and the results are compared with those obtained using a reference technique. The sample size and other required parameters of reference samples can be reduced for analysis with the HTS method. Typical limitations of the conventional analytical technique (measurement precision, relatively large sample size, need for extensive sample preparation, manual sample analysis, etc.) should be taken into consideration for a proper comparison.

Upon achieving a required correlation between the two methods, experiments can be performed to fabricate materials on the combinatorial scale with diverse properties. These materials are presented to both the HTS and reference techniques for further validation of the HTS method. To satisfy the needs of the reference technique, replicate small samples may be combined. At this point, the HTS method can be further refined or an alternative technique can be considered. Also, several alternative HTS techniques can be evaluated at once for the analysis of reference and combinatorial trial samples. This will provide time saving in method development and will permit a direct comparison of capabilities of several candidate HTS techniques. Once an HTS method has been selected, it is implemented in the combinatorial factory mode with no further alterations.

DATA MANAGEMENT AND INFORMATICS

The large amount of data that is generated in HTS requires effective data processing to identify the trends and relationships in the data. Advanced mathematical and statistical chemometric techniques are used in HTS to determine, often by indirect means, the properties of substances that otherwise would be very difficult to measure directly. In HTS, chemometrics has been successfully applied for qualitative and quantitative analysis. Examples of the qualitative analysis include visualization, pattern recognition, lead identification, and detection of abnormalities in the measured data [30,45–47]. Quantitative analysis using chemometric tools most often deals with multivariate calibration which improves analysis selectivity when measurements are performed with relatively nonselective or low-resolution instruments [33,48]. Quantitative analysis is often performed using linear and nonlinear multivariate calibration methods such as partial least-squares, locally weighted regression, neural networks, and others.

An example application of the multivariate calibration approaches is the implementation of an array of acoustic-wave sensors coated with diverse chemically sensitive films for analysis of volatile components in model combinatorial samples. Selective quantitation of analytes in complex mixtures was achieved by analyzing the dynamic data by means of locally weighted regression analysis [33].

One of the key challenges for informatics in combinatorial materials science is predictive modeling of materials performance [49]. In general, materials performance depends on intrinsic properties of starting components and process conditions during materials synthesis. However, when too many such parameters are involved, a quantitative prediction of materials performance, even using multivariate tools, is often too uncertain because of the lack of adequate descriptors. One such example is combinatorial development of protective organic coatings with high adhesion [50]. Adhesion of coatings, especially polymeric coatings, is extremely complex and poorly understood [51]. Because of the complexities of the adhesion process and a lack of a complete set of descriptors, a variety of practical direct adhesion test methods are used. Thus the combinatorial development of new organic coatings also involves direct adhesion testing [50,52,53].

OUTLOOK

Combinatorial and high-throughput methodologies are now recognized as the breakthrough technologies in materials science. Their success strongly relies on the development of analytical tools with new and improved capabilities. These new developments should focus on both new instrumentation for analysis of diverse materials properties and new mathematical approaches for processing large amounts of data [11]. An important aspect of the instrumental advances is the integration of new in situ measurement capabilities with microreactors. Recent catalyst discoveries already demonstrate [54] that high-throughput experimentation coupled with in situ reaction monitoring complement each other and significantly contribute to the success of the HTS programs. In particular, this interdisciplinary synergy generates new knowledge on the level of detail unavailable from conventional experimental methods. This knowledge can be further used for the development of new performance models. These models will relate intrinsic and performance properties and will provide transfer functions of scalability of materials from combinatorial to macroscale.

The number of areas of science that are being influenced by the combinatorial and high-throughput methods is rapidly expanding. Overall, these methods are becoming more accepted in the materials science community. Their maturity and acceptance in materials science is due to the combination of the growing number of research organizations worldwide that practice combinatorial materials science and creation of the combinatorial centers that focus on different aspects of this technology.

The area of high-throughput analytical instrumentation experiences the same growth of interest and attention from the academic, industrial, and governmental research communities, and governmental regulatory and funding agencies as that already achieved in another analytical subfield—process analytical instrumentation. The widespread acceptance of the latter has been recognized, for example, in changes in graduate and undergraduate curricula for future analytical chemists and process engineers to improve understanding of process analytical instrumentation and to make its application more successful [55]. Similarly, new educational programs that will train specialists in combinatorial methodology are being launched in Europe and the United States.

Importantly, analytical tools for accelerated materials discovery do not always require an enormous capital investment, as in the case of screening for pharmaceutically active molecules, and thus could be affordable in small industrial ventures, academia, and industrialized and developing countries. The advances in components miniaturization, electronics, computers, transducer technologies, and data analysis continue to be critical in providing new opportunities in the development of new concepts in high-throughput characterization.

OVERVIEW OF BOOK CHAPTERS

This book provides a comprehensive coverage of the current high-throughput analysis methods in materials science and chemistry beyond the pharmaceutical area. An international team of original contributors to the field has been assembled from academic, industrial, and governmental research laboratories. The book consists of five instrumental sections and a section on combinatorial libraries, data management, and scale-up of discovered materials. Five instrumental sections describe optical imaging and spectroscopy, separations and mass spectrometry, sensors and sensor arrays, analysis of mechanical, electrical, and thermal properties, and electrochemical analysis. This choice was determined by the broad applicability of these analysis methods across different materials screening applications.

ACKNOWLEDGMENTS

R.A.P. would like to acknowledge partial support of this work by the National Institute of Standards and Technology (NIST) Advanced Technology Program, Grant 70NANB9H3038, and by the US Department of Energy (DOE), Award DEFC07–01ID14093. E.J.A. acknowledges support of this work by the National Institute of Standards and Technology through the Directors Competence Fund and the Advanced Technology Program. We also acknowledge Dr Alamgir Karim and Dr Amit Sehgal for assistance in preparation of this material.

REFERENCES

1. Schlögl, R. *Angew. Chem. Int. Ed.* **1998**, *37*, 2333–2336.
2. Maier, W. F. *Angew. Chem. Int. Ed.* **1999**, *38*, 1216–1218.
3. Jandeleit, B., Schaefer, D. J., Powers, T. S., Turner, H. W., Weinberg, W. H. *Angew. Chem. Int. Ed.* **1999**, *38*, 2494–2532.
4. Sucholeiki, I. (Ed.), *High-Throughput Synthesis. Principles and Applications*, **2001**, New York: Marcel Dekker.
5. Reetz, M. T. *Angew. Chem. Int. Ed.* **2002**, *41*, 1335–1338.
6. Malhotra, R. (Ed.), *Combinatorial Approaches to Materials Development*, **2002**, Washington, DC: American Chemical Society.
7. Amis, E. J., Xiang, X.-D., Zhao, J.-C. *MRS Bull.* **2002**, *27*, 295–297.
8. Hewes, J. D., Kaiser, D., Karim, A., Amis, E. J. In *Encyclopedia of Chemical Technology Online Edition*, **2003**, Kirk-Othmer Online, Wiley InterScience.
9. Potyrailo, R. A. (Organizer), *Invited Symposium "Analytical Challenges and Opportunities in Combinatorial Chemistry,"* Pittsburgh Conference on Analytical Chemistry and Applied Spectroscopy, New Orleans, LA, **March 2000**.
10. Potyrailo, R. A. (Organizer), *Invited Symposium "Analytical Tools For High Throughput Chemical Analysis And Combinatorial Materials Science,"* Pittsburgh Conference on Analytical Chemistry and Applied Spectroscopy, New Orleans, LA, **March 2001**.
11. Borman, S. *Chem. Eng. News* **2002**, *80 (May 6)*, 49–52.
12. Ertl, G. *Angew. Chem. Int. Ed.* **1990**, *29*, 1219–1227.
13. Kennedy, K., Stefansky, T., Davy, G., Zackay, V. F., Parker, E. R. *J. Appl. Phys.* **1965**, *36*, 3808–3810.

14. Miller, N. C., Shirn, G. A. *Appl. Phys. Lett.* **1967**, *10*, 86–88.
15. Hanak, J. J. *J. Mater. Sci.* **1970**, *5*, 964–971.
16. Xiang, X.-D., Sun, X., Briceño, G., Lou, Y., Wang, K.-A., Chang, H., Wallace-Freedman, W. G., Chen, S.-W., Schultz, P. G. *Science* **1995**, *268*, 1738–1740.
17. Pirrung, M. C. *Chem. Rev.* **1997**, *97*, 473–488.
18. Potyrailo, R. A., Pickett, J. E. *Angew. Chem. Int. Ed.* **2002**, *41*, 4230–4233.
19. Takeuchi, T., Fukuma, D., Matsui, J. *Anal. Chem.* **1999**, *71*, 285–290.
20. Takeuchi, I., van Dover, R. B., Koinuma, H. *MRS Bull.* **2002**, *27*, 301–308.
21. Meredith, J. C., Smith, A. P., Crosby, A. J., Amis, E. J., Karim, A. In *Encyclopedia of Polymer Science and Technology*, **2002**, New York: John Wiley.
22. Meredith, J. C., Smith, A. P., Karim, A., Amis, E. J. *Macromolecules* **2000**, *33*, 9747–9756.
23. Meredith, J. C., Karim, A., Amis, E. J. *Macromolecules* **2000**, *33*, 5760–5762.
24. Smith, A. P., Douglas, J. F., Meredith, J. C., Amis, E. J., Karim, A. *J. Polym. Sci. B*, **2001**, *39*, 2141–2158.
25. Smith, A. P., Douglas, J. F., Meredith, J. C., Amis, E. J., Karim, A. *Phys. Rev. Lett.* **2001**, *87*, 015503-1-015503-4.
26. Schwabacher, A. W., Shen, Y., Johnson, C. W. *J. Am. Chem. Soc.* **1999**, *121*, 8669–8670.
27. Prince, B. J., Schwabacher, A. W., Geissinger, P. *Anal. Chem.* **2001**, *73*, 1007–1015.
28. Potyrailo, R. A. *Devices and Methods for Measurements of Barrier Properties of Coating Arrays*, US Patent 6,383,815 B1, **2002**.
29. Potyrailo, R. A., Wroczynski, R. J., Pickett, J. E., Rubinsztajn, M. *Macromol. Rapid Comm.* **2003**, *24*, 123–130.
30. Potyrailo, R. A. In K. H. J. Buschow, R. W. Cahn, M. C. Flemings, B. Ilschner, E. J. Kramer, and S. Mahajan, (Eds.), *Encylopedia of Materials: Science and Techonology*, Vol. 2, **2001**, Amsterdam, The Netherlands: Elsevier, pp. 1329–1343.
31. Potyrailo, R. A., Chisholm, B. J., Olson, D. R., Brennan, M. J., Molaison, C. A. *Anal. Chem.* **2002**, *74*, 5105–5111.
32. Potyrailo, R. A. Presented at *COMBI 2002—4th Annual International Symposium on Combinatorial Approaches for New Materials Discovery*, San Diego, CA, **January 2002**.
33. Potyrailo, R. A., May, R. J. *Rev. Sci. Instrum.* **2002**, *73*, 1277–1283.
34. Potyrailo, R. A. Presented at *COMBI 2002—5th Annual International Symposium on Combinatorial Approaches for New Materials Discovery*, San Jose, CA, **February 2003**.
35. Gremlich, H.-U. *Biotechnol. Bioeng. (Comb. Chem.)* **1998/1999**, *61*, 179–187.
36. Snively, C. M., Oskarsdottir, G., Lauterbach, J. *Catal. Today* **2001**, *67*, 357–368.
37. Swartz, M. E. (Ed.), *Analytical Techniques in Combinatorial Chemistry*, **2000**, New York: Marcel Dekker.
38. Su, H., Hou, Y., Houk, R. S., Schrader, G. L., Yeung, E. S. *Anal. Chem.* **2001**, *73*, 4434–4440.
39. Snively, C. M., Oskarsdottir, G., Lauterbach, J. *Angew. Chem. Int. Ed.* **2001**, *40*, 3028–3030.
40. Potyrailo, R. A., Wroczynski, R. J., Lemmon, J. P., Flanagan, W. P., Siclovan, O. P. *J. Comb. Chem.* **2003**, *5*, 8–17.
41. Warner, G. L., Flanagan, W. P., Potyrailo, R. A., Lemmon, J. P. *Method and Apparatus for Conducting Melt-Polymerization Reactions*, PCT Int. Appl. WO 01/21302 A1, **2001**.
42. King, J. A., Jr. In T. E. Long and M. O. Hunt (Eds.), *Solvent-Free Polymerizations and Processes. Minimization of Conventional Organic Solvents*, Vol. 713, **1998**, Washington, DC: American Chemical Society, pp. 49–77.
43. Jaramillo, T. F., Ivanovskaya, A., McFarland, E. W. *J. Comb. Chem.* **2002**, *4*, 17–22.
44. Potyrailo, R. A. *Trends Anal. Chem.* **2003**, *22*, 374–384.
45. Singh, J., Ator, M. A., Jaeger, E. P., Allen, M. P., Whipple, D. A., Soloweij, J. E., Chowdhary, S., Treasurywala, A. M. *J. Am. Chem. Soc.* **1996**, *118*, 1169–1679.
46. Stanton, D. T., Morris, T. W., Roychoudhury, S., Parker, C. N. *J. Chem. Inf. Comput. Sci.* **1999**, *39*, 21–27.
47. Linusson, A., Gottfries, J., Lindgren, F., Wold, S. *J. Med. Chem.* **2000**, *43*, 1320–1328.
48. Potyrailo, R. A., May, R. J., Shaffer, R. E., Lemmon, J. P., Wroczynski, R. J. *Method for High-Throughput Fluorescence Screening of Polymerization Reactions*, PCT Int. Appl. WO 02/33384 A1, **2002**.
49. Rajan, K., Rajagopalan, A., Suh, C. In *NATO Science Series, II: Mathematics, Physics and Chemistry*. Vol. 52, *Molten Salts: From Fundamentals to Applications*, **2002**, pp. 241–248. Kluwer Academic Publishers.

50. Potyrailo, R. A., Morris, W. G., Chisholm, B. J., Cawse, J. N., Brennan, M. J., Molaison, C. A., Hassib, L., Flanagan, W. P., Reitz, H., Medford, G. F. Presented at *Pittsburgh Conference on Analytical Chemistry and Applied Spectroscopy*, New Orleans, LA, **March 2002**, Paper 751.
51. Nelson, G. L. In J. V. Koleske (Ed.), *Paint and Coating Testing Manual*, **1995**, Philadelphia, PA: American Society for Testing and Materials, pp. 513–524.
52. Crosby, A. J., Karim, A., Amis, E. J. *Polym. Prepr.* **2001**, *42*, 645.
53. Crosby, A. J., Karim, A., Amis, E. J. *J. Polym. Sci. B* **2003**, *41*, 883–891.
54. Hunger, M., Weitkamp, J. *Angew. Chem. Int. Ed.* **2001**, *40*, 2954–2971.
55. Workman, J. J., Veltkamp, D. J., Doherty, S., Anderson, B. B., Creasy, K. E., Koch, M., Tatera, J. F., Robinson, A. L., Bond, L., Burgess, L. W., Bokerman, G. N., Ullman, A. H., Darsey, G. P., Mozayeni, F., Bamberger, J. A., Greenwood, M. S. *Anal. Chem.* **1999**, *71*, 121R–180R.

Chapter 2

Economic Impact of Combinatorial Materials Science on Industry and Society [1]

John D. Hewes

INTRODUCTION

Research and development (R&D) funding within the chemical process industry averages 3.5% of sales, or about $12–14 billion annually in the United States, and is focused mostly on process development and support of downstream activities. This R&D supports annual sales of chemicals of over $400 billion that in turn feed into critical US industries and account for over $1.2 trillion in product sales and service in downstream value.

Significant company profitability and broad economic benefits should have the potential to be leveraged by small improvements in the R&D productivity [2–4] as highlighted in the value chain for advanced materials and basic materials from the chemical process industry (CPI) (Figure 2.1). Heightened interest in increasing R&D productivity has been inversely correlated with private sector spending for discovery phase research, which has declined significantly as industrial R&D has increasingly been shifted toward lower-risk activities, i.e., applied research, process development, and manufacturing support services. There is increasingly less research being conducted at a time when new materials are increasing in complexity. In parallel with this trend, the chemical process industry and the advanced materials sectors are under unrelenting pressure to commercialize higher-performance products at lower costs with shorter innovation cycle times. The link between financial performance and time to market is demonstrated [5].

The advanced materials sector covers markets in metals, optical-photonic materials, polymers, ceramics, composite materials, superconducting materials, electronic materials, and magnetic materials. These markets are heavily interconnected, and there is strong reliance on strategic alliances for technology developments for sustaining a competitive edge.

Traditional chemical and materials science research is based on the iteration of one-at-a-time, or serial, synthesis, characterization, and scale-up of samples. This research methodology has become inadequate for the rapid innovation of new materials of such

John D. Hewes • National Institute of Standards and Technology, Gaithersburg, MD 20899-4700, USA
Current address: Honeywell International, Inc. Washington D.C. 20004, USA, email: John.Hewes@honeywell.com

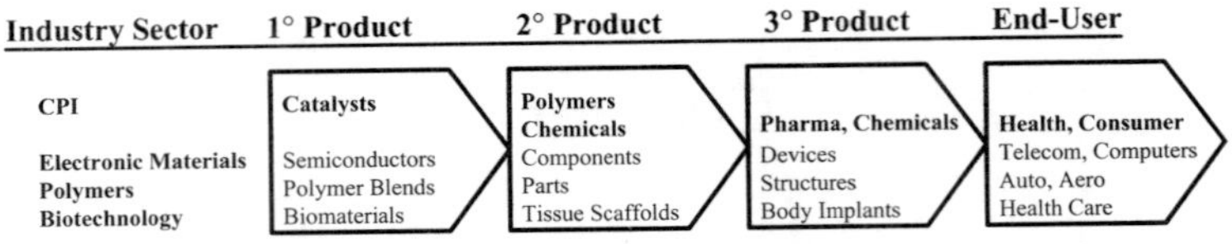

FIGURE 2.1. Value chain for materials and chemical industries.

complexity that no clear understanding or predictability of performance can be obtained. Therefore research methodology reached a "strategic inflection point" [6] where major industries were required to seek a new research paradigm to retain competitive advantage. This need coincided with the convergence of several otherwise unrelated technologies, and resulted in a step change, or *methodological discontinuity*, in the way research is conducted. In effect, high-throughput experimentation (HTE), or combinatorial methods [7], permits one to look beyond the proverbial "needle in a haystack" to the prosaic "needle in a field of haystacks." Cultural barriers have in fact been greater than technological barriers, as some scientists, companies, industries and even government agencies are skeptical about implementing such processes.

The application of HTE methodology to advanced materials discovery can significantly increase the productivity of innovation. Financial models can utilize tangibles, such as reductions in cost of capital and personnel expenses, to develop realistic cost–benefit estimates based, for example, on discounted cash flow (DCF) scenarios. On the other hand, it is difficult to estimate increased value using intangibles, such as "creativity" and "serendipity," that have been noted to arise from increasing the size of experimental space.

Significant technological leaps have already been made by several companies that implemented HTE for materials discovery and chemical process development. They have reduced the concept-to-market cycle time from a traditional 8–10 years to a paradigm-breaking 3–5 years. The economic impact of this advance is the subject of this chapter.

HIGH-THROUGHPUT EXPERIMENTATION

The fusion of three otherwise diverse technological areas have resulted in a radical departure in how researchers are approaching the discovery and innovation of new materials. New ways of thinking about research and new strategies must be embraced. This is occurring at the company level as well as on a national level. Indeed, academia is facing pressure to include high-throughput technologies in curricula.

The convergence of scientific instrumentation, chemistry and materials science, and information technologies fits the definition of both architectural (company-level) innovations [8] and discontinuities [9] on a global level that are being regarded in terms of new national competencies [10], in much the same way that combinatorial chemistry accelerated drug discovery starting in the mid-1990s [11]. For example, Malaysia has adopted high-throughput screening, computational chemistry, and informatics to catalog the diverse natural products that abound within their forests with the goal of exploiting national intellectual property assets through out-licensing or domestic drug production [12].

While one might term this new paradigm a market discontinuity, in fact it is a methodological discontinuity. Researchers are enabled greater room for exploration through the vastly increased experimental space available to them using high-throughput approaches. Rather than explore experimental space in terms of data points, researchers must now assimilate large amounts of rapidly produced multidimensional data. Visualization software has revolutionized the ability to analyze multidimensional data interactively.

The need to miniaturize and automate scientific equipment is driven by the need to reduce the time to synthesize and screen materials formulations as well as to reduce the cost of raw materials and personnel. The miniaturization of chemical reactors and sensors has allowed scientists to increase the number of formulations being screened in any given time. Laboratory automation and robotics, once the domain of the drug discovery chemistry laboratory, now reduce the cost of repetitive, time-consuming, and labor-intensive tasks for materials research. Robotic synthesizers from a variety of manufacturers address specific issues for inorganic chemistry and materials science, and competition in the area has produced relatively low-cost workstations.

The need to identify higher-performing candidates consisting of complex mixtures drives such applications as industrial catalysts and electronic materials. Combinatorial methods for drug discovery accommodated the reverse challenge—identifying single molecules from complex mixtures—using deconvolution and tagging technologies, but these methods have not translated into solid state materials. This is a stubborn, yet significant, technological challenge for HTE as it is applied to materials, which many researchers hope will be solved with computer software that enables the prediction of material properties from composition, or vice versa [quantitative structure–property relationship (QSPR)].

Informatics represents the third enabler for HTE methods in materials science: the computer software needed to collect and interpret data, manage the resulting large amounts of information, and transform it into knowledge through tools such as visualization software. While not approaching the magnitude of the needs of drug discovery and cheminformatics, materials science requires tools that would handle complex compositions and structures, and integrate simulation, modeling, experimental design and statistical engines, database query engines, and data warehousing.

In the pharmaceutical arena, the need for earlier market entry to attain profitable patent position justifies the implementation to support the risk and cost of implementing combinatorial chemistry for drug discovery and process development. The cost of developing new drugs now averages $350 million per drug, with a 12-year time to market, which is justified by potential annual sales of a blockbuster product in the range of $500 million to $1 billion or more per year. The pharmaceutical industry leads all industry sectors in R&D outlays (15% of $100 billion sales/per year), and actively pursues combinatorial methodologies to remain competitive; it is assumed to be less sensitive to large expenditures in view of the significant potential revenues gained from the sale of new drugs.

The potential payoffs are significantly less for implementing HTE in materials science, the research budgets are smaller, and the technical challenges are greater. As shown in Table 2.1, however, there exist strong motivations to implement HTE in the chemicals and materials arenas. Success stories from companies that have implemented HTE indicate that catalysts, polymers, and electronic materials are being commercialized faster than in those using traditional methods, and these are impacting products downstream.

TABLE 2.1. Technical Challenges and Economic Benefits for HTE [57]

Technical areas	Technical challenges	Impacted products	Economic benefits
Industrial chemicals and monomers	Faster catalyst screening Capability of screening extremely diverse combinations of catalyst ingredients	Industrial chemicals Engineering Thermoplastics Other plastics	Lower cost Lower energy usage New products based on newly affordable raw materials
Electronic materials	Deposition methods Interfacial mixing/diffusion	Semiconductors Dielectric materials	New products for wireless Miniaturization Simplified circuits
Polymers	Faster screening	Engineering plastics Polymer blends Thermoplastics	New products New markets (e.g., automotive glazing)
Catalysts	Process optimization		Reduce domestic energy consumption
Surface modifiers Ceramics	Deposition methods Interfacial mixing/diffusion Electronic properties Higher strength	Aircraft engines, advanced power machines Conductors, semiconductors Machine tools	Efficiency from higher engine temperatures, increased service life and reduction of downtime Higher component densities and speeds Machining of new alloys, increased productivity

TABLE 2.2. Companies with HTE Activities

3M	ABB Lummus Global
Air Products & Chemicals Inc.	Akzo Nobel Central Research
Honeywell International Inc.	Amoco Chemicals Co.
Baxter Corp.	Bayer Inc.
Boehringer Ingelheim	Cabot Corporation
Cargill Incorporated	The BOC Group Inc.
Chevron Texaco Research & Tech.	Clairol Inc.
Corning Incorporated	Dow Chemical Company
DuPont de Nemours Experimental Station	Eastman Chemical Co.
Engelhard Corporation	Exxon Mobil Technology
Exxon Mobil Technology	FMC
General Electric Global Research	Henkel Chemicals Group
Hercules Incorporated	Hoechst/Celanese
Lucent Technologies	Exxon Mobil Technology
National Semiconductor	Praxair, Inc.
PQ Corporation	Rohm and Haas Company
UOP LLC	Versicor Inc.

R&D spending against sales revenues was utilized to rank industries [13] according to their spending for innovation as an indirect measure of their sensitivity to implement a significant program in combinatorial methodologies. Figure 2.2 indicates the level of spending for research in different industries as a percentage of revenues, and indirectly

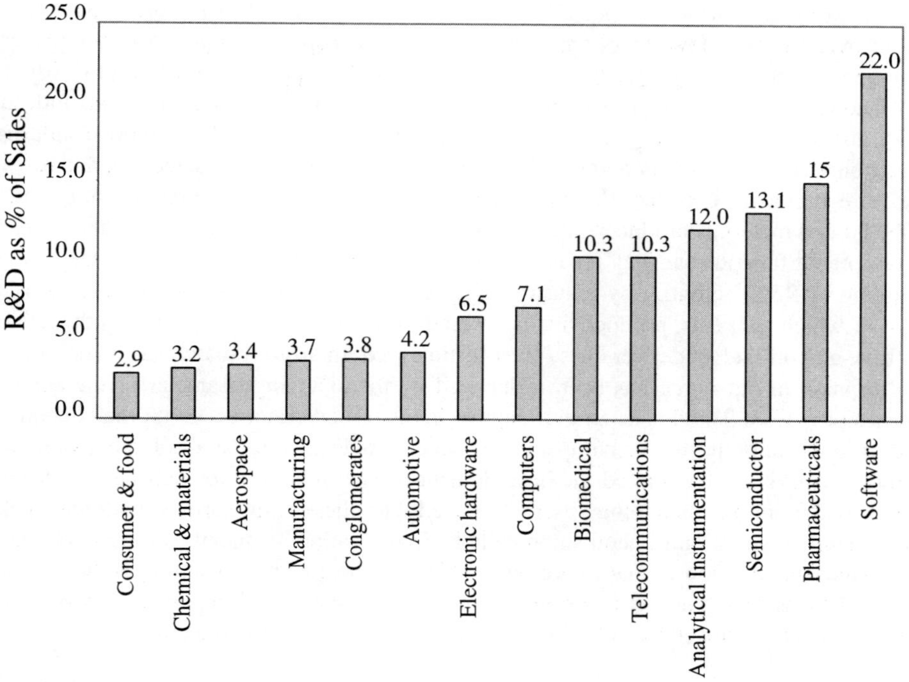

FIGURE 2.2. Technology industry spending on research.

indicates the potential market for combinatorial methods. As indicated earlier, the pharmaceutical industry leads most other sectors in the level of funding of research; although the significant portion of this spending occurs in the later stages of research (clinical trials), early time to market and a more thorough screening in preclinical studies is a strong driver. Downstream market impact could affect the aerospace, automotive, telecommunications, and other important industries in the future as new materials performance diffuses into this infrastructure. These data (Figure 2.2) indicate that chemical and materials companies (excluding petrochemical, which averages less than 1%) spent approximately 6% of revenues on R&D activities in 2002, ranking near the bottom of technology-oriented industries.

The smallest increases in R&D spending occur in the mature industries that have narrow profit margins based on commodities manufacturing, such as petroleum refining and extraction, non-ferrous metals and products, and commodity and specialty chemicals. Inadequate investment in supporting and adapting R&D at the present time may have significant consequences for the national competitiveness of commodity industries in the future.

As a result of the increasing challenges in global markets, related to downward price pressures and customer demands for higher performance, companies are changing the structure of their R&D efforts, for example, by establishing research centers in nations with lower personnel costs and throwing out ever broader nets in an effort to bring in more diverse strategic partners, especially involving smaller firms. Restructuring of industrial R&D over the past few years has resulted in a shift in the responsibility for research; suppliers are now being made to be more responsible for their own research [14]. Much of that knowledge supply chain is being outsourced in the chemicals and materials industries [15], just as it has been in the pharmaceutical [16], computer, and automotive industries. Functions such as discovery and process development are being contracted out, and research staff are being decentralized and internationalized. Original equipment manufacturers (OEMs) and materials producers are starting to outsource ("disintermediate") discovery and development functions as they focus on, and consolidate, their core business areas. The outsourcing of R&D activities by industry is being driven by a number of considerations, not least of which are costs, productivity, decreased overhead, and an emphasis on the need for on-time and on-budget deliverables. "Just-in-time" externalized research has come of age as the decision to outsource has been influenced by the changing character of the corporate research capacity. The downsizing of captive industrial laboratories—with the concomitant dedication to only those core competencies which are deemed to be absolutely essential for corporate survival—has forced the development of new relationships between the technology provider and the technology user [9]. Faced with these slim margins, materials makers are pursuing a variation of acquisition called "decaptivation"—manufacturing materials for companies that previously made their own. This shift in strategy is resulting in the formation of small firms that function as providers of services, such as discovery and development (the "discovery sector"), as well as manufacturers of chemical intermediates.

COMMERCIAL OUTLOOK

HTE will have tangible as well as intangible impacts. Tangible changes will result from different costs of capital and reduced personnel expenses, while intangible impacts will be seen in reduced time to market, defensive and offensive intellectual property positioning, and the reduction of the "risk" of discovery research and innovation.

Estimating the economic impact of HTE methodologies is difficult. On a company basis, DCF techniques indicate that HTE can provide significant value to industry, as shown below. One might also estimate the potential market in the chemical process industry and materials sectors and develop scenarios based on different levels of penetration into the R&D market. Thirdly, one might also apply real options analysis as it defines the value of R&D based on costs as well as risks.

The assumptions made in this chapter are as follows.

1. HTE is currently used for discovery and applied research in the chemical and electronics sectors. Of the $206 billion spent on R&D in the United States in 1997, 16% of combined industry and government funds were spent on basic research, 23% on applied research, and 61% on development [17]. The National Science Foundation [18] estimated that 8% of industry R&D expenditures in 2000 went to basic research ($13 billion), 20% to applied research ($33 billion), and 72% to development ($120 billion). Computers and electronics accounted for 20% of the total R&D spending in the United States ($36 billion) and the chemical industry expended $20 billion. Spending for early-stage research was reported by

Booz-Allen Hamilton to consume 33% of R&D funds in the chemical industry, 13% in the pharmaceutical industry, and around 10% for electronics, equipment, and telecommunications [19].

2. The financial value of implementing HTE is reflected in many metrics, such as profit margin, revenues, revenues gained from new products, size of R&D budget, competitive risk, market maturity, and market growth. In addition to absolute levels of and changes in R&D expenditures, another key indicator of the health of industrial science and technology is R&D intensity. There are a number of ways to measure R&D intensity, but the one used most frequently is the ratio of R&D funds to net sales. This statistic provides a way to gauge the relative importance of R&D across industries and firms in the same industry. Estimating this trend across different industries is difficult, but to a first approximation both the R&D funding levels and its ratio versus revenues indicate the financial incentive and strength of the particular industry to implement HTE.
3. Significant differences exist in the needs of companies within these sectors: mature commodities manufacturers will focus on process development or realignment of licensing agreements and may develop an internal combinatorial program; medium-sized and smaller firms may focus on new product development with external support. The risk of *not* implementing HTE may be a life-or-death business decision based on intangibles such as market position or patent assets.

HTE will be applied to advanced materials research areas where empirical or semi-empirical methods have dominated (i.e., where complexity overwhelms scientific understanding): heterogeneous catalysts and biocatalysts; "smart" materials such as high-temperature (high-T_c) superconductors and thermoelectrics; polymers and polymer additives; dyes and colorants; biomaterials; electronic, optical and optoelectronic materials; hard coats; and metal alloys [20].

The primary market for application of HTE in the advanced materials industry is likely the chemical and materials suppliers, with secondary markets in chemical intermediates. Secondary, and much smaller markets, are in polymers, biomaterials, and electronic materials. Downstream impact would probably follow on computer, telecommunications, biomedical, automotive, and aerospace sectors, as shown in Figure 2.1. The chemical process industry uses catalysts to manufacture chemical intermediates and polymers; the chemical intermediates are converted into higher-value chemicals and pharmaceuticals, which eventually feed into the health and consumer markets. Polymers produced by catalytic and non-catalytic processes feed into several secondary markets, including specialty polymers (blends, alloys, etc.) that are used in many downstream markets. Similarly, the electronic materials sector produces inorganic and organic materials that enter the semiconductor markets to produce components and devices, and eventually feed into electronics such telecommunications and computer equipment. Aerospace, automotive, biomedical, telecommunications, and computers are the principal end-user sectors driving the chemical process industry, advanced materials, semiconductors, and electronic hardware sectors; the analytical instruments and software sectors provide infrastructural support [21].

The rate-determining step in developing a world-class competency in HTE is the integration of otherwise divergent technologies into an efficient laboratory "factory" focused on specific, relevant, and qualified market opportunities. Indeed, the development of geographically dispersed "virtual laboratories" fits into this scenario quite easily, as demonstrated by commercial [22] and government [23] efforts. A strong need has been expressed by the private sector [24] to develop infrastructural technologies and standards in support of HTE. The private sector has also indicated a need for government funding to facilitate the development of new methodologies, metrics, and standards for HTE, for example, through funding of enabling technologies by the US Department of Commerce's Advanced Technology Program [25]. Other funding programs and agencies have funded specific projects utilizing HTE for advanced materials, but have not specifically targeted development of infrastructure technologies or methods.

The major commercial hurdle for companies to adopt HTE is validation that the methodologies lead to increased productivity, for example, in terms of increased return on investment (ROI) and higher DCF for new products. The chemical process industry and materials industries are characterized by R&D budgets that are typically 3–4% of sales, with profit margins in the range of 8–10%. As indicated earlier, improved return on R&D investment and faster time-to-product launch are expected to result in higher overall net present value (NPV). Several large companies have reported publicly that their HTE efforts have been technically successful in catalyst discovery and process development, at a typical entry cost of $10–20 million [26]. R&D spending for large chemical and pharmaceutical companies was evaluated on a company-wide basis (Table 2.3). Among the major industrial sectors, $111 billion was spent on R&D by the top 500 R&D firms in 1997: the information and electronics sector had the highest share at $45.8 billion (41%), followed by medical substances and devices at $20 billion (18%), basic industries and materials at $8 billion (8%), and chemicals at $7 billion (6%) [27].

THE "DISCOVERY INDUSTRY"

Companies are pursuing several business models to obtain HTE capabilities:

1. developing internal facilities (e.g., General Electric Global Research Center, DuPont Central Research, UOP LLC, Albemarle) at a cost estimated to be $5–20 million per site;
2. forming a strategic alliance with service providers having a core competency in HTE discovery methods (e.g., Dow Chemical or BASF with Symyx Technologies, W.R. Grace with Avantium, ChevronPhillips with hte Inc.), often with shared financial and/or knowledge capital;
3. developing a consortium (European CombiCat Consortium, Combichem Initiative in the United Kingdom);
4. spinning off a start-up (Shell Chemical Company and Avantium Technologies, DuPont and CombiPhos Catalysts Inc., UOP LLC and Torial LLC).

There are currently more than half a dozen start-up firms worldwide with business plans to develop and apply HTE to profitable research areas such as catalyst and electronic materials discovery and process optimization. These companies are targeting enterprises that cannot justify the costs of implementing in-house combinatorial facilities. Symyx Technologies, the current market leader, had an initial public offering on November 19, 1999 and reported 2002 revenues of $60 million [28,29] based on contracts with other companies and sales of its discovery platforms. Avantium Technologies (Amsterdam) recently announced the formation of a research alliance with Malaysia [30] and with the Industrial Technology Research Institute in Taiwan [31].

One indicator of this step change in adopting the new research methodologies is shown by the current global spending in HTE for advanced materials discovery, which was conservatively valued between $350 million [32] and $500 million [33] in the year 2000, up from an estimated spending of several million dollars between 1995 and 1997 at the Lawrence Berkeley National Laboratory, where Peter Schultz initiated the movement with the publication of research describing the high-throughput discovery of phosphors and high-T_c superconductors [34].

Adoption of HTE has resulted in successes already. Recently, UOP LLC (Des Plaines, IL) announced that a new heterogeneous catalyst discovered using HTE will enter commercial demonstration scale in 2002 [35]; UOP has spun off a company called Torial LLC that will conduct contract research in catalysis. DuPont (Wilmington, DE) recently announced that a homogenous catalyst will move into commercialization in the 2002–2003 timeframe [36] and has spun off a company called Combiphos (Princeton, NJ) to conduct contract research in the synthesis of ligands for organometallic catalysts. Symyx Technologies (Santa Clara, CA) has announced the successful introduction of other advanced materials, such as an X-ray phosphor for the Agfa-Gevaert Group (Mortsel, Belgium) and a biocompatible polymer for Applied Biosystems Inc. (Foster City, CA), discovered using HTE [37].

THE VALUATION OF HTE

DCF Analysis

An illuminating illustration of the potential to bring value to an organization was presented by Thomke et al. [38] in an economics case study on combinatorial chemistry. Using data from the combinatorial synthesis firm Pharmacopeia, the authors compared the costs associated with parallel combinatorial chemistry compared with traditional chemistry methods in a drug discovery program aimed at a specific therapeutic agent.

A financial pro forma analysis [39] was developed with a sensitivity analysis for a single-company chemical commercialization effort without HTE and one with HTE in the discovery phase only. However, "productivity" is not yet well defined in microeconomics [40] or macroeconomics [41] for the following factors which take into account the combination of concepts to define value of implementing the HTE methodology:

- cost reduction: decreased labor cost per experiment using automation
- probability of success (POS): ability to examine very large experimental spaces
- cycle-time reduction: reduction in time from idea to product launch (from a typical 10 years to 8 years).

Thomke et al. indicated that the pharmaceutical industry, where the leading companies are approaching discovery throughputs of 2 million hits per year, reports 100-fold decreases in cost per hit (Table 2.4). The pro forma yielded changes in "R&D yield" using the Thomke case study inputs of resource allocations and the time value of capital expenses. We observed an increased NPV resulting from the earlier commercial launch obtained with HTE (Figure 2.3). Without HTE, idea to launch took an estimated 9.5 years, with an NPV of $203 million at market saturation. With HTE, the idea-to-launch cycle time of 6.5 years led to an NPV of $286 million; use of HTE in the applied research and process development stages would offer additional benefits.

In an effort to define what potential value HTE methods could bring to a chemical company, Busch [42] surveyed 30 experts from the plastics industry to define the effort, costs, and potential profits in discovering a mid-level "engineering polymer" such as polycarbonate, with global annual consumption of around 500 million kg and a market opportunity of $2 billion. Busch observed that, using traditional laboratory methods, the first year for revenue generation is 12 years following the discovery phase, and financial break-even occurred on average 17 years after ideation; an NPV of $10 million was assumed at project start. HTE methods, on the other hand, reduced the time interval for both the discovery phase and scale-up by 1 year each, and this earlier market entry granted the plastic a 15% price premium; even with $6 million in depreciated capital outlays for equipment, the present value was calculated to be $37 million. Busch concluded that, at least for a high-valued polymer, HTE methods are financially viable to implement based on ROI calculations.

Neither of these studies takes into account the "prediscovery efficiency" [41,43] introduced by HTE methods. For example, the ability to scan literally a hundred to a thousand times more samples allows a trained research team to explore areas of chemistry that would not otherwise be possible under normal industrial time constraints. As indicated earlier, Symyx reported examples of ternary catalysts with substantially improved performance over binary catalyst mixtures as reported in the open literature and patents.

Pioneering researchers are showing that HTE provides a more powerful method to generate and evaluate a large number of catalysts in very short time. Recently, both UOP LLC [44] and DuPont de Nemours [45] announced the commercial utilization of new

TABLE 2.3. Traditional Methods Versus Combinatorial Chemistry in Drug Discovery [38]

Project variable	Combinatorial approach	Traditional approach
Total development time	3.5 months	5 years
No. chemists needed	4	15
No. compounds tested	Approx. 9,000	Approx. 3,750
No. of serial rounds	1	100 (250 max)
No. compounds per round	Approx. 9,000	Approx. 38
Cost of screen per round	$10,000	$10,000
Total costs—chemists[a]	$167,000	$18.75 million
Total cost per compound[b]	$19	$5,000

[a] Estimated $250,000 per full time employee.

[b] Since the capital costs for increased throughput are incremental, a 10-fold increase in productivity would equal less than $2 per sample.

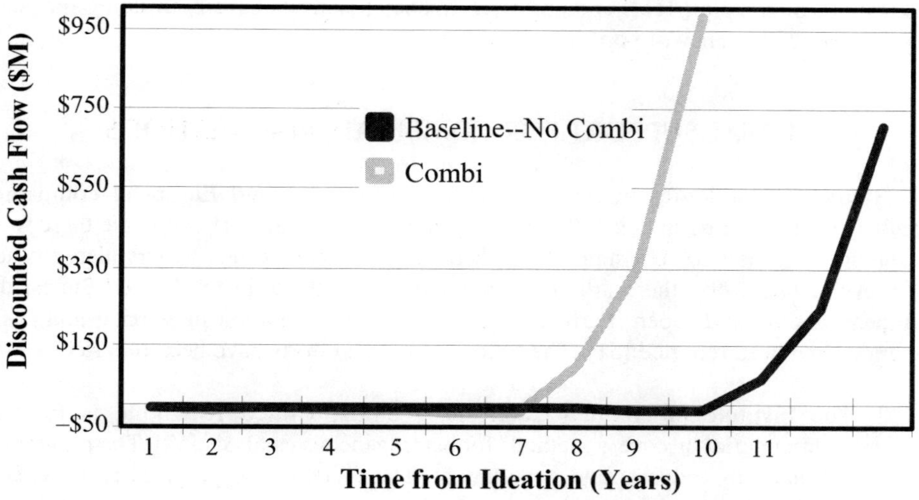

FIGURE 2.3. Time sensitivity analysis for HTE commercialization of a chemical intermediate.

catalysts developed with HTE methods in significantly less time than using classical methods. UOP confirmed that their commercialization cycle time has been reduced from 8–10 years using traditional methods to 3–5 years using HTE.

Real Options Theory

According to Boer [46], the valuation of R&D has become more sophisticated than DCF models. Tools developed in the past decade for the valuation of tangible and intangible assets, such as intellectual capital, have evolved to the point where innovation risk can be assessed.

Imagine a game of seven-card stud poker. On each deal, the player antes up to see his next card—and his opponents' cards. If he does not like what he sees, he folds. The theory defines technology investments as "options" to be pursued. As opportunities present themselves, managers can place small bets, or investments, on a variety of technology projects and see how they play out. As one becomes more likely to yield gains, the "betting strategy" becomes clearer [47].

Real options theory is being applied in the R&D management community [43,48] and commercially for the sale of intellectual property assets developed in R&D settings [49]. Sophisticated tools have been made available to the general analysts, for example, in the form of spreadsheets, informatics and data query tools for on-line databases, visualization tools such as Spotfire [13], and Monte Carlo models. The chemical process industry and materials industry have placed their bets on significantly increasing R&D productivity to improve bottom-line performance. Consolidation of experiments into large diverse libraries using HTE is being driven by risk reduction. As the risks inherent in the discovery phase of innovation decreases with implementation of HTE methods, the economic power of diversity also explains the rise of HTE [38]. By diversifying risk [50] over extremely large "portfolios" of samples, HTE is doing what financial analysts have been doing for years with increased efficiency of portfolio management.

GLOBAL SUPPORT FOR COMBINATORIAL METHODS

Symyx Technologies marketed their services to US and European companies simultaneously, starting in late 1997, and they jumped on board early on while there were no public indications of US support for their efforts. Japanese researchers also profited early. As of late 1999, there are government funding efforts in the United States, the European Union, and Japan. Early indications of implementation in Mexico and Latin America have been reported [51]. The following global efforts have been reported.

1. The Institute of Applied Catalysis, Prenton, United Kingdom, is funding combinatorial discovery methods for heterogeneous catalysis [52]. The program is cofunded by industry and government. The focus of the program is a low-cost small-scale "multichannel" catalyst testing system.
2. The European Union European Research Network for Sustainable Technologies (COST) is funding combinatorial research to several institutes such as the Universität Berlin in the area of supported catalysts. Funding levels are estimated to be in the range of $2.5 million per year.
3. The German Federal Ministry of Education, Science, Research, and Technology (BMBF) is funding basic research in applying combinatorial approaches to catalysis following a highly successful 5-year effort supporting drug discovery. The level of effort is approximately $12 million per year from 1999 to 2004.
4. The government of The Netherlands and three Dutch universities (Twente, Delft, and Eindhoven) are directly supporting the Shell Oil spin-off company Avantium.
5. A consortium of 10 research partners (three large companies, two small/medium-sized companies, four research institutions, and one university) from six European nations was grouped to fulfill a fast-track program to develop new methodologies for the development of heterogeneous catalysts using combinatorial methods. The Institut Française du Pétrole (Rueil–Malmaison) and the Institut de Recherches sur le Catalyse–CNRS (Villeurbanne) are administrators of the effort [53] which has a total budget of $5 million with funding from the European Commission ($3 million), company "equity," and the European Research Network for Sustainable Technologies (COST).
6. The Japanese Ministry of Economy, Trade and Industry (METI) initiated support for high-throughput screening beginning in 1997 with funding to the Tokyo Institute of Technology [54]. The funding was approximately $40–60 million over 5 years and was extended in 2000. The principle areas of study are in electronic materials such as zinc oxides for light-emitting diodes and new battery materials.
7. The Advanced Technology Program (ATP), administered by the U.S. Department of Commerce's National Institute of Standards and Technology, funds high-technical risk research projects in all technology areas. The ATP initiated extramural funding in HTE methods applied to catalysts, polymers, and electronic materials, starting in 1999, that now totals $40 million over 5 years.

TECHNOLOGICAL CHALLENGES

High-throughput R&D methodologies will eventually reach industries and companies that cannot make the substantial capital investment required today. Software technology must be capable of defining profitable experimental spaces and visualization of complex data relationships, and of correlating target materials with properties to permit database queries from a broad spectrum of data-mining engines and the development of structure–property relationships. New tools will have to be developed to enable the integration of this information into molecular and property modeling engines. The basic underlying hardware technologies needed for high-throughput library fabrication and analysis are enabling and emerging technologies that are finding applications in other arenas. Broad technical needs have been identified by industry: massively parallel computers and high-bandwidth networks, and micromachines and microreactor technologies (MRTs) based on microelectromechanical systems (MEMS) to address the need for higher library densities to facilitate reduced raw materials costs for library fabrication and economies of scale and modularity in laboratory instrumentation [55,56].

CONCLUSIONS

High-throughput screening and combinatorial chemistry revolutionized drug discovery in the pharmaceutical industry. Materials science, biotechnology, chemistry, physics, and other technological bases will undergo a similar transformation as more challenges are conquered. Large coordinated efforts are needed to maintain US leadership in chemicals and materials research by providing globally competitive increases in efficiency of discovery and process optimization. Estimates of increased value to commercial opportunities are being validated in the chemical industry, and these will translate into broad economic benefit. Software must be capable of defining the smallest possible experimental spaces while subsequently allowing for the visualization of complex relationships; ultimately, when enough data have been collected and efficiently assimilated, the development of structure–property relationships will facilitate even greater developments in materials science.

REFERENCES AND NOTES

1. Certain commercial equipment, instruments, or materials are identified in this paper in order to specify the experimental procedure adequately. Such identification is not intended to imply recommendation or endorsement by the National Institute of Standards and Technology, nor is it intended to imply that the materials or equipment identified are necessarily the best available for the purpose.
2. Link, A.N. Basic research and productivity increase in manufacturing: Additional evidence. *Am. Econ. Rev.* **1981**, *71(12)* 1111–1112.
3. Griliches, Z. Productivity, R&D and basic research at the firm level in the 1970's. *Am. Econ. Rev.* **1986**, *76*, 141–154.
4. Krogh, L.C. Measuring and improving laboratory productivity/quality. *Res. Manage.* **1987**, *November–December*, 22–24.
5. For example, Ellis, L.W., Curtis, C.C. 'Speedy R&D: How beneficial? *Res.-Technol. Manage.* **1995**, *July–August*, 42–51.

6. Grove, A.S., *Only the Paranoid Survive: How to Exploit the Crisis Points That Challenge Every Company*, **1999**, New York: Bantam Books,
7. Fairley, P. *Chem. Week* August 12, **1998**, 18); Fairley P., Scott, A. *Chemi. Week* August 11, **1999**, 27–32; Wood, A.; Scott, A., *Chem. Week* August 9, **2000**, 39–42; Borman, S. *Chem. Eng. News* August 27, **2001**, 49–58; Lonergan, M. *Chem. Eng. News*, March 26, **2001**, 230; Watkins, K. *Chem. Eng. News*, October 22, **2001**, 30–34.
8. Henderson, R., Clark, K. Architectural innovation: reconfiguration of existing product technologies and the failure of large firms. *Admin. Sci. Q.*, **1990**, *35*, 9–30.
9. Tushman, M., Anderson, P. Technological discontinuities and organizational environments, *Admin. Sci. Qu.*, **1986**, *31*, 439–465.
10. Miertus, S. *Trends and applications of combinatorial chemistry and combinatorial technologies*, Workshop, Budapest, Hungary, Oct. 15, **2001**.
11. Anon. The world in 1997. *Economist* **1996**; Anon. Mercky waters. *Economist* May 24, **1997**.
12. Miertus, S. (Ed.), *ICS–UNIDO Expert Group Meeting on Computational and Combinatorial Chemistry*, Trieste, Italy, January 30, **2001**.
13. Anon., *R&D Mag.* October, **1998**.
14. The amount of funding being cut from industry R&D for this trend and how much is being invested by the supply chain is not readily available. Supplier R&D figures generally do not show up as expenditures by the operating industry but are buried in the R&D of the vendors and therefore are not in the available statistics. Studt, T., Duga, J.J. Industry spends big on development while Feds focus on research, *R&D Mag.* March **2000** (http://www.rdmag.com/features/0100forecast_forecast.htm).
15. Hume, C., Schmitt, B. A dose of caution: Revising intermediates prospects. *Chem. Week*, April 26, **2000**, 35-G5.
16. Zahr, S.A. Outsourcing: A new reality in the pharmaceutical industry. *Chim. Oggi* **2000**, *May*, 18–19.
17. American Association for the Advancement of Science, Directorate for Science and Policy, http://www.aaas.org/spp/dspp/rd/fy00f5.pdf; NSF Science and Engineering Indicators 1998, http://www.nsf.gov/sbe/srs/seind98/access/c4/c4s1.htm#c4s112a
18. National Science Foundation, *Science and Engineering Indicators—2002*, http://www.nsf.gov
19. Booz-Allen Hamilton estimate in Branscomb L.M., Auerwald, P.E. Between invention and innovation: An analysis of funding for early-stage technology development. NIST GCR-02-841. November, **2002**.
20. Schultz, P.G. et al., *Combinatorial synthesis of organometallic materials*, US Patent 6,420,179, July 16, **2002**; Schultz, P. G. et al., *Combinatorial screening of inorganic and organometallic materials*, US Patent 6,410,331, June 25, 2002); Schultz, P.G. et al., *Combinatorial synthesis of novel materials*, US Patent 6,346,290, February 12, 2002; Schultz, P.G. et al., *Combinatorial synthesis of novel materials*, US Patent 6,326,090, December 4, **2001**); Schultz, P.G. et al., *Combinatorial synthesis of novel materials*, US Patent 6,004,617, December 21, **1999**; Schultz, P.G. et al., *Combinatorial synthesis of novel materials*, US Patent 5,985,356, November 16, 1999; Schultz, P.G. et al., *Giant magnetoresistive cobalt oxide compounds*, US Patent 5,776,359, July 7, **1998** (developed under contract no. DE-AC03–76SF00098 awarded by the Department of Energy and assigned to Symyx Technologies).
21. Anon., *R&D Mag.*, June **1998**.
22. Maxwell, I. *ATP/AIChE Workshop on HTE for Materials Research, Los Angeles, CA*, November 15, 2000.
23. *ICS–UNIDO Expert Working Group on Computational and Combinatorial Chemistry*, Juraquino, Mexico, May 29–31, **2000**.
24. *Vision 2020 Roadmapping Workshop*, NIST, Gaithersburg, MD, June 1, **2000**, available at http://www.nist.gov/combi.
25. *Advanced Technology Program HTE Project Review*, NIST, Gaithersburg, MD, October 19, **2001**, available at http://www.atp.nist.gov/www/ccmr/ccmr_off.htm.
26. Symyx's discovery tools sells for approximately $15–20 million. Companies with internal facilities have spent up to that amount to implement; however, other companies have adequate installations for significantly lower expenditures depending on throughput and application area.
27. Shephard, C., Payson, S. *U.S. Corporate R&D*. Vol. 2, *Company Information on Top 500 Firms in R&D*, **1999**, Washington, DC: National Science Foundation, available at http://www.nsf.gov/sbe/srs/nsf00302/pdf/nsf00302.pdf
28. Fairley, P. *Chem. Week* August 11, **1999**, 27.
29. Symyx Technologies 10-K, March 19, 2002, retrieved from http://www.sec.gov/Archives/edgar/data/1095330/000089161802001268/f79645e10-k.htm#010

30. Rouhi, A.M. Chemspec in Basel draws good crowd. *Chem. Eng. News* **2002**, *80* (29), 63–64.
31. Anon., Materials research for electronic: Avantium Technologies and Industrial Technology Research Institute, Taiwan. *Chemical Business Newsbase*, April 26, **2001**.
32. Estimate based on publicly announced contracts to combinatorial chemistry firms.
33. The Catalyst Group, *Combinatorial catalysis* (A multi-client study), February, **2000**.
34. Xiang, X.-D., Sun, X., Briceno, G., Lou, Y., Wang, K.-A., Chang, H., Wallace-Freedman, W. G., Chen, S.-W., Schultz, P. G. *Science* **1995**, *268*, 1738–1740; Jandeleit, B., Schaefer, D.J., Powers, T.S., Turner, H.W., Weinberg, W.H. *Angew. Chem.-Int. Ed.* **1999**, *38* (17), 2495–2532.
35. Holmgren, J. presented at *CombiCat 2001*, Catalyst Group, Philadelphia, PA, November **2001**; *Combi2002*, Knowledge Foundation, San Diego, CA, January **2002**.
36. Li, G. presented at *CombiCat 2001*, Catalyst Group, Philadelphia, PA, November **2001**; *Combi2002*, Knowledge Foundation, San Diego, CA, January **2002**.
37. Cohan, P. presented at *Combi2002*, Knowledge Foundation, San Diego, CA, January **2002**; http://www.symyx.com
38. Thomke, S., Von Hippel, E., Franke, R., Modes of experimentation: An innovation process and competitive variable. *Res. Policy* **1998**, *27*, 315–332.
39. Boer, P. *The Valuation of Technology*, **1999**, New York: John Wiley; *The Real Options Solution: Finding Total Value in a High-Risk World*, **2002**, New York: John Wiley.
40. Krogh, L.C. Measuring and improving laboratory productivity/ quality, *Res.-Technol. Manage.* **1987** *30* (6), 22–24; Meinhart, W.A., Pederson, J.A. Measuring the performance of R&D professionals, *Res.-Technol. Manage.* **1989** *32* (4), 19–21; Packer, M.B. Analyzing productivity in R&D organizations, *Res.-Technol. Manage.* **1983**, *26* (1), 13–20; Pappas, R.A., Remer, D.S. Measuring R&D productivity, *Res.-Technol. Manage.*, **1985** *28* (3), 15–22; Schainblatt, A.H. How companies measure the productivity of engineers and scientists, *Res.-Technol. Manage.*, **1982** *25* (3), 10–18; Szakonyi, R. Measuring R&D effectiveness—I, *Res.-Technol. Manage.* **1994**, *37* (2), 27–32; Szakonyi, R. Measuring R&D effectiveness—II, *Res.-Technol. Manage.* **1994**, 37 (3), 44–55; Whiteley, R.L., Parish, T., Dressler, R., Nicholson, G. Evaluating R&D performance using the new sales ratio, *Res.-Technol. Manage.* **1998**, *41* (5), 20.
41. Werner, B.M., Souder, W.E. "Measuring R&D performance—U.S. and German practices," *Res.-Technol. Manage.* **1997**, *40* (3) 28–32; Werner, B.M.; Souder, W.E. Measuring R&D performance—State of the art. *Res.-Technol. Manage.* **1997**, *40* (2), 34–42, and references cited therein.
42. Busch, J.V. Combinatorial analysis: How much is it worth? *Res.-Technol. Manage.*, March-April, 2001, pp. 38–45.
43. Anonymous, *Chemical Business Newsbase (Cambridge)*, Aug 20, 2001.
44. Holmgren, J. presented at *Combi 2002*, Knowledge Foundation, San Diego, CA, January **2002**; Dahl, I., Karlsson, A., Akporiaye, D.E., Wendelbo, R., Vanden Bussche, K.M., Towler, G.P., US Patent 6,368,865, April 9, **2002** (assigned to UOPLLC), and references cited therein.
45. Li, G.Y. presented at *Combi 2002*, Knowledge Foundation, San Diego, CA, January **2002**.
46. Boer, F. P. Financial management of R&D 2002, *Res.-Technol. Manage.* **2002**, *45*, 23–35.
47. Intel is piloting a method of quantifying the business value of IT investments. *Business Value Index*. http://www.intel.com/business/bss/
48. Angelis, D.I. Capturing the option value of R&D. *Res.-Technol. Manage.* **2000**, vol. 40; No 4 (July–August), p. 31–34.
49. For example, http://www.yet2com.com and competitors.
50. "Diversity" has two definitions, pertaining both to the statistical distribution of samples in a library and to the distribution of risk in a research effort.
51. *ICS–UNIDO Workshop on Combinatorial Chemistry and Molecular Modeling*, Queretaro, Mexico, May 28–June 2, **2002**.
52. http://www.iac.org.uk/rschprog.htm.
53. Mirodatos, C., Savary, L. *Combi 2000-Combinatorial Approaches to New Materials Discovery*, Knowledge Foundation, San Diego, CA, January **2000**; http://catalyse.univ-lyon1.fr and http://www.ifp.fr.
54. Koinuma, H., Koida, T., Ohnishi, T. et al. Parallel fabrication of artificially designed superlattices by combinatorial laser MBE. *Appl. Phys. A.* **1999**, *69*, S29–S31; Matsumoto, Y., Murakami, M., Jin, Z.W. et al., Combinatorial laser molecular beam epitaxy (MBE) growth of Mg–Zn–O alloy for band gap engineering. *Jpn J. Appl. Phys.* **1999**, *38*, Part 2, No 6A/B, p. L603–L605.
55. For-profit organizations (Knowledge Foundation, CHI, The Catalyst Group, etc.) as well as non-profit organizations (American Chemical Society, American Physical Society, Materials Research Society,

Association of Industrial Chemical Engineers, etc.) have held extensive public discussions of this area. Consultation of the corresponding web sites will redirect interested parties to specific events. Vision 2020 Technology Roadmap for HTE, was held June 2, 2000 at NIST, Gaithersburg, MD. In addition, the ATP has held three public workshops, the summaries of which can be found at http://www.atp.nist.gov/www/ccmr/ccmr_off.htm.

56. Jensen, K.F. et al., Wegeng, R.S. et al., Ehrhard, W. et al., Ryley, J.F. et al. Presented at 2000 National Meeting of the Association of Industrial Chemical Engineers, Houston, TX, November **2000**; US Department of Defense, Defense Advanced Research Program Agency MicroFLUMES Program, http://www.darpa.mil/MTO/mFlumes/index.html.
57. White Paper submitted to the NIST Advanced Technology Program April, 1999 by General Electric Corporate Research and Development Center.

Optical Imaging and Spectroscopy

Chapter 3

Image Analysis for High-Throughput Materials Science

Alamgir Karim, Amit Sehgal, J. Carson Meredith*, Alfred J. Crosby†, and Eric J. Amis

INTRODUCTION

Imaging plays a key role in modern day materials science by providing a high-density format of data representation that can be visually interpreted or processed by the human brain in a fraction of a second. This efficient parallel processing makes imaging naturally amenable to data acquisition and interpretation in high-throughput or combinatorial approaches. To marry imaging and high-throughput methods, four main issues must be addressed: image collection, image processing, image analysis, and informatics.

The most common route for image collection is microscopy, where a sample is placed at the focal point of a lens or beam and "imaged" in a snapshot. High-throughput imaging methods rely upon rapid or automated imaging sequences that minimize human intervention during the various steps from imaging a combinatorial library sample to the processing of the large number of image datasets. Automation in image collection can include programmed sample translation, autofocus, automated illumination control, autosetting resolution, and process controls such as temperature, pressure, humidity, etc.

Although image collection in high-throughput methods may require robotics or clever combinatorial library design, the most significant challenge in high-throughput imaging is the process of image analysis and informatics. While the "first take" from an image may be processed by human "inspection," commuting an image into data and ultimately information requires computer programs with higher-level analyses such as mathematical

Alamgir Karim, Amit Sehgal, J. Carson Meredith, Alfred J. Crosby, and Eric J. Amis • Polymers Division, National Institute of Standards and Technology, Gaithersburg, MD 20899, USA
* Current address: Assistant Professor, School of Chemical Engineering, 778 Atlantic Drive, Georgia Institute of Technology, Atlanta, GA 30332, USA
† Current address: Assistant Professor, Department of Polymer Science and Engineering, University of Massachusetts, Amherst, MA 01003, USA
Contact author: Alamgir Karim, email: Karim@nist.gov

High-Throughput Analysis, edited by Potyrailo and Amis
Kluwer Academic Publishers, New York, 2003.

transforms, correlation functions, statistical shape and size analyses, spectra, etc. Furthermore, the high-informational content of images may be expanded to analyze dynamic process by recording "movies" of data. This expansion not only modifies image collection, but also requires integrating the data analysis over an entire set of images. In general, whether a single image or a set of images is considered, the progression to analysis of individual image data, definition of descriptors, classification and data-mining, and subsequent pattern-recognition algorithms complete the high-throughput research cycle for extracting information.

In this chapter, we illustrate aspects of image analysis to the generation of descriptors in a high-throughput setting for three specific problems in polymer material science—dewetting, adhesion, and biocompatibility. The scope and difficulty of the problems discussed ranges from simple polymeric to an intermediate analysis scale involving imaging or visualizing real-time adhesion (kinetics) and on to the more complex and significant world of biomaterials science. The hope is that such illustrations will inspire common theme development for collective advancement of high-throughput imaging techniques applicable to problems in materials science involving equilibrium and non-equilibrium phenomena such as phase behavior, crystallization, fracture mechanics, and *in situ* imaging of cellular development in tissue-engineered constructs. Although the extraction of information from experiments in materials science is typically problem dependent, and ranges from the simple to the complex, it should be possible to define classes of problems in materials science where a similar approach may be applicable to different scenarios.

AUTOMATED ANALYSIS OF OPTICAL IMAGES OF THIN FILM DEWETTING FROM COMBINATORIAL LIBRARIES

Dewetting of Polymers

Polymer thin films are an important class of advanced materials that play a central role in nanoscale science and engineering. Unfortunately thin films can become unstable at thicknesses below 100 nm due to attractive van der Waals (VDW) interactions [1–3]. Dewetting must be avoided in film processing, yet robust methods for preventing VDW instability have not emerged. The lack of previous research and theory on all but model polymers leads to the exploratory high-risk nature of dewetting research. The risk is increased by the experimental difficulty in maintaining consistent substrate chemistry over the many samples needed to test hypotheses. To overcome this limitation in experimental efficiency, a methodology for preparing combinatorial libraries with thickness and temperature gradients has recently been developed [4]. A movie of the flow coating process that was used to make the thickness gradient libraries for the dewetting experiments is enclosed on the CD. Since, the details of the library preparation and screening have been presented elsewhere [4], they are not repeated here.

Approach for High-Throughput Quantitative Analysis

The focus of this discussion is on the informatic analysis necessary to process the large number of optical images resulting from a single combinatorial thickness–temperature library. For example, in our previous study, a 5×5 grid of images covering the temperature T and thickness h range of a typical library was collected every 5 min for 2 hr, using automated optical microscopy [4]. This modest data collection rate resulted in 600 optical images. In an industrial high-throughput laboratory, several such experiments might be performed daily. Each optical image reports the dewetting present at a distinct T, h, and time t on the library.

To our knowledge, systematic studies of the temperature dependence of dewetting rates have not been reported. However, through its effect on viscosity, T is expected to have a strong influence on hole drainage rates [5]. Figure 3.1 shows the 5×5 grid of images for the T–h library collected in one scan. Automated image analysis of hole area as a function of T, h, and t assays a broad range of dewetting rates in a single experiment. Figure 3.1(a)

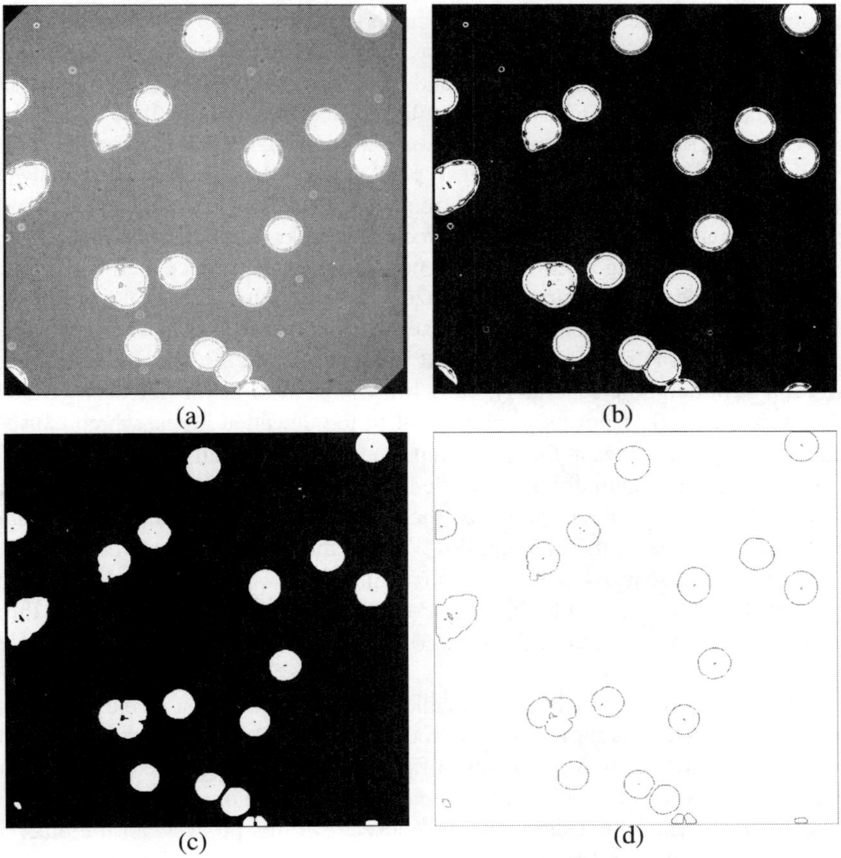

FIGURE 3.1. (a) Optical image of dewetted polystyrene film on a silicon substrate. Image width is 740 μm. (b) Thresholded black-and-white version of the original image in (a). (c) Image after three rounds of closure, followed by three rounds of opening to eliminate rim artifacts. (d) Skeletonized version of image (c), in which the edges of the holes have been detected by removing all black pixels except those bordering white areas. NIH Image software was used for all image processing.

shows a raw optical image of a dewetting poly(styrene) film on a silicon surface. It is necessary to extract from the image various parameters describing the extent of dewetting (e.g., the dewetted area fraction) and the statistical properties of the various hole sizes. In order to prevent a serious bottleneck in the combinatorial workflow, it is also necessary to develop an automated image analysis routine that can extract these parameters efficiently.

In order to extract measurable physical quantities (e.g., hole size) from the images (analysis), it is first necessary to process the gray-scale (or color) images to transform them into a format that can be easily analyzed. The first, and perhaps most fundamental, processing technique, which is almost always necessary prior to quantitative analysis, is thresholding. Thresholding, as the name suggests, involves setting a threshold gray-scale value (e.g., a number between 0 to 255) and dividing the image into white and black areas based upon whether each pixel value in the image is below or above the threshold. This transforms a gray-scale image into a true binary image, as illustrated in Figure 3.1(b), which is the result of thresholding Figure 3.1(a) using NIH Image [6]. The purpose of thresholding is to separate image features of interests (e.g., the dewetted holes in this case) from other background features. White areas in Figure 3.1(b) correspond to holes (high reflectivity of the bare silicon substrate) and black areas correspond to polystyrene. For automated analysis, the threshold value should be chosen automatically or set initially by the user. Common methods of choosing the threshold value are based upon the median or mean of the high (darkest) and low (lightest) pixel values. More sophisticated analyses will determine the histogram representing the frequency of each pixel value and then perform a weighted average or take the median between two peaks in the distribution.

However, thresholding alone is usually not sufficient to produce an image that can be analyzed accurately for hole or particle sizes and fractions. Thresholding relies upon the regions of interest having gray-scale pixel values that are significantly different than the rest of the image. For example, note that in Figure 3.1(b) the thresholded image does capture each dewetted hole accurately, but a rim can be seen just inside each of the holes. This rim results directly from the rims present in the dewetted holes, which cause slight variations in the pixel values at the hole circumference. The small circumference of white space outside the rim is actually not dewetted, but it is counted as dewetted in the thresholded image because the pixel values of the rim are too close to that of the undesired portion of the image (e.g., the hole interior). Hence the dewetted area would be overestimated if quantitative analysis of black versus white areas were performed at this stage. It is very difficult, if not impossible, in many dewetting images to choose a single threshold value that does not result in capture of the rim artifact and further processing is usually necessary.

Fortunately, robust methods are available for eliminating most of the effects of such thresholding artifacts. In order to ameliorate the problem of rim artifacts in Figure 3.1(b), we use the morphological transformations of *closing* and *opening*. These transformations, which are usually performed only on thresholded images, are one of the most basic forms of digital noise filtering. First the image is "closed". In this process, all the black image pixels are scanned and the first layer of neighboring white pixels (pixel value of zero), if present, are filled in with black. This operation has the effect of literally closing "holes" or white spaces in the image. Of course, the number of closures necessary for removing the rims is determined by half the width (in pixels) of the largest rim. After the closure operations the rims will no longer be visible, since they will be filled in with black pixels.

However, the dewetted holes, or the areas of interest, will also have been closed slightly. Hence it is now necessary to "open" the holes back to their original size by scanning the white pixels and removing the first layer of neighboring black pixels. After performing a number of openings equal to the number of closings, the dewetted hole size will be essentially equal to the interior hole size from the original gray-scale image. Figure 3.1(c) shows the closed and opened image resulting from Figure 3.1(a). Indeed, the rims have disappeared and the holes appear to retain their original size. However, there is a small uncertainty which is caused by the inevitable "digitization" that occurs during opening and closing. The number of uncertain pixels, which can be reported as a contributor to error in measurements, is equal to the number of closings (or openings) performed.

The assumption behind closing–opening as a noise-filtering routine is that the undesired or noisy artifacts are small enough that they can be closed out before significant error in the larger regions of interest are introduced. In addition, in the case opposite to the data presented in Figure 3.1, when dark regions of interest are surrounded by a white background, the process is reversed to removed dark noisy spots (e.g., opening is followed by closing). NIH Image [6], as well as most commercial image analysis packages, incorporate opening and closure as morphological transformations.

The thresholded, closed, and opened image in Figure 3.1(c) can now be used to quantify the relative area fraction of dewetted (white) versus wetted (black) surface area. The wetted fraction, for example, is measured by scanning the pixels and simply counting the number of black pixels and dividing by the total number of pixels in the image. Obviously, the remaining fraction is the dewetted portion. The other major property that usually must be determined is the size and number of holes (or particles in the case of black-on-white images). Essentially all robust image analysis packages incorporate automated subroutines for measuring the number and sizes (maximum and minimum radius) of holes or particles. Although the precise nature of these analysis algorithms is beyond the scope of this chapter, they are based upon identification of continuous "curves" of black pixels that define the perimeter of a hole or particle. Depending on the specific particle analysis algorithm, one may work from the image in Figure 3.1(c), or some analysis routines require that the edges of the holes (or particles) be clearly identified. Edge detection is another common feature of image analysis packages, and includes routines such as "skeletonization," the "watershed" procedure, and the Sobel edge-detection algorithm. An example of a skeletonized form of Figure 3.1(c) is given in Figure 3.1(d)

Imaging Packages

The power of combining automated image analysis with high-throughput experimentation is the ability to measure desired properties over large numbers of parameters in a short amount of time. In the combinatorial dewetting study mentioned above [4], the automated image analysis allows collection of dewetting area from over a large range of T and h values: (92°C $< T <$ 135°C) and (59 nm $< h <$ 86 nm). *Those data led to the discovery, for the first time, of a T, h superposition for dewetting rates.*

Each of the morphological transformations and analyses described above were performed with the freely available package NIH Image (for Macintosh) or Scion Image (for IBM-compatible machines). NIH Image incorporates a programming language that allows essentially all the NIH Image analysis and processing functions to be called automatically. In addition, image files can be opened and closed automatically, and data written to text files. The automated processing and analysis of hundreds of dewetting images, and the storage of data obtained from the images, can be carried out in a matter of minutes. For the more specialized user, details on the use of Matlab programs to analyze images of polymer dewetting can be found in a NIST Intra-Agency Report, NISTIR 6869 as Appendix A in the enclosed CD [7].

IMAGE ANALYSIS FOR HIGH-THROUGHPUT ADHESION MEASUREMENTS

JKR Adhesion Test Method

From biological implants to integrated chip technology, the interface between two contacting materials often defines the strength and reliability of the overall system. For this reason, the ability to properly characterize the interfacial strength, or adhesion, is critical to the success of many technologies. Various experimental methods have been developed to characterize adhesion for all material classes. A few common measurement methods for polymer interfaces include the peel test, the blister test, and probe-type tests [8,9]. The peel test is the most common adhesion test used in industry because of its simplicity, history, and direct analogy to many customer applications. Owing to the complicated stress distribution near the peel front, an absolute parameter that describes the adhesive strength can be difficult to determine [10].

In academia, methods such as the Johnson–Kendall–Roberts (JKR) method or the probe-tack test are often used to quantify the strength of polymer interfaces [8,11]. Although these tests are used in some industrial settings, the full quantitative collection of data is limited to academic laboratories. These data, including the applied displacement, resulting force, and contact area, are recorded throughout a typical experiment. Accordingly, these data are used to describe the formation and failure of the polymer interface through well-known fracture mechanics relationships [11,12]. Typically, this quantitative analysis is time consuming and the increase in understanding does not outweigh the inefficiency of using this process to screen numerous polymer formulations. For this reason, many industrial laboratories that use probe-type tests rely on the maximum applied force or the force history to describe the adhesive strength of the polymer interface. Although this approach can provide reliable data in certain cases, the understanding of the material system is not fully realized for future interfacial design.

High-Throughput MCAT Approach

To increase the efficiency of quantitative adhesion testing, we have recently introduced a high-throughput adhesion measurement technique. This technique builds upon the classical JKR theory [13] and uses the inherently parallel nature of optical imaging to map the adhesion of properly designed combinatorial libraries. The classical JKR theory describes the relationship between the applied force and resulting contact area for a single spherical lens contacting and separating from a flat substrate [13]. This relationship balances the elastic restoring forces and attractive adhesive forces between the two contacting bodies. Therefore, if the elastic material properties are known, the JKR expression provides a convenient means of quantifying the adhesive forces for a given interface. Although this measurement is absolute and quantitative, the classical approach can be time consuming.

To improve the efficiency of this method, we have introduced a new method called the multilens contact adhesion test (MCAT) [14,15]. For this technique, a two-dimensional array of spherical lenses is brought into contact with a complementary substrate. The array of lenses, the complementary substrate, or both can be made from the polymer of interest. During the contact and separation of this array, an automated microscope optically records the contact area of each lens while a displacement sensor monitors the relative displacement of the overall array. A movie of the MCAT adhesion test for a PDMS microlens array with a polystyrene strip is included on the CD. It shows the displacement–contact area history for the entire lens array (albeit at low magnification) in real time. With these two measurable quantities (contact area and displacement) and knowledge of the mechanical properties, the adhesion of the interface at the contact of each lens can be quantified during one test. As described in previous sections, if combinatorial processing techniques are utilized, every point in the two-dimensional array can probe the adhesion of a unique combination of conditions. Although the MCAT technique can produce sets of data equivalent to thousands of conventional adhesion tests within the same time required for one experiment (approximately 20 min), the primary challenge is processing the data and presenting it in a convenient and storable manner. For these purposes, we rely upon image-processing techniques to analyze and reformat the experimental results.

Imaging Challenges

During typical implementation of the MCAT method, optical images of the contacting array are recorded at regular time intervals by a CCD camera connected to a computer. A typical image of the entire array is shown in Figure 3.2. In a synchronized manner, the relative displacement of the two contacting bodies is recorded with an automated data acquisition system. The changes in the contact area for each lens are correlated with the displacement history for the entire array to analyze the data from a single test properly. Since a given array can contain over 1000 independent contact points and a typical MCAT test can acquire over 1000 images in sequence, automated analysis of the data is required to exploit the high-throughput experimentation that this method provides. In this section, we discuss two challenges in MCAT data analysis. The first challenge resides in isolating the contact areas of each lens, measuring the contact area, correlating this area to the appropriate displacement measurement, and subsequently tracking all changes in the data throughout the entire test. The second challenge involves developing convenient formats for storing and screening the analyzed results.

For the purposes of this discussion, we focus on a subset of a typical array used in the MCAT. The procedures discussed for this subset can be scaled up to larger arrays through simple programming steps that complicate the focus of this section. In Figure 3.3(a), three

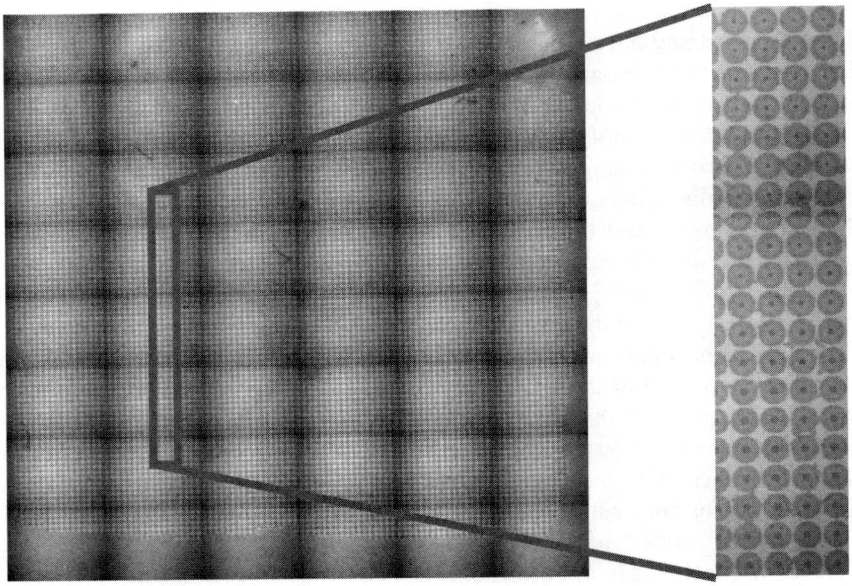

FIGURE 3.2. Image of entire array of microlenses in contact during MCAT experiment. The magnified image on the right allows individual microlenses and respective contact areas to be distinguished.

images of nine microlenses are shown in sequence. The first image in the sequence shows the microlenses before contact. In the second image, two microlenses have established contact as indicated by the slightly darker contact areas in the middle of the microlenses. For the third image, seven of the nine microlenses have contacted the underlying substrate. These three images are chosen to represent a typical progression of contact during an MCAT experiment. In total, 820 images were collected to record fully the entire process of interface formation and failure during this MCAT experiment. All 820 images are processed in an automated manner, as described below, to provide the highest dynamic resolution.

High-Throughput Image Analysis

The first goal in analyzing these results is to process the "as-recorded" images into a format where the contact areas for each microlens can be quantified and addressed according to its position within the array. As in previous discussions, to accomplish this task we use image-processing techniques that are provided in many commercially available image-processing software packages. For our analysis, we used National Instruments Labview Programming environment with the Vision software module to create a custom-designed processing routine. The processing begins by opening a background image which is a single image recorded shortly before any of the microlenses contact the complementary

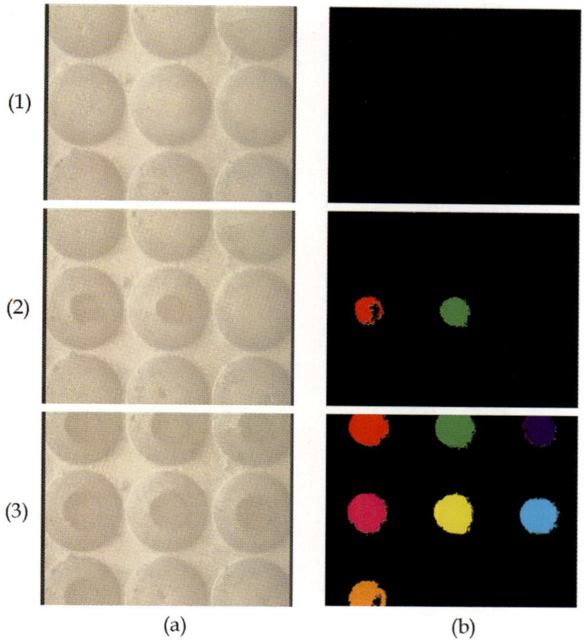

FIGURE 3.3. (a) Real images collected during the contact of nine polydimethylsiloxane microlenses with a glass slide. (b) Transformed binary images with false colors indicating that contact areas are individually labeled.

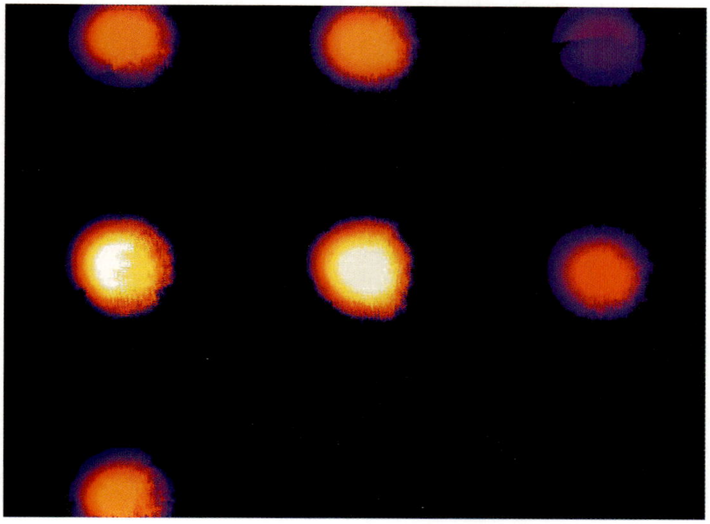

FIGURE 3.5. An example of a property map that can be used to facilitate screening and storage problems associated with the informatics step of combinatorial methods. Specifically, this image displays the contact history of seven polydimethylsiloxane microlenses contacting a glass slide. Color represents time in this image.

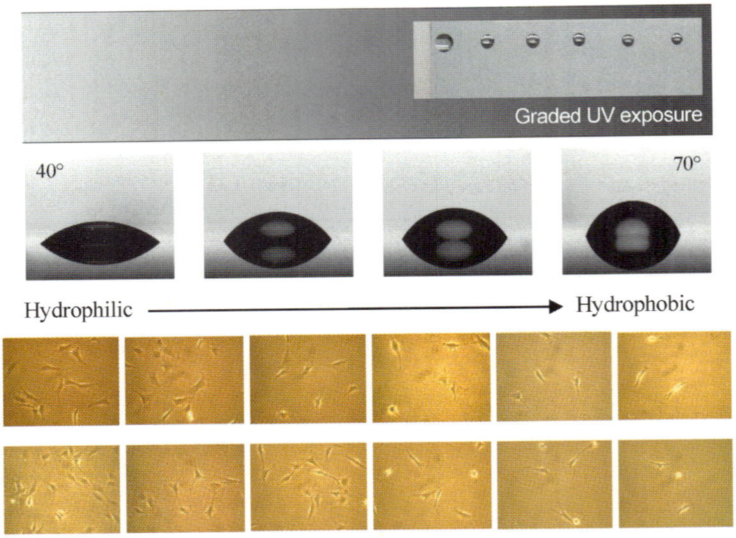

FIGURE 3.6. Cellular response on chemical gradient surfaces. The images represent the surface modification of PCL film by gradient UV-ozonolysis and response of osteoblast-like MC3T3-E1 cells. The gray-scale gradient (top) represents a linear variable neutral density filter providing a systematic change in UV dosage to the PCL film. The inset shows a digital photograph of droplets of water progressively changing from spread to beaded upon a portion of the PCL film. The film goes from hydrophilic to hydrophobic from left to right (denoted by the change in water contact angle from 40º to 70º). The osteoblast-like cell proliferation and spreading decreases dramatically with the increase in water contact angle.

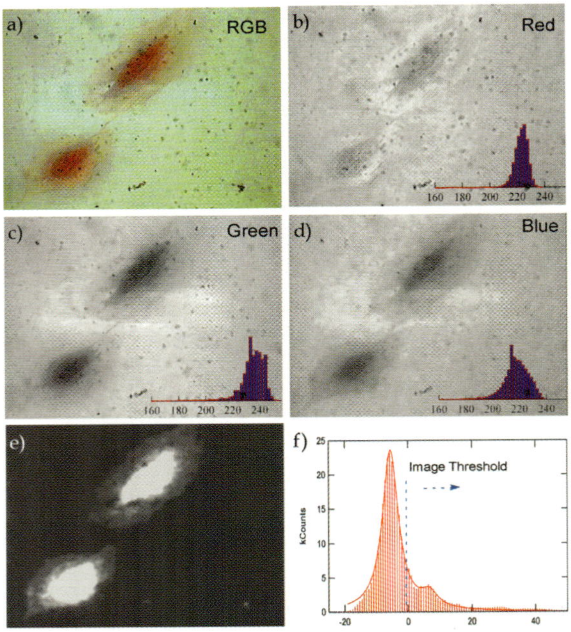

FIGURE 3.7. (a) RGB image of safranine O stained MC3T3-E1 cells on a smooth film. The image is resolved into its respective (b) red, (c) green, and (d) blue channels. Their respective image histograms (inset) show that the intensity of the blue (215.8±9.92 counts) and green (230.65±10.06 counts) channels has broader distributions about the mean than the red channel (221.6±5.02). The standard uncertainties are at 1σ about the mean. (e) The "red-hue" image obtained according to Eqn (2) is shown on the bottom left as a gray-scale image, clearly identifying the cellular regions from the noncellular background. (f) The image histogram of pixel intensities of this image shows sufficient contrast with a bimodal distribution. The arbitrarily chosen biomodal lorentzian is fit to the histogram (shown) with a standard uncertainty of 422 counts.

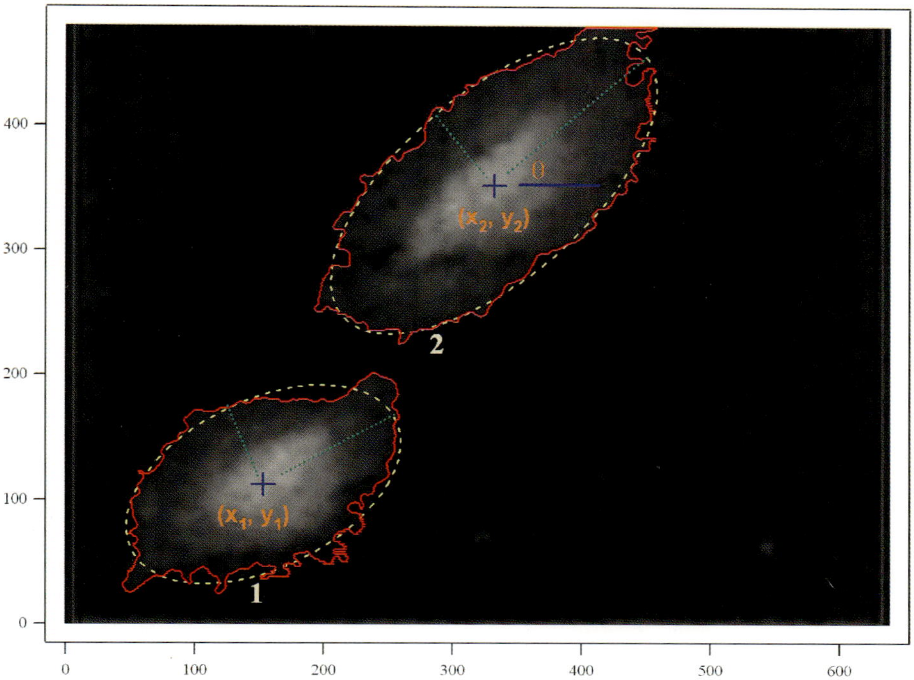

FIGURE 3.8. Particle analysis procedures for cell morphometry. Thresholding the "red-hue" image defines the cell perimeter marked in red. The centers of area (x_i,y_i) and the equivalent ellipse (dashed yellow line) for the cells are estimated from the moments analysis. The moment axis for orientation (θ) is defined as the angle between the major axis of the ellipse and the *x* axis.

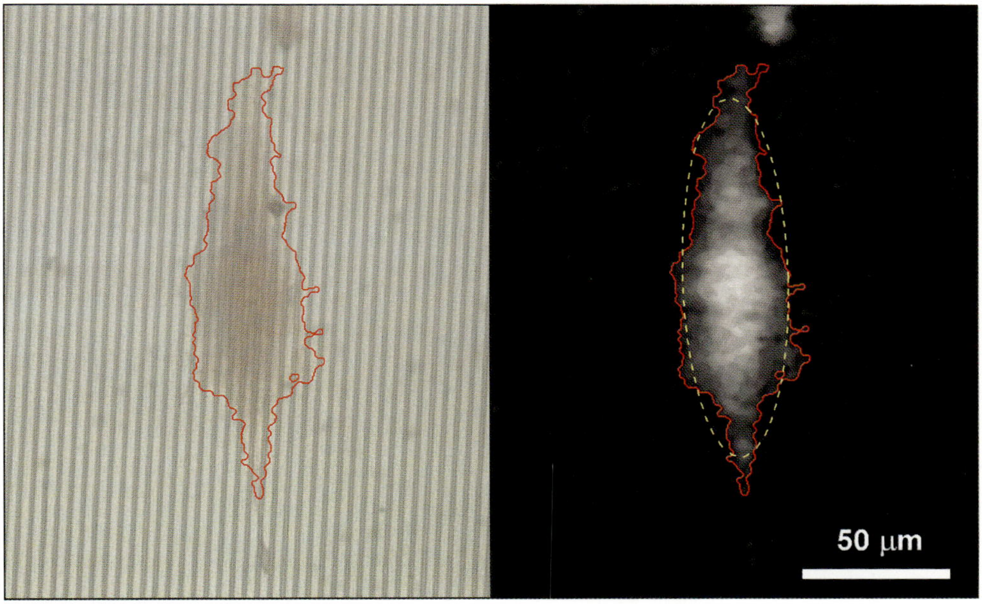

FIGURE 3.9. RGB image of a Safranine O stained MC3T3-E1 cell on an epoxy MIMIC with ridge groove topography (left). The "red-hue" image enables the background topography to be subtracted out, allowing image analysis. The overlaid cell perimeter (red) and the equivalent ellipse were estimated from particle analysis macros.

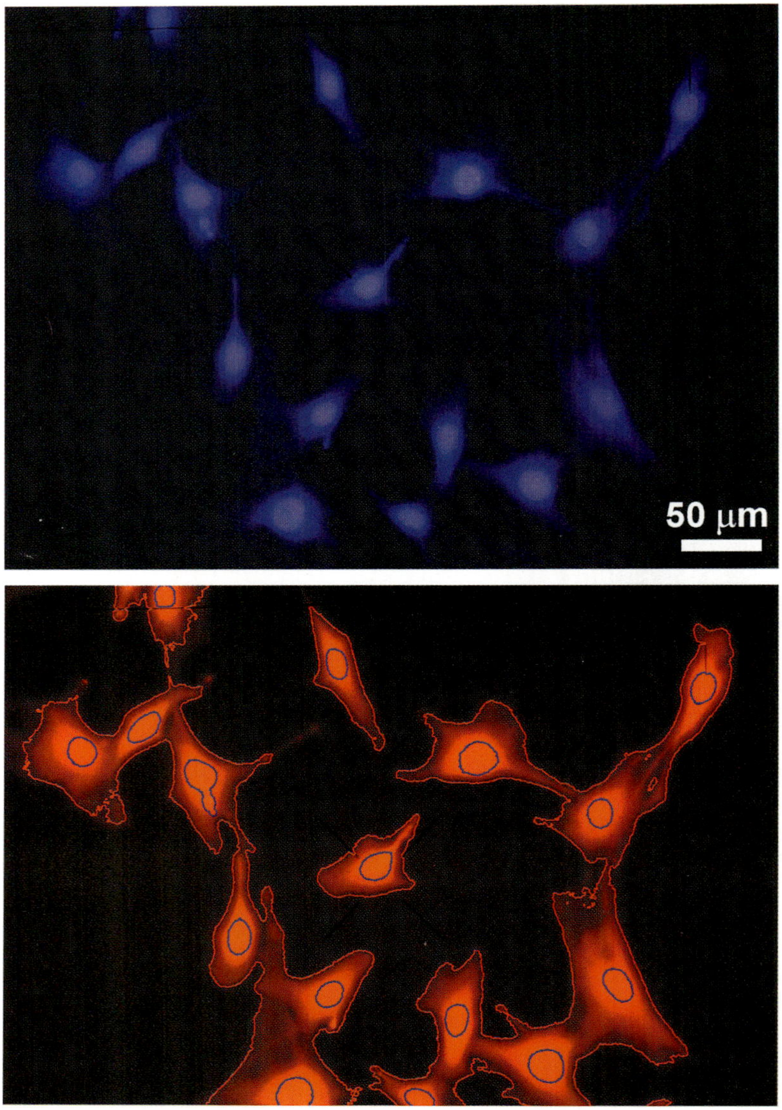

FIGURE 3.10. Fluorescence imaging of an MC3T3-E1 cell population on a test biomaterial. The double-staining procedure with DAPI (blue) and Texas red-C2-maleimide (red) stains allow the nuclear and cytoplasmic regions to be distinguished. The blue fluorescing DAPI stain in UV illumination is used to estimate cell proliferation from nuclear count per unit area (top). The red fluorescing C2 stain (bottom) allows cell-spreading areas to be calculated after nuclear correction through spatial correspondence. Thresholded nuclear and cytoplasmic perimeters estimated are overlaid in blue and red lines, respectively.

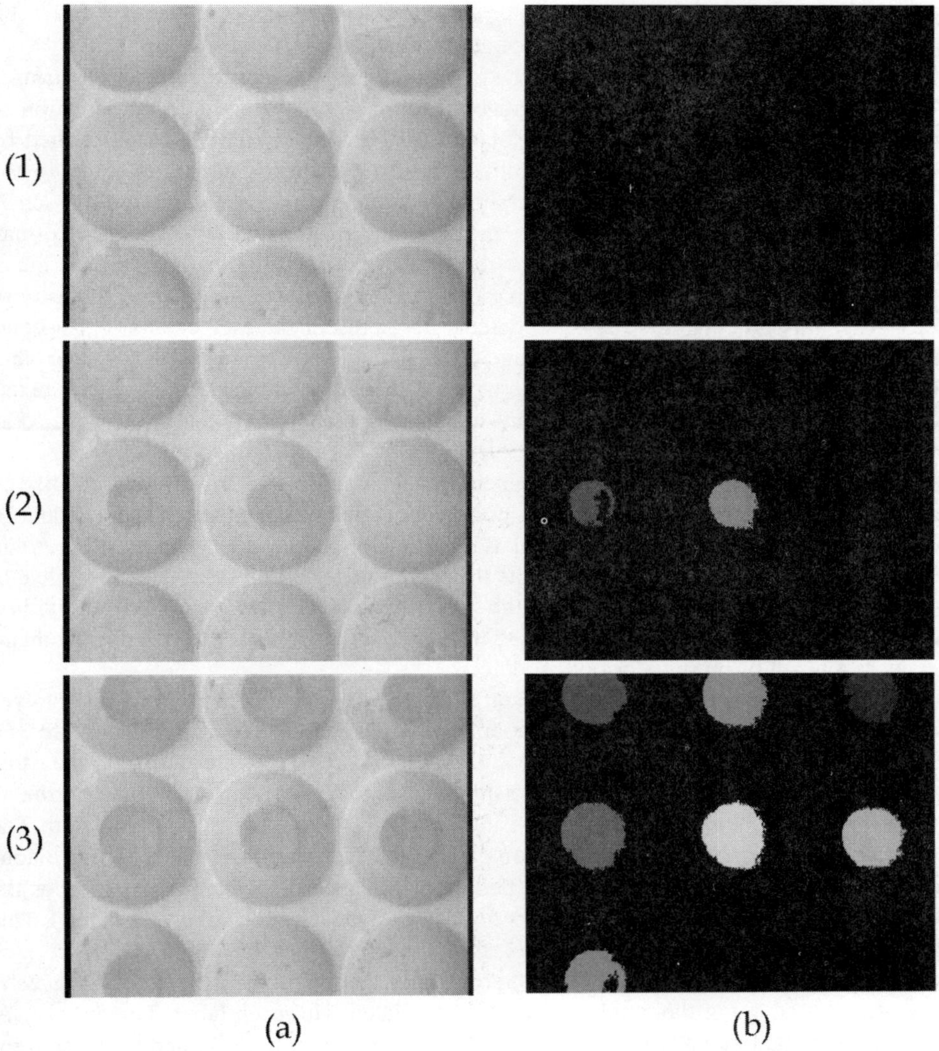

FIGURE 3.3. (a) Real images collected during the contact of nine polydimethylsiloxane microlenses with a glass slide. (b) Transformed binary images with false colors indicating that contact areas are individually labeled. A color version of this figure appears facing page 40.

substrate. We subtract this image from each of the images in the experimental image set. Specifically, the images are subtracted and the absolute value of the difference is recorded as a new image and stored as a temporary image on the hard drive. This process of taking the absolute difference allows us to correct for intensity variations associated with lighting and focusing difficulties that can vary from test to test. After taking the absolute difference, the maximum and minimum intensities over the entire image set are determined. These values establish the range of intensities that are used over the entire image set.

Accordingly, based on this range, we renormalize the entire set to use the full 8 bit intensity range and consequently enhance the contrast of our images.

By taking the absolute difference and renormalizing the entire data set based on a common intensity range, we have equalized the image set to facilitate the isolation of the contact areas for all images. The isolation of the contact areas is accomplished by applying a uniform two-point threshold, as described earlier in this chapter, to the entire image set. The values of this threshold will vary from test to test and are set by the user. As in any thresholding process, the resultant images will be in a binary format, where pixels with intensities between the threshold values will be set to a uniform intensity value of unity and pixels not meeting these criteria will be assigned an intensity of zero. After this processing step, some defects may present themselves and a "fill-hole" technique can be used to correct the image. The "fill-hole" technique essentially searches for all pixels that are set to zero and determines if they are completely surround by pixels with an intensity of unity. If any pixels are surrounded, then these pixels are reassigned an intensity of unity.

After converting every image to a binary image, a particle filter is used to identify all pixels that are connected through eight-point connectivity criteria. Each independent or non-connected group of pixels is labeled as a particle or contact area. The most important detail of this particle filter is that it limits the identification of an independent particle to particles that have diameters within the physical size constraints of our individual microlenses. This filtering process helps to minimize false positives that can present themselves as a result of physical or optical defects.

After all particles are identified on a single image, the next process involves assigning the particle an appropriate spatial label, quantifying the area of the particle, and correlating the area and label with the corresponding displacement and time data that were recorded independently. To assign a spatial label to a particle, an arbitrary coordinate system is established in the image and the centroid of the particle is calculated. The centroid values for each particle are compared with the centroid values from previous images. If a match is found, the current particle is assigned the same particle label as the particle on previous images. If a match is not found, a new particle label is assigned. This process yields a unique number, or label, that is used to track the contact areas of each microlens from image to image. After appropriately labeling the contact area, the area is quantified by counting the number of pixels associated with each label. This number of pixels is converted into a "real" quantity through the appropriate calibrated conversion for the optical system. This area and label are placed in a list with the corresponding displacement. The time of the image and the time of the displacement measurement are used to link the two measured quantities. After processing the entire image set, a data table including particle label, area, displacement, and time is configured and is ready for subsequent analysis.

A movie of transformed binary images of an MCAT test vizualizing the contact history of nine polydimethylsiloxane microlenses with a glass slide is included on the CD. The optical images are frame-by-frame processed through the particle analysis procedures described above to generate the binary images with arbitrary false colors indicating individually labeled contact areas. The time at contact is different for each lens. The displacement–contact area history in time for individual lenses may be distinctly identified in this image-processing step.

Visually Mapping Adhesion

In conventional probe-type adhesion tests, the data listed in the results table would be used to quantify the adhesion energy *G* as a function of contact radius *a* or contact velocity d*a*/d*t*. A representative adhesion energy chart is shown in Figure 3.4(a). This conventional display of adhesion data presents the second challenge that was introduced earlier: How should adhesion data from a high-throughput MCAT experiment be presented and stored? If a single chart like the one shown in Figure 3.4(b) was produced for an MCAT experiment, the required storage space would be unmanageable for most research efforts and it would be nearly impossible to screen these curves efficiently. To facilitate screening of MCAT results and to minimize the storage of data, we format the results into single images, or resultant maps. An example of this formatting is the creation of a "contact history" map to collapse an entire image sequence for one MCAT experiment into a single image. In other words, through image processing we use color or gray-scale intensities to represent a parameter. In this example, color is used to represent contact time. From these maps, technicians, engineers, scientists, or computer algorithms can efficiently search for "hot spots" in the MCAT data.

The first step in creating a "contact history" map involves creating a blank image in the image-processing software. This image, which begins as a blank template, will be referred to as the "composite" image. After initializing the "composite" image, the particle identification steps are used as described above to create a binary image where the contact areas are identified with pixels of intensity equal to unity. This binary image is then multiplied by the time interval between images. After multiplication, the resulting binary image should be added to the "composite" image. If this process is repeated for every image in the image sequence, the "composite" image will be a single image that tracks the

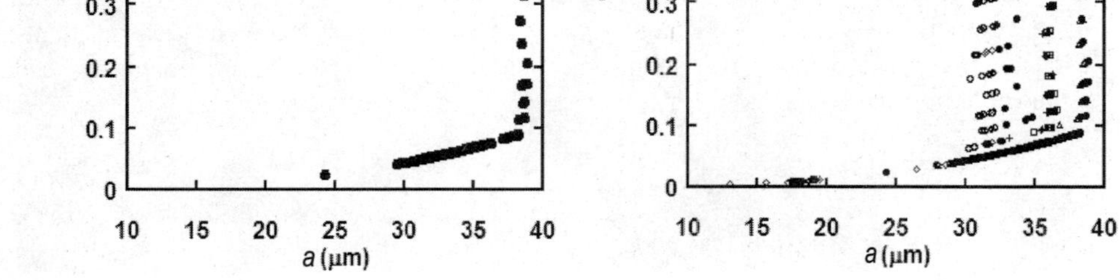

FIGURE 3.4. (a) Representative data of energy release rate *G* as a function of contact radius ***a*** for a single microlens contact sequence. (b) Seven energy release rate curves plotted on the same chart. Different data symbols distinguish data from different microlenses. The standard uncertainties in the measurements are less than the symbol size. This representation of MCAT data makes knowledge discovery difficult.

time history of the contact areas. The range of intensities in this composite image are assessed and renormalized to provide the greatest resolution and contrast in tracking the contact times. The result of this process is shown in Figure 3.5. Essentially, the brightest pixels (highest intensity) will be those pixels that were in contact the longest.

Although this example involves contact time, other properties such as displacement, energy release rate, and interfacial crack velocity can also be represented in this single-image/map format. These maps not only minimize storage space by transforming image sequences that can reach over 1 Gbyte into images of a few megabytes, but they also provide a convenient format for analyzing and screening the combinatorial data. These maps can be read by technicians, engineers, and scientists, or they can be screened by prewritten computer software. This format has been demonstrated to be conducive with high-throughput informatics in other combinatorial efforts, such as in the pharmaceutical industry.

In general, image processing not only permits an automated high-throughput collection and analysis of MCAT data, but, most importantly, it facilitates the proper formatting for subsequent storage and screening. These latter benefits provide the foundation for the informatics envelope that is critical to the success of combinatorial discovery methods.

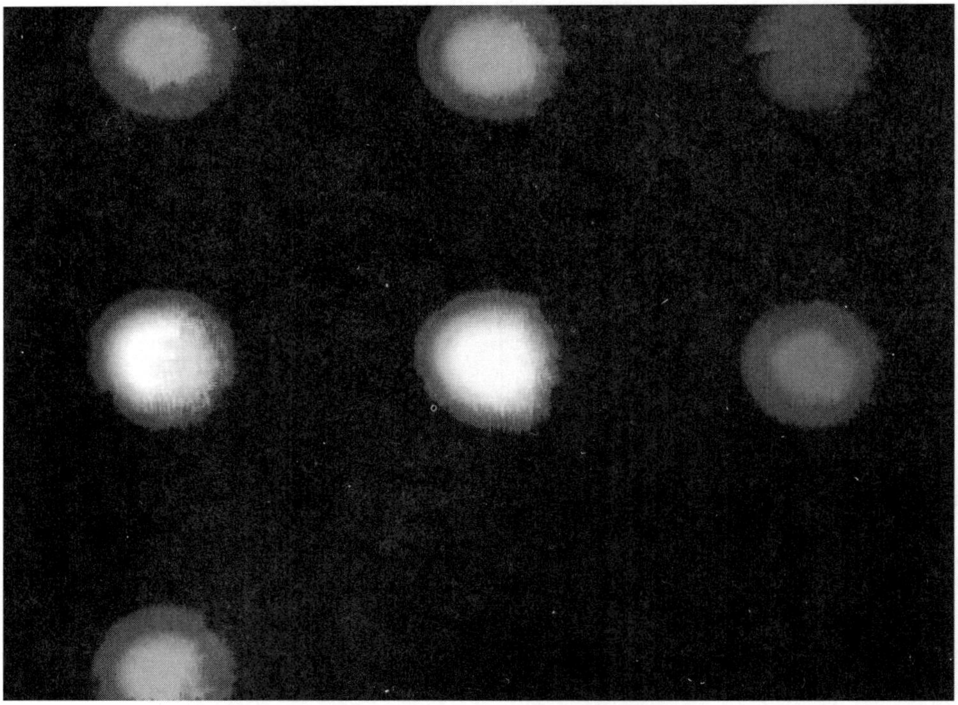

FIGURE 3.5. An example of a property map that can be used to facilitate screening and storage problems associated with the informatics step of combinatorial methods. Specifically, this image displays the contact history of seven polydimethylsiloxane microlenses contacting a glass slide. Color represents time in this image. A color version of this figure appears between pages 40–41.

MORPHOMETRIC ANALYSIS FOR HIGH-THROUGHPUT ASSAY OF CELLULAR RESPONSE

Cell–Materials Interactions

The new paradigm of "functional biomaterials" is inspired from nature. The level of complexity inherent in tissue is now increasingly being extended to the design of clinically relevant biomimetic materials. In tissue, the myriads of variables that influence cell growth and expression include soluble factors, nutrition, extent of vascularization, cell–cell interactions, and interaction with the extracellular matrix. Among all these, the extracellular matrix alone presents a heterogeneous complex environment composed of spatially distributed structural and functional components. Therefore this biomaterial interface must be designed to provide optimal biochemical [16], topographical [17], and mechanosensory [18] cues for guidance of cell adhesion, phenotypical expression, and proliferation [19]. The complexity and broad scope of these interactions suggests a need for a high-throughput investigative approach capable of mapping cellular "response functions" with stimuli from the bio-interface.

High-Throughput Test Templates

For conducting high-throughput investigations of cell–material interactions, test substrates need to be prepared with a systematic variation of the surface properties of biomaterials. A simple example for investigating the influence of surface chemistry is highlighted in Figure 3.6 where the hydrophilicity of a poly(ε-caprolactone) (PCL) film was systematically varied on a single test substrate across the film by graded UV-ozonolysis [20,21]. In another test platform described later, the effect of topographical constraint was investigated on microfabricated epoxy micromoldings in capillaries (MIMICs) [22]. The MIMICs were designed to present a range of periodic square-wave-section ridge–groove topographical patterns on a chip with no change in surface chemistry. Both these test chips (discussed in greater detail below) are model substrates for high-throughput investigation of how surface chemistry and topography independently influence cellular response. The test templates, either continuous gradients (surface energy gradients) or discrete patterned substrates (epoxy MIMICs), can then be defined in terms of a virtual array of test conditions at the bio-interface. The presentation of a range of stimuli to assay cellular response in one test is particularly significant for investigations in tissue engineering where random errors from one experiment to another (differences in cell seeding density, soluble factors, cell cycle, etc.) may be substantial. This allows for identification of subtle influences of the imposed test variable, rapid development of functional relationships, and expansion of the scope of the study to more dimensions such as cell type, serum concentration in the culture media, or cell–cell interactions. These are experimentally relevant but typically are too tedious to address in most studies. This would also allow for development of tissue-specific empirical models from in vitro high-throughput "test-chips," which are crucial for the application of biomaterials in tissue-engineered implants and biomedical devices.

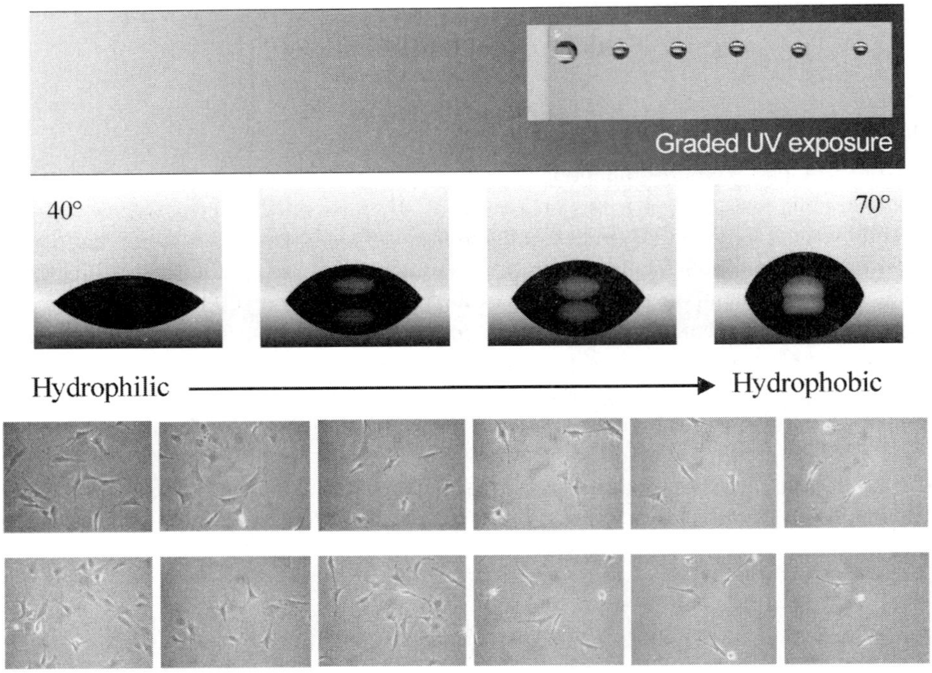

FIGURE 3.6. Cellular response on chemical gradient surfaces. The images represent the surface modification of PCL film by gradient UV-ozonolysis and response of osteoblast-like MC3T3-E1 cells. The gray-scale gradient (top) represents a linear variable neutral density filter providing a systematic change in UV dosage to the PCL film. The inset shows a digital photograph of droplets of water progressively changing from spread to beaded upon a portion of the PCL film. The film goes from hydrophilic to hydrophobic from left to right (denoted by the change in water contact angle from 40° to 70°). The osteoblast-like cell proliferation and spreading decreases dramatically with the increase in water contact angle. A color version of this figure appears between pages 40–41.

In order to correlate the systematic change in the surface properties with cell response, methodologies need to be developed for rapid measurement of these material properties as well as for quantitative determination of cellular response in terms of measurable parameters. Cellular response to this array of test stimuli is typically investigated by microscopic techniques ranging from the more rapid automated optical and fluorescence microscopy to programmable atomic force microscopy for three-dimensional imaging. All the techniques listed above measure parameters such as proliferation rates (cell count) or cell shape as descriptors for cell response. It is now established that expression of anchorage-dependent mammalian cells is mediated by cell shape [19,23–25]. Folkman and co-workers [24,26] quantified this correlation between cell spreading and DNA synthesis by bovine aortic endothelial cells with the thickness of poly(2-hydroxyethyl methacrylate) coatings. Therefore description of cell shape from images can be exploited as a measurable diagnostic tool for cell function. Cellular morphometric tools, developed for anatomic pathology [27], can be extended for prognosis of cellular reception on biomaterials. We describe image-analysis procedures and shape functionals that can be used as measurable descriptors of cell response to substrate libraries of material properties such as surface functionality and topography.

Chemical Gradient Surfaces and Cytomorphological Response Descriptors

PCL is a biodegradable semicrystalline aliphatic polyester that has been intensively studied as a non-toxic tissue-compatible material and used in relatively long-term implantable devices. It is used in biodegradable staples and sutures in Europe, in implantable contraceptive devices as Capronor, and in Duros drug-delivery platforms. However, the clinical procedure often involves UV sterilization of the device before implantation this standard procedure, though promoting clinical hygiene, can significantly alter cellular reception or protein absorption and enhance or compromise its function. The UV sterilization procedure was adapted for ozonolysis of PCL films in a gradient of UV radiation through a fused silica linear variable density filter for a few minutes [21]. A systematic change in the optical density of the filter (shown in Figure 3.6 as variation of gray scale) or the gradient in UV dosage directly translates into the progressive variation in the nominal concentration of ozone or atomic oxygen in proximity to the surface. The graded surface oxidation process results in surfaces with a systematic variation in surface chemical composition across the film. High-throughput time-of-flight secondary ion mass spectrometry (TOF-SIMS) was used to raster across the sample and provide information on the surface's chemical signature [20]. On this surface the photo-oxidation process results in cleavage of the labile ester linkage, causing an increase in the surface density of the –COOH end functional group. This increase in surface functionality is correlated with the hydrophilicity, as assessed by automated water contact angle measurements that progressively increase from 40° to 70°. The calibration of the substrate properties with respect to a spatial reference grid based on sample fiduciaries is the first step towards automated image acquisition and analysis.

The effect of surface treatment on the bioreceptivity of PCL was then assayed by in-vitro culture of osteoblast-like mouse calvarial MC3T3 cells. Cellular proliferation and spreading was monitored on a robotic *x*–*y* translation stage and an array of images as correlated with the spatial reference grid was obtained over several days. Figure 3.6 shows two panels of phase contrast micrographs of live cells in culture media. A clear trend is immediately observed with an increase in cell adhesion (population) and spreading with increasing philicity. The systematic variation of surface functionality over macroscopic distances (~ 5 cm) provides an ideal test substrate where the macroscopic distances and relatively low-motility cell line were chosen to preclude effects such as chemotaxis. Qualitatively, the extent of cell adhesion and cell spreading was found to correlate directly to the contact angle of the substrate. After 72 hr, the population of adherent cells decreased monotonically from 200 mm^{-2} to 25 mm^{-2} from the relatively hydrophilic end ($\theta H_2O \approx 40°$) to the hydrophobic end ($\theta H_2O \approx 70°$). The simple method for the creation of the test substrate allows for direct visualization of a qualitative trend as depicted in the images in Figure 3.6, which is followed by a more rigorous analysis.

Imaging a cell, which is a translucent cytoplasmic volume enclosed by a lipid membrane, has traditionally been achieved with specific and nonspecific stains that enhance optical contrast. The images of cells in Figure 3.6 highlight the typical problems encountered while imaging nonstained cells. Poor optical phase contrast through a polystyrene petri dish, nonuniform illumination (optical path through the media, petri dish, and condensation on the lid), and inability to subtract a background that is different for each element in an array are some of the concerns. One strategy to overcome these experimental difficulties is staining the cells with safranine-O (Aldrich), a deep red membrane stain that allows for visualization of the cell. The cells are fixated with 1% by mass fraction of glutaraldehyde and air dried prior to imaging. Sample data acquired as a color 640 × 480 pixel RGB image [Figure 3.7(a)] is mathematically treated as a three-dimensional matrix

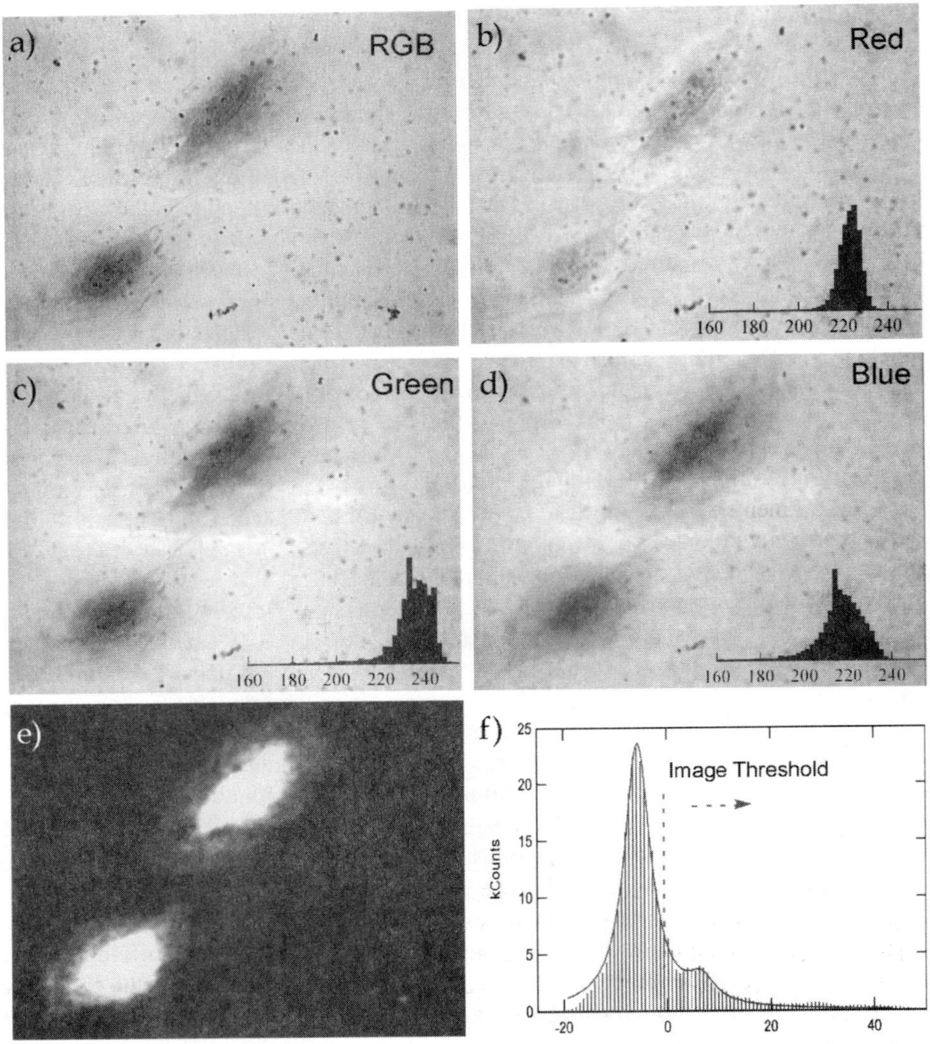

FIGURE 3.7. (a) RGB image of safranine O stained MC3T3-E1 cells on a smooth film. The image is resolved into its respective (b) red, (c) green, and (d) blue channels. Their respective image histograms (inset) show that the intensity of the blue (215.8 ± 9.92 counts) and green (230.65 ± 10.06 counts) channels has broader distributions about the mean than the red channel (221.6 ± 5.02). The standard uncertainties are at 1σ about the mean. (e) The "red-hue" image obtained according to Eqn (2) is shown on the bottom left as a gray-scale image, clearly identifying the cellular regions from the noncellular background. (f) The image histogram of pixel intensities of this image shows sufficient contrast with a bimodal distribution. The arbitrarily chosen biomodal lorentzian is fit to the histogram (shown) with a standard uncertainty of 422 counts. A color version of this figure appears between pages 40–41.

that can be resolved into the red [Figure 3.7(b)], green [Figure 3.7(c)], and blue [Figure 3.7(d)] channels as two-dimensional matrices:

$$\begin{aligned} \text{sampleRed}[] &= \text{sampleRGB}[x][y][0] \\ \text{sampleGreen}[] &= \text{sampleRGB}[x][y][1] \\ \text{sampleBlue}[] &= \text{sampleRGB}[x][y][2] \end{aligned} \tag{1}$$

Acquiring a gray-scale image or conversion of the RGB image into an 8 bit gray scale with average intensity from the red, green, and blue channels would result in loss of the "red stain" information whereby the observer's eye can discriminate the red regions of the cytoplasmic mass and the contiguous faint membranous spread. Thresholding of the gray-scale image would make discrimination based on contrast and quantitative analysis nearly impossible. The individual channels by themselves also show very little contrast with broad ill-defined unimodal image histograms of the frequency of pixel intensities (inset). In the RGB composite image, the "white" background has a high intensity *and* equivalent values in the red, green, and blue channels, while the dark spots again have equivalent values but low intensity of all three channels (red, green, and blue). The red channel has the least contrast, having a relatively invariant intensity throughout the image. The cellular regions that appear to have a red hue have a higher relative intensity of red than blue and green [dark in Figures 3.7(c) and 3.7(d)]. Therefore the specific red stain that can be discriminated by eye can be translated into a quantifiable "red hue" image [Figure 3.7(c)] by the simple transformation

$$\text{sampleHue}[] = \text{sampleRed}[] - (\text{sampleGreen}[] + \text{sampleBlue}[])/2 \tag{2}$$

The regions defined by the cell are immediately distinguishable from the background, while all the particulate debris on the substrate from the culture media and other imperfections in the image are simultaneously eliminated. This transformation fortuitously also allows for compensation of nonuniform illumination and effectively subtracts the "background." This is particularly useful in an array scan where the background may be different for each image in the array. The image histogram is better defined as a bimodal distribution that is robustly fit as a sum of two lorentzians. Convolution of the individual channels with a 3×3 gaussian smoothing filter was necessary for noise reduction from pixelation. The images can then be thresholded at the median of the two peaks as discussed before [Figure 3.7(f)]. This simple data reduction of the array of images, though performed using algorithms developed in IGOR, can be performed in most image-analysis packages. The procedure enables high-throughput assay for systematic quantification of poor-contrast images of cells, taking advantage of the specific colorimetric stain to identify the cytoplasmic boundaries from the background. The reduced images are then further processed through particle analysis algorithms to extract measurable physical parameters that are tabulated and output to text files as an array of descriptors of cell response.

The cytomorphological changes in the cell cycle [28] can be correlated with cellular response on a biomaterial surface. The cell response can be broadly characterized by processes such as cell adhesion, migration, spreading, proliferation, differentiation, and apoptosis. These processes can be put in context of the measures of merit extracted from the particle analysis procedures. The cell adhesion or initial cell attachment is directly measured as the cell count per unit area 6–8 hr before the onset of mitosis. For the surface energy gradient it may be directly correlated with the density and the favorable presentation of the ligands (e.g., arginine–glycine–aspartine or RGD peptide sequences) in absorbed proteins like fibronectin or vitronectin. Proliferation rates may be put in perspective of an evolving cell count. Cell attachment is typically followed by the growth of spike-like processes or filopodial extensions, cytoplasmic wedding, flattening of the cell mass, and ruffling morphology of the peripheral cytoplasm or lamellopodial spreading. Transmigrating cells have a spindle-like shape with directed filopodial extension (or oval cells with directed lamellopodial spreading at later stages), while the polygonal-shaped morphology with centrifugal filopodial extensions may be expected for adherent cells prior to spreading and division. Cell spreading and growth in size in the G_1 phase of the cell cycle is essential before DNA synthesis (S phase) and cell division (mitosis) can occur [28]. The extent of cell spreading, the filopodial extensions, and the aspect ratio of the cell are all important for the phenomenological development of the cell.

A schematic outline of the analyses employed is given in Figure 3.8. The edge-detection algorithms and particle analysis procedures in IGOR Pro [6] are used to determine the particle (cell) boundary (in red). All features of intensity below the threshold or "holes" within the cell perimeter are eliminated. As can been seen from the figure, the

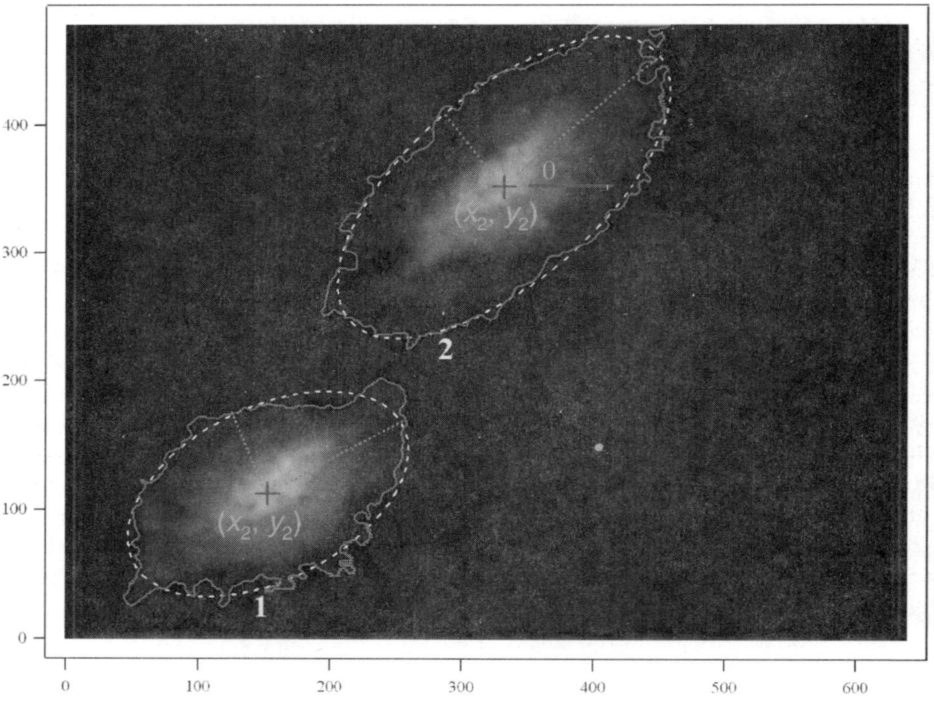

FIGURE 3.8. Particle analysis procedures for cell morphometry. Thresholding the "red-hue" image defines the cell perimeter marked in red. The centers of area (x_i, y_i) and the equivalent ellipse (dashed yellow line) for the cells are estimated from the moments analysis. The moment axis for orientation (θ) is defined as the angle between the major axis of the ellipse and the x axis. A color version of this figure appears between pages 40–41.

procedure is fairly accurate in determining the cell count n, the cell spreading area A, and the cell perimeter p. Defining constraints of minimal area (e.g., 5000 square pixels) allow for excluding artifacts and debris from the cell count. The thresholded binary mask image that is thus created from the particle analysis procedure defines the cell area. Moments analysis of this area is then used to calculate the center of area (x_0,y_0), and to fit an equivalent ellipse (shown as a yellow dashed line in Figure 3.8) with the same area as the cell. The center of area is calculated simply from

$$x_0 = \Sigma x_i / N$$

$$y_0 = \Sigma y_i / N \qquad (3)$$

The orientation or the moment axis through the center of mass for each cell is then estimated by minimizing the moment of inertia if the cell were to be rotated about this axis [29]. A brief description of the analytical development is included for the interested reader on the CD in Appendix B. The major (a) and the minor (b) axes of the equivalent ellipse (used to define the aspect ratio $\rho = a / b$) and the orientation of the major axis from the y axis (θ) are obtained from the fit. The calculation of equivalent ellipse as a shape functional is rigorous for estimation of aspect ratios of anisotropy in areal distribution of arbitrarily shaped cells. The analysis procedure for the cell area also allows for calculation of shape functionals such as the circularity [$\varphi = p^2/(4\pi A)$]. A perfect circle has $\varphi = 1$. Distortions in shape, such as filipodial extensions, only increase the perimeter with little change in area, giving circularities greater than unity. Other more sophisticated measures such as fractal dimension, lacunarity of the cell perimeter, etc. have also been performed for neuronal and glial cells [30].

Cells on Topographically Structured Surfaces

The robust nature of the procedure for contrast enhancement and background elimination of RGB images is further demonstrated by image analysis of cells on topographically patterned model surfaces. These test patterns present a range of scales of the square-wave-section-ridge–groove geometry on a single substrate. This allows systematic quantification of cytoplasmic distortion with scales of topographical constraint. A non-photolithographic processing technique by MIMIC was utilized for replicating the structured epoxy surfaces. The method first involves preparation of poly(dimethyl siloxane) (PDMS) stamps by casting vinyl terminated siloxanes (Dow Corning with Sylgard 184 base resin and curing agent) on photolithographically created silicon masters with the desired micropattern. The PDMS is cured at room temperature and then peeled of the master to yield negative molds/stamps of the pattern used subsequently for the MIMICs. The negative molds/stamps are designed with smooth regions and with multiple regions or bands with linear stripe patterns, the scale of which is varied systematically across the PDMS surface. Images acquired in an array on these patterns were processed using the procedures defined above. Clark et al. [31], and several other works in literature, have claimed analysis of optical images of cells on topographically structured surfaces to be very difficult or nearly impossible due to contributions from the substrate features. The use of a colorimetric stain and the subsequent processing of the color channels allows for

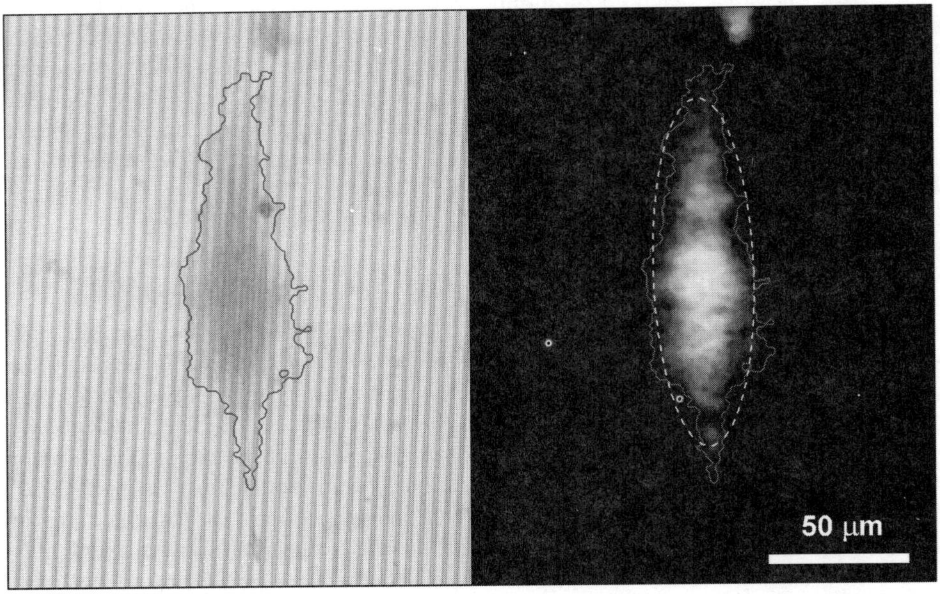

FIGURE 3.9. RGB image of a Safranine O stained MC3T3-E1 cell on an epoxy MIMIC with ridge groove topography (left). The "red-hue" image enables the background topography to be subtracted out, allowing image analysis. The overlaid cell perimeter (red) and the equivalent ellipse were estimated from particle analysis macros. A color version of this figure appears between pages 40–41.

subtracting out the background topography (Figure 3.9) and estimation of the cell shape functionals as described before. The orientation of the cytoplasmic mass, the change in aspect ratio, and the directional moments of areal spread with progressive topographical constraint are then easily estimated as descriptors of cell response.

Fluorescence Imaging Techniques

The image analysis protocols described above are at relatively low cell-seeding densities and high magnifications. The low seeding densities allow for characterization of shape functionals for individual cells. However, one of the key features of tissue development is cell–cell contact. Though the colorimetric stains do allow for a convenient way for self-consistent subtraction of the background and topography, they may not be able to account readily for cells with contiguous membranes. Specific fluorescent stains for cellular organelles may then be exploited for identification of individual cells. A schema for automated analysis using fluorescence microscopy has been outlined by Elliot and co-workers [32,33] where a DAPI fluorescent dye (4′,6-diamidino-2-diphenylindole dihydrochloride; Molecular Probes Inc.) is used to specifically stain the nuclei while a Texas red-C_2-maleimide ($C_{37}H_{36}N_4O_8S_2$; Cat. No. T-6008; Molecular Probes Inc.) cytoplasmic stain identifies the cell mass. Automated fluorescence microscopy then involves software control of the motorized stage position, autofocusing, switching of the excitation

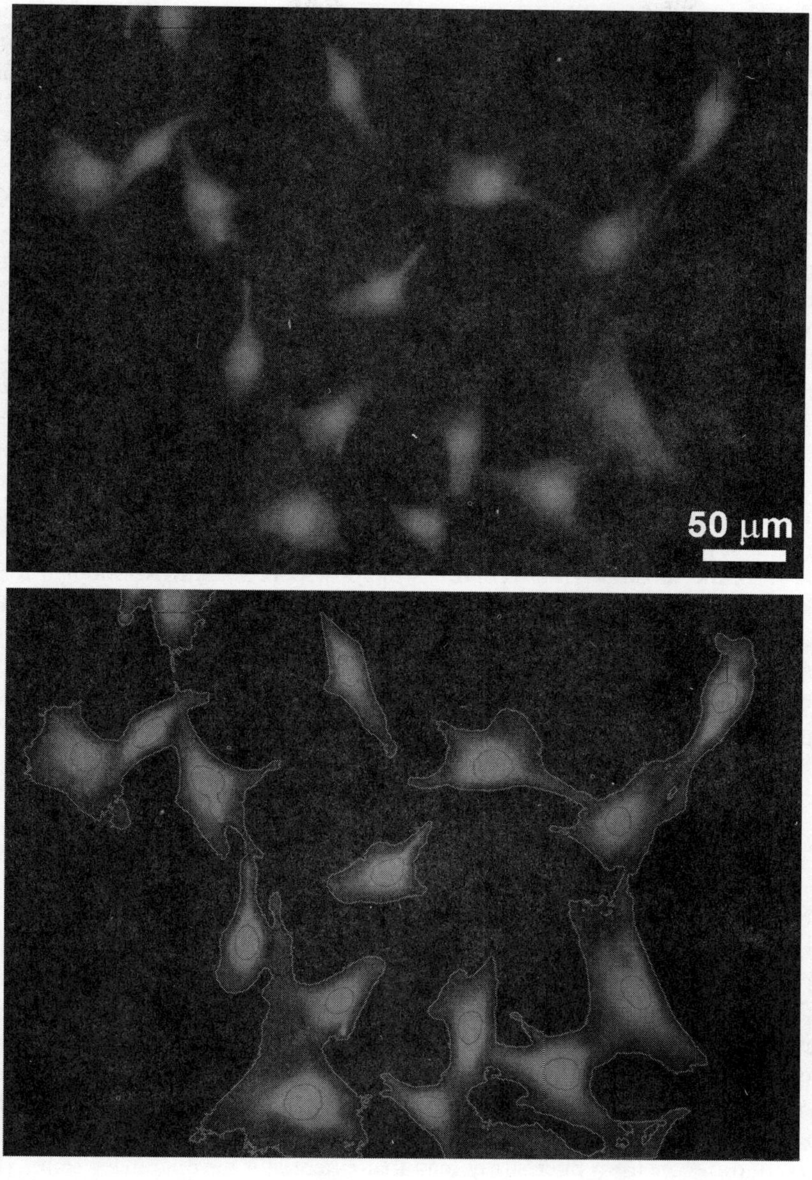

FIGURE 3.10. Fluorescence imaging of an MC3T3-E1 cell population on a test biomaterial. The double-staining procedure with DAPI (blue) and Texas red-C_2-maleimide (red) stains allow the nuclear and cytoplasmic regions to be distinguished. The blue fluorescing DAPI stain in UV illumination is used to estimate cell proliferation from nuclear count per unit area (left). The red fluorescing C_2 stain (right) allows cell-spreading areas to be calculated after nuclear correction through spatial correspondence. Thresholded nuclear and cytoplasmic perimeters estimated are overlaid in blue and red lines, respectively. A color version of this figure appears between pages 40–41.

filters and bandpass filter blocks to the appropriate wavelengths (UV excitation for DAPI—fluoresces in blue; green excitation for Texas Red stain—fluoresces in red). Figure 3.10 (top) highlights a typical image where the discrete nuclei (DAPI stain) are easily identified by particle analysis algorithms giving a cell count. Thus proliferation rates of cell may readily be generated on high-throughput test templates correlated with the substrate stimuli. Switching to the filter set for the Texas red stain (Figure 3.10, bottom) and simultaneous image acquisition in the red channel and thresholding allow for identification of the cellular regions from the noncellular background. However, "spatial correspondence" of the Texas red stained "particle" and a DAPI stained nucleus identifies an individual cell. If two or more nuclei are identified within an area outlined by the Texas red stain, further treatment is required. Elliott and co-workers outline a nuclear correction for cell spreading, whereby the cell-spreading area determined from particle analysis of the Texas red image is scaled by the number of nuclei with spatial correspondence. A more rigorous approach can be employed using the erosion algorithms described earlier to identify the cell–cell contact lines. The small but finite error associated with this analysis can also be estimated with ease.

The above description outlines methodologies to generate test substrates for high-throughput assays of cellular reception under various chemical and topographical stimuli provided from the substrate. It also outlines the protocols that can be adopted for image analysis, and some of the pitfalls commonly encountered in image analysis of cells on a biomaterial. The general protocols may be incorporated in simplified codes in various image analysis software platforms. The few simple areal distribution metrics and shape functionals introduced merely provide a framework for morphometric analysis to generate response functions as design inputs for tissue engineering constructs. Further levels of sophistication may then be adapted from these simple procedures.

CONCLUDING REMARKS

We introduced the theme of incorporating imaging as a useful high-throughput tool to investigate problems at the materials interface. Microarray imaging and data analysis [34,35] (or DNA chips) are presently being extensively exploited for detecting gene-expression patterns and in other bio-informatic high-throughput approaches. The development through this chapter highlights the distinction between imaging in materials science, and microarray-based imaging methodologies in screening for discrete hits. The imaging approach for materials explores multiparameter variable space allowing for measurement of properties that may take a finite range of values. These data may be considered to be qualitatively different from microarray data that essentially have a binary state. The data typically show gradual trends that give insight into fundamental phenomenology rather than discrete on–off responses in microarrays from genetic permutations and specific interactions. Only recently, methods in microarrays have been developed which measure absolute expression levels with a range of signal intensities [36].

In this chapter, we have illustrated a few select examples—dewetting of polymers, combinatorial adhesion, and cellular morphometrics—that dealt with the aspects of image collection, image processing, and analytical methodologies to define measurable descriptors. Some of the challenges encountered in image processing at the materials interface were discussed and simple solutions were offered. Similar challenges have traditionally been encountered with processing large volumes of image data in military reconnaissance [37] and in weather patterns. Sophisticated combinatorial algorithms for image-analysis problems such as edge detection, clustering, contrast enhancement, and target recognition developed for these applications may also be exploited for application to measurements in materials science. The measurable descriptors from the image analysis may be constituted into a relational or graph structure with the axes scaled to independent variables. This would generate an image matrix of measured properties. The adhesion measurements outline a visualization technique or generation of *property images* that would lend themselves to subsequent pattern-recognition algorithms. Once several of these higher-level property images have been generated for a host of test variables, thresholding criteria may be used to discretize the property maps based on performance criteria. The mathematical tools and algorithms that have been developed for microarrays would then be an invaluable resource for adaptation to problems in materials science. Generation of databases of measured material properties on this basis would lead to the role of "informatics" in high-throughput imaging. These databases would be amenable to data mining and local search and genetic algorithms such as simulated annealing and neural nets. The promise of "informatics" is to enhance the scientific process through better image documentation and handling, facilitate faster and more sophisticated data analysis, and provide feedback to refine subsequent library design. Indeed, the "combinatorial or high-throughput loop" in which imaging plays an important role cannot be closed without the feedback mechanisms of informatic analysis.

ACKNOWLEDGMENTS

A.K. and A.S. would like to acknowledge Professor Paula Hammond for assistance with microcontact printing. A.S. is also grateful to Sonya V. Roberson for the SIMS characterization of the PCL films, and to Newell R. Washburn and John T. Elliott for informative discussions on fluorescent labeling and visualization of cells.

REFERENCES AND NOTES

1. Brochard-Wyart, F., Redon, C., Sykes, C. Dewetting of ultrathin liquid-films. *C. R. Acad. Sci. Ser. II* **1992**, *314* (1), 19–24.
2. Sharma, A., Reiter, G. Instability of thin polymer films on coated substrates: Rupture, dewetting, and drop formation. *J. Colloid Interface Sci.* **1996**, *178* (2), 383–399.
3. Sharma, A., Khanna, R. Pattern formation in unstable thin liquid films under the influence of antagonistic short- and long-range forces. *J. Chem. Phys.* **1999**, *110* (10), 4929–4936.
4. Meredith, J. C., et al. Combinatorial materials science for polymer thin-film dewetting. *Macromolecules* **2000**, *33* (26), 9747–9756.
5. Wyart, F. B., Martin, P. Redon, C. Liquid–liquid dewetting. *Langmuir* **1993**, *9* (12), 3682–3690.
6. Certain equipment and instruments or materials are identified in the paper to adequately specify the experimental details. Such identification does not imply that recommendation by the National Institute of Standards and Technology, nor does it imply that the materials are necessarily the best available for the purpose.
7. Beichl, I., Bernstein, J. Karim, A. (Eds.) Automated image processing tools for high-throughput measurements of polymer coatings; initial report of software to quantify features, NIST Intra-Agency Reports, Vol. 6869, **2001**, Gaithersburg; MD: National Institute of Standards and Technology.

8. Creton, C. In R. W. Cahn, P. Haasen, and E. J. Kramer (Eds.), *Materials science and technology: a comprehensive treatment*, **1997**, Weinheim, Germany: VCH Verlagsgesellschaft, pp. 708–740.
9. Creton, C. and Papon, E. Guest Editors MRS Bulletin, "Materials science of adhesives: how to bond things together", Volume 28, No. 6, page 419, June **2003**.
10. Kaelble, D. H. Theory and analysis of peel adhesion: Adhesive thickness effects. *J. Adhes.* **1992**, *37*, 205–214.
11. Shull, K. R. et al. Axisymmetric adhesion tests of soft materials. *Macromol. Chem. Phys.* **1998**, *199* (4), 489–511.
12. Barquins, M., Maugis, D. Tackiness of elastomers. *J. Adhes.* **1981**, *13*, 53–65.
13. Johnson, K. L., Kendall, K., Roberts, A. D. Surface energy and the contact of elastic solids. *Proc. R. Soc. London, Ser. A* **1971**, *324*, 301–313.
14. Crosby, A. J., Karim, A., Amis, E. J. Combinatorial investigations of polymer adhesion. *Abstr. Pap. Am. Chem. Soc.* **2001**, *222*, 450-POLY.
15. Crosby, A. J., Karim, A., Amis, E. J. Combinatorial investigations of interfacial failure. *J. Polym. Sci. Part B, Poly.* **2003**, *41*, 883–891.
16. Geiger, B., et al. Transmembrane extracellular matrix-cytoskeleton crosstalk. *Nature Rev. Mol. Cell Biol.* **2001**, *2* (11), 793–805.
17. Flemming, R. G., et al. Effects of synthetic micro- and nano-structured surfaces on cell behavior. *Biomaterials* **1999**, *20* (6), 573–588.
18. Balaban, N. Q., et al. Force and focal adhesion assembly: a close relationship studied using elastic micropatterned substrates. *Nature Cell Biol.* **2001**, *3* (5), 466–472.
19. Hoch, H. C., et al. Signaling for growth orientation and cell-differentiation by surface-topography in *Uromyces. Science* **1987**, *235* (4796), 1659–1662.
20. Roberson, S. V., et al. Time of flight secondary ion mass spectrometry (ToF-SIMS) for high-throughput characterization of biosurfaces. *Appl. Surf. Sci.* **2003**, *203–204*, 855–858.
21. Roberson, S. V., et al. Multifunctional ToF-SIMS: Combinatorial mapping of gradient energy substrates. *Appl. Surf. Sci.* **2002**, *200*, 150–164.
22. Qin, D. et al. In A. Manz and H. Becker (Eds.), Microsystem Technology in Chemistry and Life Sciences, Vol. 194, **1998**, Berlin: Springer-Verlag, pp. 1–20.
23. Glowacki, J., Trepman, E., Folkman, J. Cell-shape and phenotypic-expression in chondrocytes. *Proc. Soc. Exp. Biol. Med.* **1983**, *172* (1), 93–98.
24. Folkman, J., Moscona, A. Role of cell-shape in growth-control. *Nature* **1978**, *273* (5661), 345–349.
25. Broutyboye, D., Tucker, R. W. Folkman, J. Transformed and neoplastic phenotype–reversibility during culture by cell-density and cell-shape. *Intl. J. Cancer* **1980**, *26* (4), 501–507.
26. Folkman, J., Greenspan, H. P. Influence of geometry on control of cell-growth. *Biochim. Biophys. Acta* **1975**, *417* (3–4), 211–236.
27. True, L. D. Morphometric applications in anatomic pathology. *Hum. Pathol.* **1996**, *27* (5), 450–467.
28. Alberts, B., et al. Molecular biology of the cell, 3rd ed, **1994**, New York: Garland Publishing.
29. Russ, J. C. The image processing handbook, 2nd ed, **1994**, Boca Raton, FL: CRC Press, p. 674.
30. Smith, T. G., Jr, Lange, G. D., Marks, W. B. Fractal methods and results in cellular morphology—dimensions, lacunarity and multifractals. *J. Neurosci. Methods* **1996**, *69*, 123–136.
31. Clark, P., et al. Topographical control of cell behavior .2. Multiple grooved substrata. *Development* **1990**, *108* (4), 635–644.
32. Elliott, J.T. et al. Thin films of collagen affect smooth muscle cell morphology. *Langmuir* **2003**, *19* (5), 1506–1514.
33. Mei, Y., et al. Biocompatibility of sorbitol containing polyesters. I. Synthesis, surface analysis and cell response in vitro. *Biomaterials* **2002**, submitted for publication.
34. Gershon, D. Microarray technology: An array of opportunities. *Nature* **2002**, *416*, 885–891.
35. Dopazo, J. Microarray data processing and analysis. In S. M. Lin and K. F. Johnson (Eds.), Microarray Data Analysis, **2002**, Boston, MA: Kluwer Academic, pp. 43–63.
36. Dudley, A. M., et al. Measuring absolute expression with microarrays with a calibrated sample and extended signal intensity range. *Proc. Natl Acad. Sci.* **2002**, *99* (11), 7554–7559.
37. Toet, A. (Ed.). Combinatorial optimization and image analysis: a Literature Survey, **1995**, CALMA—Combinatorial Algorithms for Military Applications, http://citeseer.nj.nec.com/toet95combinatorial.html

Chapter 4

Laser-Induced Fluorescence Detection in High-Throughput Screening of Heterogeneous Catalysts

Hui Su and Edward S. Yeung

INTRODUCTION

In the discovery of new materials with useful properties, combinatorial chemistry has established itself as one of the most important methodologies because it is capable of speeding up the discovery process hundreds to thousands of times. The application of combinatorial chemistry in solid state heterogeneous catalysis is a relatively young research area compared with its application in the pharmaceutical industry. There are three aspects of equal importance in solid state combinatorial chemistry: (1) combinatorial synthesis, (2) high-throughput screening, and (3) effective compositional analysis of library components. Success has been achieved in the combinatorial synthesis of a solid state catalyst library by thin-film deposition and liquid-dosing techniques. However, more effort is needed in developing high-throughput screening techniques as well as fast and effective approaches for compositional analysis of solid catalysts in the libraries.

We introduce laser-induced fluorescence imaging (LIFI) as a high-throughput screening technique for heterogeneous catalysts. The scheme is based on the fact that the creation or destruction of chemical bonds alters the fluorescence properties of suitably designed molecules. By irradiating the region immediately above the catalytic surface with a laser, the fluorescence intensity of a selected product or reactant can be imaged by a charge-coupled device (CCD) camera to follow the catalytic activity as a function of time and space. By monitoring the catalytic activity of vanadium pentoxide catalysts in situ for the oxidation of naphthalene, we show that LIFI has good detection performance and excellent spatial and temporal resolution for high-throughput screening. The sample packing density can reach 250×250 subunits/cm^2 for 40 μm wells. The same experimental

Hui Su and Edward S. Yeung • Ames Laboratory-USDOE and Department of Chemistry, Iowa State University, Ames, IA 50011, USA
Contact author: Edward S. Yeung. email: yeung@ameslab.gov

High-Throughput Analysis, edited by Potyrailo and Amis
Kluwer Academic Publishers, New York, 2003.

setup can simultaneously screen solid catalysts via near-IR thermography detection to provide even more details about the reactions.

We further applied the above imaging approach to the combinatorial discovery of heterogeneous catalysts and the optimization of reaction conditions for the oxidation of naphthalene and the acylation of aromatics. For naphthalene oxidation, we tested the catalytic activities of libraries of various combinations of binary and ternary vanadium-based catalysts. Sample preparation methods and reaction temperatures were screened for optimized catalytic performance. The promotional effects of different oxides—Mo(VI), Sn(V), Ti(V), and W(VI)—were revealed in a high-throughput fashion through the combined information obtained by fluorescence and IR imaging. Several combinations with superior catalytic performance compared with pure vanadium pentoxide were discovered. In the combinatorial study of the catalytic activities of zeolites in the acylation of aromatics, the effects of temperature, Si/Al ratios, and the structure of zeolites on their catalytic performance were studied.

COMBINATORIAL CHEMISTRY IN SOLID STATE MATERIALS DISCOVERY

Overview

Discovering a new material is always exciting, but the cost of time and resources is often a constraint. The traditional approach of discovery is a linear strategy: synthesize a new material with one method, and then test this material for a certain property, again and again, until one material is found with the desired property. Recently, numerous efforts have been devoted to accelerating the discovery of new materials, and combinatorial chemistry is the most focused and successful approach. An ideal combinatorial solution to discovery is to identify the target compound(s) by simultaneously testing the specific property of a large number of compounds with related compositions which are synthesized at the same time. The implementation of combinatorial methodology can make the discovery process hundreds, or even thousands, of times more efficient.

The concept of combinatorial chemistry can be traced back to 1963 with Merrifield's solid state synthesis of peptides [1], and it has achieved enormous success in drug discovery. The combinatorial discovery of solid state materials and its applications outside the pharmaceutical industry came much later, but its impact is as promising as that in drug discovery. This is because solid state materials with useful properties, such as superconductors, magnetic materials, light-emitting materials, and catalysts, are in high demand in a large variety of new technologies and have huge economical potential.

The Forerunner: "Multiple-Sample Concept"

In 1970, Hanak proposed a new concept of material research, the "multiple-sample concept," after he became increasingly impatient with the traditional methodology of material research [2]. Hanak introduced an approach that consisted of synthesizing, analyzing, testing, and evaluating multicomponent systems in integrated steps. A radio-frequency cosputtering technique was adopted and modified to codeposit two or three elements on a substrate to provide a continuum of compositions. Then, the compositions were determined by computing the thickness of a few spots on the sputtered thin film. Many new superconductors were found using this approach [3]. Hanak's approach resulted in an increase of up to 100-fold in the rate of discovering new materials. Hanak's idea was the forerunner of a concept that is now taking material science by storm—the combinatorial approach. But, because of the lack of computers in the 1970s, his idea did not become popular since compositional analysis, automated testing, and data processing cannot be done without a computer.

First Success: Thin-Film Deposition

More than 20 years after Hanak's "multiple-sample concept," inspired by the success of combinatorial chemistry in biology, materials scientists once again tackled the traditional approach of materials discovery with much more sophisticated modern technologies. Schultz and Xiang decided to extend the success of the combinatorial approach from organic and biological molecules to the remainder of the periodic table. For the first time they demonstrated that the combinatorial approach could be used to discover solid state materials with novel properties [4].

Schultz and his co-workers developed a methodology of thin-film deposition which allowed the parallel synthesis of spatially addressable arrays containing superconducting thin film. By sequentially depositing the individual precursors of interest on a suitable substrate through a series of physical masks, a spatially defined library of solid thin films could be generated. In their study, polished MgO or $LaAlO_3$ single crystals were used as substrates, and CuO, Bi_2O_3, CaO, PbO, $SrCO_3$, Y_2O_3, and $BaCO_3$ were used as precursors to make superconductive cuprates. After a series of depositions, the library array was thermally processed to produce crystalline cuprates. Then, the resistance at each site in the array was measured as a function of temperature. The sites containing the known superconductors showed large resistance drops, which were indicative of superconductivity. Superconductivity was found in thin films containing $BiCuCaSrO_x$ and $BiPbCuCaSrO_x$ in a 16-member binary library, which is consistent with what had been found by conventional one-at-a-time synthetic methods. The effects of deposition sequence and elemental composition were evaluated in a 128-member library.

The work of Xiang and Schultz showed that relatively complex solid state materials with useful properties could be made and tested in a parallel fashion, and the discovery process could be accelerated by a factor of thousands. Soon, other reports began appearing in the literature, in which combinatorial methods were used to search for a wide variety of materials including magnetoresistive materials [5], luminescent materials [6–9], and catalysts [10–13].

Solution-Phase Synthesis of Solid State Materials Libraries

Although thin-film deposition has achieved stunning success in the combinatorial discovery of solid state materials, it faces the challenging problem of the bulk synthesis of newly discovered compounds. In addition, for many materials, solid state synthesis may not be best accomplished by thin-film deposition. Solution-phase synthesis, such as sol–gel and hydrothermal synthesis, polymerization, and precipitation reaction, is free of the above problems. With the combination of solution-phase synthesis and automated ink-jet dosing system, sun and co-workers [14] prepared a microscale library of new luminescent materials from aqueous solutions of lanthanum, europium, gadolinium, and aluminum nitrates. Volumes as small as 0.5 nL could be dosed with high precision and spatial resolution. A similar ink-jet technique was also applied in the first combinatorial discovery of a new heterogeneous electrocatalysts for applications in fuel cells [12]. Mallouk and co-workers projected the three-dimensional phase diagrams of platinum, osmium, rhodium, palladium, and iridium onto planar libraries and investigated the activity of the components for methanol decomposition by use of an H^+-sensitive fluorescence indicator.

HIGH-THROUGHPUT SCREENING OF HETEROGENEOUS CATALYSTS

Overview

Catalytic gas processing is a multimillion dollar concern in the chemical industry. A minor increase in the yield, a minor decrease in undesirable by-products, the creation of novel products, or an extension of the lifetime of the catalyst are all urgent issues. An efficient approach towards understanding these complex systems is through combinatorial assays. In situ screening of these arrays under continuous flow in a microreactor system will quickly reveal the best combination of catalyst and reaction conditions.

The synthesis of a large number of solid materials is just the first step in the combinatorial discovery of solid state materials. A high-throughput screening technique is needed to test the properties of individual materials in a solid state library with the same efficiency as the synthetic process. For solid state materials, such as luminescent materials, superconductors, and magnetoresistive materials, fast screening does not seem to be a problem since only the specific physical properties of the solid materials are accessed. Superconductivity and magnetoresistivity can be tested by scanning probes, and luminescence can even be tested by the human eye. For solid heterogeneous catalysts, the situation is very different since catalytic activity is the screening target. Also, most of the heterogeneous catalytic reactions are in the gas phase, which makes high throughput even more difficult. The search for catalysts using combinatorial approaches has been under way for several years. Scientists have devised several methods for screening solid catalysts libraries, and each has its own advantages and disadvantages.

IR Thermography

In 1996 Willson and co-workers [15] reported IR thermographic imaging of combinatorial libraries, which was the first example of a highly parallel approach to the screening of heterogeneous oxidation catalysts. Given that catalytic reactions are exothermic, active catalysts reveal themselves as "hot spots" in IR images, which is the basis of IR

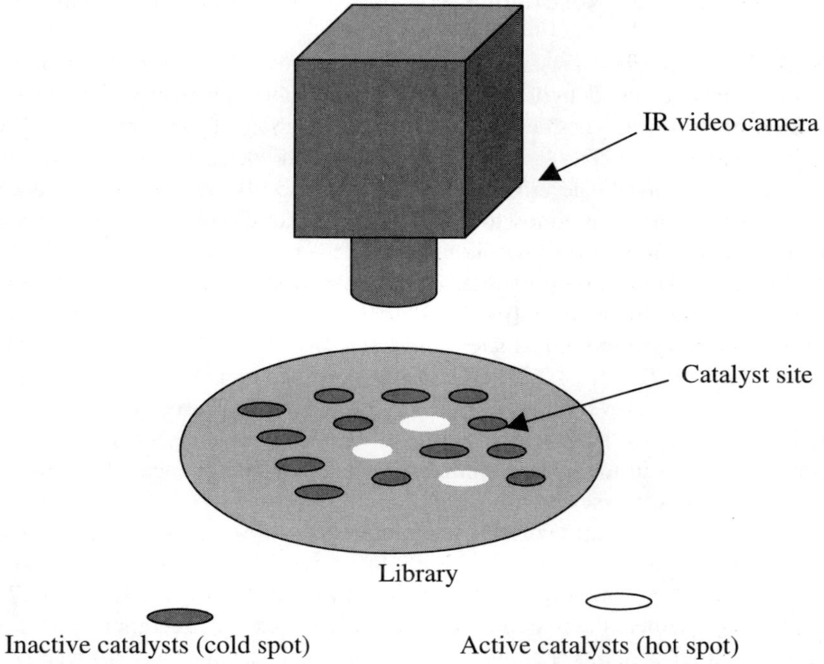

FIGURE 4.1. IR thermography screening of heterogeneous catalysts (adapted from ref. 15).

thermography (Figure 4.1). Catalysts on the solid state library can be tested simultaneously and in situ by this approach. In Willson's study, combinatorial screening of catalysts for hydrogen oxidation was described. The catalysts were placed in a reactor in predefined locations. Then they were scanned with an IR imaging camera having a temperature sensitivity of 0.025°C. Initial measurements of IR emission were made in the absence of reaction. Iridium, palladium, and platinum were found to be catalytically active in this reaction starting at 80°C, whereas the other candidates remained inactive up to 300°C.

IR thermography was also applied to the screening of potential heterogeneous catalysts by Maier and co-workers [16]. They examined different compositions of amorphous microporous mixed-metal oxides based on silica and titania. Microliter amounts of precursor solutions were pipetted into tiny wells on the surface of a slate substrate. Then the solutions were thermally processed to yield an array of solid samples. In their study, Maier and co-workers assessed a library of 37 oxides for catalytic activity in the hydrogenation of 1-hexyne at 100°C. An IR image taken during the reaction revealed four spots that were hotter than the substrate, indicating that these samples were active catalysts in the reaction.

The above studies have proven that thermal imaging can provide convenient and fast parallel testing of whole catalyst libraries. However, no chemical information about the products generated in the reaction can be obtained because only the temperature is measured. IR thermography can tell which catalyst is active, but cannot tell for which reaction the catalyst is active. The absence of chemical specificity is thus the major limitation of thermal imaging since more than one reaction is involved in most of the heterogeneous catalytic systems.

Laser-Induced Resonance-Enhanced Multiphoton Ionization

Senkan [11] described a laser-based high-throughput screening method for solid state catalysts that activates the dehydrogenation of cyclohexane to benzene. In Senkan's work, a tunable laser was used for the selective photoionization of product molecules in the vicinity of catalytic sites. The resulting photoions or photoelectrons were detected by an array of microelectrodes. The cross-section of ionization is significantly enhanced only when the laser frequency is tuned to a real intermediate electronic state of a molecule, which means that simultaneous photoionization of the reactants and other products can be avoided. To demonstrate this principle, Senkan constructed an 8×9 library containing platinum and palladium catalysts [Figure 4.2(a)]. Reactant gases containing cyclohexane were forced through the individual sites, and a UV laser beam was passed through the air space above the sites [Figure 4.2(b)]. The beam was tuned to a wavelength that selectively ionized any benzene converted to $C_6H_6^+$ and an electron. These charged species were detected by an array of microelectrodes placed above the sites near the laser beam. The signals recorded in the experiment clearly distinguished between the active and the inactive sites on the catalyst library.

Senkan's original setup evolved to a microreactor array [Figure 4.2(c)] which was used to screen a 66-member library of Pt–Pd–In compositions for catalytic activity in the cyclohexane dehydrogenation reaction [17]. The microreactor array consists of 17 narrow micromachined channels that extend through a nonporous silica ceramic slab about 3 in. long. Each channel contains a small well for the catalyst pellet and a piece of porous alumina impregnated with 1 wt.% of a Pt–Pd–In composition. The slab was heated to the reaction temperature while the reaction gas mixture was fed into each of the 17 channels in parallel. Near the exit, all 17 streams passed through the laser beam, which ionized the benzene product. The resulting benzene ions were picked up by an array of 17 microelectrodes. Each run of the microreactor array took 2 or 3 min. To screen the 66-member library, this procedure had to be repeated five times.

There are some limitations in Senkan's photonionization setup as a combinatorial screening technique. The laser beam irradiates only one line of the planar array such that the throughput is not very impressive. Also, probing the ions in the reactor with microelectrodes is invasive and hard to use in real applications. In addition, the setup is complicated because of all the mirrors and electrodes involved.

Scanning Microprobe Mass Spectrometry

Selectivity is a criterion that is as important as total activity for heterogeneous catalysts. Scanning microprobe mass spectrometry (MS) has been developed by symyx researchers into a high-throughput screening technique for both catalytic activity and selectivity of solid catalysts [Figure 4.3(a)] [13]. It was used to screen libraries of Rh–Pd–Pt compositions for activity in reactions that occur in catalytic converters, such as $CO + \frac{1}{2}O_2 \rightarrow CO_2$. In their work, $15 \times 15 \times 15$ triangular libraries containing 120 different catalysts each were generated by radiofrequency sputtering thin-film techniques or liquid-dosing robotics. A probe system of concentric tubes allows the gas stream containing the reactants (CO and O_2) to be delivered to a catalyst site and then removed by vacuum for analysis in

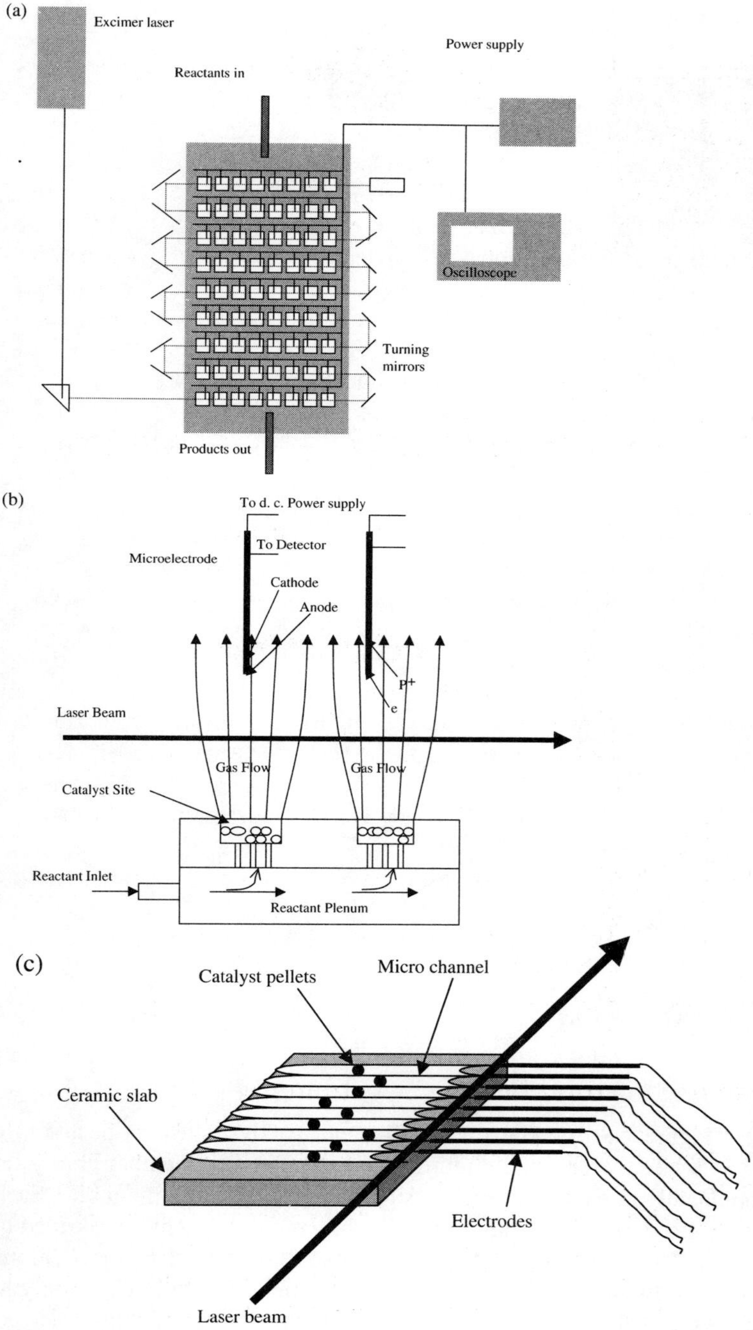

FIGURE 4.2. High-throughput screening of catalyst libraries with laser-induced resonance-enhanced multiphoton ionization (adapted from refs 11 and 17). (a) Experimental setup for the screening of a 72-member catalyst library. (b) Arrangement of individual sites on the catalyst library. (c) Array microreactor system.

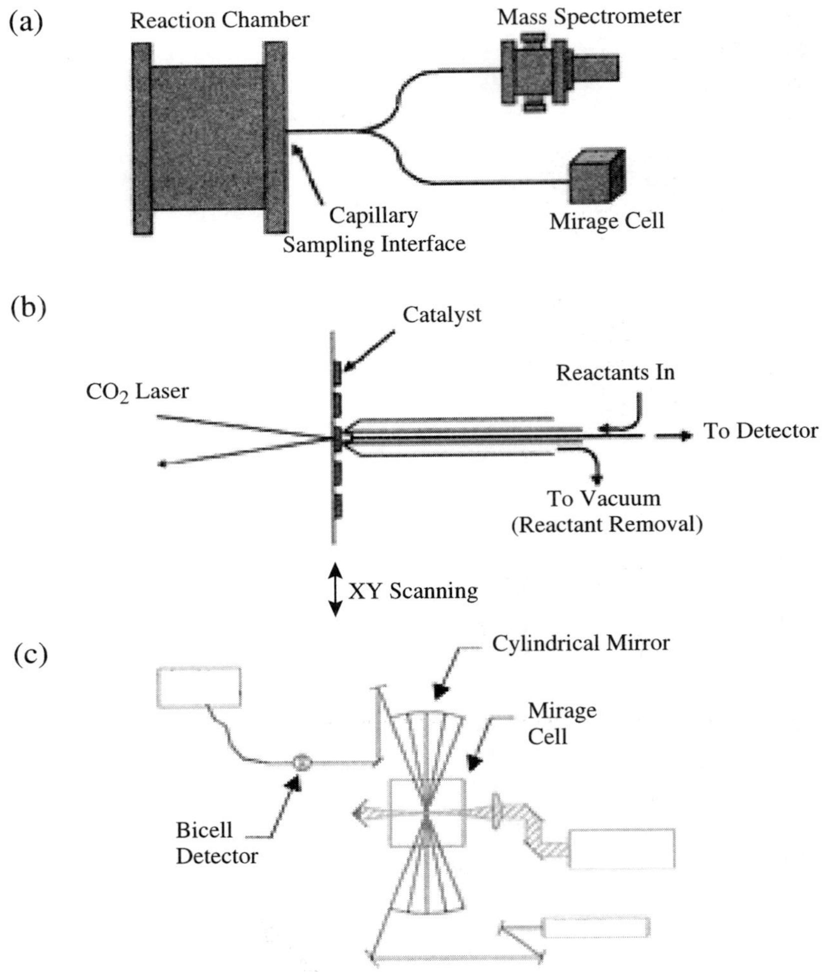

FIGURE 4.3. Scanning microprobe MS apparatus (adapted from ref. 18). (a) Experimental setup; (b) gas delivery/removal/sample nozzle; (c) photothermal deflection detector.

a mass spectrometer [Figure 4.3(b)]. The 1.5 mm diameter catalyst site was heated to a desired temperature by a laser before the sampling was done. The 120-member library sat on a stage that automatically moved each catalyst site into position under the probe. Each measurement was completed in about 1 min, so that slightly more than 2 hr was needed to screen the entire library. The screening results revealed that Pt and Rh-rich compositions were most active for producing CO_2 of the three noble metals tested. The catalytic performance of copper replacement catalysts was also evaluated by this system.

Shortly after Symyx's first report of the scanning MS high-throughput screening system, they announced a full implementation and integration of combinatorial methodologies for synthesis, screening, discovery, and optimization of multiple-component catalysts for oxidative dehydrogenation of C_2H_6 [18]. In addition to MS, photothermal deflection was developed as part of screening system [Figure 4.3(c)].

Scanning MS is capable of testing more than one product in screening the catalytic performance of heterogeneous catalysts. However, this is still a sequential testing system that cannot fundamentally improve the screening throughput.

Fluorescence Indicator

Mallouk and co-workers demonstrated that metal alloy electrocatalysts could be optimized by using the production of protons during the electrochemical oxidation of methanol in a fuel cell [12]. By converting the ions generated in an electrochemical reaction to a fluorescence signal, the active composition in a large electrode array was identified (Figure 4.4). A fluorescence acid–base indicator was used to image high concentrations of hydrogen ions, which were generated in the electro-oxidation of methanol. In a 645-member electrode array containing five elements, (platinium, ruthenium, osmium, iridium, and rhodium) 80 binary, 280 ternary, and 280 quaternary combinations were screened to identify the most active regions of phase space. Subsequent "zoom" screens pinpointed several very active compositions.

A screening scheme based on a fluorescence indicator can process a large number of catalyst combinations simultaneously with high accuracy, but its applications are limited to solution-based heterogeneous catalytic reactions in which fluorescent ions must be generated.

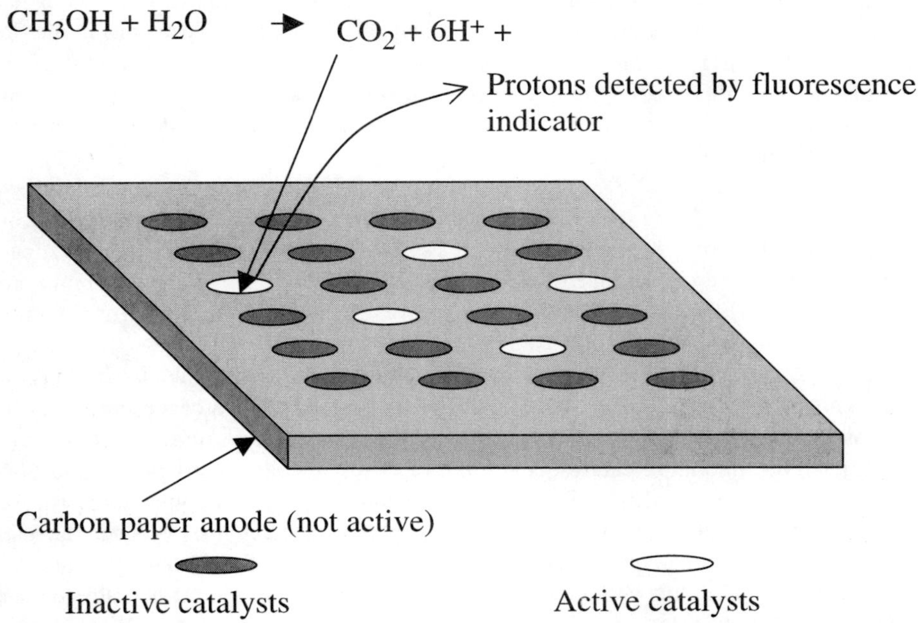

FIGURE 4.4. High-throughput screening of electrocatalysts by fluorescence indicators (adapted from ref. 12).

Laser-Induced Fluorescence Imaging

Recently, we demonstrated laser-induced fluorescence imaging (LIFI) as an alternative for high-throughput screening of heterogeneous catalysts in situ with micron-scale spatial resolution and millisecond temporal resolution [19]. This scheme is based on monitoring the fluorescence of either the reagent or a specific product molecule. By irradiating the region immediately above the catalytic surface (within 100 μm) with a laser focused into a sheet, the selected fluorescence intensity can be imaged by a CCD camera to follow the local catalytic activity. Unlike thermography [14,15] the fluorescence signal is specific and the response is linear. Interferences from variations in surface composition and the associated material emissivity are also absent. Secondary reactions do not contribute to the signal. Even thermoneutral reactions can be followed. Using this technique, dense arrays of catalyst formulations can be screened simultaneously to provide rapid access to the optimal combination of conditions.

We monitored the catalytic activity of vanadium pentoxide in the oxidation of naphthalene to naphthoquinone by oxygen, which is an important industrial process. With 488 nm excitation, naphthoquinone fluoresces, while naphthalene and the other major product, phthalic anhydride, do not fluoresce. The reaction is carried out at 330–370°C in a flow cell (Figure 4.5). Fluorescence emission at 515–545 nm is collected by a cooled scientific CCD camera.

To evaluate the spatial resolution of LIFI, we used a $4\times$ microscope objective to image a 2 mm × 2 mm area with 3 s exposure time. No naphthoquinone fluorescence is detected above the V_2O_5 surface until oxygen gas carrying 7% naphthalene is introduced into the reactor. Therefore the background signal can be recorded in the absence of the reactants and subtracted from each subsequent image. On comparing the optical image of the edge of the catalyst and the fluorescence image (Figure 4.6), a 20 μm shift in the flow direction can be observed. These experiments demonstrate that LIFI has good detection performance and the spatial and temporal resolution needed for high-throughput screening of heterogeneous catalysts. Therefore the sample packing density can reach 250×250 subunits/cm^2 for 40 μm wells.

In a larger format, a 3×2 library containing six identical V_2O_5 sample dots was prepared by pipetting 20 μg of V_2O_5–cyclohexane slurry solution into small wells (2 mm wide and 0.5 mm deep with 2 mm spacing) on a stainless steel disk and dried in air. A camera lens instead of the microscope objective was used to image the 17 mm × 17 mm area. The variation in fluorescence among the wells is $\pm$ 1.1%. Thus detection of minor differences in catalytic activity is feasible.

The 3×2 library (Figure 4.7) was then constructed with six sample dots with different amounts of V_2O_5 (10, 20, 30, 40, 50, and 60 μg). In situ fluorescence images of this library show that the relationship between catalytic activity (proportional to fluorescence intensity) and the amount of catalyst is not linear because catalytic activity is proportional to the available surface area, not the mass of the catalyst. At low amounts, there is residual contribution from the background of laser scatter and thermal emission. Then the intensity increases because the surface area increases linearly with the amount present. As a thick layer is formed in the well, material buried at the bottom of the wells is shielded and no further increase in surface area is effected. This occurs at 50 μg per well. When adapted to 40 μm wells, the amount of material need for screening will be 20 ng each.

FIGURE 4.5. Instrumentation for in situ screening of heterogeneous catalysts by fluorescence and IR imaging.

We studied the difference between LIFI and IR thermography in the system shown in Figure 4.7. At 330°C, only fluorescence can be observed by this CCD camera. However, at 350°C and above, elevated intensities can be detected in the wells above the background level (stainless steel portions of the plate) even in the absence of laser irradiation. This is near-IR emission from the catalytic material as a result of the exothermicity of the reaction (i.e., IR thermography). The camera is sensitive to radiation down to around 1000 nm. This is sufficient to record the short-wavelength tail of the blackbody radiation from the source. Because of the low dark count and low read noise of the camera, flat fielding (subtraction of the image from background levels) reveals a thermal image. The blackbody behavior is confirmed by placing short-pass filters in front of the camera at successively longer and longer wavelengths between 800 and 1000 nm.

For studies conducted between 330 and 370°C, the lower temperature favors S/N for LIFI while the higher temperature favors S/N for thermography. The spatial resolution of thermography is limited only by the pixel resolution of the imaging system, while that of

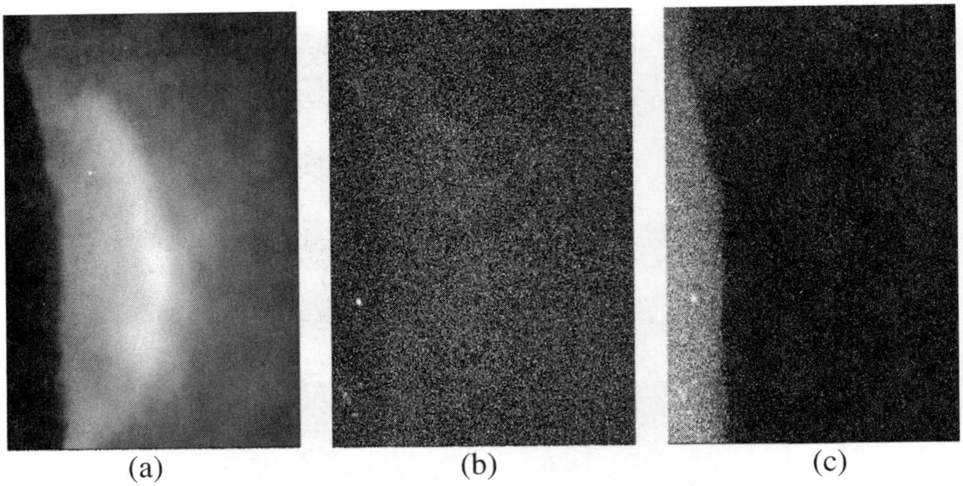

FIGURE 4.6. (a) Optical image of V_2O_5 pellet, left (darker) part of the image. Reactants flow from left to right. (b) Fluorescence image when no reactants had been introduced into reactor. (c) Fluorescence image after reactant was introduced into the reactor.

LIFI is further limited by the 20 μm flow distortion. In this setup, both sets of information can be obtained by chopping the laser irradiation or by rotating filters between the visible and near-IR. We note that CCD thermography is only useful at these high temperatures when blackbody emission begins to reach the responsive region of the device. When it does, the combination of nonspecific temperature increase and species-specific concentration maps offers unique insights into the catalytic process. Finally, the catalytic system here happens to involve fluorescent species. However, it should be possible to design model reactants for just about any catalytic reaction so that either the reactant or the product fluoresces on the breaking or making of the bond of interest.

Catalytic Screening

We used LIFI and IR thermography to screen the activity and selectivity of binary vanadium-based catalysts of V–Mo(VI)–O, V–W(VI)–O, V–Sn(IV)–O, and V–Ti(IV)–O for selective catalytic oxidation of naphthalene (NA) to naphthoquinone (NQ), which is the competing product of phthalic anhydride (PhA) [20]. Based on the screening results of binary libraries, we prepared and screened ternary 15-membered libraries of V–Sn–Mo and V–Sn–W. Composition of components and sample preparation methods are two variables in constructing the libraries. The temperature effect on the catalytic performance on catalysts is also studied by monitoring the catalytic reaction at different reaction temperatures.

Binary catalysts of V_2O_5–MoO_3, V_2O_5–WO_3, V_2O_5–SnO_2, and V_2O_5–TiO_2 with atomic ratios of 9, 2.33, 1, 0.43, and 0.11 were prepared in two ways. In method A the oxides of two elements were mixed. In method B a solution of NH_4VO_3 in oxalic acid was

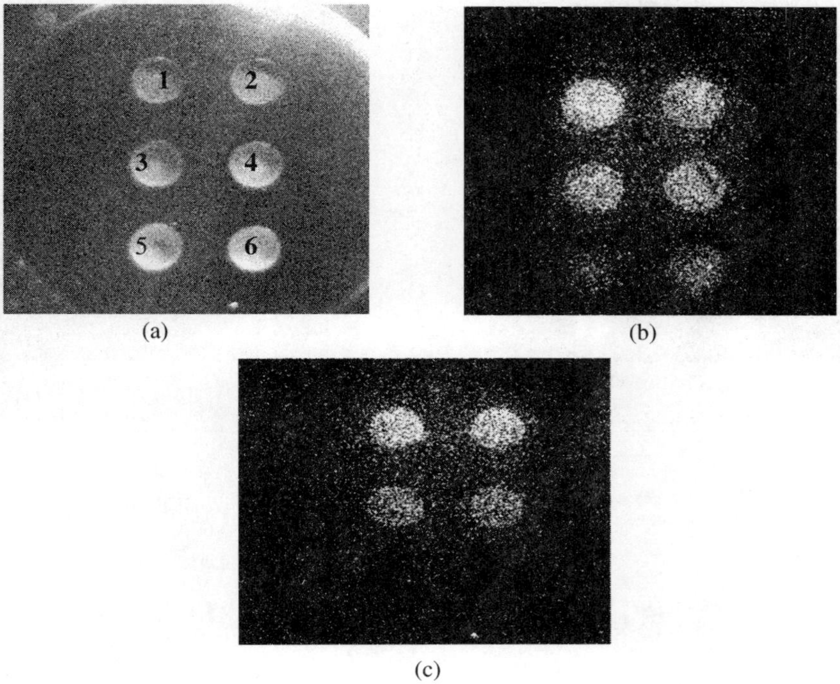

FIGURE 4.7. Catalytic screening by fluorescence imaging. (a) Optical image of six 2 mm diameter wells with 2 mm spacing between them, each containing different catalysts. (b) Fluorescence image of naphthoquinone produced above each well for (1) 60 μg, (2) 50 μg, (3) 40 μg, (4) 30 μg, (5) 10 μg, and (6) 20 μg V_2O_5. (c) Fluorescence image of naphthoquinone produced above each well: (1), (2), 50 μg pure V_2O_5; (3), (4), 50 μg 1:1 V_2O_5–TiO_2; (5), (6), 50 μg pure TiO_2.

mixed with $(NH_4)_2MoO_4$, $(NH_4)_2WO_4$, $Sn(OH)_2$, and TiO_2, respectively, followed by evaporation drying overnight and subsequent calcination in an oxygen flow at 773 K for 3 hr. Ternary catalysts of V–Sn–W and V–Sn–Mo were prepared by method B, and the compositions of these catalysts are mixture of 5%, 10%, and 15% of Mo or W with V–Sn in automatic ratios of 9, 2.33, 1, 0.43, and 0.11.

The V–Mo binary catalyst is a very interesting catalytic system. Both V_2O_5 and MoO_3 are catalytically active according to Figures 4.8(a) and 4.8(b). As shown in Figure 4.8(c), the addition of MoO_3 up to 50% decreases the catalytic activity of vanadium in the selective oxidation of NA to NQ, and further increase of molybdenum results in an increase in the activity. This result agrees well with the reported results [21]. However, the screening result from IR imaging does not support this conclusion [Figures 4.8(b) and 4.8(c)]. The addition of MoO_3 was reported to increase the overall activity as well as the activity of selective oxidation of benzene to maleic anhydride (MA). The result shown in Figure 4.8(b) goes the other way. From our experimental results, the addition of Mo(VI) decreased the selective activity of V_2O_5 in the oxidation of NA to PhA. One explanation for this result is the difference in activity of surface active sites in the production of

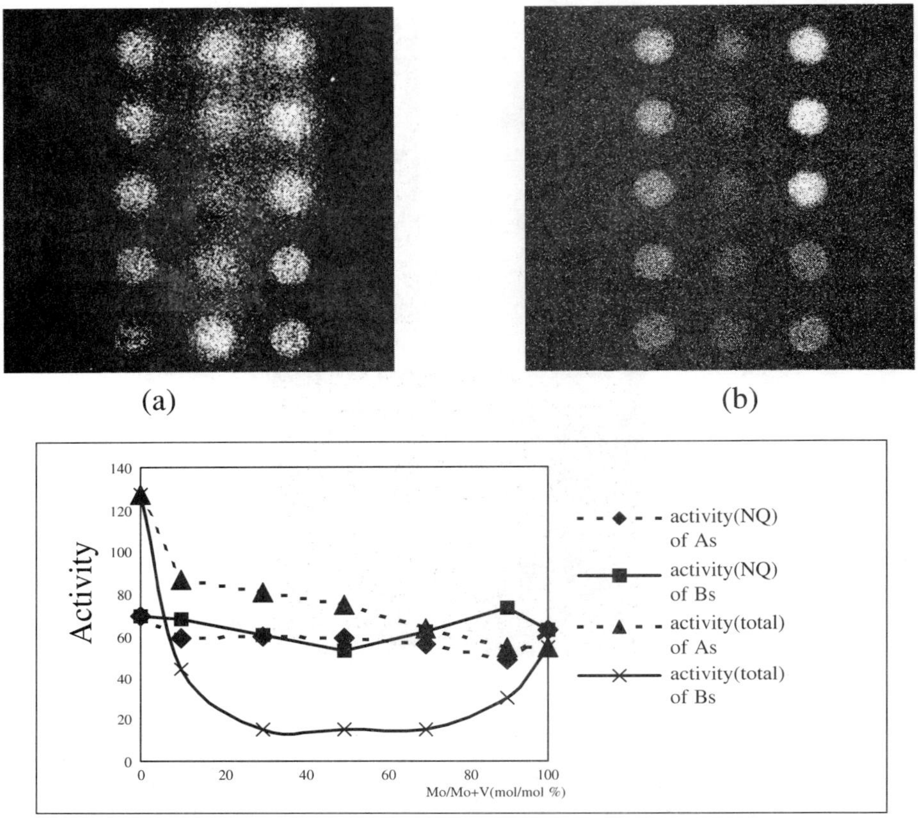

FIGURE 4.8. Catalytic performance of V–Mo–O binary system; (a) fluorescence screening; (b) near-IR thermography screening; (c) activity (NQ) and total activity of the catalysts.

MA and PhA. Samples A2 and B2 have the same sample composition and similar activity for NQ, but A2 shows higher overall activity, which means that the surface active sites of catalysts made by method B are less favorable for the formation of PhA. Catalyst B5 has a slightly higher catalytic activity than pure V_2O_5 in selective oxidation of NA to NQ while the IR detection shows that it is much less active than pure vanadium. This shows that IR thermography screening is not always informative.

Screening of Ternary Heterogeneous Catalysts

Based on our experiment results [20], Mo, Sn, and W seem to be able to modify the active sites of vanadium catalyst in their own ways, especially when prepared by method B. We prepared a library which consists of V–Sn–Mo–O and V–Sn–W catalysts in various ratios. The combinations 63% V–27% Sn–10% Mo and 45% V–45% Sn–10% Mo are the most active catalysts in selective oxidation of NA to NQ. Comparison of the selective activities of these two combinations with pure V_2O_5 shows that 45% V–45% Sn–10% Mo demonstrates a 70% increase. The interesting result here is that, in contrast, the overall activity of 45% V–45% Sn–10% Mo is depressed dramatically compared with pure V_2O_5. Therefore, 45% V–45% Sn–10% Mo is more active and more selective in the oxidation of NA to NQ than V_2O_5, the conventional catalyst used in industry.

Temperature Dependence of Catalytic Activity

It is important to identify a suitable temperature range for a heterogeneous catalyst since its activity and selectivity are temperature dependent. A ternary library was screened by LIFI at temperatures of 603 K, 613 K, 623 K, 633 K, and 643 K (Figure 4.9). The fluorescence images of the catalysts show that different materials demonstrate different sensitivities to temperature changes. Samples with 10% Mo prepared by method B are among the most active catalysts; therefore they were chosen to be screened further at various reaction temperatures with pure V_2O_5 and catalysts that have the same V–Sn composition but no molybdenum. The addition of 10% Mo boosts the activity of V–Sn (7:3), V–Sn (5:5), and V–Sn (3:7), and the activity enhancement varies as the temperature changes. At lower temperatures (up to 633 K), 63% V–27% Sn–10% Mo is more active than 45% V–45% Sn–10% Mo, but at temperatures above 633 K, 45% V–45% Sn–10% Mo is more active than 63% V–27% Sn–10% Mo. Within the temperature range studied all the

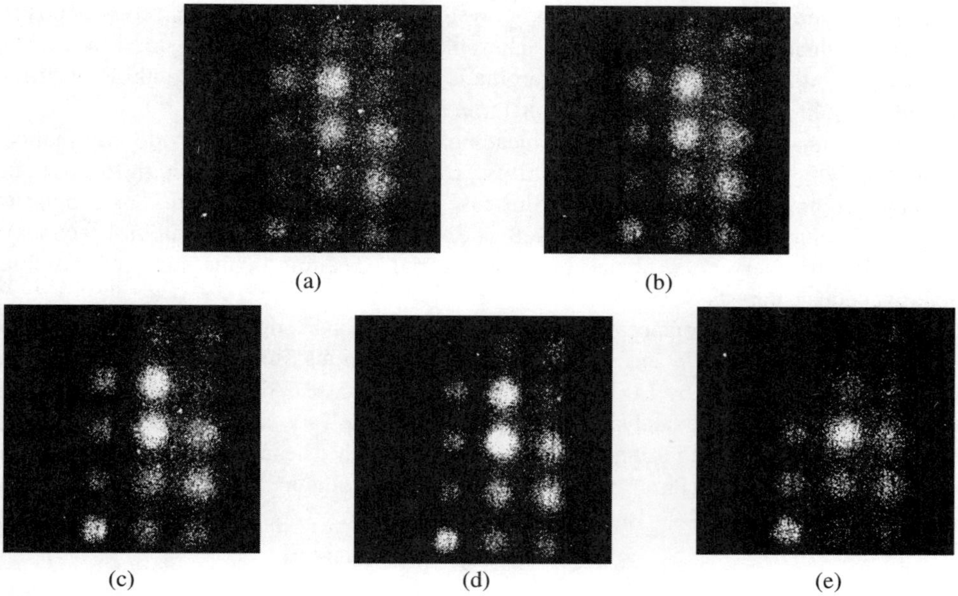

FIGURE 4.9. Temperature dependence of catalytic activity (NQ); LIFI screening of V–Sn–Mo–O catalysts at temperatures of (a) 603 K, (b) 613 K, (c) 623 K, (d) 633 K, and (e) 643 K.

catalysts in this library except 45% V–45% Sn–10% Mo reach their highest activity around 633 K.

There are several conclusions that we can make based on our screening experiments.

1. The addition of molybdenum and tungsten to V_2O_5 modifies the surface active sites, but does not increase the activity and selectivity of selective oxidation of naphthalene to phthalic anhydride while it does increase the activity and selectivity of selective oxidation of benzene to maleic anhydride as reported in the literature. SnO_2 behaves similarly to what has been reported.
2. Ternary samples of 63% V–27% Sn–10% Mo and 45% Sn–45%–10% Mo were found to have up to 70% higher activities than pure V_2O_5 as well as improved selectivity in converting naphthalene to naphthoquinone.
3. Combinatorial studies of the temperature dependence of catalytic activity have been accomplished by LIFI.
4. LIFI enables solid heterogeneous catalysts to be screened at a speed of 15 s per library. Thus the screening throughput is hundreds of times faster than conventional approaches (calculation based on a 15-member library).

Screening of Zeolites in the Acylation of Benzene

Catalyst deactivation is one of the most important problems in catalysis. The stability of catalysts in terms of activity and selectivity can sometimes last as long as years, but sometimes they become inactive in seconds. Zeolites are typically unstable but are extremely useful heterogeneous catalytic systems in industry. Carbonaceous deposits (coke) can deactivate zeolites easily. The coke structure varies with conditions, but consists essentially of ill-defined polyaromatic compounds [22]. The catalytic stability period of zeolites in oil cracking is about 1 min [23].

For an unstable catalytic system, meaningful screening should provide information about catalytic activity as well as stability. The time resolution of a high-throughout screening technique plays a key role in this case. As mentioned above, LIFI has good spatial resolution and detection limits as well as excellent time resolution (submilliseconds). This makes it a very good choice in combinatorial screening or the study of unstable heterogeneous catalysts.

The catalytic performance of a 15-member library consisting of mordenites (MORs), ferrierites (FERs), ZSM-5, and γ and β zeolites with various Si/Al ratios is screened in a high-throughput fashion by LIFI. The effects of structures, Si/Al ratio, reaction temperature, and reactants on the catalytic properties of these catalysts can be revealed in minutes instead of the days required using the conventional approach. The interaction of benzene (Bz) with phthalic anhydride (PhA) catalyzed by zeolites is studied:

$$\mathrm{Bz} + \mathrm{PhA} \xrightarrow{H^+} \text{anthraquinone}$$

All zeolite samples were pretreated by heating at 550°C in an oxygen flow for 5 hr. The catalyst libraries were constructed by depositing cyclohexane slurries of zeolites on

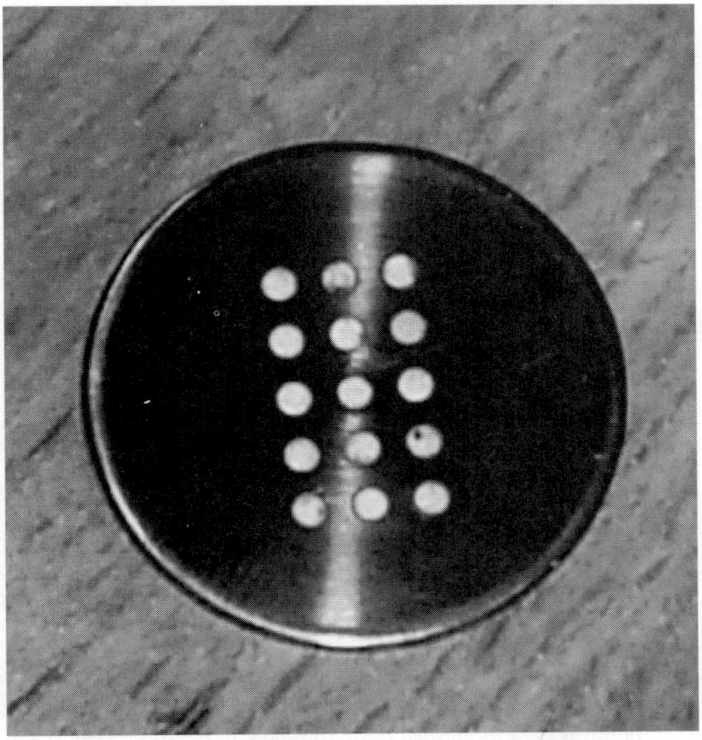

Sample Arrangement

Z1	Z5	Z9
Z2	Z6	Z10
Z3	Z7	Z14
Z4	Z8	Z15
Z11	Z12	Z13

FIGURE 4.10. Simultaneous screening of 15 different zeolite catalysts. The assignments are listed in Table 1.

sample plates. The sample arrangement is shown in Figure 4.10. The zeolites in the 15-member library are listed in Table 4.1. This library is 17 mm in diameter, and the diameter of an individual catalyst site is 1 mm. The amount of catalyst in each site is about 100 μg.

Only β zeolites were found to be active in the reaction temperature range of 190–340°C, which agrees with the previous report [24]. All the selected zeolite are well characterized, and they all have relatively large pore sizes. The structure of the MORs is a one-dimensional tube with an eight-member ring pore structure; FERs and MFIs are two-dimensional interconnecting structure with eight or ten-member rings; and FAU and β zeolite have three-dimensional networks with a 12-member ring structure. β Zeolite is the only

TABLE 4.1. Arrangement of Zeolites in the 15-member Library in Figure 4.10

Sample	Structure	Si/Al ratio
Z1	Zeolite γ	5
Z2	Zeolite γ	30
Z3	Zeolite γ	60
Z4	Zeolite γ	80
Z5	Zeolite β	25
Z6	Zeolite β	75
Z7	Zeolite β	150
Z8	Zeolite β	300
Z9	Mordenite	20
Z10	Mordenite	90
Z11	ZSM-5	30
Z12	ZSM-5	80
Z13	ZSM-5	280
Z14	Ferrierite	20
Z15	Ferrierite	55

high-silica material having a full three-dimensional network of 12-member rings, which might explain its catalytic activity in reactions involving large aromatic molecules.

Reaction temperature is another critical parameter in catalytic reactions. The screening results of activities of four β zeolites at temperatures of 190, 220, 250, and 280°C were obtained. In this temperature range, an increase in temperature causes a decrease in catalytic activity of β zeolites in the acylation of benzene. Similar results have been reported in the literature [24], and the decrease in catalytic activity is due to the selectivity change. The above screening results were obtained in 10 min by LIFI, which is at least 100 times faster (for 15 samples) than the conventional testing approach using gas chromatography.

In our experiments, β zeolites started to deactivate in less than 1 min at a reaction temperature of 280°C, while they were relatively stable at lower reaction temperatures (< 260°C). This is probably because either the coke formed at higher temperatures is of larger size than that formed at lower temperatures or the high temperatures increase the production of coke. The stability of β zeolites with various Si/Al ratios seems be affected similarly at elevated temperatures.

Oxidizing the coke with air or oxygen at high temperatures usually regenerates zeolite catalysts. β Zeolites were placed in the reactor at 300°C to catalyze the acylation of benzene, and substantial deactivation was observed by LIFI after 2 min. Then oxygen instead of reactant mixtures was introduced to regenerate the catalyst. Lower recovery was observed at lower regeneration time, but an increase in regeneration time beyond 3 s did not increase the recovery of activity for deactivated zeolites. This suggests that the coke built up on zeolites cannot be totally eliminated, or that it changed the surface conditions of zeolite in some way.

SUMMARY

The use of in situ real-time imaging can reveal details about heterogeneous reactions in an efficient manner. Both total activity and selected reaction pathways can be monitored. Dynamic processes such as catalyst deactivation and regeneration can be followed. The high data rate also allows high-throughput screening of the activity as a function of composition, structure, gas flow rate, temperature, and heterogeneity. Although the above examples show imaging using visible laser excitation, this approach can be readily modified for UV imaging. In that case, even exotic species such as C_2, CN, OH, CO, etc. can be monitored.

ACKNOWLEDGMENTS

We thank Dr Glen Schrader for help in designing the reactor and for providing the vanadium catalysts. The Ames Laboratory is operated for the U.S. Department of Energy by Iowa State University under Contract No. W-7405-Eng-82. This work was supported by the Director of Science, Office of Basic Energy Sciences, Division of Chemical Sciences.

REFERENCES

1. Merrifield, R. B. *Journal of the American Chemical Society* **1963**, *85*, 2149–2154.
2. Hanak, J. J. *J. Mater. Sci.* **1970**, *5*, 964–971.
3. Hanak, J. J., Gittleman, J. I., Pellicane, J. P., Bozowski, S. *Phys. Lett.* **1969**, *30A*, 201–202.
4. Xiang, X. D., Sun, X., Briceno, G., Lou, Y., Wang, K. A., Chang, H., Wallace-Freedman, W. G., Chen, S. W., Schultz, P. G. *Science* **1995**, *268*, 1738–1740.
5. Briceno, G., Chang, H., Sun, X. D., Schultz, P. G., Xiang, X. D. *Science* **1995**, *270*, 273–275.
6. Danielson, E., Golden, J. H., McFarland, E. W., Reaves, C. M., Weinberg, W. H., Wu, X. D. *Nature* **1997**, *389*, 944–948.
7. Wei, T., Wallace-Freedman, W. G., Schultz, P. G., Xiang, X. D. *Appl. Phys. Lett.* **1998**, *72*, 525–527.
8. Danielson, E., Devenny, M., Giaquinta, D. M., Golden, J. H., Haushalter, R. C., McFarland, E. W., Poojary, D. M., Reaves, C. M., Weinberg, W. H., Wu, X. D. *Science* **1998**, *279*, 837–839.
9. Wang, J., Yoo, Y., Gao, C., Takuchi, I., Sun, X., Change, H., Xiang, X. D., Schultz, P. G. *Science* **1998**, *279*, 1712–1714.
10. Taylor, S. J., Morken, J. P. *Science* **1998**, *280*, 267–270.
11. Senkan, S. M. *Nature* **1998**, *394*, 350–353.
12. Reddington, E., Sapienza, A., Gurau, B., Viswanathan, R., Sarangapani, S., Smotkin, E. S., Mallouk, T. E. *Science* **1998**, *280*, 1735–1737.
13. Cong, P., Doolen, P. D., Fan, Q., Giaquinta, D. M., Guan, S., Mcfarland, E. W., Poojary, D. M., Self, K., Turner, H. W., Weinberg, W. H. *Angew. Chem. Int. Ed.* **1999**, *38*, 484–488.
14. Sun, X. D., Wang, K. A., Yoo, Y., Wallace-Freedman, W. G., Gao, C., Xiang, X. D., Schultz, P. G. *Adv. Mater.* **1997**, *9*, 1046–1049.
15. Moates, F. C., Somani, M., Annamalai, J., Richardson, J. T., Luss, D., Willson, R. C. *Ind. Eng. Chem. Res.* **1996**, *35*, 4801–4803.
16. Klein, J., Lehmann, K. W., Schmidt, H. W., Maier, W. F. *Angew. Chem. Int. Ed.* **1998**, *37*, 3369–3372.
17. Senkan, S. M., Ozturk, S. *Angew. Chem. Int. Ed.* **1999**, *38*, 791–795.
18. Cong, P. J., Dehestani, A., Doolen, R., Giaquinta, D. M., Guan, S., Markov, V., Poojary, D., Self, K., Turner, H., Weinberg, W. H. *Proc. Natl Acad. Sci. USA* **1999**, *96*, 11077–11080.
19. Su, H., Yeung, E. S. High-throughput screening of heterogeneous catalysts by laser-induced fluorescence imaging. *J. Am. Chem. Soc.* **2000**, *122*, 7422–7423.

20. Su, H., Hou, Y., Houk, R. S., Schrader, G. L., Yeung, E. S. Combinatorial screening of heterogeneous catalysts in selective oxidation of naphthalene by laser-induced fluorescence imaging. *Anal. Chem.* **2001**, *73*, 4434–4440.
21. Satsuma, A., Okada, F., Hattori, A., Miyamoto, A., Hattori, T., Marukami, Y. *Appl. Catal.* **1991**, *72*, 295–310.
22. Satterfield, C. N. *Heterogeneous Catalysis in Industrial Practice*, 2nd ed., **1991**, New York: McGraw-Hill, p. 254.
23. Brito-Alayon, A., Hughes, R., Kann, E. K. T. *Chem. Eng. Sci.* **1981**, *36*, 445–453.
24. Weitkamp, J., Karge, H. G., Pfeifer, H., Hölderich, W. *Stud. Surf. Sci. Catal. (Zeolites and Microporous Crystals)* **1994**, *84*, 1904–1912.

Chapter 5

Spectroscopic Imaging in the Mid-Infrared Applied to High-Throughput Studies of Supported Catalyst Libraries

Steven S. Lasko*, Reed J. Hendershot, Yu Fu, Mark-Florian Fellmann*, Gudbjorg Oskarsdottir*, Christopher M. Snively, and Jochen Lauterbach

INTRODUCTION

The "combinatorial approach" has shown very promising results for pharmaceuticals and small organic molecules [1–3]. High-throughput (HT) screening of heterogeneous catalysts goes back at least 15 years [4], but has recently been rediscovered as a method for rapidly and efficiently identifying catalyst formulations [5–14]. The approach consists of two key steps: the systematic synthesis of a large number of potentially useful formulations (collectively referred to herein as a "library," although this term has recently been contested [15]), and the subsequent rapid testing of this library to determine the usefulness of each formulation to the specific application. Several methodologies have been developed for the rapid and efficient generation of catalyst libraries, ranging from evaporation methods [16] and robotic dispensing of catalyst precursors [17–19] to parallel hydrothermal processing [20–22]. Once generated, the libraries must be characterized and tested. For heterogeneously catalyzed reactions, this step can range from simple qualitative activity screening to the quantitative measurement of selectivity or turnover rates. Currently available analytical techniques are often incapable of keeping pace with the large numbers of compounds created. This creates a bottleneck, slowing the entire discovery process, and has created a serious demand for the development of analytical techniques specifically designed for the HT analysis of combinatorial heterogeneous catalyst libraries.

Steven S. Lasko, Reed J. Hendershot, Yu Fu, Mark-Florian Fellmann, Gudbjorg Oskarsdottir, Christopher M. Snively, and Jochen Lauterbach • Department of Chemical Engineering, University of Delaware, Newark, DE 19716, USA
* Current address: School of Chemical Engineering, Purdue University, West Lafayette, IN 47907, USA
Contact author: Jochen Lauterbach, email: lauterba@che.udel.edu

High-Throughput Analysis, edited by Potyrailo and Amis
Kluwer Academic Publishers, New York, 2003.

To date, three general types of approaches have been utilized for the analysis of combinatorial libraries of heterogeneous catalysts. The first approach is to implement currently available "one sample at a time" techniques, such as mass spectrometry (MS) or gas chromatography (GC) [23–27]. While this approach is the easiest to implement with a small number of samples, it is apparent that it is too time consuming for the screening of catalyst libraries with hundreds or more members, since the screening time is directly proportional to the number of samples to be analyzed. The second approach is based on the modification of conventional serial techniques using automation approaches in order to decrease the screening time. These include MS [28–31], gas sensors [32], and resonance-enhanced multiphoton ionization (REMPI) [5,33]. All these methods take advantage of the ability to run reactions in some parallel fashion, but the testing is still performed in a sequential manner. At this stage, sacrifice of high-quality quantitative data is often made for improved speed by utilizing reaction conditions or reactor geometries that do not resemble real-world systems. In addition, many of these methods are not necessarily quantitative.

The third approach involves the development of truly *parallel screening* techniques which gather data simultaneously from all elements in a library. This category currently includes laser-induced fluorescence imaging [34,35], IR thermography [36,37], and *Fourier transform IR (FTIR) imaging* [38–43]. Laser-induced fluorescence imaging relies on a fixed laser wavelength exciting fluorescence in molecules. The laser excites both products and reactants in the gas phase above the catalyst library [35]. Only relative activity changes among a group of catalysts can be determined [34]. IR thermography detects IR radiation emitted by catalysts and has been used to detect activity. However, thermal imaging does not possess the ability to chemically resolve product composition and therefore, it is unclear whether the apparent "activity" is due to the desired reaction or other side reactions. However, this is a key issue when studying complex reactions using supported catalysts, and highlights the necessity to gather chemically specific information from each library member. Over the past 3 years, our group has demonstrated that FTIR imaging (often referred to as hyperspectral imaging) can be successfully applied as a rapid and truly parallel HT analytical technique for a variety of combinatorial systems. FTIR imaging is a technique that couples a *focal plane array (FPA) detector* with an FTIR spectrometer to enable the collection of spatially resolved spectral information [38–60]. The basic concept of the application of FTIR imaging to the analysis of combinatorial libraries is to place many samples in the field of view of the instrument and analyze them simultaneously [38–41,43]. This provides a multiplex advantage, in which chemically specific information is simultaneously collected from many samples during a single collection.

BASICS OF FTIR SPECTROSCOPY

A basic understanding of FTIR spectroscopy is necessary before discussing FTIR imaging. For a more thorough review of the subject, *Chemical Fourier Transform Infrared Spectroscopy* by Griffiths and de Haseth [61] is highly recommended.

The Michelson interferometer was invented in 1880 by A. A. Michelson and is at the heart of all FTIR spectrometers. A broad-band IR beam is passed through a beamsplitter. One beam is sent to a fixed mirror and the other to a moving mirror. The light is then reflected back to the beamsplitter where it combines constructively or destructively, depending on the frequency and the optical path difference (retardation). The intensity at the detector can be written as

$$I'(\bar{\nu},\delta) = 0.5I(\bar{\nu})(1 + \cos 2\pi\bar{\nu}\,\delta) \qquad (1)$$

where $I'(\bar{\nu},\delta)$ is the intensity measured at the detector, $I(\bar{\nu})$ is the intensity of the source, $\bar{\nu}$ is the wavenumber, and δ is the retardation.

An *interferogram*, given as intensity versus retardation, is created from the summation at each retardation of intensities in Eqn (1) at all frequencies that can be detected. Only the modulated component is important in spectrometric measurements. A mathematical operation called a Fourier transform (FT) is used to convert the measured intensity versus retardation data into intensity versus frequency data (spectra). Additional data-processing techniques, such as zero filling, apodization, and phase correction, are routinely applied to the data but will not be discussed here.

When measuring an IR spectrum, there are several methods for the IR radiation to interact with the sample. Perhaps the most common method, and the one which will be discussed here, is *IR transmission spectroscopy*, where the IR beam is passed directly through the sample and the absorption of IR light due to specific molecular vibrations and rotations is measured. In order to absorb IR radiation, a molecule must undergo a net change in vibrational or rotational dipole moment. Homonuclear species, such as O_2, N_2, and H_2, undergo no net change in dipole moment, and therefore such compounds will not absorb in the IR and cannot be detected.

Measuring the *absorption spectrum* of a sample requires the measurement of a background in order to remove the frequency-dependent contributions of the source, detector, and elements in the beam path. A spectrum collected for a sample is ratioed against the background spectrum in order to produce a transmittance spectrum as follows:

$$T(\bar{\nu}) = \frac{I(\nu)}{I_0(\nu)} \qquad (2)$$

where $T(\bar{\nu})$ is the transmittance, $I(\bar{\nu})$ is the intensity measured with a sample in the beam, and $I_0(\bar{\nu})$ is the intensity measured with no sample in the beam (background).

Transmittance spectra are useful for determining species that may exist in a sample but, because the intensity is not linearly related to the concentration of the absorbing species, care has to be taken when performing quantitative calculations. Instead, absorbance is used and calculated from the transmittance by

$$A(\bar{\nu}) = -\log_{10} T(\bar{\nu}) \qquad (3)$$

where $A(\bar{\nu})$ is the absorbance and $T(\bar{\nu})$ is the transmittance.

Beer's law relates concentration to absorbance at a specific frequency by

$$A(\bar{\nu}) = a(\bar{\nu})bc \qquad (4)$$

where $A(\bar{\nu})$ is the absorbance, $a(\bar{\nu})$ is the absorptivity constant, b is the path length, and c is the concentration.

The peak height or integrated peak area at lower concentrations is linearly related to the concentration of the sample species being analyzed. Its measurement allows direct determination of the chemical composition of gas mixtures. Even at higher concentrations, when this relationship is often nonlinear, it is constant for a specific optical setup and therefore can be calibrated. Path length also has a linear dependence on absorbance, suggesting that increasing the path length will increase the lower detection limit. However, it must also be recognized that this can reduce the upper limit of concentrations that can be studied owing to total absorption of the IR radiation.

The spectral resolution is determined by the maximum retardation of the interferometer mirror—the greater the retardation, the greater the resolution. The frequency range is determined by the time between discrete data points during collection of the interferogram. All frequency intensities that are collected by the detector are "folded over" into the frequency range that is being examined, decreasing the signal-to-noise ratio. To prevent this interference, low- and high-pass filters should be used at the highest and lowest desired frequencies.

Since the sample spectrum is ratioed against the background spectrum, all factors which contribute to the background interferogram should be identical during collection of the sample interferogram. Three factors are especially important to maintain for producing optimum spectra: temperature, optical alignment, and gas composition in the IR beam path. Atmospheric gases, specifically water and carbon dioxide, absorb in the IR at frequencies that are often studied and therefore should be purged from the system with a noninterfering gas.

EXTENSION OF FTIR TO HIGH-THROUGHPUT SCREENING

The previous discussion assumes that the interferograms are recorded by a single detector element. If it is desired to collect spatially resolved information, so that the chemical makeup at any point can be determined, *spectral imaging* techniques must be applied. An FPA detector allows for the collection of spectral data from multiple spatial regions simultaneously, thus creating the full spectral image in one single measurement. Images at specific frequencies, or spectra at specific spatial points, are available after the Fourier transform and other processing of the interferogram data set are performed. The most common types of FPAs for the IR region are indium antimonide (InSb) and *mercury cadmium telluride* (MCT). Detectors are selected based on spectral range desired, sensitivity, response speed, pixel density, and cost.

In general, the key to the imaging technique is maintaining the spatial separation of library components without fully obstructing the IR beam path. At the same time, it is desirable to physically isolate each member of the library to prevent any cross talk between catalysts though the gas phase or thermal interaction. With a 64×64 array detector, for example, it would theoretically be possible to test 4,096 samples simultaneously. In practice, though, it is not possible to achieve a one-to-one sample-to-pixel correlation [41]. The MCT FPA used in our setup has a viewing window that is too small to image a sample library directly ($4\ mm \times 4\ mm$, individual pixel size $61\ \mu m \times 61\ \mu m$). This problem is overcome using a series of mirrors and lenses to expand, collimate, and refocus the IR beam. The proper selection of mirror and lens materials is an important consideration when designing an imaging system. Only front-coated mirrors can be used to prevent absorption and distortion of the IR beam. There are many choices of IR transparent materials for lenses and windows, but some are very expensive or hygroscopic, and some are not transparent over the entire spectral range of interest. The most common lens materials for the mid-IR region are BaF_2, CaF_2, Ge, and ZnSe. Our optical setup uses a combination of BaF_2 and ZnSe lenses, allowing access to a spectral range limited by the response of the FPA of 4,000–800 cm^{-1}.

DATA HANDLING AND MANIPULATION

A major stumbling block in HT screening is the difficulty in dealing with the enormous quantity of data that is produced. Commercial software is often not capable of handling this new experimental paradigm. For the particular measurement technique reported here, the software must be able to deal with all calculations normally involved with FTIR spectroscopy. In addition, one must be able to filter out the important information in the data, while at the same time differentiating between multiple spatial regions of interest. In other words, each pixel (detector) in the FPA must be referenced and tracked as the data processing proceeds. The following steps are routinely performed when analyzing HT data:

- zero-filling, apodization, and FT of interferograms
- averaging over multiple FT data sets
- calculation of absorbance spectra
- selection and grouping of data sets representing individual samples
- calculation of concentrations from baseline-corrected absorbance bands using previously performed calibrations
- creation of plots of temperature, feed concentration, or time versus product concentration
- calculation of orders of reaction, activation energies, or pre-exponential factors.

The limiting step using modern equipment is not the speed of the interferometer, but the speed at which the data can be collected. With a 64×64 pixel FPA, typically a collection frame rate of 315 Hz can be achieved, resulting in an 8 cm^{-1} resolution file 4.8 Mbyte in size, created in less that 2 s. Increasing resolution also increases data file size due to the dependence on the mirror retardation (time at a constant mirror speed). For example, a single 0.5 cm^{-1} resolution data set results in a 77 Mbyte file. This shows the need to accurately track, analyze, and categorize the data of interest in an efficient manner. After experimenting with several mathematical software packages, we decided to develop our own software using Borland C++ as a programming language. Our current software package can perform all the above-mentioned tasks on a timescale of seconds or less. It has a point-and-click interface, and examples of the software interface are shown in Figures 5.1 and 5.2.

EXPERIMENTAL METHODOLOGIES

Two interchangeable systems were designed for the study of heterogeneous catalytic reactions using the same gas inlet system, spectrometer, FPA, and optical elements.

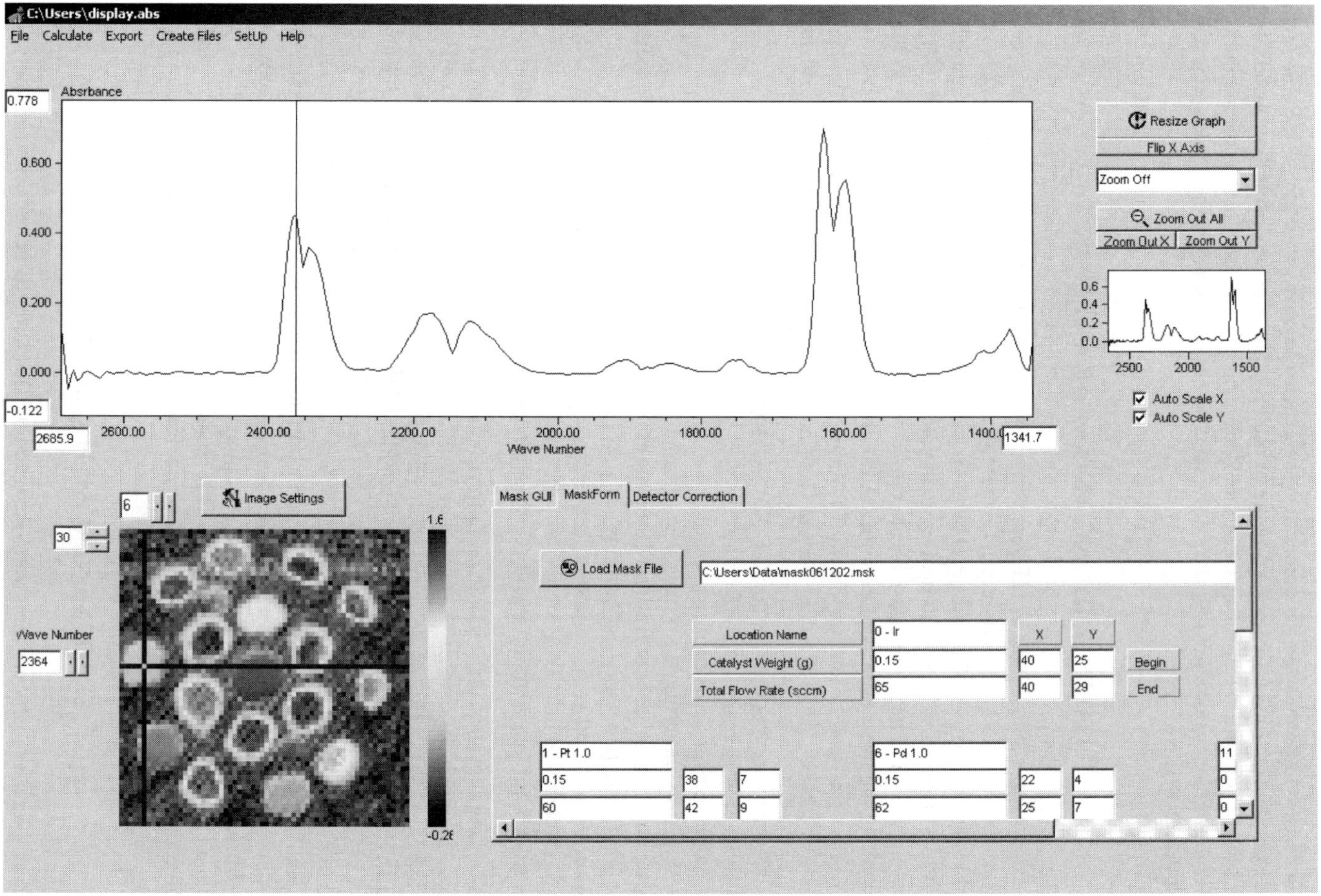

FIGURE 5.1. Screenshot of data visualization software. The image on the lower left shows the spectral intensity of the IR band selected in the main window (in this case CO_2). The main window shows the entire spectrum of the selected pixel.

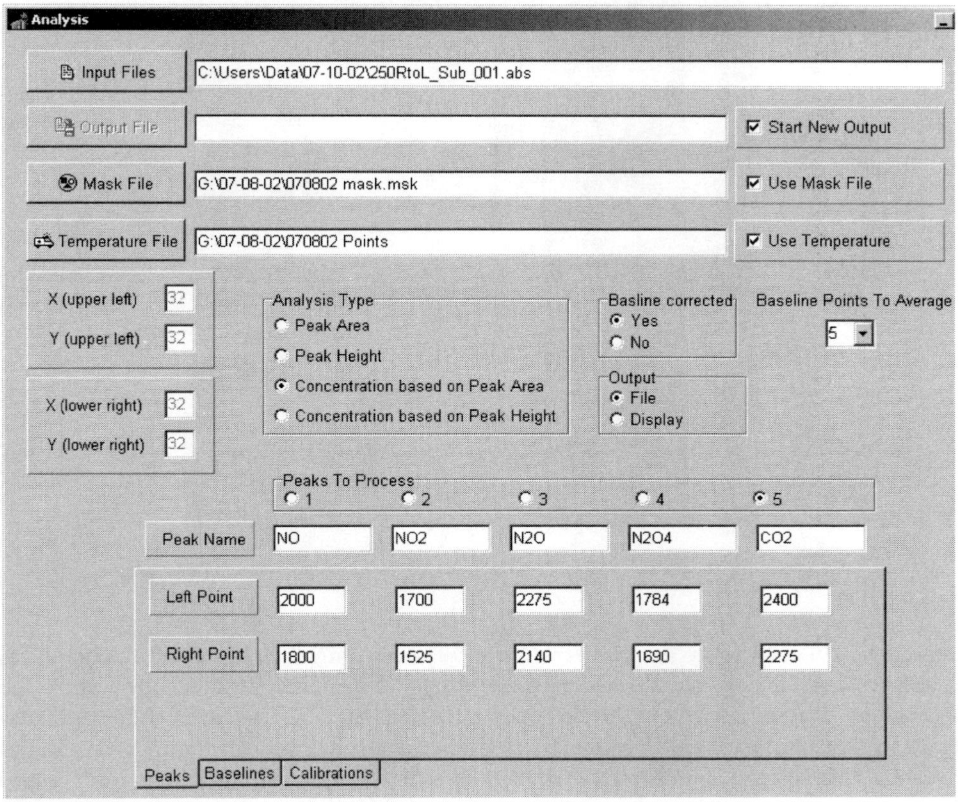

FIGURE 5.2. Screenshot of data analysis software. Multiple IR bands can be selected in the lower section of this window. These bands can then be automatically integrated or peak heights can be extracted and converted to concentrations using previously obtained calibration curves.

One system was designed to study the effluent stream of 16 reactors in parallel [38,39], while the other system allows us to study *adsorbates* on seven supported catalysts simultaneously [38,43]. The *gas phase sampling* system (setup shown in Figure 5.3) can be converted to the adsorbed surface species system by removing the 16-well reactor and replacing the gas phase array with the *IR transmission reactor*.

Study of Adsorbed Molecules on Supported Catalysts

The activity of heterogeneous catalysts is governed by the interaction of the reactants and products with the catalyst surface. Specifically, the study of adsorbed species on the catalyst surface assists in better understanding how a catalyst functions. This type of study can be performed on supported catalysts using transmission FTIR spectroscopy [62–64]. A powder composed of the supported catalyst is pressed into a self-supporting disk sufficiently thin to allow for the transmission of IR radiation. This disk is held in a reactor equipped with IR-transparent windows to allow analysis via transmission FTIR. Using this

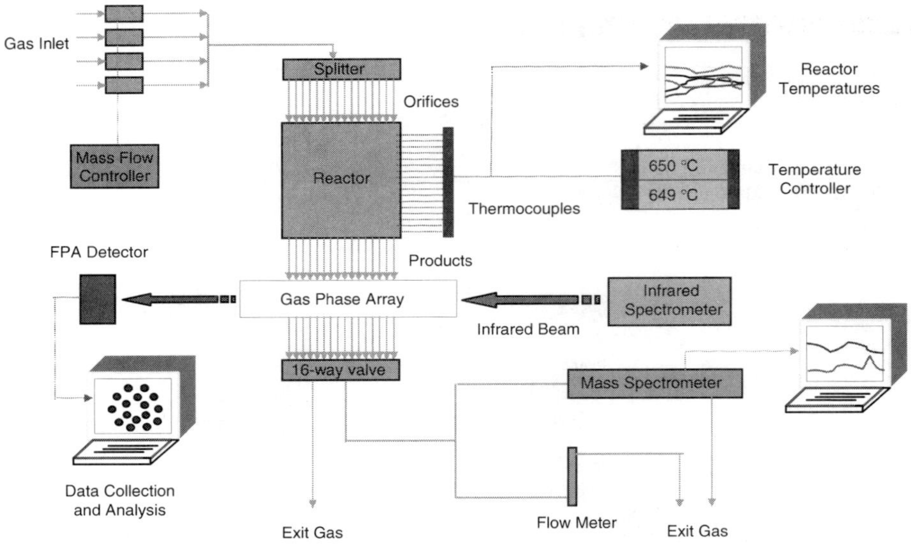

FIGURE 5.3. Schematics of the HT FTIR gas phase reaction product sampling setup.

approach, a wide variety of pressures (atmospheric to high vacuum) and temperatures (room temperature to several hundred Kelvin) can be achieved. In order to increase the throughput of this experiment, we designed a reactor that can hold seven supported catalysts so that pretreatment of catalysts, reactions, and imaging can be done in parallel [38,43]. This reactor is constructed entirely of stainless steel ultrahigh-vacuum components. At the center of the reactor is a flange that holds the supported catalysts, and also provides temperature control through built-in heating cartridges and thermocouples. Additional flanges on either side of this central flange contain a common gas inlet, common exhaust port, a pressure gauge, and IR-transparent windows. Dosing and transient studies can also be performed to show the catalyst response to differing reactant concentrations. These experiments provide various ways of testing and confirming theoretical reaction pathways.

Gas Phase Reaction Product Sampling

In order to evaluate the performance of individual catalyst samples in a library, it is desirable to determine their exit stream composition directly under the operating conditions that they will be exposed to in a production facility. It is also important to pretreat the catalysts, run the reaction, and analyze all product streams in parallel. For this purpose, we developed a *gas phase array* (GPA) sampling cell and a *parallel reactor system* for HT analysis [41]. The GPA is composed of 16 tightly packed stainless steel tubes, each with a separate gas inlet and outlet, capped at the ends with IR-transparent windows. The windows are sealed with a gasket sheet consisting of Buna rubber, Viton, or expanded polytetrafluoroethylene, depending on the specific chemical resistance required.

The array is a flow-through design, which allows both steady state and transient studies to be performed.

The GPA is connected to a home-built parallel reactor system (see Figure 5.3 for details). Premixed reactant gases are evenly divided through a 16-way splitter valve followed by individual laser-drilled orifices for each vertically aligned reactor to produce a uniform pressure drop and uniform flow conditions for each catalyst. From this point, all gas streams remain in separate channels until after the GPA, preventing the mixing of product streams. Individual vertical stainless steel reactor cartridges hold the catalysts in powder form, forcing the reaction gases to pass through the catalyst bed. Individual O-rings, which are not located in the hot zone of the reactor, seal the reactor channels. Exchanging 16 catalyst samples routinely takes 15–20 min. The setup allows for space velocities in the range of 200,000 to 3,500 ml hr^{-1} $g_{cat}{}^{-1}$, making operation in the differential regime and kinetic analysis feasible. With this design, the catalysts do not need to be pressed into pellets or prepared on a substrate and can easily be recovered for further testing or characterization. All transfer, reactor, and GPA tubing in this system are made from stainless steel and wrapped with heating tape, making studies of many condensable or polymerizable products gases possible.

The temperature of each reactor is monitored with its own thermocouple placed directly in the catalyst bed. A PID controller is used to control 16 heating cartridges (100 W each), allowing a maximum operating temperature of the reactor of over 750°C. The temperatures of the individual reactors are logged constantly by computer software written in house using LabView. The data are then appended to the spectroscopic data by the data-processing software package. In our system, it was found that typical temperature distributions among individual reactors are $\pm 1\%$ with no reaction occurring. As will be shown later in this chapter, measuring an average reactor temperature or measuring temperature distributions prior to full operation is insufficient for a real kinetic analysis of the data because of the exothermicity of many reactions. In this case, the potential for thermal cross talk among the reactors also has to be taken into consideration.

To monitor IR-inactive species, a mass spectrometer with a computer-controlled switching valve has been incorporated downstream of the GPA to analyze an individual reactor. However, it is well known that quantifying MS results can be quite difficult [65–67]. Like other MS approaches, it is a serial process with all associated disadvantages. Other serial techniques, such as GC, could also be implemented, if desired.

The lower limit of detection (LLD) is the point where the characteristic IR absorption peak is statistically insignificant to the noise in the system [68,69]. The lower limit of quantification (LLQ) is the lowest value where the characteristic transmission peak is statistically measurable to the noise in the system. These are defined as follows:

$$LLD \Rightarrow P/N = 3N \quad LLQ \Rightarrow P/N = 10N \qquad (5)$$

where P is peak height and N is root mean square of spectral noise, as shown in Figure 5.4.

The values in Eqn (5) can be determined by passing known concentrations through the GPA, measuring the peaks, and comparing them with the noise as previously calculated. The LLD varies for different molecules depending on the absorptivity of the particular molecule. These values can easily be determined by mixing various concentrations of

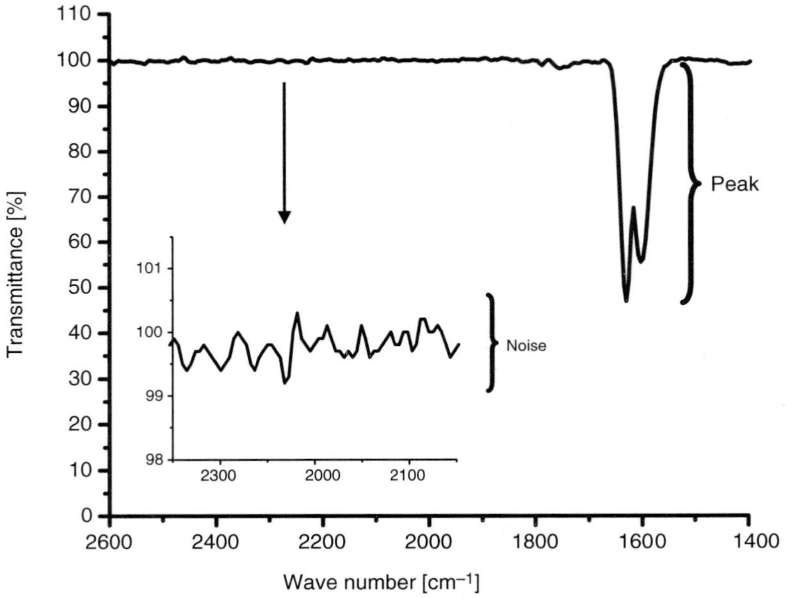

FIGURE 5.4. Transmission spectrum of NO_2 demonstrating the definitions of typical peak signal and spectral noise.

TABLE 5.1. LLD and LLQ for Various Gaseous Species

	LLD (ppm)	LLQ (ppm)
NO	370	900
N_2O	20	50
NO_2	30	70

gaseous species and passing them through the GPA. Example measurements for our system are given in Table 5.1.

To summarize, this setup allows in situ pretreated catalysts to be studied out various temperatures, feed concentrations, and flows under both transient and steady state conditions. Orders of reaction, activation energies, and pre-exponential factors can then be calculated from these quantitative data.

CASE STUDY: CO OXIDATION

In order to demonstrate the applicability of our setup for FTIR imaging of gaseous species, we present results of CO oxidation studies on supported metal catalyst at atmospheric pressure. Over the course of 2 days, 28 operating conditions were tested for each of the 16 samples for a total of 448 experiments and approximately 2.5 million unique IR spectra.

Three different amounts (0.50, 0.15, and 0.05 g) of the following catalysts on Al_2O_3 support were loaded into the reactor: 1 wt% Pt, 1 wt% Pd, 5 wt% Ir, 5 wt% Rh, and 5 wt% Ru. The catalysts were pretreated with three cycles of heating to 500°C in helium, followed by 30 min in a stream of 25 sccm oxygen and 60 sccm helium at 500°C, and cooling to 300°C, followed by 30 min in a stream of 15 sccm hydrogen and 60 sccm helium per reactor channel. Four reaction temperatures were tested (60, 105, 165, and 240°C) at seven CO concentrations (0–10.5 mol%) in 75 sccm helium and 10 sccm oxygen per reactor channel.

CO oxidation on supported catalysts is an exothermic reaction with a very simple product stream. Therefore IR thermography could be used for this reaction to give relative catalysts performance, but only because both of the preceding statements are true [37]. In our reactor setup, thermocouples in each catalyst bed can provide the same information without the complexity associated with thermal imaging (electronics, background correction, emissivity, etc.). However, more detailed and quantitative information can be gained by recording temperature and IR spectra simultaneously. This will be shown using the 2.95 mol% CO feed as an example. Figure 5.5 shows baseline corrected spectral images of the gas phase array that are obtained by using the intensity at a particular wavelength in each pixel of the detector. In this case, the P branch was used for CO (2,180 cm^{-1}) and the R branch for CO_2 (2,372 cm^{-1}). As CO is converted to CO_2 for a particular reactor, the intensity in the image for CO decreases as the CO_2 intensity increases. As can be seen from these images, our technique provides immediate visualization of the catalyst performance and enables the user to follow both reactant and product concentration in one data set.

Figure 5.6 shows the sample effluent gas phase spectra for the 0.05 g Rh catalyst (second row, second from the left in any image of Figure 5.5). Eighteen spectral images

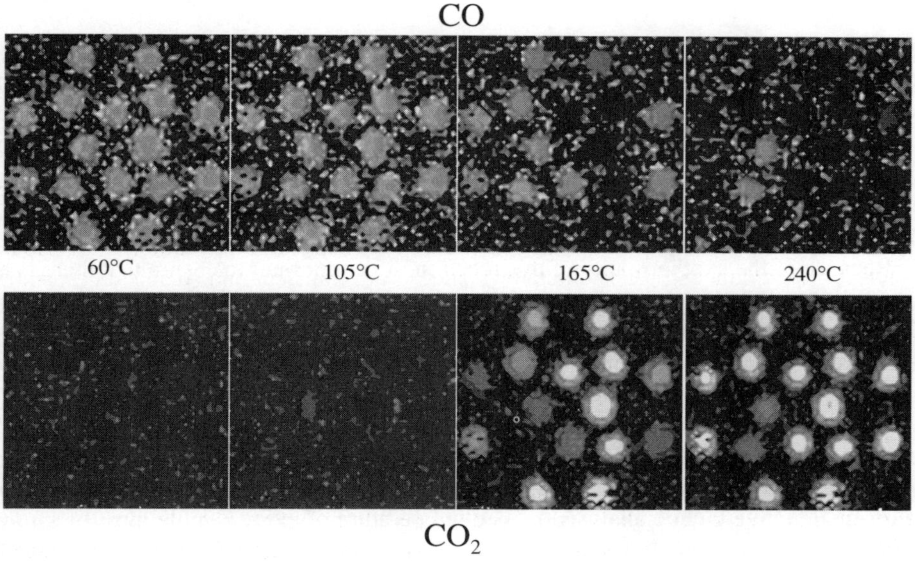

FIGURE 5.5. Spectral images of both CO and CO_2 as a function of temperature for all 16 reactor channels.

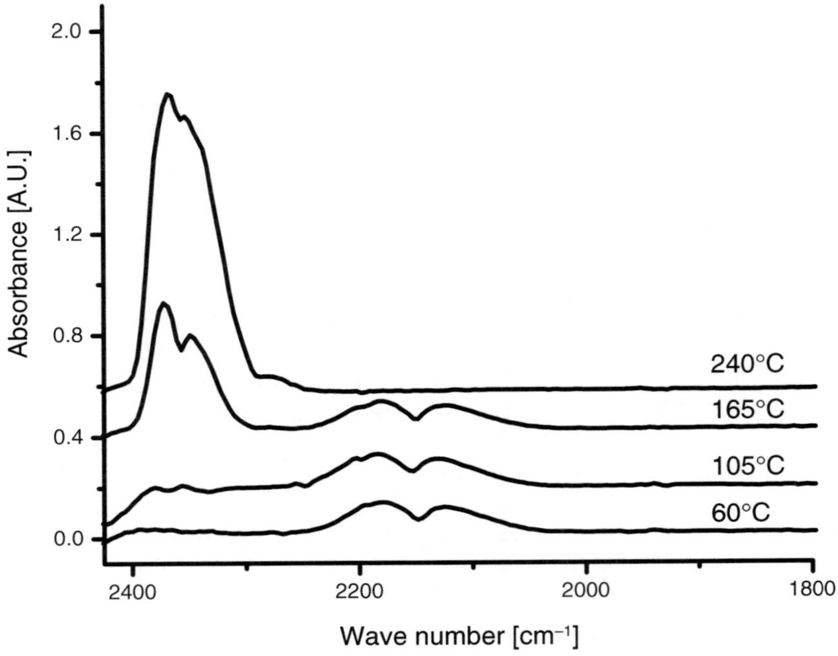

FIGURE 5.6. Representative IR spectra of the effluent gas stream for the 5 wt% Rh (0.05 g) catalyst sample at four temperatures. As expected, the increase in CO_2 production coincides with a decrease of CO in the effluent stream.

and 9 pixels for this channel for a total of 162 total spectra were averaged. The decrease in CO and the increase in CO_2 are clearly visible with increasing temperature. Because the absorptivity in the IR of CO is less than that of CO_2, peak intensities cannot be compared directly [see Eqn (4)], but must be converted to concentrations using calibrations. The results for a CO_2 calibration for all 16 reactors are shown in Figure 5.7. These data are very reproducible and absolute concentrations can easily be calculated once a set of calibrations has been obtained.

Because of the exothermicity of this reaction, it is important to ensure that there is no thermal cross talk between reactors. At full conversion, some catalyst beds heated to temperatures more than 170°C above the set-point of the reactor. However, no thermal cross talk between reactors is evident at any time during the experiments. Figure 5.8 shows the distribution of catalyst bed temperatures at a reactor set-point of 165°C and a feed of 10.5 mol% CO. Quenching the reaction by removing the reactants from the feed gas decreases the temperature to the set point within several seconds, showing that the temperature distribution is not a result of our reactor design. It is also clear from these data that for quantitative kinetic analysis the bed temperature of every sample must be known and that the frequently made assumption of uniform temperature in the reactor will lead to erroneous results.

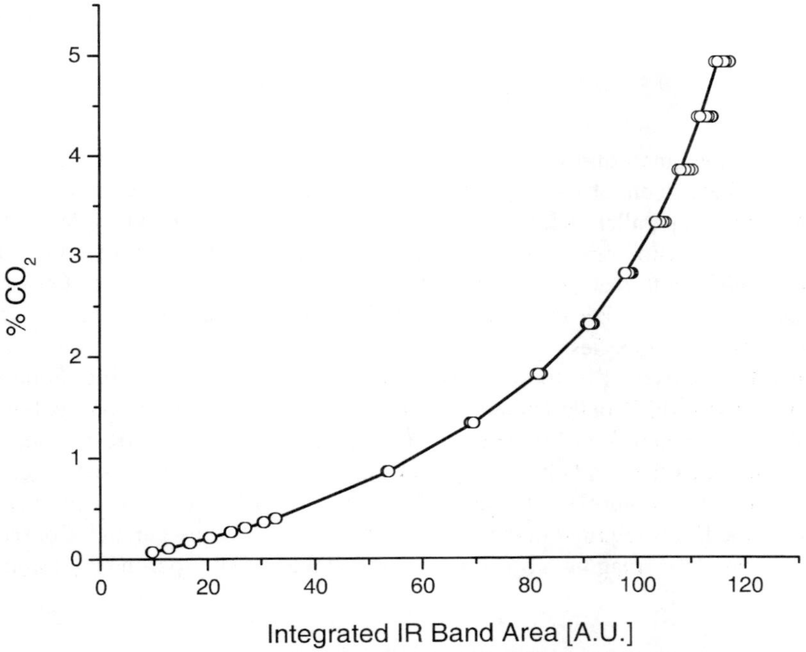

FIGURE 5.7. Concentration of CO_2 in all 16 channels of the GPA measured with FTIR imaging as a function of the integrated IR band area. The data points for all 16 channels mostly lie on top of each other.

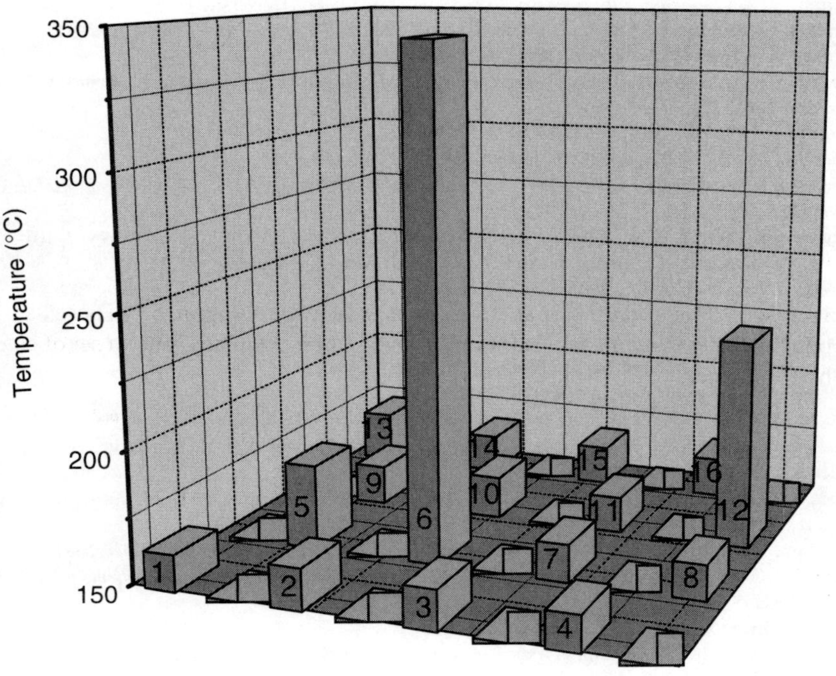

FIGURE 5.8. Distribution of catalyst bed temperature for all 16 catalyst samples at a reactor set-point of 165°C and feed concentration of 10.5 mol% of CO. The positions in the graph correspond to the locations of the individual reactors in the reactor array. The high temperature of bed 6 does not affect the surrounding channels and shows the absence of any thermal cross talk.

CONCLUSIONS

We have designed and constructed the first HT setup which allows quantitative evaluation of the effluent of 16 supported catalysts and simultaneous measurement of their bed temperature in parallel under realistic operating conditions with a temporal resolution below 2 s. It now becomes possible not only to screen catalysts rapidly for their activity, but also to follow both reactants and products in the exit stream in a quantitative manner and to determine key properties of the catalysts, such as conversion, selectivity, reaction order, and activation energies.

Every HT analytical technique has its disadvantages. For example, homonuclear species do not absorb IR radiation and cannot be detected by IR spectroscopy, fragmentation patterns of molecules in MS make peak overlaps very common, IR thermography is not chemically sensitive, and GC is very slow and careful choice of columns is often necessary to separate a complex effluent stream. The only HT methodologies that can be scaled up to the levels required in the future are those that are truly parallel. Consequently, we believe that FTIR imaging has enormous potential as a HT experimental technique.

REFERENCES

1. Szostak, J. W. *Chem. Rev.* **1997**, *97*, 347–348.
2. Terrett, N. *Drug Discovery Today* **1997**, *2*, 250.
3. Terrett, N. K. *Combinatorial Chemistry* **1998**, New York: Oxford University Press.
4. Creer, J. G., Jackson, P., Pandy, G., Percival, G. G., Seddon, D. *Appl. Catal.* **1986**, *22*, 85–95.
5. Senkan, S. *Angew. Chem. Int. Ed.* **2001**, *40*, 312–329.
6. Hagemeyer, A., Jandeleit, B., Liu, Y. M., Poojary, D. M., Turner, H. W., Volpe, A. F., Weinberg, W. H. *Appl. Catal. A* **2001**, *221*, 23–43.
7. Engstrom, J. R., Weinberg, W. H. *AIChE J.* **2000**, *46*, 2–5.
8. Weinberg, W. H., McFarland, E. W. *Tibtech* **1999**, *17*, 107–115.
9. Jandeleit, B., Schaefer, D. J., Powers, T. S., Turner, H. W., Weinberg, W. H. *Angew. Chem. Int. Ed.* **1999**, *38*, 2495–2532.
10. Jandeleit, B., Turner, H. W., Uno, T., van Beek, J. A. M., Weinberg, W. H. *Cat Tech* **1998**, *2*, 101–123.
11. Cong, P., Giaquinta, D., Guan, S., McFarland, E., Turner, H., Weinberg, W. H. *AIChE Spring National Meeting, New Orleans*, LA, **1998**.
12. Scheidtmann, J., Weiss, P. A., Maier, W. F. *Appl. Catal. A* **2001**, *222*, 79–89.
13. Maier, W. F., Kirsten, G., Orschel, M., Weiss, P. A. *Chim. Oggi–Chem. Today* **2000**, *18*, 15–19.
14. Maier, W. F. *Angew. Chem. Int. Ed.* **1999**, *38*, 1216–1218.
15. Hoffmann, R. *Angew. Chem. Int. Ed.* **2001**, *40*, 3337.
16. Liu, Y. M., Cong, P. J., Doolen, R. D., Turner, H. W., Weinberg, W. H. *Catal. Today* **2000**, *61*, 87–92.
17. Richter, A., Langpape, M., Kolf, S., Grubert, G., Eckelt, R., Radnik, J., Schneider, A., Pohl, M. M., Fricke, R. *Appl. Catal. B* **2002**, *36*, 261–277.
18. Rodemerck, U., Wolf, D., Buyevskaya, O. V., Claus, P., Senkan, S., Baerns, M. *Chem. Eng. J.* **2001**, *82*, 3–11.
19. Krantz, K., Ozturk, S., Senkan, S. *Catal. Today* **2000**, *62*, 281–289.
20. Newsam, J. M., Bein, T., Klein, J., Maier, W. F., Stichert, W. *Micropor. Mesopor. Mater.* **2001**, *48*, 355–365.
21. Bein, T., Gardner, D., Hilbrandt, N., Choi, K., Fanwick, P. *Abstr. Pap. Am. Chem. Soc.* **1999**, *218*, 143-ANYL.
22. Bein, T. *Angew. Chem. Int. Ed.* **1999**, *38*, 323–326.
23. Hoffmann, C., Wolf, A., Schuth, F. *Angew. Chem. Int. Ed.* **1999**, *38*, 2800–2803.
24. Reetz, M. T., Kuhling, K. M., Wilensek, S., Husmann, H., Hausig, U. W., Hermes, M. *Catal. Today* **2001**, *67*, 389–396.

25. Buyevskaya, O. V., Bruckner, A., Kondratenko, E. V., Wolf, D., Baerns, M. *Catal. Today* **2001**, *67*, 369–378.
26. Hagemeyer, A., Borade, R., Desrosiers, P., Guan, S. H., Lowe, D. M., Poojary, D. M., Turner, H., Weinberg, H., Zhou, X. P., Armbrust, R., Fengler, G., Notheis, U. *Appl. Catal. A* **2002**, *227*, 43–61.
27. Desrosiers, P., Guram, A., Hagemeyer, A., Jandeleit, B., Poojary, D. M., Turner, H., Weinberg, H. *Catal. Today* **2001**, *67*, 397–402.
28. Orschel, M., Klein, J., Schmidt, H.-W., Maier, W. F. *Angew. Chem. Int. Ed.* **1999**, *38*, 2791–2794.
29. Weinberg, H. W., McFarland, E. W., Cong, P., Guan, S. US Patent 5,959,297, **1999** (to Symyx Technologies).
30. Claus, P., Honicke, D., Zech, T. *Catal. Today* **2001**, *67*, 319–339.
31. Schrader, W., Eipper, A., Pugh, D. J., Reetz, M. T. *Can. J. Chem.—Rev. Can. Chim.* **2002**, *80*, 626–632.
32. Yamada, Y., Ueda, A., Zhao, Z., Maekawa, T., Suzuki, K., Takada, T., Kobayashi, T. *Catal. Today* **2001**, *67*, 379–387.
33. Senkan, S. M. *Nature* **1998**, *394*, 350–353.
34. Su, H., Hou, Y. J., Houk, R. S., Schrader, G. L., Yeung, E. S. *Anal. Chem.* **2001**, *73*, 4434–4440.
35. Su, H., Yeung, E. S. *J. Am. Chem. Soc.* **2000**, *122*, 7422–7423.
36. Maier, W. F., Holzwarth, A., Schmidt, H. W. *Angew. Chem. Int. Ed.* **1998**, *37*, 2644–2647.
37. Moates, F. C., Somani, M., Annamalai, J., Richardson, J. T., Luss, D., Willson, R. C. *Ind. Eng. Chem. Res.* **1996**, *35*, 4801–4803.
38. Snively, C. M., Oskarsdottir, G., Lauterbach, J. *Catal. Today* **2001**, *67*, 357–368.
39. Snively, C. M., Oskarsdottir, G., Lauterbach, J. *Angew. Chem. Int. Ed.* **2001**, *40*, 3028–3030.
40. Snively, C. M., Oskarsdottir, G., Lauterbach, J. *J. Comb. Chem.* **2000**, *2*, 243–245.
41. Snively, C. M., Lauterbach, J. *Spectroscopy* **2002**, *17*, 26–33.
42. Snively, C. M., Lauterbach, J. In E. G. Derouane, F. Lemos, A. Corma, and F. R. Ribero (Eds.), *Combinatorial Catalysis and High-Throughput Catalyst Design and Testing*, Vol. 560, **1999**, Kluwer Academic, pp. 437–439.
43. Snively, C. M., Katzenberger, S., Oskarsdottir, G., Lauterbach, J. *Opt. Lett.* **1999**, *24*, 1841–1843.
44. Lauterbach, J., Snively, C. M., Oskarsdottir, G. In R. Malhotra (Ed.), *Combinatorial Materials Development*, Vol. 814, **2002**, Washington, DC: American Chemical Society, 65–87.
45. Lauterbach, J., Snively, C. M., Oskarsdottir, G. *Abstr. Pap. Am. Chem. Soc.* **2000**, *219*, 6-MTLS.
46. Snively, C. M., Koenig, J. L. *Appl. Spectrosc.* **1999**, *53*, 170–177.
47. Snively, C. M., Koenig, J. L. *J. Polym. Sci. Polym. Phys. Ed.* **1999**, *37*, 2353–2359.
48. Snively, C. M. Unpublished PhD Thesis, **1999**, Department of Macromolecular Science, Case Western Reserve University; Cleveland, OH.
49. Snively, C. M., Koenig, J. L. *J. Polym. Sci. Polym. Phys. Ed.* **1999**, *37*, 2261–2268.
50. Snively, C., Lauterbach, J. US Patent, pending.
51. Snively, C. M., Koenig, J. L. *J. Polym. Sci. Polym. Phys. Ed.* **1999**, *37*, 2353–2359.
52. Snively, C. M., Koenig, J. L. *Macromol.* **1998**, *31*, 3753–3755.
53. Koenig, J. L., Snively, C. M. *Spectroscopy* **1998**, *13*, 22–28.
54. Lewis, E. N., Levin, I. W., Crocombe, R. A. *Mikrochim. Acta* **1997**, Suppl. 14, 589–590.
55. Lewis, E. N., Kidder, L. H., Pentchev, I. W., Levin, I. W., Lester, D. S. *Biophys. J.* **1997**, *72*, MPME5.
56. Lewis, E. N., Gorbach, A. M., Marcott, C., Levin, I. W. *Appl. Spectrosc.* **1996**, *50*, 263–269.
57. Kidder, L. H., Levin, I. W., Lewis, E. N., Kleiman, V. D., Heilweil, E. J. *Opt. Lett.* **1997**, *22*, 742–744.
58. Lewis, E. N., Levin, I. W. *Appl. Spectrosc.* **1995**, *49*, 672–678.
59. Lewis, E. N., Treado, P. J., Reeder, R. C., Story, G. M., Dowrey, A. E., Marcott, C., Levin, I. W. *Anal. Chem.* **1995**, *67*, 3377–3381.
60. Treado, P. J., Levin, I. W., Lewis, E. N. *Appl. Spectrosc.* **1994**, *48*, 607–615.
61. Griffiths, P. R., de Haseth, J. A. *Fourier Transform Infrared Spectrometry*, **1986**, New York: John Wiley.
62. Fanson, P. T., Delgass, W. N., Lauterbach, J. *J. Cat.* **2001**, *204*, 35–52.
63. Barshad, Y., Zhou, X., Gulari, E. *J. Cat.* **1985**, *94*, 128–141.
64. Ryczkowski, J. *Catal. Today* **2001**, *68*, 263–381.
65. Hall, J., Lundgren, S., Keck, K. E., Kasemo, B. *Int. J. Mass Spectrom. Ion Phys.* **1991**, *108*, 1–21.
66. Gaskell, S. J., Finlay, E. M. H. *Trac-Trends Anal. Chem.* **1988**, *7*, 202–208.
67. Babalievski, F. V. *Anal. Chim. Acta* **1987**, *201*, 241–252.
68. Adams, M. J. *Chemometrics in Analytical Spectroscopy*, **1995**, Cambridge: Royal Society of Chemistry.
69. Miller, J. C., Miller, J. N. *Statistics for Analytical Chemistry*, **1988**, Chichester, UK: Ellis Horwood.

Chapter 6

Linear Spatially Encoded Combinatorial Chemistry with Fourier Transform Library Analysis

Alan W. Schwabacher, Christopher W. Johnson, and Peter Geissinger

COMBINATORIAL STRATEGIES

An effective combinatorial chemical technique must facilitate all aspects of the experiment, including synthesis, assay, and association of structure with activity; it must also provide output data in a usable form. The combinatorial technique best applied to a given study depends on many factors, with the availability of a high-throughput assay critical to the choice of library size. It is also important to distinguish between library experiments where the target is a substance, and those where information is sought. A selection scheme with subsequent decoding of active compounds is appropriate when searching for a substance, while a spatially encoded procedure in which all library members are identified can potentially provide more information.

"OPTIMAL" PROCEDURES

Parallel synthesis is important to synthetic efficiency, particularly in larger libraries. Consider a library prepared by six sequential steps using 20 reagents at each step to yield 6.4×10^7 compounds. A parallel synthesis, with all reactions at each step carried out simultaneously, would require six sequential reactions to build the entire library. Such simultaneity is not sufficient for what we term "fully parallel" synthesis. For a synthesis to be fully parallel, we also require that any reagent used in a given step be handled once and delivered to a single reaction vessel per step; placing a reagent in a vessel is an operation. For instance, a parallel synthesis carried out robotically in microtiter plates, would require

Alan W. Schwabacher, Christopher W. Johnson, and Peter Geissinger • Department of Chemistry, University of Wisconsin-Milwaukee 3210 N. Cramer Street, Milwaukee, WI 53211, USA
Contact author: Alan W. Schwabacher, email: awschwab@uwm.edu

High-Throughput Analysis, edited by Potyrailo and Amis
Kluwer Academic Publishers, New York, 2003.

at least 3.8×10^8 independently addressed operations, so that while the reactions are parallel, the manipulations are not. A partially serial procedure, such as the very large scale integrated parallel synthesis (VLSIPS) method [1], which uses photolithographic masking, would entail 120 sequential reactions. A fully parallel synthesis of the same library would entail six sequential steps with a total of 120 operations.

The key to a fully parallel procedure is the ability to subdivide the library for reaction at one step, and to resubdivide the library in an orthogonal manner in subsequent steps. The split and pool synthesis [2,3] provides a particularly convenient way to do this. For investigations seeking a substance, an appropriate analytical [4,5] or tagging [6–8] method allows identification of selected library members. If identification of all or many library members is desired, efficiencies in identification with a spatially encoded method can compensate for the less efficient synthesis.

The ideal assay for a combinatorial experiment provides the activity of each library member simultaneously, in such a way as to associate connectivity with a given structure. This ideal is most closely approached using two-dimensional arrays of compounds in which an image of the array can provide activity data where the identity of a given compound is associated with that compound's position in the array. In general, synthetic methods producing such arrays are not fully parallel.

LINEAR SPATIAL ENCODED SYNTHESIS (LSES)

Solid supports of effectively zero dimensions—particles—provide optimal flexibility to allow fully parallel synthesis, while solid supports of two-dimensional contiguity are particularly appropriate for the analysis of positionally identified library members. We describe here how solid supports of effectively one-dimensional morphology (threads) can provide fully parallel synthesis with full positional identification [9]. Position in a single dimension along the thread identifies the region of solid support bearing a given compound, while the linear morphology allows reconfiguration in such a way as to provide all the combinations useful for synthesis. Other researchers have described procedures that use a thread to organize and identify library members during synthesis, though these involved substantially more manipulation during synthesis [10,11].

The challenge is to enable reaction of a set of compounds with a set of reagents such that all combinations are prepared, and to facilitate multiple steps of exponential diversification. Our approach uses a one-dimensionally organized solid support in place of effectively zero-dimensional beads or two-dimensional surfaces. A linear solid support ("thread") allows fully parallel synthesis in a simple way (Figure 6.1). The thread is wrapped around a cylinder in a single spiral layer, and impermeable barriers are placed lengthwise along the cylinder such that spaces between divisions constitute distinct reaction vessels. Coupling a different reagent to each of the regions (one reagent per reaction vessel) creates repeating domains along the thread. Each domain bears a single species, and the identity of each species is specified by its distance from the end of the thread. If three reagents were used for the first step, the sequence of compounds attached to the thread would be A B C A B C A B C A B C A B C A B C A B C A B C A B C . . . .

Before a second set of reagents can be coupled, the thread must be redivided into regions such that all reagent combinations will be equally represented. This is done by

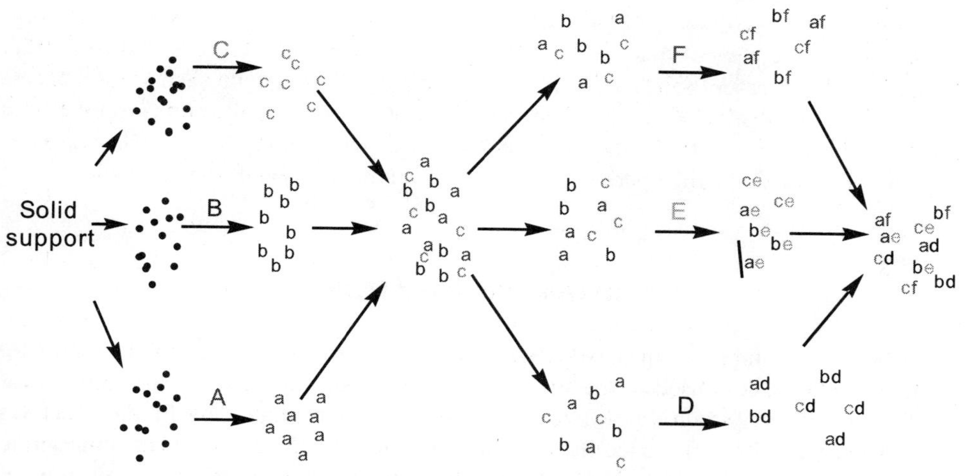

FIGURE 6.1. Split–mix Synthesis. Fully parallel synthesis on a solid support is possible because at the library is subdivided in an orthogonal manner each reaction step. Each particle has a single synthetic history; given sufficient particles, all combinations are formed. Several schemes are available to recover or retain information lost on mixing, with some cost.

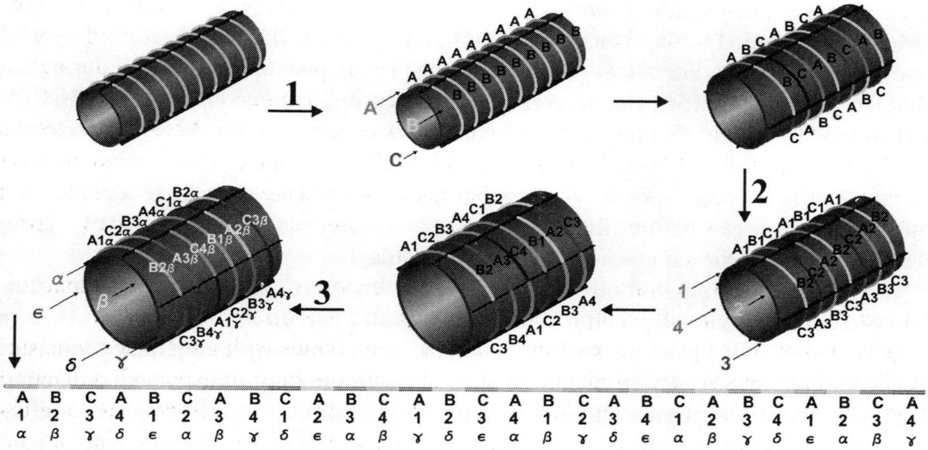

FIGURE 6.2. Linear spatial encoded synthesis. Fully parallel synthesis on domains of a solid support of linear morphology. Step 1 is carried out while the support is wrapped on a cylinder. Wrapping onto a cylinder of different appropriate diameter puts regions out of phase, so that step 2 reactions form all combinations. The process can be repeated as desired to produce libraries in a known sequence.

wrapping the thread around a cylinder of an appropriate diameter different from the first, followed by division into regions and coupling of the second set of reagents (Figure 6.2). If four reagents were used for this step, the sequence of domains would be exposed to reagents in the following pattern: 1 2 3 4 1 2 3 4 1 2 3 4 1 2 3 4 1 2 3 4 1 2 3 4 1 2 3 4.... As each of these second reagents is paired with one of the first reagents, the net result is the sequence A1 B2 C3 A4 B1 C2 A3 B4 C1 A2 B3 C4 . . . . The identity of each compound is specified by its position, and the sequence is such that if one starts anywhere and proceeds along the sequence, there are no duplicates until the entire set has been encountered. Repetition of this process allows use of as many steps as desired, with each set of reagents on a cylinder of distinct diameter. The identity of each functional group installed by a given reaction varies periodically as a function of distance along the thread, with each cylinder providing a distinct period for the corresponding portion of the molecule.

LIBRARY PREPARATION

Cotton [12] thread is an inexpensive, convenient, and appropriate solid support for peptide synthesis; it is used in the simplest variant of our technique. Cotton thread was treated with carbonyldiimidazole, generation of the intermediate acyl imidazolide was confirmed by diffuse reflectance IR, and the functionalized thread was then subjected to reaction with 1,11-diamino-3,6,9-trioxaundecane. A density of amine groups of 5×10^{-8} mol/cm was determined by ninhydrin assay [13]. Direct acylation of the cotton with a protected amino acid gave an ester-linked thread [14]. Replicate libraries were prepared on sequential domains of these two solid support materials to enable cleavage and analysis as well as solid phase binding assays. The resulting thread materials are abbreviated herein as thread~NH_2. Peptide coupling was carried out under conditions previously described for peptide synthesis on a cellulose support, using Fluorenyl Methoxy Carbonyl (FMOC) protected Hydroxy benzotriazole (HOBT) esters at 0.3 M in N-methyl pyrolidinone (NMP) [14]. Acylation was monitored by the bromophenol blue method during coupling [15] and quantitated by ninhydrin assay. Successive 30 min couplings of FMOC-Ala to thread~NH_2, with acetic anhydride end-capping after each coupling before deprotection, gave successive yields of 50%, 70%, 100%, and 100%. Thus some of the amino groups of the thread~NH_2 appear to be less accessible than others. Therefore three alanines were coupled to the thread before library synthesis to ensure that these less reactive groups were terminated in the same way throughout the thread.

Cylinders of ultrahigh-molecular-weight polyethylene (UHMW PE) were machined about 30 cm long and of precise diameter so as to have circumferences of 1 cm per compound. A winding machine analogous to those used for winding electronic tuning coils was used to wrap thread very evenly in a single layer around these cylinders. Division lengthwise along the cylinder into regions onto which distinct amino acids were to be coupled was carried out as follows. A modified hot melt glue gun was used to apply a paraffin wax barrier on parallel black lines ruled every 1 cm lengthwise along the cylinder of thread (Figure 6.3). The wax was applied in sufficient quantity that bleed-through of peptide coupling reagents did not occur over the time of coupling. Solutions of activated FMOC amino acids in NMP at 0.3 M were applied to the thread by pipette, being careful to keep each amino acid to its own space between wax lines. At this concentration, the amount of activated amino acid absorbed by the region of thread is sufficient to acylate the peptide, as has been observed with paper solid support [14]. In some cases colorimetric assay allows qualitative assessment of the evenness and completeness of coupling. Over the course of the coupling reaction, bromophenol blue adsorbed on the thread changes from blue to yellow as the amine groups are consumed. Regions that remained blue were

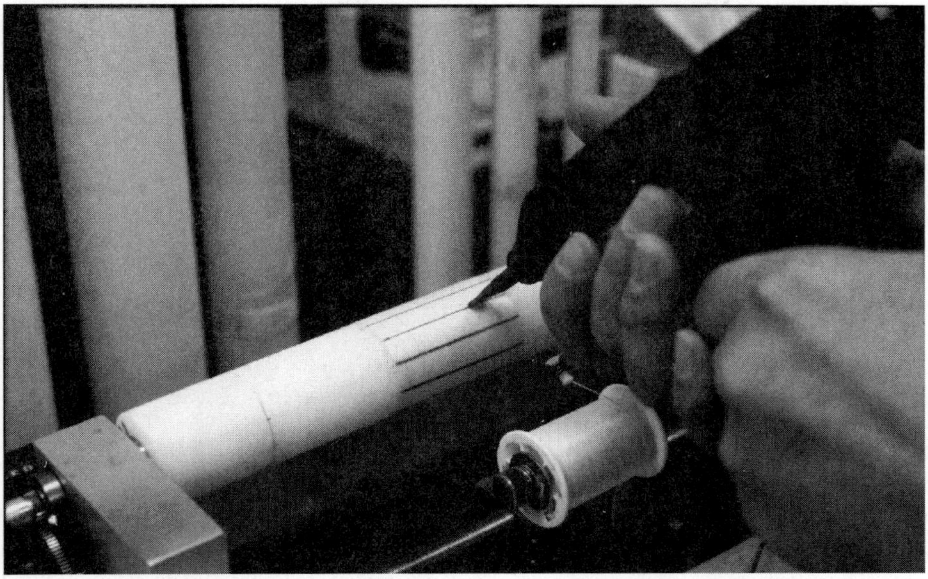

FIGURE 6.3. Application of barriers. For peptide synthesis on cotton thread as solid support, paraffin barriers are applied using a hot glue gun with a modified small orifice tip. Activated amino acids applied between barriers quickly spread to all thread regions desired, without leakage across barriers. Sufficient reagent for reaction is held by capillary action, so that immediately after application of one reagent the cylinder can be turned for application of the next. The thread winding lathe and other cylinders are also visible.

blotted to remove acylation solution, and recoupled in situ. After coupling, all regions were blotted and removed from the cylinder for end-capping, rinsing, deprotection with pyrrolidine, rinsing, bromophenol blue treatment, and then wrapping around the next cylinder for further reaction. At the end of the synthesis, side-chain deprotection was carried out with trifluoroacetic acid (TFA) in CH_2Cl_2. For simple t-Botoxycarbonyl (Boc) removal 20% TFA for 20 min is sufficient, and causes little degradation of the cellulose.

LIBRARY EVALUATION

After preparation of the peptide library, assay and analysis were conducted. First, fluorescein-conjugated streptavidin was incubated with the thread library to tag those library components which bound to streptavidin with fluorescence dependent upon binding propensity.

Wheels with sides to hold the thread were installed in ordinary audiocassette cases, along with Teflon tubes to direct the thread, which replaces the tape. An ordinary audiotape player with the record/play heads removed serves to pull the thread at a constant rate. This is convenient for libraries of moderate size, but clearly is not a limitation on the method as spools of any size driven by a stepper motor can pull thread. Teflon tubes direct thread from cassette to a cell, which was made to fit into a standard fluorescence spectrometer (Figure 6.4). An aluminum block the size and shape of a standard fluorescence

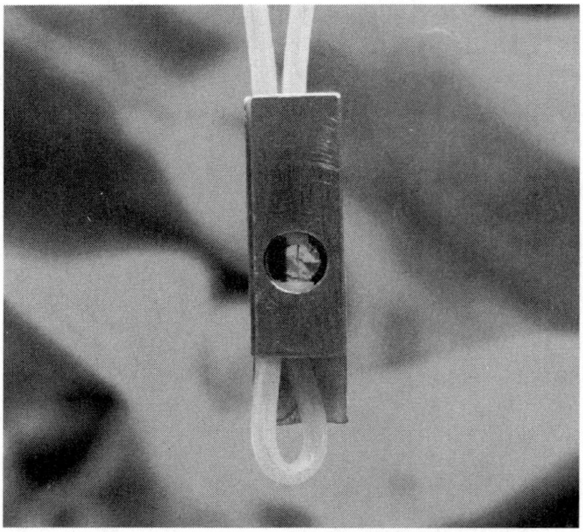

FIGURE 6.4. Thread fluorescence cell. Teflon tubes direct the thread library from the driving system (cassette player) to a spectrofluorimeter. Perpendicular lenses focus exciting light and collimate emitted light so that the time varying fluorescence output constitutes the activity of each library member in sequence.

cell was prepared, with Teflon tubes to direct the thread down the center of the block, through the light beam, and up the corner out of the light path. Perpendicular lenses mounted in the cell focus the excitation beam into a small spot (<1 mm) on the thread, and collimate the emitted light.

With the cell in a fluorescence spectrometer, the thread was pulled at a constant rate through a monochromatic beam of light focused on the thread, and the dispersed light was filtered and then detected by a photomultiplier tube (PMT). The PMT signal was fed into a computer that recorded the time course of the signal. Time corresponds to distance along the thread because of the constant speed of the thread, so the activities of each sequential library member are read out in known sequence.

The sequence of compounds facilitates chemical evaluation of library synthesis. If a subset of a large library were to be fully characterized, one would want to ensure that all the constituent reagents are represented in the subset. If one were to evaluate any contiguous set of compounds sufficient to represent the maximum number of reagents used at a given step, it would contain all the different reagents used at each step for the entire library.

FOURIER TRANSFORM DATA ANALYSIS

The sequence of compounds formed by this mode of synthesis is particularly conducive to analysis. As the identity of each library member is known, "best" activities are immediately associated with structures, and trends among compounds of various activity levels can be identified by any appropriate method. Library data obtained using an assay of large dynamic range can provide substantially more information than a simple screening threshold, but provide less confidence than careful quantitative binding or other measurements on purified and well-characterized samples. Such semiquantitative data must be handled in a responsible manner so as to extract maximal information while avoiding overinterpretation. We restrict the discussion here to the effects of substitution, rather than shape, polarity, and electronic properties that underlie activity and form the basis of a more complete analysis.

Consider the sequence of activity data for each library member to be a time-varying signal of the library as a whole (Figure 6.5). To be extracted from this signal are the trends among compounds: which features enhance or diminish activity, to what extent are these contributions linearly combinable, and groupings of library members to facilitate further analysis. We apply standard techniques of digital signal processing to the whole library signal, but emphasize here that this is quite different from the usual signal so analyzed. Our signal is not a discretely sampled representation of a continuous function, but a set of discrete measurements on distinct substances, ordered so as to correlate structural features with signal frequencies. One consequence is that the usual radix 2 fast Fourier transform (FFT) algorithms are not appropriate, and an algorithm appropriate to the number of data points present must be used. We use the powerful and convenient FFTW function library [16].

At each synthetic step, adjacent domains along the solid support react with different species. If the library members are considered to bear groups installed by each reaction (not the only possibility, but the simplest to describe), then every group in every molecular position changes on proceeding from one library member to the next in sequence.

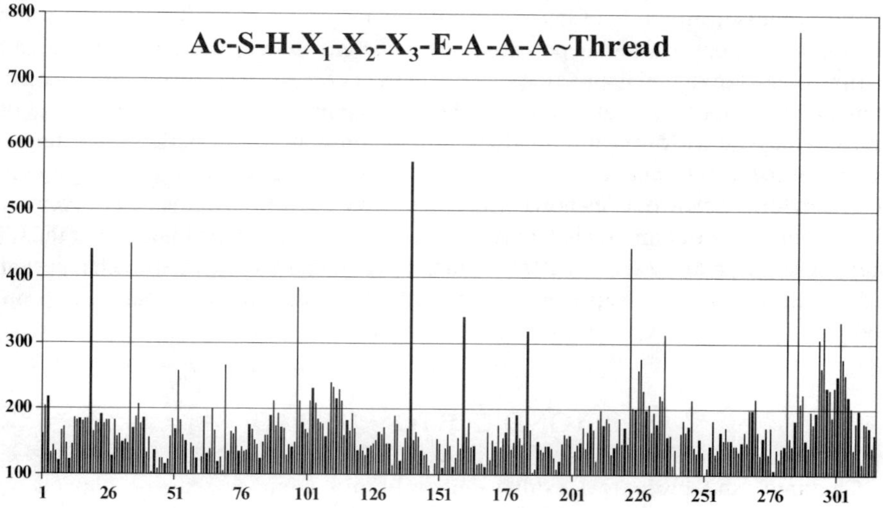

FIGURE 6.5. Thread library evaluation. Fluorescence (arbitrary units) as a function of sequential library member. Fluorescein-conjugated streptavidin bound to peptides leads to an output signal. Peptides have the structure shown, with variation in the x_1, x_2, and x_3 positions. Identities and sequences of variants at these positions are given in Figure 6.6.

As one proceeds along the sequence of compounds, the identity of the group in each position changes periodically, with the period determined by the size of the cylinder upon which the solid support was wrapped during the reaction that installed groups in that region of the molecules. Thus there is a characteristic period associated with each region of the molecule. If we look for periodicity in the activity data signal, the periods observed will indicate which molecular regions are associated with activity variation.

Fourier transformation of the library signal yields a frequency-domain spectrum with the same information as the original data in a form amenable to generalization. The significance to activity of library structural variation at a particular molecular position is indicated by a peak at a frequency equal to the reciprocal of the characteristic period defined above. These characteristic frequencies are known from the mode of synthesis, and so spectral interpretation is straightforward. As each frequency-domain peak is derived from all the (time-domain) library data, idiosyncratic or spurious activities contribute little. The time-domain waveforms are sets of repeated spikes that do not approximate sine waves, and so their Fourier transforms show peaks not only at the characteristic repetition frequency, but also at higher harmonics. These harmonics add spectrum complexity, but also contain useful information. The pattern of frequencies observed indicates those molecular regions of significance to activity, but not which functional groups are responsible for that activity. These data are present in the phases of the frequency-domain peaks. To extract them, one constructs a small data array consisting of the signal at a characteristic frequency and its harmonics, including both real and imaginary data. Transformation of this array back to the time domain yields a set of data points, equal in number to the characteristic period; each point represents the activity of a particular functional group at that molecular position, averaged over the entire library with regard to the other molecular positions. The functional group data appear in the same sequence as they do in the library sequence. We term such a data set a "fitness profile", (Figure 6.6).

Such fitness profiles are of interest, but of limited scope. In most situations, structural variation at a single molecular position is insufficient to determine substantial activity variation. If two molecular positions must interact, for example, to present functional groups to a receptor, their pairing in the library sequence will recur at the product of the characteristic periods for each individual group. Consequently, a peak in the frequency-domain spectrum at the characteristic frequency will be larger than the two single-cylinder fundamental frequencies. A fitness profile extracted at this lower frequency (larger period) will show the effect of variation in these two positions, averaged over the rest of the library (Figure 6.7). A profile of $x \times y$ data points, corresponding to two cylinder periods of x and y, will diagonally fill an x by y array, with the ith data point appearing at (i mod x, i mod y).

BEYOND COTTON THREAD

Chemical and biological evaluation of library members can be carried out in many ways. Obviously, compounds can be removed from the thread and handled in any desired manner. However, for analytical methods compatible with solid state materials, surface-bound reagents, or liquids held by capillary action within a region of thread, the linear form can be useful. A wheel is the simplest sample-changer available. We have

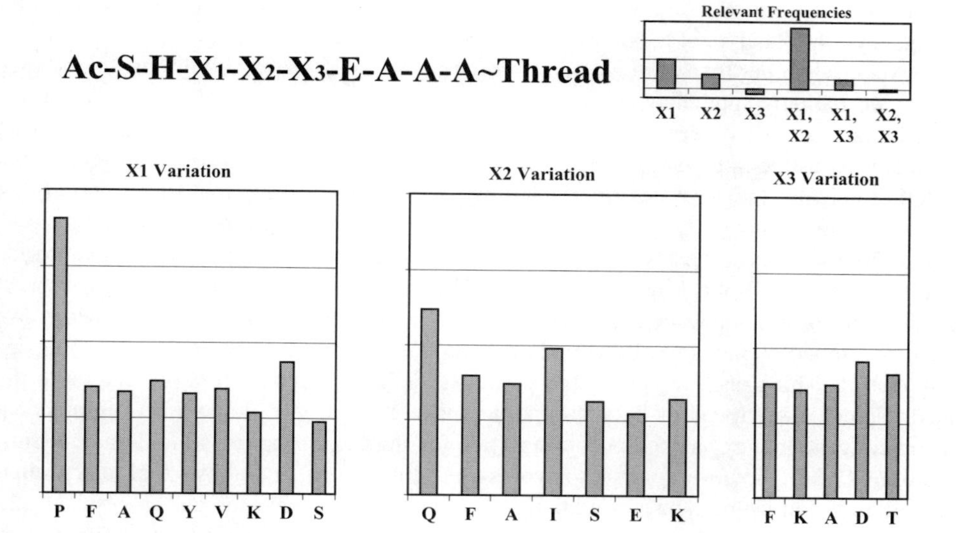

FIGURE 6.6. First-order fitness profiles. Fitness profiles, extracted from the frequency domain library spectrum as described in the text, for the three cycles corresponding to single cylinder sizes. Note that frequencies corresponding to x_1 and x_2 have larger peaks than x_3 (inset).

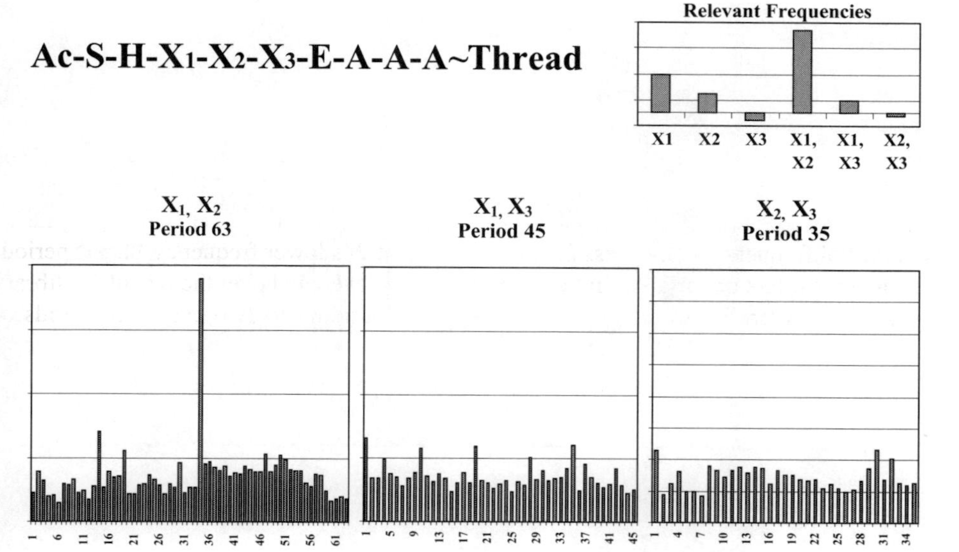

FIGURE 6.7. Second-order fitness profiles. Fitness profiles, extracted from the frequency domain library spectrum as described in the text, for the three cycles corresponding to pairings of cylinder sizes. Note that the frequency corresponding to $x_1 \times x_2$ has the largest peak in the frequency domain spectrum, and that the extracted fitness profile shows the significance of the pairing of groups in these positions.

demonstrated fluorescence measurements while the library is pulled through a detector, and many other analytical techniques are possible.

Materials other than cotton offer a broader range of reaction conditions. Our choice as the most widely applicable of the available solid supports is the SPOCC resin developed by Meldal and co-workers [17,18]. A particularly interesting possibility is the use of an optical fiber as the linear support for combinatorial synthesis. Synthesis is carried out in a SPOCC gel film that replaces fiber cladding. If the library is exposed to a solution that allows fluorescent activity evaluation, light may be guided along the optical fiber so as to excite fluorophores at the fiber–solution interface by an evanescent mechanism that avoids excitation of fluorophore in free solution. This allows simple binding assays without rinsing out excess fluorescent materials, as well as real-time kinetic measurements. We have developed methods to allow readout of an entire array as a single time-varying signal, distinguishing each region by the time it takes light to travel from one region to the next (Figure 6.8). Speed-of-light limitations make evaluation of large arrays impractical using conventional experimental configuration for fluorescent optical time-domain identification [19,20]. However, our setup avoids these limitations, and allows high discrimination of emission from intense excitation by applying a double evanescent process [21,22].

The basic idea of linearization, with multiple dimensions specified by frequency, is generalizable in many ways. Most obvious is that the solid support could be of various materials, and the method of its division into sections modified to allow other chemical reaction reagents and conditions. However, no solid support need be involved; libraries prepared on pins, in 384-well plates, or with radio frequency tags could benefit in design and analysis by application of these ideas. Experiments on multiple drug interactions could

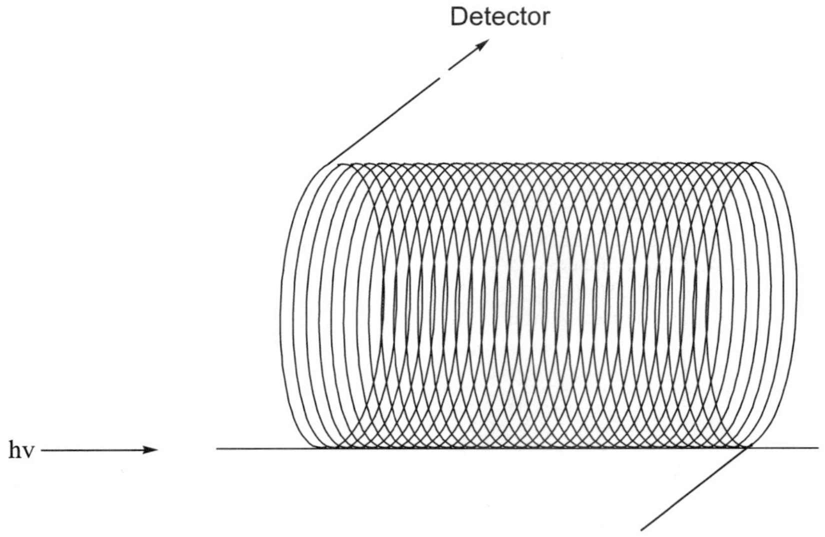

FIGURE 6.8. Apparatus for optical time-of-flight evaluation. Optical time-of-flight evaluation of fluorescence from sensor regions closely spaced on the excitation fiber can be distinguished because the detection fiber serves as an optical delay line. Other advantages of this scheme are described in the references, including Chapter 15 in this book.

be carried out by passing a dilute suspension of cells down a tube to which various sets of drugs are added in a cyclic way with different cycle times for each drug. Linear arrays of new solid state materials with useful emissive [23], magnetic [24], or conductive properties, prepared by vapor deposition, could be made with compositions varying cyclically at a different period for each component, thus increasing the dimensions of varaibility available to these gradient libraries. Enzyme specificity could be evaluated by screening an entire library of one substrate with a library of another, under a library of reaction conditions. Trends observed in the evaluation of a library of catalysts can lead not only to an excellent and specific catalyst, but also to a well-defined model of its activity and direction for modifications [25].

ACKNOWLEDGMENTS

In the Schwabacher group, this work was performed primarily by graduate students Christopher Johnson and Yixing Shen, and by Dr Maureen Prince. Contributions by undergraduates Michael Kordaszewicz and Angela Eder are also gratefully acknowledged. In the Geissinger group the main contributor to this work was Dr Barry Prince, with additional contributions from Dr Nadejda Kaltcheva and from two undergraduate students, Janelle Scharon and Tami Zembrowski.

This chapter is based upon work supported by the National Science Foundation under Grants CHE-9874241 (A.W.S.) and CHE-0078895 (P.G.).

REFERENCES

1. Fodor, S. P. A. et al. *Science* **1991**, *251*, 767–773.
2. Furka, A., Sebestyen, M., Asgedom, M., Dibo, G. *Int. J. Pept. Protein. Res.* **1991**, *37*, 487–493.
3. Lam, K. S., Lebl, M., Krchnak, V. *Chem. Rev.* **1997**, *97*, 411–448.
4. Wang, P., Arabaci, G., Pei, D. *J. Comb. Chem.*, **2001**, *3*, 251–254.
5. Lepre, C. A. *Drug Disc. Today*, **2001**, *6*, 133–140.
6. Brenner, S. et al. *Proc. Natl Acad. Sci. USA* **1992**, *89*, 5381–5383.
7. Ohlmeyer, M. H. J. et al. *Proc. Natl Acad. Sci. USA* **1993**, *90*, 10922–10926.
8. Battersby, B. J., Lawrie, G. A., Trau, M. *Drug Discover. Today*, **2001**, *6*, S19–S26.
9. Schwabacher, A. W., Shen, Y., Johnson, C. W. *J. Amr. Chem. Soc.* **1999**, *121*, 8669–8670.
10. Furka, A., Christensen, J. W., Healy, E., Tanner, H. R., Saneii, H. *J. Comb. Chem.* **2000**, *2*, 220–223.
11. Smith, J. M., Gard, J., Cummings, W., Kanizsai, A., Krchnak, V. *J. Comb. Chem.* **1999**, *1*, 368–370.
12. Stankova, M., Wade, S., Lam, K. S., Lebl, M. *Pept. Res.* **1994**, *7*, 292.
13. Stewart J. M., Young, J. D. *Solid Phase Peptide Synthesis*, **1984**, Pierce Chemical Co., Rockford, IL.
14. Frank, R. Tetrahedron **1992**, *48*, 9217–9232.
15. Krchnak, V. et al. *Collect. Czech. Chem. Commun.* **1988**, *53*, 2542.
16. Frigo, M., Johnson, S. G. *FFTW: An Adaptive Software Architecture for the FFT*, Vol. 3, **1998**, Seattle, WA, p. 1381 (software downloadable from www.fftw.org).
17. Rademann, J., Grøtli, M., Meldal, M., Bock, K., *J. Am. Chem. Soc.* **1999**, *121*, 5459–5466.
18. Grøtli, M., Rademann, J., Groth, T., Lubell, W. D., Miranda, L. P., Meldal, M. *J. Comb. Chem.* **2001**, *3*, 28–33.
19. Kvasnik, F., McGrath, A. D. Distributed chemical sensing utilising evanescent wave interactions. *Proc. of S.P.I.E. Conf. on Chemical, Biochemical, and Environmental Fiber Sensors, Boston, MA*, **1989**, Publisher: S.P.I.E- the International Society for Optical Engineering, Bellingham, WA, pp.75–82.

20. Browne, C. A., Tarrant, D. H., Olteanu, M. S., Mullens, J. W., Chronister, E. L. *Anal. Chem.* **1996**, *68*, 2289–2295.
21. Prince, B. J., Schwabacher, A. W., Geissinger, P. *Anal. Chem.* **2001**, *73*, 1007–1015.
22. Prince, B. J., Kaltcheva, N. T., Schwabacher, A. W., Geissinger, P. *Appl. Spectrosc.* **2001**, *55*, 1018–1024.
23. Danielson, E., et al. *Nature* **1997**, *389*, 944–948.
24. Briceño, G., Chang, H., Sun, X., Schultz, P.G., and Xiang, X.-D. *Science* **1995**, *270*, 273–275.
25. Rodgers, J. R., and Villars, P. *Mater. Res. Soc. Bull.* **1993**, *18*, 27–29.

Separations and Mass Spectrometry

Chapter 7

Separation Approaches Toward Rapid and Complex Molecular Characterization of Diverse Polymers

Miroslav Petro, Son Hoai Nguyen, and Han-Ting Chang

INTRODUCTION

Within the last few years, combinatorial material science has evolved into a generally accepted approach for discovery and optimization of a variety of materials, ranging from catalysts through electronic materials to polymers (see refs. 1–12 and references cited therein).

The term "material" typically does not refer to a single chemical entity but represents a complex formulation of numerous chemical individuals in a characteristic physical form. In addition, individual material components often exhibit their own structural variations. For example, synthetic polymers consist of macromolecules of variety of lengths, structures, and chemical compositions, featuring certain distribution profiles in each of the characteristics [13]. Also, all the distributions exist in a mutual relationship. That means that each fraction of one distribution shows characteristic distributions of the other parameters. The component diversity of polymeric materials, together with the structural polydispersity of its individual components, may diminish our ability to understand the material structure–performance relationship. Conversely, the lack of the understanding often results in tailoring the material properties with additional components instead of replacing the main component with a better one. The combinatorial approach breaks this cycle of step-by-step improvements, thus offering us a chance to discover materials possessing highly desirable but otherwise irreconcilable properties. Screening for molecular characteristics provides the desired link between the synthesis and application-related performance of new materials and completes the whole combinatorial discovery/optimization process.

Miroslav Petro, Son Hoai Nguyen, and Han-Ting Chang • Symyx Technologies Inc., 3100 Central Expressway, Santa Clara, CA 95051, USA
Contact author: Miroslav Petro, email: mpetro@symyx.com

High-Throughput Analysis, edited by Potyrailo and Amis
Kluwer Academic Publishers, New York, 2003.

The structural complexity of the materials may significantly complicate the actual screening process as well as data interpretation. On-line monitoring of a property change within a combinatorial space can be performed rapidly, often providing enough information on the material diversity to allow continuation of the discovery or optimization processes [3,6,9,11] and may lead to discoveries that would be hard to achieve in a classical non-combinatorial way [12]. However, the signal from a property of interest might often be affected by numerous factors related to the environment and conditions of the screening measurement as well as by a signal from unknown interferences. Adding a separation step prior to detection can minimize the complexity of the sample and eliminate some undesired interfering factors, thus significantly increasing the reliability and relevance of the screening results. Unfortunately, in contrast with measurements based on an "instant impact" (spectroscopy, microscopy, etc.), separation-based techniques require molecules to move in time and space, which is limited by diffusion and other phenomena that slow down the analysis significantly. Therefore speeding up the separation process becomes a necessity when it comes to high-throughput screening purposes.

One of the most important tasks in the polymer characterization area seems to be precise and accurate description of the molecular weight distribution (MWD), typically via determination of weight and number averages of the molecular weights and their ratio, known as the polydispersity index [13]. The most popular technique for molecular weight characterization of polymers is undoubtedly size-exclusion chromatography (SEC), also called gel-permeation chromatography (GPC) [14]. The technique separates macromolecules according to their hydrodynamic volumes based on the different permeability of macromolecules of different sizes to the pores of the separation medium. Since the speed of the size-exclusion separation is inherently limited by diffusion of macromolecules to the pores, acceleration of GPC was expected to be difficult to achieve. Nevertheless, the authors of this chapter, together with other scientists at Symyx Technologies, have approached and solved a variety of problems that prevented the size-exclusion mechanism from being applied to rapid polymer screening [15]. The resulting Rapid GPC™ system and methodology, in several variations, combinations, and forms of integration with other systems, is used widely within our polymer discovery programs [16] and accelerates the research programs of some of our partners [17], and certain components of it will soon become commercially available to researchers worldwide through Polymer Laboratories (UK) [18]. Recently, a number of GPC applications at increased speed have been reported [19], following a general trend of growing interest of polymer chemists in high-throughput experimentation [7–10].

Several other approaches to molecular weight characterization have been applied to the combinatorial screening of polymers with different levels of success. For example, flow injection techniques based on light scattering (LS) with minimal or no separation have been found to be well suited to molecular weight screening, provided that the polymer samples are available in a relatively clean form [20,21]. After implementation of a number of improvements and enhancements of both the LS system and the procedure, the speed of serial analysis can approach a few seconds [15,21]. Despite the obvious advantages of LS-based methods in high-throughput screening, other components of combinatorial polymer samples, such as monomers, initiators, catalysts, products of undesired polymerization, dust particles, etc., often severely interfere with the LS signal, thus limiting use of the technique in the combinatorial discovery process.

Separation according to molecular weight can also be achieved by injecting a polymer into a stream of a nonsolvent passing through a separation medium. The surface of the chromatographic medium accommodates the precipitating polymer. The precipitation of the polymer is followed by its elution by a thermodynamically good solvent in order of decreased solubility, i.e. small molecules first and large macromolecules later [22,23]. In this mode of separation, a chromatographic medium based on porous polymeric monoliths offers a unique combination of speed and resolution [23,24]. This separation technique has recently been applied in a combinatorial polymer project focused on identification of the best catalysts for atom-transfer radical polymerization [15] (see Chapter 9 of this book). Despite success in this particular application, the applicability of the rapid precipitation–redissolution chromatography in combinatorial screening is limited to homopolymers, since the retention of macromolecules not only depends on the molecular weights but it is also strongly affected by other molecular characteristics, namely chemical composition. Even if a deconvolution of individual distributions is possible, dealing with a wide diversity of polymer samples becomes impractical.

In contrast with GPC, which is quite insensitive to the polymer chemical composition owing to the physical nature of the size-exclusion mechanism, chromatographic separation of (macro)molecules under conditions which promote their interaction with the surface of a separation medium inherently senses the chemistry of the polymer chain [25]. The most efficient form of such separation is high-performance liquid chromatography (HPLC) or, more specifically, adsorption chromatography in both normal-phase and reversed-phase modes. The HPLC technique offers a unique flexibility that allows chromatographers to tune up the separation conditions for almost any type of polymer. Under optimized conditions, the HPLC trace can truly represent the chemical composition distribution (CCD) of the polymer sample [26]. On the other hand, interpretation of the chromatographic traces is often complicated by the mixed character of the separation mechanism [27]. Here, instead of benefiting from retention influenced by the length of the polymer chains, the molecular weight effect represents an undesired phenomenon. Again, the elimination or deconvolution of multiple retention mechanisms becomes a serious complication when it comes to a wide diversity of samples. This and several other problems need to be tackled in order to turn HPLC into a rapid, robust, and reliable copolymer CCD screen. For example, the separation needs to be sensitive enough to differentiate two copolymers of very similar chemical compositions but, on the other hand, universal enough to cover a very broad range of monomers eventually appearing in the combinatorial space. Separation media based on highly hydrophilic polymeric beads may offer the right combination of such specificity and universality [28].

The existence of several molecular distributions in a mutual relationship complicates the polymer analysis even further. The challenge of interrelated molecular distributions can be addressed by multidimensional chromatography. For example, two-dimensional (2D) chromatography, which combines a separation according to molecular weights with a separation according to a chemical composition, provides the MWD profiles for every fraction of the CCD and vice versa [29]. In the preferred 2D chromatographic system, in which the two dimensions are connected via a switching valve, the effluent from one chromatographic column is sampled and loaded continuously into the second separation mode. As a result, each of the distribution slices from one dimension is independently analyzed for distribution of another molecular parameter in the second dimension. Because of this independence, the 2D chromatography may uncover structural differences that might otherwise be undetected by any other analytical technique.

In this chaper, we give a brief overview of some of our developments in the area of polymer screening for molecular characteristics using a simple case study of a model copolymer library, characterized by rapid GPC, HPLC, and 2D chromatography, and discuss the speed and performance of these techniques.

EXPERIMENTAL PROCEDURE

Synthesis

Reactor

A Symyx 96-well glass-lined aluminum batch reactor was used. The reactor has an 8×12 well format with the industrial microtiter plate standard of 9 mm well spacing and uses disposable glass inserts as the reaction container. The glass vial insert holds up to 1 mL of total volume and is sealed against a Teflon-faced silicon gasket during polymerization. In this case, the reactor was equipped with a magnetic agitation system, heater, and timer. The synthesis station is controlled by Symyx Impressionist™ software.

Chemicals

Styrene, 2-hydroxyethyl methacrylate (HEMA), 2,2′-azobisisobutyronitrile (AIBN), *N,N*-dimethylformamide (DMF), and all other chemicals were purchased from commercial sources and used without any further purifications but were degassed by applying vacuum.

Synthesis Procedure

The library was designed using Symyx Library Studio®. The design shown in Figure 7.1 portrays the composition variation of HEMA–styrene from homo-polyHEMA to homopolystyrene in the columns and the variation of molecular weight by changing monomer-to-initiator ratio in the rows. The total volume of each polymerization was 600 μL and the total monomer concentration was kept at 20% solid. The initiator content varies from 0.1 mol% to 3.2 mol% to the total monomers. The library preparation was carried out under an argon atmosphere where all chemicals were dissolved in DMF. The stock solutions were dispensed into the reactor by a liquid-handling robot operated by Symyx Impressionist™ software. The array of the resulting reaction mixture was sealed and maintained at 70°C for 12 hr. The resulting polymer solutions were then transferred for characterization without any post-purification.

The model polymers 1 and 2 as well as a calibration library of polystyrenes, with related data shown in Figures 7.4 and 7.5, respectively, were prepared by the Symyx living polymerization process [30]. Narrow polystyrene standards for GPC calibration were purchased from Polymer Laboratories.

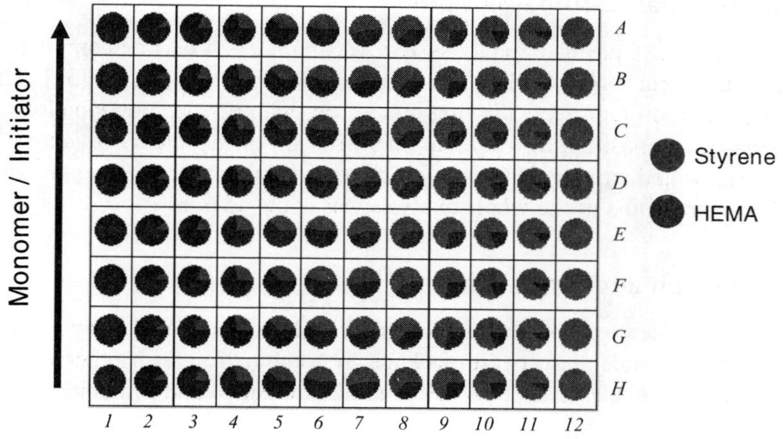

FIGURE 7.1. Compositional design of the model styrene–HEMA copolymer library (from Symyx Library Studio® library design software). The area in the pie chart is proportional to the weight fraction of individual monomers in the reaction mixture.

Screening

Molecular Weight Screen

A Symyx High-Throughput Rapid GPC™ system [15] equipped with column-selection valves that enable choice of a high-resolution separation with an 8 min injection interval (Accelerated-GPC) or high-speed separation with an injection interval of 90 or 30 s (Rapid-GPC), was used for the separations according to the molecular weights of the polymers. Symyx proprietary columns of various dimensions were packed with a mixture of polystyrene beads of different porosities obtained from Polymer Laboratories. A mixture of DMF with 0.1% trifluoroacetic acid was used as the mobile phase.

The Accelerated-GPC technique was used for the stand-alone molecular weight characterization of the model library. The Rapid-GPC technique was used as a part of the 2D chromatography system.

The molecular weight data obtained using the Symyx Accelerated- and Rapid-GPC techniques were confirmed by a conventional GPC composed of a Waters 410 Refractive Index detector, a Waters 717plus Autosampler, a Waters 515 isocratic pump, and a column set of four PL MIXED-B columns (Polymer Laboratories). The conventional GPC system was calibrated using narrow polystyrene standards (Polymer Laboratories).

Chemical Composition Screen

Separations according to the chemical composition of copolymers were performed using our Rapid HPLC™ system adapted for rapid screening and composed of a Symyx automated liquid-handling robot equipped with Waters 515 pumps, Valco injection and switching valves, and an ELSD 1000 detector (Polymer Laboratories). The Symyx HPLC column was packed with a separation material similar to that described earlier [28].

Two-Dimensional Chromatography

The Symyx 2D Chromatography system was built as an on-line combination of the essential parts of our Rapid GPC™ and Rapid HPLC™ systems [31]. The HPLC part of the separation represents the first dimension that provides fractionation according to the chemical composition distribution. Each fraction of the chemical composition distribution was injected into the GPC part, which provided the molecular weight distribution profile. The overall time of the whole analysis was about 15 min.

Data Acquisition and Processing

The whole screening process has been integrated and automated using components of the Symyx Renaissance™ software package in combination with other commercially available software, such as LabVIEW (National Instruments) and Millenium (Waters).

RESULTS AND DISCUSSION

At Symyx Technologies we have integrated parallel polymer synthesis, screening for macromolecular characteristics, and a number of screens that measure physical, mechanical, or other material properties into comprehensive and integrated discovery work flows. The performance of each part of this work flow is in a relation with the other parts and therefore cannot be fully appreciated without taking into account the complexity of the whole process. Since the challenges coming from such integration exceed the scope of this chapter, we limit our present focus to developments in the molecular characterization area.

Over recent years, the high-throughput screening approach to macromolecular characteristics has been applied at Symyx Technologies Inc. to a wide variety of random copolymers of up to four comonomers. The pool of available reactants includes about 50 different monomers categorized into several groups, such as neutral hydrophilic, neutral hydrophobic, anionic, and cationic. Some of them are proprietary to Symyx or its partners, but they are often combined with common monomers such as acrylamide, butyl acrylate, acrylic acid or 2-(dimethylamino)ethyl methacrylate. It is often the case that a proper monomer combination dials in a proper balance of molecular interactions, thus defining the performance of the resulting polymer in a specific application. The purpose of the study presented here is to demonstrate the characterization power of our screening approach on the specific example of model styrene–HEMA copolymer library.

Library Synthesis

As can be seen from the library design (Figure 7.1), the monomer/initiator variation was introduced to generate the molecular weight diversity, while the different ratios of the two monomers in the feed solution should result in a corresponding gradient of chemical composition. The chosen polymerization system, styrene–HEMA in DMF, is known to provide a statistically random copolymer since the reactivity ratios of these two monomers are practically identical [32]. Therefore we can expect that the percentage of incorporation of each monomer in the copolymer would correspond to the design. The molecular weight diversity can be generated by introducing a gradual change of relevant polymerization conditions. For example, the higher the initiator concentration the lower is the degree of polymerization and vice versa. Thus different lengths of the copolymer can be prepared by simply varying the initiator-to-monomer ratio. Although the theoretical assumption often differs from reality due to other unforeseen factors, the corresponding diversity in both molecular weight and chemical composition can be achieved.

In combinatorial polymer science, a well-understood polymerization system, like the one described above, can serve as a calibration library to validate screening protocols and performance of the screening techniques as well as to set the reference for screening of as yet unknown materials. The library used as a model for this case study represents wide chemical diversity, since it contains styrene–HEMA copolymers over the whole range of the composition. In addition, the molecular weight of each composition varies over a wide range from approximately 3×10^3 to 3×10^6.

Molecular Weight Screen

Figure 7.2 shows the results of our stand-alone molecular weight screen using the Symyx Rapid GPC™ system in an "accelerated" mode, which means an injection frequency of 8 min. The speed of separation can be increased further as discussed below. Consistency of results indicates very good reproducibility of the complete screening workflow. The results can also be considered as a validation of the complete workflow,

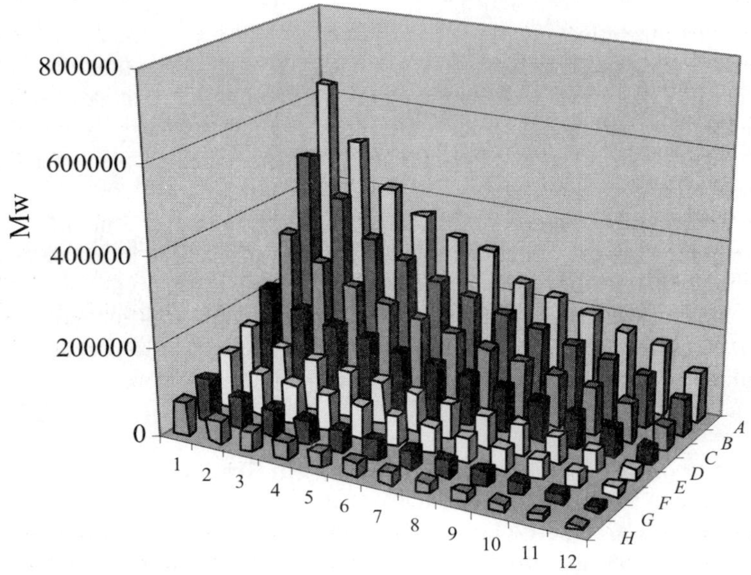

FIGURE 7.2. Molecular weight data for the model styrene–HEMA copolymer library obtained by the Rapid GPC™ analysis. See Figure 7.1 for the library design.

including polymer synthesis in a parallel reactor, sample preparation and handling protocols, screening procedures, and, finally, data processing. The fact that there are almost no outliers indicates that the screen can be applied not only to libraries featuring expected trends in molecular weights, but also to extremely diverse polymers with molecular properties that are hard to predict. Even in the case when molecular weights can be calculated from the monomer-to-initiator ratio, the actual values can be affected by numerous other factors and need to be measured to confirm the expected values. The fact that the GPC results are consistent and the trend in molecular weights matches expectation indicates reasonable precision and accuracy of the screen and demonstrates the universality of the screening approach.

The reproducibility of our high-throughput molecular weight measurements is comparable to that of conventional GPC, provided that the systems are maintained in a proper way. In our experience, injection-to-injection reproducibility within a single set of injections is typically below $\pm 2\%$ of an average value. A larger variation of up to 5% is observed on day-to-day basis. Considering the fact that the whole library of polymers is typically analyzed by Rapid GPC™ system within the same day, during which the system performance normally remains constant, as opposed to the much longer period of time needed for conventional analysis, high-throughput screening can often be more precise than conventional analysis. As described earlier [15], the accuracy of the molecular weight measurements using our high-throughput system depends, as for any GPC system, on the calibration. Typically, the accuracy in terms of the weight-average molecular weights measured within the calibrated range is comparable to that obtained by conventional GPC. However, because of higher speed of separation in the rapid mode, there is an extra band broadening that translates into a deviation in the polydispersity indices obtained by rapid versus conventional GPCs. Several other factors, such as molecular weight range, profile of the GPC calibration curve, and linearity of the detector response, also have an effect. Further studies of the effect of speed of analysis on polydispersity indices are desirable but would exceed the scope of the current study. All of the polymers within the model library used in this study have their molecular weights well within the calibrated range, thus providing a highly accurate set of average molecular weights relative to polystyrene standards.

Upon repetitive analysis of a large number of diverse samples, a number of additional occurrences such as column aging and contamination may also induce measurement errors. We have studied and optimized our separation system over a number of years and found the combination of column material and mobile phase used in this study to be very robust. For example, we have re-run the model styrene–HEMA library after about a year of extensive use of the system and were able to reproduce most of the molecular weights to within 10% of the original values, with just a few outliers that were more likely caused by contamination of the specific wells than by system deterioration.

Chemical Composition Screen

One of the reasons why we have selected the combination of styrene and HEMA for copolymerization is because the monomers have very similar reactivities. This allows us to predict that the resulting random copolymer will have the chemical composition that corresponds to the ratio of the comonomers added to the reaction mixture. It can be seen

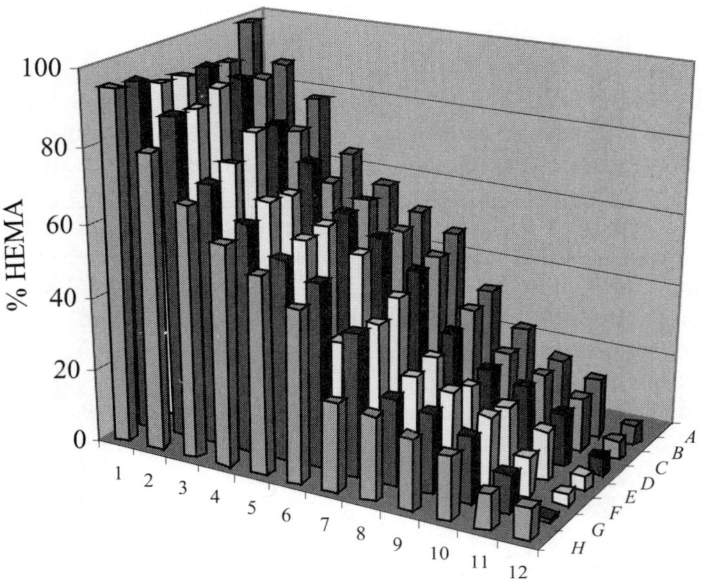

FIGURE 7.3. Chemical composition data for the model styrene–HEMA copolymer library obtained by the Rapid HPLC™ analysis. See Figure 7.1 for the library design.

in Figure 7.3 that different feed ratios between the monomers result in corresponding changes in HPLC retention, thus confirming expectations. Also, this demonstrates the ability of the screening technique to determine the composition of a copolymer made from monomers of unknown reactivities, or simply allows us to assess monomer reactivity quickly in a variety of combinations and under a variety of polymerization conditions. High-throughput screening for monomer incorporation is one of the main applications of polymer HPLC at Symyx.

GPC: Speed Versus Resolution

Figure 7.4 compares the chromatographic traces of two model polymers obtained using three different GPC techniques. The molecular weight of the polymer eluted first was about twice as high as that of the polymer eluted later. The MWDs of the polymers partially overlap. Therefore separation of the polymer mixture would be difficult to achieve. Instead, the traces of both polymers are overlaid, so that the extent of the overlap can show the resolution of the system.

Figure 7.4(a) represents the separation of the model polymers using conventional GPC with four top-quality columns 30 cm long. Single analysis using such a conventional setup takes about 50 min. Figure 7.4(b) shows the chromatograms of the same polymers separated using the Symyx Accelerated-GPC technique, which provides enhanced speed of analysis while preserving the resolution of conventional GPC. The interval between two consecutive injections using the accelerated GPC approach is 8 min. As portrayed by the

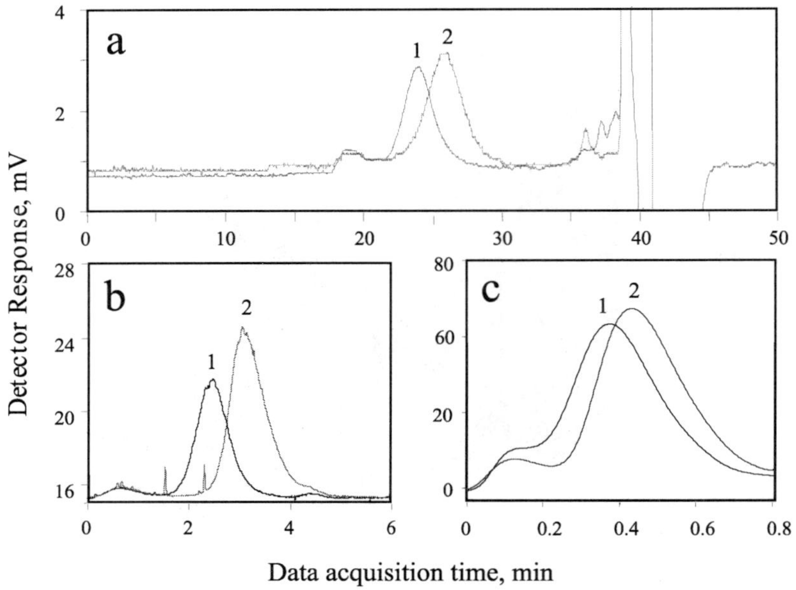

FIGURE 7.4. Comparison of separation performances of (a) conventional, (b) accelerated, and (c) rapid GPCs. The weight-average molecular weights of polymer 1 and polymer 2 are 142,000 and 71,000, respectively, as determined by conventional GPC using tetrahydrofuran as the mobile phase and calibrated using narrow polystyrene standards. The polydisperisity index values of polymer 1 and polymer 2 obtained by conventional GPC are 1.24 and 1.23, respectively.

overlaid traces, the resolution power of conventional and accelerated GPCs is about the same. The speed increase is a result of several improvements in the chromatographic system, including utilization of a void elution time between the injection and elution of macromolecules. The last set of overlaid traces [Figure 7.4(c)] depicts the resolution of the Symyx Rapid GPC™ system tuned up to an injection frequency of 90 s. Despite some loss of resolution in this case, the molecular weight changes can still be clearly identified, which satisfies most of the screening requirements. In addition, the simple overlay of the traces allows us to compare the MWD profiles and even estimate the polydispersities. For example, we can distinguish between the molecular weight increase caused by broadening of the MWD as opposed to the molecular weight increase caused by an extension of the "living" polymer chains. Such a chain-extension analysis serves successfully in our company as a screen to identify a living polymerization process [30].

The plot in Figure 7.5 compares the weight-average molecular weights (Mw) for a large number and a wide variety of narrow polystyrenes as obtained by GPC in conventional versus rapid mode of operation. This documents that rapid GPC can provide average values of the molecular weights at a precision and accuracy that are comparable to those of conventional GPC, provided that both systems use the same mobile phase and are properly calibrated in a wide range of molecular weights. Even if the weight-average molecular weights can be determined quickly at a reasonable level of precision and accuracy, complete description of the MWD by accurate determination of several of its moments still represents a significant challenge. For example, polydispersity indices can

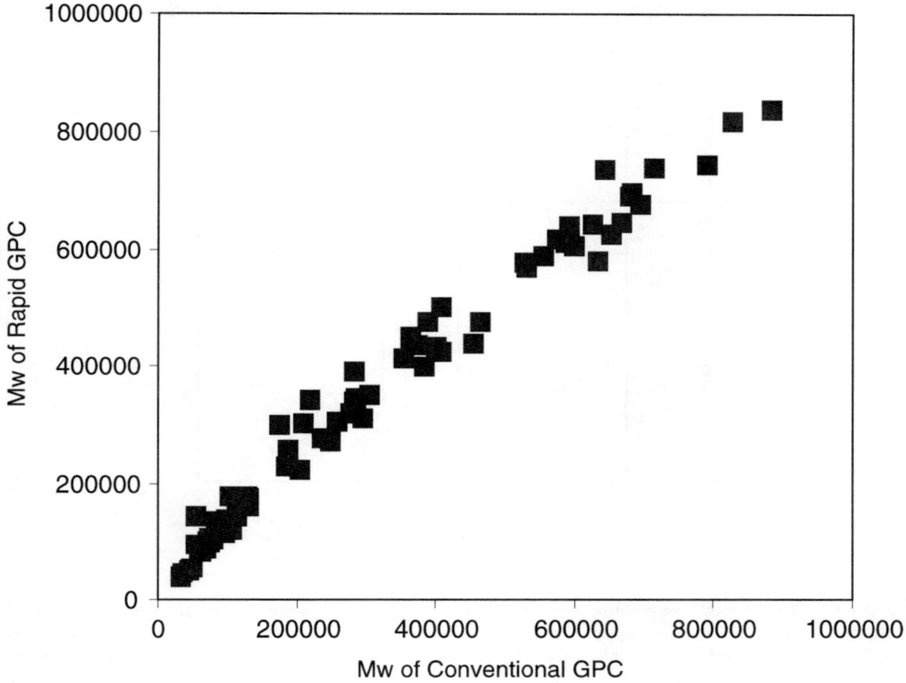

FIGURE 7.5. Correlation plot comparing the molecular weights obtained by conventional versus rapid GPCs. The model library consisted of narrow homopolystyrenes prepared at Symyx [31] and narrow polystyrene standards from Polymer Laboratories. The Rapid GPC™ system was programmed to a 90 s injection interval. Toluene was used as the mobile phase in both rapid and conventional systems. The molecular weights were calculated using polystyrene calibration.

be calculated with a satisfactory accuracy only when the polymer peak fits completely within the elution time range corresponding to the linear part of the GPC calibration curve. Any band broadening caused by diffusion or other phenomena unrelated to size exclusion that brings the macromolecules outside the calibrated range seriously complicates assignment of the correct molecular weights to those fractions. Any deviation from reality for those fractions significantly affects at least one moment of the MWD that describes the whole polymer.

For applications that require rapid monitoring of changes in MWD profiles rather than accurate determination of the molecular weight moments, the speed of the GPC separation can be increased even further. Figure 7.6 shows overlaid GPC traces obtained for a series of polystyrene standards injected into the analytical system at intervals of 30 s. Despite the obvious decrease in the resolution power at this speed of operation, the GPC calibration curve shows perfect linearity over a very wide range of molecular weights. This preserves a very good correlation between the chromatographic retention and the molecular weights for majority of diverse samples, thus enabling a sufficiently accurate assessment of the molecular weights. The rapid monitoring of changes in the MWD profiles helped us to navigate the synthesis parameters in our combinatorial discovery programs,

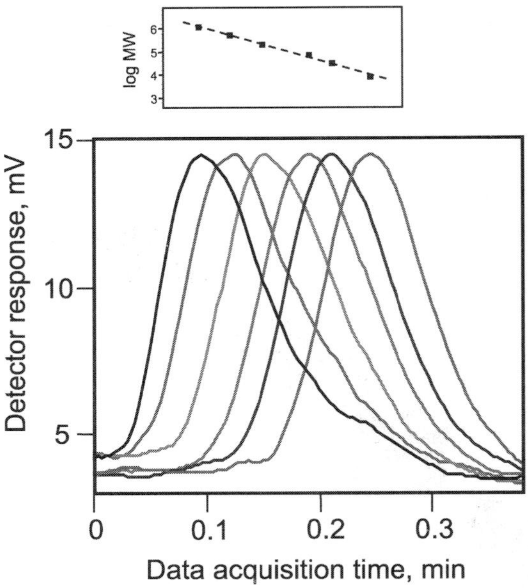

FIGURE 7.6. Calibration traces and the calibration curve for the 30 s GPC. The narrow polystyrene standards of molecular weights of 7,000, 30,000, 70,000, 200,000, 500,000, and 2,000,000 were selected from the polystyrene library described in Figure 7.5. The polydispersity indices of these standards, as obtained by conventional GPC, were in the range 1.05–1.20.

and it might be very useful in some other areas such as quality control in industrial processes.

HPLC: Chemical Sensitivity Versus Universality

As discussed earlier, the retention of macromolecules in various modes of HPLC not only reflects the composition of the polymer chain or the chain-end chemistry but it is also sensitive to the overall length of the chain as well. Typically, the larger the (macro) molecule the stronger is the retention in the interactive HPLC. As a result, the molecular weight effect may seriously complicate the interpretation of the results.

In order to minimize the undesired molecular weight effect on the separation according to the chemistry of the polymer chains, we have carefully optimized our separation conditions. In our best scenario, we have combined two separation modes, known as normal-phase HPLC and hydrophilic-interaction chromatography, in a single chromatographic procedure using an extremely hydrophilic separation medium and a relay gradient of the mobile phase from hexane to water. These conditions can easily handle a wide chemical diversity of the polymer samples while suppressing the undesired molecular weight effect very effectively. Figure 7.7 shows overlaid chromatograms of three groups of polymers differing in chemical composition, namely the three selected columns of the styrene–HEMA library described above. It is obvious that, despite the large variety in

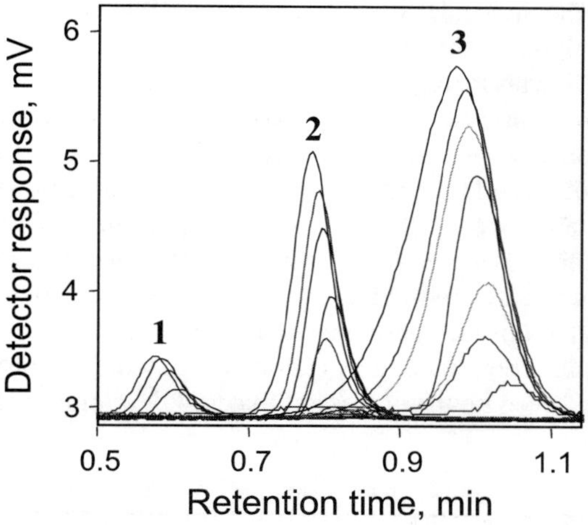

FIGURE 7.7. Traces of selected styrene–HEMA copolymers as obtained by Rapid HPLC™. Peaks in the retention time range 0.5–0.7 min (peak 1), 0.7–0.9 min (peak 2), and 0.9–1.1 min (peak 3), correspond to the styrene–HEMA copolymers made of the reaction mixture containing 9%, 45%, and 91% HEMA, respectively. The copolymers are described in the compositional map in Figure 7.1 as columns 11, 7, and 2, respectively. Molecular weights of the selected polymers varied in the range 15,000–160,000, 28,000–245,000, and 50,000–510,000 for peak groups 1, 2, and 3, respectively.

molecular weights, the polymers elute strictly in a narrow range of retention times characteristic of their chemical composition. Also, the complete polymer elution and equilibration of the separation column back to the initial state can be achieved in about 2 min. This unprecedented universality and speed of the separation allows us to use this technique for rapid screening of random copolymers for their chemical composition. A single HPLC column has been used for several thousand injections of polymers of extremely wide chemical diversity without any significant deterioration in retention properties, thus proving its durability and applicability in combinatorial chemistry.

Combination of the MWD and CCD Analyses Using 2D Chromatography

In order to accomplish a complete polymer analysis for both MWD and CCD, we have combined our screens based on GPC and HPLC into a single 2D chromatographic technique. HPLC represents the first dimension of our 2D system which fractionates a polymer sample according to its chemical composition. The 15 min long HPLC run produced 30 consecutive chemical composition fractions, which are automatically loaded at a frequency of 30 s into an on-line Rapid GPC™ system representing the second dimension of the Symyx 2D chromatograph [31]. Our data acquisition and processing software puts all the MWD profiles coming from the GPC along the time line representing the HPLC retention time, thus creating three-dimensional *x*–*y*–color maps exemplifying the polymer MWD, CCD, and concentration, respectively.

Figure 7.8 plots the molecular weight and the chemical composition data for selected styrene–HEMA copolymers obtained by GPC and HPLC individually [Figure 7.8(a)] compared with their 2D chromatographic equivalents [Figure 7.8(b)]. Even if both results reveal the same trends, only the 2D chromatography can capture the whole complexity of the MWD–CCD relationship; separately run GPC and HPLC can offer only a cross-section of the distributions and might be blind to some significant sample attributes. The 2D chromatographic map depicts the main molecular features of random copolymers in their mutual relationship and therefore it can be considered to be a molecular fingerprint of particular polymer sample. Since the fingerprint is in a certain relation with the physical, mechanical, and other properties defining the performance of the polymeric material in targeted applications, 2D chromatography can serve as a tool for the synthesis optimization that may significantly speed up the process of material development and scale-up.

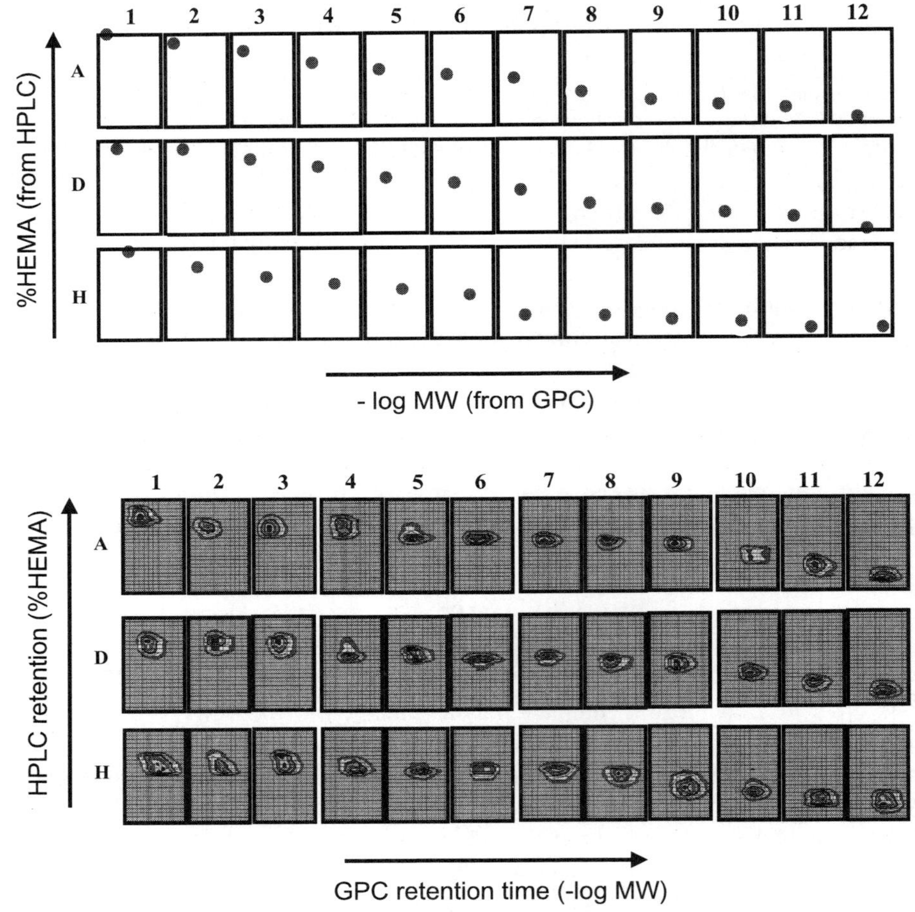

FIGURE 7.8. (a) Plot of the chemical composition and molecular weight data for the selected styrene–HEMA copolymers, as obtained by stand-alone Rapid GPC™ and Rapid HPLC™ systems. (b) Two-dimensional chromatographic maps for the selected styrene–HEMA copolymers. The row and column assignments for the selected polymers correspond to those in Figures 7.1, 7.2, and 7.3.

CONCLUSIONS

Screening for basic molecular characteristics makes an essential link between the synthesis of new polymeric materials and their performance in a final application. Therefore the rapid molecular characterization serves well as a navigational tool in the polymer discovery and optimization process.

It is also known that multiple polymer distributions and the complexity of the polymer formulation are primarily responsible for the material properties. We have shown that a more complete understanding of the material composition, including multiple polymer distributions, can be achieved using multidimensional separation approaches. The performance of an integrated system that combines the molecular weight and chemical composition characterizations was demonstrated by screening a model copolymer library. The advantages of implementation of such a system into a consistent work flow for rapid screening of combinatorial polymer libraries were discussed.

Although success in material discovery depends on many additional factors, real-time control over the synthesis parameters via monitoring the molecular characterization outputs in a robust and automated way represents a breakthrough in steering the polymer synthesis process toward a new material discovery or material optimization. Two-dimensional chromatography that can rapidly handle a large number of diverse polymers appears to be a universal tool for combinatorial screening of polymers for their multiple molecular characteristics. To our knowledge, the performance characteristics of our polymer characterization system described in this chapter are currently unprecedented.

ACKNOWLEDGMENTS

The authors of this chapter thank Professor Jean M. J. Fréchet and Dr. František Švec of the University of California at Berkeley for a gift of chromatographic beads, Dr Oleg Kolosov and Dr Dominique Charmot for helpful discussions, and Ms Isabel Galdo for performing some of the experiments.

REFERENCES

1. Schultz, P. G., Xiang, X.-D. Combinatorial approaches to materials science, *Curr. Opin. Solid State Mater. Sci.* **1998**, *3* (2), 153–158.
2. Cawse, J. N. Experimental strategies for combinatorial and high-throughput materials development, *Acc. Chem. Res.* **2001**, *34* (3), 213–221.
3. McFarland, E. W., Weinberg, W. H., Combinatorial approaches to materials discovery, *Trends Biotechnol.* **1999**, *173*, 107–115.
4. Hewes, J. D. High throughput methodologies: a new paradigm for chemicals and materials research, *Chim. Oggi* **2000**, *18* (9), 20–24.
5. Jandeleit, B., Schaefer, D. J., Powers, T. S., Turner, H. W., Weinberg, W. H. Combinatorial materials science and catalysis, *Angew. Chem. Int. Ed.* **1999**, *38* (17), 2494–2532; *Angew. Chem.* **1999**, *111*, 2648–2689.
6. Weinberg, W. H., Danielson, E., Golden, J., McFarland, E., Reaves, C., Wu, X. A combinatorial approach to the discovery and optimization of new luminescent materials, *Nature* **1997**, *389*, 944–948.
7. Griffith, L. G. Polymeric biomaterials, *Acta Mater.* **2000**, *48* (1), 263–277.

8. Reynolds, C. H. Designing diverse and focused combinatorial libraries of synthetic polymers, *J. Comb. Chem.* **1999**, *1* (4), 297–306.
9. Meredith, J. C., Smith, A. P., Karim, A., Amis, E. J. Combinatorial materials science for polymer thin-film dewetting, *Macromolecules* **2000**, *33* (26), 9747–9756.
10. Brocchini, S., James, K., Tangpasuthadol, V., Kohn, J. A combinatorial approach for polymer design, *J. Am. Chem. Soc.* **1997**, *119* (19), 4553–4554.
11. Dickinson, T. A., Walt, D. R., White, J., Kauer, J. S. Generating sensor diversity through combinatorial polymer synthesis, *Anal. Chem.* **1997**, *69* (17), 3413–3418.
12. Danielson, E., Devenney, M., Giaquinta, D. M., Golden, J. H., Haushalter, R. C., McFarland, E. W., Poojary, D. M., Reaves, C. M., Weinberg, W. H., Wu, X. D. A rare-earth phosphor containing one-dimensional chains identified through combinatorial methods, *Science* **1998**, *279* (5352), 837–839.
13. Flory, P. J. *Principles of Polymer Chemistry*, **1953**, Ithaca, NY: Cornell University Press; Stevens, M. P. *Polymer Chemistry: An Introduction*, **1998**, Oxford: Oxford University Press; Ravve, A. *Principles of Polymer Chemistry*, **1995**, New York: Plenum Press.
14. Yau, W. W., Kirkland, J. J., Bly, D. D. *Modern Size Exclusion Chromatography*, **1979**, New York: Wiley Interscience.
15. Petro, M., Safir, A. L., Nielsen, R. B. Role of high-throughput chromatography in combinatorial chemistry of polymeric materials, *Polym. Prepr.* **1999**, *40* (2), 702; Nielsen, R. B., Safir, A. L., Petro, M., Lee, T. S., Huefner, P. The preparation and rapid screening of combinatorial polymer libraries, *Polymer Materials: Science and Engineering*, Preprints **1999**, *80*, 92; Petro, M., Safir, A., Nielsen, R. B., Lee, T. S. *Automated sampling methods for rapid characterization of polymers*, US Patent 6,265,226 **2001**; Petro, M. *Indirect calibration of polymer characterization systems*, US Patent 6,294,388 **2001**; Petro, M., Safir, A. L., Nielsen, R. B., Dales, G. C., Carlson, E. D., Lee, T. S. *High-temperature characterization of polymers, US Patent 6,260,407* **2001**; Petro, M., Nguyen, S. H., Regan, J., Galdo, I., Zhou, B., DelVecchio, J. Rapid separation and detection techniques in combinatorial polymer screening, *Polym. Prepr.* **2001**, *42* (2), 655.
16. Symyx Technologies Inc., http://www.symyx.com/cu_tech-bene.html, retrieved February 19, **2002**.
17. Cohan, P. Achieving strategic change through innovation, Presented at *Combi 2000* Knowledge Foundation, San Diego, CA, January **2000**.
18. Polymer Laboratories Ltd, http://www.polymerlabs.com/press/symyx01.htm, retrieved February 19, **2002**.
19. Wu, C. S. (chair), New columns, detectors, and strategies for high throughput GPC/SEC and flow injection polymer analysis, *Pittcon 2002*, New Orleans, LA, March **2002**, Sessions 48 and 186.
20. Wyatt, P. J., Podzimek, S. High speed molar mass measurements using MALS, *Book of Abstracts, Int. GPC Symp.*, Las Vegas, NV, **2000**; Flow injection polymer analysis with triple detection, *Application note #11*, Viscotek, Houston, TX
21. Nielsen, R. B., Kuebler, S. C., Bennett, J., Safir, A. L., Petro, M. *Flow-injection analysis and variable-flow light-scattering methods and apparatus for characterizing polymers*, US Patent 6,175,409 **2001**.
22. Glöckner, G. *Pure Appl. Chem.* **1983**, *55*, 1553; Glöckner, G., Van den Berg, J. H. M. *J. Chromatogr.* **1986**, *352*, 511.
23. Petro, M., Svec, F., Gitsov, I., Fréchet, J. M. J. Molded monolithic rod of macroporous poly(styrene-co-divinylbenzene) as a novel separation medium for HPLC of synthetic polymers: "On-column" precipitation–redissolution chromatography as an alternative to size-exclusion chromatography of macromolecules, *Anal. Chem.* **1996**, *68*, 315–321.
24. Petro, M., Svec, F., Fréchet, J. M. J. Molded continuous poly(styrene-co-divinylbenzene) rod as a separation medium for the very fast separation of polymers—Comparison of the chromatographic properties of the monolithic rod with columns packed with porous and non-Porous beads in high-performance liquid chromatography of polystyrenes, *J. Chromatogr. A* **1996**, *752*, 59–66.
25. Glöckner, G. *Gradient HPLC of Copolymers and Chromatographic Cross-Fractionation*, **1991** Berlin: Springer-Verlag.
26. Pasch, H., Trathnigg, B. *HPLC of Polymers*, **1999**, Berlin: Springer-Verlag.
27. Berek, D. Liquid adsorption chromatography of copolymers: Molar mass (in)dependent retention, *Macromolecules* **1999**, *32*, 3671.
28. Petro, M., Svec, F., Fréchet, J. M. J. *Anal. Chem.* **1997**, *69*, 3131; Petro, M., Svec, F., Fréchet, J. M. J., Haque, S. A., Wang, H. C. *J. Polym. Sci. A* **1997**, *35*, 1173; Svec, F., Petro, M., Fréchet, J. M. J. *Collect. Czech. Chem. Commun.* **2001**, *66*, 1047–1061.

29. Pasch, H. Hyphenated techniques in liquid chromatography of Polymers, *Adv. Polym. Sci. Ser.* **2000**, *150*, 1–66; Kilz P., et al., *Adv. Chem. Ser.* **1995**, *247*, 223–242. Murphy, R. E. *Anal. Chem.* **1998**, *70*, 1585–1594; Pasch, H., Trathnigg, B. *HPLC of Polymers*, **1999**, Berlin: Springer-Verlag, Chapter 7.
30. Chang H.-T., et al. US Patent 6,380,335, **2002**; US Patent 6,482,909, 2002 *US Patent 6,395,850*, **2002**; *US Patent 6,518,448*, **2003**; US Patent 6,518,364, **2003**.
31. Petro M., et al. US Patent 6,475,391; US Patent 6,406,632; US Patent 6,454,947; US Patent 6,416,663; US *Patent 6,491,816*.
32. Brandrup, J., Immergut, E. H. *Polymer Handbook* **1989**, New York: John Wiley.

Chapter 8

High-Throughput Gas Chromatography and Mass Spectrometry for Heterogeneous Catalysis: Screening of Catalytic Activities and Selectivities

Pierre-Alain W. Weiss, Jens W. Saalfrank, Jens Scheidtmann, Hans-Werner Schmidt, and Wilhelm F. Maier

INTRODUCTION

Combinatorial approaches to materials and catalysis research and development are increasingly used to speed up lead discovery and optimization [1]. An efficient high-throughput experiment (HTE) requires the complete combinatorial workflow, such as automated synthesis of catalyst libraries, means of characterization, technology for desired property screening and tailored software for control of experiments, data transfer and storage, and data analysis [2]. Such complete setups are presently only available in few industrial laboratories and the development of individual building blocks and their application is typical for the present research activities.

In heterogeneous catalysis the rapid evaluation of potential catalysts is essential for such combinatorial approaches. In high-throughput (HT) applications two modes based on significantly different equipment—primary and secondary screening technologies—can be differentiated. In primary screening the emphasis is on high-throughput with a maximal number of samples; libraries often contain several hundred samples. Here, where experimental conditions may differ significantly from conventional reaction conditions, accuracy and reproducibility of conversion or activity data are lower than in secondary screening. This mode is best suited for lead discovery and exploration of large-parameter spaces.

Pierre-Alain W. Weiss, Jens W. Saalfrank, Jens Scheidtmann, and Wilhelm F. Maier • Lehrstuhl für Technische Chemie, Universität des Saarlandes, Im Stadtwald, D-66123 Saarbrücken, Germany
Hans-Werner Schmidt • Max-Planck-Institut für Kohlenforschung, Kaiser-Wilhelm-Platz 1, D-45470 Mülheim an der Ruhr, Germany
Contact author: Wilhelm F. Maier, email: w.f.maier@mx.uni-saarland.de

High-Throughput Analysis, edited by Potyrailo and Amis
Kluwer Academic Publishers, New York, 2003.

In secondary screening more emphasis is put on accuracy and reproducibility. Experimental conditions are usually very close to conventional reaction conditions; the number of samples usually varies between 10 and 50 per experiment.

A very rapid primary screening method for the activity screening of heterogeneous catalysts is emissivity-corrected IR thermography [3], where the relative heat of reaction of hundreds of catalysts can be monitored simultaneously. This method can be recommended for all processes where heat of reactions correlates with the property of interest. The method fails when parallel or sequential reactions dominate the observed heat of reaction [4]. Here, traditional methods for activity studies in heterogeneous catalysis, such as on-line mass spectrometry (MS) or gas chromatography (GC), can be applied. Equipment used for HT-MS has already been reported by several research groups almost simultaneously. Symyx has reported specially developed MS equipment which allows product composition to be monitored sequentially by quadrupole mass spectrometry (QMS) in primary screening applications. In a pressure chamber the catalyst spots on a flat library are heated from the back by a CO_2 laser, while on the opposite side the catalyst sample is supplied with the feed gas mixture from one capillary and simultaneously a small part of the feed–product mixture is transported to the MS inlet by another capillary and analyzed [5]. A simpler secondary screening device was developed by Senkan and his group [6]. A heated gas phase flow reactor with multiple channels in a ceramic block, each equipped with an individual catalyst, is sampled sequentially by QMS. Here, the whole reactor is moved by an *xyz* table to insert the fixed capillary of the MS into the exit opening of individual reactor outlets for sampling. Catalysts are prepared by impregnation of prefabricated silica or alumina tablets and inserted manually into the reactor. In another typical secondary screening setup, Claus and co-workers reported a multitubular reactor, where 15 parallel gas phase flow quartz tubes are sequentially sampled by a 16-valve multiport and analyzed by QMS [7]. In a miniaturized approach, the same group reported the use of the multiple-channel arrays of a micromixer in combination with QMS for HTE in heterogeneous catalysis [8]. Our approach, targeted to primary screening, was based on the use of plate libraries, where individual catalysts have been placed in wells and monitored in the open well by a moving capillary which is connected to the MS [9]. It was surprising to note, that even hydrogenation reactions could be carried out successfully in such open reactors.

The use of gas chromatography for HTE has also been reported. In an impressive publication, Weinberg and co-workers described the first HT development of multi-element catalysts for oxidative ethane dehydrogenation in which a 48-channel gas phase flow reactor was used for secondary screening with on-line monitoring of product compositions by GC [10]. Sequential analysis by GC was also reported by Schüth and co-workers in their description of a parallelized 49-channel reactor for the HT screening of heterogeneously catalysed gas phase reactions [11].

In this publication we describe our new system for parallel HT screening of catalyst libraries by MS and GC (primary screening technology). The system is based on the setup reported previously [9]. Sampling is achieved by moving a capillary bundle, consisting of a feed gas capillary and two capillaries feeding the MS and the GC sequentially from well to well using a modified synthesis robot. The feed gas replaces the air over the catalyst surface and catalysts usually show detectable activity within seconds. The number of samples per library has been increased from the original 33 to 207. The advantage of the original setup was its low cost and the easy exchange of libraries for screening with our IR thermography test rig. A difficulty was the lack of stability; breakdown of the system was experienced frequently for a variety of reasons. Since our original publication we have not only enhanced our test rig for HT-MS and added GC as a supplementary analysis tool, but have modified the whole setup significantly to ensure stable operation. The original QMS has been replaced by a double-focusing MS which provides much higher sensitivity and, owing to its high-resolution mode, allows the simultaneous detection of mass fragments of the same unit mass, such as N_2 and CO. Furthermore, suitable software has been developed to improve control of the HTE and facilitate analysis of the large amount of data produced in each experiment. In this chapter we describe the basic features of this new HT technique and demonstrate its capabilities with selected experimental examples.

EXPERIMENTAL SETUP

In 1999 we reported our initial setup for MS detection of reaction selectivity and catalytic activity of heterogeneously catalyzed gas phase reactions by spatially resolved MS [9]. However, the initial system was unstable and was prone to breakdowns. With increasing experience the system has been improved constantly since then and has finally reached an acceptable degree of perfection and stability. The setup is shown schematically in Figure 8.1. The catalysts are present as solids on the bottom of the wells of the library

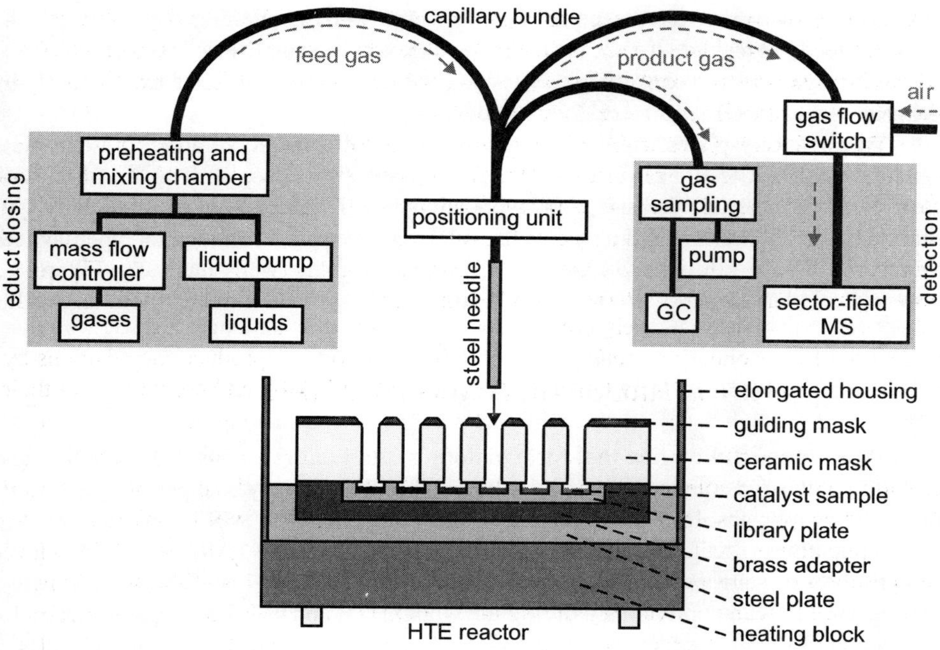

FIGURE 8.1. Schematic representation of the HT test rig.

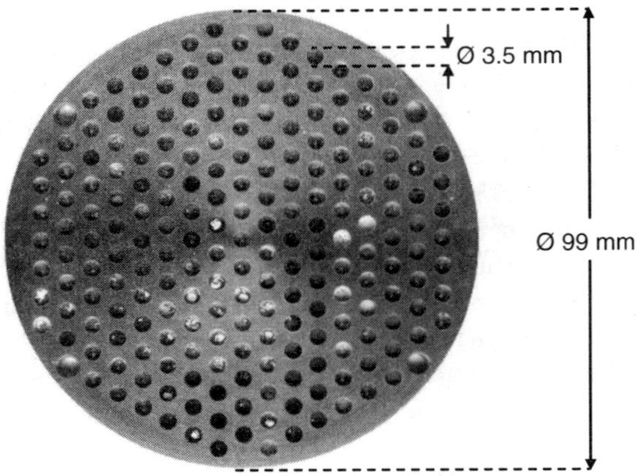

FIGURE 8.2. Combinatorial library plate with 207 wells filled with different catalysts.

plates (see Figure 8.2). The central part of the new setup is the modified capillary bundle ending in the capillary bundle needle (cb-needle), which is automatically positioned into the library wells by a modified synthesis robot for reaction product sampling. All capillaries are heated by heating wires and are thermally insulated to avoid condensation of feed and products. The gaseous and liquid reagents enter the feed gas capillary via a preheating and mixing chamber. A ceramic mask covers the library plate, thus transforming the library into an array of open "microreactors" ending in the wells containing the catalysts. The preheated feed gas replaces the air and overflows the catalyst sample in a "microreactor," and the resulting gas mixture containing educts and products can be analyzed simultaneously by the mass spectrometer and the gas chromatograph.

Typical library plates used in our group consist of a metal, ceramic, or slate plate containing a hexagonal arrangement of 207 separate 2 mm deep drill holes filled with catalysts. Plate material, well size, and layout are optional. Samples are either directly synthesized in the 207 wells by filling the wells with reagents using a commercial pipetting unit or prepared first as powders and are then filled into the plate wells manually. Figure 8.2 shows a steel plate library with catalysts (10 μmol each) synthesized directly in the wells.

HIGH-THROUGHPUT MASS SPECTROMETRY

MS is a powerful on-line tool for monitoring catalyzed reactions. Consumption and formation of desired compounds can be recorded through intensity changes of selected *m*/*z* signals. However, absolute quantification of gas components by MS is problematic. For HTE applications monitoring of selected ion intensity is usually sufficient. A problem is the definition of start and end in each well of the library plate. For stability reasons we have selected to scan continuously during the whole HTE and select associated spectra by comparison of the time tables of MS and HTE. As a single scan takes only a few seconds, several scans of the resulting gas mixture are recorded for each catalyst. Thus possible trends in activity and selectivity over the short range of the measurement time can be recognized for each sample and more precise averaged values can be calculated. The ion current of new signals after entering a "microreactor" needs 20–30 s to reach a stable level. Therefore we use a standard measurement time of 60 s for MS analysis of each "microreactor" and average the intensities obtained in the last 20 s. Together with the movement of the cb-needle a total measurement time of 70–80 s per catalyst results.

Although the data obtained with a low-cost gas analyzer (QMS) is sufficient for a variety of applications [9], two problems remain.

1. The resulting gas mixture may contain several compounds within a close *m/z* range, such as CO ($m/z = 27.9949$) and N_2 ($m/z = 28.0615$) as the products of partial oxidation of hydrocarbons by air. Such products cannot be resolved by QMS. If the intensities for both fragments are similar, the minimal mass resolution $(m/z)/\Delta(m/z)$ necessary for separation is ~ 2500. If only traces of CO are detected in a gas mixture consisting mainly of nitrogen (this corresponds to common analytic tasks), the resolution must be higher. Other compound pairs that appear in reactions of interest for catalysis and cannot be separated by QMS are, to name just a few, CH_4/O_2, N_2/C_2H_4, C_3H_6O/C_4H_{10} and CO_2/N_2O, with necessary resolutions >900, $>1{,}100$, $>1{,}600$, and $>3{,}900$, respectively.
2. The amount of reaction product formed on the small outer surface of catalysts in library plates is lower in this HTE than in conventional experiments. QMS is sensitive enough to detect small amounts of a compound, but for effective discovery of new catalysts in primary screening even traces of product should be detectable.

FIGURE 8.3. Photograph of the high-resolution SFMS.

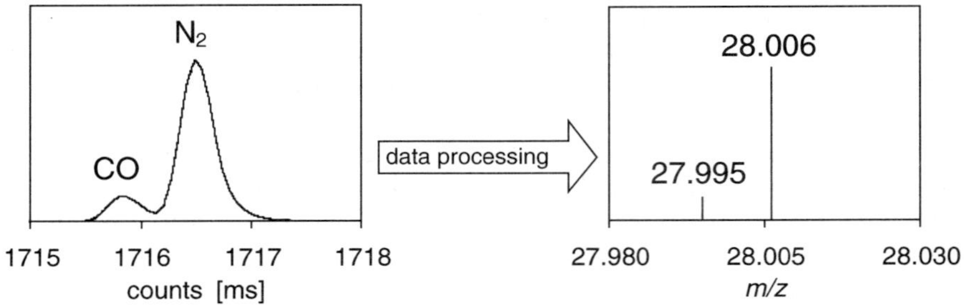

FIGURE 8.4. CO and N_2 measured with a resolution of ~2500.

For these reasons the QMS was replaced by a double-focusing sector-field MS (SFMS) (AMD-MMH1) with a resolution ranging from 400 to 10,000. The sensitivity is much higher than that of the QMS and increases with decreasing mass resolution. The upper limit for mass detection is about $m/z = 10{,}000$. The SFMS is shown in Figure 8.3 and an example for high-resolution application is shown in Figure 8.4.

HIGH-THROUGHPUT GAS CHROMATOGRAPHY

GC is used here to monitor compounds which cannot be identified clearly by MS, such as isomers or compounds of identical elemental composition and similar fragmentation pattern. Since both the MS and the GC are attached to the same test rig, they complement each other. The GC analysis time is dependent on the retention times of the compounds needed for baseline separation. This is in sharp conflict with HTE, which requires analysis times per well to be minimized. Thus measuring wide ranges of compounds with high precision should not be the subject of such a study. In contrast, separation quality should be optimized for a small number of essential compounds with the clear aim of keeping the analysis time as short as possible.

For our present study—the search for new isomerization catalysts—*n*-pentane isomerization was chosen as a model system. The sensitivity of GC with a flame ionization detector (FID) is significantly higher than that of MS. In addition, programmed temperature vaporizer (PTV) inlets can be cooled to trap traces of product in the gas mixture and thus dramatically improve the detection limit. After testing such an inlet system, we found that the sensitivity of the FID itself is sufficient for discovery even with extremely low product formation. The main disadvantage of PTV inlets for HTE is the long time delay per sample due to cooling, gas trapping, and heating for gas desorption. The best solution turned out to be a simple 180 μL gas sample loop with a sampling pump combined with a classical gas chromatograph (Agilent 6890). The problem of rapid baseline resolution for the essential components, *n*-pentane and *i*-pentane, was solved with a 15 m fused silica capillary column (Petrocol DH 50.2, Supelcoo). The separation of *n*-pentane and *i*-pentane is achieved within 30 s with optimized settings [Figure 8.5(a)]. The time needed to rinse the 100°C hot transfer capillary and the gas sample loop was investigated with a gas mixture of 2.5% *n*-pentane in hydrogen containing traces of *i*-pentane. This feed gas mixture was fed just as in a regular

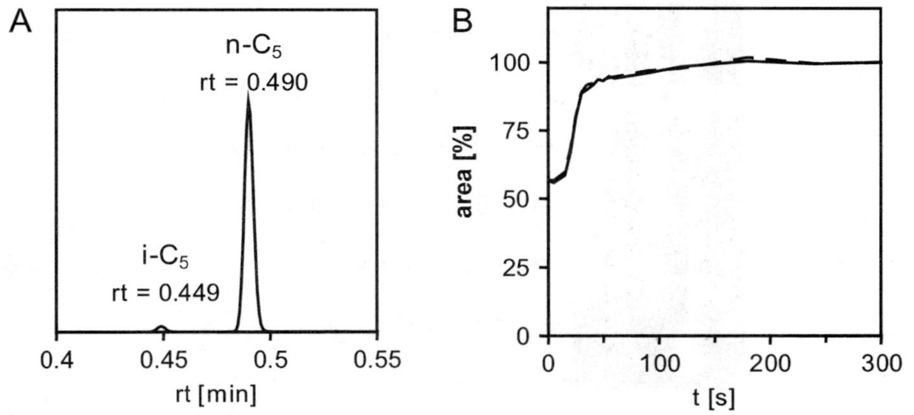

FIGURE 8.5. Optimization of the GC and overall measurement time: (A) gas chromatographic separation of *n*-pentane and *i*-pentane; (B) dependence of the signal area on the offset time before the start of GC solid curve, *n*-pentane area; dashed curve, *i*-pentane area (normalized to their respective last values).

measurement, but an offset time between entering the hole and the GC start was varied from measurement to measurement. The inlet flow to the gas sample loop was 1 mL min^{-1}. The results are shown in Figure 8.5(b); an offset time of 60 s is sufficient to detect ~ 95% of the maximum signal. There is no sampling discrimination for one of these compounds. The capillary bundle can exit the position immediately after the GC start and enter the next hole while the GC still runs. This leads to an overall measurement time of 80 s per well or ~ 4.5 hr per library plate with 207 positions with simultaneous MS and GC analysis. The high sensitivity of the method is demonstrated later in this chapter.

Micro-GC has successfully been introduced recently for fast on-line monitoring and may present a useful alternative to our present GC system in such an HTE setup, especially since sampling pump and capillary interface are standard equipment here. However, products with higher boiling points are a subject of concern, since condensation in the heated injection system may cause damage. Therefore at present we rely on the robustness of our conventional GC system.

CAPILLARY BUNDLE

The actual screening probe in our test rig is the cb-needle holding the ends of the capillary bundle. The positioning of the cb-needle is controlled by a commercial pipetting robot (Miniprep 50, Tecan), whose pipetting unit is replaced by the capillary bundle. Heat-sensitive parts of the moving robot arm had to be replaced. The capillary bundle consists of a capillary for the feed gas supply and two capillaries directing the resulting gas mixture to the GC and the MS. The GC capillary is connected to a pump, which ensures the necessary product gas flow for the sampling valve. The flow into the MS capillary is brought about by the pressure gradient between the normal pressure at the inlet end and the low pressure of the MS inlet chamber over the length of the capillary (length, 2–3 m;

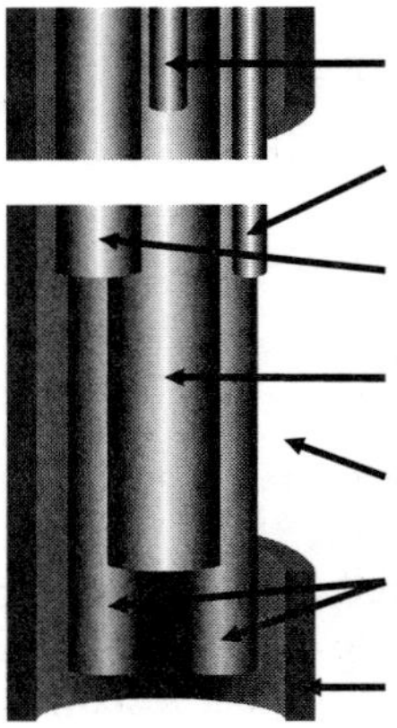

FIGURE 8.6. Detailed scheme of the capillary bundle in the cb-needle.

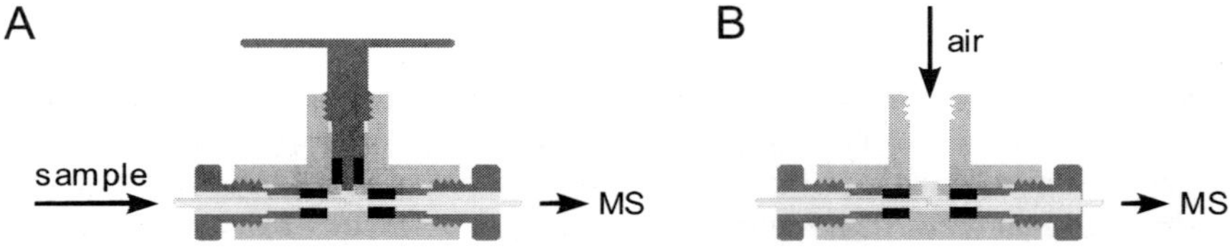

FIGURE 8.7. Function of the gas flow bypass: (A) sample flow to MS; (B) air flow to MS.

inside diameter, 75–100 μm). The wall of the cb-needle has openings on the side of its end for better gas exchange, which substantially improved the sensitivity of the method. A detailed scheme of the capillary bundle in the cb-needle is shown in Figure 8.6. As common polyimide-coated fused silica capillaries broke regularly, they were replaced by stainless steel tubings. The capillary system must be heated to prevent condensation of products with higher boiling points. However, this heating, must not constrain the flexibility of the capillaries. Commercial heating tubes commonly used for capillaries do not comply with this requirement. Our solution has been to thread the capillaries into thin silicone-coated glass-fiber tubes together with heating wires (Thermocoax). Capillaries can now be heated up to 200°C. A low-cost alternative for heating wires is realized with the alienated usage of thin thermocouples (Ø 0.5 mm). The temperature of the capillaries was set by applying appropriate constant voltage. In the presence of compounds with high boiling points regulated extra heating is necessary for the cb-needle owing to its higher heat emission. The temperature is monitored with tenuous thermocouples (Ø 0.25 mm) at different points of the system.

In some situations it may be desirable to turn off the gas flow to one of the analytic devices. This is readily done for the GC by switching off the pumping system of the gas sampling loop, but is more complicated with the MS where the pressure gradient across the sampling capillary is used for sampling. Here a simple means for disconnecting the MS from the sampling capillary was required. This was achieved using a small gas flow bypass constructed in our mechanics shop by adaptation of a commercial capillary connector shown in Figure 8.7 (Graphpack-3D/2 connector, Gerstel), which allows the sample flow to be replaced by air.

EDUCT DOSING

The amount of catalyst in our HT approach is quite small, ranging from 10 to 300 μmol depending on the synthesis procedure. The feed gas flow has to be adjusted to these small catalyst amounts, especially since the feed does not flow through the catalyst bed but is reflected by its surface. The final flow rate is a compromise—the slower the flow, the higher the conversion—but the flow has to be higher than the sampling flows of GC (here $1\,mL\,min^{-1}$) and MS (here ~ $50\,\mu L\,min^{-1}$) and high enough to enter the detector through the capillary during the sampling time of a single well. In our experience a flow rate between 4 and $10\,mL\,min^{-1}$ worked very well. Such low flow rates are readily obtained with gaseous feeds by commercial electronic mass flow controllers.

Precise and continuous liquid dosing is more demanding, as usually only few microliters per minute of liquid reagent must be mixed with the carrier gas. Since turbulence is difficult to achieve with such low flow rates, a combined preheating–mixing chamber was constructed in our machine shops: the liquid- and gas-carrying steel capillaries are wound around a heating cartridge and end in a small cavity (volume ~ 300 μL), pointing towards each other separated by a distance of less than 1 mm. The exit capillary (actual feed gas capillary) is positioned far from the ends of the inlet capillaries.

A commercial syringe pump (Infors) was employed for such small liquid flows with satisfying results. One of the problems observed, especially with liquids of high vapor

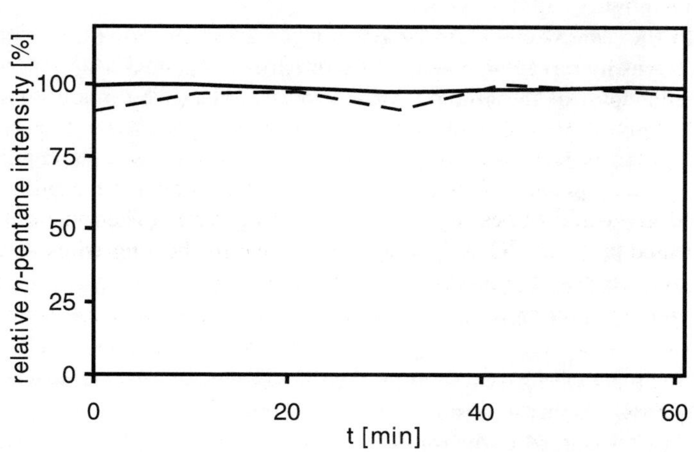

FIGURE 8.8. Reliability of microliter dosing of liquid educts: dashed curve, liquid *n*-pentane with syringe pump ($1.1\,\mu L\,min^{-1}$ *n*-pentane for $10.2\,mL_n\,min^{-1}$ gas mixture with 2.15 mol% *n*-pentane in hydrogen); solid line, *n*-pentane in gas mixture ($5.0\,mL_n\,min^{-1}$ gas mixture with 2.5 mol% *n*-pentane in hydrogen).

pressure, was the formation of gas bubbles in the syringe, leading to false dosing. This could be corrected by positioning the pump vertically with the dosing needle pointing to the top, so that gas or vapor resulting from outgassing could leave the dosing system. The most precise dosing is obtained with smallest syringe volumes; however, for dosing times of 4–10 hr large syringe volumes may be required or syringes have to be refilled frequently, which would result in undesired interruption of the screening process. Therefore selection of suitable syringe volumes is essential. A more precise but less flexible alternative consists in the use of special gas mixtures containing all necessary educt compounds. Nevertheless, the composition of such gas mixtures is restricted by the partial pressure of the compound with the lowest boiling point. The continuity of the flow of *n*-pentane in hydrogen was compared for dosing with a syringe pump and from a special gas mixture (Messer-Griesheim) (Figure 8.8).

HTE REACTOR

The reactor should heat the library plate uniformly and isolate the head spaces above the catalysts from each other and from air. A principal reactor cross-section is shown in Figure 8.1; detailed drawings are shown in Figure 8.9. For better heat transport the library plate is embedded in a brass adapter, which is positioned in a steel plate. This stack is placed over a commercial heating element for household stoves. Radiative losses from the top are reduced by a ceramic mask 15 mm thick that has the same drill hole pattern as the library plate. Furthermore, this mask separates the gas spaces above the catalysts, thus transforming the library into an array of open "microreactors" ending in the wells containing the catalysts. An additional brass mask 5 mm thick is positioned on top of the ceramic mask to guide the capillary bundle into the microreactors. The diameter of its holes is only 2.3 mm, so that the brass masks and the cb-needle form a lid for the respective reaction compartments. The drill holes of the ceramic are larger (Ø 3.5 mm) to avoid abrasion by the moving capillary bundle.

Owing to this sandwich setup the reactor can easily be modified by the user for special applications by replacing single layers or parts of the stack. All upper components from the brass adapter to the guiding mask can be replaced by a similar stack for the screening of differently structured libraries, thus increasing the range of application of the same reactor block. Very active catalysts in the library (e.g., a reference sample) may contaminate the porous ceramic mask, which cannot be cleaned properly. Subsequent measurements may show false positives due to such contaminations, as will be demonstrated later. These could lead to misinterpretations, especially in first-generation libraries of low activity. In such cases an additional steel mask is used as a separator between the library plate and the ceramic mask or as a complete replacement of the ceramic and the brass mask.

The heating plate (Ø 140 mm) and its covering steel plate (Ø 190 mm) are larger than the library plate (Ø 99 mm) to ensure a low temperature gradient across the library. The reactor housing is elongated to the top to protect from air currents and additionally insulated (see Figure 8.1). The temperature is controlled with a thermocouple in the brass adapter and checked with a second thermocouple positioned in an empty library "microreactor." Our reactor was designed to work at temperatures up to 800°C. Temperature is

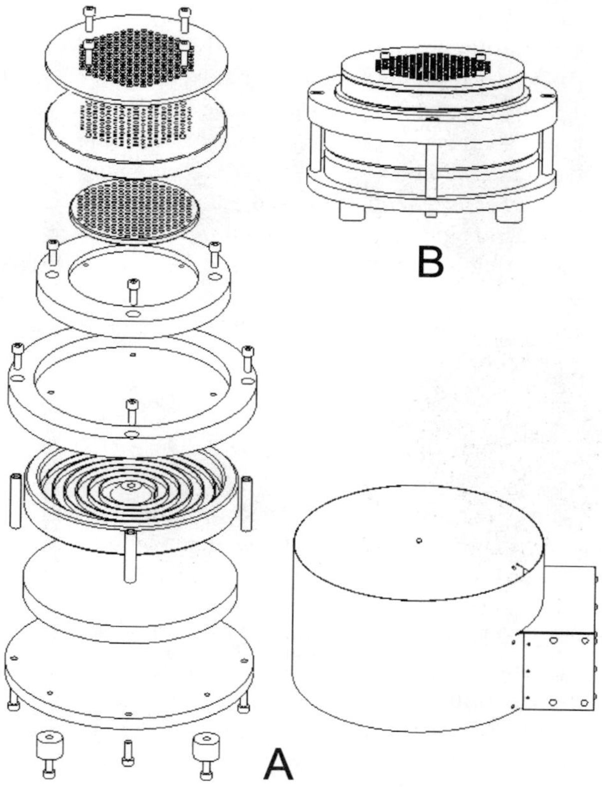

FIGURE 8.9. Detailed drawings of the parallel reactor: (A) exploded view showing all components; (B) illustration of the mounted inner parts.

controlled with a commercial controller with temperature self-optimization. The maximal temperature variation in a single "microreactor" is better than ± 1°C. The temperature distribution over the whole library, shown for different temperatures in Figure 8.10, is sufficient for such a primary screening device.

Since the reactor is open, the catalysts are in contact with air during the whole experiment. Although most of the air is replaced by the feed gas flow, complete replacement cannot be achieved. This may be an advantage in oxidation reactions, but is a clear drawback in reactions with atmospheres other than air, such as reduction and isomerization reactions. Gas exchange with air is strongly increased by thermal convection effects, especially at outer library positions. The dependence of the oxygen content in the air in individual "microreactors" on the educt flow was analysed using MS (Figure 8.11). As absolute intensities may vary from one measurement to the next, values are displayed in a relative scale obtained from division of the ion intensity of *m/z* 32 by total ion current (O_2 MS%). For general understanding, the uncorrected oxygen value for pure air varies between 17 and 18 O_2 MS%. Figure 8.11 shows the increasing relative oxygen exposure of the catalysts with decreasing flow.

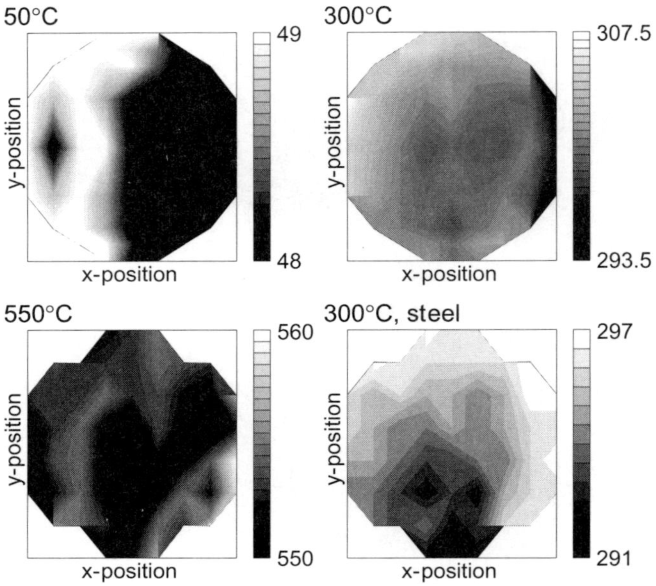

FIGURE 8.10. Temperature distribution over a slate library: 50°C: with ceramic and brass mask, $T_{mean} = 321.6\,K$, $\sigma = 0.2\%$, $\Delta_{max} = 0.3\%$. 300°C: with ceramic and brass mask, $T_{mean} = 574.6\,K$, $\sigma = 0.5\%$, $\Delta_{max} = 2.4\%$. 550°C: with ceramic and brass mask, $T_{mean} = n26.2\,K$, $\sigma = 0.3\%$, $\Delta_{max} = 1.2\%$. 300°C, steel: only with steel mask, $T_{mean} = 567.9\,K$, $\sigma = 0.3\%$, $\Delta_{max} = 1.1\%$.

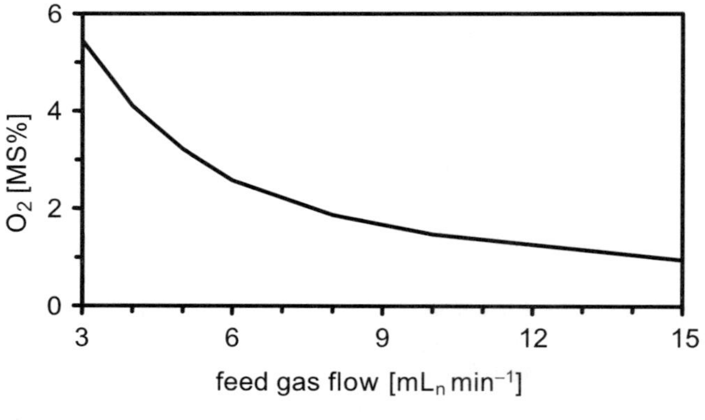

FIGURE 8.11. Dependence of oxygen content on feedgas flow at 300°C in the middle position of the library reactor.

In an attempt to reduce the amount of air in the reactor, a 100 mm thick steel ring was welded onto the steel plate around the brass adapter reducing horizontal air flow (Figure 8.12). Six uniformly distributed steel capillaries welded to the ring allow the passage of additional inert gas into the reactor close to the top of the library. To reduce convection

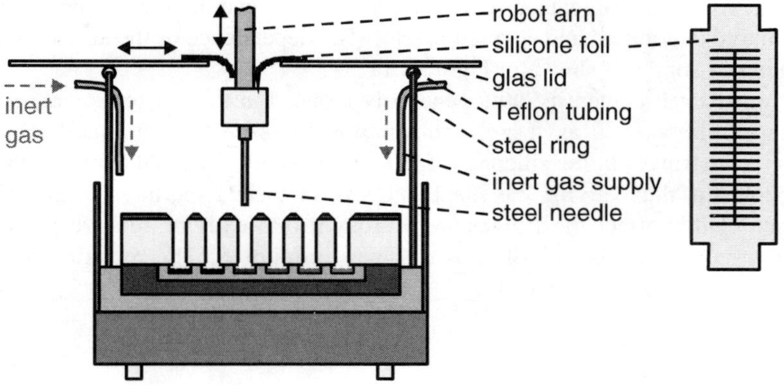

FIGURE 8.12. Scheme of the modified reactor with added steel ring and glas lid. The silicone foil with cut-in fringes is used to seal the groove in the glas lid during cb-needle movement.

FIGURE 8.13. Reactor with upper steel ring, inert gas supply, and glass lid.

a lid, consisting of a glass plate sliding on Teflon tubing affixed to the steel ring, was placed on top of the upper steel ring. The lid is moved by a frame fixed at the robot arm suspension bar moving in the x direction. For y and z movement of the robot arm a slit was cut into the glass lid. An attached fringed silicone foil further reduces convection through the trench. A picture of the optimized reactor with the moving glass lid is shown in Figure 8.13.

Air exclusion experiments have been carried out by feeding nitrogen and argon as the inert gas into the modified and covered reactor. The dependence of the air oxygen content on the inert gas supply is shown in Figure 8.14.

The air oxygen content is homogeneously reduced in the reactor gas atmosphere at room temperature as well as at high temperature (Figure 8.15, left panel). The relative positions of the fringes in the silicone film are permanently changed owing to the motion of the robot arm, thus varying the shape and size of the lid aperture. To investigate the importance of this effect the fringes were brought into extreme relative positions and the oxygen content in the middle position was monitored. The resulting variation is negligible (Figure 8.15, right panel).

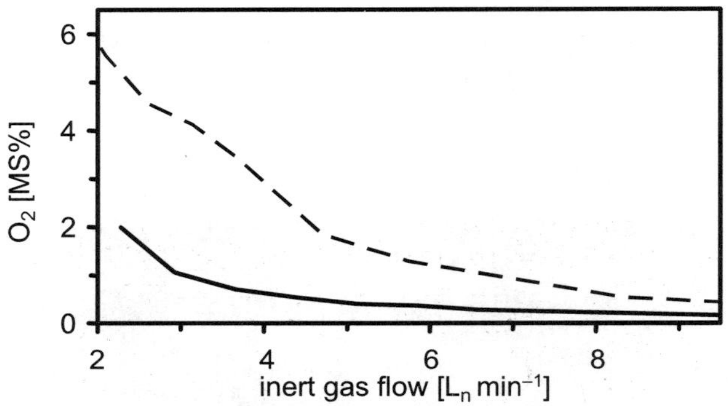

FIGURE 8.14. Oxygen dependence on the inert gas supply at 300°C in the "microreactors" of the parallel reactor (middle position; 5 mL$_n$ min^{-1} H_2 as feed gas): dashed curve, nitrogen as inert gas; solid curve, argon as inert gas.

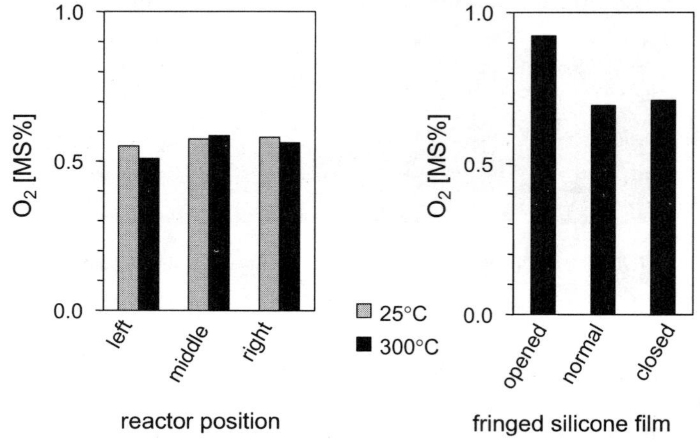

FIGURE 8.15. Oxygen content dependence on position on the library plate (left) and silicone film (right) (5–6 L$_n$ min^{-1} argon as inert gas; 5 mL$_n$ min^{-1} hydrogen as feed gas).

The oxygen is successfully displaced even at higher temperatures, but the argon flow must be at least $3\,L_n\ min^{-1}$ to reach contents lower than 1 vol%. The costs may be justifiable, taking the high sample quantity measured into account; nevertheless a more efficient alternative seems advantageous for frequent use. The reactor and the moving components could be placed into a glovebox, perhaps with an adequate cooling system, if a real protective atmosphere is required. It would probably be sufficient to have heat-conducting aluminum walls to compensate the reactor radiation.

SOFTWARE

Software is of central importance in HTE. It is concerned with the preparation of libraries with large numbers of samples, the control of the screening experiment, the evaluation of the resulting large quantities of data, their summation and visualization, and data storage. Since commercial software is not available to meet the requirements for our tasks, an extensive software package has been and still is being developed in our group.

Library design and synthesis are assisted by a custom-designed software package named Plattenbau, which reduces the time necessary for library synthesis by a factor of up to 20 [12].

The software necessary for a complete screening experiment (HTE) with the high-throughput test rig described here has to meet several requirements. It has to control the movement of the capillary bundle and it has to start and end analysis and data accumulation for each "microreactor." The cb-needle must be able to enter the "microreactors" in an arbitrary sequence. The sequence to be measured should be easily edited and should be saved in a file describing the planned run. Three definite time settings must be variable: an offset time (t_{offset}) between entering a hole and starting the measurement, a measurement time (t_{meas}) in the hole after the start, and finally a waiting time (t_{wait}) between exiting a hole and entering the next one. The cycle carried out for each sample is illustrated in Figure 8.16. For the true time between two measurements the robot movement time t_{travel} must also be taken into account. The real GC measuring time may be longer than t_{meas}

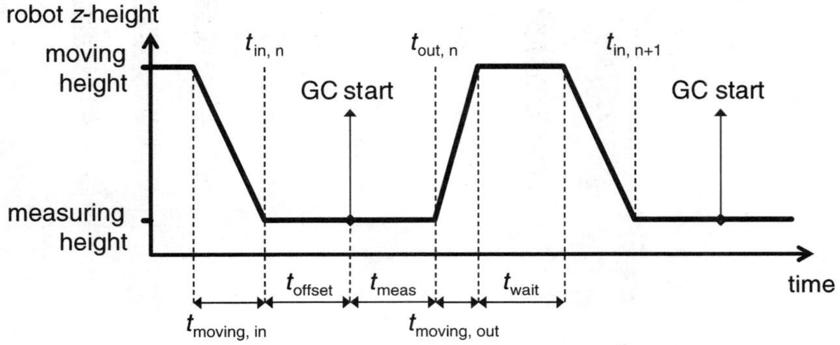

FIGURE 8.16. Cycle carried out for each sample in a measurement sequence.

but should be shorter than $t_{meas} + t_{wait} + t_{travel}$. As an additional precaution the next starting signal must not be generated if a "ready" signal from the GC has not been issued. The entering and exiting times (t_{in} and t_{out}) should be recorded together with the coordinates of the sampled position and be saved permanently into a file, thus describing the measured sequence. The starting signal is fundamental with GC because a single measurement is recorded for each sample, detection with GC is not a permanent on-line scan as is the case with MS.

The commercial software supplied with the pipetting robot (Logic v 1.94, Tecan) does not meet these requirements. The main problems are the missing start option and the requirement that the library pattern exactly matches the robot's coordinate axes, leading to time-consuming reactor positioning. For this reason a custom-made and user-friendly solution had to be programmed as part of our controlling software TestRig. A screenshot of the software is presented in Figure 8.17. The software displays the structure of the well pattern on the library plate (quadratic or hexagonal); positions that do not exist on the actual plate are faded out. It allows the user to freely define the whole sequence of the HTE experiment by simple pointing or marking. The automatic start and end and the GC start and end at each library position can be selected. GC measurements are only collected while the cb-needle is inside a "microreactor." MS, when used, samples continuously during the HTE. Individual library positions can be measured several times, for example, to repeat the

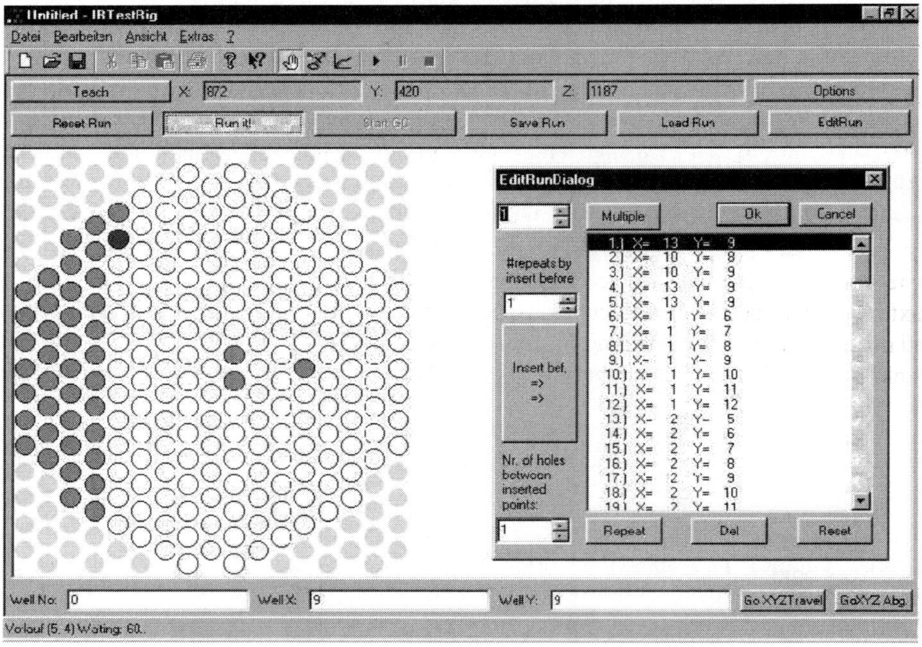

FIGURE 8.17. Screenshot of the software TestRig controlling the HTE.

measurement of a reference material after a certain number of measurements. A panel at the bottom of the window provides information about the status of the HTE. The HTE sequence is created in the EditRunDialog window and is saved in ASCII format. This file comprises the position coordinates sorted by a consecutive run number. Repetitions of segments can be selected by setting a repetition. A special feature for subsequent insertion of repeated blank or reference measurements is implemented. The number of samples to be measured between two blank or reference segments can be selected conveniently, and several identical runs of a library can be measured automatically for reproduction. The status of the sequence is permanently saved in an ASCII file containing t_{in} and t_{out} and the position coordinates for each measurement.

During such an HTE controlled by TestRig, huge amounts of data (MS or GC) are recorded sequentially. The software provided with the mass spectrometer (MASPEC Data System for Windows 14.0f) and with the gas chromatograph (GC ChemStation Rev.A.08.03, Agilent) both include data conversion into ASCII format.

The file generated for GC contains the integrated peaks for each single chromatogram. Each peak is defined by a retention time (rt) corresponding to a compound and by an area value (area) corresponding to its abundance. Each chromatogram is given a run number (run#) that can be assigned to a library position or catalyst. Thus each data point is well defined by three variables that are listed by run# in the GC file (run# $\leftarrow$ rt/area). The user is usually interested in one or more specific products. Thus all gas chromatograms have to be sorted by rt (rt $\leftarrow$ run#/area). A single HTE produces about 220 discrete gas chromatograms, including blanks and reference samples. Each gas chromatogram contains many different signals, identified by rt. The sorting of gas chromatograms can be achieved within the Chemstation software, but only for signals that have been previously defined in the internal calibration.

A typical evaluation of HTE data from GC for n-pentane isomerization with commercial data processing and visualization software is now described. Once the data are converted to ASCII format using Chemstation, the run# and area values of the preselected *i*-pentane and *n*-pentane rt are inserted into an Excel or Origin spreadsheet. The run# and an area of zero must be completed manually for the missing lines. Signal fluctuation can generally be reduced by normalizing to an internal standard; this means division by a signal representative for the measurement. As no internal standard is available for division in the gas mixture used, i-pentane area fluctuations are reduced by division by the sum of the i-pentane and n-pentane areas. The resulting fraction is now defined as figure of merit "i-C_5 yield." The blanks measured at regular intervals during the HTE are averaged and the standard deviation (σ_{blank}) is calculated. The yield values are corrected for background by subtraction of this average blank value. Only samples with a yield bigger than $<blank> + 2\sigma_{blank}$ are considered as being active (hit criterion). Thus the probability that a hit is true and not an artifact coming from blank scattering is 97.7% if the respective sample has only been measured once. The relative i-pentane yield of a sample is obtained with further standardization with a reference catalyst. This standardization only makes sense if the activity of the reference catalyst is very reproducible from one library to the next. As a run# is assigned even to blanks and other measurements of minor importance after data processing, a sample number (sample#) is introduced that is not assigned to those measurements. The dependence of the figure of merit on the sample# is displayed in Excel or Origin, and its dependence on the x

and *y* position coordinates is visualized in Spotfire or Origin. The attribution of an activity value to its respective catalyst is done manually. This mainly manual evaluation with commercial data processing and visualization software is a time-consuming and unpleasant task and is restricted to few preselected signals of special interest. The evaluation is demonstrated below for a concrete screening experiment (see Figure 8.19).

MS evaluation is done in an analogous way. Argon is generally used as an internal standard. Approximately 5000 mass scans are recorded in an ordinary sequence instead of one discrete measurements per sample. Assigning and averaging these scans for each "microreactor" without special software assistance is a time-consuming manual task. Furthermore, without the files generated from the robot control the assignment is only possible if some ion currents permit the monitoring of the sequence, showing significant differences in and out of the drill holes.

Data evaluation from GC and MS measurements with commercial software is achievable with a reasonable amount of effort. Nevertheless, evaluation and visualization software tools are being created in our group to achieve real high-throughput and more effective data extraction, thus significantly reducing the evaluation time. Solutions for the storage and management of the processed data, and accordingly data bases in general, are being developed in collaboration with our partners.

VALIDATION AND OPTIMIZATION OF SETTINGS

Validation of the test rig is demonstrated here with n-pentane isomerization. A standard parameter set has been chosen for comparison of different experiments. The optimized conditions generally used for n-pentane isomerization are described here. A premixed gas mixture of 2.5% *n*-pentane in hydrogen (purity 3.0) was used as feed gas. The mixture contains less than 20 ppm (v/v) of i-pentane. Temperatures are 100°C for the preheating in the mixing chamber, 100°C for the capillaries, and 300°C for the reactor, measured in the library plate. Experiments are carried out at normal pressure as the reactor is not closed; no additional inert gas is used for air exclusion. The cb-needle is always positioned in the "microreactors" at 2.1 mm above the bottom of the 2 mm deep wells in the library plate to guarantee a minimal safety distance of 0.1 mm between the cb-needle end and the catalyst surface. The resulting gas mixture is simultaneously monitored by MS for detection of oxidation products and GC for hydrocarbon detection. The treatment of the catalysts and libraries has been standardized as well; catalyst reduction steps have been abandoned. Samples are always calcined at least at reaction temperature before mounting into the reactor. Additional settings are described above in the appropriate sections of this chapter.

Two reference catalysts have been used for validation and activity comparison: a mordenite containing 1 wt% platinum and calcined at 500°C (refcat-1) and an industrial pentane isomerization catalyst (refcat-2). Both catalysts were milled manually to form fine powders. Approximately 5 mg of each of these catalysts was used to fill wells of the library plate.

The feed gas flow has a strong effect on the product response. The maximal product response was achieved with a flow of 5 mL_n min^{-1} (standard flow rate) for several different reactions and catalysts. The representative results for the pentane isomerization with refcat-1 are shown in Figure 8.18.

For background calibration of the reproducibility and sensitivity of the GC analysis an empty library plate was measured. The plate was prepared and measured in the same way as catalyst libraries in order to make the same errors. The central drill hole was chosen as the blank position; all other empty positions were considered to be samples. All 206 samples were sampled once in the experiment; the blank was chosen as the first measurement and repeated after each 30 samples, and the sequence was finished by a blank again ($run\#_{blank} = 1, 32, 63, 94, 125, 156, 187, 214$). The "product gas mixture" of the library was analyzed under standard conditions by GC and MS simultaneously. The total experiment with 214 single measurements took 4.7 hr. The general data evaluation procedure was performed as described in the previous section. The raw and processed data detected in this screening experiment are displayed in Figure 8.19. The blanks are indicated by diamonds in the upper linear graph and are excluded in the lower surface plot. The linear presentation helps us to recognize linear trends such as intensity changes with time. The lower presentation helps us to recognize position-dependent trends, such as higher activities in a certain region of the plate and to assign activities to specific catalysts. The results demonstrate that no trend was observable. The maximum difference in *i*-pentane yield measured is 10.1 ppm, the hit criterion $<blank> + 2\sigma_{blank}$ is 0.52 ppm, and $<x> + 2\sigma_x$ for all measurements in the sequence is 2.4 ppm. In this case hits are those with values higher than 0.52 ppm. The two positions with the highest values were found to be contaminated with traces of reference catalyst. The hit criterion is redefined for every screening experiment as it depends on the respective blank values. Values of $<blank> + 2\sigma_{blank}$ greater than 50 ppm have never been recorded for this reaction, thus providing the sensitivity of this high-throughput screening. The probability for blank scattering artifacts is only 2.3%; these artifact values are mostly close to the hit criterion and therefore much lower than real catalyst values. Consequently traces of isomerization activity can be detected very precisely with this high-throughput test rig, proving its potential for catalyst discovery! The sensitivity of

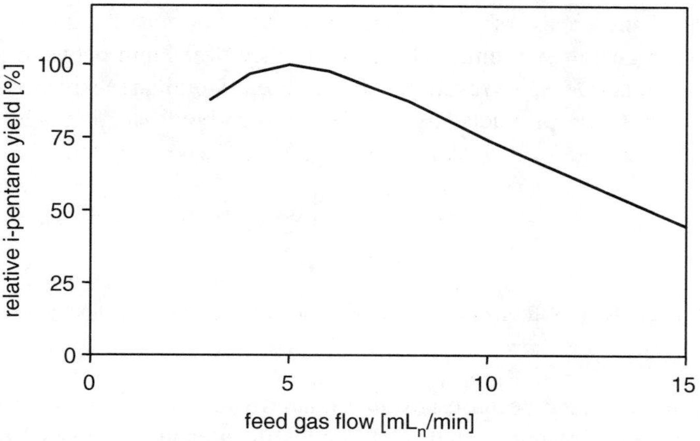

FIGURE 8.18. Optimization of the feed gas flow for pentane isomerization.

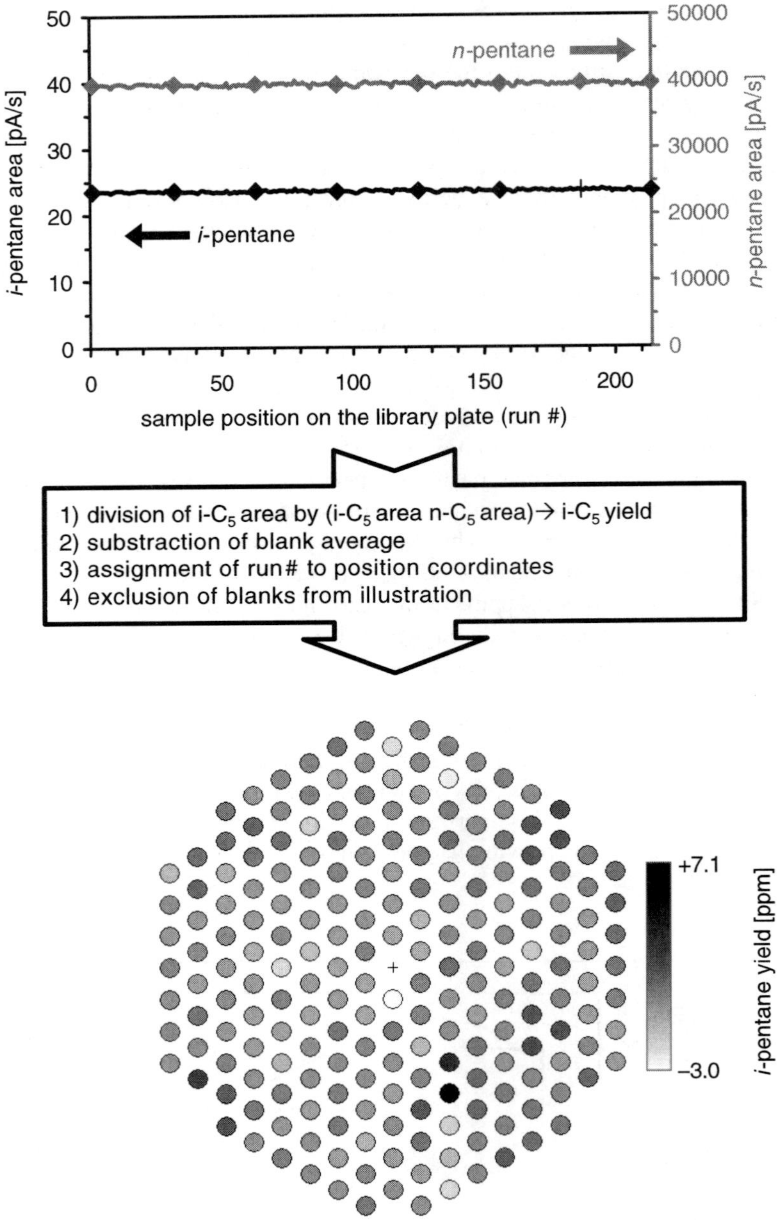

FIGURE 8.19. Validation procedure demonstrated using an empty library plate.

the MS was not calculated because absolute quantification is not possible with common mass spectrometers. However, small differences in intensity changes can be detected reproducibly with the SFMS. The total oxidation products CO_2 and H_2O were present in all positions as they were constant parts of the background signal. These compounds arise from air entering the drill holes and from desorption from reactor and library components. For this reason it is advisable to dry the catalyst library before screening if it is required to detect traces of activity down to few parts per million.

The total activity and long time performance of refcat-1 is shown in Figure 8.20. Although the total yield of 1.2% may seem low, it is orders of magnitude higher than the hit criterion of 0.52 ppm!

The distance between the end of the cb-needle and the catalyst surface plays an important role for measured activities. This distance changes with the fill height of the catalysts, while the cb-needle is always positioned at the same height. The effect of the fill height of refcat-1 on isomerization is shown in Figure 8.21. Consequently, it is essential to maintain the same fill height for manually filled powder libraries to minimize errors. This

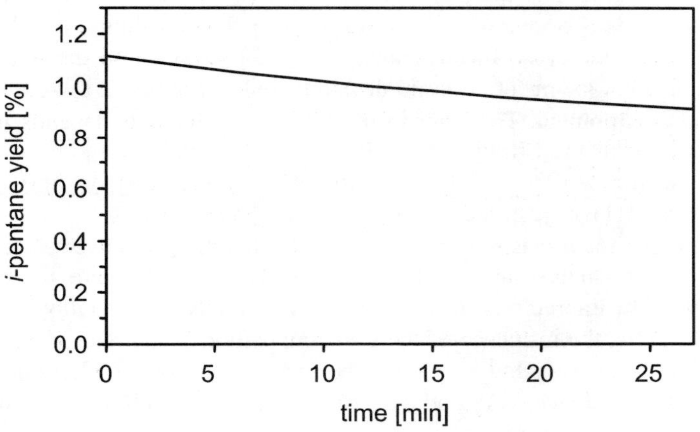

FIGURE 8.20. Isomerization yield and deactivation with time of refcat-1 under standard reaction conditions.

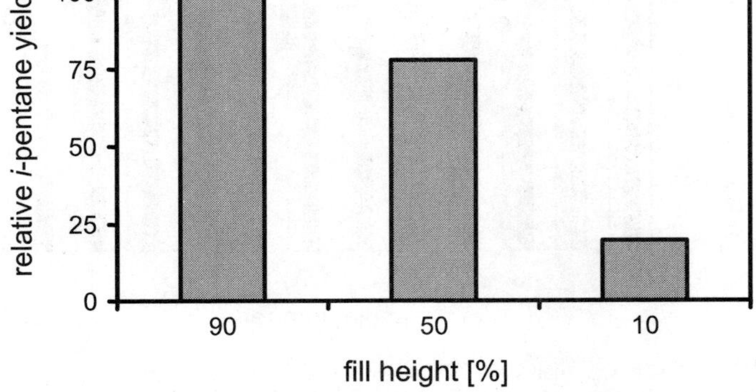

FIGURE 8.21. Effect of the fill height on product formation.

parameter cannot be controlled as readily with libraries synthesized automatically by a robot, because there is no way to predict the final catalyst volume in the well after calcination. The samples may also show very different textures, such as solid glass films or fluffy powders. This can results in large errors and is a major drawback of direct synthesis in library wells. However, even such a potentially severe disadvantage of directly synthesized libraries does not really affect the potential of this setup for discovery of new catalysts. Active samples will show statistically significant activity because of the high sensitivity of the method, so the decision as to whether a sample is active or not can still be made, independent of the error in absolute signal height. If reliability of relative activity of the catalysts is important, libraries using milled and manually filled catalyst powders should be chosen.

The preparation of libraries with powdered catalysts starts with the library design using the Plattenbau software, followed by automated synthesis into small glass vials with a pipetting unit. The samples are then dried and calcined, followed by manual crushing of the solid in the vials if necessary (visual inspection). The crushing of 200 solids in this manner takes approximately 1.5 hr. Transfer of the powders into the library plate has been done manually (3–4 hr for 200 samples), but could be replaced by robotic systems in a professional environment. The error in the HTE measurements depends in part on the precision achieved in the filling step. Additional processing steps, such as calcination, impregnation with new dopants, etc., may follow. The increased synthesis time of such a library plate for HTE is justified by the gain in data reliability.

Another experiment was designed to test for reliability of activity detection and for the effect of air on catalyst activity. Refcat-1 powder was filled into several drill holes manually and rather imprecisely, filling the drill holes only to a roughly similar level. In two complete HTEs, the influence of the gas composition in the reactor on the measured *i*-pentane yield was investigated with and without an argon flow of $5\,L_n\,min^{-1}$ through the inlets of the additional steel ring under standard sampling conditions (Figure 8.22). The

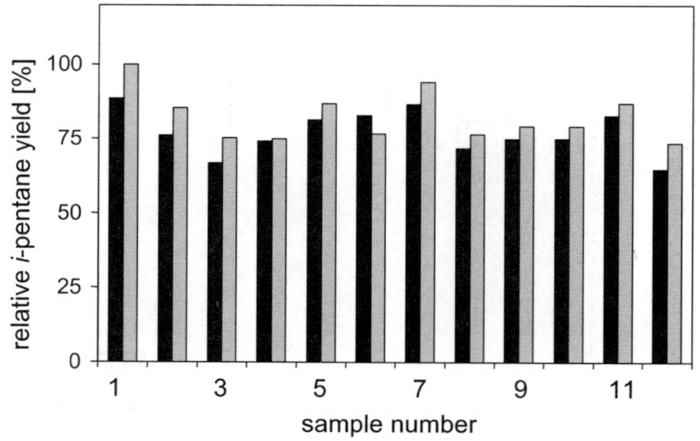

FIGURE 8.22. Reproduction experiments and influence of the reactor atmosphere: black bars, without additional inert gas; gray bars, with $5\,L_n\,min^{-1}$ argon.

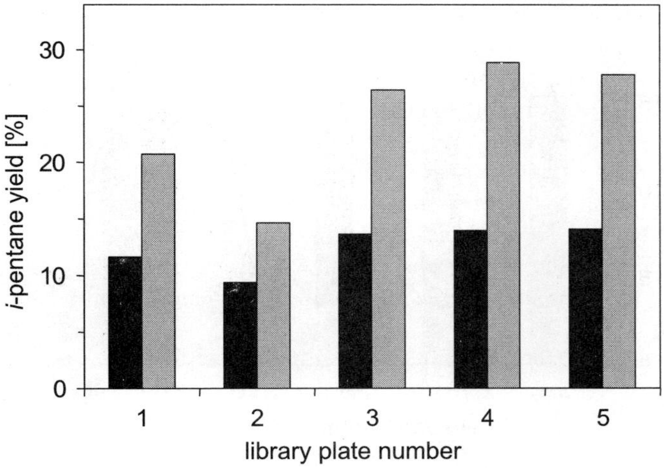

FIGURE 8.23. Influence of the library plate pretreatment on the activity: black bars, refcat-1; gray bars, refcat-2.

averaged relative *i*-pentane yield and the absolute standard deviation for the experiment without argon are 77.2% and 7.6%, respectively. The corresponding values for the experiment with argon are 82.5% and 8.4%, respectively. The errors can be reduced further by using more precise plate filling. The results show that activity and differences in activity are reliably detected and that the presence of air (oxygen) does not significantly influence the production of *i*-pentane. A protective gas atmosphere in the HTE seems unnecessary for isomerization activities. The platinum-containing zeolite samples were active, although the common catalyst reduction for optimal performance was not carried out.

In Figure 8.23 the activities of the two reference catalysts on different library plates are compared. The plates were measured under the same conditions in different experiments. Heating the reactor to 300°C always took ~ 0.5 hr, but a parameter that was unaccounted for in the screening of the two first plates was the time between reaching the reaction temperature and starting the test sequence. Another source of error is the time between the HTE start and the measurement of individual catalysts (up to 4.5 hr here). This leads to additional "calcination" times that obviously play a role on the absolute catalyst activity. The time between heating to reaction temperature and start of HTE was set to a standard value of 2 hr for libraries 3–5 in Figure 8.23, leading to better reproducibility. The relative errors from plate to plate are smaller for refcat-1 than for refcat-2.

EXAMPLES OF TEST APPLICATIONS

The following applications demonstrate the power of our test rig for discovery and for comparison of heterogeneous catalysts. Background correction was always obtained from the data collected in empty reference "microreactors."

Ruthenium-containing mixed-oxide catalysts with compositions Ru_5Zr_{95}, $Ru_{10}Zr_{90}$, $Ru_{15}Zr_{85}$, $Ru_{20}Zr_{80}$, and $Ru_{25}Zr_{75}$ were tested for CO oxidation at 100°C. This reaction is

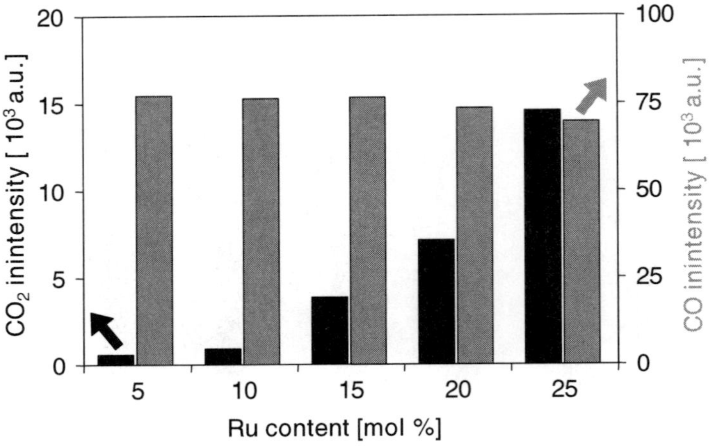

FIGURE 8.24. Dependence of CO oxidation activity on the ruthenium content of the catalyst (Ru_xM oxides, 0.65 $mL_n\ min^{-1}$ CO, 4.35 $mL_n\ min^{-1}$ synthetic air, 100°C): black bars, product CO_2 signal (left axis); gray bars, educt CO signal (right axis).

an advantageous model reaction, because only CO_2 is likely to be produced. The feed gas containing 15 vol% of CO in synthetic air was not preheated. The resulting gas mixture was monitored with the SFMS in high-resolution mode. For more accuracy the mass calibration necessary for high-resolution measurements was performed with the detected educt and product gas mixture itself. The intensities for CO and CO_2 given in Figure 8.24 are background-corrected. The activity of the powder catalysts increases with the ruthenium content, as was confirmed with additional catalytic tests using different methods. The precision of the high-throughput MS in high-resolution mode is sufficient to detect such relationships.

In a second experiment a library plate containing a large variety of mixed-oxide powder catalysts was screened for CO oxidation with high-throughput MS. The detection of CO_2 was realized in low-mass resolution to gain more sensitivity with the SFMS. Consequently CO was not detected. The reaction was carried out at two temperatures, 50°C and 100°C. The feed gas was not preheated and contained 3% of CO in a mixture consisting of 20% O_2 and 80% Ar. The argon signal at $m/z = 40$ was used as an internal standard to improve stability of the intensities over the full HTE. The background-corrected CO_2 intensities are shown in Figure 8.25. As expected, the activities of the samples increase with temperature. As the powders have similar fill heights, a ranking list of the catalysts was obtained through sorting by relative activities. The same sample was the most active for both temperatures.

If a certain catalyst property is of special interest, it may be useful to not only display the respective activities but also an additional figure of merit permitting conclusions for this property. As an example, a property of interest might be the temperature dependence of the catalyst activity as a first indicator of differences in the activation barrier. By comparison of the two plots in Figure 8.25 it can be seen that some samples exhibit a stronger gain in activity with increasing temperature than the most active sample, but it is hard to

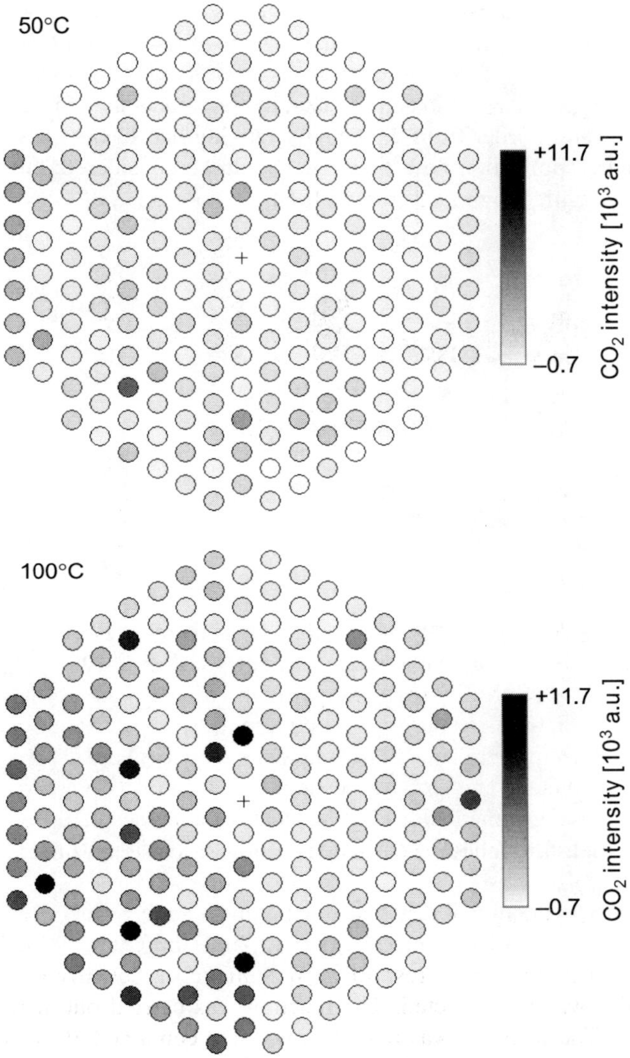

FIGURE 8.25. CO oxidation activity of a library plate at different reaction temperatures (0.15 mL$_n$ min^{-1} CO, 0.97 mL$_n$ min^{-1} O_2, 3.88 mL$_n$ min^{-1} Ar).

decide for which sample this property is strongest. For this reason a function describing the request is introduced using the available data sets. The activities of both plots were each normalized to the sample which was the most active in both experiments. The lowest value was set to zero and the highest to 100, thus giving relative CO_2 formation values from zero to 100% for each plot. The difference between the values at 100°C and at 50°C was then determined, resulting in an additional figure of merit Δ(relative CO_2 formation) [%]. Logically, the difference value for the reference sample is zero. A positive value signifies that the relative increase in activity with temperature was higher for this sample than for the reference, which is actually the required property. A plot of this figure of merit is given in Figure 8.26; the reference sample is identified by an asterisk. Many catalysts show a higher increase in activity than the reference, thus forming a second ranking list with a different order from the activity ranking list. In the quest for catalysts for higher-temperature application both rankings should be considered; the most interesting catalysts are those with a high rank in both lists.

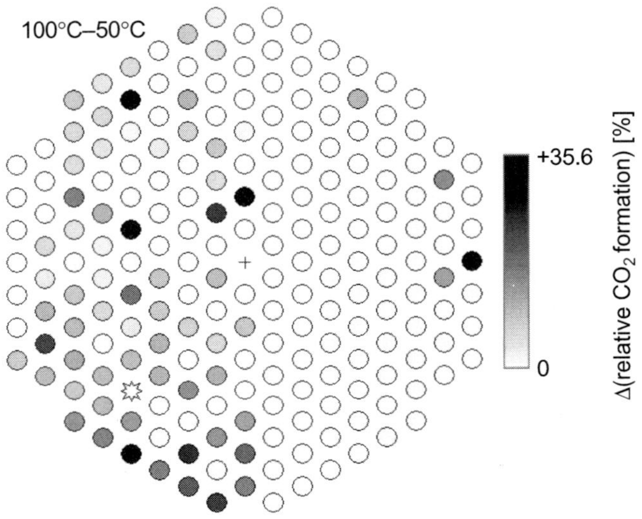

FIGURE 8.26. Increase in CO oxidation activity with temperature (see text).

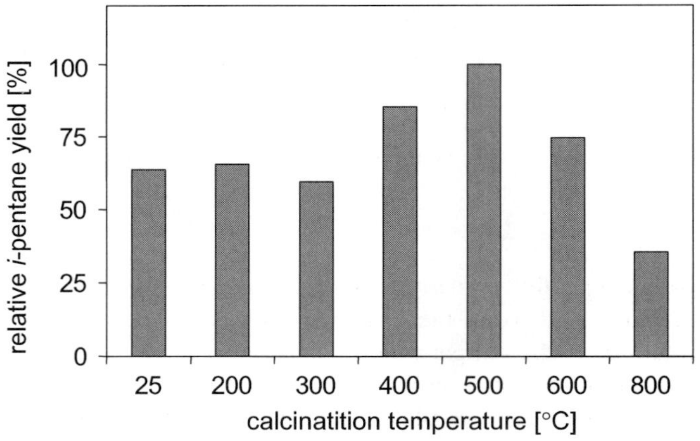

FIGURE 8.27. Optimization of the calcination temperature of mordenite doped with 1 wt% Pt.

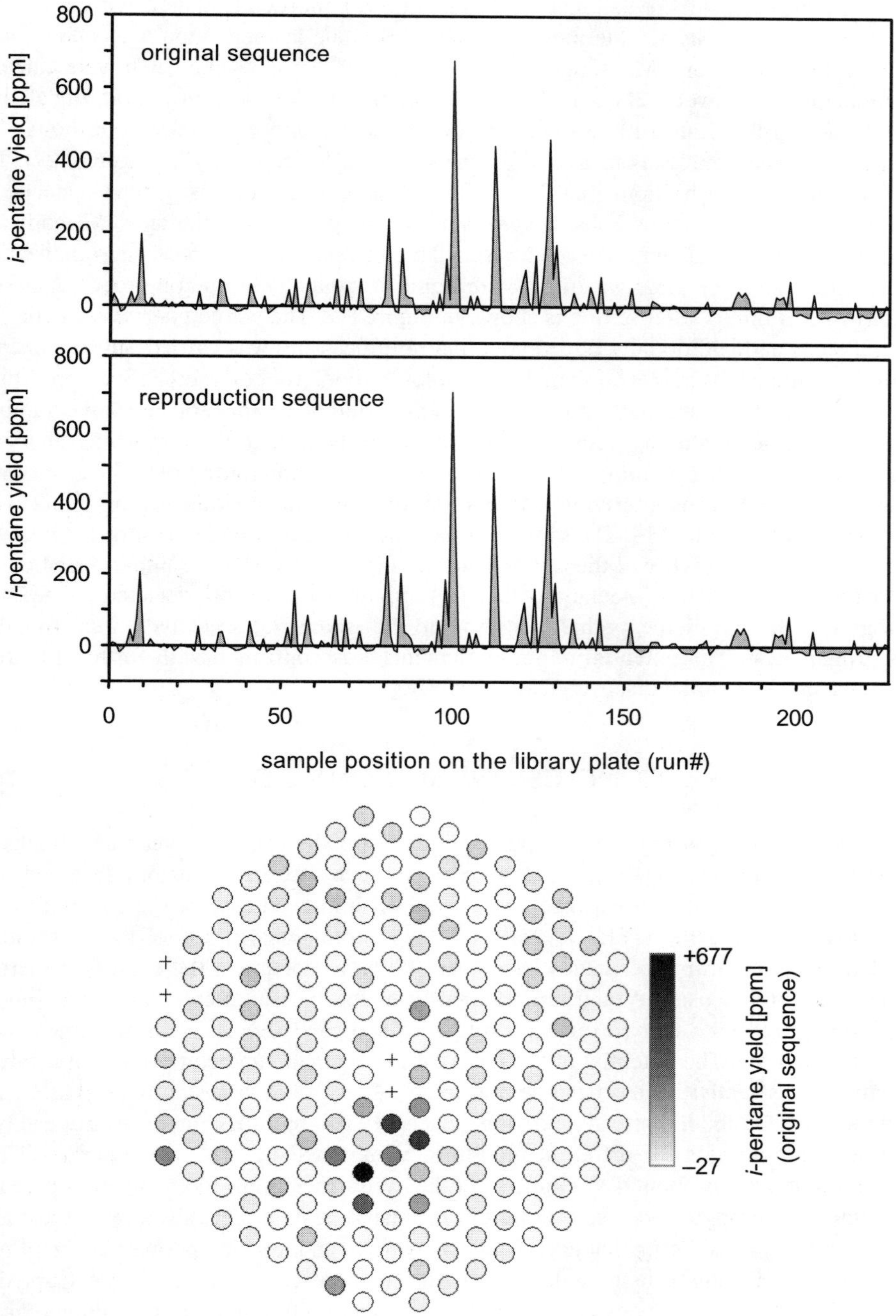

FIGURE 8.28. Isomerization activities on a directly synthesized library plate.

The influence of the calcination temperature of mordenite doped with 1 wt% of platinum on its activity for the more demanding pentane isomerization was studied with high-throughput GC and MS simultaneously. Samples from the same batch were calcined at temperatures between 200 and 600°C. A sample of each portion and an uncalcined sample were filled into a library support and measured under standard conditions. The highest *i*-pentane signal was measured for the sample calcined at 500°C (Figure 8.27). The results gained with high-throughput GC are in good agreement with those gained in analogous conventional tests. This proves that the errors made with powder samples are sufficiently low to allow trends to be recognized, even for a challenging reaction like *n*-pentane isomerization.

A steel substrate plate was filled with liquid reagents by a pipetting robot. A photography of this library after drying is shown in Figure 8.2. The whole plate was calcined in several steps and additionally doped later, but no processing was carried out to make the sample texture or fill height similar. The samples do not have common bulk compositions but are highly divergent and consequently very different in appearance. Some samples consist of brittle powder agglomerates almost filling the hole to the top, while others are films just covering the bottom of the drill hole, small loose monoliths, etc. The library was tested for *n*-pentane isomerization under standard conditions and the products were monitored with GC and MS. The screening sequence was repeated afterwards to check the convincing reproducibility of the test once more (Figure 8.28). Forty samples satisfied the hit criteria and produced *i*-pentane with a yield higher than 43 ppm detected in the chromatograms. Several elements that contribute to the isomerization activity were found in this primary test. The potential of these elements was confirmed with further libraries. Active samples could be discovered successfully.

CONCLUSIONS AND OUTLOOK

The setup of new high-throughput equipment for the primary screening of catalyst libraries by simultaneous MS and GC analysis has been described. Custom-designed control and evaluation software has been developed for improved ease of use and significant time savings during the HTE. Several possible sources of errors have been excluded. Validation and control experiments have been shown to document the equipment performance. The high sensitivity and reproducibility of the measurements have been shown. High-throughput tests carried out with reference samples were in good agreement with conventional tests. The potential of the setup for the detection and comparison of catalytic activities and selectivities has been demonstrated. The system is presently used as a standard tool in HTE and has already led to the discovery and optimization of several catalytically active materials for different reactions, which will be reported separately. The present throughput is about 400 samples per day. With such a fast method catalysts requiring time-consuming activation may be discriminated against. Yet this is not necessarily a disadvantage, considering that good catalysts without an induction period may be of preferred interest. Owing to its modular design, the setup described can easily be adapted to other tasks. Other analytic devices can be connected and different types of samples can be investigated. In general this system can be modified for each type of spatially resolved analysis in sampling, such as coupled GC–MS, capillary electrophoresis, photo-acoustics, IR, UV–VIS, and Raman spectroscopy, and many others, by moving capillaries and fibers.

ACKNOWLEDGMENTS

This research was funded by the German Ministry for Education and Research (BMBF) under contract number 03C0307A. We thank Bayer AG for partial support of our mass spectrometer. Thanks go to R. Richter for his engineering help and for the technical drawings, to the machine shops at the University of the Saarland for the production of the parts, to B. Gretschel and C. Thome for programming, to H. Höltzen for her assistance during the experimental period, and to S. Thomson, MPI Mülheim, and the Süd-Chemie AG for the reference isomerization catalysts. J.S. thanks the BMBF for support under grant 03C0311.

REFERENCES

1. Jandeleit, B., Schaefer, D. J., Powers, T. S., Turner, H. W., Weinberg, W. H. *Angew. Chem. Int. Ed.* **1999**, *38*, 2476.
2. Scheidtmann, J, Weiss, P. A., Maier, W. F. *Appl. Cat. A* **2001**, *222*, 79.
3. Holzwarth, A., Schmidt, H.-W., Maier, W. F. *Angew. Chem. Int. Ed.* **1998**, *37*, 2644.
4. Maier, W. F., Kirsten, G., Orschel, M., Weiss, P. A. *Chim. Oggi.* **2000**, *18*, 15.
5. Cong, P., Doolen, R. D., Fan, Q., Giaquinta, D. M., Guan, S., McFarland, E. W., Poojary, D. M., Self, K., Turner, H. W., Weinberg, W. H. *Angew. Chem. Int. Ed.* **1999**, *38*, 483.
6. Senkan, S., Krantz, K., Ozturk, S., Zengin, V., Onal, I. *Angew. Chem. Int. Ed.* **1999**, *38*, 2794.
7. Rodemerck, U, Ignaszewski, P., Lucas, M., Claus, P. *Chem. Eng. Technol.* **2000**, *23*, 413.
8. Claus, P., Hönicke, D., Zech, T. *Catal. Today* **2001**, *67*, 319.
9. Orschel, M., Klein, J., Schmidt, H.-W., Maier, W. F. *Angew. Chem. Int. Ed.* **1999**, *38*, 2791.
10. Cong, P., Dehestani, A., Doolen, R., Giaquinta, D. M., Guan, S., Markov, V., Poojary, D., Self, K., Turner, H., Weinberg, W. H. *Proc. Natl. Acad. Sci. USA*, **1999**, *96*, 11077.
11. Hoffmann, C., Schmidt, H.-W., Schüth, F. *J. Catal.* **2001**, *198*, 348.
12. Scheidtmann, J., Saalfrank, J. W., Maier, W. F. Science and Technology in Catalysis 2002, *Studies in Surf. Sci. and Catalysis* (**2003**) *145*, 13–20.

Chapter 9

Rapid Determination of Molecular Parameters of Synthetic Polymers Using Precipitation–Redissolution HPLC and a "Molded" Monolithic Column

David Sýkora, Frantisek Svec, Jean M. J. Fréchet, Miroslav Petro, and Adam L. Safir

INTRODUCTION

Synthetic polymers present many unique separation challenges because they consist of macromolecules featuring a distribution of structurally different chains that can vary in both chain length and end-groups. Separation of copolymers is even more difficult since, in addition to the previous two variables, they also may differ in ratios of individual repeat units and their sequence distribution. Therefore a variety of methods have been developed for the characterization of molecular parameters of both natural and synthetic polymers.

SIZE-EXCLUSION CHROMATOGRAPHY

Synthetic polymers are currently primarily characterized by size-exclusion chromatography (SEC). Within the last four decades, SEC has almost completely replaced

David Sýkora • Department of Analytical Chemistry, Institute of Chemical Technology, Technická 5,166 28 Prague, Czech Republic
Frantisek Svec and Jean M. J. Fréchet • University of California, Department of Chemistry, Berkeley, CA 94720, USA
Miroslav Petro and Adam L. Safir • Symyx Technologies Inc., 3100 Central Expressway, Santa Clara, CA 95051, USA
Contanct author: Frantisek Svec, email: svec@uclink4.berkeley.edu

the tedious fractionation methods that were used for the determination of the molecular weight distribution of synthetic macromolecules [1,2]. SEC is based on differences in the accessibility of pores of the separation medium to molecules of varying hydrodynamic volume. This method provides accurate, reliable, and reproducible data on the molecular weight distribution of soluble polymers within a reasonably short period of time. In general, the separation medium for SEC must be inert, because all interactions between the solutes and the stationary phase have to be avoided. Modern SEC employs small rigid polymeric or silica-based beads of controlled pore-size to separate molecules differing in hydrodynamic volume, and to calculate average molecular weights and molecular weight distribution information for desired polymers.

Although SEC is a very powerful method for the characterization of polymers, it also has some weaknesses, such as the decrease in resolution which is observed for very high molecular weight polymers and the high shear forces within the pores that may degrade the separated macromolecules [3,4]. Moreover, as follows from the nature of SEC, this method is not able to efficiently separate macromolecules of the same hydrodynamic volume that differ in chemical composition, physical structure, or even molecular weight. The peak capacity is rather limited in the SEC mode [5], and the separation range of SEC columns is determined by the porosity of the packing. Therefore it cannot be improved significantly by an optimization of the separation conditions [6]. Typical SEC separation run times are often in the range of several tens of minutes since the flow rate is low (typically 1–2 mL min^{-1}) to achieve the highest column efficiency, and a series of rather large columns (typically two to four columns, each 300 mm $\times$ 8 mm i.d.) with large void volumes are often used to achieve the desired resolution. Long run times are ill suited for applications such as the high-throughput screening of large number of polymers produced by combinatorial techniques or on-line control of polymerization processes in which the speed of analysis is the most important issue. Despite recently reported success [7], substantial acceleration of true SEC separations remains difficult to achieve.

In contrast with SEC separations, which run under isocratic conditions and typically at an optimal flow rate that should not be changed, several variables of the gradient elution process in interactive chromatography can be adjusted without decreasing resolution. For example, the speed of the separation can be increased considerably by using high flow rates and steep gradients. However, high flow rates are not feasible for columns packed with small microparticles as they often exhibit prohibitively high flow resistance. In contrast, monolithic columns tolerate very high flow velocities since their resistance to flow is very low.

ALTERNATIVE METHODS FOR SEPARATION OF SYNTHETIC POLYMERS

Several other methods that complement SEC, such as matrix-assisted laser desorption/ionization—mass spectrometry (MALDI-MS), electrophoresis, field-flow fractionation (FFF), and hydrodynamic liquid chromatography have also been used for characterization of the molecular parameters of polymers [8–10]. In addition to molecular weight characteristics, some of these methods may also provide information on the chemical composition of the polymers.

In consideration of the application of monolithic stationary phases for the separation of synthetic polymers the most important alternative to SEC is interactive high-performance liquid chromatography (HPLC). The literature contains several mechanisms for the chromatographic separation of synthetic polymers in interactive modes. Snyder and co-workers [11] have indicated that only two of them appear to be really relevant. The first is the mechanism including the conventional retention resulting from adsorption at the surface of the stationary phase that is typical of both normal- and reversed-phase chromatography and constitutes the essence of adsorption HPLC. This mechanism holds for systems that involve low sample loads and separation media with large pores [11,12]. The second mechanism concerns separations using precipitation–redissolution processes in a gradient of the mobile phase. This mobile phase or eluent typically consists of two solvents that differ significantly in their solvating strength with respect to the polymer. Use of columns packed with small-pore media that cannot be permeated by the polymer molecules facilitates the separation. This chromatographic technique is also called precipitation–redissolution HPLC [12–14].

In contrast with the separation of small molecules, which can easily be achieved in an isocratic mode, the separation of macromolecules generally requires a gradient of the mobile phase with an increasing solvating power in order to achieve the elution of all of the macromolecular components, which may have significant differences in their retentions.

ADSORPTION HPLC

HPLC, which was introduced in the late 1960s, has sparked an enormous interest in chromatographic separation process. In particular, HPLC in the reversed-phase mode is used most frequently because it is particularly suitable for the separation of small hydrophobic molecules and molecules of medium polarity. Reversed-phase silica-based column packings afford very high column efficiencies and reproducible results. However, the use of this chromatographic mode for the separation of polymers is still marginal compared with SEC, and currently is used mainly in the separations of biopolymers [6]. Interactive HPLC can also be used for the selective separation and characterization of synthetic polymers. These separations rely on differences in enthalpy resulting from the interaction of a polymer with the stationary phase such as adsorption and partition [12,13,15] as well as solubility typical for the precipitation–redissolution mode (see below) [11–14,16].

As a rule, the retention factor in a fixed solvent increases exponentially with increasing molecular weight of the polymer, and the retention time for large molecules might be too long for practical use. Therefore the use of isocratic elution with a single solvent is limited to separations of oligomers [12,17,18], although the isocratic separation of high molecular weight polymers has also been attempted [19–21]. In contrast, elution using a mobile phase with increasing solvency [14,22–25] or a temperature gradient [26–29] enables separations of both oligomers and polymers in a much shorter period of time. In addition to the determination of molecular weight, interactive HPLC can also separate polymers according to their chemical composition [30]. In contrast with SEC, interactive HPLC commonly offers much higher resolution and a larger number of variables such as stationary phase properties, mobile phase composition, flow rate, and gradient profile, and temperature can also be used to control the chromatographic process and fine tune the separation.

Because of the specific requirements, including the need to optimize separation conditions for each individual polymer, re-equilibration between consecutive runs, and method validation for each chromatographic system, HPLC currently only complements the more popular SEC. It should also be emphasized that in real separations of synthetic polymers, a mixed retention mechanism usually occurs in which retention is a result of the combined effects of both adsorption and precipitation–redissolution. The extent of these interactions depends on a number of factors such as molecular weight and chemical structure of the solute (polymer), as well as the composition of the mobile phase. Since the concepts of these two mechanisms differ significantly and, in addition, a typical polymer is composed of molecules that may vary in length, end-groups, and composition, a rigorous description of the separation process in interactive HPLC is more difficult and parameter dependent than that in SEC.

PRECIPITATION–REDISSOLUTION HPLC

The concept of using polymer precipitation followed by subsequent redissolution was first used for the fractionation of polymers in the early 1950s [31]. However, this approach has only recently been exploited for the HPLC of polymers [16,30,32]. In this technique, the polymer solution is injected into a stream of the mobile phase in which the polymer is insoluble. Therefore the macromolecules precipitate and form a separate gel phase that adsorbs at the surface of the packing and does not move along the column. The solvency of the mobile phase is then increased gradually until it reaches a point at which some of the macromolecules start to dissolve again and travel with the stream. If the column is packed with a support that contains pores smaller than the size of the polymer molecules, the mobile phase penetrates the pores, while the dissolved molecules move with the stream only through the interstitial voids. As a result, the polymer solution moves forward faster than the solvent gradient and reaches a point at which the solvency of the mobile phase is no longer sufficient to hold the polymer in solution, and the polymer precipitates again. This newly formed precipitated gel phase will redissolve only when the solvent strength is enhanced. A multitude of such precipitation–redissolution steps is repeated over and over again until the macromolecule finally leaves the column. The solubility of each polymer molecule in the mobile phase depends on both molecular weight and composition. As a result, separation of macromolecules differing in these properties is achieved. Under ideal conditions, the column packing should serve only as a support for the precipitated phase and the separation should not depend on its chemistry. However, this is often not the case and, in reality, interactions often play a significant role, even in the precipitation–redissolution separation processes [32,33].

COLUMN TECHNOLOGIES FOR THE SEPARATION OF POLYMERS

In SEC the stationary phase usually has a format of rigid porous polymeric or silica-based beads. The separation results from the distribution of the polymer molecules between the moving mobile phase and the stagnant portion of the mobile phase retained within the porous structure of the stationary phase. Stationary phases used for interactive

HPLC may also have the same shape as porous beads. However, a slow mass transfer within the pores is a serious problem in chromatographic separations of macromolecules such as synthetic polymers using packed columns [30,34]. The diffusion coefficients of macromolecules are at least two orders of magnitude smaller than those of low molecular weight compounds [31]. As a result, the diffusion of a polymer molecule through the stagnant mobile phase within the pores of a particulate separation medium is slow and may significantly contribute to band spreading. Reducing the particle size of porous beads can offset the negative effect of diffusion. The theory suggests that beads with a size of less than 1 μm would be ideal to reduce the effect of diffusion on the separation of polymers with a molecular weight over 10^5 [35]. Provided that such porous particles were available, they would currently be less practical because of difficulties with their packing in columns and the very high back pressure they would generate. As a consequence of the slow mass transfer, low flow rates are necessary in both SEC and interactive HPLC to operate at the optimal flow velocity of the mobile phase and obtain acceptable resolution.

Obviously, nonporous separation media are advantageous for gradient HPLC of large molecules in adsorption modes because the restricted diffusion in pores that is characteristic of porous media is completely eliminated, and mass transfer in pores is no longer a problem [36–38]. Consequently, much faster separations can be achieved easily. Nonporous beads are currently popular for the separation of proteins, oligonucleotides, and DNA fragments. However, they have seldom been used for the characterization of synthetic polymers [33,39]. Small nonporous particles (1.5–4 μm) are preferred because they have sufficient surface area for the separation of *detectable* amounts of components in the polymer sample. However, owing to the lack of porosity, the surface area remains rather limited, often not exceeding 1 m^2/column. As a consequence of small particle size, only relatively short columns (3–5 cm) can be used to keep the back pressures within limits tolerated by typical instrumentation. The limited surface area available for accommodation of precipitated macromolecules can also lead to coprecipitation and subsequent coelution of macromolecules with different molecular weights, resulting in deterioration of the separation [39].

Another way to solve the slow diffusion of macromolecules is to accelerate mass transfer by convection. Regnier designed a technique called perfusion chromatography, in which a small fraction of the mobile phase is driven through the very large pores (600–800 nm) of chromatographic particles, thus increasing the diffusivity of large molecules [40,41]. Even the convection of only a small part of the mobile phase, typically less than 5%, has a large effect as demonstrated for the very fast separation of proteins and nucleic acids. However, only the recent development of monolithic columns has enabled the potential of rapid determination of molecular parameters of synthetic polymers using precipitation–redissolution separations to be fully exploited.

MONOLITHIC COLUMNS

Preparation of Rigid Polymer Monoliths

The preparation of the new class of rigid macroporous organic polymers produced by a "molding" process is simple and straightforward (Figure 9.1). The mold, typically

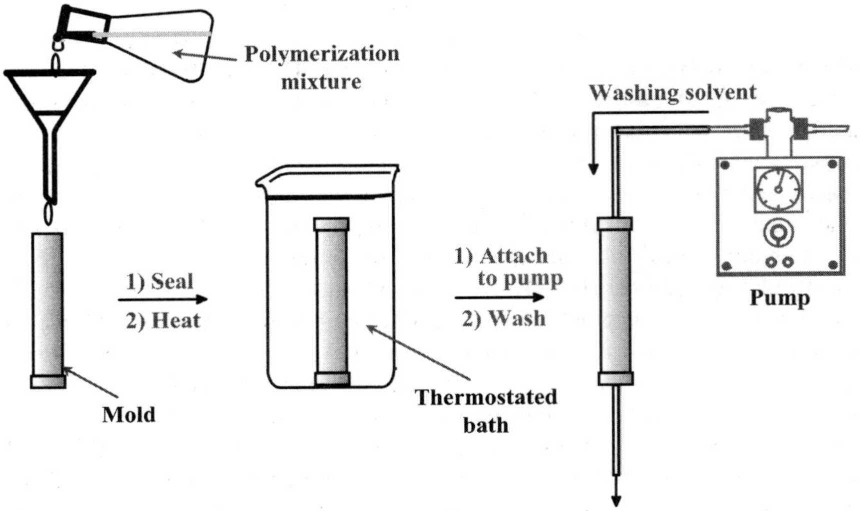

FIGURE 9.1. Preparation of a macroporous monolith by a "molding" process.

a tube, is sealed at one end, filled with a polymerization mixture, and sealed at the other end. The polymerization is then triggered, frequently by heating in a bath at a temperature of 55–80°C. After the polymerization is completed, the seals are removed, the tube is provided with fittings and attached to a chromatographic pump, and a solvent is pumped through the column to remove the porogens and any other soluble compounds that remained in the polymer monolith. A broad variety of tube sizes and materials, such as stainless steel and poly(ether-ether-ketone) (PEEK), as well as glass tubes, fused silica capillaries, and microfluidic chips, have been used as molds for the preparation of monoliths [42–47].

While the preparation of cylindrical monoliths with a homogeneous porous structure and diameter up to about 10–25 mm is readily achieved in a single-step polymerization, larger monoliths are somewhat more difficult to prepare. Dissipation of the heat of polymerization is frequently slow, and the exotherm may be sufficient to increase the reaction temperature substantially, accelerate the polymerization dramatically, and cause a rapid decomposition of the initiator. If this process is not controlled, monoliths with unpredictable radial and axial gradients of porosity are obtained. However, the slow and gradual addition of the polymerization mixture to the reaction vessel in which the polymerization reaction proceeds continuously minimizes the exotherm and allows the preparation of very large diameter monoliths with homogeneous porous structures [47]. Podgornik et al. [48] have analyzed a heat release during the polymerization and have derived a mathematical model for the prediction of the maximal thickness of the monolithic annulus having a uniform structure. On the basis of their calculations, they demonstrated an alternative solution of the heat-dissipation problem by preparing annular monoliths of different diameters and merging them into a single monolithic unit with a large volume.

Control of Porous Properties

Many applications of porous materials such as catalysis, adsorption, ion exchange, solid phase synthesis, and chromatography, rely on intimate contact with a surface of a support. In order to obtain a high surface area value, a large number of smaller pores should be incorporated into the polymer. The most substantial contribution to the overall surface area comes from the micropores, with diameters smaller than 2 nm, followed by the mesopores ranging from 2 to 50 nm. Macropores with a size exceeding 50 nm make only an insignificant contribution to the overall surface area. However, these pores are essential to enable liquid to flow through the material at a reasonably low pressure. This pressure, in turn, depends on the overall porous properties of the material. Therefore the pore-size distribution of the monolith should be adjusted properly to fit each type of application.

The pore-size distributions of the molded monoliths are quite different from those observed for "classical" macroporous beads. An example of a pore-size distribution curve is shown in Figure 9.2. An extensive study of the types of pores obtained during polymerization both in suspension and in an unstirred mold has revealed that, in contrast to conventional wisdom, there are some important differences between the suspension polymerization used for the preparation of beads and the bulk-like polymerization process utilized for the preparation of molded monoliths. In the case of polymerization in an unstirred mold, the most important differences are the lack of interfacial tension between the aqueous and organic phases, and the absence of dynamic forces that are typical of stirred dispersions [49].

The porosity and flow characteristics of macroporous polymer monoliths intended for use as separation media for chromatography, flow-through reactors, catalysts, or supports for solid phase chemistry have to be adjusted during their preparation. Key variables such

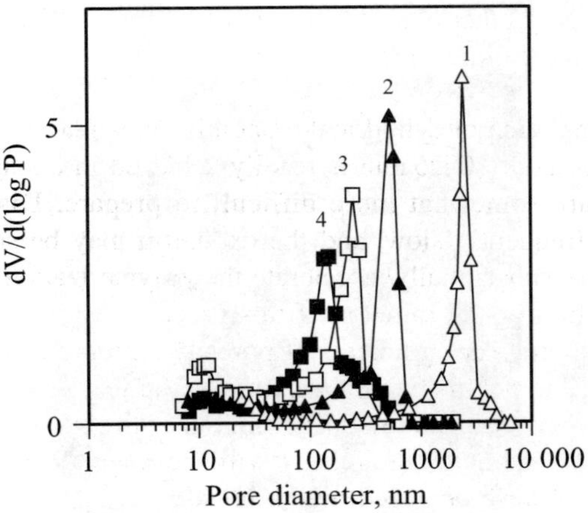

FIGURE 9.2. Effect of dodecanol in the porogenic solvent on differential pore-size distribution of poly(glycidyl methacrylate-*co*-ethylene dimethacrylate) monoliths. Conditions: polymerization time, 24 hr; temperature, 70°C; polymerization mixture, 24% glycidyl methacrylate, 16% ethylene dimethacrylate, and cyclohexanol/dodecanol contents in mixtures of (1) 60/0, (2) 57/3, (3) 54/6, and (4) 45/15 vol%. (Reprinted with permission from ref. 51. Copyright 1996 American Chemical Society.)

as temperature, composition of the pore-forming solvent mixture, and content of cross-linking divinyl monomer allow the tuning of the average pore-size within a broad range spanning at least two orders of magnitude from tens to thousands of nanometers.

The polymerization temperature, through its effects on the kinetics of polymerization, is a particularly effective means of control, allowing the preparation of macroporous polymers with different pore-size distributions from a single composition of the polymerization mixture composition. The effect of temperature can readily be explained in terms of the nucleation rates, and the shift in pore-size distribution induced by changes in the polymerization temperature can be accounted for by the difference in the *number of nuclei* that result from these changes [50,51].

The choice of pore-forming solvent is another tool that may be used for the control of porous properties without changing the chemical composition of the final polymer. In general, larger pores are obtained in a poorer solvent due to an earlier onset of phase separation. The porogenic solvent controls the porous properties of the monolith through the *solvation of the polymer chains* in the reaction medium during the early stages of the polymerization [51].

In contrast, increasing the proportion of the cross-linking agent present in the monomer mixture affects the chemical composition of the final monoliths. At the same time, it also decreases their average pore-size as a result of early formation of highly cross-linked globules with a reduced tendency to coalesce. The experimental results imply that, in this case, the pore-size distribution is controlled by limitations in *swelling of cross-linked nuclei* [51].

Morphology of Monoliths

The morphology of the monoliths is closely related to their porous properties, and is also a direct consequence of the quality of the porogenic solvent as well as the percentage of cross-linking monomer and the ratio of the monomer to the porogen phases. The

FIGURE 9.3. Scanning electron micrograph of the inner part of the porous poly(styrene-*co*-divinylbenzene) monolith.

presence of interactive effects between these reaction conditions was verified using multivariate analysis [52].

In general, the morphology of macroporous materials is rather complex. The scanning electron micrograph shown in Figure 9.3 reveals the details of the globular internal structure of a molded monolith of porous poly(styrene-*co*-divinylbenzene). Although this morphology featuring individual microglobules and their irregular clusters is similar to that found for beads [53], the size of both the clusters and the irregular voids between clusters is much larger.

Hydrodynamic Properties

For practical reasons, the pressure needed to drive the liquid through any system should be as low as possible. Because all the mobile phase must flow through the monoliths, the first concern is their permeability to liquids, which depends *prima loco* on the size of their pores. A monolith with pores only of the size found in typical macroporous beads would be crushed by the extremely high pressures required for flow. Obviously, lower flow resistance can be achieved with materials that have a large number of broad channels. However, many applications also require a large surface area in order to achieve a large capacity. This high surface area is generally a characteristic of porous material that contains smaller pores. Therefore a balance must be found between the requirements of low flow resistance and high surface area, and an ideal monolith should contain both large pores for convection and a connected network of shorter and smaller pores for high interactive capacity.

Figure 9.4 shows the back pressure per unit of flow rate (flow resistance) as a function of the mode pore-size for two monoliths of different chemistries. Typically, the pressure

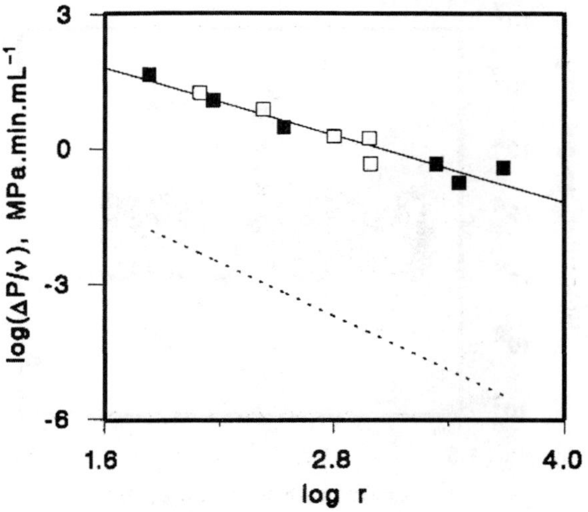

FIGURE 9.4. Flow resistance (back pressure per unit of flow rate) of molded poly(glycidyl methacrylate-*co*-ethylene dimethacrylate) (open squares) and poly(styrene-*co*-divinylbenzene) (solid squares) monoliths of dimensions 100 mm × 8 mm i.d. as a function of pore diameter at the mode of pore-size distribution curve (solid line) and the flow resistance calculated from the Hagen–Poiseuille equation (dotted line). (Reprinted with permission from ref. 51. Copyright 1996 American Chemical Society.)

needed to sustain even a very modest flow rate is quite high for materials that have a mean pore diameter of less than about 500 nm, while high flow rates can readily be achieved at low pressures with materials that have pores larger than 1000 nm. Although the shape of the pores within the monoliths is very different from those of a tube, the Hagen–Poiseuille equation essentially holds also for the flow through the molded porous poly(glycidyl methacrylate-*co*-ethylene dimethacrylate) and poly(styrene-*co*-divinylbenzene) monoliths because, as expected, flow does not depend on the chemistry of the material [51].

CHROMATOGRAPHY

In contrast with the standard beaded separation media with typical porous properties, the monolith provides almost no separation in the size-exclusion mode. SEC experiments confirm the lack of separation of polystyrene standards in the range of molecular weights between 10^4 and 10^6 and clearly document the absence of medium-sized pores (Figure 9.5). However, this is a substantial advantage if the monolithic column is to be used for separations that rely on the interaction of the solutes with the stationary and mobile phases rather than on their sizes, because the size-exclusion effect does not interfere with the separation.

The most frequently used poly(styrene-*co*-divinylbenzene) monoliths are sufficiently hydrophobic for use in reversed-phase chromatography without any further modification. A comparative study of separations of alkylbenzenes demonstrated that these monolithic columns exhibit surface polarity equal to that of poly(styrene-*co*-divinylbenzene) beads [54]. The majority of the separations using monolithic columns performed so far have been carried out with polystyrene standards since these model polymers are available in a broad range of well-characterized molecular weights.

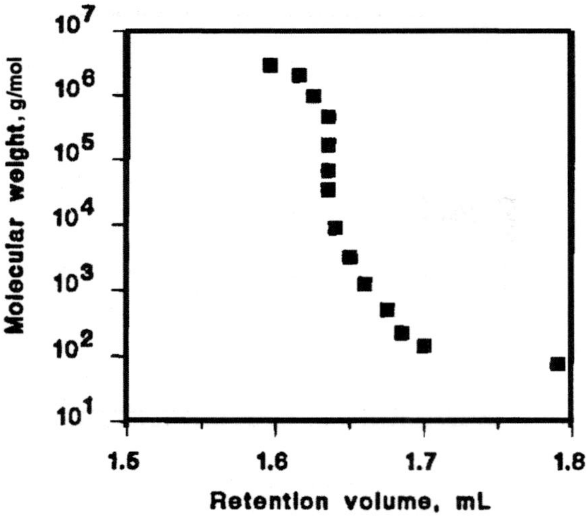

FIGURE 9.5. SEC calibration curve for the monolithic poly(styrene-*co*-divinylbenzene) rod column. Conditions: column 50 mm × 8 mm i.d.; mobile phase THF; flow rate 1 mL min^{-1}; analytes polystyrene standards and alkylbenzenes; injection volume 20 μL; UV detection at 254 nm. (Reprinted from ref. 56. Copyright 1996 American Chemical Society.)

Rapid Separations

In order to achieve a good separation, classical chromatographic columns are typically packed with micrometer-sized particles. It is well known that resolution increases with a decrease in the particle size of the sorbent. However, the reduction of particle diameter leads simultaneously to a rapid build-up of a back-pressure in the column. Consequently, a trade-off is required to achieve a reasonable resolution under acceptable flow rate and analysis time. Hence column packings of particle size 3–10 μm are the most common at present in analytical HPLC.

One of the most significant advantages of specifically designed monolithic columns is their capability to facilitate fast separations using very high flow rates of the mobile phase but still keeping a moderate back-pressure. For example, Figure 9.6 shows fast separations of three polystyrene standards using a poly(styrene-*co*-divinylbenzene) monolithic column at a flow rate of 20 mL min^{-1} and two different very steep gradients of the mobile phase. The back-pressure of the 50 mm × 8 mm i.d. column using pure tetrahydrofuran (THF) was only 2.6 MPa despite such a high flow rate. The separation was excellent at a gradient time of 30 s, and three well-resolved peaks were obtained within 8 s [Figure 9.6(a)]. An application using an even faster gradient of 12 s still leads to a good separation, which is achieved within mere 4 s [Figure 9.6(b)].

Figure 9.7 shows the chromatograms obtained using a column packed with porous monodispersed 7 μm poly(styrene-*co*-divinylbenzene) beads. The separation conditions were the same as those used for with the monolithic column. In contrast, a back-pressure of 9.6 MPa at 20 mL min^{-1} was significantly higher than that observed for the monolithic column of the same dimensions. The separations using 30 s gradients are comparable

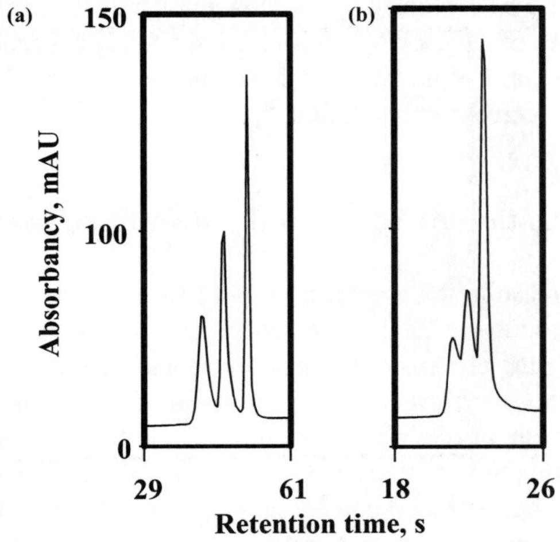

FIGURE 9.6. Rapid separations of polystyrene standards on a monolithic column. Conditions: column, 50 mm × 8 mm i.d. poly(styrene-*co*-divinylbenzene) monolith; mobile phase, linear gradient from 100% to zero methanol in THF within (a) 30 s and (b) 12 s; flow rate, 20 mL min^{-1}; analytes, molecular weights of (1) 9,200, (2) 34,000, and (3) 980,000, 3 mg mL^{-1} of each standard in tetrahydrofuran; injection volume, 20 μL; UV detection at 254 nm. (Reprinted from ref. 39. Copyright 1996 Elsevier.)

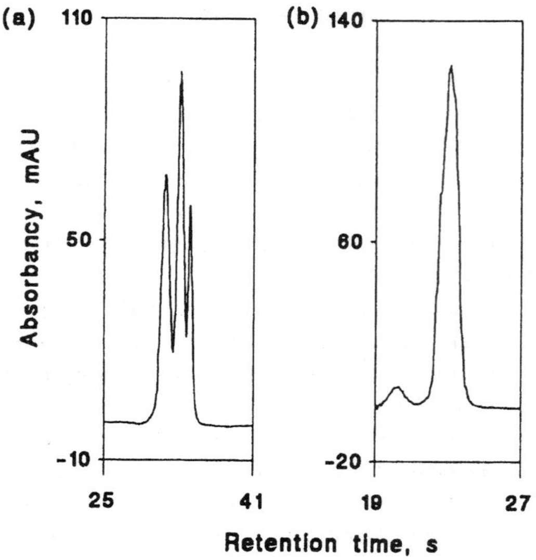

FIGURE 9.7. Rapid separation of polystyrene standards on a conventional particle packed column. Conditions: column, 50 mm × 8 mm i.d., porous poly(styrene-*co*-divinylbenzene) beads; other conditions as in Figure 9.6. (Reprinted from ref. 39. Copyright 1996 Elsevier.)

for columns of both types [Figures 9.6(a) and 9.7(a)]. However, the polystyrene standards are coeluted from the packed column in only a single peak if a 12 s gradient is used [Figure 9.7(b)]. Obviously, an increase in the gradient steepness, while keeping the flow rate unchanged, leads to a rapid deterioration of the separation. This result is not completely unexpected and can easily be explained by the slow mass transfer within the stagnant mobile phase located in the pores of the packing. In contrast, the total convection of the entire mobile phase through the large pores of the monolithic stationary phase facilitates very fast separations with only a slight decrease in resolution.

Optimization of Speed and Solvent Consumption

In general, an increase in the resolution of an SEC system can be achieved by using better column packing and/or a longer column, which represent a "hardware approach," rather than by changing the chromatographic conditions such as the flow rate. In contrast with SEC, which is always performed in an isocratic mode using an optimal flow rate that cannot be changed without an adverse effect on the separation efficiency, several variables controlling the resolution are available for the gradient elution technique. For example, the speed of separation can be increased considerably by using high flow rates and steep gradients. However, high flow rates are not possible for columns packed with typical small microparticles as they often exhibit prohibitively high flow resistance. In contrast, molded-rod columns allow the application of very high flow rates at a reasonable back pressure, thus making very fast chromatographic runs possible and optimization of the method more flexible.

The average retention factor $\bar{k}$ in the gradient elution, depends on the gradient time t_G, the volume flow rate F, the change $\Delta\phi$ in composition of the mobile phase, the column dead volume V_m, and the constant S, which is calculated for each solute from the retention data in isocratic systems and characterizes the strength of the interaction between the solute and the stationary phase [34]:

$$\bar{k} = (t_G F)/(\Delta\phi\ V_m S) \tag{1}$$

For a constant range of mobile phase composition and a specific column and solute, the denominator in Eqn (1) remains constant and the average retention factor $\bar{k}$ depends only on the gradient time and the flow rate. Because the product $t_G F$ of these variables is the gradient volume V_G, equal peak capacities that are independent of flow rate and gradient steepness should be achieved within the same gradient volume. Figure 9.8 shows separations of styrene oligomers obtained on a 50 mm × 8 mm i.d. monolithic column with gradient times of 200 and 20 min and flow rates of 1 and 10 mL min^{-1}, respectively. Because the gradient volume of 200 mL is the same for both separations, no significant differences can be seen between these two chromatograms. This result supports the validity of Eqn (1) and clearly shows that remarkable acceleration of the separation process is feasible using the monolithic stationary phase.

In addition, the effect of the absolute value of the gradient volume on the resolution in gradient elution should also be mentioned. This is demonstrated in the separation shown in Figure 9.9(a) in which the gradient volume was decreased 10-fold with respect to the chromatographic conditions of Figure 9.8. The significant deterioration of the separation is a direct implication of a gradient volume which is too low.

The speed of the analysis itself is only one of a number of important parameters concerning optimization of the gradient separations. Several other aspects are just as important. For example, much effort is presently focused on the reduction of solvent consumption in fast gradient techniques because of both environmental protection and waste solvent disposal costs. The original separations of synthetic polymers were carried out

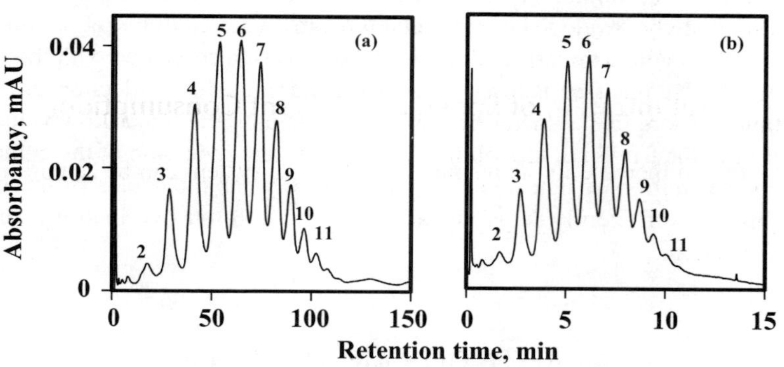

FIGURE 9.8. Effect of flow rate and gradient time on the separation of styrene oligomers using a monolithic poly(styrene-*co*-divinylbenzene) column. Conditions: column 50 mm × 8 mm i.d.; mobile phase (a) linear gradient from 60 to 30% water in THF within 200 min, flow rate 1 mL min^{-1}, and (b) linear gradient from 60 to 30% water in THF within 20 min, flow rate 10 mL min^{-1}; analyte 15 mg mL^{-1} in THF; injection volume 20 μL; UV detection at 254 nm. Peak numbers correspond to the number of styrene units in the oligomer. (Reprinted from ref. 56. Copyright 1996 American Chemical Society.)

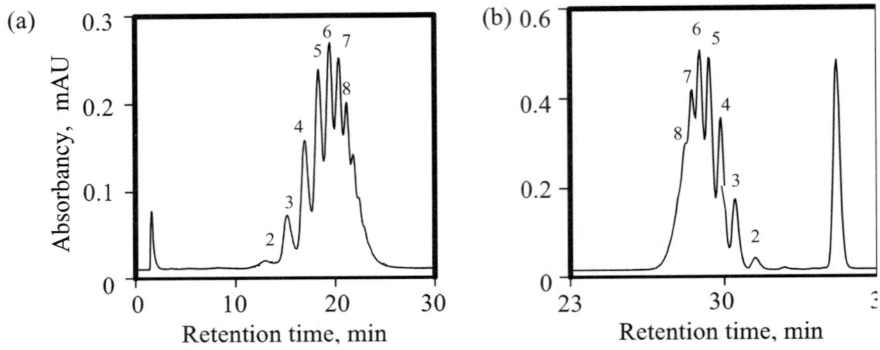

FIGURE 9.9. Separation of styrene oligomers by (a) reversed-phase chromatography and (b) SEC. (a) Conditions: monolithic poly(styrene-*co*-divinylbenzene) column 50 mm × 8 mm i.d.; mobile phase, linear gradient from 60 to 30% water in THF within 20 min; flow rate 1 mL min^{-1}; injection volume 20 μL; UV detection at 254 nm. (b) Conditions: series of four 300 mm × 7.5 mm i.d. PL Gel columns (100 Å, 500 Å, 10^5 Å and Mixed C); mobile phase, THF; flow rate 1 mL min^{-1}; injection volume 100 μL; toluene added as a flow marker; UV detection at 254 nm; temperature 25°C. Peak numbers correspond to the number of styrene units in the oligomer. (Reprinted from ref. 56. Copyright 1996 American Chemical Society.)

using columns with, by today's standards, a rather wide internal diameter of 8 mm. Under these conditions, the requirement to achieve sufficiently large gradient volume to maintain the desired resolution requires high volumetric flow rates typically in the range of 10–20 mL min^{-1}. Consequently, the daily consumption of the mobile phase can easily reach up to 6–12 L just for a single 8 mm i.d. column. In addition, these high flow rates are also very close to the upper limit of the typical analytical HPLC pump, which is 10 mL min^{-1}. Thus the application of semipreparative HPLC pumps would be needed.

These problems imposed a strong motivation for reduction of the size of the monolithic columns while retaining their excellent separation characteristics. This has also led to the recent transfer of the concept of the rapid separations of synthetic polymers to the smaller monolithic column format [55]. Almost identical chromatographic results were obtained using newly developed commercial 4.6 mm i.d. monolithic rod columns (ISCO Inc., Lincoln, NE) with a three-fold decreased cross-section compared with the previously used 8 mm i.d. columns (see below). This modification enables reduction in the solvent consumption by more than a factor of 3 while maintaining the linear flow velocity at the same level as in the 8 mm i.d. monoliths. Although further reduction of the column size is feasible, the separation of detectable amounts of polymer molecules in the interactive mode requires a sufficiently large surface area and column overloading can become a serious limitation.

RETENTION MECHANISM

Oligomers

The separation of styrene oligomers by HPLC on reversed-phase octadecyl silica columns in a gradient of the mobile phase was studied as early as 1983 [12].

These experiments demonstrated that the separation followed the normal course for reversed-phase chromatography of small molecules and that retention depended on both the composition of the mobile phase and the number of repeat units in the oligomer. Larger oligomer molecules, having longer hydrophobic chains, exhibited longer retention times or required a higher percentage of good solvent in the mobile phase used for their elution. This means that shorter oligomers eluted before longer ones, quite unlike SEC for which large molecules elute first.

Figure 9.9 shows the separations of a commercial sample of polystyrene oligomers with a number average molecular weight of 630 in a short monolithic column using the gradient HPLC mode and compares it with the separation in a series of four high-performance SEC columns. Despite a very narrow polydispersity of 1.07, both the chromatograms exhibit a number of peaks that can be assigned to the individual styrene oligomers. Addition of a small amount of the dimer (2,4-diphenyloctane) to this sample led to an increase in the height of peak 2, thus clearly confirming its identity. As expected, because of the opposite order of elution, the chromatograms in Figure 9.9 are almost mirror images. The resolution achieved with the monolithic column is good and the chromatogram in Figure 9.9(a) even indicates the presence of an undecamer. Unlike the SEC separation in Figure 9.9(b), the resolution of gradient HPLC shown in Figure 9.9(a) can be further improved simply by increasing gradient volume (Figure 9.8).

A simple calculation [56] provides the composition of the mobile phase in which the oligomers are eluted. For alkylbenzenes this composition is a linear function of molecular weight which, for a process driven by hydrophobic interaction, obeys the equation [12]

$$C_{\mathrm{THF}} = a + b\,M \tag{2}$$

where a and b are constants and M is the molecular weight.

In contrast, the data obtained for styrene oligomers is a better fit to the equation [56]

$$C_{\mathrm{THF}} = a' + b' \log M \tag{3}$$

This indicates that selectivity for the separation of styrene oligomers decreases as the molecular weight of the oligomers increases, and therefore the rules typical of reversed-phase chromatography of small molecules no longer apply completely.

Earlier literature often reports the large effect of sample load on peak shapes and retention for the gradient elution of macromolecules [13,14,16,57,58]. Monolithic columns tolerate much higher sample injections than packed columns. Figure 9.10 shows the separations of 0.05–6.25 mg of styrene oligomers using a 50 mm × 8 mm i.d. poly(styrene-*co*-divinylbenzene) monolith. The retention times for all the oligomers remain almost constant over the entire load range, while the resolution decreases only slightly for injections larger than 0.5 mg. This indicates a very broad dynamic range of the monolithic separation media.

Polymers

Reversed-phase separation of polystyrenes differs from that of small molecules (alkylbenzenes) and oligomers [11–13]. In particular, the dependence of the logarithm of

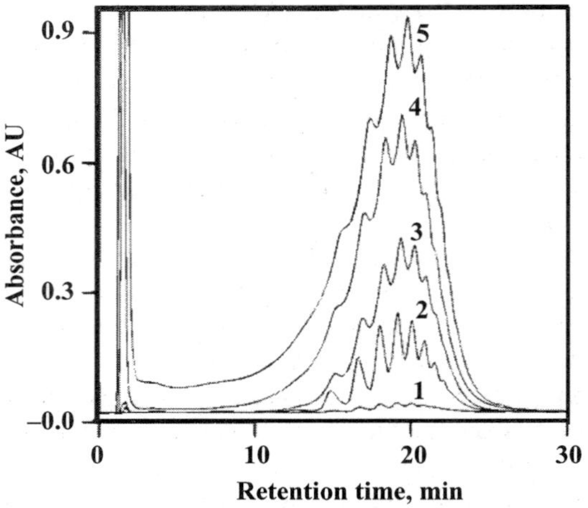

FIGURE 9.10. Effect of sample load on the separation of styrene oligomers in a monolithic poly(styrene-*co*-divinylbenzene) column. Conditions: column 50 mm × 8 mm i.d.; mobile phase, linear gradient from 60 to 30% water in THF within 20 min; flow rate 1 mL min^{-1}; analytes 25 mg mL^{-1} in THF; injection volumes of (1) 2 μL, (2) 20 μL, (3) 50 μL, (4) 100 μL, and (5) 250 μL. UV detection at 254 nm. (Reprinted from ref. 56. Copyright 1996 American Chemical Society.)

the retention factor k on temperature is much steeper, and it is not a linear function of molecular weight. However, the general elution pattern remains unchanged and, once again, the species with lower molecular weights elute prior to those with higher molecular weights.

THF was found to be a suitable good solvent for the gradient elution of polystyrene standards. Various poor solvents such as water, methanol, and acetonitrile were evaluated as precipitants [56]. Provided that the contents of the relevant solvents were properly adjusted, similar separations were obtained for all three precipitants. The amount of tetrahydrofuran in the different mobile phases used to elute the peaks was then plotted against $M^{-1/2}$ to give straight lines that were fitted to the equation [56]

$$C_{\mathrm{THF}} = a'' + b'' M^{-1/2} \tag{4}$$

where a'' and b'' are constants and M is again the molecular weight of the polymer. This finding confirms that precipitation of the polymer in the mobile phase is the driving force for the separation. Water was found to be the most powerful precipitant, but the selectivity of the tetrahyrofuran–water system was poor. Acetonitrile affords the highest selectivity but it was the poorest precipitant, and up to about 80% of acetonitrile had to be added to the mobile phase in order to observe retention of polystyrene with a molecular weight of 9,200. An extrapolation revealed that styrene oligomers with a molecular weight of less than about 2,600 could be completely soluble in pure acetonitrile. This disqualifies acetonitrile from use for the chromatographic analysis of unknown polymers or broad molecular weight samples because the low molecular weight part of the sample may not precipitate. Methanol proved to be the precipitant of choice as it provides sufficient selectivity and still precipitates even the oligomers.

Mixed Mechanism

The results obtained from the separations of alkylbenzenes, styrene oligomers, and polymers discussed above revealed different retention behavior for each of these three families of compounds [56]. The percentage of good solvent needed for the elution is a linear function of molecular weight for low molecular weight compounds, of the logarithm of molecular weight for oligomers, and of the square root of molecular weight for polymers. The appropriate S-shaped curve that depicts the percentage of THF required for the elution of solutes that have similar chemical composition but differ in their molecular weights can be clearly divided into three sections as shown in Figure 9.11. This indicates that the separation mechanism is different for each class of compounds and changes from a typical reversed-phase mechanism to one involving solubility differences, i.e. precipitation–redissolution. Thus, for the separation mechanism of polystyrenes on monolithic columns, it can be concluded that indeed two different retention mechanisms, adsorption and precipitation–redissolution, simultaneously control the separation of high molecular weight solutes.

It follows from the extensive study by Quarry et al. [13] that the molecular weight of the polymer, its solubility, and the amount injected are the most important variables in determining which retention mechanism is operative in the specific separation. For example, the composition of the mobile phase required for elution remains constant within a broad range of amount of sample injected for the adsorption mode, while the percentage of good solvent necessary for elution increases in the case of the precipitation–redissolution mechanism. Measurements of isocratic retention also provide insight into the retention mechanism. Thus in the adsorption mechanism the retention time depends on the composition of the mobile

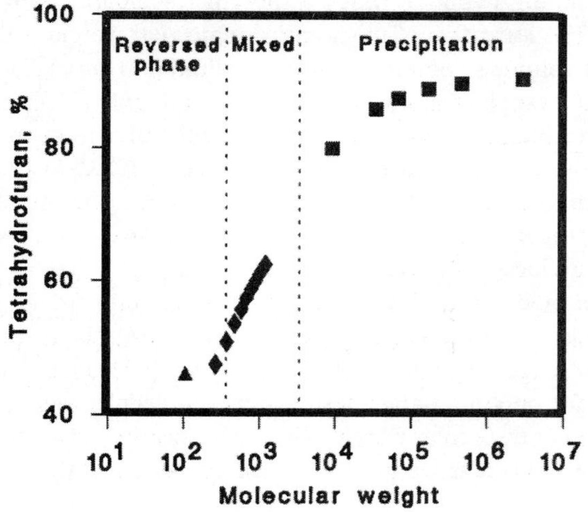

FIGURE 9.11. Composition of the mobile phase for the elution of ethylbenzene, styrene oligomers, and polystyrenes from the monolithic poly(styrene-*co*-divinylbenzene) column as a function of their molecular weight. Conditions: column 50 mm × 8 mm i.d.; mobile phase, linear gradient from 60 to 30% water in THF within 20 min; flow rate 1 mL min^{-1}; analytes, injection volume 20 μL; UV detection at 254 nm. (Reprinted from ref. 56. Copyright 1996 American Chemical Society.)

phase and polymer analytes can be always eluted as a distinct near-Gaussian peak. In contrast, it is impossible to elute peaks at retention times exceeding that of the standard under nonretained conditions in the precipitation–redissolution mode. Experiments with poly(styrene-*co*-divinylbenzene) monolithic columns clearly indicated the strong effect of precipitation on the separation process of polystyrenes even for a standard with a relatively low molecular weight of 9,680 in the THF–methanol system [39]. Although the adsorption mechanism of retention cannot be excluded completely, the experimental data suggest that the precipitation–redissolution effect is quite strong and contributes considerably to the retention of synthetic polymers in the monolithic column. This is expected as the monolithic columns have a limited surface area in the large pores available for the interactions [51] and, in contrast with some of the silica-based separation media, these columns also contain very small pores that are not permeable by polymer solutes. Therefore during the gradient elution these solutes migrate through the column faster than through the mobile phase zone capable of dissolving them. As a result, the polymer molecules precipitate as they move from the zone of stronger solvent into the zone of poorer solvent and dissolve only when the solvent strength is again sufficient to dissolve them.

SEPARATIONS USING NONLINEAR GRADIENTS

Polystyrenes

The precipitation–redissolution technique using monolithic columns has a great potential for high-throughput characterization of molecular parameters of synthetic polymers. However, precipitation–redissolution HPLC is a relative method and therefore must be calibrated for determination of actual molecular weights. Typically, the calibration is performed by determining retention times of polymer standards with narrow molecular weight distribution. The logarithm of the nominal molecular weight is then plotted against the retention time or volume to obtain the desired calibration curve. The calibration curve should span as broad a range of molecular weights as possible to ensure that the molecular parameters of the future "unknown" samples will fall within this range. Since no "universal" calibration currently exists for the gradient HPLC technique, standards of the same chemical composition as the analyzed samples must be used for the calibration. This requirement restricts the precipitation–redissolution analysis to polymers for which standards with well-defined molecular weights are available.

Use of the linear gradient of THF in methanol discussed in previous sections provides a nonlinear calibration curve with an exponential shape for polystyrene standards within the molecular weight range 3,000–980,000 (Figure 9.12) [55]. This shape is not favorable since the resolution for polymer molecules of higher molecular weight is lower than that for molecules with lower molecular weight. This "nonlinearity" also affects the peak shape of samples of higher molecular weight. The peak shape is not Gaussian and appears to exhibit fronting.

In order to avoid this undesired effect, a nonlinear gradient was developed with a profile that compensates for the decreasing resolution in the high molecular weight region and leads to a linear calibration curve [55]. The gradient shape can be inferred from cloud-point curves that are obtained by titration of a THF solution of polystyrene standards with methanol. These curves were found to be well suited as a guide to the design of the

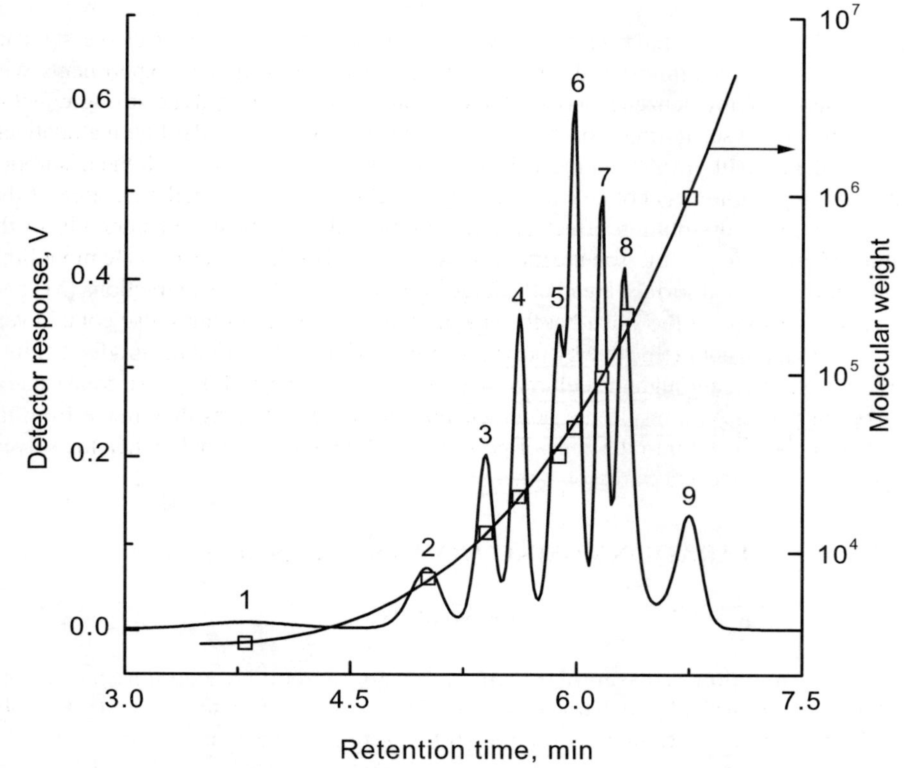

FIGURE 9.12. Separation of a mixture of nine polystyrene standards on a monolithic 50 mm × 4.6 mm i.d. column using a linear gradient and the corresponding calibration curve obtained from the elution data. Separation conditions: linear gradient 0–60% THF in methanol in 5 min; flow rate 1 mL min^{-1}; sample volume 10 μL; overall sample concentration 18 mg mL^{-1} (2 mg mL^{-1} of each standard) in THF; evaporative light-scattering detector (ELSD). Molecular weights of polystyrene standards: (1) 3,000, (2) 7,000, (3) 12,900, (4) 20,650, (5) 34,800, (6) 50,400, (7) 96,000, (8) 214,500, and (9) 980,000. (Reprinted from ref. 55. Copyright 2000 John Wiley.)

gradient shape. The final optimization has led to a gradient composed of five linear ramps that are steeper at the beginning and shallower at the end. An excellent separation of a mixture of nine polystyrene standards [Figure 9.13(a)] was achieved using this "tailored" nonlinear gradient within 6.5 min at a flow rate of only 1 mL min^{-1}. In contrast, Figure 9.13(b) shows the SEC separation of the same mixture of polystyrene standards using a tandem of two columns at the same flow rate which required 18 min. This figure also depicts calibration curves for both methods. The HPLC calibration is linear over the entire range of measured molecular weights.

The accuracy of this HPLC method featuring nonlinear gradient elution for molecular weight determinations was compared with that of HPLC with a linear gradient and SEC using binary mixtures of two narrow molecular weight polystyrene standards. These binary mixtures were used to validate the absence of mutual effects of polymer molecules of different molecular weights on both precipitation and redissolution processes.

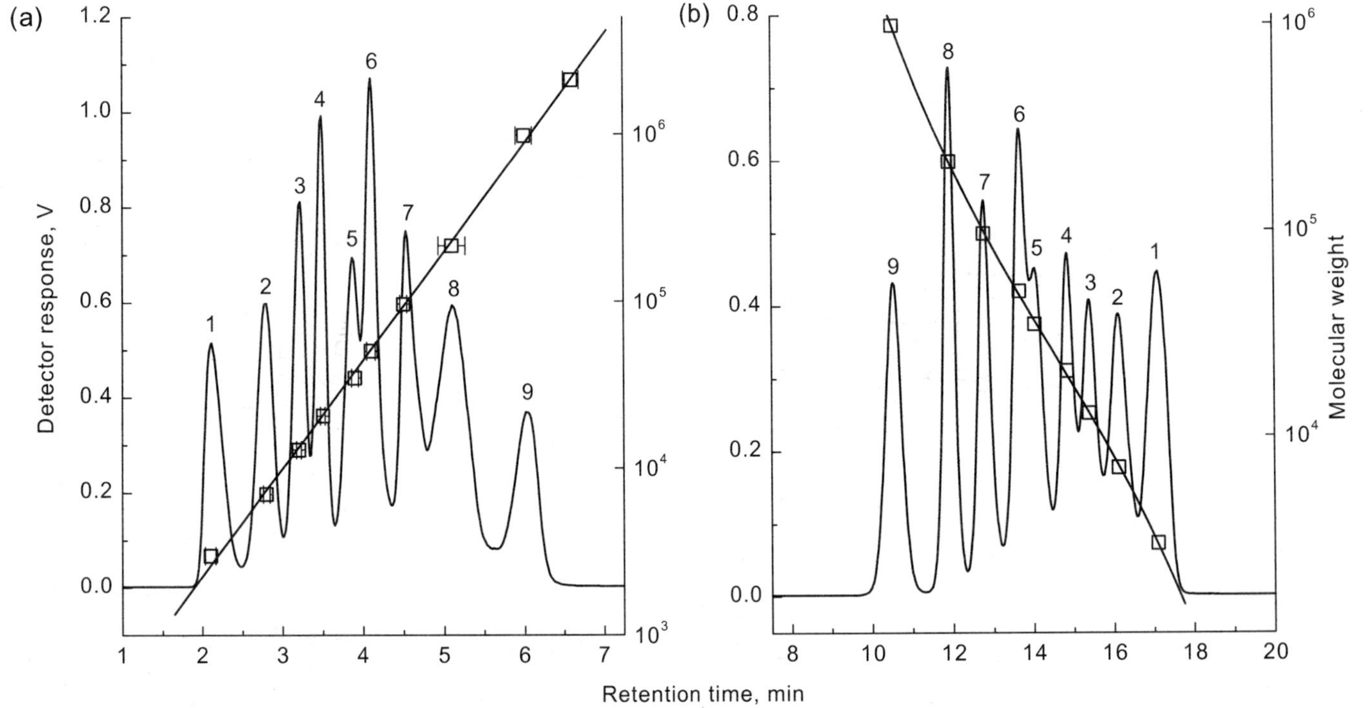

FIGURE 9.13. Separation of a mixture of nine polystyrene standards using (a) a nonlinear gradient in a monolithic 50 mm × 4.6 mm i.d. poly(styrene-*co*-divinylbenzene) column and (b) SEC in two 300 mm × 7.5 mm i.d. PL gel columns, together with the corresponding calibration curves obtained from the elution data. Separation conditions: flow rate 1 mL min^{-1}; overall sample concentration 18 mg mL^{-1} (2 mg mL^{-1} of each standard) in THF; ELSD. (a) HPLC eluent: nonlinear gradient consisting of 0–35% THF in methanol in 0.5 min, 35–50% in 1.5 min, 50–55% in 1 min, 55–59% in 1 min, and 59–60% in 1 min. (b) SEC eluent: THF. Molecular weights of polystyrene standards: (1) 3,000, (2) 7,000, (3) 12,900, (4) 20,650, (5) 34,800, (6) 50,400, (7) 96,000, (8) 214,500, and (9) 980,000. (Reprinted from ref. 55. Copyright 2000 John Wiley.)

Table 9.1 summarizes the experimentally obtained results. Using the linear gradient approach, both M_w and M_n calculated for standards with lower molecular weight were slightly higher than those calculated from SEC. Simultaneously, the molecular weight parameters of high molecular weight standards were generally much smaller because of the peak tail compression effect resulting from the nonlinearity of the calibration curve. In contrast, the agreement of M_w and M_n obtained using the nonlinear gradient HPLC elution and SEC was much better; their chromatographic traces were very close and matched well the nominal molecular weights provided by manufacturer. This demonstrates that the optimized precipitation–redissolution chromatography can afford characteristics that are in a good agreement with those obtained using SEC.

The use of nonlinear gradients is also a means of accelerating the separation. For example, Figure 9.14 shows the rapid and efficient separation of polystyrene standards with molecular weights ranging from 3,000 to 980,000 using the poly(styrene-*co*-divinylbenzene) monolithic column at only a moderate volumetric flow rate of 2 mL min^{-1}.

Similar to the styrene oligomers, overloading the monolithic columns with polystyrenes is of much less concern than in conventional columns containing particulate packings. Figure 9.15 shows the separations of mixture of five polystyrene standards on a 50 mm × 4.6 mm i.d. monolithic column at overall concentrations ranging from 2.5 to 50 mg mL^{-1}. The retention times for all the peaks remain almost constant over the entire load range and demonstrate the broad dynamic range of the monolithic column. This also suggests that even small monolithic columns may be useful for the preparative separation of polymers. Higher column loading can also be achieved by injecting larger volumes of samples. However, comparative experiments document that this approach leads to a larger shift in retention times for high molecular weight polymers. In addition, larger injection

TABLE 9.1. Molecular Parameters of Individual Polystyrene Standards with Narrow Molecular Weight Distribution as Calculated for Each Peak After the Separations of Their Binary Mixtures Determined by HPLC and SEC

Binary polymer mixture[a] (kDa)	$M_w \times 10^{-3}$			$M_n \times 10^{-3}$			M_w/M_n		
	HPLC		SEC	HPLC		SEC	HPLC		SEC
	Linear	Stepwise		Linear	Stepwise		Linear	Stepwise	
18.1	18.9	17.7	17.4	18.5	17.6	17.2	1.01	1.01	1.01
150	139.8	153.4	148.5	137.8	152.1	146.1	1.01	1.01	1.02
30	32.6	28.9	30.1	32.3	28.7	29.9	1.01	1.01	1.01
200	168.9	211.4	211.6	166.5	209.4	209.9	1.01	1.01	1.01

HPLC conditions: 50 mm × 4.6 mm i.d. monolithic poly(styrene-*co*-divinylbenzene) column (ICSO Inc., Lincoln, NE); ELSD (Polymer Laboratories); conditions of linear and five-step gradient as described in Figures 9.12 and 9.13(a), respectively.
SEC conditions: two 300 mm × 7.5 mm i.d. 10^3 Å and 10^5 Å PL gel columns (Polymer Laboratories); THF; flow rate 1 mL min^{-1}; injections of 20 μL of polymer solution, concentration 1–2 mg mL^{-1}.
[a]Values provided by manufacturer.

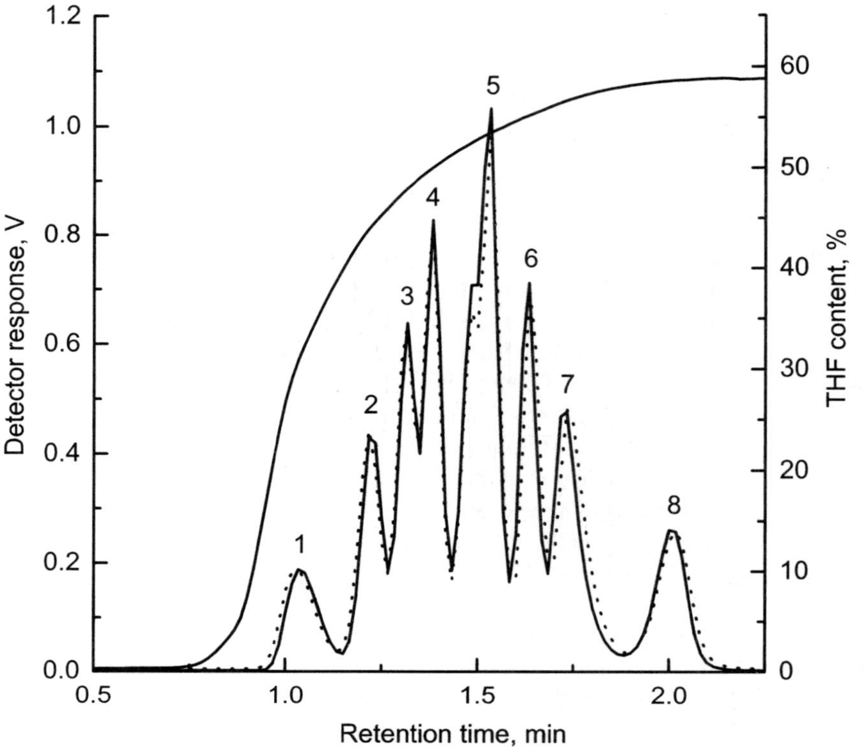

FIGURE 9.14. Rapid separation of a mixture of eight polystyrene standards using a monolithic 50 mm × 4.6 mm i.d. poly(styrene-*co*-divinylbenzene) column and the corresponding gradient profile monitored by the UV detector. Separation conditions: flow rate 2 mL min^{-1}; 1.25 min nonlinear gradient of THF in methanol consisting of 0–35% THF in methanol in 0.12 min, 35–50% in 0.38 min, 50–55% in 0.25 min, 55–59% in 0.25 min, and 59–60% in 0.25 min; overall sample concentration 16 mg mL^{-1} (2 mg mL^{-1} of each standard) in THF; ELSD. Molecular weights of polystyrene standards: (1) 3,000, (2) 7,000, (3) 12,900, (4) 20,650, (5) 50,400, (6) 96,000, (7) 214,500, and (8) 980,000. The solid and dotted lines represent the chromatograms of the same polymer mixture recorded about 400 injections apart. (Reprinted from ref. 55. Copyright 2000 John Wiley.)

volumes introduce larger volumes of the good solvent into the system thereby triggering the undesired pre-elution [14,30,39].

Poly(methyl methacrylates)

Methacrylate polymers are more polar than polystyrene and therefore their solubilities are also different. In order to characterize poly(methacrylates) it is again necessary to perform all the steps including cloud-point titration, optimization of the gradient shape, and calibration with standards. These experiments indicated that a simple two-step gradient of THF in methanol was sufficient too "linearize" the calibration curve for six poly(methyl methacrylate) standards with molecular weights in the range 5,750–910,000 [55].

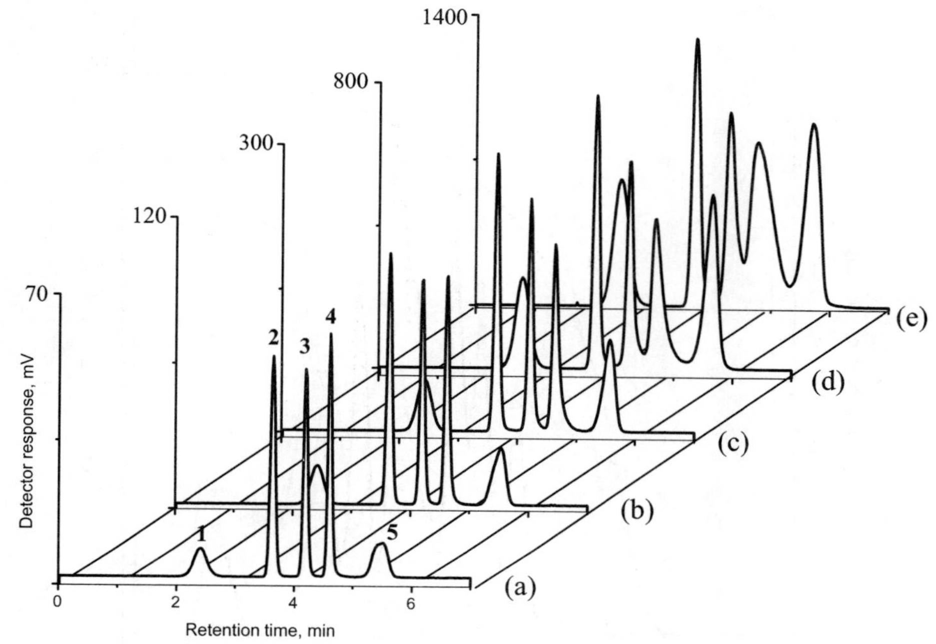

FIGURE 9.15. Effect of sample concentration on retention times of polystyrene standards. Column: 50 mm × 4.6 mm i.d. poly(styrene-*co*-divinylbenzene) monolith; 5 min nonlinear gradient consisting of 0–35% THF in methanol in 0.5 min, 35–50% in 1.5 min, 50–55% in 1 min, 55–59% in 1 min, and 59–60% in 1 min; flow rate 1 mL min^{-1}; sample volume 10 μL; overall concentration (a) 2.5, (b) 5.0, (c) 10.0, (d) 25.0, and (e) 50 mg mL^{-1} in THF; ELSD. Molecular weights of polystyrene standards: (1) 4,950, (2) 30,300, (3) 66,000, (4) 135,000, and (5) 629,000. (Reprinted from ref. 55. Copyright 2000 John Wiley).

Rapid separation of this mixture accomplished within 1.2 min at a flow rate of only 1 mL min^{-1} is demonstrated in Figure 9.16.

Poly(vinyl acetates)

In contrast with polystyrene and poly(methyl methacrylate), methanol is a solvent for poly(vinyl acetate). Therefore a series of solvent-precipitant pairs were tested [55]. Molecular-weight-dependent elution was achieved with THF–water, dichloromethane–n-hexane, and THF–n-hexane. The last pair appeared to give the best results in these tests. Thus optimization of the gradient elution was carried out with this pair. Figure 9.17 shows the HPLC traces for "secondary" standards with a broader molecular weight distribution and compares them with SEC. The HPLC separation appears to be more sensitive to changes in molecular weights than the SEC separation. Unlike the separations of polystyrene and poly(methyl methacrylate), an acceleration of the poly(vinyl acetate) elution could not be achieved using means such as changes in flow rate and gradient steepness without a concomitant rapid deterioration in the selectivity.

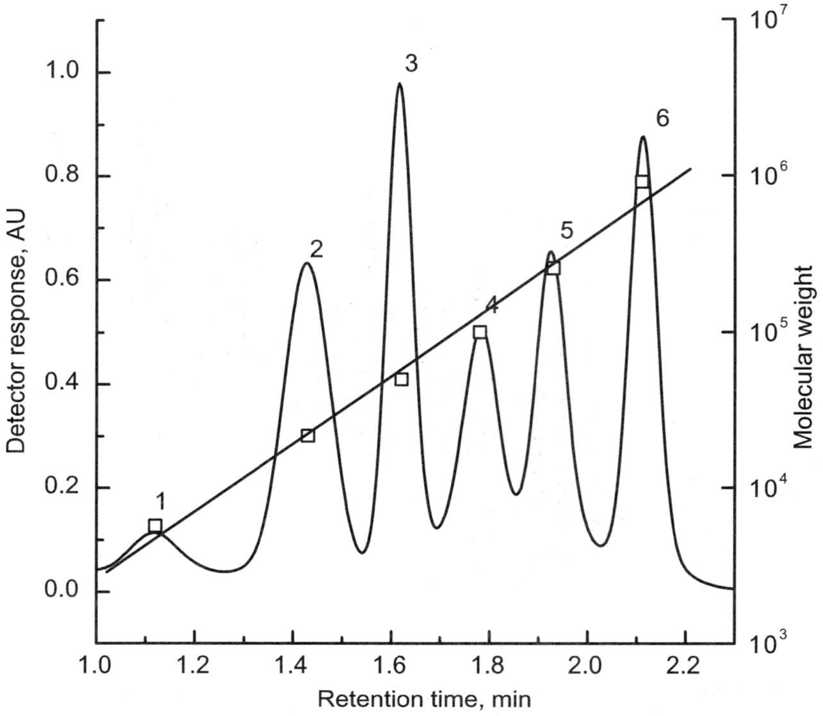

FIGURE 9.16. Rapid separation of a mixture of six poly(methyl methacrylate) standards using a monolithic 50 mm × 4.6 mm i.d. poly(styrene-*co*-divinylbenzene) column and the corresponding calibration curve obtained from the elution data. Separation conditions: 2 min gradient of THF in methanol consisting of 0–35% THF in methanol in 1 min and 35–50% in 1 min; flow rate 1 mL min^{-1}; sample volume 10 μL; overall sample concentration 30 mg mL^{-1} in THF; UV detection at 220 nm. Molecular weights of polymer standards: (1) 5,750, (2) 21,650, (3) 50,000, (4) 100,000, (5) 254,000, and (6) 910,000. (Reprinted from ref. 55. Copyright 2000 John Wiley.)

Polybutadiene

Since THF–methanol did not perform well in the separation of polybutadienes in the HPLC mode, dichloromethane was used as the solvent in combination with methanol (precipitant) to achieve the molecular-weight-dependent separation [55]. The optimized five-step gradient profile again provides a linear calibration curve for all five polybutadiene standards that were available. The separation of these polybutadienes is shown in Figure 9.18. The run time was about 6 min at a flow rate of 1 mL min^{-1}. However, doubling the flow rate to 2 mL min^{-1} and halving the gradient time enabled the separation to be achieved within 3 min. A further acceleration was possible by running the gradient for only 1.25 min. Although the resolution of peaks 2 and 3 is slightly lower in this rapid run, complete separation of the mixture of polybutadiene standards is achieved within 1 min.

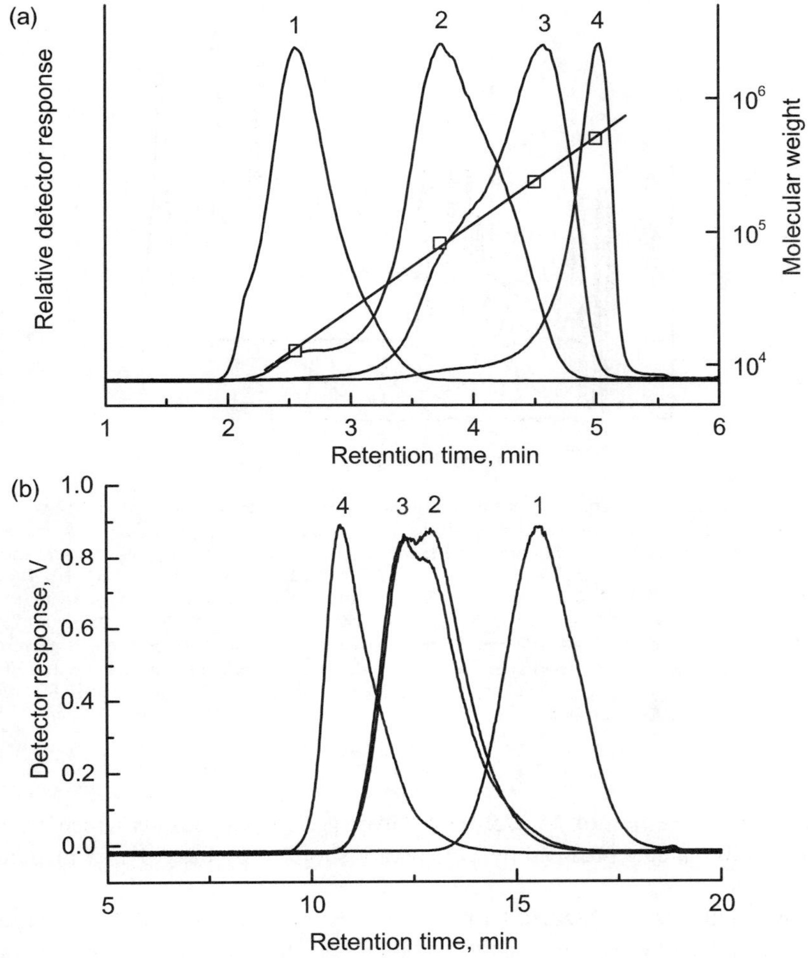

FIGURE 9.17. Elution traces of individually injected secondary poly(vinyl acetate) standards using (a) HPLC in a monolithic 50 mm × 4.6 mm i.d. poly(styrene-*co*-divinylbenzene) column and (b) SEC in two PL gel columns (300 mm × 7.5 mm i.d.), and the corresponding HPLC calibration curve obtained from the elution data. Separation conditions: 5 min gradient consisting of 0–46% THF in hexane in 1 min, 46–60% in 1.2 min, 60–67% in 1.8 min, and 67–100% in 1 min; flow rate 1 mL min^{-1}; sample volume 10 μL; sample concentration 5 mg mL^{-1}; ELSD. Molecular weights: (1) 12,800, (2) 83,000, (3) 129,000, and (4) 500,000. (Reprinted from ref. 55. Copyright 2000 John Wiley.)

DETERMINATION OF MOLECULAR PARAMETERS OF "REAL-LIFE" POLYMERS

A number of different polymers can be subjected to rapid separation using monolithic columns. Typical examples include polystyrene, poly(methacrylates), poly(vinyl acetate), and poly(butadiene). Moreover, as has already been described, a rigorously designed gradient profile affords a linear calibration curve for well-characterized standards that

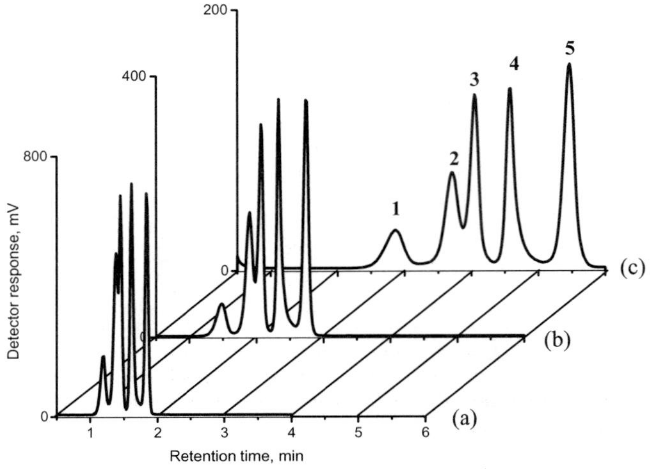

FIGURE 9.18. Effect of gradient time and flow rate on the separation of a mixture of five polybutadiene standards using a 50 mm × 4.6 mm i.d. monolithic poly(styrene-*co*-divinylbenzene) column. Separation conditions: nonlinear gradients of dichloromethane in methanol in overall gradient times of (a) 5.5 min consisting of 0–75% in 3 min, 75–79% in 0.5 min, 79–81% in 0.5 min, 81–84% in 1 min, and 84–100% in 0.5 min, (b) 2.75 min consisting of 0–75% in 1.5 min, 75–79% in 0.25 min, 79–81% in 0.25 min, 81–84% in 0.5 min, and 84–100% in 0.25 min, and (c) 1.38 min consisting of 0–75% in 0.75 min, 75–79% in 0.13 min, 79–81% in 0.12 min, 81–84% in 0.25 min, and 84–100% in 0.13 min; flow rates (a) 1 mL min^{-1} and (b,c) 2 mL min^{-1}; sample volume 10 μL; overall sample concentration 25 mg mL^{-1} (each standard 5 mg mL^{-1}) in THF; ELSD. Molecular weights of polymer standards: (1) 520, (2) 2,820, (3) 6,000, (4) 24,800, and (5) 215,800. (Reprinted from ref. 55. Copyright 2000 John Wiley.)

enables the determination of M_w, M_n, and polydispersity with results which are in good agreement with the data obtained by SEC and/or supplied by the polymer manufacturers (Table 9.1).

Although the results obtained for narrow molecular weight polystyrene standards had been encouraging, the real test of the suitability of the method was the separation of common polymers with a broader molecular weight distribution. For example, a variety of polystyrenes from different sources were analyzed using both precipitation–redissolution HPLC and SEC and their molecular characteristics were calculated [55]. Table 9.2 compares results obtained using these two chromatographic modes. Despite the differences in separation mechanisms, an excellent match was obtained. Lower polydispersity values calculated from HPLC data may result from peak compression in the HPLC mode and/or band broadening in the SEC mode. Moreover, it is important to emphasize that the determination using interactive HPLC in a gradient mode is completed in about 6 min, with a provision for a further acceleration, while the complete SEC trace is eluted only after 20 min.

Similar determinations of molecular parameters for two isotactic poly(methyl methacrylates) using the optimized gradient are summarized in Table 9.3 [55]. Although the values of molecular weight and polydispersity determined by HPLC and SEC do not match exactly, they are again very close. However, the HPLC data are obtained in about one-tenth the time required to record the complete SEC trace.

TABLE 9.2. Molecular Parameters of Various Polystyrenes as Determined by HPLC and SEC

Origin	$M_w \times 10^{-3}$		$M_n \times 10^{-3}$		M_w/M_n	
	HPLC	SEC	HPLC	SEC	HPLC	SEC
Rhodia	73.4	72.5	67.2	67.7	1.09	1.08
Sp²	211.3	208.1	159.8	147.3	1.32	1.41
Aldrich 1	266.1	285.1	202.3	202.6	1.32	1.41
Bayer 1	290.2	297.5	213.4	206.5	1.36	1.44
Kodak	293.1	311.8	227.9	221.2	1.29	1.41
Aldrich 2	294.0	330.5	204.5	211.4	1.44	1.56
Bayer 2	302.3	319.9	225.7	223.2	1.34	1.43

HPLC conditions: 50 mm × 4.6 mm i.d. monolithic poly(styrene-*co*-divinylbenzene) column (ICSO Inc., Lincoln, NE); ELSD (Polymer Laboratories), conditions of five-step gradient as described in Figure 9.13(a).
SEC conditions: 300 mm × 7.5 mm i.d. 10^3 Å and 10^5 Å PL gel columns (Polymer Laboratories); THF; flow rate 1 mL min^{-1}; injections of 20 μL of polymer solution, concentration 1–2 mg mL^{-1}.

TABLE 9.3. Molecular Parameters of Isotactic Poly(Methyl Methacrylates) Determined by HPLC and SEC

Polymer	$M_w \times 10^{-3}$		$M_n \times 10^{-3}$		M_w/M_n	
	HPLC	SEC	HPLC	SEC	HPLC	SEC
PMMA 1	31.2	33.0	29.1	29.4	1.07	1.12
PMMA 2	128.5	125.7	111.0	95.7	1.16	1.31

HPLC conditions: 50 mm × 4.6 mm i.d. monolithic poly(styrene-*co*-divinylbenzene) column (ICSO Inc., Lincoln, NE); UV detection; two-step gradient of THF in methanol (0–35% in 1 min, 35–50% in 1 min); gradient time 2 min; flow rate 1 mL min^{-1}.
SEC conditions: two 300 mm × 7.5 mm i.d. 10^3 Å and 10^5 Å PL gel columns (Polymer Laboratories); THF; flow rate 1 mL min^{-1}; injections of 20 μL of polymer solution, concentration 1–2 mg mL^{-1}.

HIGH-THROUGHPUT DETERMINATION OF MOLECULAR PARAMETERS

Perhaps the best demonstration of the performance of the monolithic columns is the rapid screening of a polystyrene library for differences in molecular weights. This library has been prepared using the living free radical polymerization (LFRP) technique. The molecular weight of polymers prepared using LFRP is affected by a number of factors including the polymerization conditions and the composition of polymerization mixture. In addition to a monomer, this mixture typically consists of a LFRP control agent and possibly other additives. A series of polymerizations, varying polymerization recipe, was carried out at Symyx Technologies to provide a library of polymers. Screening of this library enabled selection of the most active control system. By using a heated 96-well glass-lined aluminum reactor stirred using an orbital shaker, a large number of polystyrenes could be created in a short period of time. Obviously, a high-throughput characterization technique was required to determine the molecular weights of all of these polymers at the rate at which they were prepared.

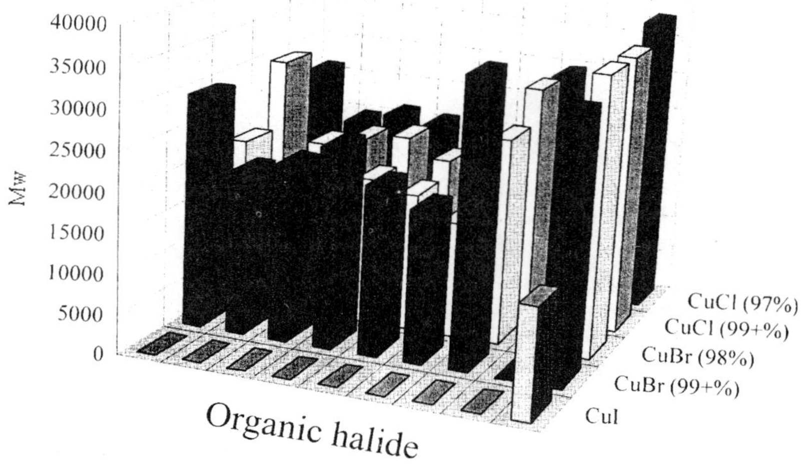

FIGURE 9.19. Molecular weights of polymers prepared by LFRP upon simultaneous variation of control agent and additive determined using high-throughput precipitation–redissolution HPLC. Separation conditions: 50 mm × 8 mm i.d. monolithic poly(styrene-*co*-divinylbenzene) column; mobile phase gradient of 0–30% tetrahydrofuran in methanol in 45 s, followed by 30–0% gradient in 6 s and washing with methanol for 30 s; injection frequency 110 s.

Precipitation–redissolution HPLC was a good candidate for achieving this goal. A robotic device was used to dilute the homogeneous polymer solution formed in each well with THF and inject it directly into the chromatographic system. Alternatively, the polymer was first purified using an automated routine. This has been achieved by precipitation triggered by addition of methanol to the mixture, separation of solid polymer by decanting, and dissolution in THF. The chromatographic procedure included automatic filling of a loop from which the sample was injected into a 50 mm × 8 mm i.d. monolithic poly(styrene-*co*-divinylbenzene) column. Since the mobile phase at this stage was pure methanol, polystyrene sample precipitated and was dissolved only after application of a 45 s long gradient of 0–30% THF. The system was equilibrated after each analysis by returning to pure methanol in 6 s, which was pumped through the column for the next 30 s. Using this approach, injections took place every 110 s. This frequency enables analysis of more than 30 samples per hour and the complete screening of all polymers created in the 96-well polymerization reactor is achieved in about 3 hr. The bar diagram shown in Figure 9.19 demonstrates the results of this rapid screening.

STABILITY OF MONOLITHIC COLUMNS

The monolithic columns are surprisingly rugged. The solid and the dotted lines in Figure 9.14 represent the chromatograms of the same polymer mixture recorded more than 2 months and about 400 injections apart [55]. Even after such a long time, the very small difference that can be observed between these two runs lies within the experimental error of the measurements. Moreover, it should be considered that in each gradient run one component of the mobile phase is a good swelling agent for the monolithic material of the column while the other is a precipitant. Although the high level of cross-linking does not allow extensive swelling of the monolith, even small volumetric changes of the matrix constitute a periodic stress for the column. However, this repeated stress had no effect on long-term column performance. Excellent column stability was also demonstrated by the use of several different solvents such as THF, dichloromethane, methanol, hexane, and water with repeated changes in gradient composition without any adverse effects on the separations. An occasional low-flow-rate flushing with THF was the only "maintenance" carried out on the column. Similarly, frequent changes of the flow rate, which is necessary mainly in the method optimization step, have no detrimental effect on long-term column performance and no change in back-pressure was observed.

CONCLUSIONS

The simplicity of the preparation, the unique flow properties, and the enhanced mass transport ability of monolithic columns makes them attractive as an alternative to particulate column packing for the separation of macromolecules. The rapid and efficient determination of molecular weight parameters of synthetic polymers in a gradient elution HPLC mode facilitated by the use of small monolithic poly(styrene-*co*-divinylbenzene) columns constitute a viable, less expensive, and much faster alternative to the SEC separations using costly sets of sophisticated columns. Although the use of monolithic columns has so far been reported for separations of only limited number of types of polymers, the short molded monoliths are likely to be useful for high-throughput separations of many polymers using optimized gradients of suitable pairs of solvents and non-solvents. The sample loading affects the separation much less than in the case of packed beds. This is a promising feature for future high-throughput separations of polymers on the preparative scale.

ACKNOWLEDGMENTS

Support of this work by a research grant from the Ministry of Education of the Czech Republic (CB MSM 223400008) and by a grant from the National Institute of General Medical Sciences, National Institutes of Health (GM–48364) is gratefully acknowledged. Thanks are also due to ISCO Inc. for the gift of the monolithic columns.

REFERENCES

1. Cantow, M. *Polymer Fractionation*, **1967**, New York: Academic Press.
2. Francuskiewicz, F. *Polymer Fractionation*, **1994**, Berlin: Springer.
3. Revillon, A. Alternatives to size-exclusion chromatography. *J. Liq. Chromatogr.* **1994**, *17*, 2991–3023.
4. Chubarova, E. V., Nesterov, V. V. Degradation of high-molecular-mass polystyrene during chromatography on porous and nonporous silicate beads—Correlations with solvent quality. *Vysokomol. Soed. A* **1994**, *36*, 1373–1378.

5. Poole, C. F., Poole, S. K. *Chromatography Today*, **1995**, Amsterdam, The Netherlands: Elsevier.
6. Gooding, K. M., Regnier, F. E. *HPLC of Biological Molecules*, **1990**, New York: Marcel Dekker.
7. Petro, M., Safir, A. L., Nielsen, R. B. Role of high throughput chromatography in combinatorial chemistry of polymeric materials. *Polym. Prepr.* **1999**, *40*, 702.
8. Radota, B. J. *Analytical Isotachophoresis*, **1988**, Weinheim, Germany: VCH.
9. Janča, J. *Field-Flow Fractionation*, **1984**, New York: John Wiley.
10. Stegeman, G., Kraak, J. C., Poppe, H., Tijssen, R. Hydrodynamic chromatography of polymers in packed-columns. *J. Chromatogr. A* **1993**, *657*, 283–303.
11. Stadalius, M. A., Quarry, M. A., Mourey, T. H., Snyder, L. R. Conventional chromatographic theory versus "critical" solution behavior in the separation of large molecules by gradient elution. *J. Chromatogr.* **1986**, *358*, 17–37.
12. Larmann, J. P., DeStefano, J. J., Goldberg, A. P., Stout, R. W., Snyder, L. R., Stadalius, M. A. Separation of macromolecules by reversed-phase high-performance liquid-chromatography—Pore-size and surface-area effects for polystyrene samples of varying molecular-weight. *J. Chromatogr.* **1983**, *255*, 163–189.
13. Quarry, M. A., Stadalius, M. A., Mourey, T. H., Snyder, L. R. General model for the separation of large molecules by gradient elution. Sorption versus precipitation. *J. Chromatogr.* **1986**, *358*, 1–16.
14. Shalliker, R. A., Kavanagh, P. E., Russel, I. M. Reversed-phase gradient elution behaviour of polystyrenes in a dichloromethane methanol solvent system. *J. Chromatogr.* **1991**, *543*, 157–169.
15. Kaczmarski, K., Prus, W., Kowalska, T. Adsorption/partition model of liquid chromatography for chemically bonded stationary phases of the aliphatic cyano, reversed-phase C-8 and reversed-phase C-18 types. *J. Chromatogr. A* **2000**, *869*, 57–64.
16. Glöckner, G. Characterization of co-polymers by means of liquid-chromatography. *Pure Appl. Chem.* **1983**, *55*, 1553–1562.
17. Jandera, P., Holčapek, M., Kolářová, L. Retention mechanism, isocratic and gradient-elution separation and characterization of co-polymers in normal-phase and reversed-phase high-performance liquid chromatography. *J. Chromatogr. A* **2000**, *869*, 65–84.
18. Pasch, H., Trathnigg, B. *HPLC of Polymers*, **1998**, Berlin: Springer.
19. Lochmuller, C. H., McGranaghan, M. B. Isocratic elution of high molecular-weight monodisperse polystyrenes. *Anal. Chem.* **1989**, *61*, 2449–2455.
20. Northrop, D. M., Martire, D. E., Scott, R. P. W. Liquid-chromatographic retention behavior of polystyrene homopolymers on a C4, bimodal pore diameter, reversed-phase column. *Anal. Chem.* **1992**, *64*, 16–21.
21. Lochmuller, C. H., Jiang, C., Elomaa, M. Retention behavior of poly(methyl methacrylate) and poly(ethylene glycol) in reversed-phase liquid-chromatography. *J. Chromatogr. Sci.* **1995**, *33*, 561–567.
22. Armstrong, D. W., Bui, K. H. Nonaqueous reversed-phase liquid chromatographic fractionation of polystyrene. *Anal. Chem.* **1982**, *54*, 706–708.
23. Lochmuller, C. H., Jiang, C., Liu, Q., Antonucci, V., Elomaa, M. High-performance liquid chromatography of polymers: Retention mechanisms and recent advances. *Crit. Rev. Anal. Chem.* **1996**, *26*, 29–59.
24. Rissler, K. Separation of polyester oligomers by gradient high-performance liquid chromatography. *J. Chromatogr. A* **1997**, *786*, 85–98.
25. Klumperman, B., Philipsen, H. J. A. Coupling matters: Gradient polymer elution chromatography as a versatile tool in polymer characterization. *LC-GC* **1999**, *17*, 118–130.
26. Lochmuller, C. H., Moebus, M. A., Liu, Q., Jiang, C. Temperature effect on retention and separation of poly(ethylene glycol)s in reversed-phase liquid chromatography. *J. Chromatogr. Sci.* **1996**, *34*, 69–76.
27. Lee, W., Lee, H. C., Chang, T. Y., Kim, S. B. Characterization of poly(methyl methacrylate) by temperature gradient interaction chromatography with on-line light scattering detection. *Macromolecules* **1998**, *31*, 344–348.
28. Lee, H. C., Chang, T. H., Harville, S., Mays, J. W. Characterization of linear and star polystyrene by temperature-gradient interaction chromatography with a light-scattering detector. *Macromolecules* **1998**, *31*, 690–694.
29. Lee, W., Lee, H. C., Park, T., Chang, T., Chang, J. Y. Temperature gradient interaction chromatography of low molecular weight polystyrene. *Polymer* **1999**, *40*, 7227–7231.
30. Glöckner, G. *Gradient HPLC of Copolymers and Chromatographic Cross-Fractionation*, **1991**, Berlin: Springer.
31. Desreux, V., Spiegels, M. C. Fractionation of polythene by extraction. *Bull. Soc. Chim. Belg.* **1950**, *59*, 476–489.

32. Glöckner, G., van den Berg, J. H. M. Precipitation and adsorption phenomena in polymer chromatography. *J. Chromatogr.* **1986**, *352*, 511–522.
33. Bullock, J. High-temperature reversed-phase high-performance liquid-chromatographic analysis of a synthetic copolymer on a nonporous support. *J. Chromatogr.* **1995**, *694*, 415–423.
34. Snyder, L. R., Stadalius, K. A., Quarry, M. A. Gradient elution in reversed-phase HPLC separation of macromolecules. *Anal. Chem.* **1983**, *55*, 1412A–1430A.
35. Snyder, L. R., Stadalius, M. A. In C. Horváth (Ed.), *High-Performance Liquid Chromatography, Advances and Perspectives* **1986**, New York: Academic Press Vol. 4, pp.195–218.
36. Colwell, L. F., Hartwick, R. A. Nonporous silica supports for high-performance liquid-chromatography. *J. Liq. Chromatogr.* **1987**, *10*, 2721–2744.
37. Lee, D. P. Chromatographic evaluation of large-pore and non-porous polymeric reversed phases. *J. Chromatogr.* **1988**, *443*, 143–153.
38. Huber, C. G., Oefner, P. J., Preuss, E., Bonn, G. K. High-resolution liquid-chromatography of DNA fragments on nonporous poly(styrene-divinylbenzene) particles. *Nucleic Acid Res.* **1993**, *21*, 1061–1066.
39. Petro, M., Svec, F., Fréchet, J. M. J. Molded continuous poly(styrene-*co*-divinylbenzene) rod as a separation medium for the very fast separation of polymers—Comparison of the chromatographic properties of the monolithic rod with columns packed with porous and non-porous beads in high-performance liquid chromatography of polystyrenes. *J. Chromatogr. A* **1996**, *752*, 59–66.
40. Regnier, F. E. Perfusion chromatography. *Nature* **1991**, *350*, 634–635.
41. Afeyan, N. B., Gordon, N. F., Maszaroff, I., Varady, L., Yang, S. P., Regnier, F. E. Flow-through particles for the high-performance liquid-chromatographic separation of biomolecules—Perfusion chromatography. *J. Chromatogr.* **1990**, *519*, 1–29.
42. Wang, Q. C., Svec, F., Fréchet, J. M. J. Macroporous polymeric stationary phase rod as continuous separation medium for reversed phase chromatography. *Anal. Chem.* **1993**, *65*, 2243–2248.
43. Svec, F., Fréchet, J. M. J. "Molded" rods of polymer for preparative separations of biological products. *Biotech. Bioeng.* **1995**, *48*, 476–480.
44. Svec, F., Fréchet, J. M. J. Modified poly(glycidyl methacrylate-*co*-ethylene dimethacrylate) continuous rod columns for preparative-scale ion-exchange chromatography of proteins. *J. Chromatogr. A* **1995**, *702*, 89–95.
45. Svec, F., Fréchet, J. M. J. Continuous rods of macroporous polymer as high-performance liquid chromatography separation media. *Anal. Chem.* **1992**, *64*, 820–822.
46. Svec, F., Fréchet, J. M. J. New designs of macroporous polymers and supports: From separation to biocatalysis. *Science* **1996**, *273*, 205–211.
47. Peters, E. C., Svec, F., Fréchet, J. M. J. The preparation of large diameter "molded" porous polymer monoliths and the control of pore structure homogeneity. *Chem. Mater.* **1997**, *9*, 1898–1902.
48. Podgornik, A., Barut, M., Štancar, A., Josic, D., Koloini, T. Construction of large volume monolithic columns. *Anal. Chem.* **2000**, *72*, 5693–5699.
49. Svec, F., Fréchet, J. M. J. Kinetic control of pore formation in macroporous polymers—Formation of molded porous materials with high-flow characteristics for separations or catalysis. *Chem. Mater.* **1995**, *7*, 707–715.
50. Svec, F., Fréchet, J. M. J. Temperature, a simple and efficient tool for the control of pore-size distribution in macroporous polymers. *Macromolecules* **1995**, *28*, 7580–7582.
51. Viklund, C., Svec, F., Fréchet, J. M. J., Irgum, K. Monolithic, "molded", porous materials with high flow characteristics for separations, catalysis, or solid-phase chemistry: Control of porous properties during polymerization. *Chem. Mater.* **1996**, *8*, 744–750.
52. Viklund, C., Pontén, E., Glad, B., Irgum, K., Horstedt, P., Svec, F. "Molded" macroporous poly(glycidyl methacrylate-*co*-trimethylolpropane trimethacrylate) materials with fine controlled porous properties: Preparation of monoliths using photoinitiated polymerization. *Chem. Mater.* **1997**, *9*, 463–471.
53. Pelzbauer, Z., Lukáš, J., Svec, F., Kálal, J. Morphology of polymeric sorbents based on glycidyl methacrylate copolymers. *J. Chromatogr.* **1979**, *171*, 101–107.
54. Wang, Q. C., Svec, F., Fréchet, J. M. J. Reversed-phase chromatography of small molecules and peptides on a continuous rod of macroporous poly(styrene-*co*-divinylbenzene), *J. Chromatogr. A* **1994**, *669*, 230–235.
55. Jančo, M., Sýkora, D., Svec, F., Fréchet, J. M. J., Schweer, J., Holm, R. Rapid determination of molecular parameters of synthetic polymers by precipitation–redissolution high-performance liquid chromatography using "molded" monolithic column, *J. Polym. Sci. A* **2000**, *38*, 2767–2778.
56. Petro, M., Svec, F., Fréchet, J. M. J. Molded monolithic rod of macroporous poly(styrene-*co*-divinylbenzene) as a separation medium for HPLC of synthetic polymers: "On-column" precipitation–redissolution

chromatography as an alternative to size-exclusion chromatography of styrene oligomers and polymers. *Anal. Chem.* **1996**, *68*, 315–321.

57. Engelhardt, H., Czok, M., Schultz, R., Schweinheim, E. Sample size and retention values in high-performance liquid chromatography of biological and synthetic polymers. *J. Chromatogr.* **1988**, *458*, 79–92.
58. Glöckner, G., Engelhardt, H., Wolf, D., Schultz, R. Sample size and retention values in gradient liquid chromatography of synthetic polymers. *Chromatographia* **1996**, *42*, 185–190.

Chapter 10

Ion Mobility/Time-of-Flight Analysis of Combinatorial Library Mixtures

Catherine A. Srebalus Barnes and David E. Clemmer

INTRODUCTION

Recent developments in combinatorial synthesis and high-throughput screening (HTS) have emphasized parallel strategies for the generation and characterization of large libraries of synthetic molecules. In the era following the sequencing of the human genome [1], proteomics has been developed as a high-throughput approach to the discovery of biological target molecules that are up- or downregulated in diseased cells [2]. The identification of such targets will certainly increase the demand for combinatorial libraries of potential binding ligands that are chemically diverse, well characterized, and appropriate for screening against multiple target molecules.

Automated high-throughput purification, characterization, and screening strategies have been used for the processing of isolated components in compound libraries that are present in relatively high purity. However, high-throughput component characterization based on analytical technologies that involve costly instrumentation (spectroscopy, mass spectrometry, etc.) is inherently limited due to the requirements for serial detection. A potential solution to the analytical bottleneck of high-throughput strategies involves the simultaneous characterization of components from library mixtures. Early efforts to identify new therapeutic molecules from combinatorial library mixtures were plagued by the lack of available analytical methods for characterization of the library mixtures prior to screening. Screening results obtained from uncharacterized mixtures often provide unreliable estimates of binding affinity because the relative concentrations of mixture components are highly variable and some components may be missing entirely.

The development of analytical methods for the analysis of mixtures containing large numbers of isomeric components is important for maximizing the advantages afforded by combinatorial synthesis. The combination of high-throughput parallel automation with strategies for mixture analysis would facilitate the characterization of microtiter plate

Catherine A. Srebalus Barnes • Eli Lilly and Company, Indianapolis, IN 46285, USA
David E. Clemmer • Indiana University, Department of Chemistry, Bloomington, IN 47405, USA
Contact author: Catherine A. Srebalus Barnes, email: SREBALUS_BARNES_CATHERINE_A@Lilly.com

High-Throughput Analysis, edited by Potyrailo and Amis
Kluwer Academic Publishers, New York, 2003.

library arrays containing hundreds to thousands of components in each well. These same approaches for the characterization of library mixtures would be readily applicable to proteomics investigations that require technologies for the analysis of biological mixtures.

Mass spectrometric (MS) techniques [3] or combined separation–MS methods [4] have been used to characterize small library mixtures. However, these strategies are not suited to assigning mixtures of isomers. We have recently developed a nested high-resolution ion mobility–time-of-flight mass spectrometry (IMS–TOFMS) approach for the characterization of biological mixtures [5]. IMS–TOFMS strategies have several advantages over conventional mass spectrometry. The incorporation of a high-efficiency gas-phase mobility separation prior to mass-to-charge analysis allows many isomeric mixture components to be distinguished. Using IMS–TOFMS technologies it is possible to assess the composition and purity of combinatorial peptide mixtures generated using mix and split methods. We have applied these methods to the characterization of large library mixtures containing several hundreds to thousands of peptides.

This chapter will focus on the principles of IMS–MS technologies for the gas phase separation of isomeric combinatorial peptide library components. Developments in IMS–MS instrumentation and the optimization of fundamental experimental parameters such as resolving power and duty cycle will be discussed. Several applications of IMS–TOFMS strategies for library characterization will be presented which detail the analysis of peptide library mixtures before and after affinity screening. Recent developments in the coupling of condensed phase HPLC separations with IMS–TOFMS and the identification of mobility-separated ions using a parallel collision induced dissociation (CID) sequencing approach will also be included.

EXPERIMENTAL APPROACH

General

Several reviews have detailed recent developments in IMS [6]. The separation of gas phase ions using IMS results from differences in the time required for ions to move through a buffer gas under the influence of an electric field. In the mobility experiment, ions are accelerated by a uniform electric field and decelerated by collisions with buffer gas atoms resulting in constant experimental drift velocities. Drift velocities for polyatomic ions are dependent on both the overall charge state of the ion and the average collision cross-section (or shape) of the ion. Ion mobility distributions are recorded using drift tubes of variable design (described below) by measuring the time required for a pulse of ions to travel across a well-defined drift region under the influence of a uniform electric field.

The coupling of ion mobility with MS provides additional information regarding the mass-to-charge (m/z) ratios of the ions in the mobility distribution. Various instrument configurations have been reported in which mass analyzers are placed either before or after the drift tube. Traditionally, quadrupole mass filters have been used to select ions of specific m/z ratio following the mobility separation [7–9]. In these systems, ion mobility distributions are recorded sequentially for each ion of interest. This chapter will focus on the use of time-of-flight mass spectrometry (TOFMS) for simultaneous acquisition of mobility and m/z data for mixtures of ions. Incorporation of a time-of-flight analyzer at the exit of the drift tube permits acquisition of mass spectra for all ions in a mixture simultaneously. The IMS–TOFMS approach provides increased spectral resolution in the MS dimension, higher sensitivity, and increased throughput.

A variety of different ion sources have been used for the generation of ions for introduction into the drift tube. Early ion mobility studies utilized electron impact (EI) sources. EI sources are commonly used in modern IMS instruments designed to detect explosives, chemical warfare agents, and environmental pollutants [10]. Pulsed laser vaporization sources have been utilized in ion mobility studies for the examination of cluster and atomic metal ions [7]. More recent work involving the study of biomolecules has employed soft ionization techniques such as matrix-assisted laser desorption (MALDI) [8] and electrospray ionization (ESI) [9]. This chapter will focus on the use of ESI for the direct generation of ions from solutions of combinatorial library mixtures.

Injected-Ion Drift Tubes

Injected-ion drift tube configurations require the transfer of ions from the ion source through a high-vacuum region followed by "injection" of the ions into a drift tube containing several torr of buffer gas. Figure 10.1 shows a schematic diagram of the injected ion drift tube developed in the Clemmer laboratory [11]. Similar injected ion drift tubes have been described previously [7,8,12]. In the apparatus shown in Figure 10.1, ions are formed by electrospraying solutions into a variable-temperature differentially pumped desolvation region (~2–10 Torr). Ions exit this desolvation region through an orifice into a high-vacuum chamber (10^{-5}–10^{-4} Torr) where they are focused into a beam and accelerated to the desired kinetic energy using a series of electrostatic lenses. Short pulses of

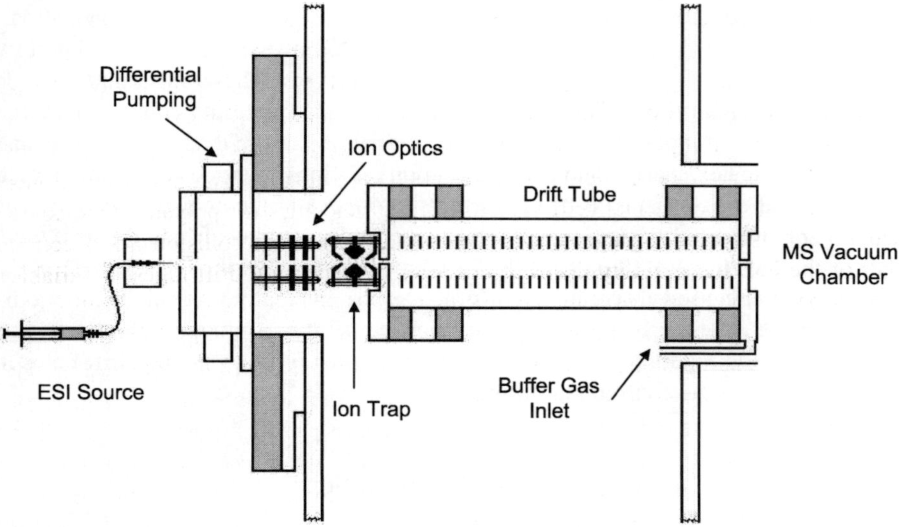

FIGURE 10.1. Schematic diagram of the injected-ion drift tube equipped with the ion trap interface. (Adapted from ref. 14.)

ions (20–50 μs in duration) are injected through a small aperture into the drift tube which contains ~1–5 Torr of helium buffer gas. Ion pulses are generated by variation of the potential on a deflector in the high-vacuum lens system.

An ion trap mounted on the front of the drift tube can be used to generate shorter pulses of ions (0.6–1 μs in duration) that are both spatially and temporally focused [13]. Ejection of ions from the ion trap into the drift tube is accomplished by tuning the radio-frequency (rf) field and supplying a pulse to one of the trap end-caps. In addition, the ion trap is used for the accumulation of ions from the continuous ESI source between mobility pulses. A bias voltage (~30 V) applied to the trap prevents ions from leaking out during accumulation. Ions enter the trap continuously (except during the ejection pulse) and are trapped using a 1.1 MHz rf field (~2–3 kV, peak-to-peak) applied to the ring electrode. Conventional drift tube designs result in ion losses of up to 99.99% because ions are discarded between mobility pulses. Accumulation of ions in the ion trap results in duty cycles that approach 100% and improvements in signal-to-noise ratio (S/N) by factors of ~10–30.

In the injected-ion configuration, the injection energy is defined by the ion charge state and the voltage difference between the exit of the high-pressure ion source and the drift tube entrance flange. As ions enter the drift tube, they are rapidly heated as their kinetic energies are thermalized by collisions with the buffer gas. Further collisions cool the ions to the temperature of the buffer gas. This heating–cooling cycle can be varied by changing the injection energy. Low injection energies can be used to study ion conformations formed in the ion source and weakly bound non-covalent complexes. Higher injection energies can be used to study temperature induced structural transitions and ion dissociation [14]. Practical limitations on injection energies are imposed by the need to overcome the counterflow of buffer gas out of the drift tube entrance aperture. Typically, ion signals decrease rapidly at injection energies below ~50 V. Operating pressures for injected ion drift tubes are typically below ~10 Torr because, at higher pressures, the injection energies required to overcome the buffer gas flow cause excessive ion fragmentation.

Several designs of the injected-ion drift tube body have been used in our laboratory to provide a sealed cavity that can be filled with buffer gas. These variations include a stainless steel design with Teflon spacers at each end to isolate the entrance and exit plates. An alternative modular design, made from alternating 2.5 cm thick copper rings and 0.9 cm thick Myokcalex spacers and sealed by Viton® O-rings, provides a variable length drift tube body. The insertion of cartridge heaters into holes in the copper rings provides uniform heating of the buffer gas for temperature-dependent studies. Inside the drift tube, equally spaced BeCu rings (0.025 cm thick) connected by a series of 5.00 MΩ high-vacuum resistors are used to create a uniform electric field along the drift tube axis. Limitations in the applied field strength are imposed by the discharge potential of the buffer gas at particular operating pressures and temperatures. Typical applied drift fields in the injected-ion drift tube configuration range from ~3 to 15 V cm^{-1}.

High-Pressure Drift Tubes

In high-pressure drift tube configurations, the ion source is coupled directly to the front of the drift tube. The injection process is not required because the ions do not enter

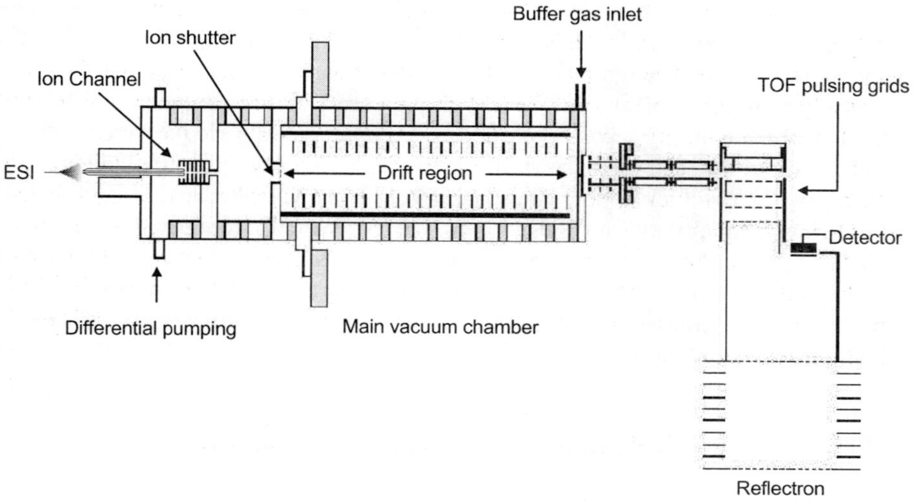

FIGURE 10.2. Schematic diagram of the high-resolution IMS–TOFMS instrument. (Reproduced from ref. 40.)

a high-vacuum region between the source and the high-pressure drift tube. Figure 10.2 shows a schematic diagram of the high-pressure drift tube–orthogonal reflectron geometry TOFMS designed in the Clemmer laboratory [14]. This instrument is discussed in greater detail below; however, the high-pressure drift tube component will be the focus of the present section. The drift tube is similar to a configuration described by Dugourd et al. [15]. In this design, ions are electrosprayed at atmospheric pressures directly into a differentially pumped desolvation region through a capillary tube which extends into the center of a series of electrostatic lenses. Ions are forced through the tube by the pressure gradient across the capillary. We estimate that the pressure in the differential pumping region is ~200–250 Torr. The lenses are connected by a series of resistors in order to create a uniform electric field and the voltage drop across these lenses is ~1000–2000 V. This field guides the ions from the source into the body of the drift tube against the counterflow of helium buffer gas, while preventing neutrals from entering the drift region.

Once inside the body of the drift tube, the ions drift through the buffer gas under the influence of a field provided by the external copper ring electrodes until they reach an ion gate. The drift region is defined as the region between this ion gate and the exit of the drift tube. A variety of gating configurations can be used to generate ion pulses to initiate the mobility experiment. One of the designs implemented in our laboratory uses a wire mesh maintained at a potential exceeding the potential of the first lens of the drift region. This voltage increase prevents the transport of ions into the drift region. Short pulses of ions are introduced into the drift region by dropping the potential on the ion gate for a period of ~50–300 μs. Recently, Wyttenbach et al. [16] have described the use of an ion funnel interface (similar to the design of Shaffer et al. [17]) to store ions at intermediate pressures between mobility pulses. The use of an ion funnel between the atmospheric ESI source and high-pressure drift tube is analogous to the ion trap interface described above for the ion trap–injected-ion drift tube configuration; however, the ion funnel should effectively store ions at higher pressures. Although the ion funnel interface has been reported to increase the duty cycle to nearly 100%, it has currently only been tested at low pressures (4–5 Torr). Operating pressures in the high-pressure drift region of the instrument shown in Figure 10.2 are typically between ~150 and 200 Torr. High-pressure drift tube configurations have been reported in which mobility separations are carried out at atmospheric pressures [18].

The body of the drift tube shown in Figure 10.2 is variable in length and comprised of alternating 2.5 cm thick copper rings and 0.9 cm thick Mycalex spacers and sealed by Viton O-rings. Equally spaced BeCu rings (0.025 cm thick) connected by a series of 5.00 MΩ high-vacuum resistors are used to create a uniform electric field inside the drift tube body from the ion gate to the exit of the drift tube. Higher operating pressures permit the use of increased drift fields which typically range from ~150–280 V cm^{-1} in the high-pressure configuration.

Fundamentals of Mobility Separations

The resolving power of the mobility separation is given by [19]

$$\frac{t_D}{\Delta t} \approx \left(\frac{LEze}{16\, k_b T \ln 2}\right)^{1/2}$$

where Δt is the peak width at half maximum, L is the length of the drift tube, T is the temperature of the buffer gas, ze is the ionic charge, and k_b is Boltzmann's constant. Typical resolving powers in the low-pressure injected-ion drift tube range from 20 to 30. The higher buffer gas pressures used in the high-pressure drift tube (~150–250 Torr) relative to the injected ion drift tube (~2–5 Torr) enable the use of higher field strengths (~150–280 V cm^{-1}) which favorably impact the experimental resolving power. Substantially higher resolution mobility separations can be obtained using the high-pressure drift tube. Typical experimental resolving powers in this system range from ~100 to 150 (relative to ~20–30 for the low-pressure injected-ion drift tube). In many cases, high-resolution mobility separations are capable of completely resolving isomeric mixture components in combinatorial libraries.

The resolving power of mobility separations also increases for ions of higher charge state. This is due to the amount of time that the ions spend in the drift region and the associated diffusion of the ion packet as it travels the length of the drift region. Doubly charged ions experience twice the drift force of singly charged ions with similar collision cross-sections. Higher experimental resolving powers are obtained for mobility distributions of highly charged ions because of the decreased peak broadening due to diffusion. Experimental resolving powers in excess of 200 have been reported for ion mobility distributions of highly charged peptide ions obtained using high-pressure drift tubes [20,21].

When multiple isomers or ion conformations are expected at a given m/z ratio, it is helpful to compare experimental peak widths with peak shapes that are calculated for transport of a single geometric structure through the drift tube given by the following relation [22]:

$$\Phi(t) = \int P(t_p)dt_p \frac{C}{(Dt)^{1/2}}\left(v_D + \frac{L}{t}\right)\left[1 - \exp\left(\frac{-r_o^2}{4Dt}\right)\right]\exp\left(\frac{-(L - v_D t)^2}{4Dt}\right) \quad (2)$$

where $\Phi(t)$ is the intensity of ions passing through the exit aperture as a function of time, r_o is the entrance aperture radius, v_D is the measured drift velocity, C is a constant related to the exit aperture area, $P(t_p)dt_p$ is the distribution function for the pulse of ions entering the drift tube, and D is the diffusion constant. Peaks that are significantly broader than calculated distributions suggest the presence of multiple isomers (or conformations) with similar mobilities. Increased peak widths may also indicate interconversion of ion conformers during the mobility experiment or fluctuations in experimental variables such as pressure and temperature.

Determination of Experimental Cross-Sections

Collision cross-sections can be obtained from experimental drift times using the following relation [22]:

$$\Omega = \frac{(18\pi)^{1/2}}{16} \frac{ze}{(k_bT)^{1/2}} \left[\frac{1}{m_I} + \frac{1}{m_B}\right]^{1/2} \frac{t_D E}{L} \frac{760}{P} \frac{T}{273.2} \frac{1}{N} \tag{3}$$

where t_D is the experimental drift time (corrected for time spent in regions of the instrument other than the drift region), E is the electric field, T and P are the temperature and pressure of the buffer gas, L is the length of the drift region, N is the neutral number density, k_b is Boltzmann's constant, and m_I and m_B are the mass of the ion and the buffer gas, respectively. The experimental parameters in Eqn (3) can be measured precisely and therefore experimental collision cross-section measurements are expected to be very accurate. Any two measurements of collision cross-section for a given ion using different instruments in our laboratory typically differ by less than 1% (relative uncertainty). When mobility experiments are carried out at low E/N values, drift velocities are linearly dependent on E and ions are not expected to align in the drift region. Under these conditions, experimental collision cross-sections reflect average values for all possible orientations of the ion [23].

Assignment of Mobility-Separated Isomers

Experimental collision cross-sections provide a rough estimate of the average shape of the ion as it travels through the drift tube. Significantly more detailed information can be obtained from comparisons between experimental collision cross-sections and calculated values obtained for model conformers. Meaningful comparisons require that structures obtained in the molecular modeling simulations are representative of the structures observed experimentally. Structures from the molecular modeling simulations are considered as candidate structures if their calculated collision cross-sections agree to within $\pm 2\%$ of the experimental cross-section. Candidate structures that satisfy this criterion and represent the lowest energy structure obtained in a given simulation are considered representative structural types.

Cross-sections for low-energy structures obtained in the simulations can be calculated using one of the following theoretical approaches. The projection approximation developed by Bowers [24], Jarrold [25], and their co-workers simplifies the molecular collision integral by placing spheres the size of atomic integrals at the locations of each atom. The molecule is then projected onto a randomly chosen plane in space to generate circles (with radii corresponding to the collision radii) in the plane at the positions of each projected atom. For a series of randomly selected points within an area enclosing the projection, the ratio of hits relative to the number of attempts multiplied by the area defines the cross-section for a given projection. Orientationally averaged cross-sections are obtained for many random projections until the average cross-section converges to a value within acceptable error limits.

The exact hard-sphere scattering method (EHSS) [26] developed by Jarrold and co-workers includes contributions to the collision cross section from the scattering angle between buffer gas atoms and the drifting ion. The most rigorous method to date is the trajectory method [27], also developed by Jarrold and co-workers. Trajectory methods use a realistic potential to account for long-range potential interactions between the ion and the buffer gas.

Nested Measurements of Drift (Flight) Times

The coupling of ion mobility separations with mass spectrometry has traditionally involved the use of quadrupole mass filters to select ions of known *m/z* either before or after the mobility separation in the drift tube [7–9]. In these systems, ion mobility distributions must be recorded for each *m/z*-selected ion of interest. Ion mobility–time-of-flight approaches, in which complete mass spectra are recorded for all ions simultaneously, have inherently higher throughput and permit the direct measurement of *m/z* and conformer-resolved information for the total ion system [5]. Initial IMS–TOFMS developments in the Clemmer laboratory involved the coupling of an injected-ion drift tube and a linear-geometry TOF mass spectrometer. However, subsequent configurations utilizing a high-pressure drift tube and orthogonal geometry reflectron TOF region have been demonstrated to provide significantly higher resolution in both the mobility and TOF dimensions [28]. Recently, additional reports have been published describing the coupling of ion mobility separations with TOF mass analyzers [29,30]. The high-pressure drift tube–orthogonal reflectron TOF instrument developed in the Clemmer laboratory and shown in Figure 10.2 will be the focus of the discussion here.

In this instrument, the mobility separation occurs as described above for the high-pressure drift tube. The mobility of an ion through the high pressure of helium buffer gas is dependent on both its orientationally averaged collision cross-section and charge state. For a given charge state, compact ions undergo fewer collisions with the buffer gas atoms and thus have higher mobilities (shorter drift times) than more elongated conformations. For ions with similar collision cross-sections, higher charge states experience a greater drift force than lower charge states and thus have higher mobilities.

As ions exit the drift tube, they are focused into the source region of a TOF mass spectrometer. High-frequency pulses, synchronous with lower-frequency mobility pulses, are used to initiate the TOF experiment. Because drift times through the high pressure of helium buffer gas are substantially longer (milliseconds) than flight times in the evacuated TOF region (microseconds), hundreds of TOF mass spectra can be acquired for a single pulse of ions into the drift tube. Arrival times at the TOF detector are known with respect to both the mobility and TOF pulses. This permits the accumulation of drift-time and flight-time information for mixtures of ions without changing any experimental parameters. Even small relative differences in mobility for different mixture components are highly reproducible.

We refer to the IMS–TOFMS experiment as a nested experiment, because for each ion we obtain both a drift time t_D and a flight time t_f. Positions of peaks in the nested drift (flight) time data are reported using the t_D (t_f) nomenclature described previously[5]. Unless otherwise noted, values of t_D and t_f are given in units of milliseconds and microseconds, respectively. Flight times are converted to *m/z* ratios using a multipoint calibration method. Typically, the resolving power of the TOF measurement ranges from ~1500 to 2500 [$m/\Delta m$ definition, where Δm is the full width at half maximum (FWHM) of a peak]. The resolving power of the high-pressure drift tube (defined as $t/\Delta t$, where Δt is the FWHM) is typically ~100–150 as described above.

APPLICATIONS OF ION MOBILITY–TIME-OF-FLIGHT TECHNIQUES

Determining Synthetic Failures in Combinatorial Libraries

Figure 10.3 shows a TOF mass spectrum for a tripeptide library of the general form (D)-Glu-Xxx-Xxx-$CONH_2$, where Xxx represents one of 26 different residues [10 naturally occurring L-amino acids, 5 (D)-amino acids, and 11 synthetic residues] [31]. Although 676 peptides are expected in the mixture, only 300 of these have unique molecular weights [32]. A comparison of the experimental mass spectrum and a calculated spectrum generated by assuming that all components are present in equal abundance and have equal ionization efficiencies is also shown in Figure 10.3. The comparison suggests that, in general, library synthesis must have proceeded as expected. In several regions, experimental peaks are substantially larger (or smaller) than the calculated intensities. From the mass spectrum alone it is unclear whether these variations arise because there are more (or fewer) components present within the *m/z* peak than expected from the synthesis, or because the relative abundance of one or more components is higher (or lower) than expected. In several cases, expected peptides appear to be missing altogether. For example, the calculated spectrum contains a peak at $m/z = 378.4$ that corresponds to the expected (D)-Glu-*D*-Ala-TIC, (D)-Glu-TIC-(*D*)-Ala, (D)-Glu-β-Ala-TIC and (D)-Glu-TIC-β-Ala peptides [31]; however, only a very small peak is observed at this *m/z* ratio in the experimental spectrum. In all, nine peaks at *m/z* ratios that are unique to peptides containing tetrahydroisoquinoline-3-carboxylic acid (TIC) are expected but only three are observed experimentally. These results suggest that problems arise in the synthetic coupling steps or workup of some TIC peptides. Many other peaks associated with TIC peptides are found at *m/z* ratios where other peptides are also present. Slight reductions in the intensities of some of these peaks would be consistent with missing peptides; however, small variations in abundance or ionization efficiency would also explain these results [33].

Substantially more information about the number of different isomers (and true relative abundances of different isomers) can be obtained by mobility separation of the ion mixture prior to *m/z* analysis. Figure 10.4 shows a narrow range ($m/z = 414$–462) of the

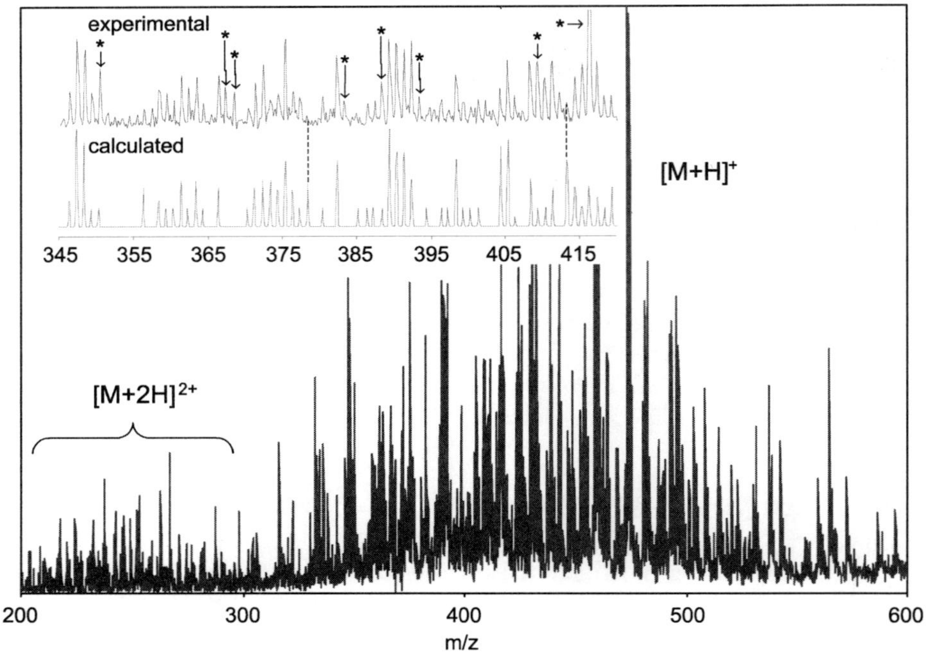

FIGURE 10.3. TOF mass spectrum recorded upon electrospraying the (D)-Glu-Xxx-Xxx-$CONH_2$ tripeptide library which is expected to contain 676 different components. The inset shows a comparison of an expanded region of the experimental data with a mass spectrum that was calculated assuming that all peptides from the synthesis are present in equal abundance and have identical electrospray ionization efficiencies. The dashed lines show regions where several expected peptides appear to be missing. The arrows and asterisks indicate regions where experimental ion intensities are substantially greater than expected based on the number of isobaric species. The less intense distribution of peaks in the $m/z = 200$–300 region primarily corresponds to doubly protonated forms of the tripeptides. (Reproduced from ref. 28.)

two-dimensional IMS–TOFMS dataset obtained for the (D)-Glu-Xxx-Xxx-$CONH_2$ mixture. Ion mobility slices at specific m/z ratios provide a direct measure of the number and relative abundance of isomers and permit the synthesis to be examined in more detail. At $m/z = 460.4$, four peaks with similar intensities are observed at 20.2, 21.4, 22.1, and 22.9 ms. In addition, a small shoulder is observed in the mobility distribution at 22.7 ms. At this m/z ratio seven peptides are expected, including a pair of (D)-Glu-TIC-Chex and (D)-Glu-Chex-TIC sequence isomers. The observation of only five resolved peaks in the ion mobility slice is consistent with the absence of two peptides (possibly the TIC sequence isomers). Because the other peptides that are expected at 460.4 contain residues that are generally represented as expected throughout the library, we suspect that TIC-containing peptides are missing. The ion mobility slice obtained for $m/z = 459.3$ suggests that the peptides in this region are represented in the library as expected. In this mobility distribution there are four peaks with nearly equal intensities. In this case, the ion mobility data provide clear evidence for the two expected TIC sequence isomers [(D)-Glu-*3*-Ala-TIC and (D)-Glu-TIC-*3*-Ala], even in the presence of the [(D)-Glu-(D)-Asn-*N*-Ala and (D)-Glu-*N*-Ala-(D)-Asn] peptides that are also expected at this m/z ratio. In many of the cases involving apparent deletions associated with the TIC residue, problems with

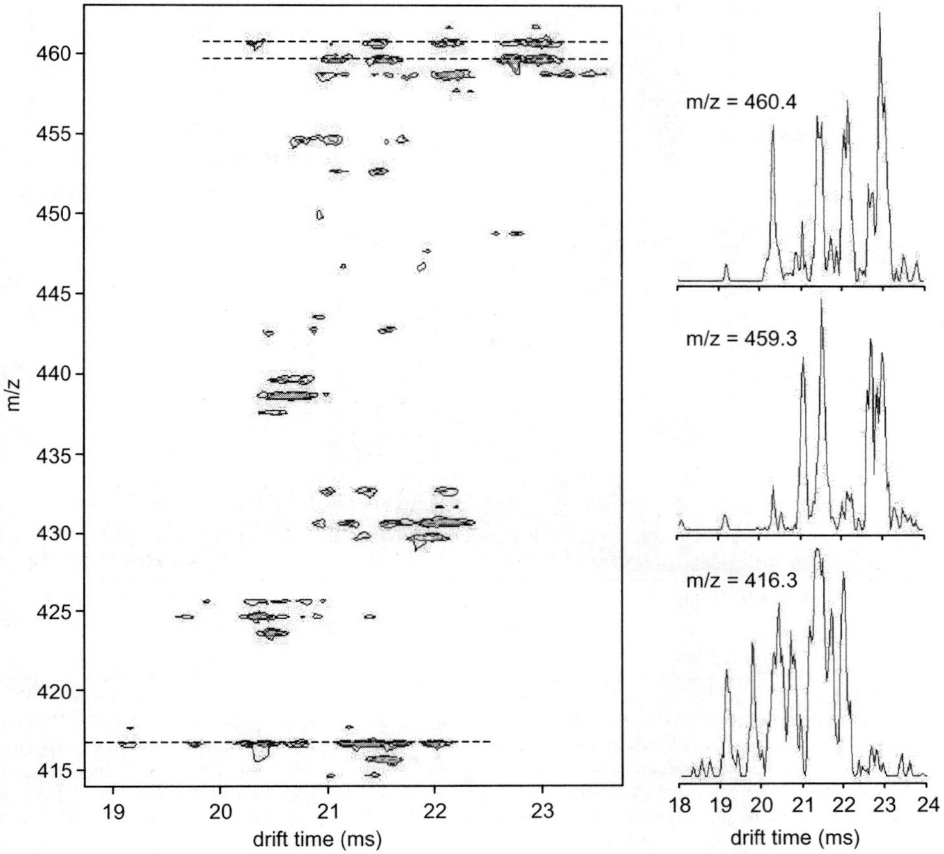

FIGURE 10.4. Expanded region of the nested IMS–TOFMS data for the (D)-Glu-Xxx-Xxx-CONH$_2$ library over an *m/z* range of 414–462. The contours are shown on a nine-point scale; many features with fewer than nine counts are not apparent in this plot. The horizontal dashed lines show the positions of slices through the contours at *m/z* = 416.3, 459.3 and 460.4 which correspond to the ion mobility distributions that are shown (right). (Reproduced from ref. 28.)

synthetic steps can be rationalized by considering side reactions that are known in the chemistry of cyclic amino acids (e.g., Pro, which should display similar reactivity). Residues such as TIC and Pro are prone to numerous side-chain and backbone cyclization and cleavage pathways that would compete with desired products [34].

The ion mobility slice for *m/z* = 416.3 suggests that eight different peptides are present; seven features are resolved and one is substantially broader and more intense than is expected for a single peptide. Only six peptides are expected at this *m/z* ratio in the library. In addition, there is a large discrepancy between the relative peak intensities at this *m/z* ratio in the experimental and calculated mass spectra. The eight peaks in the ion mobility distribution suggest the presence of at least two synthetic contaminants. In this region (and others) it appears that a primary source of contamination is the incomplete removal of side-chain protecting groups. The additional peaks at *m/z* = 416.3 are consistent with the presence of (D)-Glu-Pro-Asp(OtBu) and (D)-Glu-Asp(OtBu)-Pro species ($m/z = 416.3$) that arise from incomplete removal of the *t*-butyl protecting group (OtBu) bound to the carboxylic acid side chain of the Asp residue. Peaks in regions of the spectrum corresponding to deprotected forms of these Asp peptides have lower intensities than expected. Other residues that are often observed in their protected forms in the library include hydroxyproline, serine, and, to a lesser extent, tyrosine. All of these residue side chains are protected using a *t*-butyl group during synthesis. Removal of the *t*-butyl protecting group from hydroxyl-containing side chains is known to be a relatively difficult step [34], consistent with our experimental findings. Our results suggest that a simple means of improving the purity and equalizing the relative abundances of at least 30 peptides (which are now found as *t*-butyl ethers) would be to utilize longer deprotection times or a different protection strategy.

Identification of RNase S-Protein Binders

Although it is possible to find some peptides that bind to a given target by screening large uncharacterized libraries, screening results which involve the analysis of only positive "hits" should be interpreted cautiously [35]. It is unknown whether components from these libraries that are found to bind have binding affinities that are truly greater than all other expected components, or whether the hits simply correspond to components that were present in much greater concentrations owing to variations in solubility or synthetic yield.

We have used IMS–TOFMS to examine libraries that contain numerous peptide sequences that have been reported to bind to the RNase S-protein [36,37]. Following the initial characterization, the library was screened against the S-protein using an affinity chromatography approach. Binding was assessed by analysis of the components that pass through the column, as well as the bound species that were released by washing the column with dilute acid. Four nonisomeric 24-component octapeptide libraries were synthesized, characterized using IMS–TOFMS, and then combined to create a well-defined mixture of 96 related peptide sequences. The 96 components in the larger library can be described by the general sequence NH_2-$GX_1X_2FX_3X_4X_5G$-CO_2H, ($X_1 =$ L, F, V, and Y; $X_2 =$ N and F; $X_3 =$ E, V, and T; $X_4 =$ V and L; $X_5 =$ V and L).

Figure 10.5 shows the IMS–TOFMS data for the 96-component library mixture. Several distinct features are readily apparent in the data, including evidence for peptide ion monomers $[M + H]^+$ and $[M + 2H]^{2+}$ (where M stands for the distribution of eight residue peptides that are expected to be present in the library), doubly charged peptide dimers $[2M + 2H]^{2+}$, triply charged peptide trimers $[3M + 3H]^{3+}$, and families of peptides that arise from a failure of the combinatorial synthesis which leads to two- or three-residue deletion peptides $[M - 2X + H]^+$ and $[M - 3X + H]^+$, respectively [38]. From comparisons of calculated m/z ratios with the experimental values we find that the family of $[M + H]^+$ peaks in Figure 10.5 shows evidence for all 96 expected peptides. It is important to note that even for sequences where some fraction of the expected peptides are observed as deletion peptides, most of the synthetic steps have proceeded to near completion. That is, all of the expected eight residue peptides (that would be influenced by deletion peptides) are present in the $[M + H]^+$ family, albeit in some cases in slightly lower abundance (relative to the majority of sequences that show no evidence of deletion peptide peaks).

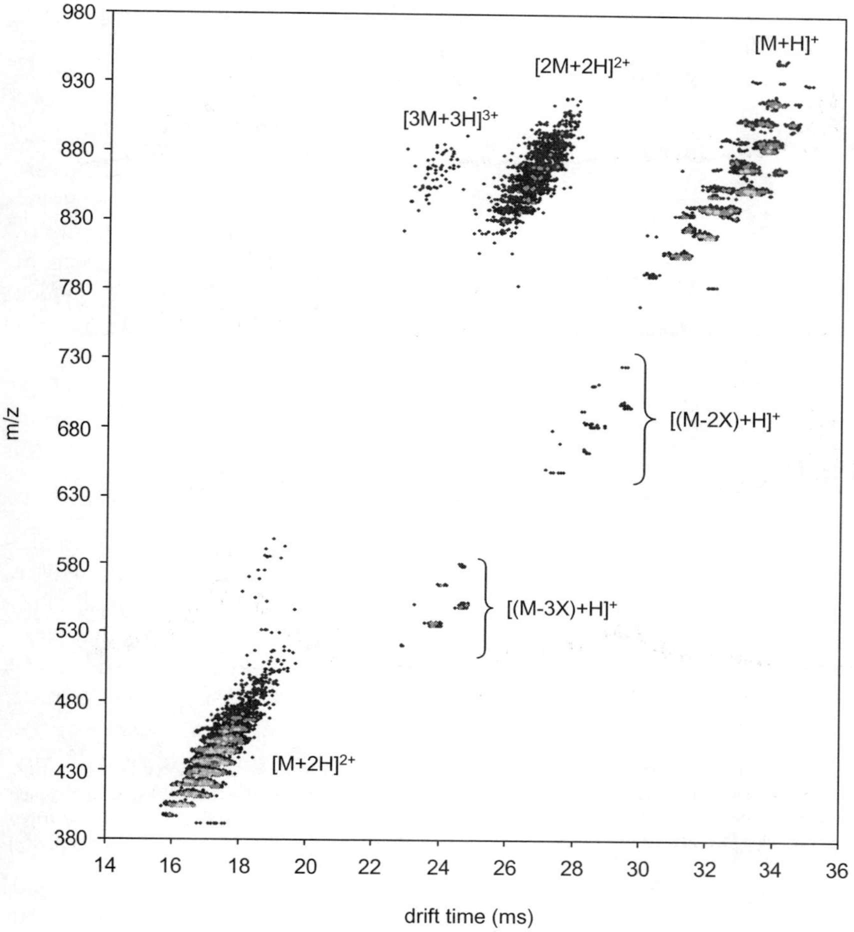

FIGURE 10.5. Nested IMS–TOFMS data for a 0.25 mg mL^{-1} (total peptides) solution of the NH_2-$GX_1X_2FX_3X_4X_5G$-CO_2H library mixture in 49 : 49 : 2 water : acetonitrile : acetic acid (equivalent amounts of libraries 1–4). These data were collected using a helium buffer gas pressure of 158.64 Torr and a drift field of 137 V cm^{-1}. Monomeric peaks corresponding to $[M+H]^+$, $[M+2H]^{2+}$ parent peptide ions, and $[M-nX+H]^+$ ions for two and three amino acid deletion peptides are observed. In addition, lower intensity distributions are present that correspond to $[2M+2H]^{2+}$ and $[3M+3H]^{3+}$. See text for discussion. (Reproduced from ref. 40.)

Examination of the relative peak intensities in the various charge-state families of different datasets reveals some variability that appears to depend on the ESI conditions. However, for a given set of conditions, the relative peak intensities within a family and the relative distributions of the different families are very reproducible. This can be confirmed by examining the mass spectral intensities obtained by integrating the two-dimensional data sets over specified regions, such as those shown in Figure 10.6. Integration of a region including only singly charged ions removes contributions to intensities from doubly charged dimers and triply charged trimer ions that overlap along the *m/z* dimension.

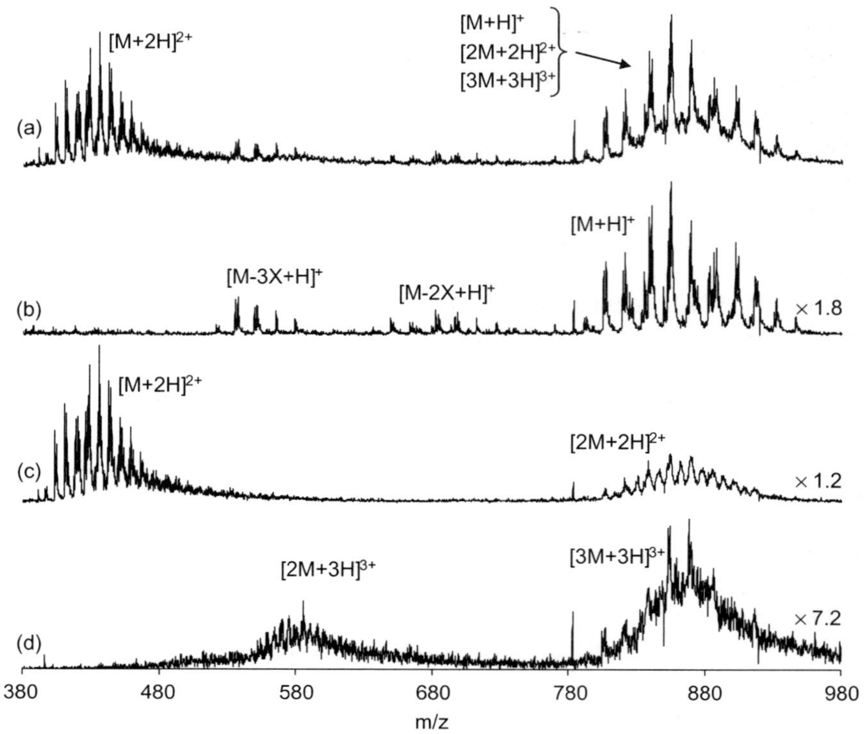

FIGURE 10.6. Integrated mass spectra obtained by summing various regions of the data shown in Figure 10.5. (a) Integration across all drift times—equivalent to the complete mass spectrum. (b), (c), (d) Mass spectra obtained by integrating the ion intensity over regions of the nested data set that includes only singly, doubly, or triply charged ions, respectively. The intensities of these data have been scaled as indicated. (Reproduced from ref. 40.)

All 96 expected library peptides that are observed in the $[M + H]^+$ distribution are also observed as $[M + 2H]^{2+}$ ions.

Characterization of the 96-component mixture prior to screening is critical to the binding analysis that follows. Analysis of normalized peak intensities in the ion mobility data for the 24 component libraries provides an assessment of the relative abundance of each isomer in the larger 96-component mixture (data not shown). In addition, the mobility separation is important for removing contributions to mass spectral peak intensities from multiply charged multimers.

Binding of the library peptides to the S-protein was assessed via affinity chromatography [39]. The screening of a mixture of components provides information regarding relative binding affinities because the fraction of bound ligand should increase according to ligand–target binding affinities. Other important parameters that affect ligand binding include equilibration time, column geometry, and flow rate [39]. These should be identical for each component of the mixture when screening is performed in a single experiment.

A comparison of the integrated mass spectrum for the $[M + H]^+$ ion family in the 96-component mixture obtained prior to affinity screening with the spectra for the retained

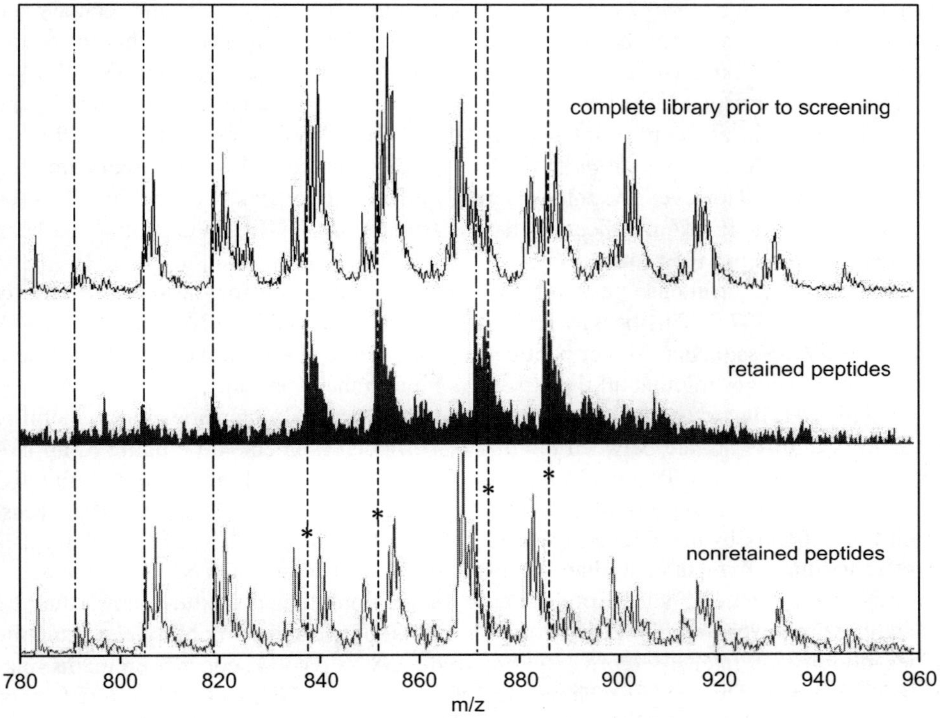

FIGURE 10.7. Integrated mass spectra for the $[M + H]^+$ family obtained from the IMS–TOFMS data of the 96-component library mixture prior to screening (top), the fraction of peptides retained on the affinity column during the screening (middle), and the nonretained peptides which passed directly through the affinity column (bottom). The dashed lines indicate selected regions where the data for the retained and nonretained fractions differ. Peptide sequences that are present in the retained peptide fraction and completely depleted in the nonretained fraction are indicated by both a dashed line and an asterisk. Sequences observed in both the retained and nonretained peptide fractions are indicated by a chain line. (Reproduced from ref. 40.)

and nonretained peptide fractions is shown in Figure 10.7. The IMS–TOFMS analysis of all three screening fractions provides an important cross check of retained and nonretained peptides against the original peptide library. In many cases, peptide sequences that bind to the S-protein are observed in the spectrum for the retained peptides but are completely depleted in the nonretained peptide fraction. Often, the depleted regions correspond to *m*/*z* values where multiple isomeric peptides are expected. The complete absence of a peak in the nonretained fraction indicates that all sequence isomers must have bound to the S-protein. Peaks at *m*/*z* values of 837.9, 852.0, 852.9, and 886.0 are significantly depleted in the data set for the nonretained peptides (relative to the data for the entire library). In all cases, high-intensity peaks at these *m*/*z* values are observed in the data set for the retained peptides.

The data for the different affinity fractions suggests three general categories of peptide binders: low-affinity peptides that are partially retained; groups of isomeric peptides where only particular isomers within a group bind; and groups of isomers that all bind strongly. Peaks at $m/z = 790.7$ are observed in the mass spectra for both the retained and nonretained fractions. Only the $[\text{GVNFVVVG} + \text{H}]^+$ ion is expected at this m/z ratio. Because peaks are observed at this m/z ratio in both affinity fractions, the GVNFVVG sequence must correspond to a low-affinity sequence that is only partially retained by the S-protein. The ratio of the peak intensity for the GVNFVVVG peptide at $m/z = 790.7$ to that for the peptide sequence at $m/z = 792.7$ is approximately 1:1 in the library mixture prior to screening. However, the intensity ratio for these two peaks is roughly 1:2 in the nonretained peptide fraction, indicating that the amount of GVNFVVVG peptide has been depleted by binding to the protein.

Alternatively, an intense peak is observed in the spectrum for the retained affinity fraction at $m/z = 837.9$. This peak, which corresponds to the GVFFVLVG, GLFFVVVG, and GVFFVVLG sequence isomer peptides, is completely absent in the nonretained peptide spectrum, suggesting that all three isomers bind to the S-protein.

Figure 10.8 illustrates the use of normalized relative peak intensities in the mobility distribution to differentiate between binding of particular isomers and binding of all isomers at a given m/z ratio. Two peaks are observed at 30.7 and 32.1 ms in the mobility distribution obtained for the 96-component mixture at $m/z = 820.8$. The peak at 32.1 ms, which corresponds to the GLNFTLVG, GVNFTLLG, and GLNFTVLG sequences, is about four times more intense than the peak at 30.7 ms for the GVNFEVVG peptide. Analysis of the same m/z slice from the data for the nonretained mixture shows that the intensity of the lower-mobility peak is only twice that of the higher-mobility peak, and the higher mobility ion intensity (corresponding to the GVNFEVVG sequence) remains relatively unchanged. This suggests some depletion of the GLNFTLVG, GVNFTLLG, or GLNFTVLG peptides following screening against the S-protein; the GVNFEVVG peptide abundance does not appear to change.

The ion mobility distribution at $m/z = 820.8$ for the mixture of retained peptides contains a low-intensity peak at 32.1 ms. This supports the idea that at least one of the GLNFTLVG, GVNFTLLG, and GLNFTVLG isomers binds to the S-protein. It is also possible that the observed peak-intensity variations are caused by low-level binding of all three isomers.

The ability to analyze the library prior to screening and to identify both retained and nonretained peptides provides checks that appear to be very important for interpretation of library screening results. Surprisingly, we find that the previously reported sequences bind with lower affinities than a series of related peptides that have significantly more nonpolar character [40]. These results suggest that these high-affinity nonpolar peptides were not present (or present in very small abundance) in the initial screens of larger libraries [37] and illustrate the importance of approaching binding results from large uncharacterized mixtures with caution.

HPLC–ION MOBILITY–TIME-OF-FLIGHT TECHNIQUES

General

Gas phase separation of peptide isomers in combinatorial libraries using an IMS–TOFMS approach provides valuable information for characterizing small to moderately sized library mixtures. However, as the number of library components increases, the

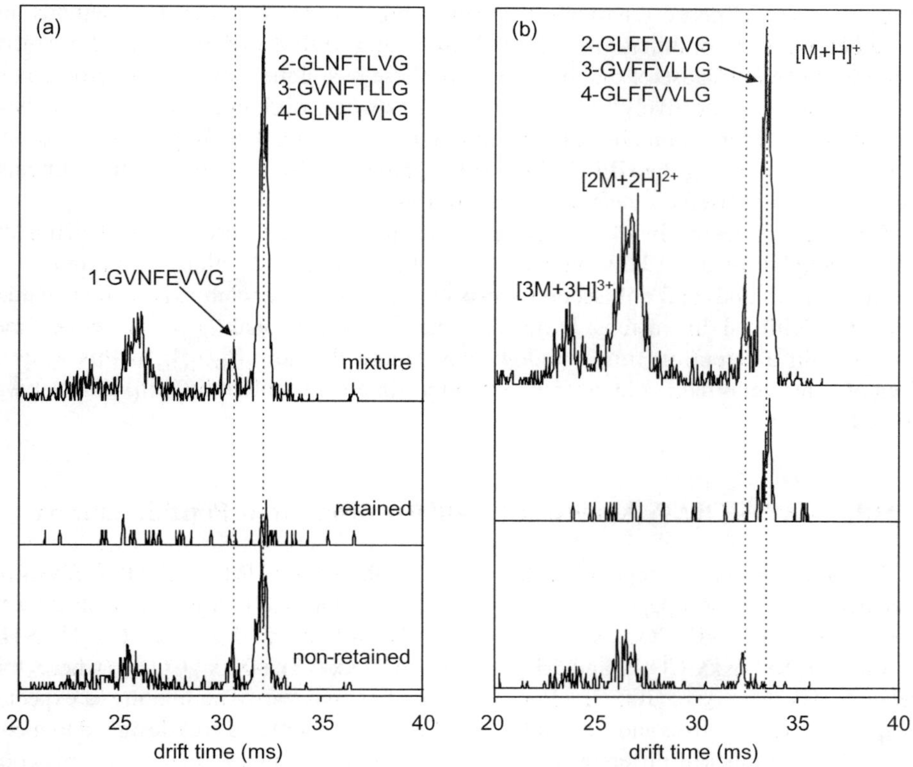

FIGURE 10.8. Ion mobility distributions obtained from the nested data sets for the library mixture prior to screening (top), the retained peptide fraction (middle), and the nonretained peptide fraction from the affinity screening (bottom). (a) Spectra generated by integrating the ion intensity for $m/z = 820.8$ over all drift times. Drift times from the different data sets have been normalized to 160 Torr. Peak intensities have been normalized based on the total ion intensity of each nested IMS–TOFMS data set. Drift times for peaks observed in the ion mobility distributions for the 96-component library mixture prior to screening are indicated using vertical dashed lines. Peak assignments given in the figure were obtained from the ion mobility distributions of the nonisomeric 24-component libraries. Based on these assignments, the higher-mobility peak at 30.6 ms corresponds to the $[M + H]^+$ ion of the GVNFEVVG peptide and the lower-mobility peak at 32.1 ms corresponds to a mixture of $[M + H]^+$ ions of the three sequence isomers GLNFTLVG, GVNFTLLG, and GLNFTVLG. (b) Ion mobility distributions at $m/z = 852.0$ for the various library screening fractions described above. The peak at 32.4 ms corresponds to an unidentified library impurity and the peak at 33.4 ms corresponds to the $[M + H]^+$ ions for the GLFFVLVG, GVFFVLLG, and GLFFVVLG peptide sequences. (Reproduced from ref. 40.)

likelihood that isomeric library components will have similar or undistinguishable mobilities also increases. For complex mixtures of closely related components the incorporation of orthogonal condensed phase separations into the IMS–TOFMS sequence can significantly increase the overall peak capacity of the approach [41].

In the multidimensional experiment, library peptide ions are electrosprayed directly from the HPLC effluent into the entrance of a high-pressure drift tube that is coupled to a reflectron geometry TOFMS. Because of the relatively low density of the buffer gas (compared with condensed phase separations), mobilities of ions in the gas phase are high; thus separation can be carried out on millisecond time-scales, making it possible to incorporate the gas phase separation between condensed phase separation methods and MS detection. The combination of HPLC, IMS, and TOFMS (demonstrated recently) [42,43] is feasible because of the time-scales of these techniques (minutes, microseconds, and milliseconds, respectively). We find many cases where different peptide isomers have identical liquid chromatography (LC). retention times, making them impossible to resolve by LC–MS alone. Using the HPLC–IMS–TOF approach, it is often possible to distinguish these components based on their separation as ions in the gas phase.

Positions of peaks in the nested drift (flight) time data are reported using the $t_D(t_f)$ nomenclature described above; unless otherwise noted values of t_D and t_f are given in microseconds and milliseconds, respectively. The nomenclature is easily extended to include additional dimensions. In this case the two-dimensional $t_D(t_f)$ data set becomes nested within a retention time window. We denote this as $t_r[t_D(t_f)]$. In this system, values of t_r are reported as the average retention time over which the $t_D(t_f)$ data set was acquired.

HPLC–IMS–TOFMS Analysis of a 4000-Component Peptide Library

Condensed phase LC separations have been combined with IMS-TOFMS analysis for the characterization of a peptide library expected to contain 4,000 different peptides with the general form: NH_2-Xxx-Xxx-Xxx-CO_2H, NH_2-Ala-Xxx-Xxx-Xxx-CO_2H, NH_2-Ser-Ala-Xxx-Xxx-Xxx-CO_2H and NH_2-Leu-Ser-Ala-Xxx-Xxx-Xxx -CO_2H (where Xxx represents Ala, Arg, Asp, Glu, Gly, Leu, Lys, Phe, Ser, and Val). The mixture is expected to contain three, four, five, and six residue peptides (1000 each) and was designed to incorporate a large number of different sequence and structural isomers. Analysis of this complex library emphasizes the application of the combined HPLC–mobility dimensions for separation of up to ~20–40 isomers at each m/z ratio. For analysis of the library, individual nested $t_D(t_f)$ datasets were obtained for 147 different 1 min retention time windows during a 170 min chromatographic run.

Figure 10.9 shows a nested $t_D(t_f)$ dataset for direct ESI analysis of the 4000-component peptide library. There are three general families of ions in the distribution. Based on previous work [44] and the isotopic spacing we assign the series of peaks from $m/z = 190$ to $m/z = 750$ as $[M+H]^+$ ions. A family of higher-mobility $[M+2H]^{2+}$ ions is observed from m/z values of ~280 to 400. A significantly lower intensity series of peaks that corresponds to ions with relatively high mobilities is observed from m/z values of ~ 350 to 750. This family can be assigned to multiply charged multimers of the general form $[2M+2H]^{2+}$. These $[2M+2H]^{2+}$ peaks appear at m/z ratios that are identical with many of the $[M+H]^+$ peptide ions; however, it is possible to distinguish them based on differences in their experimental drift times.

A typical mobility distribution is shown in Figure 10.9 for $m/z = 533.6$. From $m/z = 533.5$ to $m/z = 533.7$ it is expected that 39 different peptides will be present. Experimental drift times for the $[M+H]^+$ ions at $m/z = 533.6$ vary from ~17.5 to 19.2 ms—a range of ~10%. The broad feature suggests that many peptides with similar mobilities are present but are not resolved in the ion mobility separation.

Figure 10.9 shows several mass spectral slices across the different regions of the $t_D(t_f)$ data set. Integration across all drift times yields the mass spectrum for all ions. In this plot,

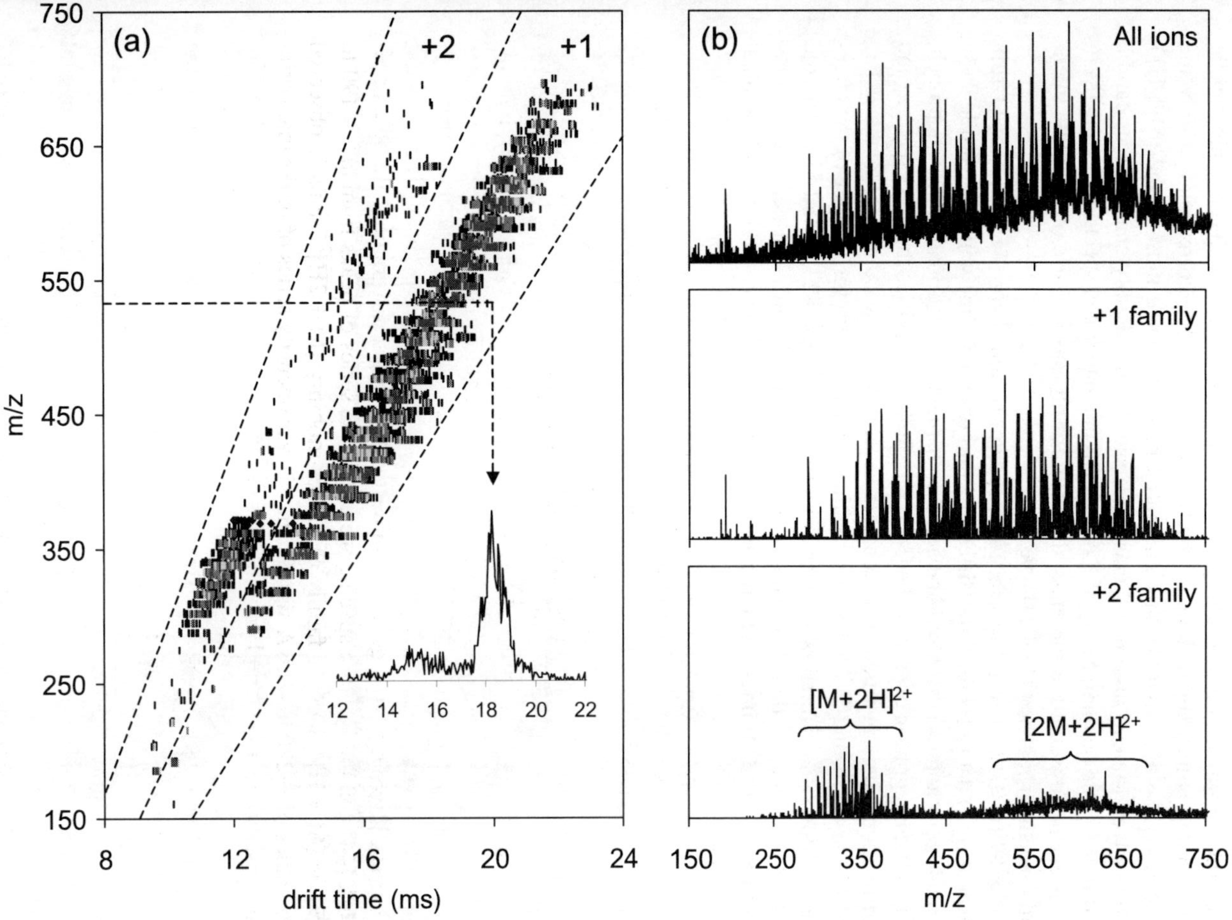

FIGURE 10.9. (a) The nested $t_D(t_f)$ data obtained for direct ESI of the 4000-component peptide library mixture. These data were recorded using a helium buffer gas pressure of 154.2 Torr and a drift field of 170.97 V cm^{-1}. Diagonal dashed lines indicate regions of the data set corresponding to low-mobility singly charged and higher-mobility doubly charged peptide ions. A sample ion mobility distribution obtained by integrating the $t_D(t_f)$ data across all drift times is also shown for $m/z = 533.6$. Drift times for library isomer ions at this m/z ratio vary from ~17.5 to 19.2 ms. (b) TOF mass spectra obtained by integration over the diagonal drift (flight) time regions indicated on the $t_D(t_f)$ plot for the +1 and +2 ion families. Integration over all drift times yields an all-ion mass spectrum that is similar to spectra obtained in conventional ESI–MS experiments (top). Integration over the specified regions of the $t_D(t_f)$ data indicated in the figure gives TOF mass spectra for the singly charged ion family at longer drift times (middle) and the higher-mobility doubly charged ion family (bottom). (Reproduced from ref. 43.)

peaks for the $[M + H]^+$ ions are superimposed on the lower-intensity $[2M + 2H]^{2+}$ peaks, making it difficult to resolve individual ions. In addition many $[M + 2H]^{2+}$ peaks are obscured by more intense features associated with $[M + H]^+$ ions. Contributions to the mass spectrum from the doubly charged $[M + 2H]^{2+}$ and $[2M + 2H]^{2+}$ ions can be removed by integrating over regions of the nested data set that correspond to only the lower-mobility $[M + H]^+$ ions. In this integrated mass spectrum most of the $[M + H]^+$ peaks are resolved to near baseline. A mass spectrum for the doubly charged ions can be generated in the same manner. Many of the peaks in this integrated spectrum for the $[M + 2H]^{2+}$ ions are not observed in the mass spectrum for all ions because of their low intensity relative to superimposed $[M + H]^+$ peaks.

Figure 10.10 shows several typical mass spectra obtained for $[M + H]^+$ ions (from integration of $t_D(t_f)$ datasets) for three different 1 min retention time windows. While these data are analogous to those that would be acquired using a conventional LC–MS approach, we note that the integration over only the $[M + H]^+$ family simplifies the mass spectra (and the interpretation). It is instructive to consider several peaks as examples. Peaks at *m/z* values of 423.4, 452.5, and 517.6 are observed in the mass spectra for all three of the retention time windows that are shown. Over *m/z* ranges of 423.4–423.5, 452.5–452.6, and 517.6–517.7 we calculate that there should be 24, 12, and 24 different peptides, respectively. Over the 147 LC frames that have been recorded, we observe an *m/z* = 423.4 ion at

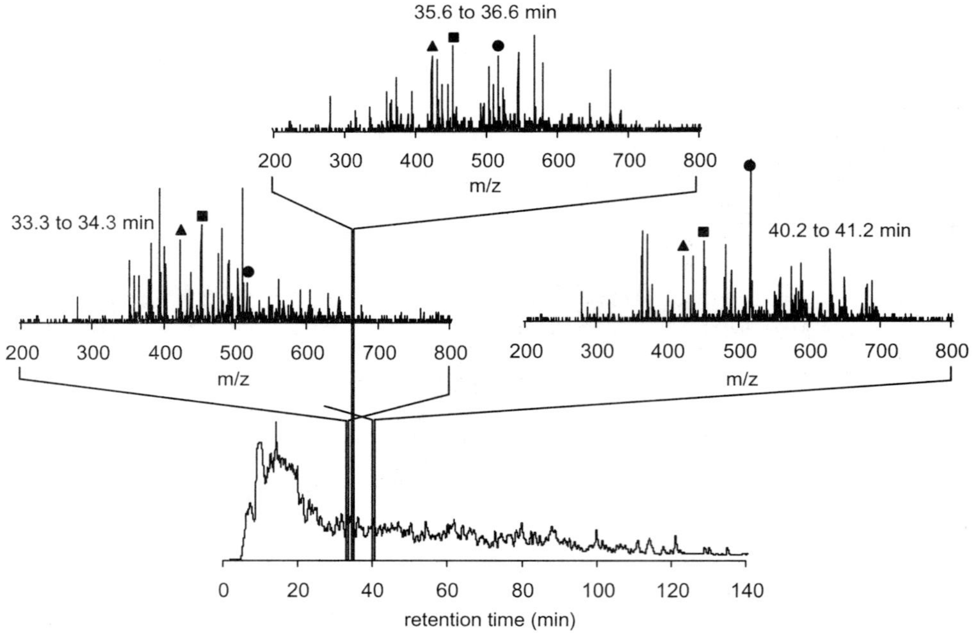

FIGURE 10.10. LC–MS data for three different 1 min retention time windows of the reversed-phase HPLC separation of the 4000-component library. The mass spectra were obtained by integration of the $t_D(t_f)$ data for the following retention time windows: t_r = 33.3–33.4 min (left), t_r = 35.6–36.6 min (center), and t_r = 40.2–41.2 min (right). Peaks at 423.4 (▲), 452.5 (■) and 517.6 (●) are indicated on the plots. The location of these t_r windows with respect to the full retention time range is indicated on the UV chromatogram. (Reproduced from ref. 43.)

10 different retention times. The $m/z = 452.5$ and $m/z = 517.6$ ions are observed at seven and eight different retention times, respectively. The observation that we resolve peaks at the same m/z ratio in multiple LC windows indicates that at least some of the sequence isomers must be separated in the HPLC. However, we note that fewer than 50% of the peptides are resolved on the basis of retention time differences alone.

An example $t_D(t_f)$ distribution recorded from $t_r = 43.7$ min to $t_r = 44.7$ min is shown in Figure 10.11. The additional peak capacity of the two-dimensional IMS–TOFMS approach is illustrated by comparing an integrated mass spectrum for the $[M + H]^+$ peptides with the two-dimensional $t_D(t_f)$ distribution. In the mass spectrum only approximately 23 peaks are resolved; however, the $t_D(t_f)$ data show evidence for 35 resolved $[M + H]^+$ peaks. In a number of cases, it appears that it is possible to separate isomers that were not resolved in the HPLC separation. For example, an ion mobility distribution at $m/z = 510.5$ shows evidence of two different isomers with drift times of 17.2 and 18.2, respectively. A very low intensity spike at $t_D = 17.7$ may also correspond to a peptide. It is generally possible to differentiate low-intensity peaks in the mobility distributions from noise based on experimental peak widths. Peaks due to library peptides should have widths

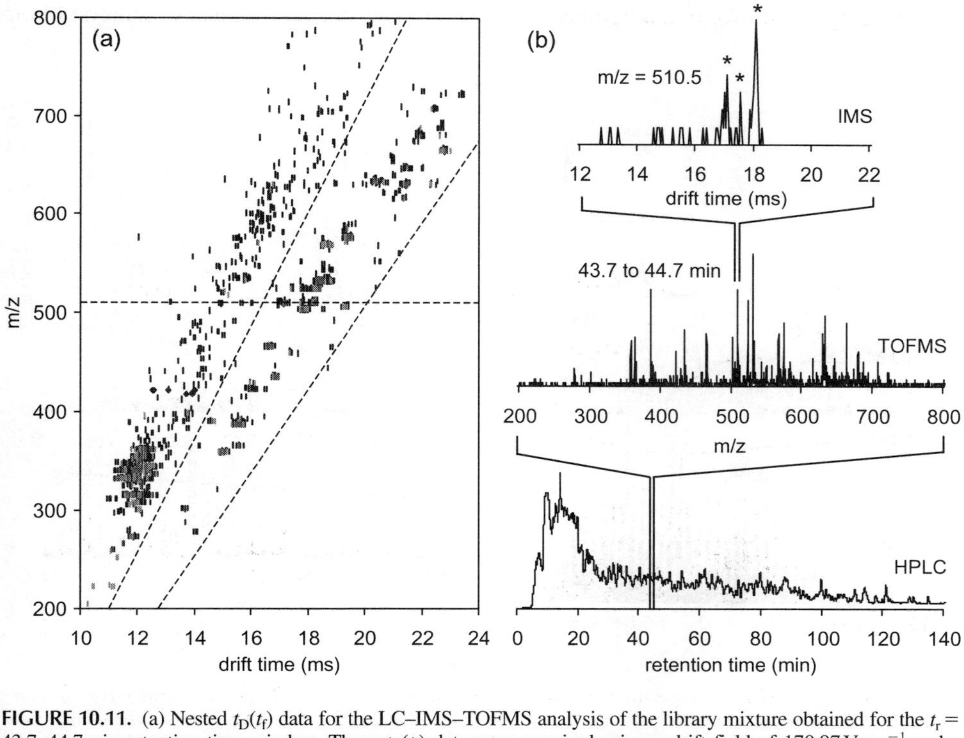

FIGURE 10.11. (a) Nested $t_D(t_f)$ data for the LC–IMS–TOFMS analysis of the library mixture obtained for the $t_r =$ 43.7–44.7 min retention time window. These $t_D(t_f)$ data were acquired using a drift field of 170.97 V cm^{-1} and a helium buffer gas pressure of 156.11 Torr. (b) Integrated data obtained from the nested $t_D(t_f)$ dataset for the individual IMS and TOF dimensions. An integrated TOF mass spectrum for the $[M + H]^+$ ions is shown with respect to the $t_r = 43.7$–44.7 min retention time window of the chromatographic separation over which the $t_D(t_f)$ data set was acquired (middle). In addition, an example drift time distribution for $m/z = 510.5$ is shown for this retention time window to illustrate the mobility separation of isomeric peptide ions (top). (Reproduced from ref. 43).

consistent with the typical resolving power of the high-pressure drift tube used in these studies ($t/\Delta t = 100$–150). Analysis of this single $t_D(t_f)$ window provides evidence for multiple $[M + H]^+$ peptides at m/z values of 510.5, 524.6, 532.6, and 631.8. Additionally, multiple peptides appear to be resolved for several $[M + 2H]^{2+}$ peaks. For example, at $m/z = 333.0$, three distinct peaks are resolved.

It is also interesting to examine the simplification of the mass spectra and ion mobility that results from incorporation of the LC separation. Figure 10.12 shows a comparison of the $t_D(t_f)$ data obtained for direct ESI of the library mixture and the data from the $t_r[t_D(t_f)]$ dataset taken at $t_r = 43.7$–44.7 min. The integrated mass spectrum for the $[M + H]^+$ ions from the direct ESI IMS–TOFMS data is significantly more complex than the mass spectrum for the single LC window. Additionally, ion mobility slices across the data sets show that peaks are much sharper in the $t_r[t_D(t_f)]$ dataset. This comparison corroborates the idea presented above—that in the direct $t_D(t_f)$ analysis, the broad features in the ion mobility data were due to many unresolved peptides.

The incorporation of the mobility separation provides several advantages over standard LC–MS experiments. One important feature of the technique involves the separation of mixture components into distinct charge-state families. This separation makes it possible to obtain integrated mass spectra for specific families of ions and permits identification of ion charge states for unambiguous mass assignment. In addition, many isomeric library

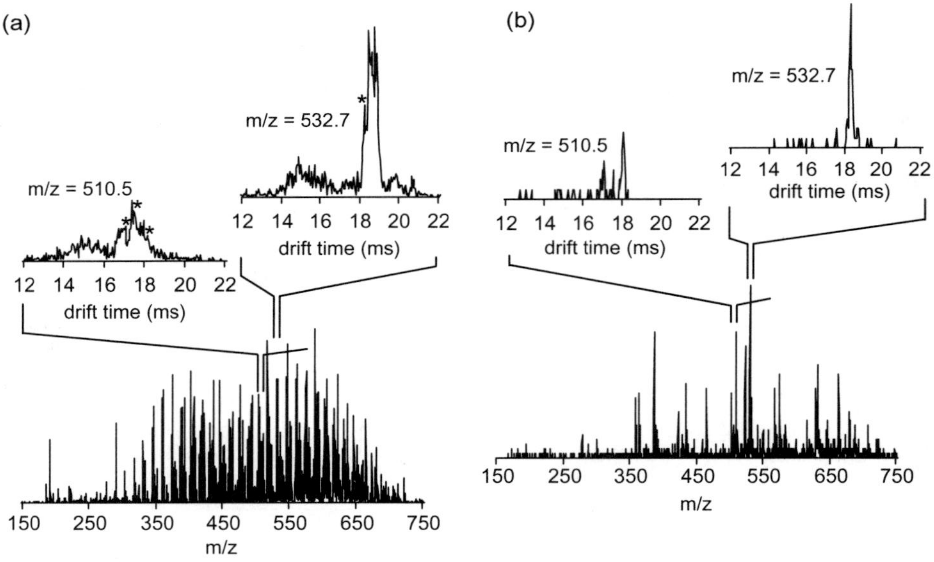

FIGURE 10.12. (a) TOF mass spectrum obtained by integration of the direct ESI $t_D(t_f)$ data ($[M + H]^+$ ions only) for the 4000-component mixture (bottom). Ion mobility distributions for the direct ESI analysis of the library mixture are also shown for $m/z = 510.5$ and $m/z = 532.7$ (top). These mobility slices show broad unresolved features that are consistent with the presence of numerous isomeric peptides with very similar mobilities. (b) Integrated $[M + H]^+$ mass spectrum obtained from the $t_r[t_D(t_f)]$ data for the $t_r = 43.7$–44.7 min window. Also shown are the corresponding mobility distributions at $m/z = 510.5$ and $m/z = 532.7$ for this t_r window (top). The features indicated with asterisks in the mobility distributions in (a) correspond to the sharp features observed in the distributions for the $t_r = 43.7$–44.7 min window shown in (b). (Reproduced from ref. 43.)

peptides that have identical LC retention times can be separated based on differences in their gas phase mobilities in the $t_r[t_D(t_f)]$ data.

MOBILITY LABELING FOR PARALLEL COLLISION-INDUCED DISSOCIATION

General Overview

One of the current limitations of IMS–TOFMS and HPLC–IMS–TOFMS strategies for the analysis of libraries involves the identification of mobility-separated isomers. We have previously used comparisons between experimental collision cross-sections and calculated values for model conformers to match peaks in the mobility distributions to specific peptide sequences [45]. This approach requires the use of extensive molecular modeling simulations to generate energy-minimized structures for each peptide expected at a given *m/z* ratio. For large libraries, these theoretical approaches are computationally expensive. In addition, the forcefields employed in the simulations have been developed from limited parameter sets because of the lack of gas phase peptide structural data.

Traditional experimental MS sequencing approaches for peak identification involve mass selection of precursor ions. Because other *m/z* ions are discarded during this *m/z* selection, these strategies are inherently inefficient. Conversely, it is difficult to obtain valuable sequencing information from direct dissociation of mixtures without *m/z* selection because the daughter ions are not uniquely associated with parent ions. A more direct experimental approach involves the use of ion mobility labeling for parallel collision-induced dissociation (CID) of mixtures. Recently, this strategy has been demonstrated in the Clemmer laboratory by introducing an octopole collision cell between the drift tube and the TOFMS [46]. When this configuration is used, peptide fragments created via CID in the collision cell are labeled according to the mobilities of their respective parent ions.

Figure 10.13 shows a schematic diagram of the ion mobility–collision cell–TOF instrument. The instrument incorporates the ion trap injected-ion drift tube described

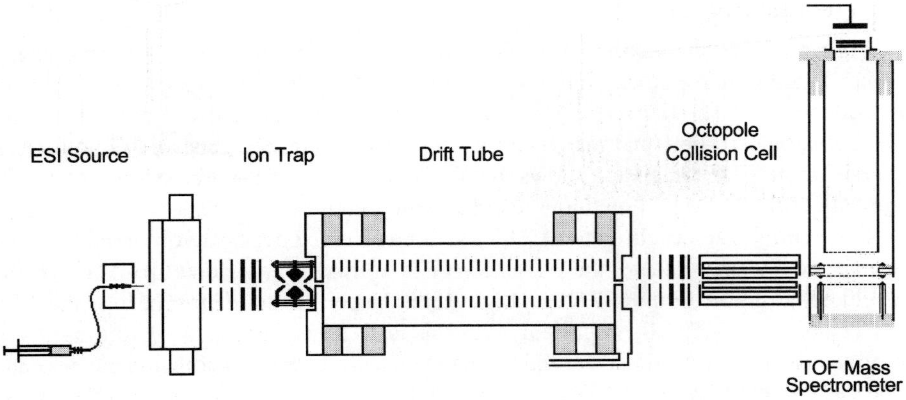

FIGURE 10.13. Schematic diagram of the ion mobility–collision cell–TOFMS instrument used for mobility-labeling CID experiments. (Reproduced from ref. 46.)

above. An octopole collision cell is inserted between the drift tube and an orthogonal linear TOF region. Fragment ions generated via CID in the octopole collision cell are labeled with respect to their precursor ions using the following experimental sequence. Ions formed by ESI are accumulated in the ion trap and injected into the drift tube as described above. Ions that exit the drift tube are focused into the octopole collision cell. The octopole is typically operated at 1.7 MHz and ~540 V (peak-to-peak) and filled with ~0.5 mTorr of argon during CID experiments. Parent and fragment ions exit the collision cell directly into the source region of the TOFMS where high-frequency pulses, synchronous with the mobility pulse, are used to initiate the TOF measurement.

Mobility Labeling of [D-Ala^{2,3}] Methionine Enkephalin Parent and Fragment Ions

This approach is demonstrated by examination of the CID patterns of the $[M+H]^+$ parent and a series of a-, b-, and y-type fragments of [D-Ala^{2,3}] methionine enkephalin that were generated prior to mobility separation. Dissociation of these fragments formed upon injection into the drift tube yields a mixture of several fragment ions that have identical *m/z* ratios. The origin of each fragment ion formed in the octopole collision cell is unambiguous because the experimental drift time for each fragment is nearly identical with that of its parent ion precursor.

Figure 10.14 shows the IMS–TOFMS data for a distribution of [D-Ala^{2,3}] methionine enkephalin ions generated by ESI. These data were obtained using an injection energy of 100 eV (for singly protonated ions) to introduce the ions into the drift tube. In addition to the intact $[M+H]^+$ ion, Figure 10.14 shows a series of a-, b-, and y-type fragment ions generated by dissociation of [D-Ala^{2,3}] methionine enkephalin upon injection into the drift tube. As discussed above, when ions are injected from the low-pressure vacuum chamber into the drift tube, they undergo a rapid heating–cooling cycle. These fragment ions formed at the front of the drift tube are separated from one another based on mobility and *m/z* differences. Also shown in the figure is a TOF mass spectrum obtained by integrating the ion intensity across the drift time dimension. This spectrum is equivalent to a conventional MS–MS spectrum.

Figure 10.15 shows the IMS–TOFMS data and integrated mass spectrum for [D-Ala^{2,3}] methionine enkephalin obtained when ~0.5 mTorr of argon is added to the octopole collision cell. The IMS–TOFMS data obtained at high injection energy with argon gas in the collision cell show that many additional peaks are present (relative to Figure 10.14). Analysis of the TOF dimension shows that the new peaks can also be assigned to expected fragments of the [D-Ala^{2,3}] methionine enkephalin peptide.

For example, at a flight time of 7.85 μs, three resolved peaks are observed at 3.01 ms, 3.24 ms, and 3.94 ms. Each of these ions can be identified as an a_2 ion of [D-Ala^{2,3}] methionine enkephalin based on the 7.85 μs flight time. Experimental drift times for these fragment peaks are essentially identical with the drift times recorded for the a_2, b_2, and unresolved b_3–y_2 pair (3.01, 3.20, and 3.94 ms) observed when no collision gas was added to the gas cell (Figure 10.14). The a_2 ion at 3.01 ms is generated by the injected-ion fragmentation process; however, the a_2 ions at 3.20 ms and 3.94 ms must arise from dissociation of the b_2 ion and the unresolved b_3–y_2 pair, respectively. Every fragment in

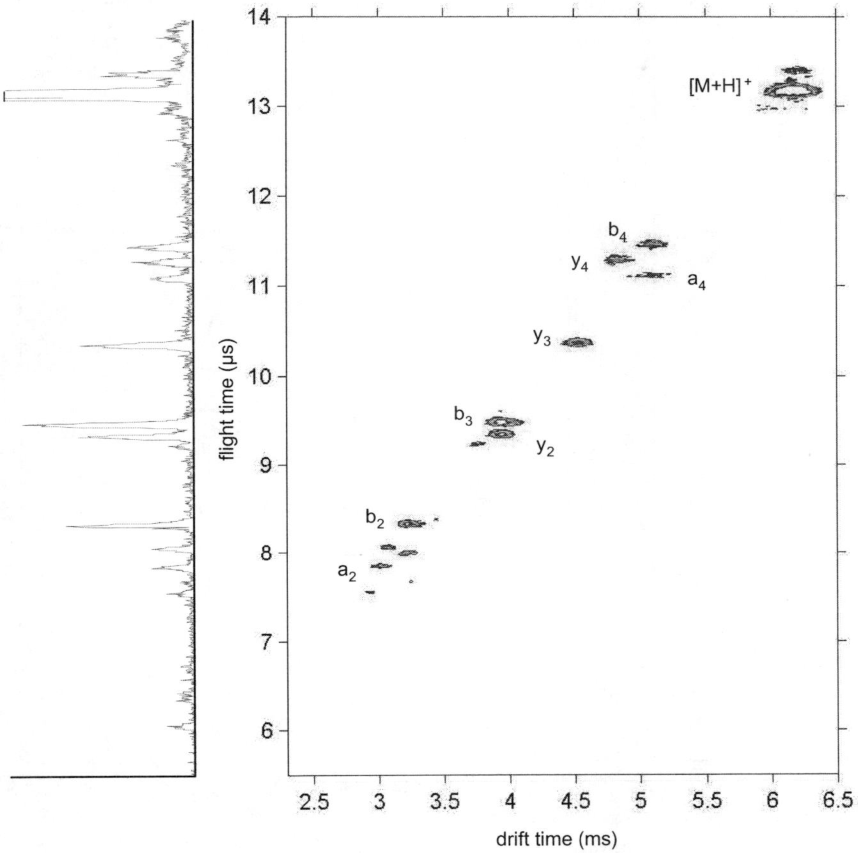

FIGURE 10.14. Two-dimensional contour plot of the nested drift (flight) time data for a mixture of ions formed from electrosprayed [D-Ala^{2,3}]-methionine enkephalin. These data were recorded with no gas in the octopole collision cell. The contours are shown on a 10-point scale; this removes all features that contain fewer than 10 counts. The distribution includes a series of a-, b-, and y-type fragment ions that are formed upon high-energy injection into the drift tube (see text). The drift time axis has been normalized to a helium pressure of 2.70 Torr. Also shown (left) is the TOF mass spectrum obtained by integrating the flight time data over the experimental drift time range. Data are plotted on a normalized intensity scale so that peak heights can be compared directly with the data in Figure 10.15. The resolving power along the flight time axis is typically 200 ($m/\Delta m$ for a singly charged ion, where Δm is determined at half maximum). Along the drift time axis, the resolving power ($t/\Delta t$, where Δt is determined at half maximum) is between 20 and 30 for all peaks. (Reproduced from ref. 46.)

Figure 10.15 corresponds to either a precursor (that did not undergo CID in the gas cell) or a CID product that is correlated with a precursor ion at a specific mobility. By using the mobility of the ions as a label for parallel sequencing, it is possible to sequence the ion mixture unambiguously even when identical fragment ions are produced from different parent ions. The integrated mass spectrum shows that the relative magnitudes of fragment peaks are substantially different from the mass spectrum in Figure 10.14. In particular, smaller fragment ions are more abundant (e.g., $m/z < 250$); an increase in intensities of small fragment peaks is consistent with formation of fragments in the octopole collision cell.

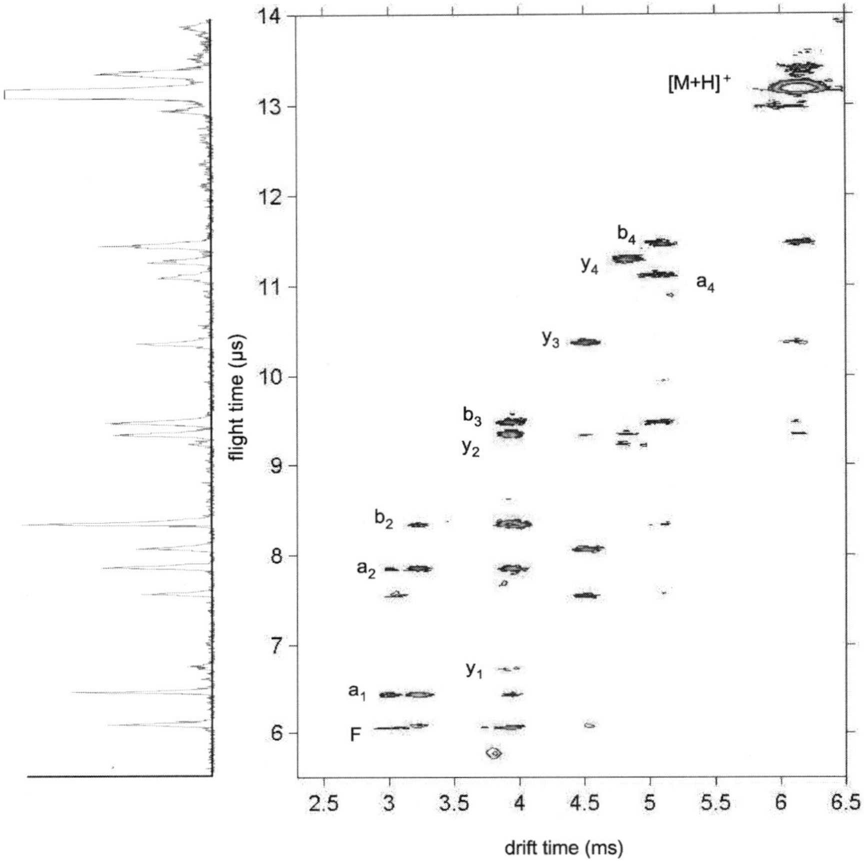

FIGURE 10.15. Two-dimensional contour plot of drift (flight) time data recorded when $\sim 5 \times 10^{-4}$ Torr of argon is added to the octopole collision cell. Fragment ions formed in the collision cell are observed at the drift times of the antecedent parent ions. All other conditions are identical with those in Figure 10.14. (Reproduced from ref. 46.)

Incorporation of an octopole collision cell between the drift tube and TOF regions of the instrument is feasible because drift times in the drift tube are substantially longer (milliseconds) than both the time required for ions to dissociate in the collision cell and flight times in the TOF region (microseconds). The coincident drift times of fragment ions generated by CID in the octopole collision cell and the associated precursor ions can be used to establish the origin of peaks. The parallel sequencing instrument used in these studies utilized a low-resolution injected-ion drift tube configuration. Development of a parallel CID instrument that incorporates a high-resolution drift tube with a high-resolution reflectron geometry time-of-flight mass spectrometer should permit parallel CID analysis of significantly larger mixtures. This configuration would be well suited for the sequencing of mobility-separated isomeric peptide components of combinatorial library mixtures.

A similar approach has recently been described by Stone and co-workers in which mobility-separated peptides are dissociated via surface-induced dissociation (SID) [47] prior to TOF analysis [48]. Recently, our group has extended the parallel-sequencing CID strategy to incorporate *m/z* selection of ions prior to CID in the octopole collision cell [49]. The addition of a quadrupole mass filter between the drift tube and the octopole collision cell enables the selection of ions based on their mobilities and *m/z* ratios prior to CID. Parent and fragment ions that exit the collision cell are analyzed using TOFMS. The approach facilitates mobility labeling of mixture components that have overlapping drift times.

FUTURE DIRECTIONS

The coupling of gas phase ion mobility separations with TOFMS provides a novel approach for the high-throughput characterization of synthetic combinatorial library mixtures. Variations in ion sources and drift tube designs permit the optimization of separations for a variety of different ion types. This chapter emphasizes the use of ESI for characterization of peptide libraries; however, IMS–TOFMS strategies have also been demonstrated for simple small-molecule mixtures. We have recently used an IMS–TOFMS approach to differentiate *ortho*-substituted diarylamines from *meta*- and *para*-substituted isomers in a diarylamine library mixture [50]. In addition, a number of other groups have examined small-molecule mixtures, such as those containing synthetic polymers [51], amphetamines [52], and primary metabolites [53], using IMS.

Key advantages of IMS/TOF strategies include the mobility separation of different charge-state families and isomeric mixture components. The separation of charge states makes it possible to simplify ESI mass spectra by removing signal contributions from common sources of chemical noise, such as highly charged multimers. Specific regions of the two-dimensional $t_D(t_f)$ data can be integrated to emphasize low intensity peaks that would otherwise be buried by signals from higher-abundance ions. Integration of mobility-resolved isomer peaks within a given charge state provides valuable relative abundance information that is not accessible using MS alone.

Incorporation of an octopole collision cell (or other means of dissociation such as an orifice skimmer cone at the back of the drift tube [54] or SID [48] between the drift tube and TOF analyzer) generates characteristic peptide fragments that can be used to sequence mobility-separated ions. The coincidence in drift time between the parent and daughter ions permits unambiguous confirmation of the origin of the fragment peaks. Because the precursor–fragment ion relationship is well defined, it is possible to sequence entire mixtures simultaneously without prior *m/z* selection.

A current limitation of the mobility-labeling parallel-sequencing technology is the loss of sensitivity associated with the multidimensional analysis. Sensitivity issues arise because of several factors. First, the pulsed nature of the ion mobility experiment results in decreased duty cycles. Coupling of an ion trap interface to the injected ion drift tube increases ion intensities by storing ions between mobility pulses. Spatial and temporal focusing in the trap increases the number of ions that enter and exit the drift tube, yielding ion signals that are sufficiently high for subsequent CID. Second, the separation of ions into multiple dimensions results in apparent signal losses, relative to single-dimensional experiments, because the ions are dispersed into multiple bins. To some extent, this constraint will be apparent in any multidimensional technique and can be compensated for by integration of the signal across one or more of the dimensions. We are currently developing mathematical protocols to identify peak positions for low intensity ions in complex multidimensional data sets.

A final consideration for improvement of the mobility-labeling parallel-sequencing strategy involves increasing the resolving power of the drift tube. As noted in Eqn (1), the ion mobility resolving power increases in proportion to $(E/T)^{1/2;}$. Experimental configurations that extend the accessible values of *E*/*T* will have a significant impact on the ability to resolve similar ion structures. Significantly higher mobility resolving power is required for the mobility separation and sequencing of isomeric mixture components. Work is currently under way in our laboratory to couple high-pressure high-resolution mobility separations with parallel-sequencing strategies. Implementation of high-pressure ion storage devices such as the ion funnel interface developed by Wyttenback et al. [16] is promising for optimization of both sensitivity and resolving power.

ACKNOWLEDGMENTS

Support was provided by a grant from the NIH (1R01GM-59145-01) and the Camille and Henry Dreyfus Research Foundation.

REFERENCES AND NOTES

1. Venter, J. C. et al., *Science* **2001**, *291*, 1304.
2. Pandey, A., Mann, M. *Nature* **2000**, *405*, 837; Harry, J. L., Wilkins, M. R., Herbert, B. R., Packer, N. H., Gooley, A. A., Williams, K. L. *Eletrophoresis* **2000**, *21*, 1071; Aebersold, R., Goodlett, D. R. *Chem. Rev.* **2001**, *101*, 269.
3. Yates, N., Wislocki, D., Roberts, A, Berk, S., Klatt, T., Shen, D-.M., Willoughby, C., Rosauer, K., Chapman, K., Griffin, P. *Anal. Chem.* **2001**, *73*, 2941; Lambert, P. H., Bertin, S., Fauchere, J. L. *Comb. Chem. High Throughput Screen.* **2001**, *4*, 317; Loo, J. A. *Eur. Mass Spectrom.* **1997**, *3*, 93; Wigger, M., Nawrocki, J. P., Watson, C. H., Eyler, J. R., Benner, S. A. *Rapid Commun. Mass Spectrom.* **1997**, *11*, 1749; Metzger, J. W., Kempter, C., Wiesmuller, K. H., Jung, G. *Anal. Biochem.* **1994**, *219*, 261; Andrews, P. C., Boyd, J., Ogorzalek Loo, R., Zhao, R., Zhu, C. Q., Grant, K., Williams, S. In J. W. Crabb (Ed.), *Techniques in Protein Chemistry*, Vol. 5, **1994**, San Diego, CA: Academic Press, p. 485.
4. Lambert, P-.H., Boutin, J. A., Fauchère, J-.L., Volland, J.-P. *Rapid Commun. Mass Spectrom.* **1997**, *11*, 1971.
5. Hoaglund, C. S., Valentine, S. J., Sporleder, C. R., Reilly, J. P., Clemmer, D. E. *Anal. Chem.* **1998**, *70*, 2236.
6. St. Louis, R. H., Hill, H. H. *Crit. Rev. Anal Chem.* **1990**, *21*, 321; Clemmer, D. E., Jarrold, M. F. *J. Mass Spectrom.* **1997**, *32*, 577; Liu, Y., Valentine, S. J., Counterman, A. E., Hoaglund, C. S., Clemmer, D. E. *Anal. Chem.* **1997**, *69*, 728A.
7. Jarrold, M. F. *J. Phys. Chem.* **1995**, *99*, 11; Radi, P. P., von Helden, G., Hsu, M. T., Kemper, P. R., Bowers, M. T. *Int. J. Mass Spectrom. Ion Proc.* **1991**, *109*, 49.
8. von Helden, G., Wyttenbach, T., Bowers, M. T. *Science* **1995**, *267*, 1483; Wyttenbach, T., von Helden, G., Bowers, M. T. *J. Am. Chem. Soc.* **1996**, *118*, 8355.
9. Clemmer, D. E., Hudgins, R. R., Jarrold, M. F. *J. Am. Chem. Soc.* **1995**, *117*, 10141.
10. Hill, H. H., Siems, W. F., St. Louis, R. H., McMinn, D. G. *Anal. Chem.* **1990**, *62*, 1201A.
11. Valentine, S. J., Clemmer, D. E. *J. Am. Chem. Soc.* **1997**, *119*, 3558; Valentine, S. J., Anderson, J., Ellington, A. E., Clemmer, D. E. *J. Phys. Chem. B* **1997**, *101*, 3891; Valentine, S. J., Counterman, A. E., Clemmer, D. E. *J. Am. Soc. Mass Spectrom.* **1997**, *8*, 954.
12. Valentine, S. J., Clemmer, D. E. *J. Am. Chem. Soc.* **1997**, *119*, 3558; Liu, Y., Valentine, S. J., Counterman, A. E., Hoaglund, C. S., Clemmer, D. E. *Anal. Chem.* **1997**, *69*, 728A.

13. Hoaglund, C. S., Valentine, S. J., Clemmer, D. E. *Anal. Chem.* **1997**, *69*, 4156.
14. Counterman, A. E., Valentine, S. J., Srebalus, C. A., Henderson, S. C., Hoaglund, C. S., Clemmer, D. E. *J. Am. Soc. Mass Spectrom.* **1998**, *9*, 743.
15. Dugourd P., Hudgins, R. R., Clemmer, D. E., Jarrold, M. F. *Rev. Sci. Instrum.* **1997**, *68*, 1122.
16. Wyttenbach, T., Kemper, P. R., Bowers, M. T. *Int. J. Mass Spectrom.* **2001**, *212*, 13.
17. Shaffer, S. A., Prior, D. C., Anderson, G. A., Udseth, H. R., Smith, R. D. *Anal. Chem.* **1998**, *70*, 4111.
18. Wu, C., Siems, W. F., Asbury, G. R., Hill, H. H. *Anal. Chem.* **1998**, *70*, 4929.
19. Revercomb, H. E., Mason, E. A. *Anal. Chem.* **1975**, *47*, 970.
20. Srebalus, C. A., Li, J., Marshall, W. S., Clemmer, D. E. *Anal. Chem.* **1999**, *71*, 3918.
21. Wu, C., Siems, W. F., Klasmeier, J., Hill, H. H. *Anal Chem.* **2000**, *72*, 391.
22. Mason, E. A., McDaniel, E. W. *Transport Properties of Ions in Gases*, **1988**, New York: Wiley.
23. Carr, T. W., *Plasma Chromatography*, **1984**, New York: Plenum Press.
24. Wyttenbach, T., von Helden, G., Batka, J. J., Carlat, D., Bowers, M. T. *J. Am. Soc. Mass Spectrom.* **1997**, *8*, 275.
25. Jarrold, M. F., Constant, V. A. *Phys. Rev. Lett.* **1991**, *67*, 2994.
26. Shvartsburg, A. A., Jarrold, M. F. *Chem. Phys. Lett.* **1996**, *261*, 86.
27. Mesleh, M. F., Hunter, J. M., Shvartsburg, A. A., Schatz, G. C., Jarrold, M. F. *J. Phys. Chem. B* **1996**, *100*, 16082.
28. Srebalus, C. A., Li, J., Marshall, W. S., Clemmer, D. E. *J. Am. Soc. Mass Spectrom.* **2000**, *11*, 352.
29. Gillig, K. J., Ruotolo, B., Stone, E. G., Russell, D. H. *Anal. Chem.* **2000**, *72*, 3965.
30. Steiner, W. E., Clowers, B. H., Fuhrer, K., Gonin, M., Matz, L. M., Siems, W. F., Schultz, A. J., Hill, H. H. *Rapid Commun. Mass Spectrom.* **2001**, *15*, 2221.
31. Mix and split strategies have been described previously (e.g., Lebl, M., Krchnak, V. *Methods Enzymol.* **1997**, *289*, (336–392). Briefly, the approach uses a gas (nitrogen) agitated manifold system and standard Fmoc chemistry protocols (Wellings, D. A., Atherton, E. *Methods Enzymol.* **1997**, *289*, 44–67). Appropriately protected library inputs (and abbreviations for less common synthetic residues) are isoleucine, tyrosine, tryptophan, serine, histidine, proline, aspartic acid, arginine, lysine, glutamine, D-asparagine, D-glutamic acid, D-alanine, D-phenylalanine, D-lysine, hydroxyproline, 3-aminopropionic acid or β-alanine (as β-Ala), aspartic (*O*-benzyl) acid, glutamic (*O*-benzyl) acid, benzyl histidine, β-(1-napthyl)-alanine (*N*-Ala), β-(3-pyridyl)-alanine (as 3-Ala), p-nitrophenylalanine, β-cyclohexylalanine (Chex), γ-aminobutyric acid, and tetrahydroisoquinoline-3-carboxylic acid (TIC).
32. The value of 300 unique molecular weights comes from considering that, of the 26 variable residues, 24 have unique masses. Thus there will be $(24 + 2 - 1)!/2!(24 - 1)! = 300$ unique peptide masses. For a discussion of combinatorial statistics see Demirev, P. A., Zubarev, R. A. *Anal. Chem.* **1997**, *69*, 2893–2900.
33. TIC peptides could be absent because they are preferentially formed as doubly protonated ions; examination of the mass spectrum indicates that this is not the case.
34. Details of peptide synthesis can be found in Bodanszky, M. *Principles of Peptide Synthesis*, 2nd ed., **1993**, Berlin: Springer-Verlag; Bodanszky, M. *Peptide Chemistry, A Practical Textbook*, 2nd ed., **1993**, Berlin: Springer-Verlag.
35. Dunayevskiy, Y., Vouros, P., Carell, T., Wintner, E. A., Rebek, Jr, J. R. *Anal. Chem.* **1995**, *67*, 2906.
36. Smith, G. P., Schultz, D. A., Ladbury, J. E. *Gene* **1993**, *128*, 37.
37. Huang, P. Y., Carbonell, R. G. *Biotechnol. Bioeng.* **1999**, *63*, 633.
38. The distribution of $[M + 2H]^{2+}$ ions is somewhat surprising since there are no arginine, lysine, or other basic residues that are normally assumed to be the charge carriers of electrosprayed ions. Presumably our peptides are protonated at the basic amino terminus and a second proton associates with N–H (or C = O) groups along the polypeptide backbone.
39. Turková, J. *Bioaffinity Chromatography*, 2nd ed., **1993**, Amsterdam: Elsevier.
40. Srebalus Barnes, C. A., Clemmer, D. E. *Anal. Chem.* **2001**, *73*, 424.
41. Giddings, J. C. *J. High Resolut. Chromatogr.* **1987**, *10*, 319.
42. Valentine, S. J., Kulchania, M., Srebalus Barnes, C. A., Clemmer, D. E. *Int. J. Mass Spectrom.* **2001**, *212*, 97.
43. Srebalus Barnes, C. A., Hilderbrand, A. E., Valentine, S. J., Clemmer, D. E. *Anal. Chem.* **2002**, *74*, 26.
44. Valentine, S. J., Counterman, A. E., Hoaglund, C. S., Clemmer, D. E. *J. Am. Soc. Mass Spectrom.* **1998**, *9*, 1213.
45. Hilderbrand, A. E., Myung, S., Srebalus Barnes, C. A., Clemmer, D. E., submitted to *J. Am. Soc. Mass Spectrom.* in preparation.

46. Hoaglund-Hyzer, C. S., Li, J., Clemmer, D. E. *Anal. Chem.* **2000**, *72*, 2737.
47. Cooks, R. G., Ast, T., Mabud, A. *Int. J. Mass Spectom. Ion Proc.* **1990**, *100*, 209.
48. Stone, E., Gillig, K. J., Ruotolo, B., Fuhrer, K., Gonin, M., Schultz, A., Russell, D. H. *Anal. Chem.* **2001**, *73*, 2233.
49. Hoaglund-Hyzer, C. S., Clemmer, D. E. *Anal. Chem.* **2001**, *73*, 177.
50. Srebalus Barnes, C. A., Clemmer, D. E., Harris, M. C., Buchwald, S. L. unpublished data.
51. Gidden J., Wyttenbach, T., Jackson, A. T., Scrivens, J. H., Bowers, M. T. *J. Am. Chem. Soc.* **2000**, *122*, 4692.
52. Steiner, W. E., Clowers, B. H., Fuhrer, K., Gonin, M., Matz, L. M., Siems, W. F., Schultz, A. J., Hill, H. H. *Rapid Commun. Mass Spectrom.* **2001**, *15*, 2221.
53. Matz, L. M., Hill, H. H. *Anal. Chem.* **2001**, *73*, 1664.
54. Lee, Y. J., Hoaglund Hyzer, C. S., Taraszka, J. A., Zientara, G. A., Counterman, A. E., Clemmer, D. E. *Anal. Chem.* **2001**, *73*, 3549.

Sensors and Sensor Arrays

Chapter 11

Acoustic Wave Sensors for High-Throughput Screening of Materials

Radislav A. Potyrailo, William G. Morris, and Ronald J. Wroczynski

INTRODUCTION

Sensors are well-accepted analytical instruments for the measurement of a diverse range of chemical and physical parameters [1–6]. For combinatorial and high-throughput (HT) screening, sensors are almost an ideal screening tool for a variety of material applications. Indeed, sensors are already miniaturized and do detect quantitatively small amounts of materials. Also, sensors are often multiplexed into arrays for improvement of selectivity and overall performance. In addition, data analysis from sensor arrays is well understood. Finally, different types of sensor transduction mechanisms are known and available to fit particular HT applications. Thus it is relatively straightforward to adapt advances in sensor development for combinatorial screening applications. Table 11.1 compares the capabilities of an ideal high-throughput screening (HTS) tool with the reported capabilities of sensors for a variety of combinatorial and other applications and demonstrates the growing potential of sensors for combinatorial screening.

A discussion of the operating principles of all known sensors is well beyond the scope of this chapter and book. This chapter covers only the operating principles and implementation of acoustic wave sensors for HT materials screening. More particularly, from a wide variety of acoustic wave sensors, we discuss only thickness shear mode (TSM) sensors, also known as quartz crystal microbalance (QCM) sensors. The latter name is less descriptive, but is widely used for historical reasons [7].

From a variety of acoustic wave sensor types, TSM sensors are the most widely used for chemical and other types of sensing. TSM devices are commercially available in different configurations and types. This availability arises from their applications as

Radislav A. Potyrailo, William G. Morris, and Ronald J. Wroczynski • General Electric Company, Global Research Center, Schenectady, NY 12301, USA
Contact author: Radislav A. Potyrailo, email: Potyrailo@crd.ge.com

High-Throughput Analysis, edited by Potyrailo and Amis
Kluwer Academic Publishers, New York, 2003.

TABLE 11.1. Capabilities of an Ideal HTS Tool Versus Reported Capabilities of Sensors

Characteristic of an ideal HTS tool	Characteristic of sensors	Reference
Analysis of small samples	Picogram mass sensitivity of acoustic wave sensors in HTS	This work
Minimal or no sample preparation	Real-time analysis of reactions	Chapter 12
Rapid turnaround in serial analysis of multiple samples	Serial analysis of volatiles from combinatorial samples	[16]
Rapid turnaround in parallel analysis of multiple samples	Analysis of solubility of materials using parallel 6×4 array	This work
Analysis during or after reaction or testing	HTS system for catalytic hydrogen-producing materials	[62]
High precision and accuracy of measurements	Prediction of concentrations of individual components in multicomponent combinatorial samples	[16]
Automated data acquisition and instrument operation and communication with other instruments	Unattended operation of sensors and communication	[63]
Data analysis handles large amounts of data	Data from thousands of sensor spots	[64]
Application of univariate or multivariate analysis	Analysis of volatiles from combinatorial samples	[16]
Developed transfer functions for the property of interest	Determination of barrier properties of materials	Chapter 14
Well-ordered data management	Analysis of signals from fiber-optic probes from individual wells in 960-well plate	[65]

frequency filters in electronic equipment (wireless communications etc.). Important advantages of using TSM sensors are the relative simplicity of the associated electronics for signal collection and good signal-to-noise performance. The latter is important for the determination of small differences between materials properties. While the sensitivity of acoustic wave devices improves as the square of their operating frequency, the noise level of highly sensitive devices is also increased. For example, other types of available sensors—surface acoustic wave (SAW) devices—have operating frequencies of up to 2 GHz, while typical operating frequencies of TSM sensors range from ~1 to ~50 MHz. Unfortunately, the increased noise in SAW devices rarely provides an overall signal-to-noise advantage. Indeed, while noise in a polymer-coated 10 MHz TSM can be as low as 0.03–0.05 Hz [8], typical noise in 433 MHz SAW device is about 2–3 Hz [9]. However, owing to the nature of the operation of SAW devices, they are somewhat more sensitive than TSM devices to changes in the viscoelastic properties of the environment.

In our HT applications, we use TSM sensors in two modes of detection—gravimetric and viscoelastic. In the gravimetric mode, mass loss or gain, caused by a variety of sources (e.g., vapor uptake, material evaporation, material removal, etc.) is measured. In the

TABLE 11.2. Attractive features of TSM sensors for HT applications

Sensors are very inexpensive and are available in different configurations
Robust sensor material (quartz) makes possible their use in harsh environments
Simple relatively inexpensive electronics for data acquisition is available
High (picogram) mass sensitivity
Determination of viscoelastic properties in acoustically thick films
Ease of multiplexing of sensors into arrays
In remote sensor-array applications, long distance between sensor array and electronics does not noticeably increase noise

viscoelastic mode we measure the changes in moduli of deposited films due to thermodynamic transitions, weathering, etc. Another operation mode of TSM sensors utilizes liquid-loading detection. Overall, a summary of the attractive features of TSM sensors for HT applications is provided in Table 11.2.

Recently, arrays of TSM sensors have been demonstrated for HTS applications in probing bioaffinity interactions in liquids [10]. In those studies, 8.5 mm diameter TSM crystals were directly mounted at the bottom of an array of wells with a format of a standard microtiter plate. Other acoustic wave sensor systems for HT analysis of properties of liquids for materials applications have been proposed [11,12]. This chapter highlights the theoretical aspects of the TSM sensor performance that form the basis for their HT applications followed by three representative examples of applications of TSM arrays in HT materials science from our laboratories.

THEORY OF OPERATION

Thickness shear mode sensors are characterized by a shear displacement in response to an applied electric field. A thin slice of a piezoelectric crystal with applied electrodes is illustrated in Figure 11.1. When a voltage is applied to the electrodes, the crystal responds by deforming slightly with the top electrode shifting laterally with respect to the bottom electrode. If the polarity of the voltage is reversed, the shear deformation takes place in the opposite direction. The shear motion gives rise to a shear displacement wave which starts at one surface and propagates through the thickness of the crystal. If an alternating voltage is applied, there will be a frequency at which the transit time of the shear wave through the crystal equals the time for a half-cycle of the voltage. Under these conditions, a standing wave, called the crystal resonance, is set up which has maxima at the electrode surfaces and a minimum (node) midway through the thickness of the crystal.

For HT applications, several types of environmental perturbations can be utilized that cause a detectable and reproducible signal change in the TSM device. The analytical signal in a TSM device can be produced by a change in the mass of the oscillating crystal (gravimetric or mass sensitivity), a change in the properties of a bulk liquid in contact with the crystal (liquid viscosity and density sensitivity), or a change in the viscoelastic properties of a film deposited onto the crystal (viscoelastic sensitivity).

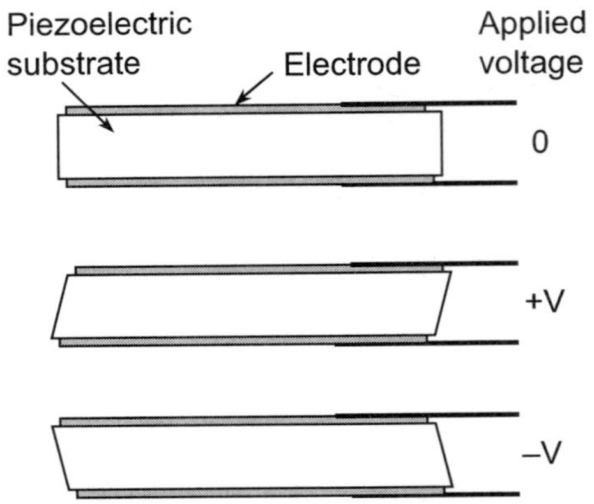

FIGURE 11.1. Operating principle of a TSM device. Electrodes (e.g., gold) are deposited on the top and bottom of a piezoelectric resonator (e.g., an AT-cut quartz crystal). An applied voltage causes a shear deformation in one direction. Reversing the voltage reverses the shear direction. An alternating voltage gives rise to shear waves that propagate through the crystal.

Mass Sensitivity

The conditions for gravimetric measurements occur when a film applied onto the surface of the sensor crystal is acoustically thin and thus remains effectively rigid during the lateral displacements [13]. Typically, a film can be considered acoustically thin if its thickness is less than 2% of the quartz crystal thickness. The frequency shift should also be less than 2%, or 200 kHz for a 10 MHz crystal. As the mass of material on the electrode surface of the crystal increases, the frequency decreases. This effect is analogous to an oscillating mechanical system consisting of a mass attached to a spring; increasing the mass will cause a decrease in the frequency.

In the gravimetric mode, it is possible to detect very small changes in the crystal mass if a film is deposited onto the crystal surface. Such a film induces a change in the resonant frequency f_r of the sensing crystal given by [14]

$$\Delta f_F = -2 f_0^2 (m_F/A)(\mu_Q \rho_Q)^{-1/2} \tag{1}$$

where f_0 is the fundamental resonant frequency of an unloaded device, μ_Q is the shear modulus of the piezoelectric substrate, ρ_Q is the substrate density, m_F is the total mass of the coating deposited to both faces of the crystal, and A is the active surface area of one face of the crystal. Parameters of a given piezoelectric substrate can be combined into a constant $C = 2f_0^2 (\mu_Q\rho_Q)^{-1/2}$. For an AT-cut quartz typically used for TSM sensors, $\mu_Q = 2.947 \times 10^{11}\,\mathrm{g\,cm^{-1}\,s^{-2}}$, $\rho_Q = 2.648\,\mathrm{g\,cm^{-3}}$, and $C = (2.264 \times 10^{-6}) f_0^2\,\mathrm{Hz\,cm^2\,g^{-1}}$. Equation (1) can be rewritten as

$$\Delta f_F = -2C\,\rho_F\,d_F \tag{2}$$

where ρ_F and d_F are the density and thickness of the film, respectively.

Equations (1) and (2) demonstrate that mass sensitivity increases as a square of the operating frequency f_0. However, to resolve this mass change, it is important to have a low noise level in the detection system. For example, with an optimized noise level of 0.03–0.05 Hz [8] in our 10 MHz sensors, the mass resolution is better than 100 pg.

The film deposited onto the crystal can be of different natures, depending on the HT application. For the development of materials for chemical sensor applications, screening of the materials involves the exposure of films to different vapors and determination of the vapor-sorption properties of the films. The best materials are those that have dramatically different sorption properties to the different vapors of interest. These materials are further used in sensor arrays.

When the analyte partitions into the film, the resonant frequency of the crystal decreases as a function of analyte mass as follows:

$$\Delta f_A = -C\, m_A/A \tag{3}$$

where Δf_A is the change in resonant frequency associated with exposure of the sensor to analyte and m_A is the mass of analyte partitioned into the film when both sides of the crystal are coated with the film and are exposed to the analyte. Mass m_A is governed by the ability of the film to sorb the analyte and by the thickness d_F of the film. The equilibrium distribution of analyte molecules between the ambient environment (gas or liquid phase) and the sorbing film is described by the partition coefficient K. This thermodynamic parameter is defined as the ratio of the concentration C_F, of the analyte in the film material to the concentration C_O of the analyte outside the film [15]:

$$K = C_F/C_O \tag{4}$$

Combining Eqns (2), (3), and (4) gives the dependence of the change in resonant frequency upon exposure of the sensor to analyte and the concentration of the analyte outside the film as

$$\Delta f_A = (\Delta f_F\, K/\rho_F)C_O \tag{5}$$

As shown by Eqn (5), selectivity of the steady state sensor response is provided by a variable partition coefficient K toward different analyte vapors. However, sensor selectivity can be improved by the temporal modulation of the sensor response. This modulation enhances selectivity of individual sensors in the array to different components in a mixture as follows:

$$\Delta f_A(t) = \Delta f_F\, C_F(t)/\rho_F \tag{6}$$

where $C_F\,(t)$ is the temporally changing vapor concentration in the sensing film.

The temporal modulation of the sensor response described by Eqn (6) can be provided by a specific design of a sample introduction system into the sensor that is responsible for $C_F(t)$. For example, as described later in this chapter, for analysis of combinatorial samples, we introduce these samples as small amounts of liquid into the sensor system followed by their evaporation at room temperature and transport of the vapors to the sensor array [16]. Thus, in this HT sensor system, the temporal modulation of the sensor response described by Eqn (6) is provided by two groups of dynamic parameters of the system responsible for $C_F(t)$. The first group is related to the introduction of a liquid sample into the gas stream. This group includes the evaporation rate of a given component of the liquid sample with a time constant τ_1 and the exponential dilution of the generated vapor in the gas carrier line with a time constant τ_2. The second group is related to the interactions of the generated vapor with a chemically sensitive material deposited onto a TSM resonator. These interactions are described by time constants τ_3 and τ_4 for sorption and desorption interactions, respectively, of the vapor with the chemically sensitive material. In general, under constant experimental conditions, different volatile components of the sample will have different time constants τ_1, τ_3, and τ_4, while τ_2 could be very similar for different vapors.

Thus the temporal distribution of vapor concentration $C_F(t)$ in a sensing film is described by a convolution (indicated by *) of temporal profiles of the generated vapor plug and dynamic interactions of the vapor plug with the sensing film:

$$C_F(t) = C_1(t, \tau_1) * C_2(t, \tau_2) * C_3(t, \tau_3) * C_4(t, \tau_4) \tag{7}$$

where $C_i\ (t, \tau_i)$ are contributions of vapor concentration in the sensing film from different processes. The differences in $C_1(t, \tau_1)$, $C_3(t, \tau_3)$, and $C_4(t, \tau_4)$, for sample components with different boiling points and kinetics of interactions with the chemically sensitive film provide an additional selectivity in sensor response.

Liquid Viscosity/Density Sensitivity

The TSM device can be operated in liquid to measure either the accumulation of mass onto the surface from the liquid phase or the properties of the contacting liquid itself [7]. Detection of mass changes is described by Eqn (1). Assuming that the change in the fundamental resonant frequency f_0 is directly proportional to the change in the series resonant frequency f_S, the decrease Δf_L in the resonant frequency upon contact with liquid is given by [7,17,18]

$$\Delta f_L = -f_0^{3/2} (\rho_L \eta_L / \pi \mu_Q \rho_Q)^{1/2} \tag{8}$$

where ρ_L is the density of the liquid and η_L is the shear viscosity of the liquid. As shown by Eqn (8), changing the viscosity and density of the liquid produce similar effects to the frequency change. However, it is possible to separate these contributions by using a combination of resonators with different surface roughness. If the surface roughness is of the order of the penetration depth of the shear wave into the liquid, an additional mass-loading effect by the picked up liquid will occur [19].

Viscoelastic Sensitivity

Environmental exposures of coated TSM crystal during HT evaluations can cause changes in the viscoelastic properties of the deposited film. These changes can be a result of a crystallization or softening of the film upon exposure to fluids or operating the coated crystal below and above the glass transition or melting temperatures. If the film is sufficiently acoustically thick (>2% of the quartz crystal thickness or >2% of the fundamental frequency), the change in the viscoelastic properties is easily detected as a change in the fundamental frequency [14]. For determination of viscoelastic properties of acoustically thin films, other types of acoustic wave devices can be more preferable. These can include SAW and flexural-plate-wave devices [7].

When TSM devices are used, in addition to measurements of fundamental frequency with a frequency counter, it is advantageous to measure additional parameters of the TSM sensors with a network analyzer or a vector voltmeter. These can include the magnitude and phase of the reflected signal (admittance) as a function of the excitation frequency f. From the measured admittance phase, it is possible to extract the information about the series and parallel resonant frequencies. The resonant behavior of the coated TSM device will depend on the polymer film thickness d_F and the shear modulus G of the film. In general, G will be composed of real and imaginary parts, G' and G'', known as the storage modulus and the loss modulus, respectively. A network analyzer is often used to measure the impedance Z_0 of the uncoated quartz crystal as a function of frequency in the vicinity of the resonant frequency, yielding the equivalent circuit parameters such as capacitance C, inductance L, and resistance R. The impedance due to the film is obtained by subtracting the uncoated impedance from the coated impedance:

$$Z_E = Z_F - Z_0 \tag{9}$$

Thus the viscoelastic properties of the coated film can be related to Z_E by [7]

$$Z_E = (1/8K^2 f_S\, c)(G\rho_F/\mu_Q\, \rho_Q)^{1/2} \tan h(\gamma d_F) \tag{10}$$

where K is the quartz electromechanical coupling coefficient, f_S is the series resonant frequency, c is the parallel static capacitance, G is the complex shear modulus of the deposited film ($G = G' + iG''$), and γ is a parameter dependent on the excitation frequency f, the film density ρ_F, and the complex shear modulus G of the film [$\gamma = i\, 2\pi f(\rho_F/G)^{1/2}$].

EXAMPLES OF APPLICATIONS TO HT MATERIALS EVALUATIONS

We describe here several examples of application of the developed TSM sensors for HTS. These examples can be broadly divided into two categories. In the first category, each sensor is used individually to determine the properties of materials in contact with the sensor. In the second category, individual sensor responses are combined for a more selective analysis of properties of complex materials.

Design and Capabilities of a 24-Channel Sensor System

A 24-channel TSM-sensor system was designed, fabricated, and tested for several required applications. The sensor system was arranged on two circuit boards. The design features of the boards are shown in Figure 11.2. The sensor system with assembled 24-sensor crystals and associated electronics is shown in Figure 11.3. One of the printed circuit boards [Figures 11.2(a) and 11.3(a)] accommodated a 6×4 array of sensor crystals. The layout of this array was compatible with available 24-well plates, which are

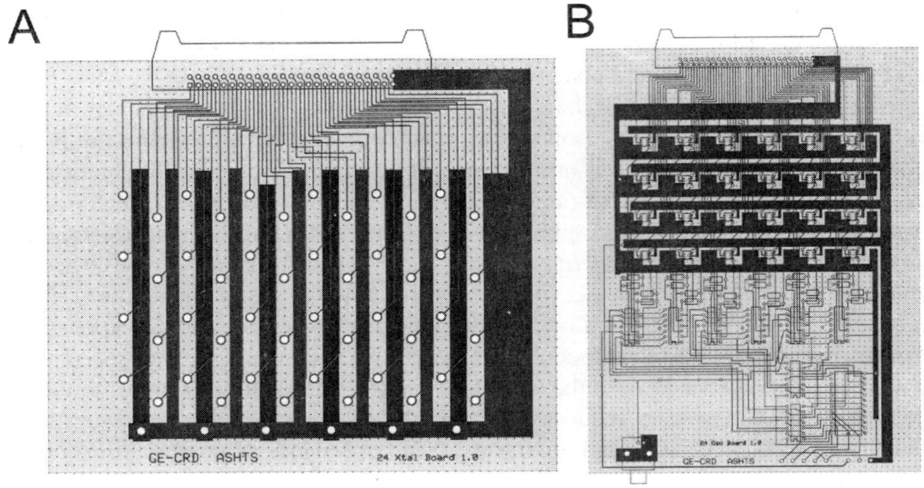

FIGURE 11.2. Design of printed circuit boards for a 24-channel TSM sensor system: (A) printed circuit board for a 6 × 4 sensor array; (B) printed circuit board for 24 oscillators, six multiplexers, and associated electronics. The scale in (A) is 1.5 times larger than that in (B).

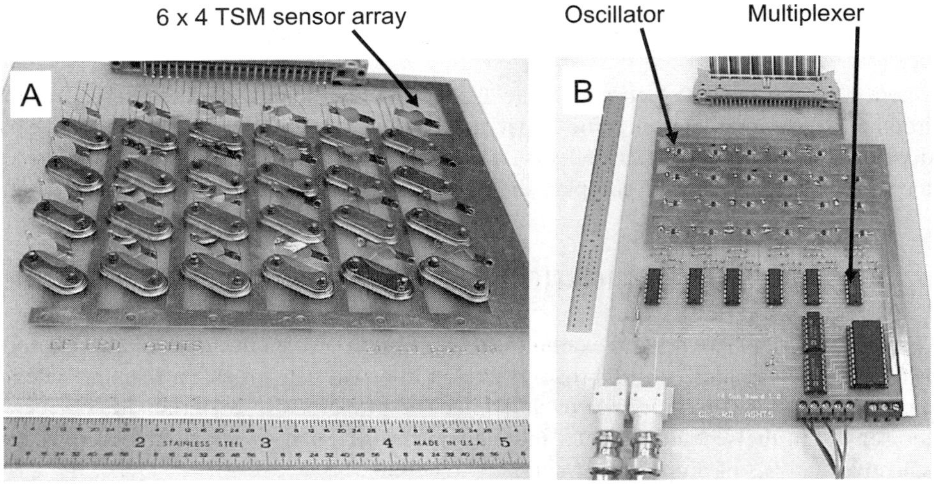

FIGURE 11.3. General view of the 24-channel TSM sensor system: (A) 6 × 4 sensor array; (B) printed circuit board with 24 oscillators, six multiplexers, and associated electronics.

in turn a derivative of a 96-well mictrotiter plate format. Connections were made from this printed circuit board to a second printed circuit board which contained 24 integrated circuit oscillators giving transistor–transistor logic (TTL) level outputs [Figures 11.2(b) and 11.3(b)]. The oscillators were able to support oscillation frequencies of different TSM crystals in the range from ~4 to ~20 MHz. The 24 oscillator signals were selected one at a time using a TTL multiplexer integrated circuit and were connected to a time-interval analyzer circuit card installed in a desktop personal computer. Data acquisition was achieved using a program written in LabVIEW (National Instruments, Austin, TX). The operator interface panel of the developed program is shown in Figure 11.4.

The main requirements for the development of the sensor system were electronic and system design requirements. Electronic requirements included low noise of operation and rapid acquisition time. System design requirements included an adequate sensor enclosure design for diverse environmental exposures and modular design for diverse applications. The electronic requirements were met by optimizing a variety of parameters to provide a desired sensor performance. These parameters included integration time, signal averaging, and absence of cross talk between channels. The optimization results are provided in Figure 11.5. These data show that under the optimized data acquisition conditions, the noise of the sensor system was less than 0.1 Hz. Such high signal quality was adequate for HTS and was applicable to different HTS projects. System design requirements were met by developing several types of sensor enclosures that included a gas flow though cell for vapor sorption studies and a platform for immersion of crystals into 24-well plate arrays. In addition, the cable design and length were optimized for the operation of the 24-channel sensor array [Figure 11.3(a)] remotely from the oscillator array [Figure 11.3(b)] to decouple environments of the sensor array exposure (e.g. high temperature, pressure, chemicals) and environment of operation of the oscillator array.

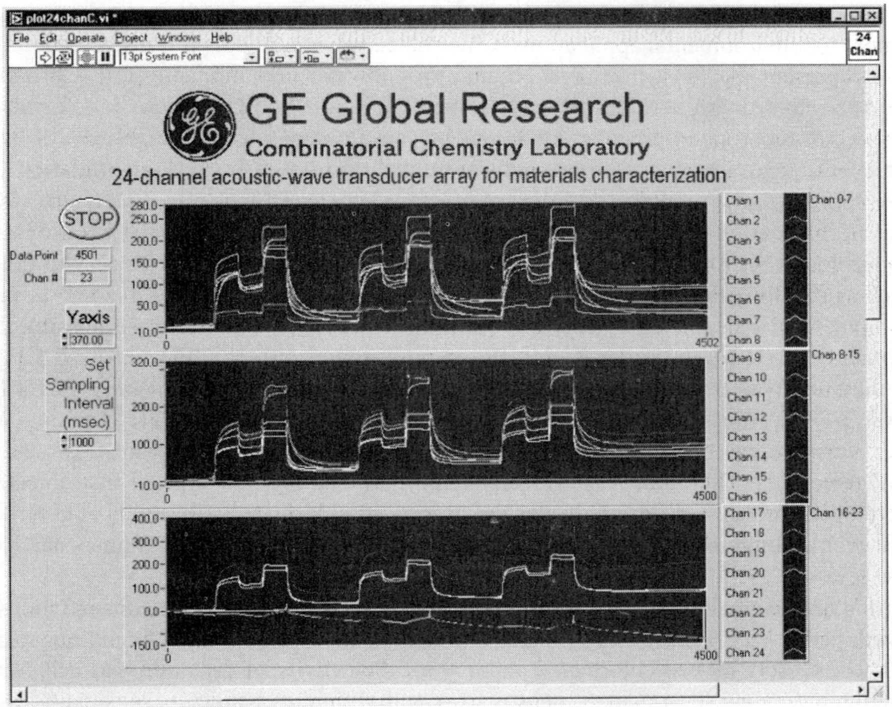

FIGURE 11.4. The operator interface panel of the developed program for data acquisition of the 24-channel sensor array.

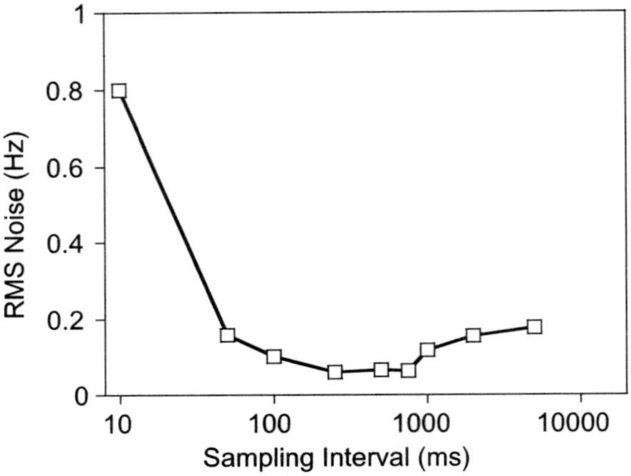

FIGURE 11.5. Results of minimization of noise in sensor signal.

Application Example 1: Development of New Sensor Materials

One of the applications of the developed 24-channel sensor system was for discovery of new sensor materials for sensing part per billion levels of chlorinated organic solvent vapors for environmental applications. Chlorinated solvents and their transformation products represent the most prevalent organic groundwater contaminants [20]. Chlorinated aliphatic hydrocarbons can be transformed biologically [21]. Such transformations can occur under environmental conditions present in an aquifer in the absence of human intervention in a process termed natural attenuation or intrinsic biodegradation [22]. Recalcitrant under aerobic conditions, tetrachloroethylene [or perchloroethylene (PCE)] and trichloroethylene (TCE), the chlorinated solvents of interest, have been observed under anaerobic conditions to be reductively dechlorinated to less chlorinated ethenes such as dichloroethylene (DCE) isomers, and vinyl chloride (VC), and in some instances completely dechlorinated to ethylene [23] and ethane [24]. Of the three possible DCE isomers, 1,1-DCE is the least significant intermediate [23], while *cis*-1,2-DCE predominates over *trans*-1,2-DCE [25–27]. Five of the six species, namely PCE, TCE, *cis*-1,2-DCE, *trans*-1,2-DCE, and VC, are regulated under the 1986 Safe Drinking Water Act Amendments because they pose a threat to public health as they are toxic and carcinogenic [23]. Therefore it is critically important that any new type of instrumentation developed for real-time groundwater monitoring of chlorinated solvents is able to detect and even distinguish PCE and TCE from their daughter products, DCE isomers and VC, at low levels.

With the aim of detecting these chlorinated solvent vapors at trace concentrations, we developed a TSM sensor array [28,29] and demonstrated discrimination and quantitation of these closely related chlorinated vapors [30]. Selectivity of detection was achieved by coating individual sensors in the sensor array with different chemically sensitive films and analyzing the response pattern of the array. The results of this discrimination are presented in Figure 11.6. It shows the response of a four-sensor array to increasing concentrations

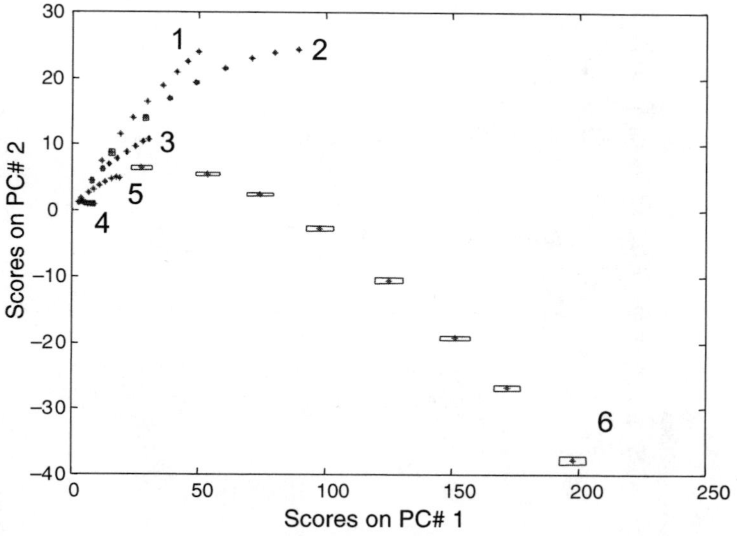

FIGURE 11.6. Plot of the first two principal components for the data set obtained from the response of the four-sensor array to six chlorinated organic vapors at different concentrations. (1) *cis*-1,2-DCE; (2) TCE; (3) *trans*-1,2-DCE; (4) VC; (5) 1,1-DCE, (6) PCE. Each data point is the mean of three measurements; squares represent 1 SD. Maximum vapor concentrations are around 100 ppm. (Adapted from ref. 30.)

(0–100 ppm) of six vapors (PCE, TCE, *cis*-1,2-DCE, *trans*-1,2-DCE, 1,1-DCE, and VC) which were analyzed using principal components analysis (PCA) [31]. The increasing vapor concentrations are indicated by the increase in the distance between the response to the blank gas (point 0,0) and the respective vapor concentration. With the increase in vapor concentration, the discrimination ability of the sensor array increases. Clearly, as shown in Figure 11.6, the six closely related chlorinated organic vapors are well separated from each other using the first two principal components of the data set obtained from the response of the four-sensor array.

While these determinations were performed for concentrations of volatiles in the part per million range, selective determinations at part per billion levels were impossible because of the lack of chemically sensitive materials with sufficiently high partition coefficients [31]. Thus the focus of our combinatorial developments in this area was to find materials with improved diversity in response to the chlorinated organic solvent vapors. For material screening, we produced diverse materials families with potential diversity of sensor response [32].

Our strategy for screening sensor materials for practical applications is presented in Figure 11.7. It included three levels of screening. The primary screen was the discovery screen where materials were exposed to a single analyte concentration. The secondary screen was the focused evaluation where the best subset of these materials was exposed to analytes and interferences. Finally, the tertiary screen involved evaluation of material performance under conditions mimicking the long-term application.

The first two screens were performed with the system shown in Figure 11.8. The 6×4 sensor array was coated with individual films and arranged in a gas flow-through

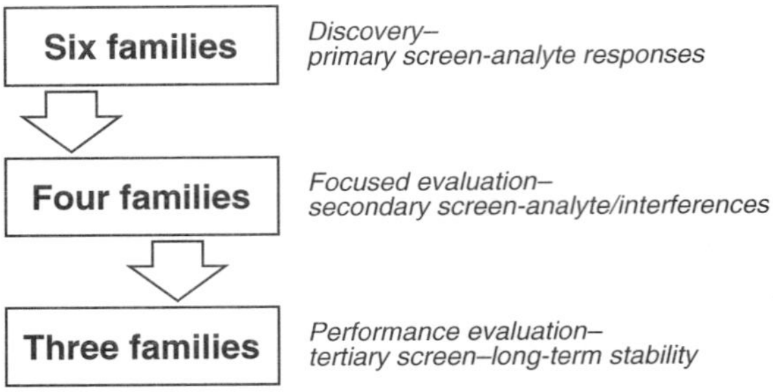

FIGURE 11.7. Strategy and materials spacing for screening of sensor materials for practical applications.

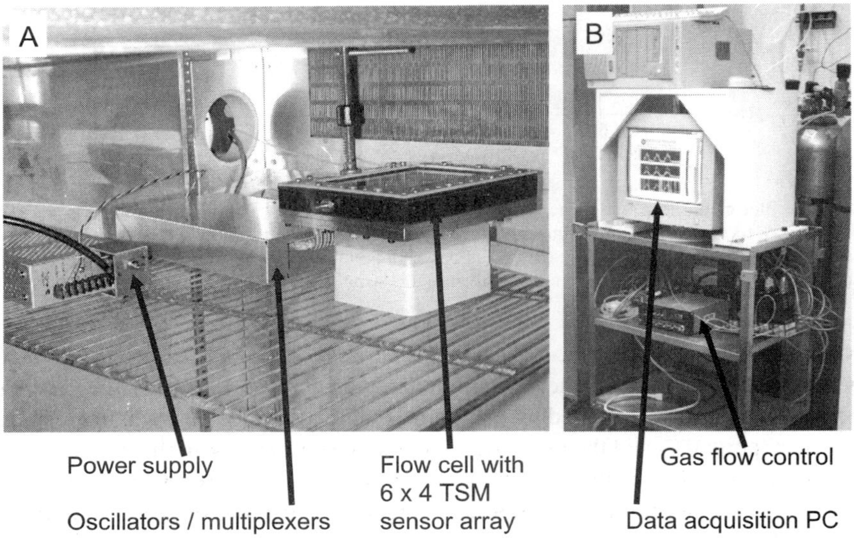

FIGURE 11.8. Operation of the HT system for primary and secondary screens: (A) system components positioned in a temperature-controlled environment; (B) computer control and vapor delivery system components.

cell. The whole arrangement was kept in an environmental chamber that provided adequate (0.1°C) temperature control. Gases were introduced into the gas cell using a four-channel computer-controlled gas flow system (nitrogen, vapor 1, vapor 2, and vapor 3). Three vapors (1–3) were introduced in sequence with no blank gas (nitrogen) in between. Blank gas was introduced between three replicates of three-gas exposures. The sampling interval was selected from the results of the noise reduction studies (see Figure 11.5). Typical exposure time per vapor was 200 data points (~80 min).

The data processing steps in the primary and secondary screens are summarized in Figures 11.9(a) and 11.9(b), respectively. The steps in the primary screen [Figure 9(a)] included visualization of individual responses for their reversibility and determination of the relative sensitivity of each material with respect to one of three vapors. Such analysis compensated for any variability in the absolute response of the replicates of the material due to thickness differences, yet demonstrated the reproducibility of the sensor response

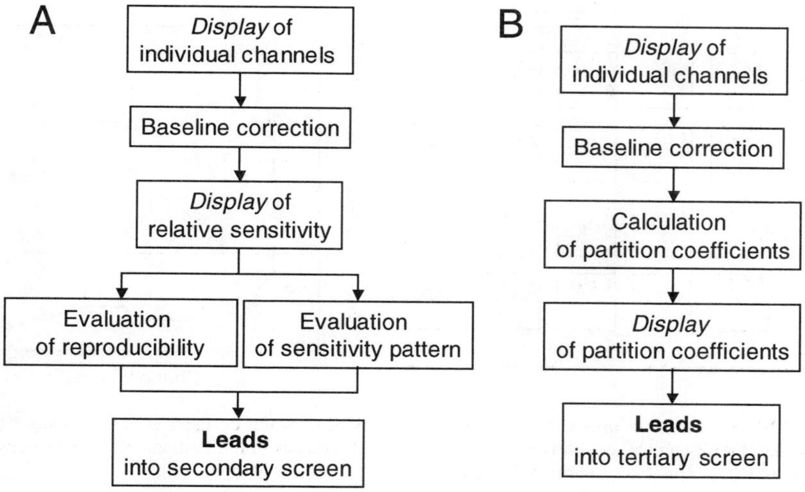

FIGURE 11.9. The data processing steps in (A) the primary and (B) the secondary screens.

and the sensitivity pattern. As a result of the primary screen, materials with good reproducibility and statistically different responses to vapors 1–3 were passed onto the secondary screen. The steps in the secondary screen [Figure 11.9(b)] included visualization of individual responses and the determination of partition coefficients of the materials with respect to analyte and interference vapors. Calculated partition coefficients were displayed for comparison. This comparison was performed using bar charts and multivariate statistical tools. As a result of the secondary screen, materials with different responses to analytes and interferences were passed onto the tertiary screen.

For development of materials for part per billion detection of chlorinated solvent vapors, we identified and made six families of materials based on the best material that we had discovered earlier [33]. This material had a partition coefficient about 1000 times higher than that of conventional materials (e.g., polyisobutylene). However, our base material alone did not provide the required discrimination for detection of different analytes and interfering species such as chloroform, toluene, and carbon tetrachloride. Thus new families of materials were developed in order to induce diversity in sensor response.

Results of the primary screening of several families of these materials are presented in Figure 11.10. The 6×4 array of sensors was comprised of a variety of materials that were deposited as films of different thickness in replicates of two. Also, two reference sealed crystals were used [channels 18 and 19 in Figure 11.10(a)]. As shown in Figure 11.10(a), the responses of all sensor materials were reversible on the timescale of testing. Results of the evaluation of the response pattern upon exposure to three analyte vapors are shown in Figure 11.10(b). These analyte vapors were PCE, TCE, and *cis*-DCE, all at 10 ppm. These data showed two important aspects of the primary screening. First, reproducibility of responses was easily quantified by the magnitude of the variation between replicate material films and between replicate measurements in individual films. For example, material B showed poor run-to-run ($n = 3$) and film-to-film ($n = 2$) reproducibility, while material E demonstrated good reproducibility for both types of

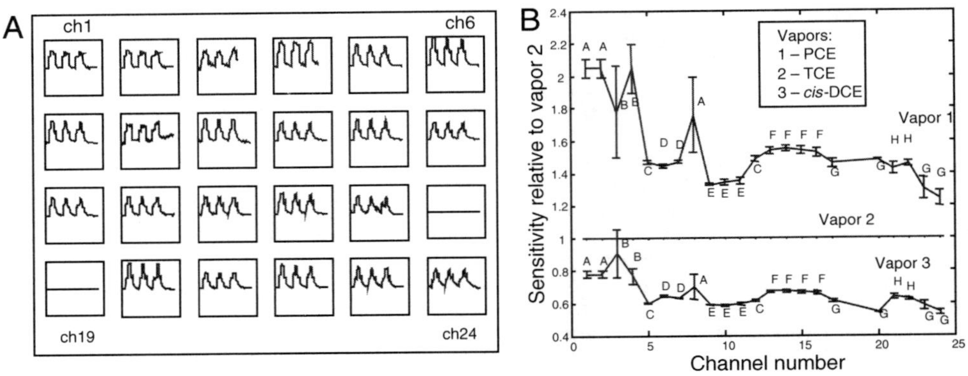

FIGURE 11.10. Results of the primary screening of several families of sensor materials: (A) visualization results of the baseline stability and reversibility of sensor response; (B) results of the evaluation of the response pattern upon exposure to three analyte vapors (PCE, TCE, and *cis*-DCE), all at 10 ppm.

measurements as indicated by the small run-to-run error bars for each material ($n = 3$) and similar relative sensitivity for replicate films with material E ($n = 3$).

The diversity of material response to the three analyte vapors was evaluated by plotting the sensor responses relative to the response of sensors to TCE. The information of interest here was the combination of the individual responses of materials. For example, materials C and E had the same relative response to *cis*-DCE, while material C had a significantly larger relative response to PCE than material E [see Figure 11.10(b)].

In the secondary screen, experiments were done repetitively over several days. The calculated partition coefficients for the materials from different days are presented in Figure 11.11. The acceptable run-to-run reproducibility (error bars of each column, $n = 3$) and day-to-day reproducibility (different columns of the same color) was observed as shown in the inset to Figure 11.11. As a result of our screening efforts, we have identified materials with improved partition coefficients (e.g., materials Q, R, and T) and with diverse responses to analytes.

Results of material performance with respect to the differences in partition coefficients to analytes (PCE, TCE, and *cis*-DCE) and interfering vapors (carbon tetrachloride, toluene, and chloroform) is summarized in Figure 11.12. Normalization of the partition coefficients for the different vapors to the partition coefficient of TCE revealed a noticeable difference in response pattern of materials to analyte and interfering vapors. A more quantitative analysis can be performed using PCA tools as shown in Figure 11.13. In this analysis, we used data from all replicates of eight types of different films. The capability for discriminating six vapors using the eight types of polymers was evaluated using a score plot [Figure 11.13(a)]. It demonstrated that these six vapors are well separated in the PCA space when these eight types of polymers are used for the determinations. To understand which materials induce the most diversity in the response, a loadings plot was constructed [Figure 11.13(b)]. The greater the distance between the films of the different types, the better was the discrimination ability of the sensor array that includes these films. For example, films of materials T and X introduce the most diversity in the recognition ability of the sensor array. In contrast, materials Q and R have very similar response

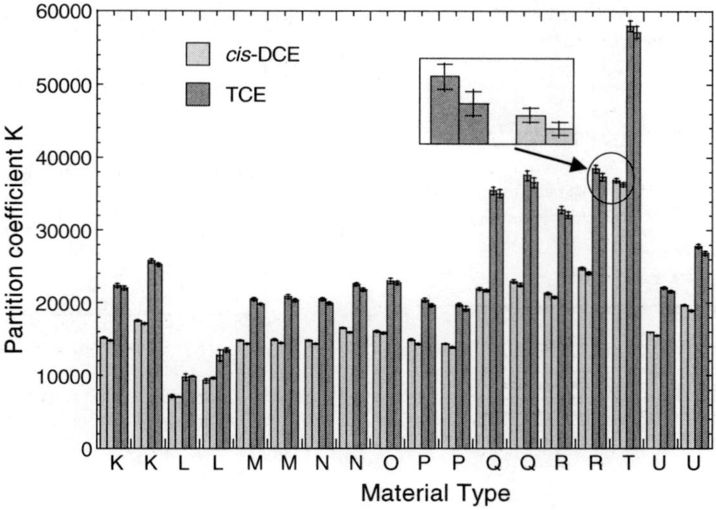

FIGURE 11.11. Results of the calculation of partition coefficients of the materials during different days of experiments: Error bars of each column, ($n = 3$) and day-to-day reproducibility (different columns of the same gray scale).

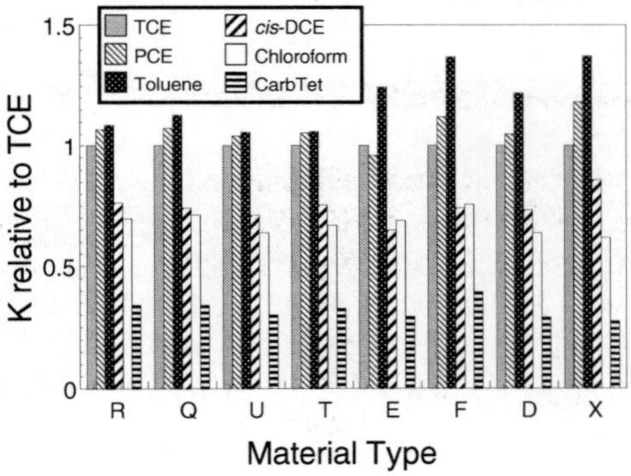

FIGURE 11.12. Results of analysis of materials performance with respect to response to analytes and interfering vapors. Visualization of relative material sensitivities as normalized partition coefficients for different vapors to the partition coefficient of TCE.

patterns to the six vapors and their combination in the sensor array will be less useful. The loadings plot also demonstrates the reproducibility of the response of replicate films of the same materials. Such information served as an additional input into the materials selection for the tertiary screen.

Using the experimentally obtained partition coefficients and assuming to a first approximation a linear response of the sensors over the concentration range studied, the response of the sensor array with some of the lead materials identified in the secondary screen (R, T, D, and X) was modeled. This analysis was performed for vapor concentrations ~100 times smaller than those used earlier (cf. Figure 11.6). There were four elements in the sensor array, the same as in the earlier array. Results of these simulations

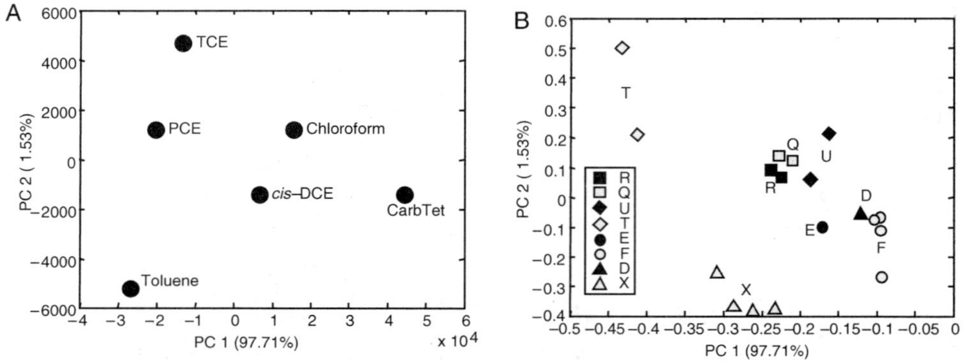

FIGURE 11.13. Application of PCA tools to the determination of differences in the response pattern of the sensor materials toward analytes (PCE, TCE, and *cis*-DCE) and interfering vapors (carbon tetrachloride, toluene, and chloroform): (a) score and (b) loading plots of the first two principal components.

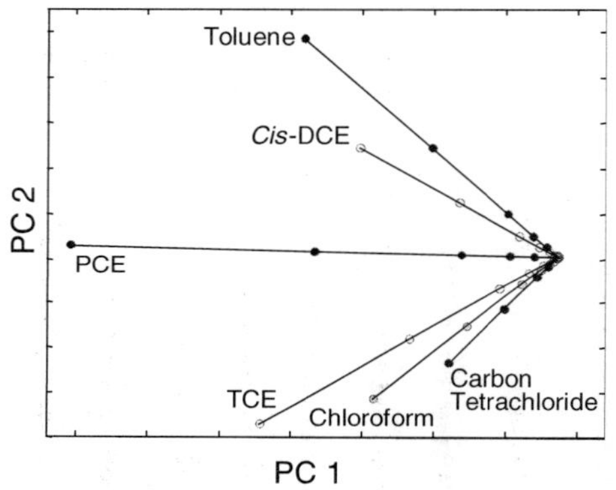

FIGURE 11.14. Results of simulations of sensor responses to increasing concentrations of analyte and interfering vapors. Maximum vapor concentrations are ~1 ppm.

are presented in Figure 11.14. The results demonstrated that the new materials developed using HT experimentation were capable of quantifying selectively several closely related chlorinated vapors in the presence of interfering vapors at part per billion concentrations.

The feasibility of applications for the tertiary screen has been also demonstrated. The aims of these screening studies were to develop transfer functions between environmental effects and performance loss and to improve understanding of the aging mechanisms of chemically sensitive materials upon different environmental exposures. In these experiments, certain materials were coated onto the sensors of the 24-channel sensor array and were aged in the flow-through cell. In brief, Figure 11.15 demonstrates replicate ($n = 3$) responses of one of the materials in the array to a nonpolar analyte vapor (1), a polar interference vapor (2), and their mixture ($1 + 2$) before and after the aging study. Several important findings followed from this preliminary study. First, the magnitude of the response to a nonpolar analyte vapor decreased upon aging. Second, the magnitude of the response to a polar interference vapor increased upon aging. Third, because of these

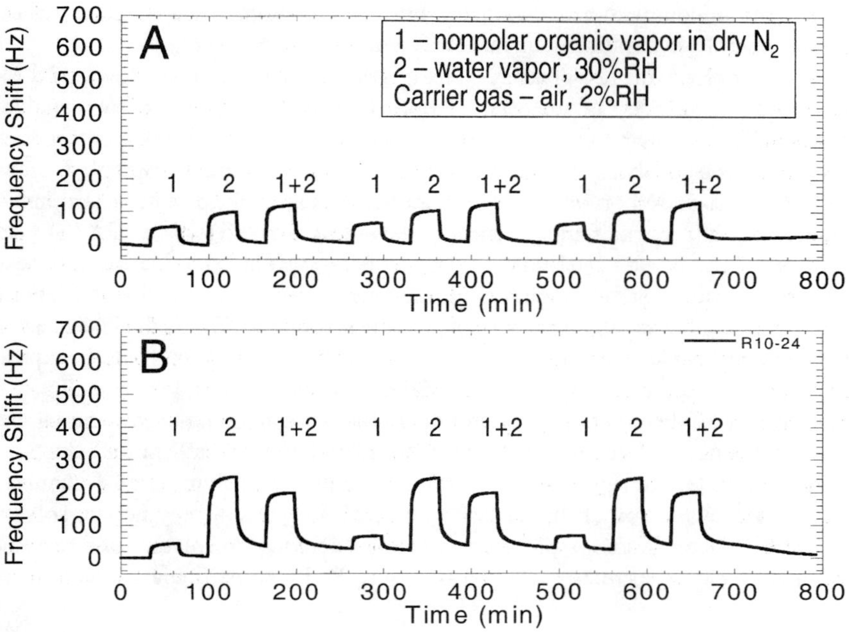

FIGURE 11.15. Example of a tertiary screen for the evaluation of the long-term stability of sensor materials. Replicate responses of one of the materials in the array to (1) a nonpolar analyte vapor, (2) a polar interfering vapor, and (1 + 2) their mixture (A) before and (B) after the aging study.

changes, the overall response pattern of the sensor performance was altered. Thus the long-term stability studies reported briefly here demonstrate their importance in the development of materials for practical sensor applications.

Application Example 2: Screening of Materials Properties—Solvent Interactions

Another application of the developed 24-channel sensor system was for the rapid screening of materials for their solvent resistance and extractability. The solvent resistance of a variety of polymers and their blends, which are used in demanding engineering and consumer applications, is of increasing interest. The extractability of additives is another important aspect of the development of polymers such as polyolefins, silicones, polycarbonates, polyetherimide resins, and many others. Among the solvents whose solubility and extractability are of interest are a variety of fuels, alkaline and acidic solutions, water, organic solvents of different polarities [34–39], and solvent mixtures.

Conventional methods of measuring extractables often require exposure of a relatively large amount of sample to a solvent for extended periods of time. Thus, conventional methods for evaluation of the solubility and extractability of materials are too time consuming to be applicable to the rapid screening of a large number of materials such as those developed in combinatorial chemistry. Further, traditional measurement techniques are difficult to apply to simultaneous measurements of multiple samples.

We have applied the 24-channel sensor system that we have developed to the evaluation of the solubility of materials. Application of this system allows analysis of picogram quantities of materials deposited or removed due to the dissolution process and permits simultaneous analysis of multiple samples. The measurement principle is shown in Figure 11.16. It was based on two modes of operation that involved rapid determination of minute quantities of material removed from or deposited onto the surface of a TSM sensor. In the first mode, a crystal is exposed to a polymer–solvent combination and residual dissolved or extracted material is quantified after sensor removal and solvent evaporation. The mass increase in the crystal is proportional to the amount of dissolved or extracted material. In the second mode, a crystal is coated with a polymer of interest and is exposed to a solvent. Upon solvent interaction, the material can be dissolved and/or crystallized. Thus the mass decrease of the crystal is proportional to the amount of removed material and the viscoelastic property is related to the solvent-induced polymer crystallization effects.

We demonstrated the applicability of this technique by applying the 24-channel sensor system to the evaluation of the solubility of different types of polymers in solvents of interest. The 6 × 4 configuration of the array enables a variety of solvents and materials to be tested in a single experiment as shown in Figure 11.17. In particular, solvent–material

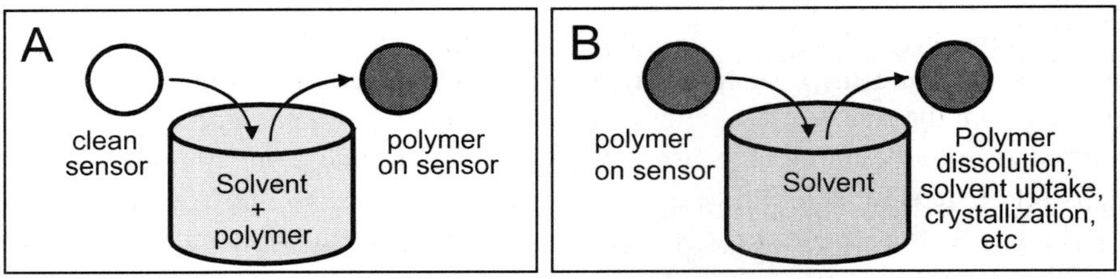

FIGURE 11.16. Principles of measurement of solvent–polymer interactions: (A) determination of minute quantities of material deposited onto the surface of a TSM sensor; (B) determination of minute quantities of material removed from the surface of a TSM sensor.

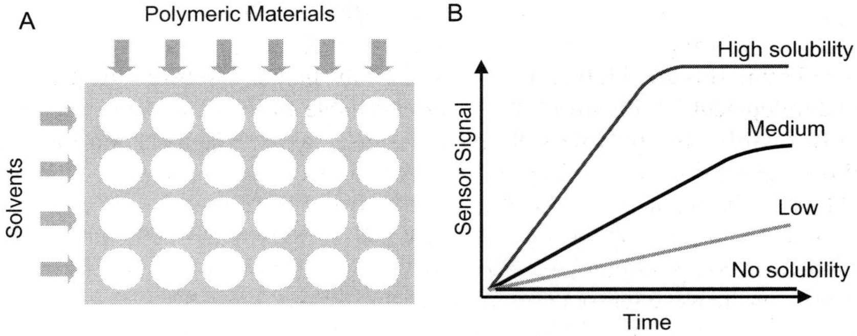

FIGURE 11.17. Experimental protocol for a 6 × 4 configuration of the array for evaluation of solvent–polymer permutations: (A) plate layout for evaluation of 6 × 4 polymer–solvent permutations; (B) conceptual response of the HT system for the evaluation of solubility amount and rate.

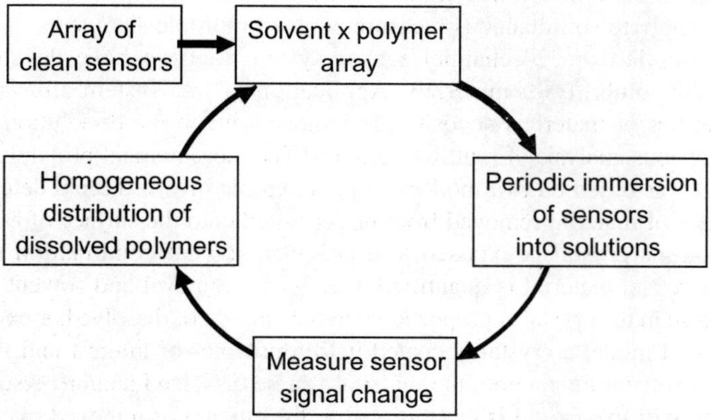

FIGURE 11.18. Process map for determinations of polymer solubility.

permutations can be studied simultaneously [Figure 11.17(a)]. Such studies are advantageous not only because they are performed in parallel but also because they reduce the variability between the measurements by reducing possible uncontrolled environmental variations that may affect individual kinetic experiments (laboratory temperature, atmospheric pressure, etc.). To study the solubility of the materials, measurements of the amount dissolved and the dissolution rate can be performed. The amount dissolved is determined from the sensor signal after a predetermined experimental time, while the rate is determined from the multiple measurements at different time points [Figure 11.17(b)].

The set of steps for solubility/extractability determinations is shown schematically in Figure 11.18. In this experiment, an array of initially clean sensors is periodically immersed into polymer–solvent systems. The measurements of the frequency changes are performed when the sensors are withdrawn from the solvents and the solvents are evaporated. In this way, the measured signal change is indicative of the amount of the deposited material from the solution. To insure a homogeneous distribution of dissolved polymer in each well, the wells can be stirred.

A general view of the sensor array system and the associated mechanical configuration is presented in Figure 11.19. For these measurements, we used sensors of smaller diameter (7 mm) [Figure 11.19(a)] than the 12 mm diameter sensors used for sensor material development [cf. Figure 11.3(a)]. Thus available 24-well glass plates could easily be used to dissolve the polymers. The 6×4 sensor array was arranged on a Z-axis translation stage as shown in Figure 11.19(b), and the whole system is shown in Figure 11.19(c). During the experiments, the array with the polymer–solvent compositions was covered to reduce evaporation of solvents. Before each immersion, the array of solvent–polymer compositions was stirred. A video demonstrating the operation of this 24-channel acoustic wave sensor array for the evaluation of the solubility of different types of polymers in solvents of interest is included on the CD (video *Sensor screening of solubility of polymers*).

Typical results of HT solubility experiments in which several types of polymers were exposed to the same solvent (chloroform) are presented in Figure 11.20. In

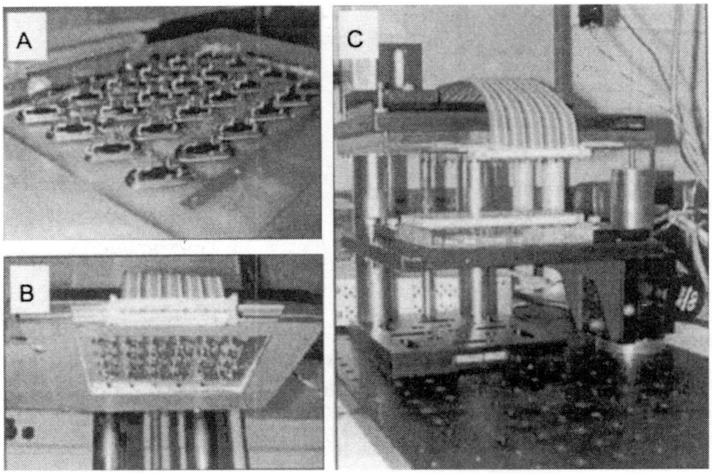

FIGURE 11.19. Twenty-four-channel sensor system for the evaluation of polymer solubility: (A) 6 × 4 array of 7 mm diameter TSM sensors; (B) sensor array arranged on a *Z*-axis translation stage for immersion experiements; (C) general view of the screening system.

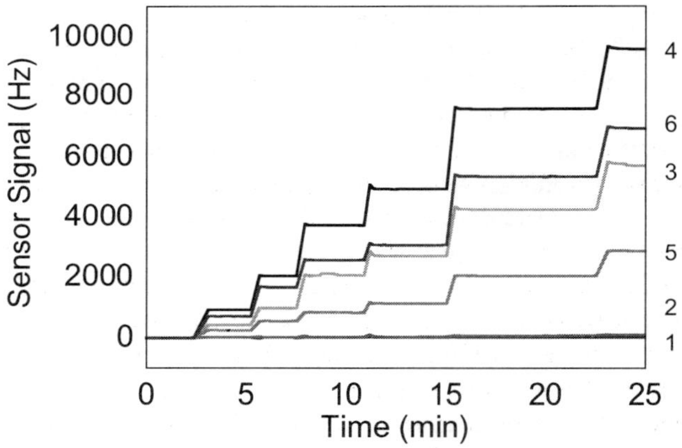

FIGURE 11.20. Typical results of the HT solubility experiments. Several types of polymers were exposed to same solvent (chloroform). In this experiment, the 6 × 4 array was periodically immersed in the solvents and the frequency change was measured upon solvent evaporation.

this experiment, the 6 × 4 array was periodically immersed in the wells and the frequency change was measured upon removal of the sensor from the solvent and solvent evaporation. The stable horizontal regions in the sensor responses are the signals in air upon deposition of different types of materials. To assess the reproducibility of the determinations, four replicate solvent–polymer solutions were made for six different polymers which were copolymers of polycarbonate. Figure 11.21 shows the data with these statistical figures of merit. The error bars for each data point indicate one standard deviation from the mean of four measurements obtained from individual crystals.

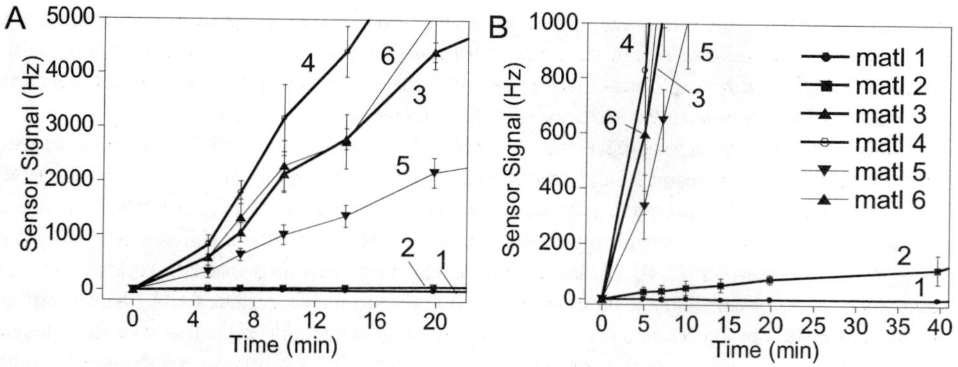

FIGURE 11.21. Reproducibility of the polymer solubility determinations. Six different polymers with four replicate solvent–polymer solutions were analyzed. The error bars for each data point indicate one standard deviation from the mean of four measurements obtained from individual crystals. The discrimination ability of the developed system is shown for (A) highly soluble and (B) poorly soluble polymers.

These data indicate the simplicity and effectiveness of this determination for both highly soluble [Figure 11.21(a)] and sparingly soluble [Figure 11.21(b)] polymers.

Application Example 3: Identification and Quantitation of Combinatorial Samples

Another sensor-based HTS system was developed and demonstrated for identification and quantitation of volatile substances in combinatorial chemical libraries. HTS of volatiles is required for a broad range of combinatorial chemistry applications ranging from the discovery of new catalysts [40,41] to chemically sensitive materials [42]. A variety of methods have been used including gas chromatography, mass spectrometry, laser-induced fluorescence, and resonance enhanced multiphoton ionization [40,41, 43–46]. Chemical sensors can be an appealing alternative for these and other applications because these instruments potentially afford many attractive features for screening of combinatorial libraries as indicated in Table 11.1. Also, by combining several broadly selective chemical sensors into an array and analyzing the data using multivariate techniques the selectivity of the response is dramatically improved [47,48].

Our sensor-based HTS system has been used for the identification and quantitation of volatile substances from multiple samples such as those created in combinatorial libraries for screening catalyst candidates in the arene oxidation process [49–52]. This industrially important process involves the catalytically supported production of oxygen-containing arene derivatives such as cresol, benzoquinone, phenol, dihydric phenols, and some others. The catalyst activity is evaluated based on the amount of oxygen-containing arene derivatives generated in the reaction using time-consuming chromatographic methods.

Our measurement method employs a combination of a periodic introduction of a minute amount (1–2 μL) of a liquid reaction sample into the HTS system, rapid evaporation of different volatile components in the sample at room temperature, and dynamic measurement of a generated vapor pulse using an array of TSM sensors. The temporal modulation of the concentration of analyte vapor and measurement of both the temporal profile and the magnitude of the response improves sensor selectivity and makes possible robust identification and quantitation of individual components in multicomponent combinatorial mixtures with a reduced number of sensors in the array.

The approach described here has several important advantages for HTS compared with other types of temporal separation and analysis of vapor mixtures. For example, one of the ways of producing a temporal separation of chemical compounds is to use a preconcentration of mixture components onto a sorptive material followed by thermal desorption of vapors [53–55]. However, while this approach improves the selectivity and sensitivity in response of a sensor array, it also considerably reduces the throughput of analysis, introduces additional baseline drift, and can thermally alter the analytes. These limitations are caused by the need for a large amount of sample to be drawn through the preconcentrator for extended periods of time (60–840 s) followed by heating to 160–200°C for desorption. Another method utilizes conventional gas chromatographic (GC) separation and a single sensor or a sensor array as a GC detector [56,57]. This approach requires a significant amount of time for all vapors to elute from a specially made GC column coated with a stationary phase, a certain type of a carrier gas (typically helium, argon, nitrogen, or hydrogen), and operation at a constant very low (<100 mL/min) flow rate in order to achieve the required separation.

In contrast, our detection method has none of the limitations of the existing methods outlined above because it does not rely on the sorption–desorption characteristics of supporting system elements such as preconcentrators or GC columns. Further, our method of dynamic sensor-based measurements allows instrument miniaturization by reducing the number of sensors in an array and increases the usefulness of the instrument in space-limited applications.

For this application, an array of four TSM sensors was adapted from our environmental monitoring applications [8,30]. This sensor system has been optimized for low noise (RMS 0.03–0.05 Hz) operation [8]. The crystals were coated with a variety of chemically sensitive polymer films [8] responding to polar and nonpolar vapors and arranged in a flow-through gas cell with a small dead volume. Upon exposure to the vapor pulse, the frequency of oscillation of the coated crystals varied as a function of both the composition of the sample stream and the concentration of the chemical species of interest [1,56]. A temporally modulated combination of responses from each of the sensors was recorded and provided signature patterns indicating the presence of certain chemicals.

A schematic diagram of the sensor-based HTS system developed is presented in Figure 11.22. A library of samples was arranged on a single-axis translation stage. An auto-injector was used to introduce a 2 μL volume of solution from each library element into the gas delivery line through an injection port. The injection port, gas delivery lines, sensor array, and other system components operated at room temperature. Each sample was evaporated upon injection into the system and the vapor was directed into the gas cell containing the sensor array. The injection port and the gas cell were connected with a 10 cm section of 1.6 mm i.d. Teflon tubing. Dry air was used as a carrier gas at a flow rate of 0.45–0.5 L min^{-1} to deliver the vapor to the sensor array. To reduce the recovery time of the sensor array, a purge line was used with a flow rate of 4.5–5.0 L min^{-1}. This line operated with the same gas for 90–250 s after a 150–380 s delay from injection of the sample into the HTS system. The flows in the carrier and purge lines were automatically

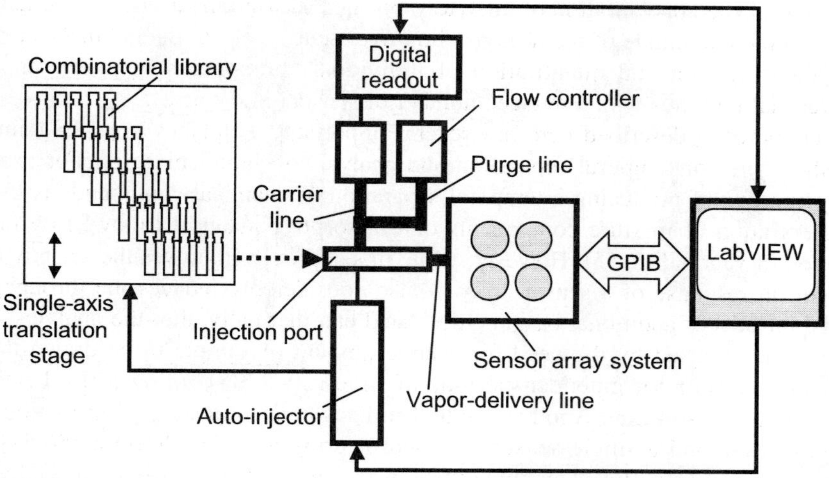

FIGURE 11.22. Schematic diagram of the sensor-based HTS system for identification and quantitation of volatile components in combinatorial mixtures. The diagram shows the four-element acoustic wave TSM sensor array system for the dynamic identification and quantitation of volatile components in combinatorial mixtures. (Adapted from ref. 16.)

regulated using individual mass-flow controllers via a four-channel readout unit. Determination of volatile arene oxidation products (cresol, benzoquinone, and phenol) was demonstrated using a 32-element model library of these analytes in different solvent mixtures containing benzene, toluene, acetonitrile, and water. The types of arene oxidation products generated are governed by the starting material [49–52].

Typical dynamic responses of four sensors are illustrated in Figure 11.23. The data demonstrate the diverse responses of different sensors to various vapors present in the samples. The vapors of solvents are more volatile (have lower boiling points) than the arene oxidation products. Thus the response of sorption-based sensors is faster for the solvent vapors and has a pronounced time delay for arene oxidation products. Specifically, sensor A could be useful for analysis of chemical compounds with high boiling points as indicated by the sensor response approximately 50 s after sample injection. Sensor B could be useful for analysis of chemical compounds with both low and high boiling points approximately 20 s and 50 s after sample injection, respectively. Sensor C could be useful for analysis of chemical compounds with low boiling points approximately 20 s after sample injection. Sensor D could be useful for analysis of chemical compounds with both low and high boiling points approximately 20 s and 50 s after sample injection, respectively.

Under the conditions used, the highest sensitivity for analysis of chemical compounds with low boiling points was provided by sensor D, where the maximum signal change for these types of compounds was about 140 Hz. The highest sensitivity for the analysis of chemical compounds with high boiling points was provided by sensor B, where the maximum signal change for these types of compounds was about 110 Hz. These initial experiments demonstrated that generation of vapor pulses and their dynamic analysis permits determinations of vapor components with reduced number of sensors. In particular, it was found that, based on the need for evaluation of arene oxidation products with

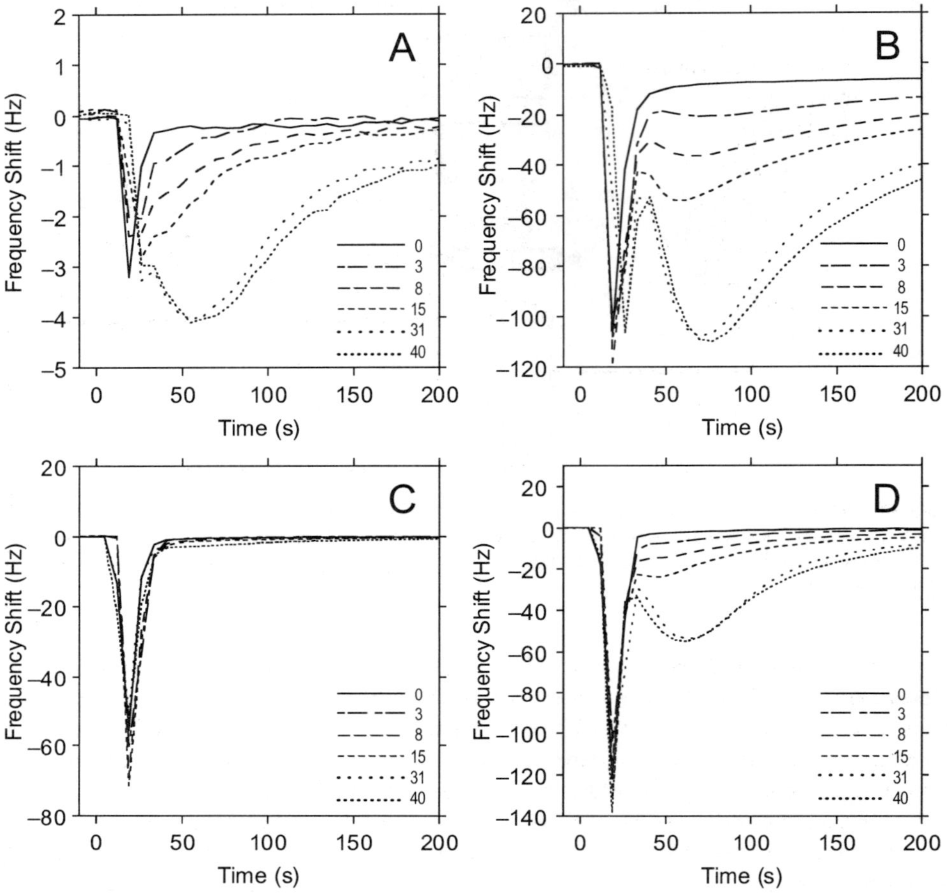

FIGURE 11.23. Time-dependent frequency changes of individual TSM sensors upon exposure to vapor pulses introduced into the HTS system. Vapor pulses are generated by evaporation of 2 μL samples of cresol solutions in toluene–acetonitrile (1:1) at room temperature. Amounts of cresol (0, 3, 8, 15, 31, and 40 μg) are indicated with different line styles. Time $t = 0$ s indicates sample introduction into the HTS system. (Adapted from ref. 16.)

high boiling points in combinatorial mixtures, sensors A, B, and D could be useful, while sensor C does not respond to these analytes. Further, because each sensor had a comparable noise level of less than 0.05 Hz RMS noise [8], a sensor with the largest response would be able to provide the best signal-to-noise performance. Sensor B had the largest response and thus was selected for further applications in the quantitation of arene oxidation products.

The speed of analysis of the sensor system was further evaluated with maximum flows of carrier and purge lines of 0.5 L min^{-1} and 5.0 L min^{-1}, respectively, available from the gas-flow controllers. At the conditions used, maximum sensor response from the arene oxidation component was observed 10 s after sample injection and can be monitored for 50 s before the operation of the purge line. Such a short response time is an important advantage of the developed instrument over conventional GC instruments that typically require 10–15 min for analysis of arene oxidation products. It is important to note that in the present design of the HTS system, the minimum time delay period was limited by the requirements to rinse the injector and purge the sensor. If necessary, the duration of purge line operation can easily be decreased by using a higher purging flow. In addition, flow of the vapors into the sensor cell be can easily increased by using a faster rate of acquisition signals from the acoustic-wave sensors.

The time-induced selectivity of sensor response predicted by Eqn (7) was used for quantitation of cresol and benzoquinone in mixtures. In these determinations, the temporal profiles of sensor response were analyzed using multivariate regression analysis tools such as locally weighted regression [30, 58, 59]. This method is especially useful for modeling nonlinear systems such as response of the sensor to cresol. Data collected with the sensor were preprocessed before constructing and validating the calibration model. Preprocessing involved taking the first derivative of the temporal profiles to improve the reproducibility. For analysis, a time window between 50 and 150 s after sample injection was utilized. Results of the multivariate analysis are shown in Figure 11.24 as a plot of predicted versus known concentrations for the validation samples of complex combinatorial mixtures containing variable concentrations of cresol, benzoquinone, and interfering vapors. The samples in Figure 11.24 cluster about the ideal line of the slope of unity, which indicates a good prediction ability of the developed model [60]. To quantify the prediction accuracy of the developed locally weighted regression model, the root mean squared error of prediction (RMSEP) was determined [60,61]. The RMSEPs for cresol and benzoquinone were calculated to be 1.5 μg and 1.8 μg, respectively.

This study demonstrated that the developed chemical sensor system was an attractive HT measurement alternative to conventional GC instruments in determinations of volatiles in complex mixtures. In addition, our detection method allowed a reduction in the number of sensor elements necessary for effective monitoring by measuring both the temporal profile of the sensor response and the magnitude of the signal change at a given time.

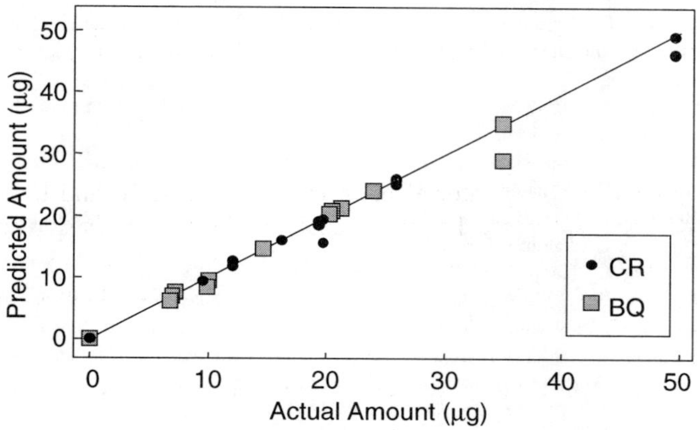

FIGURE 11.24. Multivariate validation plot between actual and predicted concentrations of cresol (CR) and benzoquinone (BQ) in complex mixtures using a developed locally weighted regression model. The solid line with unit slope indicates perfect predictive ability of the model. (Adapted from ref. 16.)

Reduction of the number of sensor elements in a sensor array led to a decrease in the physical size of the array. The demonstrated high signal-to-noise ratio in the analysis of 2 μL samples provides a straightforward possibility for analyzing sample volumes at least an order of magnitude smaller when robust nanoliter sampling systems become available. This method of dynamic sensor-based measurement allowed instrument miniaturization and increased the usefulness of the instrument in space-limited applications. Upon operation of multiple sensors in parallel, effective analysis of dense combinatorial libraries of materials is possible.

ACKNOWLEDGMENTS

We thank our co-workers, whose names are given in the references, for their contributions and Douglas Forman for preparing the video. This work has been supported by the US Department of Energy (Grants DE-AC26-01NT41188 and DEFC07-01ID14093).

REFERENCES

1. Mandelis, A., Christofides, C. *Physics, Chemistry and Technology of Solid State Gas Sensor Devices*, **1993**, New York: John.
2. Baltes, H., Göpel, W., Hesse, J. (Eds.) *Sensors Update*, **1996**, Weinheim, Germany: VCH.
3. Janata, J., Josowicz, M., Vanysek, P., DeVaney, D. M. *Anal. Chem.* **1998**, *70*, 179R–208R.
4. Potyrailo, R. A., Hobbs, S. E., Hieftje, G. M. *Fresenius J. Anal. Chem.* **1998**, *362*, 349–373.
5. Albert, K. J., Lewis, N. S., Schauer, C. L., Sotzing, G. A., Stitzel, S. E., Vaid, T. P., Walt, D. R. *Chem. Rev.* **2000**, *100*, 2595–2626.
6. Soloman, S. *Sensors Handbook*, **1999**, New York: McGraw-Hill.
7. Ballantine, D. S., Jr., White, R. M., Martin, S. J., Ricco, A. J., Frye, G. C., Zellers, E. T., Wohltjen, H. *Acoustic Wave Sensors: Theory, Design, and Physico-Chemical Applications*, **1997**, San Diego, CA: Academic Press.
8. Potyrailo, R. A., Sivavec, T. M., Bracco, A. A. *Proc. SPIE—Int. Soc. Opt. Eng.*, **1999**, *3856*, 140–147.
9. Dickert, F. L., Forth, P., Tortschanoff, M., Bulst, W. E., Fischerauer, G., Knauer, U. IEEE International Frequency Control Symposium, 28–30 May, **1997**, Orlando, FL: The Institute of Electrical and Electronics Engineers, pp. 120–123.
10. Brecht, A., Burckardt, R., Rickert, J., Stemmler, I., Schuetz, A., Fischer, S., Friedrich, T., Gauglitz, G., Goepel, W. *J. Biomol. Screen.* **1996**, *1*, 191–201.
11. Matsiev, L., Bennett, J., McFarland, E. *Method and apparatus for characterizing materials by using a mechanical resonator*, World Patent Appl. WO 99/18431, **1999**.
12. Matsiev, L., Bennett, J., McFarland, E. *Method and apparatus for characterizing materials by using a mechanical resonator*, US Patent 6,336,353, **2002**.
13. Lucklum, R., Behling, C., Hauptmann, P. *Anal. Chem.* **1999**, *71*, 2488–2496.
14. Thompson, M., Stone, D. C. *Surface-Launched Acoustic Wave Sensors: Chemical Sensing and Thin-Film Characterization*, **1997**, New York: John Wiley.
15. Grate, J. W., Abraham, H., McGill, R. A. In E. Kress-Rogers (Ed.), *Handbook of Biosensors and Electronic Noses. Medicine, Food, and the Environment*, **1997**, Boca Raton, FL: CRC Press, pp. 593–612.
16. Potyrailo, R. A., May, R. J. *Rev. Sci. Instrum.* **2002**, *73*, 1277–1283.
17. Herrmann, F., Jakoby, B., Rabe, J., Büttgenbach, S. In H. Baltes, J. Hesse, and J. G. Korvink, (Eds.), *Sensors Update*, Vol. 9, **2001**, Weinheim, Germany: VCH, pp. 105–160.
18. Kraus, P. R., Cooper, F. L., Emmons, D. H., Ferguson, S., McClain, R. D. In *Sicon'01 Sensors for Industry Conference*, **2001**, Rosemont, IL: IEEE pp. 241–245.

19. Benes, E., Gröschl, M., Seifert, F., Pohl, A. IEEE International Frequency Control Symposium, 28–30 May **1997**, Orlando, FL: The Institute of Electrical and Electronics Engineers, pp. 5–20.
20. Westerick, J. J., Mello, J. M., Thomas, R. F. *J. Am. Water Works Assoc.* **1984**, *76*, 52–59.
21. DeWeerd, K. A., Flanagan, W. P., Brennan, M. J., Principe, J. M., Spivack, J. L. *Biorem. J.* **1998**, *2*, 29–42.
22. Harkness, M. R., Bracco, A. A., Franz, T., Tsentas, C., Becker, M., Huber, W. A., Orient, J. P., Rich, C. A., Figura, M. In G. B. Wickramanayake and R. E. Hinchee (Eds.), *Natural Attenuation. Chlorinated and Recalcitrant Compounds*, **1998**, Columbus, OH: Battelle Press, pp. 231–236.
23. Freedman, D. L., Gossett, J. M. *Appl. Environ. Microbiol.* **1989**, *55*, 2144–2151.
24. de Bruin, W. P., Kotterman, M. J. J., Posthumus, M. A., Schraa, G., Zehnder, A. J. B. *Appl. Environ. Microbiol.* **1992**, *58*, 1996–2000.
25. Barrio-Lage, G. A., Parsons, F. Z., Nassar, R. S., Lorenzo, P. A. *Environ. Sci. Technol.* **1986**, *20*, 96–99.
26. Barrio-Lage, G. A., Parsons, F. Z., Nassar, R. S., Lorenzo, P. A. *Environ. Toxicol. Chem.* **1987**, *6*, 571–578.
27. Parsons, F. Z., Wood, P. R., DeMarco, J. *Am. Water Assoc.* **1984**, *76*, 56–59.
28. Potyrailo, R. A., Sivavec, T. M., Stokes, E. B., Mackenzie, P. D., Salvo, J. J., Presented at *Pittsburgh Conference on Analytical Chemistry and Applied Spectroscopy, Orlando, FL*, **1999**, Paper 724.
29. Potyrailo, R. A., Sivavec, T. M. Presented at *Federal Remediation Technologies Roundtable, St. Louis, MO*, **June 1999**.
30. Potyrailo, R. A., May, R. J., Sivavec, T. M. *Proc. SPIE—Int. Soc. Opt. Eng.* **1999**, *3856*, 80–87.
31. Potyrailo, R. A., May, R. J., Sivavec, T. M. *Recognition and quantitation of closely related chlorinated organic vapors with acoustic wave chemical sensor arrays*, GE CRD Tech. Rep. 2000CRD093, **April 2001**, Schenectady, NY: GE CRD.
32. Potyrailo, R. A., Sivavec, T. M. In *Abstracts of 9th International Meeting on Chemical Sensors, Boston, MA, July 7–10, 2002*, **2002**, Amsterdam: Elsevier Science, p. 119.
33. Sivavec, T. M., Potyrailo, R. A. *Polymer coatings for chemical sensors*, US Patent 6,357,278 B1, **2002**.
34. Hara, H. *Compounds for solvent resistant silicone gels*, US Patent 5,519,096, **1996**.
35. Hofmann, G. H. *Polymer blends*, US Patent 5,981,008, **1999**.
36. Ueno, R., Kitayama, M., Minami, K., Wakamori, H. *Azo compounds and process for producing the same*; US Patent 5,973,126, **1999**.
37. Yamaya, M., Kizaki, H., Furuya, M., Yamamoto, A. *Silicone resin-containing emulsion composition, method for making same, and article having a cured film of same*, US Patent 5,973,068, **1999**.
38. Enokida, T., Akimoto, H., Tatsu, H. *Fluoroelastomer and its cross-linkable composition*, US Patent 5,969,066, **1999**.
39. Matsuda, T., Sakamoto, T., Sato, S., Koike, N., Tarumi, Y., Kimura, T., Yamamoto, Y., Arai, M. *Room temperature curable fluoropolymer composition; and fluorine-containing organosilicon compounds, a method of producing the same, and room temperature curable silicone composition containing the same*, US Patent 5,919,886, **1999**.
40. Senkan, S. M. *Nature* **1998**, *394*, 350–353.
41. Su, H., Yeung, E. S. *J. Am. Chem. Soc.* **2000**, *122*, 7422–7423.
42. Matzger, A. J., Lawrence, C. E., Grubbs, R. H., Lewis, N. S. *J. Comb. Chem.* **2000**, *2*, 301–304.
43. Sigman, M. S., Jacobsen, E. N. *J. Am. Chem. Soc.* **1998**, *120*, 4901–4902.
44. Newcomb, W. S., Deegan, T. L., Miller, W., Porco, J., J. A. *Biotechnol. Bioeng. (Comb. Chem.)* **1998**, *61*, 55–60.
45. Newsam, J. M., Schuth, F. *Biotechnol. Bioeng. (Comb. Chem.)* **1998–1999**, *61*, 203–216.
46. Cong, P., Doolen, R. D., Fan, Q., Giaquinta, D. M., Guan, S., McFarland, E. W., Poojary, D. M., Self, K. W., Turner, H. W., Weinberg, W. H. *Angew. Chem. Int. Ed.* **1999**, *38*, 484–488.
47. Hierlemann, A., Schweizer-Berberich, M., Weimar, U., Kraus, G., Pfau, A., Göpel, W. In H., Baltes, W., Göpel, and J. Hesse (Eds.), *Sensors Update*, Vol. 2, **1996**, Weinheim, Germany: VCH, pp 119–180.
48. Gardner, J. W., Hines, E. L. In E. Kress-Rogers, (Ed.), *Handbook of Biosensors and Electronic Noses. Medicine, Food, and the Environment*, **1997**, Boca Raton, FL: CRC Press, pp. 633–652.
49. Kitano, T., Kuroda, Y., Itoh, A., Li-Fen, J., Kunai, A., Sasaki, K. **1990** *J. Chem. Soc., Perkin Trans. 2*, 1991–1995.
50. Barnhart, T. M., Hughes, A. W. *Method for hydroxylating aromatic compounds*, US Patent 5,912,391, **1999**.
51. Ishida, M.-A., Masumoto, Y., Hamada, R., Nishiyama, S., Tsuruya, S., Masai, M. **1999** *J. Chem. Soc., Perkin Trans. 2*, 847–853.
52. Barnhart, T. M. *Method and composition for hydroxylation of aromatic substrates*, US Patent 6,232,510, **2001**.

53. Kindlund, A., Sundgren, H., Lundstrom, I. *Sens. Actuators* **1984**, *6*, 1–17.
54. Groves, W. A., Zellers, E. T., Frye, G. C. *Anal. Chim. Acta* **1998**, *371*, 131–143.
55. Grate, J. W., Rose-Pehrsson, S. L., Venezky, D. L., Klusty, M., Wohltjen, H. *Anal. Chem.* **1993**, *65*, 1868–1881.
56. Guilbault, G. G. In C. Lu and A. W. Czanderna (Eds.), *Applications of Piezoelectric Quartz Crystal Microbalances*, **1984**, Amsterdam: Elsevier, pp. 251–280.
57. Breimer, M. A., Masila, M., Sadik, O. A. Presented at *Pittsburgh Conference on Analytical Chemistry and Applied Spectroscopy, Orlando, FL*, **1999**, Paper 402.
58. Cleveland, W. S., Devlin, S. J. *J. Am. Stat. Assoc.* **1988**, *83*, 596–610.
59. Naes, T., Isaksson, T., Kowalski, B. *Anal. Chem.* **1990**, *62*, 664–673.
60. Beebe, K. R., Pell, R. J., Seasholtz, M. B. *Chemometrics: A Practical Guide*, **1998**, New York: John Wiley.
61. Qu, J. Y., Shao, L. *Rev. Sci. Instrum.* **2001**, *72*, 2717–2723.
62. Jaramillo, T. F., Ivanovskaya, A., McFarland, E. W. *J. Comb. Chem.* **2002**, *4*, 17–22.
63. McAllister, M., Walt, D. R. *Proc. SPIE—Int. Soc. Opt. Eng.* **1994**, *2068*, 213–215.
64. Michael, K. L., Taylor, L. C., Schultz, S. L., Walt, D. R. *Anal. Chem.* **1998**, *70*, 1242–1248.
65. Hessling, M., Ihlemann, J., Marowsky, G. *Rev. Sci. Instrum.* **2000**, *71*, 2201–2205.

Chapter 12

High-Throughput Screening of Oxidation Catalysts with Gas Sensors

Yusuke Yamada and Tetsuhiko Kobayashi

INTRODUCTION

The rapid evaluation of catalysis has been an important technology for accelerating the investigation of heterogeneous catalysts because the catalysis evaluation process often seems to be a bottleneck in experiments. For example, the volatiles in a reaction gas have been analyzed by gas chromatography. It takes from several minutes to hours for the gas analysis of each catalyst by gas chromatography. On the other hand, it is not difficult to prepare dozens of catalysts at a time by conventional catalyst preparation methods such as impregnation and the sol–gel technique. For the rapid screening of a series of catalysts, the rapid screening technique should be applied instead of conventional gas chromatography.

Several rapid evaluation technologies have been reported depending on the volatility of a reaction gas. An IR imaging camera has been used to evaluate the complete oxidation catalysts of hydrogen, isooctane, and toluene and the hydrogenation catalysts of 1-hexyne [1,2]. The reaction heat of each catalyst was directly correlated with its activity. When information on the catalytic products is required, gas sensors [3,4], mass spectrometry [5–9], resonance-enhanced multiphoton ionization [10,11], or IR spectroscopy have been used for the identification of the volatiles [12–14]. A gas analyser with a more selective, highly sensitive, and rapid response and smaller size is required owing to the progress in the miniaturization of the catalyst library and its reactor. A certain type of gas sensor has characteristics which meet these requirements [15].

Much effort has been devoted to improving the selectivity of gas sensors [16]. The selectivity of gas sensors has been improved [17] sufficiently for the screening of heterogeneous catalysts, although the selectivity is not yet better than that of conventional gas analyses such as gas chromatography in general, because the volatiles in a reaction gas are

Yusuke Yamada and Tetsuhiko Kobayashi • Special Division for Green Life Technology, National Institute of Advanced Industrial Science and Technology (AIST), 1-8-31 Midorigaoka, Ikeda, Osaka 563-8577, Japan
Contact author: Yusuke Yamada, email: Yusuke.YAMADA@aist.go.jp

High-Throughput Analysis, edited by Potyrailo and Amis
Kluwer Academic Publishers, New York, 2003.

usually known and quantification of each volatile is often the only purpose of the product analysis. When the selectivity of a single-gas sensor is not sufficient to determine the concentration of volatiles, a gas sensor system with multiple gas sensors of different types can be used. The outputs from the sensor system are analyzed by principal component analysis or an artificial neural network [18,19]. Improvements have been made not only to the gas sensor itself but also to the chemo-informatics. These technologies improve the selectivity of the gas sensors sufficiently to quantify the volatiles in the reaction gas if appropriate sensors are selected.

In order to select suitable sensors for each catalytic system, an understanding of their principles is indispensable. This understanding helps us to analyze the data obtained correctly. For example, the presence of ethane does not interfere with the estimation of the ethylene concentration using an IR-based gas sensor, but detection of ethylene in the presence of ethane is almost impossible with a combustible gas sensor. Both gas sensors are probably satisfactory for determining the ethylene concentration in air, and the combustible gas sensor has a higher sensitivity for ethylene than the IR-based gas sensor. Gas sensors should be chosen carefully on the basis of considerations of their principles of operation.

The basic principles of a semiconductor quartz oscillator IR-based sensors, combustible gas sensors, and solid electrolyte gas sensors are described in the next section. This is followed by a review of reported examples of gas sensor systems for the evaluation of heterogeneous catalysts. These examples indicate how gas sensors should be used for the rapid screening of heterogeneous catalysts.

BASIC PRINCIPLES OF GAS SENSORS

Semiconductor Sensors [20,21]

There are two types of semiconductor sensors: the surface-controlled type and the bulk-controlled type. Of these, the surface-controlled type is frequently used while the bulk-controlled type is only used as an oxygen sensor at high temperatures. Advantages of semiconductor-type gas sensors are their simple structure, high sensitivity, rapid response, and long-term stability. Recently, their selectivity of volatiles has been dramatically improved. This type of sensor seems to be useful for the evaluation of catalysts in many systems.

The surface-controlled type of sensor detects the change produced in its electrical resistance by adsorption of volatiles at the semiconductor surface. Metal oxides such as SnO_2, ZnO, In_2O_3, Fe_2O_3, WO_3, V_2O_5, etc. are generally used as the base semiconductor. They are usually heated at in air 100–500°C. Under these conditions, molecular oxygen is absorbed on the semiconductor surface as O_2 or the O^{2-} ion as shown in Figure 12.1. These oxygen atoms receive electrons from the semiconductor surface owing to their high electron negativity. This phenomenon induces a lower electric conductivity in the semiconductor. The anionic oxygen has a high reactivity toward flammable volatiles. After the oxygen reacts with the volatiles, the electron trapped by the oxygen atoms is released and the electron conductivity of the semiconductor increases. The change in electron conductivity is recorded as the output of the semiconductor-type sensor.

Loading of catalytic materials on a semiconductor enhances the gas sensor sensitivity and/or selectivity. For example, palladium loading on SnO_2 greatly enhances its sensitivity because palladium promotes the dissociative adsorption of a flammable volatile. The dissociatively adsorbed species easily reacts with adsorbed oxygen on the SnO_2 surface. Coverage of SnO_2 with TiO_2, WO_3, etc. improves the selectivity of the sensor. These covered metal oxides are well known as partial oxidation catalysts. They probably modify the depth of oxidation of the flammable volatile. These selective gas sensors are used in odor discrimination systems that are commercially available. Modification of the catalytic layer will improve the selectivity of the semiconductor gas sensor for the evaluation of catalysts.

Acoustic Wave Devices [22,23]

Two device types have been used as sensors: thickness-shear mode (TSM) [also known as quartz crystal microbalance (QCM)] resonators, and bulk acoustic wave (BAW) and surface acoustic wave (SAW) resonators. The difference between the two types is in the direction of acoustic wave propagation; the working principles are the same.

A quartz crystal coated with an organic film has the potential to detect various kinds of volatiles with a high selectivity and sensitivity. Usually, an organic film is used as an absorbent and provides selectivity to the gas sensor. Table 12.1 summarizes the detectable gas and the organic adsorbent on quartz oscillator sensors. It includes some volatiles whose detection seems to be difficult with other kinds of gas sensors.

The operating principle of this gas sensor is often based on the gravimetric detection as described by [24]

$$\Delta F = k\,\frac{F^2 \Delta W}{A} \tag{1}$$

Where F is the resonance frequency, ΔF is the decrease in F, ΔW is the increase in mass accompanying adsorption, and A is the surface area of the adsorbent.

The quartz oscillator has a specific resonance frequency with an oscillating circuit. Its frequency decreases when volatiles are absorbed on the surface of the quartz because of the increasing mass. The shift in frequency on exposure to the volatile depends on the amount absorbed. The decrease in frequency for a 9 MHz quartz oscillator, is 400 Hz per microgram of compound absorbed.

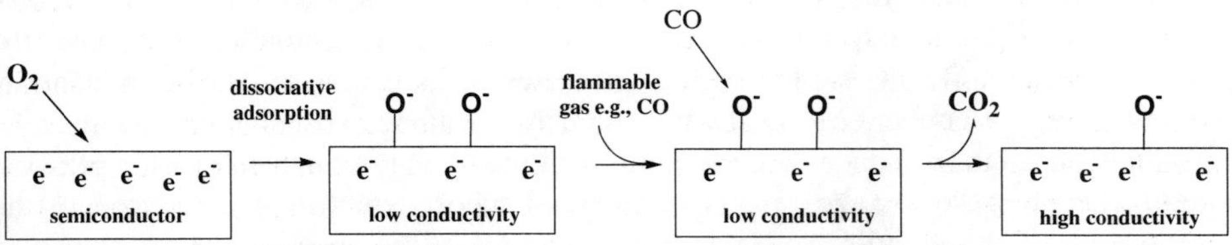

FIGURE 12.1. Schematic drawing of the working mechanism of a semiconductor.

TABLE 12.1. With acoustic wave device gas sensor

Detectable gas	Organic absorbent	Detectable concentration
NH_3	Ascorbic acid + silver nitrate	1 ppb–10 ppm
NO_2	Ucon75-H-90000	1 ppb–1 ppm
HCl	Trimethylamine hydrochloride	1 ppb–100 ppm
H_2S	Carbon	1–60 ppm
SO_2	Triethanol amine	10 ppb–30 ppm
Toluene	Carbowax 550	30–300 ppm
CO	Gold	5 ppb–50 ppm
H_2	Palladium	0–1%
H_2O	Polystyrene + polyamide	1 ppm–10%
Odorant	Bilayer lipid	
Insecticide	3-PAD + Triton X-100 + NaOH	1 ppb–15 ppm

The pioneering work on this kind of gas sensor was done by King in 1964 [25]; however, few TSM or SAW sensors are commercially available at present. The problem of the long-term stability on an organic adsorbent still remains. When an organic adsorbent with a long-term stability and high volatile selectivity is developed, this kind of gas sensor would provide the most promising method of analysis for the purpose of catalyst evaluation because of its high sensitivity and small size. The design features and several applications of the acoustic wave sensor arrays for high-throughput screening of materials are described elsewhere in this book (Chapter 11).

IR Gas Sensors

Non-dispersive IR sensors are suitable for the detection of IR-sensitive materials such as CO_2 and CO. The sensor consists of a nondispersive light source, a filter, and a photon detector as shown in Figure 12.2. The light energy transmitted through the volatiles is compared with that transmitted through the filter, usually the target compound sealed in a cell. The difference in light energy is detected as pressure difference and is converted to gas concentration. Interference from a nontargeted compounds can easily be removed by another filter.

Photoacoustic spectroscopy (PAS) has the advantage of high sensitivity. Commercially available PAS instruments using IR radiation detect the various chemicals to the order of parts per million [26]. The cell volume into which the gas is introduced is as small as 2 mL. The PAS utilizes the photon absorption properties of the volatiles. When a voltaile is irradiated by light of a specific wavelength, the molecules became excited. After the radiation is turned off, the excited molecule relaxes to its rest state. During relaxation, the excess energy is converted to heat which induces a volume expansion of the molecule. Intermittent irradiation induces a continuous volume change which induces a pressure change. The pressure change results in a sound whose volume is converted to the concentration of the volatile. A diagram of a photoacoustic spectrometer is shown in Figure 12.3.

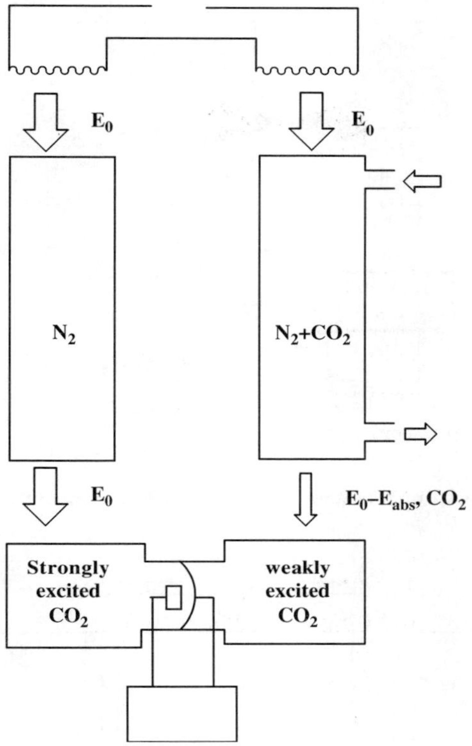

FIGURE 12.2. Structure of a nondispersive IR gas sensor.

Combustible Sensor

A combustible gas sensor is very sensitive and stable over the long term because of its very simple structure. Usually, a platinum wire is coated with a palladium or rhodium catalyst. The elements are heated to 250–400°C. When a flammable gas contacts the wire, it burns and its reaction heat induces an increase in resistance which is correlated with the concentration of the flammable volatiles. The selectivity of the gas sensor is not very high for flammable volatiles. When the catalytic system includes only one kind of flammable volatile, the combustible gas sensor is useful for detecting volatility.

Solid Electrolyte Sensors [17,27]

Solid electrolyte gas sensors have mainly been used as oxygen detectors in, for example, control of the ratio of air to fuel in an engine room, the burning control of an industrial gas burner, and detection of oxygen contamination during the refining of iron. Usually, stabilized zirconia with 8% yittria is used for the solid electrolyte at high temperature. Its gas selectivity is very high because of its working principle described below.

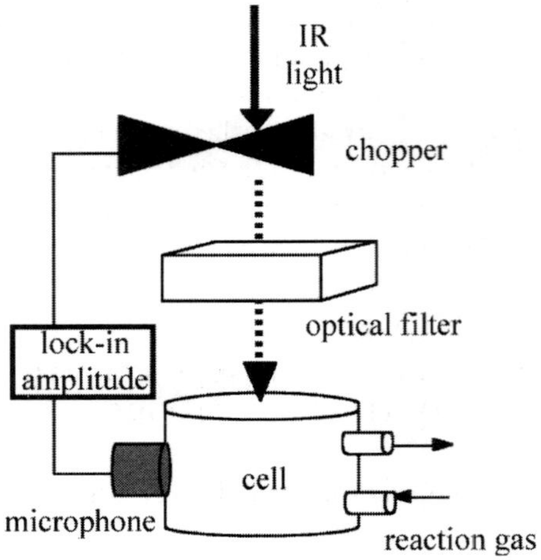

FIGURE 12.3. Structure of a photoacoustic spectrometer.

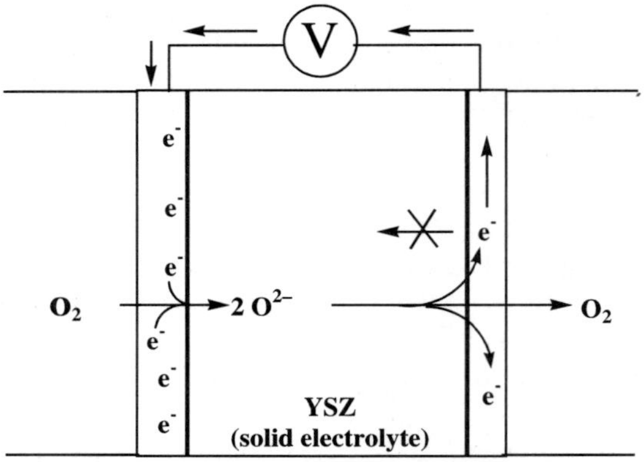

FIGURE 12.4. Structure of a solid electrolyte gas sensor.

Figure 12.4 indicates the working principle of the oxygen detector. A stabilized zirconia plate has platinum electrodes on both sides. When passed over the platinum electrode, molecular oxygen easily decomposes to anionic O^{2-}. When the concentration of O_2 or O^{2-} is higher than that on the opposite side of the electrode, a voltage occurs because the O^{2-} on the side with higher O_2 concentration penetrates into the zirconia and produces O_2 and electrons. The strength of the voltage is given by [28]

$$E = \frac{RT}{4F}\ln\left[\frac{P_{O_2}(II)}{P_{O_2}(I)}\right] \tag{2}$$

where E is the voltage, R is the gas constant, T is the absolute temperature, F is the Faraday constant, and P_{O_2} (X) is the partial pressure of oxygen in cell X. The difference in oxygen concentration is estimated by the voltage level. The voltage depends on the difference in the concentration on both sides of a logarithmic scale. These gas sensors are suitable for determining a wide range of gas concentrations.

If a suitable solid electrolyte is available, the concentrations of hydrogen, halogens, CO_2 [29], hydrogen, SO_2 [30,31] and NO_x [32] can be determined. Silver bromide is used as a solid electrolyte for the determination of bromide concentration. K_2CO_3, Sb_2O_5–$2H_2O$, Li_2SO_4–Ag_2SO_4, and $Na_3Zr_2Si_2PO_{12}$–$NaNO_x$ are used for the detection of CO_2, hydrogen, SO_2, and NO_x, respectively.

ANALYSIS OF PRODUCTS FROM CATALYTIC REACTION WITH GAS SENSORS

CO Oxidation Reaction

The evaluation of CO oxidation in air has been performed using gas sensors. Hoffmann et al. [33] determined the concentrations of both CO and CO_2, which are the reactant and product in the reaction. Nondispersive IR (ND-IR) gas sensors were used for both volatiles. Yamada et al. [3] used potentiometric CO gas sensors for CO concentration monitoring. Usually, the products of the catalytic reactions should be detected; however, only monitoring of the reactant concentration is meaningful in the reaction. The potentiometric CO sensors are sufficiently small to allow parallel sensor systems to be made.

The reaction conditions for the validation of the ND-IR gas sensor are as follows. The catalysts tested were Au/TiO_2 and Au/Co_3O_4 which are well known as the best catalysts of CO oxidation at low temperature [34]. The catalysts were placed in 16 holes (1 cm i.d.) in a brass block with nine heaters to ensure homogeneous heating. A 46 mg sample of each catalyst was placed separately in each hole. Air containing 1% CO was passed through the catalyst bed at 250 mL min^{-1} (15.6 mL per well; SV = 20,000 mL h^{-1} g_{cat}^{-1}). This condition is typical for the CO oxidation reaction. The superior catalysis of CO oxidation by Au–TiO_2 and Au–Co_3O_4 was confirmed by the gas sensors [33].

Potentiometric gas sensors have also been used to evaluate the complete oxidation catalysis of CO [3]. Catalysts composed of precious metal films on metal oxide surfaces were evaluated. Figure 12.5 shows the typical output signal change during the methane oxidation reaction over Rh/TiO_2. The output signal decreased on heating the catalysts. This means that the CO concentration was decreased by the catalytic oxidation.

Five CO gas sensors were placed in parallel as analysis tools, which dramatically shortened the analysis time. To heat the five catalysts at the same time, an oven for the gas chromatograph was used. Gas sensors enable continuous monitoring of the CO concentration. The reaction temperature was increased continuously without a plateau step.

Combustion of hydrogen, methane, and volatile organic compounds can be also detected with a similar sensor system.

Selective Oxidation of Ethane and Propane to Oxygenate Compounds

The selective oxidation catalysis of the lower alkanes has been evaluated by gas sensors with high selectivity for the desirable products. An IR gas sensor is suitable for the

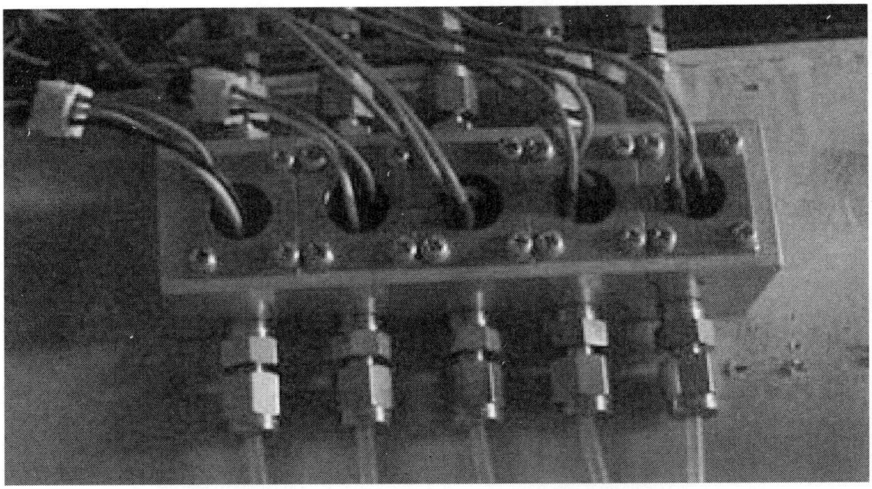

FIGURE 12.5. Five parallel combustible gas sensors for the evaluation of CO oxidation catalysts.

discrimination of multiple products because of its relatively high selectivity. The PAS instrument detects five types of reaction products during ethane oxidation [4]. Recent advances in semiconductor sensors enable oxygenates with keto groups to be distinguished from other flammable gases such as CO and propane. Yamada et al. [3] used semiconductor sensors for the detection and quantification of useful oxygenate compounds. These examples of the selective oxidation of ethane and propane are described below.

PAS was used for screening the selective oxidation of ethane. The catalysts tested were NO_2-absorbed metal oxides supporting various metal ions. Ethylene, ethanol, acetic acid, acetaldehyde, formaldehydes, and carbon oxides are formed in the reaction over these catalysts. It takes only 1 min to determine each gas concentration. The most important information collected is the concentration of C_2 oxygenate products (i.e., acetic acid, ethanol, and acetaldehyde). The sum of the concentration of acetic acid and ethanol was recorded because the reaction mechanism study indicates that acetic acid forms via ethanol in the catalytic system. Figure 12.6 shows the total yield of ethanol and acetic acid estimated by PAS in each catalytic reaction. It only took a week to obtain the data shown in Figure 12.6.

For the selective oxidation of propane, semiconductor sensors, an ND-IR sensor and a potentiometric gas sensor with an active carbon filter were used to determine the volatiles in the reaction gas. The reaction gas includes unreacted propane, carbon monoxide, carbon dioxide, aldehydes, and acetone. The concentration of the oxygenates without the carbon oxides was determined using a semiconductor sensor. The CO and CO_2 concentrations were determined using a potentiometric CO sensor and an ND-IR CO_2 gas sensor, respectively. In a typical case, the relative concentrations are almost 90% for propane, 4% for the carbon oxides, and 6% for the oxygenates. These relative concentrations can be converted to absolute concentrations of 18% for propane, 0.8% for carbon oxides, and 1% for oxygenates. These concentrations were too high for the sensors; the effluent was diluted to 1/20 with dry air.

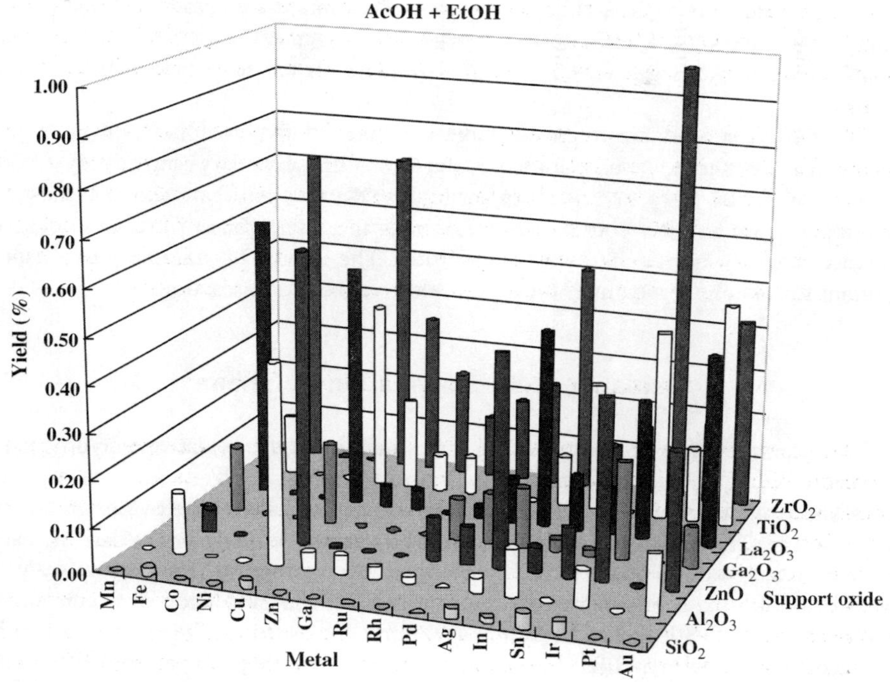

FIGURE 12.6. Total concentration of acetic acid (AcOH) and ethanol (EtOH) on ethane oxidation over supported metal oxide catalysts.

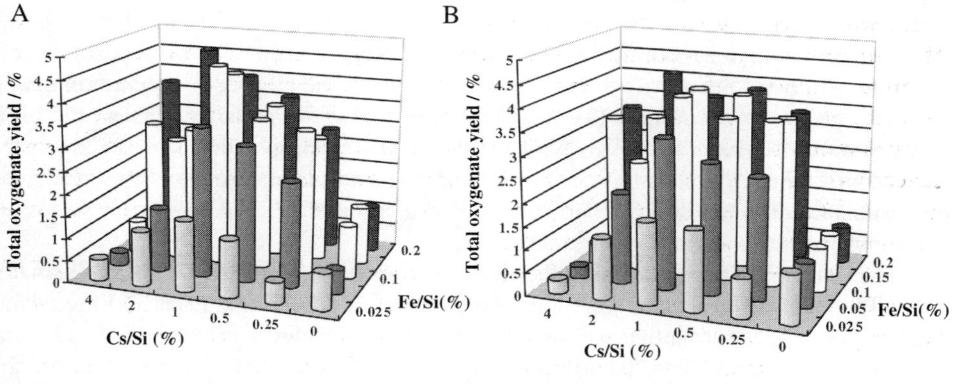

FIGURE 12.7. Oxygenate yields of propane oxidation over xCs–yFe–SiO_2 determined by (A) gas chromatography and (B) a gas sensor system (Cs/Si ranged from 0 to 4% and Fe/Si ranged from 0.025 to 0.20%).

Figure 12.7 shows a comparison of the oxygenate concentrations determined by a gas sensor system and a gas chromatograph. A difference of 0.1% yield of the oxygenate products was detected by the gas sensor system. The 0.1% difference in yield corresponds to a 30 ppm difference in absolute concentration. These gas sensors are sensitive enough for the 1/20 reduction of the gas concentration. The miniaturized scale is suitable for the primary stage catalysts. There is no selectivity for oxygenates including the keto group, also indicating that this gas sensor system is suitable for the screening of primary stage catalysts.

The IR gas sensors are relatively large compared with semiconductor gas sensors. It seems that they should be used for the evaluation of the secondary-stage catalyst library. The merit of the IR gas sensor is its rather high selectivity. Also, we can easily estimate interference from another compound. Sometimes the interference from an undesirable product can easily be reduced with a gas filter. The choice of suitable gas sensors is important for obtaining meaningful data on selective oxidation reactions.

Oxidative Dehydrogenation of Ethane

Two examples have been reported for the evaluation of the oxidative dehydrogenation of ethane with a gas sensor including gas analysis. One is an IR gas sensor system with a mass spectrometer which has difficulty with the quantification of ethylene in the presence of ethane. The sensor is highly selective and sensitive enough to evaluate the catalysis of the mixed oxides deposited on the substrate on a microgram or milligram scale. The second example is a semiconductor sensor system for the evaluation of a conventional reaction system with 300 mg of catalyst in a 10 mm i.d. glass tube. The semiconductor sensor discriminated and quantified the odorless ethylene despite the presence of flammable volatiles. Both the IR-based sensor and the semiconductor sensor are useful for the detection of ethylene in the presence of ethane.

An IR-based gas sensor (PTD detector), called the "mirage cell" in the literature, has been used for the evaluation of oxidative dehydrogenation. Cong et al. [5,6] used a mass spectrometer and the mirage cell to evaluate oxidative ethylene oxidation [5,6]. The concentration of ethylene was mainly determined by the mirage cell for the library during the primary catalytic screening. The amount of catalyst is 100–200 μg deposited on a substrate. Although the gas flow rate was not described, the ethylene concentration is of the order of parts per million. The mass spectrometer is not suitable because the high concentration of ethane interferes with the ethylene detection. The mass spectrometer detects only the CO_2 concentration. In fact, mass spectrometers are widely used for the discrimination of volatiles; however, this method is not perfect. The gas sensor was used as a supplement to the mass spectrometer.

The semiconductor sensor used in this evaluation was 13 wt% Al_2O_3–SiO_2 sensitized SnO_2. The catalysts evaluated were the mixed oxides of Ni–Fe–Ox. Each 300 mg of the catalyst placed in a conventional glass tube was passed through a mixture of N_2, O_2, and C_2H_6 at 120 mL/min ($SV = 20{,}000$ mL $h^{-1}g_{cat}^{-1}$). The effluent, diluted to 1 L with dry air, was introduced into the gas sensor system. Figure 12.8 shows the output signal of the gas sensor for CO, ethylene, and ethane. Although the sensitivity to CO is negligible compared with that to ethylene, the sensitivity to ethane was not negligible. Figure 12.9 shows the output signal of the gas sensor for a mixture of ethylene and ethane. When the concentration of ethylene ranged from zero to 10%, the output signal increased linearly in proportion to the ethane concentration. The selectivity of the gas sensor to ethylene is not perfect; however, it was selective enough for the evaluation of the ethylene concentration in the effluent.

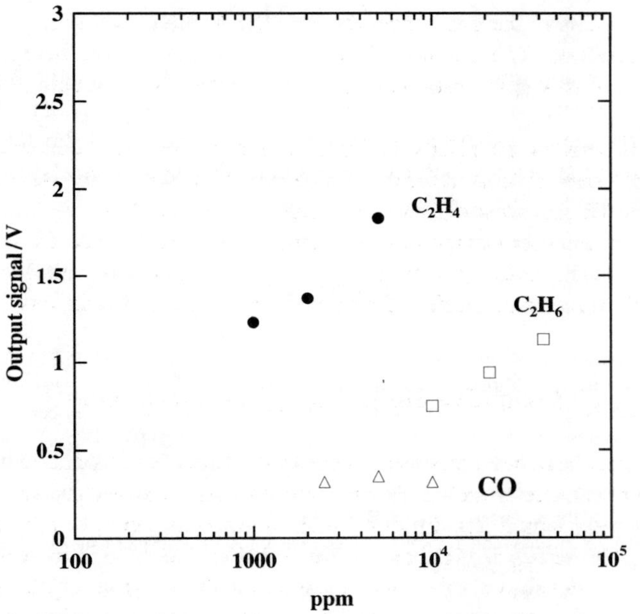

FIGURE 12.8. Output signals of a semiconductor gas sensor exposed to ethylene, ethane, and CO in air.

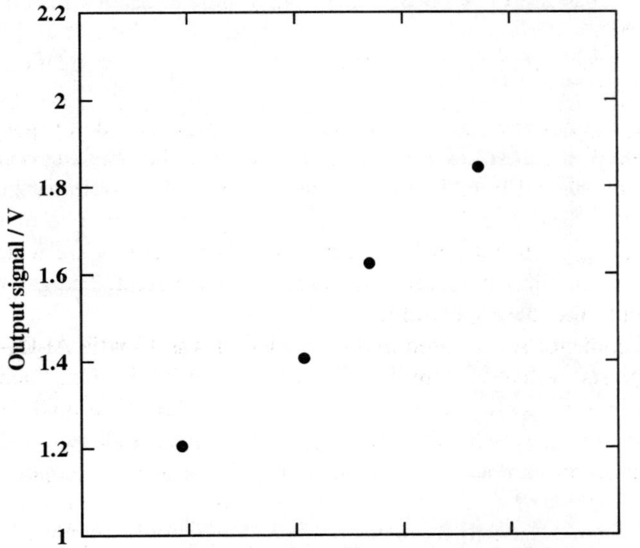

FIGURE 12.9. Output signal of the ethylene-selective gas sensor toward ethylene in ethane.

CONCLUSION

The small size and lower cost of gas sensors makes them suitable for parallel arrays of detectors that would dramatically shorten the evaluation time. The high sensitivity of these detectors matches that for the evaluation of a miniaturized catalytic library. The number of catalytic systems which can be evaluated by a gas sensor system will increase with the development of gas sensors.

REFERENCES

1. Moates, F. C., Somani, M., Annamalai, J., Richardson, J. T., Luss, D., Wilson, R. C. Infrared thermographic screening of combinatorial libraries of heterogeneous catalysts. *Ind. Eng. Chem. Res.* **1996**, *35*, 4801–4803.
2. Holzwarth, A., Schmidt, P. W., Maier, W. E. Detection of catalytic activity in combinatorial libraries of heterogeneous catalysts by IR thermography. *Angew. Chem. Int. Ed.* **1998**, *37*, 2644–2647.
3. Yamada, Y., Ueda, A., Zhao, Z., Maekawa, T., Suzuki, K., Takada, T., Kobayashi, T. Rapid evaluation of oxidation catalysis by gas sensor system: total oxidation, oxidative dehydrogenation, and selective oxidation over metal oxide catalysts. *Catal. Today* **2001**, *67*, 379–387.
4. Ueda, A., Yamada, Y., Kobayashi, T. Novel catalysts having NOx-adsorption sites for the selective oxidation of ethane. *Appl. Cat. A.* **2001**, *209*, 391–399.
5. Cong, P., Dehestani, A., Doolen, R., Giaquinta, D. M., Guan, S., Marcov, V., Poojary, D., Self, K., Turner, H., Weinberg, W. H. Combinatorial discovery of oxidative dehydrogenation catalysts within the Mo-V-Nb-O system. *Proc. Natl Acad. Sci. USA* **1999**, *96*, 11077–11080.
6. Cong, P., Doolen, R. D., Fan, Q., Giaquinta, D. M., Guan, S., McFarland, E. W., Poojary, D. M., Self, K., Turner, H. W., Weinberg, W. H. High-throughput synthesis and screening of combinatorial heterogeneous catalyst libraries. *Angew. Chem. Int. Ed.* **1999**, *38*, 483–488.
7. Liu, Y. M., Cong, P. J., Doolen, R. D., Turner, H. W., Weinberg, W. H. High-throughput synthesis and screening of V–Al–Nb and Cr–Al–Nb oxide libraries for ethane oxidative dehydrogenation to ethylene. *Catal. Today* **2000**, *61*, 87–92.
8. Holzwarth, A., Denton, P., Zanthoff, H., Mirodatos, C. Combinatorial approaches to heterogeneous catalysis: strategies and perspectives for academic research. *Catal. Today* **2001**, *67*, 309–318.
9. Senkan, S., Krantz, K., Ozturk, S., Zengin, V., Onal, I. High-throughput testing of heterogeneous catalyst libraries using arrays microreactors and mass spectrometry. *Angew. Chem. Int. Ed.* **1999**, *38*, 2794–2799.
10. Senkan, S. M. High throughput screening of solid-state catalyst libraries. *Nature* **1998**, *394*, 350.
11. Senkan, S. M., Ozturk, S. Discovery and optimization of heterogeneous catalysts by using combinatorial chemistry. *Angew. Chem. Int. Ed.* **1999**, *38*, 791–795.
12. Snively, C. M., Lauterbach, J. In *Combinatorial Catalysis and High Throughput Catalyst Design and Testing, Algarve, Portugal*, **1999**, pp. 311–313.
13. Snively, C. M., Oskarsdottir, G., Lauterbach, J. Parallel analysis of the reaction products from combinatorial catalyst libraries. *Angew. Chem. Int. Ed.* **2001**, *40*, 3028–3030.
14. Snively, C. M., Oskarsdottir, G., Lauterbach, J. Chemically sensitive parallel analysis of combinatorial catalyst libraries. *Catal. Today* **2001**, *67*, 357–368.
15. Tuller, H. L., Mlcak, R. Inorganic sensors utilizing MEMS and microelectronic technologies. *Curr. Opin. Solid State Mater. Sci.* **1998**, *3*, 501–504.
16. Thompson, M., Frank, M. D., Stone, D. C. *Interfacial Des. Chem. Sen.*, **1994**, *561*, 155–161.
17. Liu, C. C. Development of chemical sensors using microfabrication and micromachining techniques. *Mater. Chem. Phys.* **1995**, *42*, 87–90.
18. Jurs, P. C., Bakken, G. A., McClelland, H. E. Computational methods for the analysis of chemical sensor array data from volatile analytes. *Chem. Rev.* **2000**, *100*, 2649–2678.
19. Osbourn, G. C., Bartholomew, J. W., Ricco, A. J., Frye, G. C. Visual-empirical region-of-influence pattern recognition applied to chemical microsensor array selection and chemical analysis. *Acc. Chem. Res.* **1998**, *31*, 297–305.

20. Williams, D. E. Defect chemistry and gas sensing properties of semiconducting ceramic oxides. *Cryst. Lattice Defects Amorfph. Mater.* **1987**, *14*, 341–349.
21. Seker, F., Meeker, K., Kuech, T. F., Ellis, A. B. Surface chemistry of prototypical bulk II–VI and III–V semiconductors and implications for chemical sensing. *Chem. Rev.* **2000**, *100*, 2505–2536.
22. Grate, J. W. Acoustic wave microsensor arrays for vapor sensing. *Chem. Rev.* **2000**, *100*, 2627–2647.
23. Ricco, A. J., Crooks, R. M., Osbourn, G. C. Surface acoustic wave chemical sensor arrays: new chemically sensitive interfaces combined with novel cluster analysis to detect volatile organic compounds and mixtures. *Acc. Chem. Res.* **1998**, *31*, 289–296.
24. Sauerbrey, G. Z. *Phys* **1959**, *155*, 206–222.
25. King, W. H., Jr. *Anal. Chem.* **1964**, *36*, 1735–1739.
26. Web site http://www.innova.dk/genpages/WallChar/Wall.htm
27. Fisher, G. Ceramic sensors: providing control through chemical reactions. *Ceram. Bull.* **1986**, *65*, 622–629.
28. Kiukkola, K., Wagner, C. *J. Electrochem. Soc.* **1957**, *104*, 308–379.
29. Cote, R., Bole, C. W., Gauthier, M. K_2CO_3 solid electrolyte as a CO_2 probe: decomposition measurements of $CaCO_3$. *J. Electrochem. Soc.* **1984**, *131*, 63–67.
30. Singh, K., Bhoga, S. S. SO_x solid state gas sensors. *Bull. Mater. Sci.* **1999**, *22*, 71–83.
31. Skeaff, J. M., Dubreuil, A. A. Electrochemical measurement of SO_3–SO_2 in process gas streams. *Sens. Actuator B* **1993**, *10*, 161–168.
32. Menil, F., Coillard, V., Lucat, C. Critical review of nitrogen monoxide sensors for exhaust gases of lean burn engines. *Sens. Actuator B* **2000**, *67*, 1–23.
33. Hoffmann, C., Wolf, A., Schuth, F. Parallel synthesis and testing of catalysts under nearly conventional testing conditions. *Angew. Chem. Int. Ed.* **1999**, *38*, 2800–2803.
34. Haruta, M., Yamada, N., Kobayashi, T., Iijima, S. Gold catalysts prepared by coprecipitation for low-temperature oxidation of hydrogen and of carbon-monoxide. *J. Catal.* **1989**, *115*, 301.

Chapter 13

Optical Sensing Methods in High-Throughput Screening

Kerstin Kröger, Michael Seidel, and Günter Gauglitz

INTRODUCTION

In many fields of life sciences, pharmaceutical research, and chemical research and development, application of methods for validated search for active compounds using high-throughput screening (HTS) techniques is rapidly increasing. Time and costs limit assays with many compounds [1]. The aim is to establish high-throughput analysis supplying a high information density.

The explosion in the number of potential drug candidates has been enhanced by combinatorial chemistry. These compounds are pooled in substance libraries which often contain 10,000–300,000 compounds of high structure diversity. Their bioactivity is screened by detecting the reactions with biological targets. Many new targets for pharmaceutical investigation, which have not yet been detected, will be understood through advances in molecular biology, for example, high-throughput automated sequencing and rapid hybridization techniques as well as the successful efforts of the Human Genome Project. Functional genomics and proteomics will help us to understand reactions at the molecular level and will produce an increasing number of molecular targets available for therapeutic intervention.

HTS techniques are also being applied in other areas of chemical research, such as agrochemicals, adhesives, and catalysts. Extensive and diverse libraries of functional materials have been produced and screened for chemical, biological, or physical properties. In all these fields it is essential to increase productivity and reduce the costs and the time per assay. For this reason fully automated HTS will be controlled by highly sophisticated robotics in the laboratory operating round the clock. The test formats developed have a high degree of parallelization, automation, and miniaturization.

In this chapter we concentrate on improvements in optical detection methods, especially those using direct optical biosensing principles developed for screening of protein–protein interactions in HTS applications. In addition, fluorescence techniques, which are mainly used in HTS investigation of target–compound binding. Homogeneous

Kerstin Kröger, Michael Seidel, and Günter Gauglitz • Institute of Physical and Theoretical Chemistry, Auf der Morgenstelle 8, 72076 Tübingen, Germany gg@ipc.uni-tuebingen.de
Contact author: Günter Gauglitz, email: gg@ipc.uni-tuebingen.de

High-Throughput Analysis, edited by Potyrailo and Amis
Kluwer Academic Publishers, New York, 2003.

fluorescence bioassays focus on developments in high-density sample carrier and automated liquid handling. Following the establishment of microarray technology in the 1990s, there was significant progress in high-throughput analytics in genomic and proteomic research. Low-density arrays for diagnostic applications and high-density arrays with 250,000 data points/cm^2 are now commercially available [2].

Direct optical detection of biomolecular interaction on sensitive layers is important for understanding the functionality of biomolecules. Biacore instrumentation opened the market for optical biosensors in the field of bioanalysis. At present a number of instruments based on a variety of transduction principles are available, including interferometers, grating couplers, resonant mirrors, surface plasmon resonance, ellipsometry, and reflectometric interference spectroscopy.

In this chapter we will summarize the principles of optical sensors and discuss the advantages of the various techniques. Examples of applications in biomolecular interaction analysis (BIA) are given for receptor–ligand, protein–protein, DNA–protein, or DNA–DNA affinity binding using real-time measurements. Application of direct optical sensors allows determination of kinetic and thermodynamic constants of molecular reactions and quantification. Intensive work has been directed toward parallelization and miniaturization. Time- and space-resolved measurements have been achieved by two different principles. In the first, the transducer area is illuminated with monochromatic light (e.g., laser) and then monitored with a charge-coupled device (CCD) camera. In the second, many separated biosensors are used in a high-density arrangement.

TEST FORMATS

Assays can be classified according to different principles, for example, by distinguishing between assays in a homogeneous or a heterogeneous phase or by comparing direct binding assays with binding inhibition assays using either label-free or labeled molecules. Some of these assays are suitable for modern detection techniques and even allow direct monitoring. Figure 13.1 shows a combination of the classification schemes for some of the suitable assays. Heterogeneous assays use specific binding to the surface; non-bound compounds are separated physically by washing. The well-known enzyme-linked immunoassay (ELISA) is used routinely in laboratories. This assay format achieves small limit of detections and a large dynamic range of signal. However, ELISA has limited suitability for miniaturization because of the unavoidable washing steps and it is time consuming [3].

Whereas assays in the homogeneous phase usually require at least one labeled component, those in the heterogeneous phase allow direct optical detection without the use of labeled ligands or receptors. However, direct optical techniques usually require the immobilization of one of the binding partners at the surface, necessitating the use of a heterogeneous phase assay.

There is strong interest in direct monitoring of the binding of free ligands to an immobilized receptor molecule in the direct test format [Figure 13.1(a)], thus avoiding problems with loss or reduction of bioactivity. The effects of this binding process on refractometric and reflectometric principles increase in proportion to the molecular weight of the ligands. Therefore direct optical detection proves to be rather difficult and requires a high affinity,

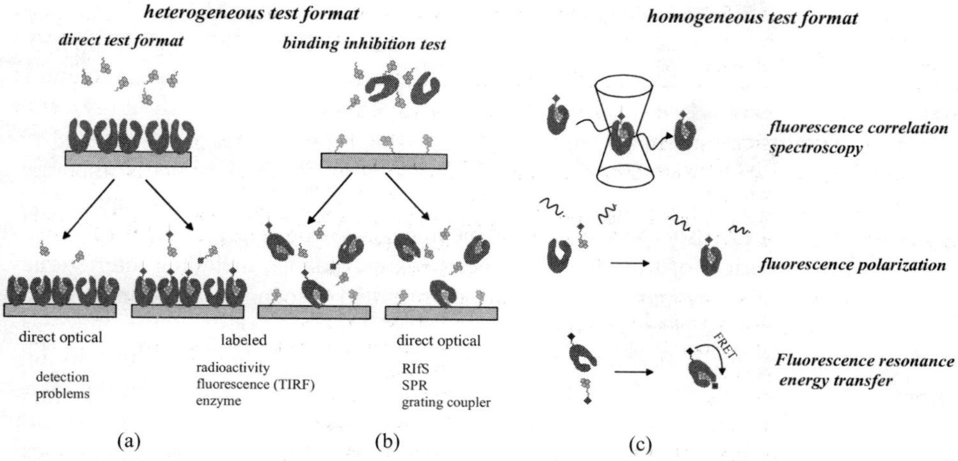

FIGURE 13.1. Survey of assays using a homogeneous or heterogeneous phase with a direct binding or binding inhibition approach using direct optical or labeled partners: (a) direct assay for ligands; (b) binding inhibition format; (c) homogeneous phase assay.

a rather high concentration, and a very large number of receptor sites immobilized at the surface, especially in the case of low molecular weight ligands. For this reason, the ligands are labeled in order to measure radioactivity, fluorescence (TIRF), or enzyme activity [Figure 13.1(a)]. However, labeling is not only a costly process which requires careful synthesis, but it can also affect biological activity, especially in the case of low molecular weight ligands.

For this reason another approach, binding inhibition assay, is chosen [Figure 13.1(b)]. In this case, ligand derivatives are immobilized at the transducer surface, and receptor and ligand molecules are mixed during a preincubation phase such that they bind to each other. In a second step, this solution is brought into contact with the immobilized ligand derivatives at the surface. Only receptors which are not bound to ligands during preincubation will be able to bind to the ligand derivatives at the surface. In the binding inhibition assay the larger receptor molecule is detected leading to better signal-to-noise ratio in the case of label-free techniques. Signal amplification can be performed with labeled receptors.

Homogeneous fluorescence assays are widely used in HTS [4,5]. The detection takes place in solution. In addition to radioactivity-based methods, fluorescence detection methods are known to have a high sensitivity. This type of test format has the advantage that no sophisticated surface chemistry is necessary and no washing steps are included. Thus background fluorescence might reduce the dynamic range of the system [6]. A variety of fluorescence spectroscopy methods can be used to perform an assay (polarization, lifetime, resonance energy transfer, intensity, and quenching).

HOMOGENEOUS FLUORESCENCE ASSAY

Common therapeutic targets for HTS are enzymes, cell surface receptors, nuclear receptors, ion channels, and signal transduction proteins (Table 13.1) [7]. In 1995, 45% of

TABLE 13.1. Homogeneous Fluorescence Assay Used in HTS

Method	Biomolecular reaction	Target	Reference
FCS	DNA–protein	Topoisomerase	[11]
	Receptor–ligand	7-Transmembrane receptor	[11]
	Enzymatic phosphorylation	CD45 phosphatase	[11]
	Protease	Cytomegalovirus protease	[11]
	Restriction endonuclease	EcoRI, BamHI, HindIII, SspI	[18]
FRET	Protease	Hepatitis C virus proteinase	[19]
	Protease	Serine proteinase	[20]
	5′-Exonuclease	Genomic DNA	[21]
FP	Receptor–ligand	Estrogen receptor	[22]
	Phosphodiesterase	Cyclic nucleotide phosphodiesterase	[23]
	Kinase	Tyrosine Kinase	[24]
TRET	Protease	HSV-I protease	[25]
	Kinase	Tyrosine kinase	[24]
	Antigen–antibody	γ-Interferon antibody	[26]

the targets in HTS were based on biomolecular binding reactions. The remainder were enzymatic reactions [8]. In 2002 only 28% of the known targets were enzymes; assays including analysis of biomolecular interactions had increased [9]. Therefore binding assays are increasing in importance and we review them as well as enzymatic assays.

Fluorescence Anisotropy/Polarization (FA/FP) [10]

The exciting light is typically linearly polarized and emission is measured in parallel and perpendicular orientation. Depending on the binding reaction, an increase in fluorescence anisotropy is detected because of decreasing rotational diffusion of the labeled ligand/receptor "complexes". This assay is preferred if small molecules contain a fluorescent label to generate a high anisotropic signal. The advantages are that only one fluorescent label is necessary, and a homogeneous assay can be performed in presence of background fluorescence.

Fluorescence Correlation Spectroscopy (FCS) [11]

FCS is another recently developed detection format that is eminently suitable for HTS. The fluorescence of molecules in a very small confocal detection volume is excited by a laser beam. Conventional FCS takes advantage of the reduced translatory diffusion (Brownian motion) of large versus small molecules. The time domain for passing the spot depends on the molecule size. Thus receptor–ligand binding reduces their translatory motion. This method can only be used even for small numbers of molecules.

Fluorescence Resonance Energy Transfer (FRET) [12]

FRET occurs when two different chromophores with an optimum spectral overlap are in proximity. The emission spectrum of the donor has to overlap the excitation spectrum of the acceptor, Radiationless energy transfer from excited donor fluorophores to acceptor molecules occurs via dipole–dipole interactions at distances of the order of 1–10 nm. If labeled receptor and donor molecules are very close, energy transfer will occur, with an efficiency dependent on $1/r^6$. As a result, fluorescence quenching of the donor is observed. Recent applications have mainly been in screening for enzyme inhibitors of cleaving reactions.

Time-Resolved Energy Transfer (TRET) [13]

TRET is a dual-labeling approach that is based upon the excitation of rare earth elements (e.g., lanthanides or europium) which are complexed by cryptates. Energy transfer occurs between the fluorescent Ln^{3+} or Eu^{3+} complexes and a suitable acceptor such as the fluorescent protein allophycocyanin (APC). This assay exhibits a large Stoke's shift of a long-lived donor fluorescence. The Förster pair has a large effective distance ($R_0 = 9$ nm). Thus TRET is very suitable for measurements of protein–protein interaction. The advantage of this assay is the discrimination between sample signal and background fluorescence based on different lifetimes of the long-lived donor and the background fluorescence. Thus TRET assays have a high sensitivity.

MINIATURIZATION AND PARALLELIZATION

In recent years there has been enormous progress in miniaturization and parallelization of fluorescence-based assays. The main research subjects are high-throughput analysis in the bioanalytical fields of clinical diagnostics, genomics, proteomics, and pharmaceutical screening. Time and costs are the limitations on assays of large numbers of analytes [14]. The aim is to establish high-throughput analysis supplying a high information density.

Two different approaches to parallelization and miniaturization have been adopted. On the one hand, there is microarray technology on flat surfaces with a high density of spots. However, only one sample can be analyzed [15]. On the other hand, there are assays in high-density low-volume wells, where many samples can be screened in parallel. The traditional 96-well microtiter plate is currently being replaced by 384- and 1,536-well plates, and assays of volumes below 1 μl performed on plates with larger numbers of wells (>3,456) are under investigation [12,16,17].

Applications in nanotiter plates in which sample volumes of 70 nL can be handled are also being developed. FRET assays are preferable for these small volumes because of the easy use of fluorescence scanning methods [27].

Miniaturization implies the integration of liquid handling in nanoliter volumes. Dispensing in the pico- to nanoliter range is manageable since non-contact microdispensing devices can be used for biological samples [28]. A variety of micropipetting techniques have been established based on ink-jet principles (thermal [29], piezoelectric-actuated [30], syringe-solenoid [31]). Newer techniques are high-parallel microdispensing devices like the TopSpot technology which is based on piezo-triggered short pneumatic pressure pulses [32].

Binding assays in the homogeneous phase are particularly suitable for miniaturization and parallelization. These assays can handle volumes up to 1 μl corresponding to the volume of a 1536-well plate. The smaller the volume, the higher is the surface-to-volume ratio and nonspecific binding to the surfaces is more likely. Another problem in open wells is the evaporation of liquids [33]. The concentration of reagents becomes higher than expected and the binding to the surface occurs. These effects reduce the signal dynamics. The traditional 96-well microtiter plate is currently being replaced by 384-well plates. Plates with 1,536 well are available, and plates containing up to 9,600 wells are under development.

Since the microarray technology was established in the 1990s, genomic and proteomic research has progressed. Low-density arrays for diagnostic applications and high-density arrays with 400,000 data points/cm^2 are commercially available. These techniques have been shown to be useful for gene expression analysis [34], sequencing [35], single nucleotide polymorphism analysis [36], and protein arrays [37, 38]. The chip technology uses mainly glass slides where probes are spotted in a microarray. Fluorescent-labeled biomolecules bind specifically to the immobilized probe and the fluorescent image is detected using a fluorescence scanning device. The main problems are the robustness of the test system for applications in quantitative bioanalysis. Fluorescence is detected at the surface with a non competitive assay format. Nonspecific binding reduces the signal-to-noise ratio.

The miniaturization of optical detection systems for low-cost diagnostic biochips in the field of medical applications is under investigation. With integrated electro-optic systems the excitation of fluorophores and the detection of the emitted light can be integrated on a sensitive transducer [39].

Surface-sensitive fluorescence techniques have been developed for simultaneous analysis of DNA–DNA, antigen–antibody, receptor–ligand interactions. Optical structuring allows parallel setups using total internal reflection fluorescence (TIRF). Microarray biosensors provide the opportunity to observe time-resolved binding reactions in parallel. They use the evanescent field of a two-dimensional waveguide to excite the fluorescent-labeled binding partner. Laser light beam is coupled into a planar optical transducer and creates an evanescent field which decays exponentially within a few 100 nm into the medium. One binding partner is immobilized at the surface and the complementary partner is labeled with a fluorophore. Non-bound biomolecules in the bulk medium are outside of the evanescent field. They are not excited, whereas the receptors recognize specifically the binding partners immobilized on a patterned surface, which can be detected. The emission light is detected with a cooled CCD or a photomultiplier tube (PMT).

Several TIRF devices have been realized (Table 13.2). One distinguishing feature is the coupling into the waveguide to generate an evanescent field along the area of the microarray. Thin-film planar waveguides (PWGs) as used by Zeptosens GmbH consist of a thin film (150–300 nm) of a material with a high refractive index (e.g., Ta_2O_5 or TiO_2), deposited on a transparent support with lower refractive index (e.g., glass or polymer). A parallel laser light beam is coupled into the waveguiding film by a diffractive grating that is etched or embossed into the substrate. The light propagates within this film and creates a strong evanescent field perpendicular to the direction of propagation into the adjacent medium [40]. It has been shown that the intensity of this evanescent field can be enhanced dramatically by either increasing the refractive index of the waveguiding layer or by decreasing the layer thickness [41]. The field strength decays exponentially with the

TABLE 13.2. Overview of Array Biosensors Based on TIRF

Optical DNA sensor chip for real-time detection of hybridization events	[42]
Novel bioaffinity sensors for trace analysis based on luminescence excitation by planar waveguides	[46]
Array biosensor for detection of biohazards	[47]
Development of a multichannel fluorescence affinity sensor system	[48]
Real-time detection of DNA hybridization and melting on oligonucleotide arrays by using optical waveguides	[49]
Planar waveguides as high-performance sensing platforms for fluorescence-based multiplexed oligonucleotide hybridization assays	[40]
Array biosensor for simultaneous identification of bacterial, viral, and protein analytes	[50]
Fiber-optic evanescent wave biosensor for the detection of oligonucleotides	[51]

distance from the waveguide surface, and its penetration depth into the solution is limited to about 400 nm. This effect can be utilized for the selective excitatiion of fluorophores located at or near the surface of the waveguide. By taking advantage of the high field intensity and the confinement of this field to the close proximity of the waveguide, PWG technology combines highly selective fluorescence detection with a very high sensitivity.

Katerkamp and co-workers have described a disposable sensor chip [42]. The chip is constructed from a prism and a cover plate, both made from poly (methylmethacrylate) (PMMA). The biomolecules are directly immobilized on the prism surface. A laser beam enters the prism at an angle such that there is total internal reflection at the interface between the prism and the flow channel. The optical system for the detection of fluorescent light is arranged below the sensor chip and consists of two lenses, an interference filter, and a photomultplier tube which is moved by a stepping motor.

Brandenburg and co-workers generated the evanescent field at a PWG by focusing an extended laser beam on the side face of a prism. The light spreads inside the sensor chip through total internal reflection so that the fluorescent material is excited by the evanescent field. The emitted light is focused via an optical system and a filter to a CCD camera. The two-dimensional readout is obtained by moving the transducer very fast [43].

The array biosensor constructed by Ligler and co-workers uses a two-dimensional array of graded index (GRIN) lenses in the imaging optics. Parallel light is launched from a diode laser into the end of a waveguide (microscope slide), resulting an evanescent field. A series of lenses are utilized to provide uniform excitation in the sensing area of the waveguide [44]. A two-dimensional array of GRIN lenses image the fluorescent array in the sensing region of the waveguide onto a CCD camera and one bandpass filter is installed to eliminate excitation and scattered light prior to imaging.

Wilkinson and co-workers describe an integrated optical (IO) TIRF transducer. Light from a linearly polarized He–Ne laser is launched into a channel waveguide which branches into two arms. An array of detection spots is produced by spotting the ligand derivatives on the channel waveguides. The excited fluorescence is collected by an array of plastic fibers of high numerical aperture located opposite the sensor position. The collected light is filtered to reject any collected pump radiation, detected with photodiodes, and acquired by lock-in detection [45].

TECHNICAL PRINCIPLES OF LABEL-FREE DIRECT OPTICAL DETECTION

Recent improvements in instrument hardware, experimental design, and data processing have allowed direct optical sensors to be used as screening platforms. In the past, label-free optical sensors were fabricated as single-channel to four-channel systems to examine biomolecular interactions. The great advantage of these sensor types for biomolecular interactions is that they allow real-time measurement of the kinetics of interactions of non-labeled compounds. Labeling can affect interactions and in some cases, especially in the measurement of protein–protein interactions, the binding partner cannot be labeled. Another advantage of label-free systems is the time required for the assay and for assay development. At present, a limited amount of patents and patents applications of screening platforms of these sensors have been published, and a commercial platform will soon be available. Such a screening platform will combine the advantages of an optical sensor with the possibility of measuring many biomolecular interactions in parallel. Affinity-based sensors using direct optical detection can be generally classified into two groups (Table 13.3):

1. Sensors based on micro-*refractometric* effects detect the binding of molecules to the surface by monitoring the change in refractive index adjacent to the sensor surface. For example binding of molecules to the surface increases the refractive index in the direct environment of the waveguide used as sensor surface. Because of the evanescent field of the light-coupling waveguide the velocity of propagation of the guided wave will be reduced. This results in a change of the actual wavelength.
2. Sensors based on *reflectometric* effects detect the change in the medium surrounding the transducer by measuring the change of reflectivity of incident radiation at an interface between the fluid and the transducer (e.g., water and glass) as a function of incidence and state of polarization.

TABLE 13.3. Types of Direct Optical Sensor and the Detection Principles used in them

	Principles	Limit of detection
Refractometric sensors		
Integrated optical interferometry	Detection of the phase shift by interference	
Mach Zehnder interferometer		1 pg/mm^2
Young interferometer		5 pg/mm^2
Fabry–Pérot interferometer		9 fg/mm^2
Grating couplers	Shift of the coupling angle of an optical grating	5 pg/mm^2
Resonant mirror (Prism coupler)	Change of the coupling conditions of an isolated waveguide layer	8 pg/mm^2
Surface plasmon resonance	Shift of the resonance energy of a surface plasmon	0.3 pg/mm^2
Reflectometric sensors		
Ellipsometry	Change of the polarization of the reflectivity	0.5 pg/mm^2
Reflectometric interference spectroscopy	Change of the interference pattern of a thin layer	1 pg/mm^2

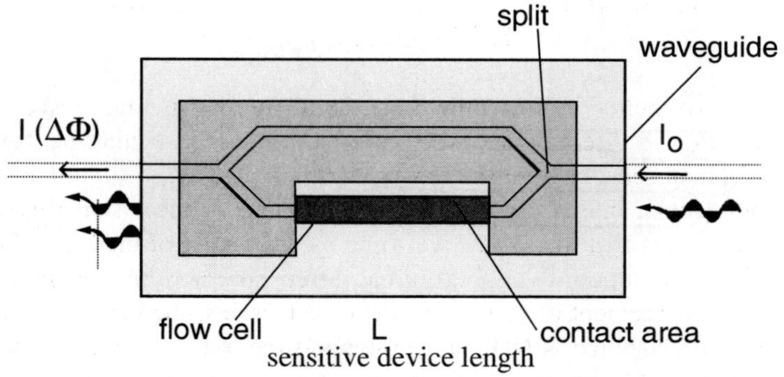

FIGURE 13.2. The Mach–Zehnder interferometer.

Integrated Optics, Interferometry, and Refractometry

Mach-Zehnder Interferometer

The Mach–Zehnder interferometer is a monomode waveguide which is split into two arms. After a distance of a few millimeters the arms are recombined (Figure 13.2). The device is covered by an isolated layer and the fluid only contacts one arm. Mostly Transverse electric (TE) polarized light from a He–Ne laser is coupled in the waveguide by a microscope lens. Binding of biomolecules to the surface in the contact area increases the average refractive index adjacent to the surface interrogated by the evanescent field. Thus the effective refractive index of the waveguide is changed. This leads to a phase shift $\Delta\phi$ between the interferometer arms and to a change in the output intensity.

$$\frac{I\,(\Delta\phi)}{I_0} = \frac{I + \cos\,(Lkn_{eff})}{2} \tag{1}$$

where L is the sensitive device length, k is the wavevector of the laser light used, and n_{eff} is the effective refractive index [52,53].

Young Interferometer

The Young Interferometer is a special application based on differential refractometry [54]. The principle of operation is shown in Figure 13.3.

A semiconducting laser is fixed on the waveguide and the light is coupled in the waveguide. Then the beam is split into two branches. Two flow cells are located at the endface of the chip. Light beam from the two exit ports is combined and radiated through the flow cells. A cylindrical lens focuses the light onto a CCD camera. The interference pattern observed is produced in the same way as diffraction at a double slit. The phase difference between the light beams from the two arms of the Y branch determines the position of the fringes in the interference pattern. If molecules are bound to the surface of one arm of the branch there is a change in the effective refractive index and thus a change in the phase shift which alters the position of the fringes in the interference pattern.

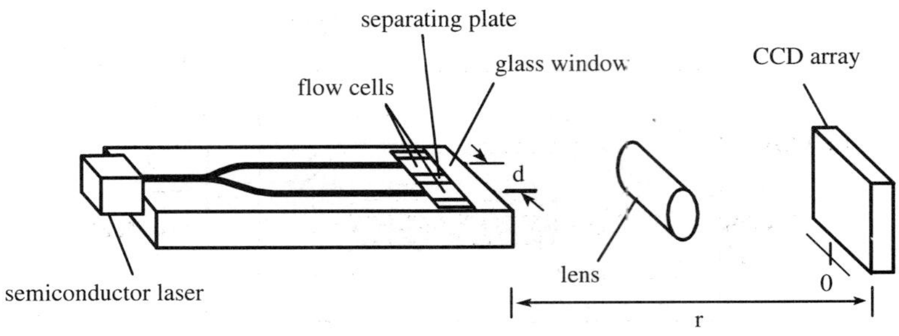

FIGURE 13.3. The Young interferometer.

The correlation between phase shift $\Delta\phi$ and the change in the effective refractive index n_{eff} is given by

$$\Delta\phi = \frac{2\pi}{\lambda} L \Delta n_{\text{eff}} \tag{2}$$

where L is the length of the flow cell and λ is the wavelength. The limit of detection is in the range of $\Delta n_{\text{eff}} = 10^{-8}$ [55].

Fabry–Pérot Interferometer

The main components in this type of interferometer are two closely spaced plates which are partially silvered. Light is incident on the transparent plates at a constant angle. Multiple reflections occur between the two plates and a small part of the transmitted light forms interference fringes (concentric circles when focused by a lens). The principle follows the interference conditions for thin films and is shown in Figure 13.4.

The condition for a maximum in the interference pattern is given by

$$2d \cos\alpha = m\lambda \tag{3}$$

where d is the distance between the plates, α is the angle of incidence of the light, and m is the order of interference. Two effects can be measured [56]:

1. the substitution of sample by water causes an increase in the refractive index;
2. a change in the concentration of the charge carrier by the formation of a "double-layer capacitance-induced dielectric change" leads to a decrease of the refractive index.

An advantage of this type of interferometer is the very low detection limit of about 9 fg mm^2. The disadvantages are the handling and the small frequency range in which this interferometer can be used. Despite these problems, a new direct optical biosensor for multi analyte detection based on a Fabry–Pérot Interferometer has been described. [57] The technique, called the spectral correlation method (SCM), measures a correlated signal from two interferometers: a Fabry–Pérot interferometer and a glass biochip interferometer (recognition spots). During measurement the distance between the two plates of the Fabry–Pérot interferometer changes periodically. The device has been developed for 96 channels.

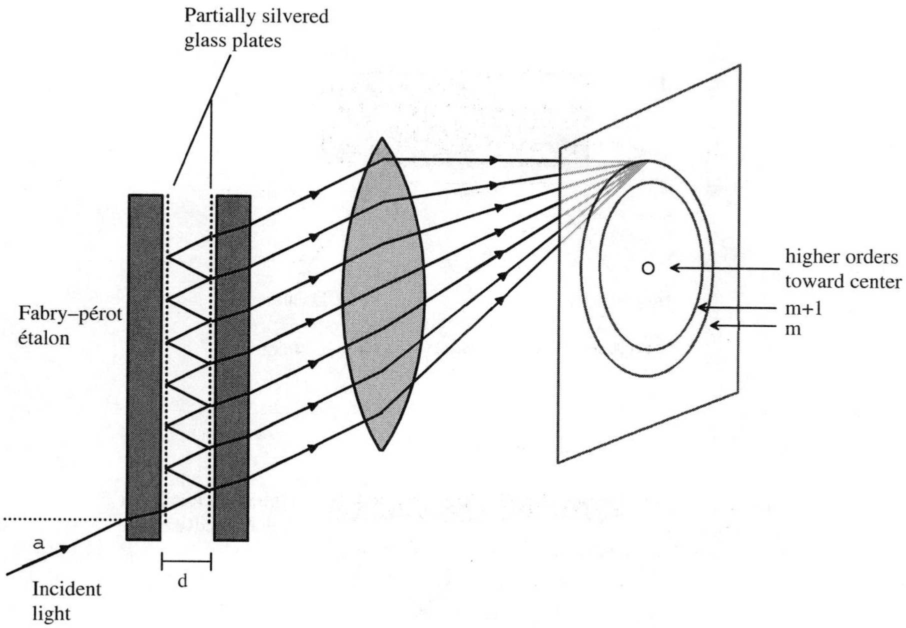

FIGURE 13.4. Fabry–Pérot interferometer.

Other Refractometric Sensors

Grating Coupler

The grating coupler is another technique which uses the principle of the evanescent field monitoring. The first applications of grating couplers used as chemical sensor were described by Lukoz and Tiefenthaler in 1983 [58]. Further development led to a reflection grating coupler in which the reflected beam is detected. This approach differs from the early models of input and output grating couplers in which the detector had to be adjusted on the thin layer of the waveguide [58,59,60]. Thus these models do not require mechanical parts. The set up of a reflection grating coupler with the in- and out-coupling light rays is shown in Figure 13.5.

Part of the radiation couples into the waveguide via the grating. This is missing in the reflected light beam, resulting in a minimum signal detected by a CCD camera. The refractive index changes when biomolecules are bound to the surface of the chip. This affects the evanescent field and the intensity of the reflected light, leading to a shift in the position of the minimum in the detected signal. The minimum position is determined and the change Δn_{eff} in the effective refractive index is calculated from

$$n_{\text{eff}} = n_0 \sin\alpha + m(\lambda/g) \quad (4)$$

where n_0 is the refractive index of the waveguide cover, α is the angle of incidence, m is the mode number, λ is the wavelength, and g is the grating number.

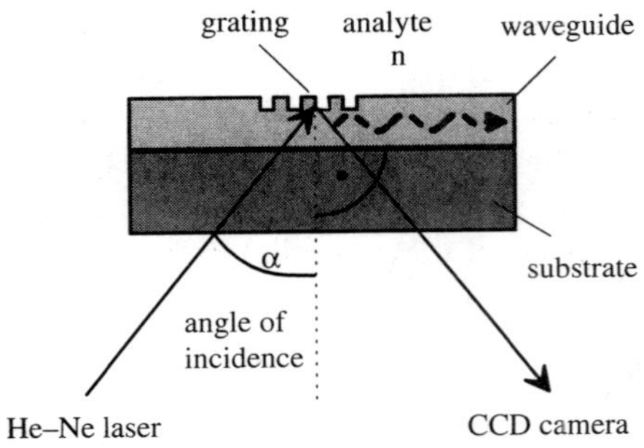

FIGURE 13.5. The reflection grating coupler.

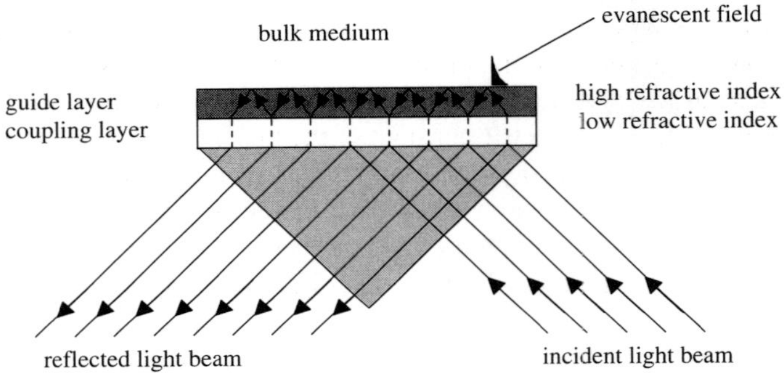

FIGURE 13.6. Arrangement of the resonant mirror in the IAsys system.

A grating coupler system BIOS™ is commercially available from Artificial Sensing Instruments (ASI), Zürich, Switzerland.

Resonant Mirror (Prism Coupler)

Two thin layers are coated on the base of a highly refractive prism resulting in a change from high to low to high refractive index layer (Figure 13.6). At the low-index layer radiation through the prism is totally reflected. This low-index layer is thin enough for light to couple into the resonant high refractive index layer (the sensing surface) via an evanescent field. Efficient coupling occurs only at certain angles of incidence where phase matching between the incident beam and the resonant modes of the high-index layer is achieved.

At this resonance angle, radiation couples into the high-index layer and propagates some distance along the sensing layer interface before coupling back into the prism via the evanescent fields in the low-index layer. Resonance can be achieved for transverse electric (TE, electric field perpendicular to the optical plane) and transverse magnetic (TM, electric vector in the optical plane) polarization. Both polarization states can be observed as fine structures in the intensity distribution of the reflected light. The angle of excitation is sensitive to changes at the sensing layer caused by changes in the refractive index during interactions at the surface. Light is reflected at many angles of incidence, but the shift in the phase of reflected light during resonance can be sensitively detected [62]. This technique was developed for a robust sensor system called IAsys (Affinity Sensors, Cambridge, UK).

A new system, IAsys Plus, uses stirred liquid in a cuvette with two cells. Problems can arise with this system if the cuvettes are not completely filled with liquid, if the cells are not filled simultaneously, and if evaporation occurs during prolonged equilibrium measurements (there is no cover for the two cells) [63].

Surface Plasmon Resonance

Surface plasmons are the quanta of collective oscillations of surface charges at the interface between metal and dielectric layers. These oscillations are correlated with electromagnetic fields which decay exponentially from the metal/dielectric interface. The fields are referred to as evanescent fields. The resonance condition for exciting surface plasmons is satisfied for light in the visible range. At this point the energy of the incident radiation is transferred to the plasmon oscillations by the evanescent field.

Excitation of surface plasmons is only possible if the angle of incidence is sufficiently large and the real part of the dielectric constant of the metal (e_2) is smaller than the dielectric constant of the prism (e_1) and the dielectric constant e_3 of the dielectric (e.g., buffer solution). For this reason a prism with a high refractive index ($e_{1\ \mathrm{air}} \ll e_{1\ \mathrm{prism}}$), is used so that the impact of the radiation (wave correlates with particle) fits the impact of the surface plasmon:

$$\sin \phi [e_1(\lambda)]^{1/2} = \left[\frac{e_{2,\mathrm{re}}(\lambda) \cdot e_3(\lambda)}{e_{2,\mathrm{re}}(\lambda) + e_3(\lambda)} \right]^{1/2} \tag{5}$$

Either gold or silver is used as the metal. Normally a gold layer is used even though the silver surface has better characteristics for the excitation of the plasmon. However, the disadvantage of silver is the oxidation of the surface. The advantage of gold is a simple thiol surface chemistry [self-assembly monolayer (SAM)].

If the dielectric constant of the ambient medium or a sensitive layer is changed, the conditions for plasmon oscillations also change.

The most common surface plasmon resonance (SPR) setup is that due to Kretschmann and Raether [64–66]. SPR can be realized in two ways, the variation of either the angle of incident or the wavelength.

In the first approach parallel (p) polarized white light (TM mode) is irradiated at a constant angle onto a prism with a high refractive index. The prism is coated with a thin layer (~50 nm) of silver or gold. The evanescent field of the totally reflected radiation (Figure 13.7) can excite surface plasmons on the opposite surface of the metal layer. Any change in the dielectric (changes in the refractive index) will vary the resonance condition and shift the resonance wavelength. The resonance wavelength of the surface plasmons is detected as a minimum of reflectivity in the reflected light (Figure 13.8). In the second approach the angle of incidence of the monochromatic light used to excite the surface plasmons is varied. With this method the resonance angle is determined (Figure 13.8). The realization is according to the first setup.

Another possibility of increasing the impact of the radiation is periodic structuring of the metal surface (see grating coupler) instead of using a prism. In this case the light is

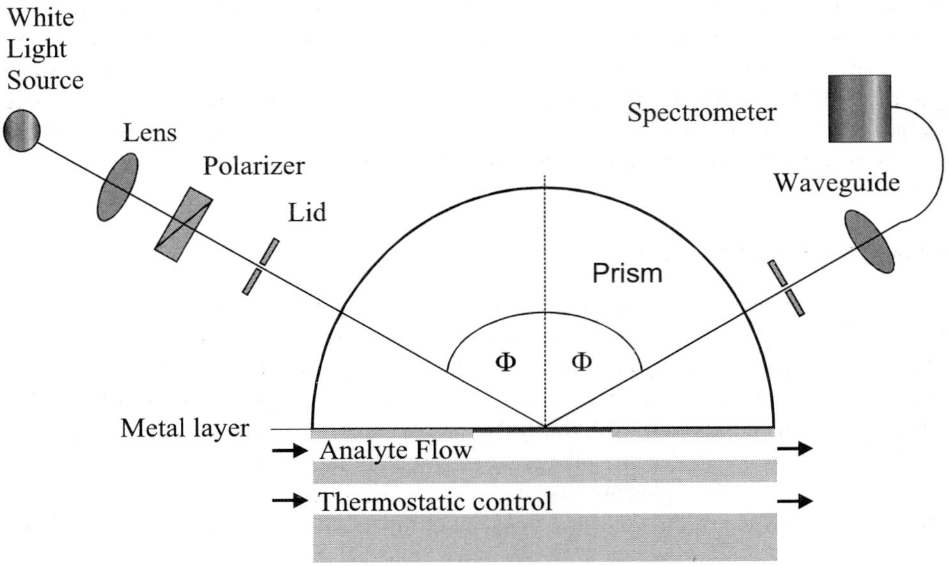

FIGURE 13.7. Kretschmann arrangement for SPR.

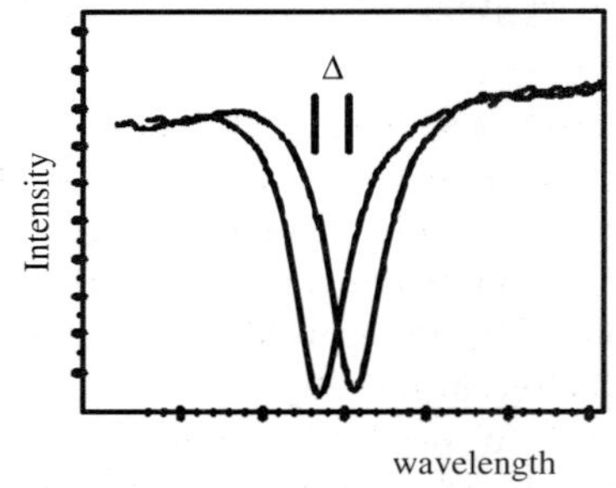

FIGURE 13.8. Reflection spectrum depending on wavelength or angle.

irradiated directly through the sensitive layer to the grating, which makes this arrangement difficult to handle in bio-applications.

BIAcore instrumentation (Fa. Pharmacia Biacore, Uppsala, Sweden) opened the market for direct optical detection of biomolecular interaction. The BIAcore 2000 series and upgrade instrumentations have a very small dead space and ensure a highly precise exchange of sample in the flow cavity by using a microflow cartridge. Sequential filling of the cell means that the different measuring spots ("channels") are reached one after the other. However, this marginal offset is almost eliminated by the software supplied. At present a system with four channels in parallel is available. In the commercial BIAcore equipment monochromatic light is irradiated sequentially at different angles and the resonance angles are determined by a spatial resolving detector first described by Malmquist [67]. In this device [68] no movable parts are used, resulting in a very robust system. This development has led to a high signal-to-noise ratio in the new device (BIAcore 3000). A combination of both detection methods, in which different wavelengths and angles are sequentially incident and detected, is used for the exact determination of the refractive indices of metals [69,70].

If a parallelized beam of sequentially chosen wavelengths is irradiated and detected at a constant angle, spatially resolved detection is possible using a CCD camera [71]. Parallel detection is also possible using defined wavelength and angle, and measuring the change of reflectivity in the proximity of the resonance peak [72]. Under these conditions the refractive indices of the various sensitive spots may not differ substantially and the sensitive layers must be very thin. Then the resonance wavelengths of the surface plasmons depend mainly on the medium to be analyzed and the resonance conditions for all layer areas are approximately satisfied at the same angle.

An increase in the sensitivity of the SPR sensors can be achieved by detecting the phase of the monochromatic light incident at an fixed angle [73,74]. Because just one angle and monochromatic light are used, this arrangement can easily be parallelized [75].

The first parallelized SPR technique was a surface plasmon imaging system [76,77]. In this method light from a 632 nm He–Ne laser is incident on a prism at a fixed angle. Binding of biomolecules to the surface causes a change in the reflectivity. The images of an array of differential reflectivity are detected using a CCD camera. Because of the occurrence of laser fringes from the coherent light source, Nelson and co-workers changed to a combination of collimated white light with a near-IR narrow band pass filter. The image is detected at 830 nm. Near-IR SPR imaging using excitation at 800–1,152 nm results in sharper SPR reflectivity curves and thus in better image contrast. This method allows the measurement of thicker films in applications involving bacteria or large microspheres, for example because of the sharper resonance and longer wavelengths. The longer decay length of the electromagnetic field results in a larger penetration depth.

Another imaging SPR sensor has been described by Johansen and co-workers [78]. The setup is based on the Kretschmann configuration with a fixed incident angle and a gold metal film. Multicolor imaging is realized by a collimated light beam from a white-light or multiple-wavelength source, a polarizer, a gold-layered prism, and a multicolour detecting camera [79].

Homola and co-workers have described a multichannel SPR sensor based on different sensing channels. The incident polychromatic light beam excites the surface plasmons in the different sensing channels at different frequencies. The encoded signal of the optical spectrum is achieved either by irradiating the light beam sequentially onto different sensing areas at different angles of incidence or by modifying the sensor surface by including a thin high refractive index overlay, which locally alters SPR frequency [80].

Graffinity Technique

The first company to announce a parallel SPR array was Graffinity Pharmaceutical Design GmbH (Heidelberg, Germany) with their label-free screening platform for chemical microarrays called Plasmon Imager R. It can detect 4,608 spots per array with a CCD camera detecting the change of the SPR resonance wavelength. With this technique it is possible to detect up to 1 mM affinity interactions between proteins.

The technique, which is based on waveguides covered with a gold layer was described for the first time in two patents [81]. However, it was difficult to construct a parallel and miniaturized sensor array. These problems were solved in a later development [82]. A waveguide array is manufactured using semiconductor technology. It consists of a comb-like structure of many waveguides. The distance between these waveguides is of the dimensions of a microarray format.

The waveguides are covered by a thin layer of metal (gold or silver) which can exhibit SPR. A monochromator or a filter wheel selects from a white-light source wavelengths which are irradiated onto the comb array passing through a polarizer for TM modes. The light beam is incident on the whole SPR array at an angle of 70°–90° (necessary for the excitation of plasmons). After diffuse scattering the radiation passing the waveguide array is detected by a CCD camera. By scanning the wavelength in steps of 1 nm an intensity spectrum is achieved for all waveguides within 16 min covering a range of 200 nm. Normally, interactions of biomolecules are faster than the time necessary to aquire such an image. However, this technique allows comparison of the shift at one extreme of the intensity spectrum before and after biomolecular interaction (imaging process). If a xenon light source is used and the wavelength scan is reduced to 60 nm, the array can be monitored in 1 min. Thus the setup comes close to real-time measurements of biomolecular interactions.

HTS Biosystems Technique

An interesting technique using grating couplers for SPR in the Kretschman–Raether configuration were reported in 1999 by Schröter and Heitman [83]. Another publication describes the principle of using gold gratings for SPR [84]. HTS Biosystems (Hopkinton, MA) in collaboration with Ciencia Inc. (East Harford, CT), offers chips for label-free proteome analysis called the FLEXchip Kinetic Analysis System™. The system is based on grating-coupled surface plasmon resonance (GC-SPR) [85]. Instead of a prism, a grating is used to couple the light into the surface plasmons. The chip consists of a plastic optical grating with a thin layer of gold. Collimated monochromatic light from a 867 nm light-emitting diode is incident on the gold array. A CCD camera measures the shift in the SPR angle for each position on the chip. Thus real-time measurements of binding events at each position of the chip are possible.

The advantage of this system, in addition to miniaturization and parallelization, is the use of plastic chips with a thin gold film. Plastics are very inexpensive and are commonly used for HTS in the field of diagnostics. By using a gold rather than a silver layer for SPR, standard coupling chemistry can be used to immobilize biomolecules. The spots are 250 μm in diameter [86].

BIAcore Technique

In April 2002 BIAcore presented the proof of principle of their array chip technology. In collaboration with some pharmaceutical companies (e.g., Millennium Pharmaceuticals) they demonstrated biological applications of SPR chip technology in the field of functional proteomics [87]. This study demonstrates the feasibility of using the SPR chip in complex assays to achieve kinetic information about binding events. In March 2002 BIAcore predicted that the SPR chip array technology would be available commercially in 2004 [88].

A possible setup of the parallel BIAcore equipment is described in a patent issued in 1998 [89]. As in the non-parallel apparatus (see earlier), monochromatic light is used at different angles of incidence. Radiation is irradiated via a collimator optic and an interference filter to a first flat scanning mirror, and is then incident on a second scanning mirror after which it is coupled via a prism or a grating into the sensor surface. The reflected light beam passes a polarizer and produces a real image on the first part of the detector. Changes in the angle of incidence are achieved by moving the mirrors. An image is produced for each angle of incidence. The angle of the incident light is determined with a reference point on the array. The light for this area is transmitted through a separate optics and illuminated to a second part of the detector. By using these optics the angle of incidence can be associated to the image of the whole array. Thus the angle of irradiation can be associated with the image on the first part of the detector. With this setup, real-time measurements of biomolecular interactions on thousands of spots on one chip can be monitored in parallel.

Further Developments

In most of the developments described above miniaturization was achieved by modifying conventional techniques. In the future, the technique based on optical label-free detection will be used to detect thousands of proteins in one run, for example, for measuring of protein–protein interactions in functional proteomics or in protein–drug investigations for pharmacological studies. Therefore the optical techniques must be parallelized and miniaturized. At present, protein chips are used where thousands of proteins are printed on one chip and the sample is poured over the chip. The protein–protein interactions can be detected using chemoluminescence or fluorescence. The next step will be the coupling of such chips with mass spectrometry for direct analysis of the protein of interest. However the surface chemistry involved in these techniques causes problems in the immobilization of an active protein structure onto a chip. Therefore some companies are developing a new approach. For example, Zyomyx (Havard, CA) has developed a technique where the proteins are not immobilized but are positioned in miniature wells on the chip. Some companies (e.g., Ciphergen) have developed label-free detection methods in which the mass spectrometer is used directly as a detector [90].

In the area of functional proteomics it is necessary to aquire information not only about the binding event but also about the protein that is involved. For this reason BIAcore have developed a system in which the nonparallel BIAcore setup is coupled with mass spectroscopy for direct detection of the mass of proteins on the chip. A cooperation with Bruker Daltonics Inc. provides a good starting point for functional proteomics [91,92].

In general, the SPR technique replaces conventional methods such as ELISA. Thus SPR is used not only for screening experiments or the determination of binding kinetics but also for diagnostics in the medical laboratory. For example, a virus or antibody in blood can be detected in 1 min in comparison with the ELISA technique which requires 4 hr for one assay.

Reflectometric Techniques

Spectral Ellipsometry

Spectral ellipsometry (SE) is a nondestructive optical method for determining the refractive index and the physical thickness of thin layers separately. Light at a specific angle of incidence of a well-defined wavelength and polarization state is radiated onto the sample, and the polarization state and the phase of the reflected light is measured (Figure 13.9). At the interface between the layer and substrate and the layer and ambient media the light obeys Fresnel's laws of reflection and transmission. These equations for radiation, polarized transverse or parallel to the plane of incidence, are important in ellipsometry. The fundamental equation of ellipsometry is

$$(\tan \overline{\underline{\psi}})e^{i\Delta} = (R_s / R_t) \tag{6}$$

where $\overline{\underline{\psi}} = |R_s| / |R_t|$ is the ratio of the absolute values of the reflectivity of the light polarized parallel (R_s) and transverse (R_t) to the plane of incidence and $\Delta = \delta_1 - \delta_2$ is the change in phase shift between the two polarized parts of the light incident onto (δ_1) and passing (δ_2) the sample. By measuring the change of state of polarization, the refractive index and the physical thickness of the layer can be determined. Taking into account multiple reflections, multilayer systems can also be characterized.

The main advantage of ellipsometry is that it is possible to determine the physical thickness and the real refractive index of a sample separately. Another advantage of ellipsometry over the techniques described above is the detection of the polarization state. Whereas changes in intensity of s and p polarized light due to slight changes in thickness are very small when measured using reflectance, in ellipsometry the phase change is 100 times larger. Thus it is possible to determine thicknesses in the range of an atomic layer using ellipsometry.

Types of Ellipsometers

Nulling Ellipsometer The monochromatic light source, polarizer, analyzer, and compensator are brought into an optical position at which the intensity of the reflected

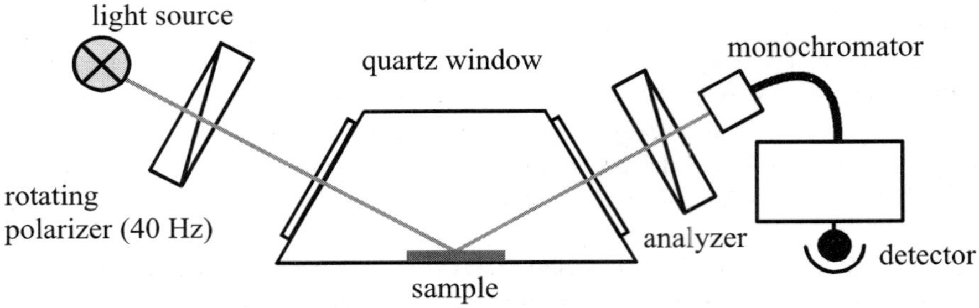

FIGURE 13.9. Setup for spectral ellipsometry with a rotary polarizer and analyzer.

light is completely extinguished. The angular positions of the optical components are determined.

Polarization Modulated Ellipsometer (PME) This type of ellipsometer uses a monochromatic light source. The polarization is modulated at a favorable point along the optical pathway by a turnable polarizer or compensator. The change of intensity is evaluated.

Rotary Analyzer/Polarizator/Compensator Ellipsometer (RAE/RPE/RCE) A spectral or monochromatic light source is used. A sinusoidally modulated intensity is detected and evaluated using a Fourier series.

Interferometric Ellipsometer (IF) This setup, using a monochromatic light source, is similar to a Michelson interferometer. The advantage is the short measuring time.

Applications of Spectral Ellipsometry

The advantage and disadvantages of ellipsometry in sensor applications have been described [92]. The advantages are label-free technique, high phase sensitivity, two types of output data, nonelectric readout, photon-based true reflection technique, possibility of using thin sensing layer, and applicability in any transparent ambient. The disadvantages are the indirect measurement and the complicated theory and instrumentation. An ellipsometric multisensing instrumentation based on multiple beams and multiple samples has been proposed [92], which demonstrates future application of ellipsometry as a screening device.

Two companies have developed imaging ellipsometers within the last decade. Beaglehole Instruments Ltd (Wellington, New Zealand) sells a device for producing imaging in 1 s using a fixed polarizer and a continuously rotating retarder with a microscope zoom lens and a CCD detector. This ellipsometer can also be used as a Brewster angle microscope. Applications of this ellipsometer are direct monitoring of surface chemical reactions, spreading of drops onto surfaces, and following dynamic phenomena in real time.

Nanofilm Technology GmbH (Göttingen, Germany) also offers an imaging ellipsometer, called the EP^3 Ellipsometry platform which is based on the null ellipsometry with single or multiple wavelengths. The applications include quality control on biochips because a three-dimensional profile of spots on surfaces or of an antibody layer on a dextran surface can be achieved [94]. In this context, it is interesting, that they offer an SPR cell for the imaging ellipsometer. This technology can be used to measure the thickness of multispots in HTS for examination of the surface structure of biological macromolecules on glass, gold, or silicon substrates.

Reflectometric Interference Spectroscopy

Principle of Reflectometry

The principle of reflectometric interference spectroscopy (RIfS) is based on the spectral distribution of multiple reflection at the interfaces of thin layers on a glass substrate (Figure 13.10).

In the high-resolution single-channel system, white light from a tungsten light source (20 W) is guided to the transducer using a polymer fiber coupler (1 mm PMMA fiber, 50/50 coupling ratio) (Microparts Dortmund, Germany). The light is reflected at the interfaces from the thin transparent interference layer (glass, SiO_2, Ta_2O_5) of the transducer (Figure 13.10) producing a reflectance pattern (Figure 13.10 right panel).

The reflectance pattern changes when the optical thickness *nd* (refractive index *n*, physical thickness *d*) changes as a result of the binding of biomolecules to the surface of the transducer. The reflected light is collected by the same fiber and spectrally detected by a diode-array spectrometer module (Zeiss SPEKOL 1100, modified as described by Schmitt et al. [95]). Monitoring of real-time binding is obtained by plotting the wavelength position of an extremum of the reflectance pattern versus time. The transducer is mounted in a flow cell of volume approximately 200 nl (50 μm high, 1 mm wide, and 4 mm long). Sample handling was carried out by flow injection analysis using an ASIA system (Ismatec Zürich, Switzerland). The setup is described in detail elsewhere [95]. In addition to the normal applications in the area of DNA antibodies, enzyme activity, and proteins (Table 13.4) this setup allows direct optical detection of the biomolecular interaction of small ligands [96].

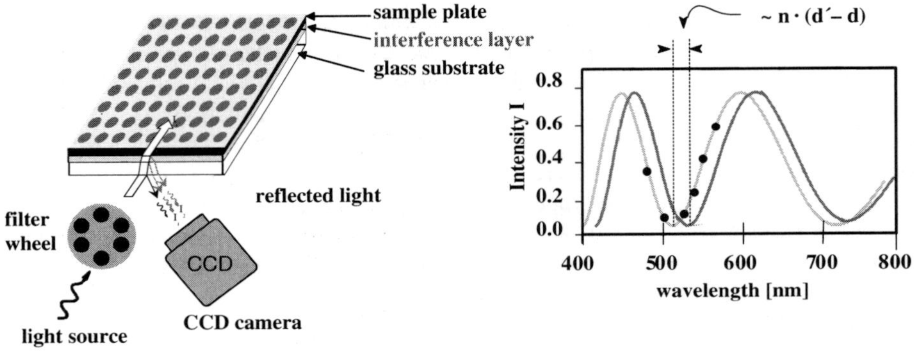

FIGURE 13.10. Principle of RIfS.

TABLE 13.4. Applications Classified with Respect to Sensor Types and Biomolecular Interactions

Method	Single/ HTS	BIA	Reference
Mach–Zehnder interferometer	S	Sugar–protein	[108]
	S	Antigen–antibody	[109]
	S	Biotin–streptavidin	[110]
	S	Receptor–ligand	[111]
	S	Antigen–antibody	[112]
	S	Antigen–antibody	[113]
	S	Antigen–antibody	[114]
Young Interferometer	S	Antigen–antibody	[115]
	S	Antigen–antibody	[116]

(continued)

TABLE 13.4. (continued)

Method	Single/ HTS	BIA	Reference
Fabry Pérot–interferometer	S	DNA–DNA Antibody–antigen Receptor–ligand	[117]
	S	DNA–protein	[118]
	S	Protein–protein	[119]
Grating coupler	S	DNA–protein	[120]
	S	DNA–DNA	[121]
	S	Antibody–antigen	[122]
	S	Antibody–antigen	[123]
	S	Antibody–antigen	[124]
	S	Receptor–ligand	[125]
	S	Antibody–antigen	[126]
	S	Antibody–antigen	[127]
Resonant Mirror	S	Clinical diagnostic Antibody–antigen	[128]
	S	Protein–protein	[129]
	S	Antibody–antigen	[130]
	S	Protein–sugar	[131]
	S	Protein–toxin	[132]
	S	DNA–ligand	[133]
	S	Protein–protein	[134]
	S	DNA–DNA	[135]
Ellipsometry	S	DNA–ligand	[136]
	S	DNA–DNA	[137]
	S	Antibody–antigen, DNA–DNA	[138]
	S	Antigen–antibody	[139]
	S	Protein–antibody	[140]
	S	Process control	[141]
	S	Biopolymers	[142]
Ellipsometry imaging	HTS	Protein–receptor	[143]
	HTS	Antigen–antibody	[144]
	HTS	Polymer architectures	[145]
SPR	S	Protein–protein	[146]
	S	Protein–peptide	[147]
	S	DNA–DNA	[148]
	S	DNA–PNA	[149]
	S	Molecular imprinting polymers– Theophylline–caffeine–xanthine	[150]
	S	Microporous sensitive layer Methanol–Ethanol	[151]
SPR image	HTS	DNA–DNA	[152]
	HTS	Antibody–antigen	[153]
	HTS	Protein–protein	[154]
	HTS	DNA–DNA	[155]
	HTS	Peptide–protein	[156]
RIfS	S	DNA–DNA	[157]
	S	PNA–DNA	[158]
	S	Antibody–antigen	[159]
	S	DNA–ligand	[160]
	HTS	Antibody–antigen	[101]
	HTS	Receptor–ligand	[100]
	HTS	Antibody–peptide	[102]
	HTS	DNA–DNA	[103]

FIGURE 13.11. Prototype of the RIfS setup for microplate format screening.

Recently, the RIfS experimental setup was parallelized, enabling HTS in 96- and 384-well microplates to be performed. This was accomplished by using a halogen light source, a rotable filter wheel with seven interference filters for light of different wavelengths in the range of 516–707 nm, and a 12-bit CCD camera (EEV, type CCD05-20) as the detection unit. Seven wavelengths are sequentially filtered from the white light source and illuminate the RIfS microplate from below. The light passes through the bulk carrier glass slide and is partially reflected at the bottom and the top of a thin interference layer consisting of a silicon dioxide layer of thickness 330 nm. The reflected light beam ($I_1 + I_2$) results in an intensity-dependent interference pattern. The intensities of the reflected beams are recorded by the CCD camera and allow calculation of the optical thickness of the layer [97,98] (Figure 13.11) [99].

To achieve compatibility with common test formats for combinatorial compound libraries, the transducer was made by gluing the bottomless scaffold of a 96-well plastic microplate onto the transducer slide. Liquid handling is accomplished by a 96-pipette robot.

Applications of the parallel RIfS system in antibody–peptide and antibody–pesticide screening and drug screening have been demonstrated [100,101,102].

A parallel miniaturized RIfS setup using the biochip format has recently been developed. It uses planar surfaces and a high density of spots. One sample can be analyzed in high-density and low-volume wells.The screening of many samples in parallel is under development. The principle of operation has been proved for DNA–DNA hybridization [103].

KINETIC ASPECTS OF BIOMOLECULAR INTERACTIONS IN HETEROGENEOUS ASSAYS

The advantage of the label-free technique presented here is the direct measurement of binding kinetics for biomolecular interactions (e.g., protein–protein interactions).

Fitting the binding curves allow both kinetic and affinity constants to be determined. Association and dissociation rate constants are obtained as described briefly below.

Association Rate Constants

The binding of a dissolved analyte B to an immobilized species A to form a complex obeys the following kinetics in the simplest case (1:1 kinetics):

$$[A]+[B] \underset{k_d}{\overset{k_a}{\rightleftharpoons}} [AB]$$

where k_a is the association rate constant ($M^{-1}s^{-1}$) and k_d is the dissociation rate constant (s^{-1}).

The rate of formation of the complex AB depends on the free concentrations of components A and B:

$$d[AB]/dt = k[A][B] - k_d[AB] \tag{7}$$

Measurement under constant flow with the concentration of the species B kept constant leads to a pseudo-first-order kinetic. Then the change of signal R with time can be expressed as

$$dR/dt = \text{const} - (k_a A_0 + k_d) \cdot R \tag{8}$$

Solving Eqn (8) gives the signal measured at time t, as

$$R_t = R_{eq} \cdot [1 - e^{k_{obs} t}] \tag{9}$$

where

$$k_{obs} = k_a C + k_d \tag{10}$$

R_t denotes the response at time t and R_{eq} is the response at equilibrium. A_0 is the initial molar concentration of the component binding to the surface.

The kinetics described here are only obtained when the diffusion of the biomolecules to the surface is not limited. Mass transport is a particular problem with large macromolecules. The limitations of the pseudo-first-order analysis are discussed elsewhere [104,105]. In order to ensure the binding rate and not the mass transport is detected the derivative of the signal dR/dt is plotted against the signal R. According to Eqn (7) the plotted curve should be linear when the curve shows single-exponential behavior [106]. Second-order kinetics and the behavior when one component is constantly replenished by laminar flow or stirred solution, resulting in pseudo first order kinetic for binding reactions to a surface, are well described in the literature [104,105,106].

Dissociation Rate Constants

Dissociation rate constants are obtained by measuring the dissociation of the bound analyte in buffer flow. The rate equation is

$$dR_t/dt = -k_d R \tag{11}$$

and the dissociation rate constant can be obtained by an exponential fit of desorption data to the integrated rate equation

$$R_t = R_{eq} e^{-k_d t} \tag{12}$$

as described in the literature [103–105]. As explained above, in order to ensure that the binding rate and not the mass transport is detected, the curve of the derivative of the signal dR/dt plotted versus signal R should be linear when the curve shows single-exponential behavior.

The RIfS, SPR, and resonant mirror methods are compared in the literature [107] with a special reference to the kinetics.

APPLICATIONS TO LABEL-FREE DIRECT OPTICAL SENSORS

Direct optical sensing methods have been described in this chapter. Some applications have been described to show the advantages of some sensor types, for example the possibility of coupling SPR to mass spectroscopy.

Normally an analytical chemist or biochemist has a problem to be solved. In this case it is useful to know which techniques have been used to solve a comparable problem. Table 13.4 gives an overview of applications classified with respect to sensor types and biomolecular interactions. The sensor applications for interferometers, grating couplers, and resonant mirrors are only available for a single channel or spot. Other applications are available in the HTS form.

REFERENCES

1. Houston, J. G., Banks, M. *Curr. Opin. Biotechnol.* **1997**, *8*, 734–740.
2. Lemieux, B., Aharoni, A., Schena, M. *Mol. Breeding* **1998**, *4*, 277–289.
3. Mendoza, L. G., McQuaey, P., Mongan, A., Gangadharan, R., Brignac, S., Eggers, M. *Biotechniques* **1999**, *27*, 778–788.
4. Hertzberg, R. P., Pope, A. J. *Curr. Opin. Chem. Biol.* **2000**, *4*, 445–451.
5. Burbaum, J. J., Sigal, N. H. *Curr. Opin. Chem. Biol.* **1997**, *1*, 72–78.
6. Pope, A. J., Haupts, U. M., Moore, K. J. *Drug Discov. Today*, **1999**, *4*, 350–362.
7. Broach, J. R., Thorner, J. *Nature* **1996**, *384* (suppl.), 14–16.
8. GlaxoWellcome *Nature* **1996**, *384* (suppl.), 1–5.
9. Müllner, S. *Transkript* **2002**, *7*, 35–37.
10. Pope, A. J., Haupts, U. M., Moore, K. J. *Drug Discov. Today* **1999**, *4*, 350–362.
11. Auer, M., Moore, K. J., Meyer-Almes, F. J., Guenther, R., Pope, A. J., Stoeckli, K. A. *DDT* **1998**, *3*, 457–465.
12. Mere, L., Bennett, T., Coassin, P., England, P., Hamman, B., Rink, T., Zimmerman, S., Negulescu, P. *Drug Discov. Today* **1999**, *4*, 363–369.
13. Turconi, S., Bingham, R. H., Haupts, U., Pope, A. J. *Drug Discov. Today* **2001**, *6*, 27–39.
14. Houston, J. G., Banks, M. *Curr. Opin. Biotechnol.* **1997**, *8*, 734–740.
15. Marshall, A., Hodgson, J. *Nat. Biotechnol.* **1998**, 16, 27–31.
16. Schullek, J. R., Butler, J. H., Ni, Z. J., Chen, D., Yuan, Z. *Anal. Biochem.* **1997**, *246*, 20–29.
17. Burbaum, J. J. *Drug Discov. Today* **1998**, *3*, 313–322.
18. Winkler, T., Kettling, U., Koltermann, A., Eigen, M. *Proc. Natl. Acad. Sci. USA* **1999**, *96*, 1375–1378.
19. Kakiuchi, N., Nishikawa, S., Hattori, M., Shimtohno, K. *J. Virol. Methods* **1999**, *80*, 77–84.
20. Grahn, S., Kurth, T., Ullmann, D., Jakubke, H. *Biochim. Biophys. Acta* **1999**, *1431*, 329–337.
21. Morin, P. A., Saiz, R., Monjazeb, A. *Biotechniques* **1999**, *27*, 538–552.

22. Li, Z., Mehdi, S., Patel, O., Kaooya, J., Judkins, M., Zhang, W., Diener, K., Lzada, A., Dunnington, D. *J. Biomol. Screen.* **2000**, *5*, 31–38.
23. Huang, W., Zhang, Y., Sportsman, J. R. *J. Biomol. Screen.* **2002**, *7*, 215–222.
24. Sills, M. A., Weiss, D., Pham, Q., Schweitzer, R., Wu, X., Wu, J. J. *J. Biomol. Screen.* **2002**, *7*, 191–214.
25. Kolb, J. M., Yamanaka, G., Manly, S. P. *J. Biomol. Screen.* **1996**, *1*, 203–210.
26. Enomoto, K., Aono, Y., Mitsugi, T., Takahashi, K., Suzuki, R., Preaudat, M., Mathis, G., Kominami, G., Takemoto, H. *J. Biomol. Screen.* **2002**, *5*, 263–268.
27. Schobel, U., Coille, I., Brecht, A., Steinwand, M., Gauglitz, G. *Anal. Chem.* **2001**, *73*, 5172–5179.
28. Seidel, M., Gauglitz, G. TrAC **2003**, *22*, 385–394.
29. Okamoto, T., Suzuki, T., Yamamoto, N. *Nat. Biotechnol.* **2000**, *18*, 438–441.
30. Schober, A., Günther, R., Schienhorst, A., Döring, M., Linemann, B. F. *Biotechniques* **1993**, *15*, 324–329.
31. Rose, D. *Drug Discov. Today* **1999**, *4*, 411–419.
32. Ducree, J., Gruhler, H., Hey, N., Muller, M., Bekesi, S., Freygang, M., Sandmaier, H., Zengerle, R. *Proc. 13th Annual International Conference on Micro Electro Mechanical Systems, Miyazaki, Japan* **2000**, *23–27*, 317–322.
33. Sundberg, S. A. *Curr. Opin. Biotechnol.* **2000**, *11*, 47–53.
34. Schena, M., Shalon, D., Davis, R. W., Brown, P. O. *Science* **1995**, *270*, 467–470.
35. Fodor, S. P. A. *Science* **1997**, *277*, 393–395.
36. Gilles, P. N., Wu, D. J., Foster, C. B., Dillon, P. J., Chanock, S. *J. Nat. Biotechn.* **1999**, *17*, 365–370.
37. Arenkov, P., Kukhtin, A., Gemmell, A., Voloshchuk, S., Chupeeva, V., Mirzabekov, A. *Anal. Biochem.* **2000**, *278*, 123–131.
38. Templin, M. F., Stoll, D., Schrenk, M., Traub, P. C., Vohringer, C. F., Joos, T. O. *Trends Biotechnol.* **2002**, *20*, 160–166.
39. Vo-Dinh, T. *Sens. Actuators B* **1998**, *51*, 52–59.
40. Budach, W., Abel, A. P., Bruno, A. E., Neuschäfer, D. *Anal. Chem.* **1999**, *71*, 3347–3355.
41. Duveneck, G., Bopp, M., Pawlak, M., Ehrat, M., Marowsky, G. World Patent WO 200107 **2001**.
42. Peter, R., Meusel, M., Grawe, F., Katerkamp, A., Cammann, K., Börchers, T. *Fresenius J. Anal. Chem.* **2001**, *371*, 120–127.
43. Brandenburg, A. World Patent WO 0125759 **2001**.
44. Golden, J. P. US Patent 5827748 **1998**.
45. Brecht, A., Klotz, A., Barzen, C., Gauglitz, G., Harris, R. D., Quigley, G. R., Wilkinson, J. S., Sztajnbok, P., Abuknesha, R., Gascon, J., Oubina, A., Barcelo, D. *Anal. Chem. Acta* **1998**, *362*, 69–79.
46. Duveneck, G. L., Pawlak, M., Neuschäfer, D., Baer, E., Budach, W., Pieles, U., Ehrat, M. *Sens. Actuators B* **1997**, *38*, 88–95.
47. Rowe-Taitt, C. A., Golden, J. P., Feldstein, M. J., Cras, J. J., Hoffman, K. E., Ligler, F. S. *Biosens. Bioelectron.* **2000**, *14*, 785–794.
48. Schuderer, J., Akkoyun, A., Brandenburg, A., Bilitewski, U., Wagner, E. *Anal. Chem.* **2000**, *72*, 3942–3948.
49. Stimpson, D. I., Hoijer, J. V., Hsieh, W. T., Jou, C., Gordon, J., Theriault, T., Gamble, R., Baldeschwieler, J. D. *Proc. Natl. Acad. Sci. USA* **1995**, *92*, 6379–6383.
50. Rowe, C. A., Tender, L. M., Feldstein, M. J., Golden, J. P., Scruggs, S. B., MacCraith, B. D., Cras, J. J., Ligler, F. S. *Anal. Chem.* **1997**, *1*, 3846–3852.
51. Abel, A. P., Weller, M. G., Duveneck, G. L., Ehrat, M., Widmer, H. M. *Anal. Chem.* **1996**, *68*, 2905–2912.
52. Ingenhoff, J., Drapp, B., Gauglitz, G., *Fresenius J. Anal. Chem.* **1993**, *346*, 580–583.
53. Brosinger, F., Freimuth, H., Lacher, M., Ehrfeld, W., Gedig, E., Katerkamp, A., Spener, F., Cammann, K. *Sens. Actuators B* **1997**, *44*, 350–355.
54. Brandenburg, A. *Sens. Actuators B* **1997**, *39*, 266–271.
55. Brynda, E., Houska, M., Brandenburg, A., Wikerstal, A. *Bionsens. Bioelectron.* **2002**, *17*, 665–675.
56. Lin, V. S. Y., Motesharei, K., Dancil, K. P. S., Sailor, M. J., Ghadiri, R. M. *Science* **1997**, *278*, 840–843.
57. Nikitin, P., Valeiko, M., Gorshkov, B. Presented at EUROPTRODE VI, Manchester April 2002.
58. Lukosz, W., Tiefenthaler, K. *2nd ECIO, Florence 1983, IEE Conf. Pub.* **1983**, *227*, 152.
59. Nellen, P. M., Lukosz, W *Sens. Actuators B* **1990**, *1*, 592–596.
60. Nellen, P. M., Lukosz, W. *Biosens. Bioelectron.* **1993**, *8*, 129–147.
61. Brandenburg, A., Gomberg, A. *Sens. Actuators B* **1993**, *17*, 35–40.
62. Cush, R., Cronin, J. M., Stewart, W. J., Maule, C. H., Molloy, J., Goddard, N. *Biosens. Bioelectron.* **1993**, *8*, 347–354.

63. Iasys Methods Guide, **1996,** Cambridge, UK: Affinity Sensors.
64. Raether, H. *Phys. Thin Films* **1997**, *9*, 145–261.
65. Raether, H. *Surface Polaritons Eds, Agranovic and Mills*, **1982** North Holland Publ. Comp. Amsterdam.
66. Stemmler, Brecht, A., Gauglitz, G., *Sens. Actuators B* **1999**, *54*, 98–105.
67. Malmquist, M. *Nature* **1993**, *361*, 186–187.
68. Fa. Pharmacia Biacore, Uppsala, Sweden, World Patent WO 90/05925, **1990**.
69. Innes, R. A., Sambles, R., J. *J. Phys. F* **1987**, *17*, 277–287.
70. Sambles, J. R., Hibbins, A. P., Jory, M. K., Azizbekyan, H. *Journal of modern optics* **2000**, *47*, 1227–1235.
71. Johansen, K., Arwin, H., Lundström, I., Liedberg, B. *Sci. Instrum.* **2000**, *71*, 3530–3538.
72. Steiner, G., Sablinskas, V., Hübner, A., Kuhne, C., Salzer, R. *J. Mol. Struc.* **1999**, *509*, 265–274.
73. Nelson, S. G., Johnston, K. S., Yee, S. S. *Sens. Actuators B* **1996**, *35/36*, 187–191.
74. Nikitin, P. I., Beloglazov, A. A., Kochergin, V. E., Valeiko, M. V., Ksenevich, T. I. *Sens. Actuators B* **1999**, *54*, 43–50.
75. Nikitin, P. I., Grigorenko, A. N., Beloglazov, A. A., Valeiko, M. V., Savchuk, A. I., Savchuk, O. A., Steiner, G., Kuhne, C., Huebner, A., Salzer, R. *Sensors and Actuators A* **2000**, *85*, 189–193.
76. Nelson, B. P., Grimsrud, T. E., Liles, M. R., Goodman, R. M., Corn, R. M. *Anal. Chem.* **2001**, *73*, 1–7.
77. Nelson, B. P., Frutos, A. G., Brockman, J. M., Corn, R. M. *Anal. Chem.* **1999**, *71*, 3928–3934.
78. Johansen, K., Arwin, H., Lundström, I., Liedberg, B. *Rev. Sci. Instrum.* **2000**, *71*, 9.
79. Institutet Polymerutveckling AB Djursholm, Sweden, World Patent WO 01/69201, **2001**.
80. Homola, J., Dostalek, J., Piliarik, S., Yee, S. Presented at EUROPTRODE VI, April **2002**.
81. Graffinity Pharmaceutical Design GmbH, Heidelberg, Germany, World Patents WO 94/16312 A1, **1994**; WO 98/32002 A1, **1998**.
82. Graffinity Pharmaceutical Design GmbH, Heidelberg, Germany, World Patents WO 00/31515 A1, **2000**; WO 01/63257, **2001**.
83. Schröter, U., Heitmann, D. *Phys. Rev. B* **1999**, *60*, 4992–4999.
84. Kreiter, M., Mittler, S., Knoll, W., Sambles, J. R. *Phys. Rev. B* **2002**, *65*, 125415.
85. World Patents WO 99/30135; **1999**.
86. Brockman, J. M., Fernandez, S. M. Am. Lab. 2001, *33*, 37–40.
87. Biacore Journal, April **2002**.
88. *BIAcore, Instrumenta 6th March* **2002** *18*, 22.
89. Fa Pharmacia Biacore, Uppsala, Sweden, World Patent WO 98/34098, **1998**.
90. Mitchell, P. *Nature biotechnol.* **2002**, *20*, 225–229.
91. Soenksen, C. P., Nordhoff, E., Jansson, O., Malmqvist, M., Roepstorff, P. *Anal. Chem.* **1998**, *70*, 2731–2736.
92. Lin, S., Long, S., Ramirez, S. M., Cotter, R. J., Woods, A. S. *Anal. Chemist.* **2000**, *72*, 2635–2640.
94. van Noort, D., Rumberg, J., Jager, E., Mandenius, C. F. *Meas. Sci. Techol.* **2000**, *11*, 801–808.
95. Schmitt, H. M., Brecht, A., Piehler, J., Gauglitz, G. *Biosens. Bioelectron.* **1997**, *12*, 809–816.
96. Piehler, J., Brecht, A., Gauglitz, G., Maul, C., Grabley, S., Zerlin, M. Biosens. Bioelectron. **1997**, 12, 531–538.
97. Rothmund, M., Schütz, A., Brecht, A., Gauglitz, G., Berthel, G., Gräfe, D. *Fresenius J. Anal. Chem.* **1997**, *359*, 15–22.
98. Gauglitz, G. *Cur. Opi. Chem. Biol.* **2000**, *4*, 351–355.
99. Patent DE 196 15 366 A1, **1996**.
100. Birkert, O., Gauglitz, G. *Anal. Bioanal. Chem.* **2002**, *372*, 141–147.
101. Birkert, O., Tuennemann, R., Jung, G., Gauglitz, G. *Anal. Chem.* **2002**, *74*, 834–840.
102. Kröger, K., Bauer, J., Fleckenstein, B., Rademann, J., Jung, G., Gauglitz, G. *Biosens. Bioelectron.* **2002**, *17, (11–12)* 937–944.
103. Wegner, M., Jung, A., Hänel, C., Gauglitz, G. in preparation.
104. O'Shannessy, D. J., Winzor, D. J. *Anal. Biochem.* **1996**, *249*, 457–468.
105. Karlsson, R., Michaelsson, A., Mattsson, L. *Anal. Biochem.* **1991**, *236*, 275–283
106. Karlsson, R., Michaelsson, A., Mattsson, L. *J. Immunol. Methods* **1991**, *145* 229–240.
107. Hänel, C., Gauglitz, G. *Anal. Bioanal. Chem.* **2002**, *372*, 91–100.
108. Busse, S., DePaoli, M., Wenz, G., Mittler, S. *Sens. Actuators B* **2001**, *80*, 116–124.
109. Drapp, B, Piehler, J., Brecht, A., Gauglitz, G., Luff, B. J., Wilkinson, J. S., Ingenhoff, J. *Sens. Actuators B* **1997**, *39*, 277–282.
110. Busse, S., Scheumann, V., Menges, B., Mittler, S. *Biosens. Bioelectron.* **2002**, *8*, 704–710.

111. Weisser, M., Tovar, G., Mittler-Neher, S., Knoll, W., Brosinger, F., Freimuth, H., Lacher, M., Ehrfeld, W. *Biosens. Bioelectron.* **1999**, *14*, 405–411.
112. Ingenhoff, J., Drapp, B., Gauglitz, G. *Fresenius J. Anal. Chem.* **1993**, *346*, 580–583.
113. Boiarski, A. A., Busch, J. R., Bhullar, B. S., Ridgway, R. W., Miller, L. S., Zulich, A. W. *Proc. SPIE. Int. Soc. Opt. Eng.* **1993** *1886* 15–26.
114. Heideman, R. G., Kooyman, R. P. H., Greve, J. *Sens. Actuators B* **1993**, *10*, 209–217.
115. Brynda, E., Houska, M., Brandenburg, A., Wikerstal, A. *Biosens. Bioelectron.* **2002**, *17*, 665–675.
116. Ymeti, A., Kanger, J. S., Wijn, R., Lambeck, P. V., Greve, J. *Sens. Actuators B* **2002**, *83*, 1–7.
117. Lin, V. S. Y., Motesharei, K., Dancil, K. P. S., Sailor, M. J., Ghadiri, R. M. *Science* **1997**, *278*, 840–843.
118. Pecora, R. *Biochem. Soc. Trans.* **1991**, *19*, 498.
119. Ryu, C. S., Cho, S. M., Kim, B. W. *Biotechn. Letters* **2001**, *23*, 653–659.
120. Bier, F. F., Scheller, F. W. *Biosens. Bioelectron.* **1996**, *11*, 669–674.
121. Bier, F. F., Kleinjung, F., Scheller, F. W. *Sens. Actuators B* **1997**, *38*, 78–82.
122. Orban, M., Katerkamp, A., Renneberg, R., Spener, F., Cammann, K. *J. Immunol. Methods* **1998**, *215*, 17–26.
123. Piehler, J., Brandenburg, A., Brecht, A., Wagner, E., Gauglitz, G. *Appl. Optics* **1997**, *36*, 6554–6562.
124. Polzius, R., Diessel, E., Bier, F. F., Bilitewski, U. *Anal. Biochem.* **1997**, *248*, 269–276.
125. Clerc, D., Lukosz, W. *Biosens. Bioelectron.* **1997**, *12*, 185–194.
126. Brandenburg, A., Polzius, R., Bier, F. F., Bilitewski, U., Wagner, E. *Sens. Actuators B* **1996**, *30*, 55–59.
127. Bernard, A., Bosshard, H. R. Eur. *J. Biochem.* **1995**, *230*, 416–423.
128. Dougan, T., Levy, J. B., Salama, A., George, A. J. T., Pusey, C. D. *Clin. Experiment. Immunol.* **2002**, *128*, 555–561.
129. Blacque, O. E., Worrall, D. M. *J. Biol. Chem.* **2002**, *277*, 10783–10788.
130. Dmitriev, D. A., Massino, Y. S., Segal, O. L., Smirnova, M. B., Pavlova, E. V., Gurevich, K. G., Gnedenko, O. V., Ivanov, Y. D., Kolyaskina, G. I., Archakov, A. I., Osipov, A. P., Dmitriev, A. D., Egorov, A. M. *J. Immunol. Methods* **2002**, *261*, 103–118.
131. Fernig, D. G. *Methods in Mol. Biol.* **2001**, 171, 505–518.
132. Puu, G. *Anal. Chem.* **2001**, *73*, 72–79.
133. Grunden, A. M., Self, W. T., Villain, M., Blalock, J. E., Shanmugam, K. T. *J. Biol. Chem.* **1999**, *274*, 24308–24315.
134. Hirmo, S., Artursson, E., Puu, G., Wadstrom, T., Nilsson, B. *J. Microbiol. Methods* **1999**, *37*, 177–182.
135. Koval, V. V., Gnedenko, O. V., Ivanov, Y. D., Fedorova, O. Z., Archakov, A. I., Knoore, D. G. *IUMBM Life* **1999**, *48*, 317–320.
136. Eskilsson, K., Leal, C., Lindman, B., Miguel, M., Nylander, T. *Langmuir* **2001**, *17*, 1666–1669.
137. Mandenius, C. F., Chollet, A., Mecklenburg, M., Lundstroem, I., Mosbach, K. *Anal. Lett.* **1989**, *22*, 2961–2973.
138. Ostroff, R. M., Maul, D., Bogart, G. R., Yang, S., Christian, J., Hopkins, D., Clark, D., Trotter, B., Moddel, G. *Clin. Chem.* **1998**, *44*, 2031–2035.
139. Tengvall, P., Askendal, A., Lundstrom, I. *Colloids Surface B* **2001**, *20*, 51–62.
140. Robers, M., Rensink, I. J. A. M., Hack, C. E., Aarden, L. A., Reutelingsperger, C. P. M., Glatz, J. F. C., Hermens, W. T. *J. Biophys.* **1999**, 76, 2769–2776.
141. Simpson, L. J., Joshi, B. S., Gonzales, L. A., Verley, J., Furtak, T. E. *Mater. Res. Soc. Symp. Proc.* **2000**, *616*, 9–14.
142. Mutschler, T., Kieser, B., Frank, R., Gauglitz, G. *Anal. Bioanal. Chem.* **2002** *374*, 658–664.
143. Kim, S. R., Abbott, N. L. *Adv. Mater.* **2001**, *13*, 1445–1449.
144. Jin, G., Meng, Y., Xing, J., Zhao, Z. *Proc. SPIE-Int. Soc. Opt.I Eng.* **1998**, *3548*, 131–135.
145. Beyerlein, D., Kratzmuller, T., Eichhorn, K. J. *Vibration. Spectrosc.* **2002** *29*, 223–227.
146. Nelson, R. W., Nedelkov, D., Tubbs, K. A. *Anal. Chem.* **2000**, *72*, 404–411.
147. Nelson, R. W., Jarwik, J. W., Taillon, B. E., Tubbs, K. A. *Anal. Chem.* **1999**, *71*, 2858–2865.
148. Mariotti, E., Minunni, M., Mascini, M. *Sensors and Microsystems, Proc. 6th Italian Conference* **2002**, pp. 5–7.
149. Jensen, K. K., Oerum, H., Nielsen, P. E., Norden B. *Biochemistry* **1997**, *36*, 5072–5077.
150. Lai, E. P. C., Fafara, A., Vandernoot, V. A., Kono, M., Polsky, B. *Can. J. Chem.* **1998**, *76*, 265–273.
151. Kieser, B., Dieterle, F., Gauglitz, G. *Anal. Chem.* **2002**, *74*, 4781–4787.
152. Jin-Lee, H., Goodrich, T. T., Corn, R. M. *Anal. Chem.* **2001**, *73*, 5525–5531.

153. Dutra, R., Fireman, M. N., Otavio, J., Fontana, E. *Proc. SPIE-Int. Soc. Opt. I Eng.* **2001**, *4254*, 128–135.
154. Smith, E. A., Erickson, M. G., Ulijasz, A. T., Weisblum, B., Corn, R. M. *Langmuir* **2003**, *19*, 1486–1492.
155. Lee, H. J., Goodrich, T., Corn, R. M. Micro Total Analysis Systems 2001, *Proc. 5th* TAS 2001 *Symp.* Monterey, CA, **2001**, pp. 441–443.
156. Houseman, B. T., Huh, J. H., Kron, S. J., Mrksich, M. *Nature Biotechnol.* **2002**, *20*, 270–274.
157. Sauer, M., Brecht, A., Charisse, K., Stemmler, I., Gauglitz, G., Bayer, E., *Anal. Chem.* **1999**, *71*, 2850–2857.
158. Kröger, K., Jung, A., Reder, S., Gauglitz, G. *Anal. Chim. Acta* **2002**, *469*, 37–48.
159. Piehler, J., Brecht, A., Giersch, T., Kramer, K., Hock, B., Gauglitz, G. *Sens. Actuators B* **1997**, *39*, 432–437.
160. Piehler, J., Brecht, A., Gauglitz, G., Maul, C., Zerlin, M., Thiericke, R., Grabley, S., *Anal. Biochem.* **1997**, *249*, 94–102.

Chapter 14

Combinatorial Study and High-Throughput Screening of Transparent Barrier Films using Chemical Sensors

Jaime C. Grunlan, Dennis Saunders, Jay Akhave, Mark Licon, Marcel Murga, Aaron Chavira, and Ali R. Mehrabi

INTRODUCTION

The combinatorial study of materials has already proven its value in the areas of biotechnology [1] and the discovery of medicinal compounds [2,3]. More recently, this methodology has moved into applications such as discovery of organometallic catalysts with special activity [4], optimization of polymer processing [5], and composite design [6]. The primary advantage of the combinatorial technique is the speed at which different materials can be synthesized, formulated, and tested for particular application [sometimes referred to as high-throughput screening (HTS)]. In addition to speed, the amount of material needed for a combinatorial study is far less than that required for conventional methods, which makes combinatorial material discovery more affordable when the materials are expensive. The need for speed in the combinatorial science magnifies the necessity for automation of different steps in the material discovery process. Consequently, a tremendous effort is focused on automating the formulation, synthesis, and screening steps in discovering new materials. Many of these methods are applicable to the screening of any type of functional material, while others are targeted for specific functionality in a particular area of application.

Materials with different levels of gas (or diffusion) barrier are used in a wide range of packaging applications, from food [7] and medical packaging [8] to the protection of

Jaime C. Grunlan, Dennis Saunders, Jay Akhave, Mark Licon, Marcel Murga, Aaron Chavira, and Ali R. Mehrabi • Avery Dennison Corporation, Avery Research Center, 2900 Bradley Street, Pasadena, CA 91107, USA
Contact author: Ali R. Mehrabi, email: Ali_Mehrabi@averydennison.com

High-Throughput Analysis, edited by Potyrailo and Amis
Kluwer Academic Publishers, New York, 2003.

electronic materials [9]. Significant levels of research and development effort have focused on the discovery of inorganic and polymeric materials for barrier applications for decades [10–12]. In this chapter the first combinatorial methodology and experimentation for discovering new diffusion barriers is presented. This study encompasses the development of new tools and methods for the entire combinatorial process, such as formulation, coating, and screening in a high-throughput manner. The screening of diffusion barriers includes measurement of their gas barrier properties (e.g., oxygen and moisture vapor) and also their optical clarity (or haziness). A variety of HTS methods are described that quickly screen many films simultaneously for the desired property. Each of these methods rely on a change in the optical property (e.g., absorbance or fluorescence of light). The screening process is also automated using a variety of image-processing techniques. In some instances, additional mathematical modeling is also used to improve the accuracy of the experimental measurements. In the following section a general method for preparation of laminate sensors is described.

LAMINATE SENSORS

Laminate sensors are multilayer structures of different materials, in which one or more layers have the desired sensing functionality. Other layers in the laminate protect the sensor layer and facilitate application and testing. In this section, different variations of laminate sensors are described and their advantages and disadvantages are discussed.

Sensor Design

In the first method for the fabrication of a laminate sensor, a sensor dye formulation is coated on a substrate film (2 mil PET film). After drying the sensor layer, a second layer of pressure-sensitive adhesive is coated over the sensor layer. The film is then covered and protected by a release paper. After applying and coating the barrier materials using the spin-casting method on the other side of the substrate film, the release paper is removed and the whole construction is laminated to a glass plate (Figure 14.1). This construction ensures a proper seal between interfaces and also uniform adhesion of the laminate to the glass plate.

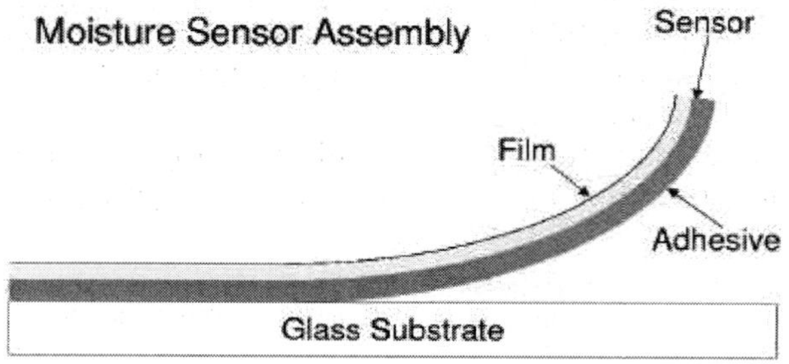

FIGURE 14.1. Sealing the laminate sensor to the glass substrate.

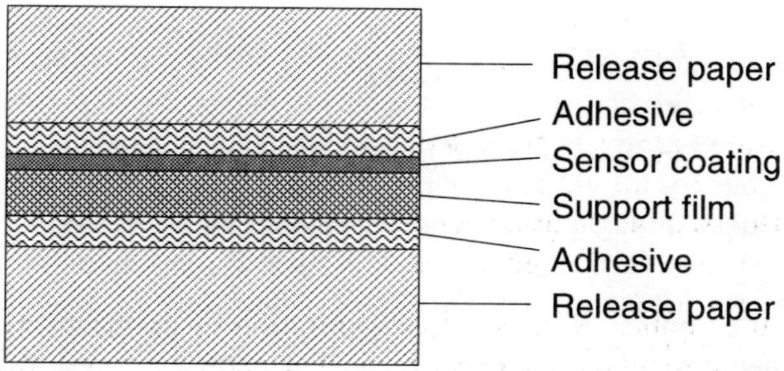

FIGURE 14.2. Sensor sandwiched between two transfer adhesives.

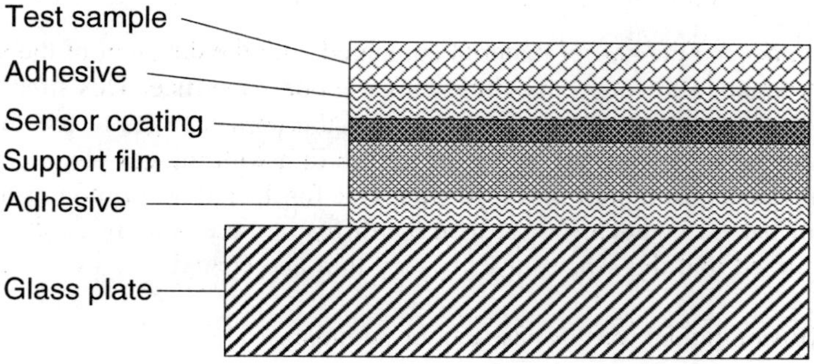

FIGURE 14.3. Construction cross-section of an MHTS or OHTS sample before edge sealing.

A modified version of the laminate shown in Figure 14.1 sandwiches a sensor-coated support film between two pieces of transfer adhesive as shown in Figure 14.2. First, a sensor solution containing a special polymer and an oxygen- or water-sensitive dye is coated onto a piece of 2 mil PET. A pressure-sensitive adhesive, coated onto release paper (transfer adhesive), is then laminated to both sides of the sensor-coated film.

Samples to be tested for moisture or oxygen barrier, referred to as moisture high-throughput screening (MHTS) or oxygen high-throughput screening (OHTS), are prepared by removing the top release paper and laminating the sample on the exposed adhesive layer. A circular piece of the sample laminate is then cut with a half-inch punch. The bottom release paper is removed from the circle and the construction is laminated to a glass plate. Samples are positioned on the plate in a 6×8 array corresponding to the positions read by the fluorescence or UV–vis spectrum plate reader. Figure 14.3 shows a cross section of an individual sample on the plate. The actual function of these films and their use in testing barrier properties is described in more detail in the following sections.

Mathematical Model

In a typical high-throughput experiment, a multilayer consisting of a laminate sensor material and a barrier film is constructed. The barrier–sensor construction is then placed

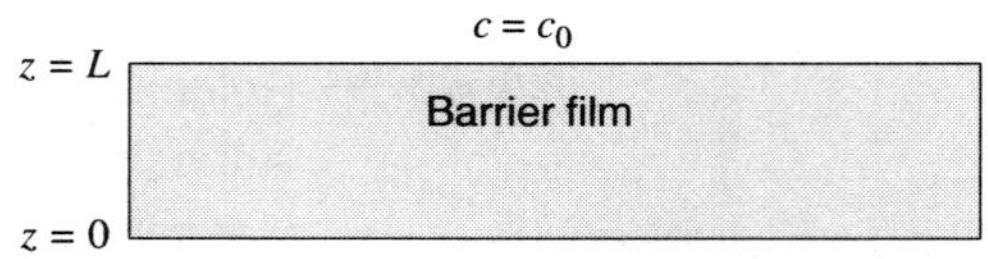

FIGURE 14.4. One-dimensional diffusion through a barrier film.

in an environmental chamber. A constant concentration of permeant gas (moisture or oxygen) is maintained throughout the experiment. The permeant gas diffuses through the barrier material and reaches the sensor layer causing a given response (e.g., color change, fluorescence, etc.), that is measured with respect to time. Substrate films with a relatively high rate of gas transmission are chosen, which allows the barrier layer to control the gas transmission rate.

The thickness of the barrier material compared with the diameter of the sample to be tested is very small, allowing a one-dimensional model to suffice. This simplified case is represented in Figure 14.4. The concentration of the permeant gas at the surface of the barrier is assumed to be constant and is in equilibrium with the gas phase concentration. This is a reasonable assumption since the timescale for the gas adsorption from the bulk to the surface of barrier is much shorter than the diffusion timescale inside the barrier film. It is also assumed that the sensor layer is very thin and coated at $z = 0$ in the schematic Figure 14.4.

The diffusion model for this case can be summarized [13]

$$\frac{\partial c}{\partial t} = D\frac{\partial^2 c}{\partial z^2} \tag{1}$$

where the initial conditions are

$$c(L, t) = c_0$$

$$\frac{\partial c}{\partial z} = 0, \quad z = 0$$

$$c(z, 0) = 0$$

and D and L are the diffusion coefficient of the permeant gas in the barrier material and the thickness of the barrier layer, respectively. The solution to Eqn (1) is obtained analytically by using the separation of variables:

$$\frac{c}{c_0} = 1 - 2\sum_{n=0}^{\infty}\frac{(-1)^n}{(n+\frac{1}{2})\pi}\exp\left[-\left(n+\frac{1}{2}\right)^2\pi^2\frac{D}{L^2}t\right]\cos\left(n+\frac{1}{2}\right)\frac{\pi z}{L}$$

Therefore, at the sensor layer (i.e., $z = 0$) the concentration takes the form

$$\frac{c}{c_0} = 1 - 2\sum_{n=0}^{\infty}\frac{(-1)^n}{(n+\frac{1}{2})\pi}\exp\left[-\left(n+\frac{1}{2}\right)^2\pi^2\frac{D}{L^2}t\right]$$

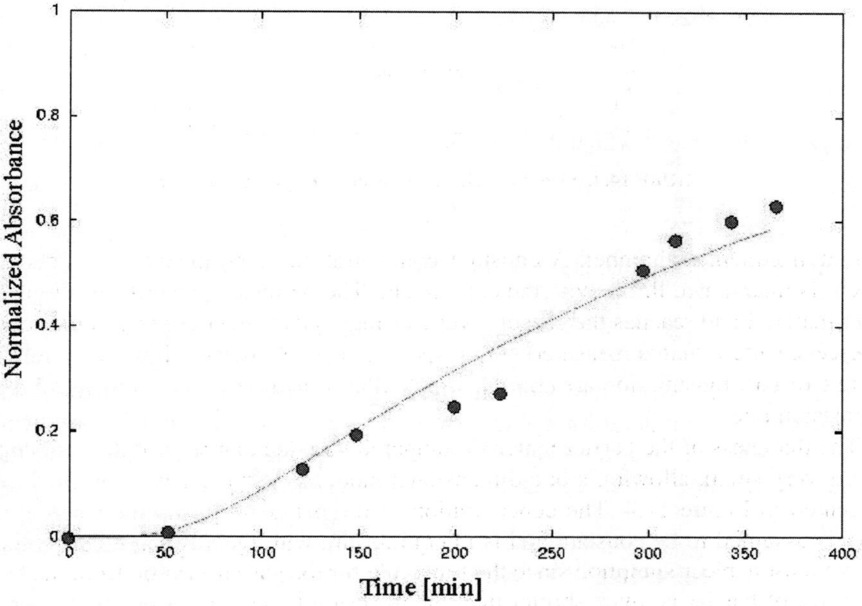

FIGURE 14.5. Typical fit of Eqn (2) to the experimental data.

Assuming that $\alpha = D/L^2$, where α^{-1} is the diffusion timescale, and taking into the account that the absorbance of the sensor dye is proportional to the moisture concentration, we have

$$\varphi(t) = \frac{A}{A_0} = 1 - 2\sum_{n=0}^{\infty} \frac{(-1)^n}{\left(n+\frac{1}{2}\right)\pi} \exp\left[-\left(n+\frac{1}{2}\right)^2 \pi^2 \alpha t\right] \quad (2)$$

where A is the absorbance at the desired wavelength. The values for A_0 and α are unknown and need to be calculated. The absorbance for the dye at the maximum concentration (A_0) is measured by reading the asymptotic limit for the absorbance of the dye where no barrier coating is present. The value for α is then calculated by a nonlinear fit of Eqn (2) to the experimental values measured for each of the barrier coatings.

Figure 14.5 shows a typical fit to the experimental curve and the values for α are calculated. The errors and variations in the experimental values are due to the variations in the relative humidity caused by ambient temperature change and possibly diffusion of moisture through the interface between the barrier coating and the substrate film.

Keeping only the first two terms in the series expansion given in Eqn (2), we can write

$$\varphi(t) = \left[1 - \frac{4}{\pi}\exp\left(-\frac{\pi^2}{4}\alpha t\right) + \frac{4}{3\pi}\exp\left(-\frac{9\pi^2}{4}\alpha t\right)\right]$$

To find the inflection point we set

$$\frac{\partial^2 \varphi}{\partial^2 t} = 0$$

and therefore the time at which the curvature changes is

$$t_i = \frac{3\ln(3)}{2\pi^2 \alpha}$$

Finally, the slope of the curve at the inflection point is given by:

$$\left.\frac{\partial \varphi}{\partial t}\right|_{t_i} = \pi \alpha C_1$$

where $C_1 \approx 0.5887$. Since the slope at the inflection point changes linearly with α, it can be used as a measure for gas transmission rate through the film. Consequently, in the rest of measurements, the slope of the absorbance curve is correlated with the gas transmission rate through the film in order to simplify the calculations.

SENSING CHEMISTRY AND MECHANISMS

Moisture Sensor

Optochemical sensors for humidity have been widely studied in a variety of applications and conditions [14–19]. In particular, sensors based on a mixture of Nafion® and crystal violet (CV) have shown promise [14,18,19]. This system changes color from yellow to green to dark blue depending on the ambient humidity (or moisture uptake). The mechanism of the Nafion®–CV (CVN) sensor appears to be a simple acid–base reaction. Nafion® (a sulfonic acid substituted fluoropolymer) protonates the CV (a pH indicator), turning it yellow. Protonated CV is a stronger acid than water, which causes protons to be transferred from it to form H_3O^+. As the CV deprotonates, the color turns from yellow to blue and eventually to violet (see proposed reaction scheme in Figure 14.6). In a CVN sensor film, green is observed prior to blue owing to the presence of fully protonated (CVH_2^{3+}) and partially protonated (CVH^{2+}) species. The speed and extent of color change depends on the thickness of sensor material and the ratio of the components (i.e., Nafion® to CV) [14].

This material can be used as a moisture sensor when a thin sensor film, attached to a substrate, is coated with a barrier film of interest (as described for laminate sensor fabrication) [20]. Such moisture-sensing approach was co-developed by Avery and GE as a part of the joint NIST ATP project [21]. By placing this construction into a temperature-controlled chamber with 100% relative humidity and measuring its absorbance at 630 nm as a function of time, a curve related to moisture uptake is obtained. Absorbance in the yellow region of the visible light spectrum increases with time due to the color change in the sensor film as the CV becomes deprotonated (Figure 14.6). Figure 14.7 shows the effect of absorbance with time in a 100% humid environment for a 5 mil PET film as the moisture barrier. The absorbance–time curve can then be further processed in order to calculate the moisture transmission through the sample.

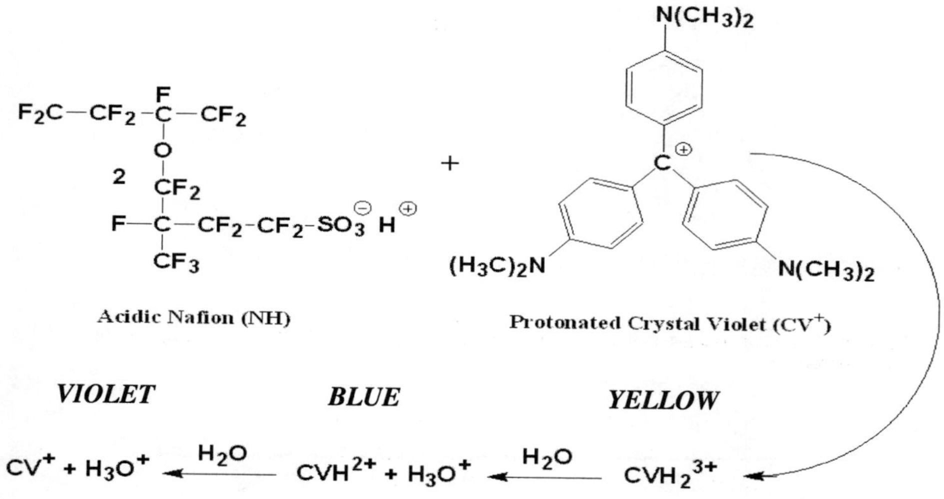

FIGURE 14.6. Proposed scheme for color change during water uptake by CVN sensor film.

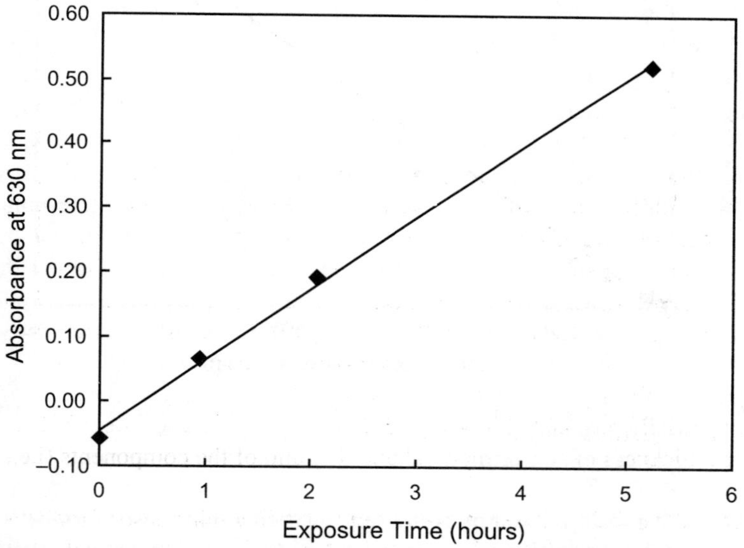

FIGURE 14.7. Visible light absorbance as a function of exposure time of a CVN sensor construction covered by 5 mil PET in a 100% humid atmosphere.

Oxygen Sensor

Fluorescence quenching has proven to be a successful method for measuring oxygen concentration [22–26]. Platinum octaethylporphyrin (PtOEP) embedded within poly [1-(trimethylsilyl)-1-propyne] (PTMSP) or poly(styrene-*co*-trifluoroethyl-methacrylate) (polystyrene-*co*-TFEM) has shown extremely high sensitivity to oxygen, with an I_0/I_{100}

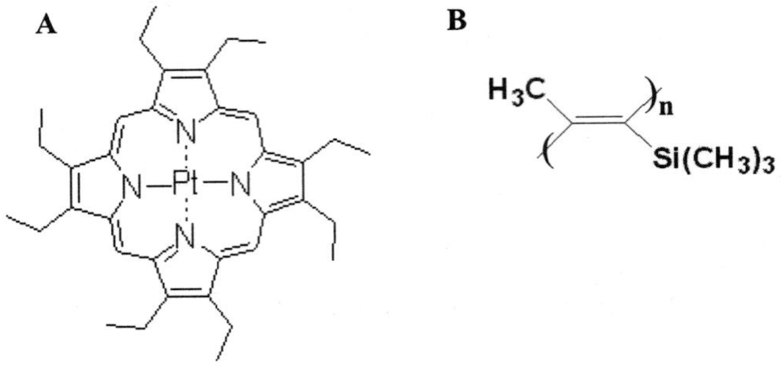

FIGURE 14.8. Chemical structures of (a) platinum octaethylporphyrin (PtOEP) and (b) poly[1-(trimethylsilyl)-1-propyne] (PTMSP).

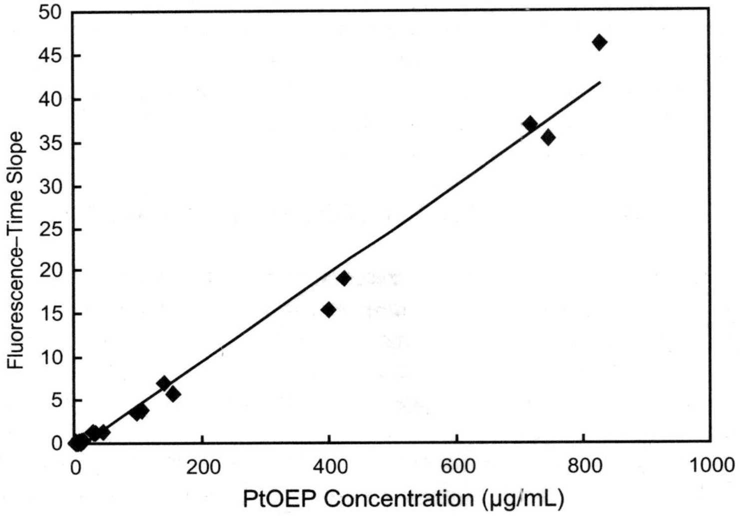

FIGURE 14.9. Effect of PtOEP concentration on oxygen sensitivity of sensor film, as evidenced by increasing fluorescence–time slope. A 175 μm thick PET film was used as the barrier for these data.

ratio of over 200 [23,24]. Thus it was suggested that such fluorescence quenching chemistries could be used for studies of barrier films [27]. The chemical structures of both of these molecules are shown in Figure 14.8. The process of using a PtOEP–PTMSP sensor film for measuring the oxygen transmission rate (OTR) through a barrier film of interest is very similar to that already described for the CVN sensor (used for MVTR measurement).

As will be described in more detail, OTR is determined using the slope of fluorescence intensity as a function of time (see Figure 14.10 as an example) which is then compared with the slopes of known standards. Sensor film thickness and PtOEP content determine the sensitivity of this form of barrier measurement. Figure 14.9 illustrates the effect of increasing the PtOEP concentration, within PTMSP, on the slope of fluorescence

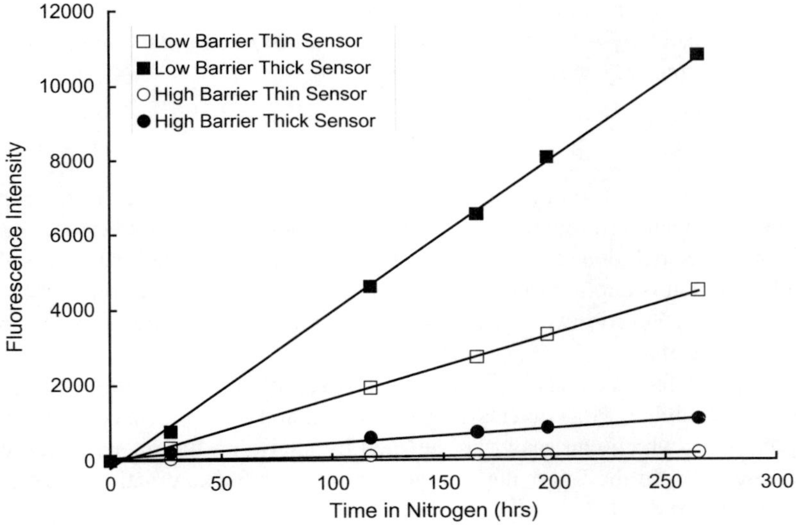

FIGURE 14.10. Effect of PtOEP–PTMSP film thickness on oxygen sensitivity, as evidenced by differences in fluorescence–time slope.

intensity as a function of time. Steeper fluorescence–time slopes can typically be interpreted as offering greater oxygen sensitivity for the same test film. Figure 14.10 shows the effect of changing the sensor film thickness on the fluorescence intensity as a function of the time curve. The fluorescence–time curves for low (OTR $\approx 8.5\,\text{cm}^3\,\text{m}^{-2}$) and high (OTR $<$ $0.005\,\text{cm}^3\,\text{m}^{-2}$ per day) barrier films are shown here for thick (6 μm) and thin (1 μm) sensor films. Thicker sensor films yield larger fluorescence–time slopes, which increases sensitivity and allows very high barrier films to be detected. The thin sensor film is unable to provide a measurable slope for the high barrier film in Figure 14.10.

Coating Defects

In a combinatorial study of coating materials, several different formulations and compositions are prepared and tested for evaluation. It is expected that many of these formulations will fail during dispensing and preparation. Failure may also happen at other stages of sample plate preparation such as mixing, coating, and screening steps. The failure in the formulation step can be a result of the incompatibility of the ingredients that causes further coagulation, gel formation, opacity, and/or phase separation. These types of failure usually propagate down to other preparation steps such as mixing, coating, and screening. On the other hand, it should be noted that these types of coating failures contain valuable information about compatibility and miscibility of the ingredients, and help to optimize the formulations further.

It is necessary to develop a procedure for identification of defects in each stage of sample plate preparation. A defect detection procedure needs to be automated in an effort to minimize human errors and unify the data interpretation. Consequently, efforts have been made to develop an automatic defect detection system. A computer code has been created that automatically identifies defects in the final combinatorial daughter plates and marks them as "defective" coatings.

Defect Categories

Different types of defects may arise in the final coatings. Some of these can produce deterioration in the barrier properties of these films, which would lead to erroneous barrier measurement. One such defect is the uneven coverage of the coating material in the well plates. Delamination is another major defect resulting from poor adhesion of the material to the substrate. Nonuniformity in the coating material, bubbles, and pinholes are other common types of defect.

In any case, defects can be observed after the daughter plates are subjected to testing in the humidity chamber. Most defects cause abnormal and nonuniform color variations in the daughter plate, which can easily be detected by analyzing the intensity variations at certain wavelengths. In the following section, a procedure for automatic defect detection is explained in more detail.

Image Enhancement

The heart of any image-processing algorithm is the image enhancement. Most as-captured images are of poor quality and make it difficult to apply a further detection algorithm. Captured images need further enhancement to correct nonuniform lighting, improve brightness and contrast, and better align the edges of the daughter plate within the image. In the first step, the images are subjected to enhancement and the variations caused by nonuniform illumination are reduced. The original image is loaded into the program and subjected to a sequence of so-called "erosion" and dilation steps [28]. As a result a background image is generated that contains the information due to improper and nonuniform lighting conditions. This image is then subtracted from the original image to obtain the background-enhanced image. It is found that for the images taken from the daughter plates having an MHTS sensor, the contrast between low and high moisture content regions is higher if the red channel is selected from the color image. For this reason the background-enhanced image is decomposed to RGB channels and only the red channel is processed in the detection algorithm. To increase or stretch the contrast, the R-channel image is first converted to the gray-level image. The histogram distribution for the gray-level image is then calculated. The contrast is increased by manipulation of the histogram automatically and stretching the gray-level range for the image. Once these steps are completed, the contrast-enhanced image is ready for the pattern-recognition algorithm. Orientational and misalignment errors do not need any correction because the detection algorithm is invariant to the rotation and translation of the coatings as long as the image is taken perpendicular to the plane where the daughter plate is positioned.

Feature Detection

Feature detection algorithms are primarily designed to operate on binary images (i.e., images that contain only 0s and 1s). The gray-level image can easily be converted to a binary image by assigning a threshold value for the gray level. The value for the threshold is varied to detect different features in the image. After thresholding and creating a binary image, the edges (borders between light and dark regions) are detected. It is crucial in this algorithm to detect the location of the coatings automatically. Automated locating reduces the effort needed for aligning the image and rescaling calculations. Since the coatings are supposed to be circular, it is useful to detect circular objects in the thresholded image. This is easily achieved by least-squares fitting a bilinear form to the edge contours resulting from the edge-detection algorithm. The coefficients in the bilinear form are constrained to ensure the fit of an ellipse to the contours. The resulting elliptical objects are screened according to their size and eccentricity. The eccentricity of an ellipse defines how it is elongated in one direction. If the eccentricity is unity the resulting curve is a circle. Therefore each ellipse with an eccentricity larger than a prescribed value (close to unity) is discarded. If a coating has an irregular shape or the red-channel intensity varies greatly in the radial direction, it would be screened out as a "defective" coating.

At this stage, all the coatings that have a circular shape are detected and their centers and radii are stored. This will reduce the image-processing effort considerably because the search for other types of defects can now be limited to the regions inside these circles. For the next defect test, the red-channel intensity variation inside each circle is measured. High-intensity variations can be a result of defects and/or thickness variations in a coating. If the intensity variations are larger than a prescribed threshold, the coatings are screened out. It is possible to define further procedures to detect other types of defects. Consequently, it may be necessary to capture images in higher resolution and classify the defects by their topological properties. However, preliminary experiments show that the current algorithm is sufficient for its intended purpose.

EXPERIMENTAL

Moisture High-Throughput Screening

MHTS Sensor Preparation

Reagent-grade CV from Aldrich (Milwaukee, WI) is dissolved in methanol to create a 0.61 wt% solution. Dissolution of CV takes approximately an hour using a magnetic stirrer. This CV solution is then combined with a 5 wt% Nafion solution (Liquion 1100, Ion Power, Bear, DE) to yield a Nafion : CV (CVN) solution ratio of 1 : 46. The CVN solution is typically stirred for 15 min to achieve homogeneity of components. Supported CVN sensor film is then prepared by coating the CVN solution onto 2 mil (50 μm) thick PET film (Mylar D, GE, Charlotte, NC) using the 4 mil opening (approximately 100 μm) on a blade-coater. After drying in a hood for 1 hr, the visible spectrum of this solid CVN film is measured using a μQuant spectrometer from Bio-Tek Instruments Inc. (Winooski, VT). The absorbance of the sensor films is measured at 435, 630, and 800 nm and the net absorbance ratio, which should be 19.6 ± 5.0, is determined from

$$\text{ratio} = \frac{A_{630} - A_{800}}{A_{435} - A_{800}} \tag{3}$$

When the ratio is too low, indicated by a greenish film color, a small amount of Nafion solution (around 5% of total CVN solution volume) is added and stirred for 15 min. The newly diluted solution is again coated onto PET and dried, and the absorbance ratio is retested. This procedure is repeated until the appropriate absorbance ratio is achieved. The final adjusted sensor solution is then filtered through a G6 glass filter (Fisher Scientific Inc., Hampton, NH). Both sides of the supported CVN sensor film are then laminated with 3M 467 transfer adhesive (Minnesota Mining and Manufacturing, St. Paul, MN). Finally, the supported sensor film is laminated to a plate of window glass measuring 3.375 in. × 5 in. × 0.125 in.

Moisture Barrier Sample Plate Preparation

In general, barrier films to be tested in MHTS fashion are prepared using blade, spin, and dip coating techniques. Test films are either coated onto a 2 mil PET substrate and later laminated to the sensor construction or they are directly coated onto the sensor film prior to lamination to the glass plate as described above (see Figure 14.1). Figure 14.11 shows an example test panel with 48 barrier films of varying thickness and composition. Barrier solutions are applied to this plate using a robot-controlled pipette that is programmed to deposit varying volumes and/or compositions of test solutions. In other instances, rectangular strips of film are simply laminated to the sensor construction. In this case, each test strip occupied a column of six spots in the 48-well framework shown in Figure 14.11.

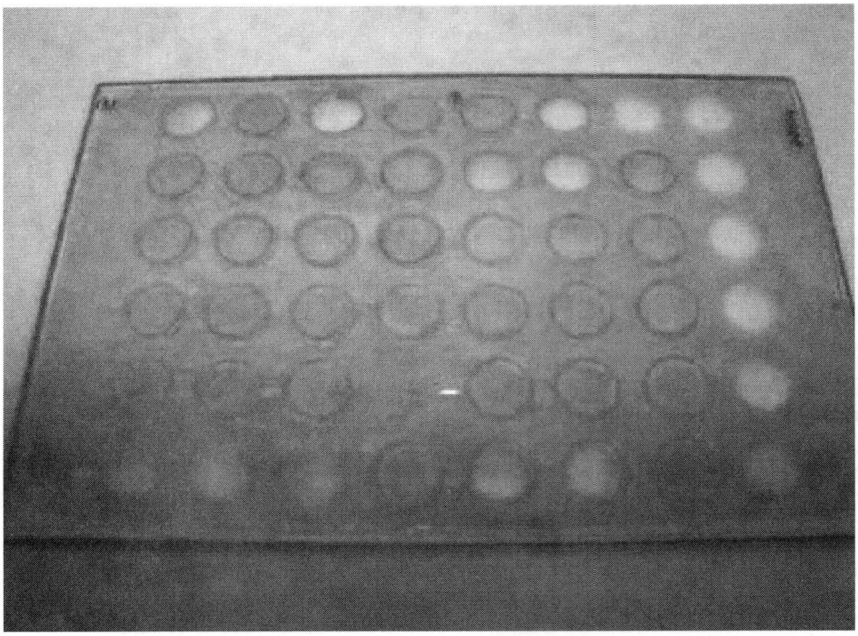

FIGURE 14.11. CVN sensor film coated with 48 spots, each having a different thickness and/or composition for MHTS testing.

Once prepared, test plates are typically stored in a desiccator, followed by 1 hr under ambient conditions to equilibrate prior to testing.

MHTS Measurement Operation

Absorbance at 630 nm (in the yellow region of the visible spectrum) is measured periodically in a 48-point array like that shown in Figure 14.11. These measurements are made using the same UV–vis spectrometer as described before. When not being tested, plates are held in temperature-controlled chambers (23 or 40°C) at a relative humidity (RH) of 98–100%. Initially, absorbance increases as a function of time, but eventually the CVN sensor becomes saturated with moisture and the absorbance increase begins to level. Slopes of absorbance at the inflection point can be calculated within 2 or 3 days with six to eight data points. Films with lower moisture permeability (i.e., lower MVTR) require a longer time for data collection, as the change in absorbance with time is much slower.

MHTS Sensor Calibration

Slopes of absorbance as a function of time are converted to MVTR values using a series of known reference films that are attached to each plate. References are established using a Permatran-W 3/31 (MG) instrument (Mocon, Minneapolis, MN) that conforms to ASTM Method F1249. Figure 14.12 shows a typical calibration curve fit with a second-order polynomial used to convert unknown absorbance–time slopes to MVTR values. The reference films shown in Figure 14.12 are summarized in Table 14.1 with respect to absorbance–time (*A*–*t*) slope, MVTR, and manufacturer. The accuracy of this technique can be further

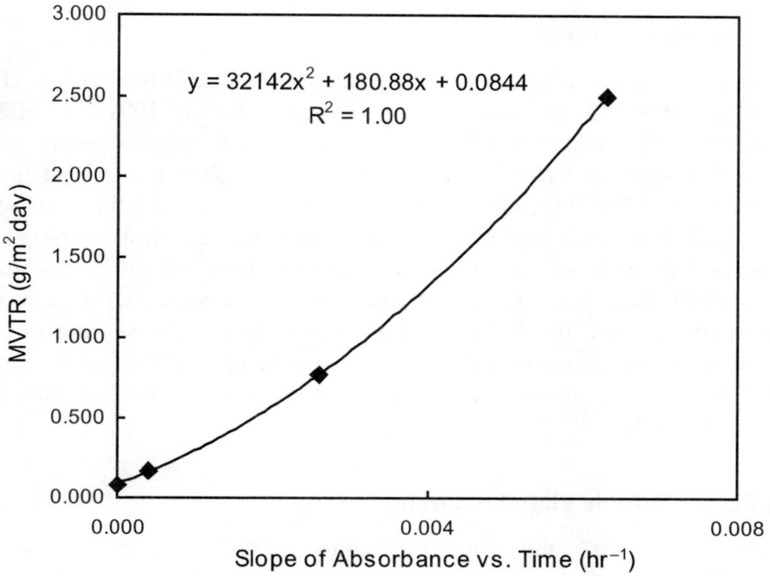

FIGURE 14.12. Calibration curve used to convert absorbance–time slopes to MVTR values during MHTS testing.

TABLE 14.1. MHTS and OHTS Reference Materials

Material	Form[a]	Manufacturer	Thickness (μm)	MVTR-23[b]	MVTR-40[c]	OTR[d]
Daran SL158	Latex	W. R. Grace	22	0.16	1.34	
PET	Film	DuPont	127	2.34	5.94	
Composaclean	Film	Cleanfilm	110	0.77	3.28	
Aclar	Film	Honeywell	38	0.31	0.38	78.7
Saran 560	Film	Dow Chemical	155	0.16	1.14	3.13
Super Escal	Film	Mitsubishi Gas	123	0.078		
Escal	Film	Mitsubishi Gas	118			<0.005
EVAL EF-F15	Film	EVALCA	16			0.24
Capran OEB	Film	Honeywell	26			0.56
Saran HB	Film	Dow Chemical	28			1.26
EVAL EF-E20	Film	EVALCA	18			2.35
ST505–PET	Film	Dupont Teijin	176			8.31

[a]Latexes are made into films by blade coating onto 2 mil PET followed by oven drying to remove water.
[b]MVTR, in units of g m^{-2} per day, was performed using Mocon at 23°C and 100% RH.
[c]MVTR, in units of g m^{-2} per day, was performed using Mocon at 40°C and 100% RH.
[d]OTR, in units of cm^3 m^{-2} per day, was performed using Mocon at 23°C and 0% RH.

improved by adding more reference films, especially ones that are close to the MVTR range of interest. The MHTS test method can produce MVTR values in excellent quantitative agreement with better-established measurement techniques (i.e., ASTM F 1249).

Oxygen High-Throughput Screening

OHTS Sensor Preparation

A sensor solution is prepared by adding 2.00 g of PTMSP (Gelest Inc., Tullytown, PA) and 37 mg of PtOEP (Porphyrin Products, Logan, UT) to 100 mL of HPLC grade toluene (Aldrich, Milwaukee, WI). This mixture is rolled for approximately 16 hr to dissolve the solids completely. After rolling, the sensor solution is stored in the dark. Supported PtOEP–PTMSP sensor film is then prepared by coating the solution onto the same 2 mil PET film that is used for the CVN sensor using the 6 mil opening (~150 μm) on a Byrd bar. The coated film is left to dry for 20 min under ambient conditions and then placed into a 70°C oven for an additional 20 min. Both sides of the sensor-coated film are then laminated with 3M 467 transfer adhesive, leaving the release liner in place. The film is again stored in complete darkness prior to use. Sample testing plates are created by laminating the PET side of the sensor-coated film to a plate of window glass measuring 3.375 in. × 5 in. × 0.125 in.

Oxygen Barrier Sample Plate Preparation

Sample testing plates are created by laminating 0.75 in. squares of oxygen barrier film to the PET side of the sensor-coated film. Next, using a punch, a 0.5 in. circle of this barrier–sensor film combination is laminated to a plate of window glass measuring

FIGURE 14.13. OHTS plate with a variety of barrier film disks laminated to PtOEP–PTMSP sensor film that is laminated to the glass and sealed around the edges with poly(vinyl acetate) latex.

3.375 in. × 5 in. × 0.125 in. These sample circles are placed on the glass plate according to a 48-well template, which is the measurement scheme employed by the spectrometer used to measure fluorescence. Oxygen leakage from the edges of these small samples can cause significant measurement errors; therefore the edges of the circles are sealed by applying a bead of Vinac XX-230 poly(vinyl acetate) latex (Air Products, Allentown, PA). Figure 14.13 shows an example OHTS test plate containing sealed barrier film disks. The sealant is coalesced for 16 hours in a 50°C oven with a dark interior. Once dry, the sample plate was maintained in complete darkness prior to testing.

OHTS Measurement Operation

Fluorescence measurements are obtained periodically using a Biotek FL×800 fluorescence plate reader. The sample plate, prepared as described, is assumed to be saturated with oxygen prior to testing (i.e., fluorescence is at a minimum). After the initial fluorescence measurement, the sample plate is kept in an oxygen-free dark chamber. Each plate is read daily over a 7–10 day period, being sure to minimize the time that the plate is exposed to ambient air and light. A positive slope for fluorescence as a function of time was produced for each plate.

OHTS Sensor Calibration

Sensor calibration is accomplished by running OHTS on a set of reference films with known OTRs. The OTR, expressed in $cm^3\ m^{-2}$ per day, is measured with a Mocon Ox-Tran 2/20 (ML System) instrument, which conforms to ASTM F1927. The reference films are prepared exactly like sample films and the fluorescence–time (F–t) slope is calculated in exactly the same manner. A calibration curve is prepared by plotting the slopes of the reference films as a function of their known OTRs. The calibration data are fitted to an equation (linear or second-degree polynomial) which is then used to calculate the OTRs

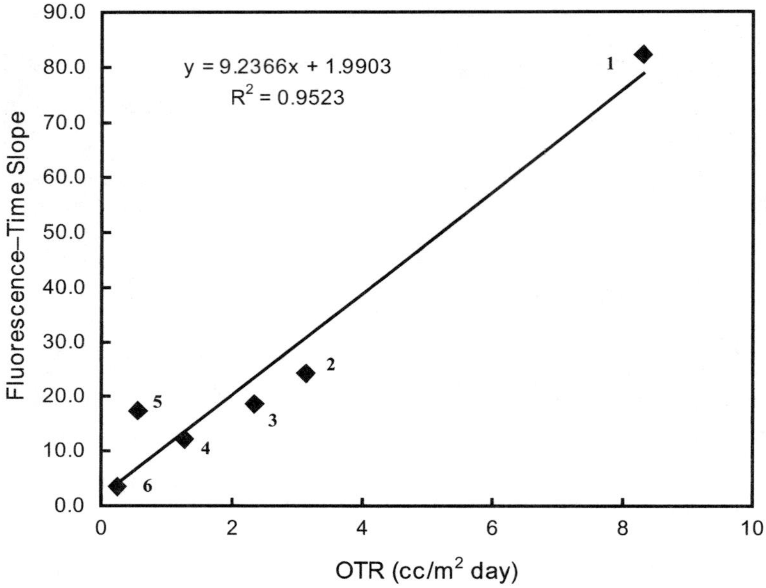

FIGURE 14.14. Example OHTS calibration curve using six reference films: (1) 7 mil PET; (2) Saran 560; (3) EVAL EF-E 20 u; (4) Saran HB; (5) 1 mil OEB; (6) EVAL EF-F 15 u.

of the samples. Table 14.1 provides a list of the current reference films along with their corresponding OTR, *F–t* slope, and manufacturer. OHTS can be done at other temperatures and humidity, provided that the OTRs of the reference films are measured under the same conditions. Figure 14.14 shows an example of an OHTS calibration curve.

Haze High-Throughput Screening

Haze Measurement Setup

Haze is "the cloudy ... appearance of an otherwise transparent specimen caused by light scattered from within the specimen or from its surfaces" (ASTM D 1003 – 92). Polymeric films are made less permeable to water vapor [29], or oxygen gas [30], by the addition of clay platelets. An increased level of haze resulting from the light scattered by the clay particles often offsets this improvement in barrier performance. High throughput haze measurement is performed for several samples simultaneously by capturing an image of the coating laid over a black-and-white texture. The degree in contrast reduction is then used as a measure for haze. Figure 14.15 shows the experimental setup for haze measurements.

In a typical haze measurement experiment, the combinatorial sample plate is placed over a high-contrast image of black and white lines. The image of the black and white lines is scanned into software to determine the maxima *W* (brightest part of the white lines) and the minima *B* (darkest part of the black lines). Two scans of each line are made, either through the hazy material or directly. The initial contrast CN_0 is simply the difference

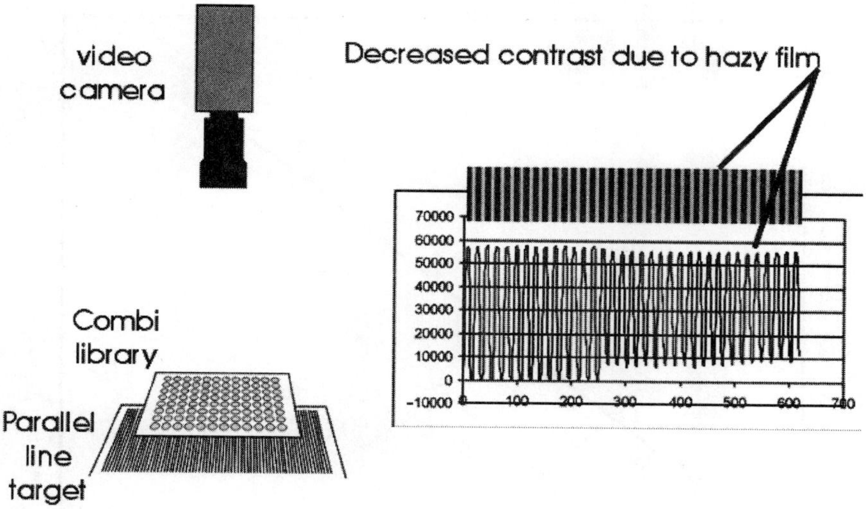

FIGURE 14.15. Haze method. The left panel shows the equipment setup, and the right panel is a photograph showing the target covered on the right-hand side with hazy film and the numerical data associated with scanning this image. The small distance between maxima and minima corresponds to the decrease in contrast.

between the maxima W_0 and the minima B_0 from the image having no film in place. The contrast with the film in place CN_x is the difference between W_x and B_x, the maxima and minima of the image taken through the film. The logarithm of the contrast ratio (CN_0/CN_x) is plotted versus the actual haze value to form a calibration line. Thus

$$CN_0 = W_0 - B_0$$

$$CN_x = W_x - B_x$$

and the value for haze is

$$haze = a \log\left(\frac{CN_0}{CN_x}\right)$$

Haze Measurement Calibration

The haze high-throughput screening (HHTS) method is calibrated by comparing the haze values with commercially available haze measurement equipment. The haze value of one or more layers of a somewhat hazy film material is measured using a Gardner Haze Gard Plus hazemeter. It is found that the haze data plotted versus the number of layers fitted a straight line. Figure 14.16 shows the linear relationship between $\log(CN_0/CN_x)$ and the number of layers, which in turn was shown to be linear with haze. This proves the concept of using contrast to measure haze is viable. The distance between the film to be measured and the target can be used to enhance the sensitivity. Figure 14.17 shows the contrast measurement CN_x as a function of distance between film and target. As the distance is increased, the contrast decreases and thus the sensitivity of the measurement is magnified.

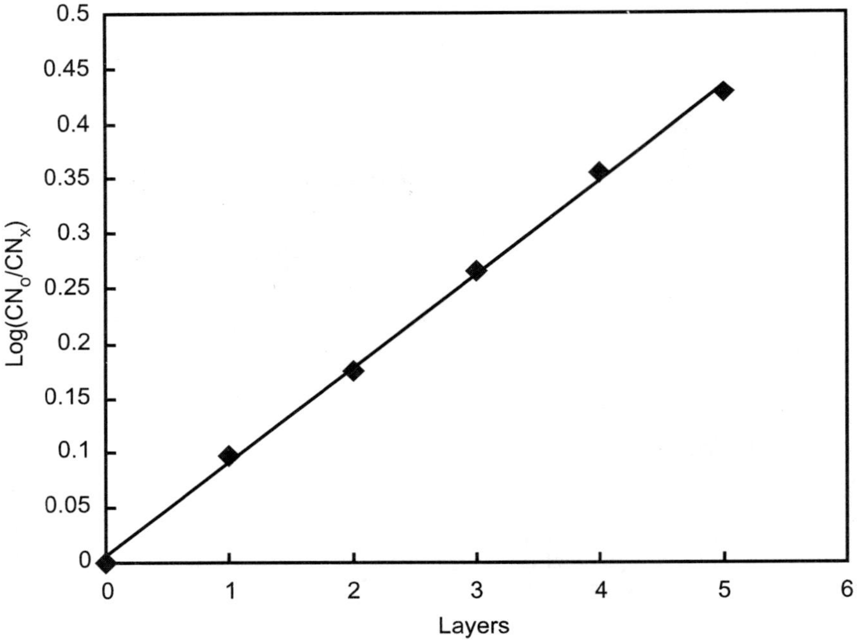

FIGURE 14.16. Haze calibration using relative contrast change.

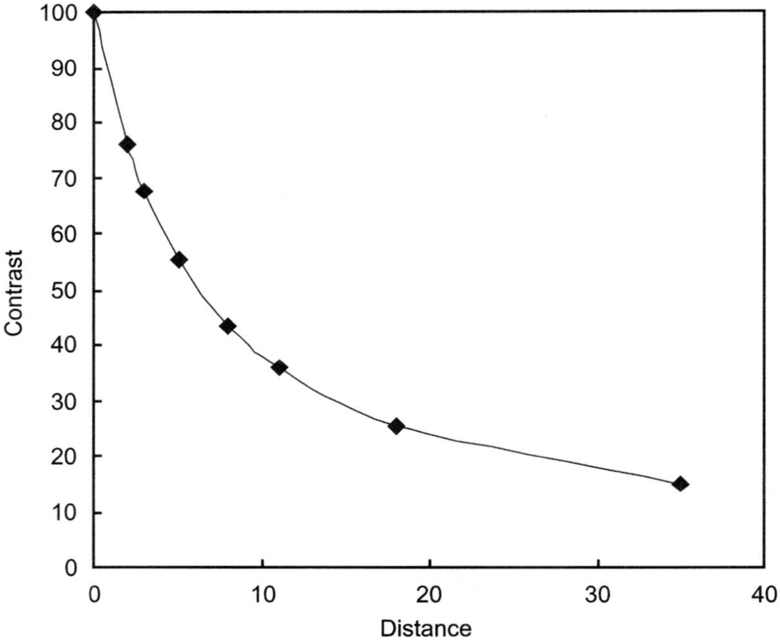

FIGURE 14.17. Change in contrast as a function of the distance between target and film.

Defect Detection Process

The following steps can be performed to detect defects automatically. After making the daughter plates and applying them to the MHTS sensor, it is necessary to prepare them for image capture. The fresh daughter plates are usually yellow in color and it is very difficult to detect any defects in the clear coatings when there is very low contrast in the pictures. If a defect is defined as an entity that deteriorates the barrier performance of the coating material, it is logical to look for a rapid color change in the MHTS sensor in the coated regions to locate the defects. To achieve this, the daughter plates can be kept in the humidity chamber for a short period of time in order to enhance the color contrast in the defective regions. Keeping daughter plates in the humidity chamber for 2–3 hr tends to make most defects visible with a reasonable contrast. Next, the daughter plates are taken out of the humidity chamber and placed over a light table. Their images are captured using a digital camera placed on a tripod to minimize the blurring effect due to camera motion. Digital images are fed into the image-processing computer code. The computer program performs several steps of image enhancement and feature detection as described before. The feature detection step can be modified and tuned to detect certain types of defects in the coatings. Finally, the data obtained from the detection algorithm are interpreted and translated in such a way that the defective coatings are marked for the user for further investigation.

RESULTS AND DISCUSSION

MHTS Results

Recent improvements in barrier film preparation, handling, and storage, along with improvements in sensor film design, have allowed the MHTS test method to provide quantitative agreement with MVTR values that are obtained using the Mocon instrument. Quantitative agreement between the Mocon and MHTS tests has been found for poly(vinylidene chloride) (PVdC) latex-based barrier films. PVdC has been widely used due to its ability to create relatively low MVTR films ($< 5\,g\,m^{-2}$ per day at 23°C and 100% RH) [31] and OTR films ($< 5\,cc\,m^{-2}$ per day at 23°C and 0% RH) [32] unlike most other polymers. Two example MHTS experiments are described here together with corresponding Mocon results to provide validation of this method of barrier film testing. A third experiment, involving MHTS measurements performed at 40°C and almost 98% RH, further illustrates the robust nature of the MHTS sensor film, as the MVTR values calculated from these slopes continue to match those of the Mocon instrument. All these experiments use some form of PVdC latex as the source of the tested barrier film.

Dilution of PVdC Latex

A series of PVdC latex-based films are created using Daran SL 158 (see Table 14.1 for more information) and diluted to 10, 20, 30, 40, 50 and 60% of the original concentration (~54 wt% solids). These dilutions show the expected trend of decreasing MVTR with increasing thickness, both visually (Figure 14.18) and based on MHTS results (Figure 14.19). Figure 14.18 shows a standard combinatorial (or "daughter") plate prepared by spin

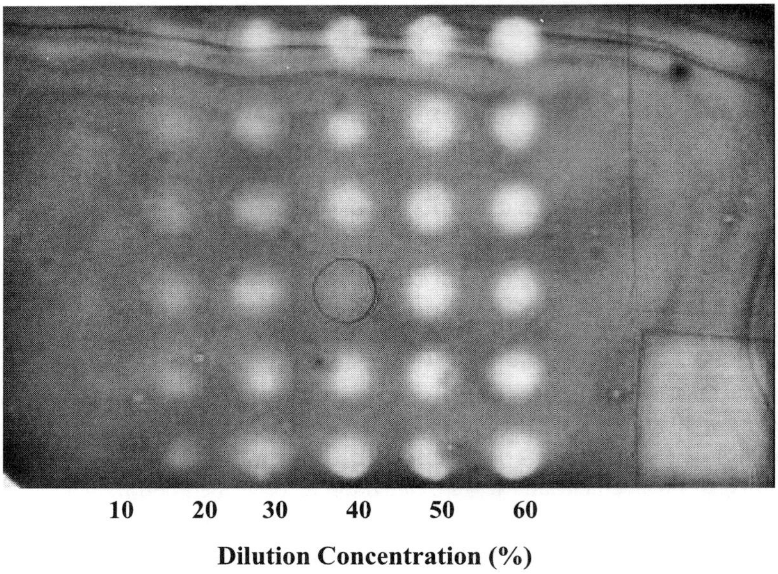

FIGURE 14.18. Image of PVdC-latex-coated daughter plate showing the influence of thickness on the moisture barrier. A higher intensity corresponds to a higher moisture barrier.

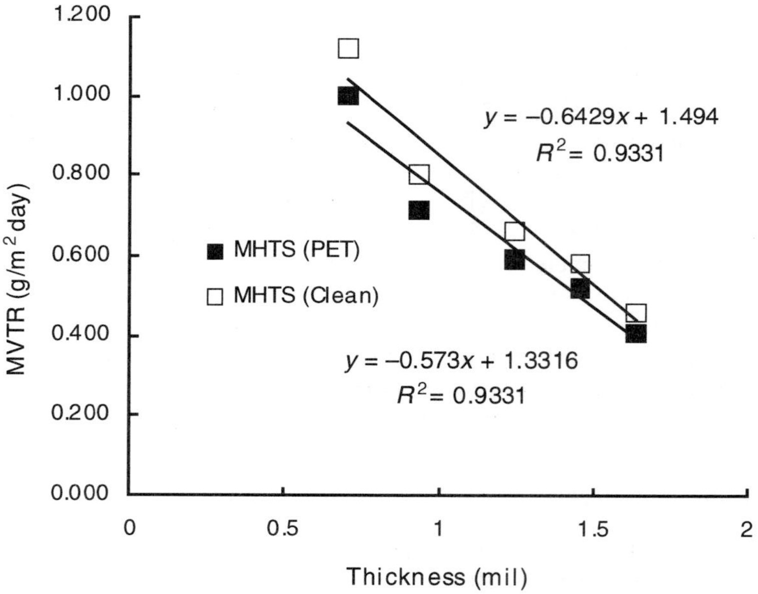

FIGURE 14.19. Calculated MVTR values based on MHTS measurements plotted as a function of spot thickness for the 12 μL daughter plate (Figure 14.18). Linear fit to the data is shown for values calculated based upon 5 mil PET (filled squares) and Composaclean (open squares) reference films.

casting of latex droplets using a special mask. Each spot is created with the same volume of liquid, which allows a thickness gradient to be created (i.e., thickness is increasing with latex solids moving left to right across the plate). Table 14.2 provides thickness and MVTR values for three of the dilution-based samples prepared on a PET substrate using a Byrd bar (i.e., blade coating). Sample sizes required for Mocon measurement (at least 4 in.^{2}) are too large to be prepared via the spin-casting method used in Figure 14.18. Blade-coated, samples can be cut for Mocon measurement and for lamination to an MHTS daughter plate from one film for direct comparison. Samples 1 and 2 (Table 14.2) show quantitative agreement between MHTS and Mocon (~2.5 g m^{-2} per day). It is assumed that this high level of agreement stems from the nearness to the 5 mil PET reference used in the MHTS test [33]. Sample 3 has a Mocon MVTR of 2.07, which is somewhat further away from the 2.5 g m^{-2} per day PET reference and yields an MHTS of 1.5 as a result. Although the qualitative trend remains intact, quantitative disagreement between Mocon and MHTS for Sample 3 results from the linear extrapolation used to calculate an MVTR value:

$$\text{MVTR (sample)} = \frac{\text{MVTR(reference)} \times \text{slope(sample)}}{\text{slope(reference)}} \tag{4}$$

where slopes are those of the absorbance versus time curves generated during MHTS testing. The relationship between slope and MVTR is not linear, but the linear extrapolation [Eqn (4)] assumes that it is. The results for each sample include the number of measurement spots, the number of points used for the slope, and the fit of each slope (expressed as an R^2 value). Unless otherwise noted, these supplementary pieces of information are averages of the total number spots used.

Blade-Coating Thickness Variation

Owing to the unpredictable thickness variation of the diluted drawdowns in the preceding example, a series of blade-coated samples were prepared using concentrated (55.4 wt%) SL 158 latex. Six films were coated using the 6, 4, 3, 2, 1, and 0.5 mil Byrd bar openings. A nice trend of increasing thickness with increasing Byrd bar opening is established with these films. Table 14.3 summarizes the results of Mocon and MHTS testing on these films. The MHTS values shown here agree very well with Mocon down to MVTR values of approximately 0.1. It is believed that the relationship between MHTS slope and MVTR is nearly linear at values much less than unity. Figure 14.20 shows MHTS and Mocon

TABLE 14.2. MHTS and Mocon Data for Diluted SL-158 Drawdowns

Sample	Formula	Duplicates	Points	R^2	Slope	CV	MHTS	Mocon
1	10% SL 158	4	4	0.9923	0.0345	33	2.54	2.55
2	20% SL 158	4	4	0.9939	0.0329	4	2.465	2.43
3	30% SL 158	4	5	0.9932	0.0200	10	1.5	2.07
4	30% SL 158							1.98
5	30% SL 158							2.22
6	30% SL 158							2.03
Ref. 1	5 mil PET	4	6	0.9972	0.0339	20	2.5	2.5
Ref. 2	Composaclean	2	7	0.9973	0.0059	27	0.77	0.77

TABLE 14.3. MHTS and Mocon Data for Concentrated SL-158 Drawdowns

Sample	Thickness	Duplicates	Points	R^2	Slope	CV	MHTS	Mocon
4	1.385	6	6	0.9911	0.0008	18	0.13	0.11
3	0.8							0.17
2	0.625	6	8	0.9960	0.0016	3	0.3	0.26
1	0.265	6	8	0.9985	0.0035	11	0.67	0.54
0.5	0.155							0.73

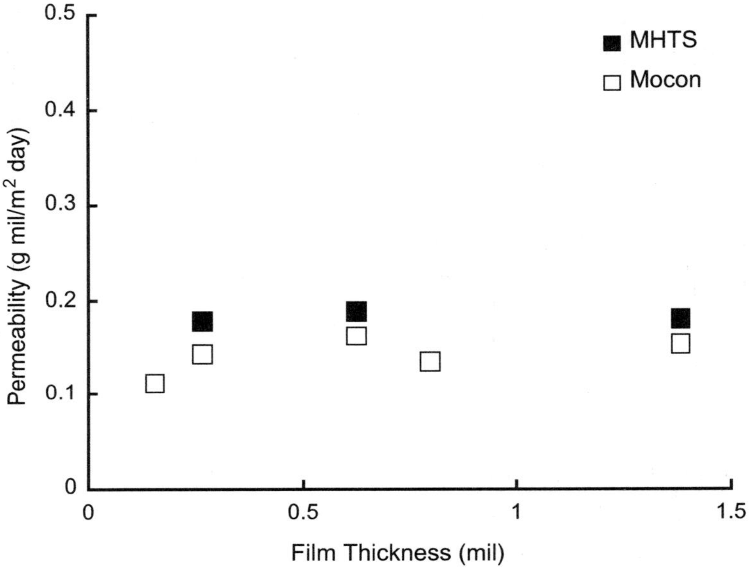

FIGURE 14.20. Permeability (or MVTR multiplied by film thickness) as a function of film thickness for MHTS and Mocon measurement techniques.

MVTR values multiplied by thickness, i.e. permeability. Mocon gives a permeability of approximately 0.15 g mil m^{-2} per day and MHTS gives an approximate value of 0.18 g mil m^{-2} per day.

Cross-linked PVdC Latex

In an effort to improve MVTR stability at 40°C, cross-linking of PVdC latex is proposed. The initial pilot cross-linking experiment does not produce good results for the intended purpose, but samples submitted for both Mocon and MHTS testing show nice agreement. Results from three of the samples containing 0, 2, and 10 wt.% cross-linker in polymer are shown in Table 14.4. The cross-linker, Astromel NW-3A from Borden Chemicals, is a water-soluble, melamine formaldehyde resin that is expected to have poor moisture resistance in the absence of a cross-linking reaction. This lack of cross-linking appears to be the case when looking at the early results in Table 14.4, which show

TABLE 14.4 MHTS and Mocon Data for Cross-linked SL-143

Sample	Formula	Spots	Points	R^2	Slope	CV	MHTS	Mocon
0	0% Crslnk	4	4	0.9977	0.0009	17	0.24	0.24
2	2% Crslnk	4	4	0.9960	0.0014	17	0.26	0.26
10	10% Crslnk	4	4	0.9985	0.0061	33	0.51	0.51
Ref. 1	5 mil PET	4	5	0.9991	0.0111	20	2.5	2.5
Ref. 2	Composaclean	2	5	0.9868	0.0016	27	0.77	0.77

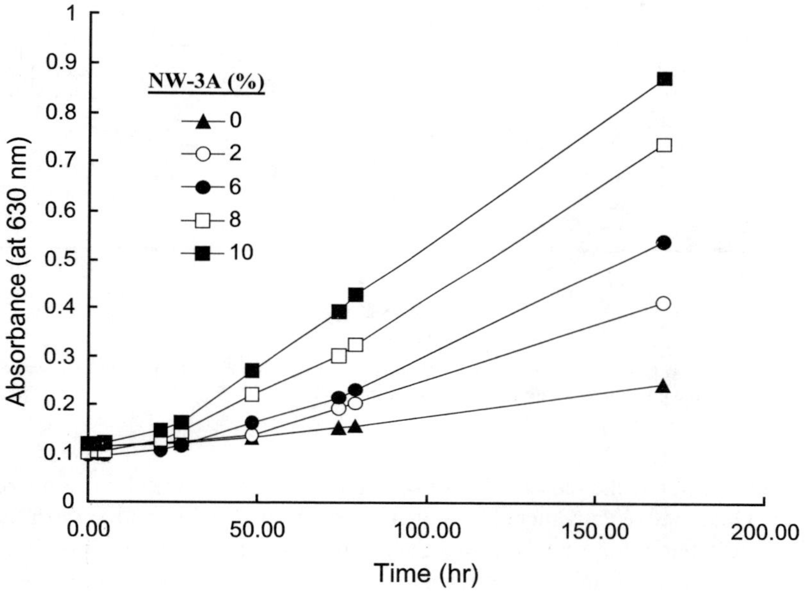

FIGURE 14.21. Absorbance of 630 nm light as a function of time (in hours) for films of SL-143 with varying concentration (in weight percent) of NW-3A melamine formaldehyde cross-linker.

increasing MVTR with NW-3A concentration. Figure 14.21 shows absorbance at 630 nm plotted as a function of time to further highlight this trend of increasing MVTR with cross-linker concentration.

OHTS Results

Much like MHTS, gradual improvements in sensor design and sample preparation have made OHTS a reasonably quantitative technique. OHTS not only provides the ability to test numerous samples simultaneously, but also allows relatively small sample sizes (< 1 in.^{2}) to be used, unlike more conventional tests (e.g., Mocon). Furthermore, film defects can easily be pinpointed with a fluorescence photograph of an OHTS sample. Fluorescence is rapidly quenched in areas of a film that have pinholes or other defects that

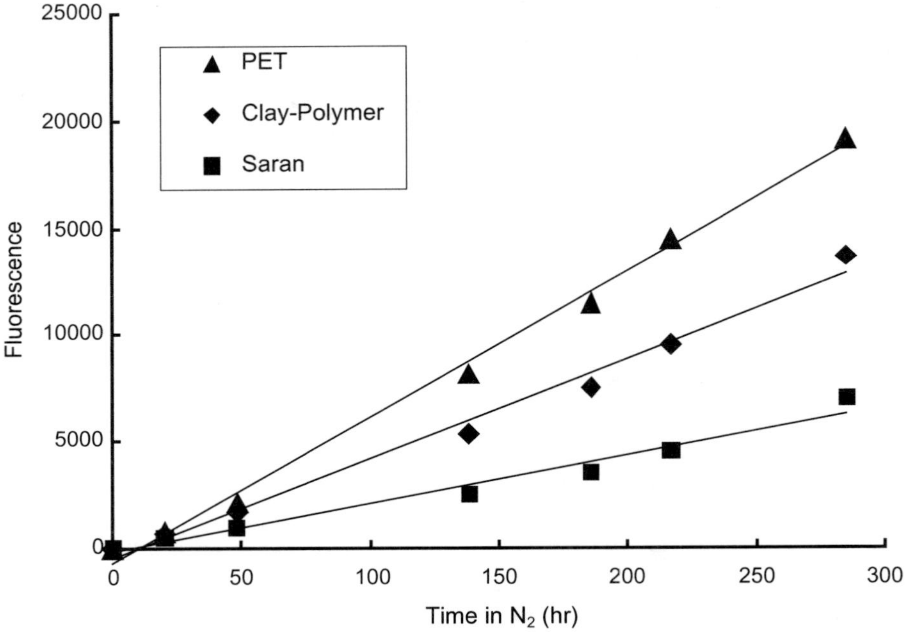

FIGURE 14.22. Standard OHTS data (fluorescence intensity as a function of time in hours) for an ESA-based oxygen barrier, shown in comparison with two reference films.

allow oxygen to infiltrate the sensor, which would appear as a different color in a fluorescence photograph. Quantitative agreement between the Mocon and OHTS test has been found for a clay-filled polymer composite film. Relatively low oxygen transmission rates have been achieved with clay-filled polymer composites due to the "tortuous" path created by the oxygen-impenetrable clay sheets [34].

Figure 14.22 shows the OHTS data collected for an oxygen barrier film comprised of polymer and clay, along with PET and Saran as reference films. As oxygen is removed from the OHTS sensor film, achieved by placing the test plate into a nitrogen atmosphere, fluorescence intensity increases. Films with a relatively high oxygen permeability (e.g., PET) have steeper slopes, as shown in Figure 14.16. In this case a two-point calibration slope created using the PET and Saran references (see OTR values in Table 14.1) yielded an OTR of $5.7\,cm^3\,m^{-2}$ per day, while the Mocon instrument produced a value of $5.0\,cm^3\,m^{-2}$ per day. A more thorough calibration curve, using seven of the references shown in Table 14.1, produced an OTR of $5.1\,cm^3\,m^{-2}$ per day, in better agreement with the Mocon result. Even the OHTS result based upon only two references is within 15% of the Mocon value.

Results of Coating Defect Detection

The results of a defect detection experiment applied to a very defective daughter plate are shown to demonstrate the process of defect detection. This daughter plate is coated with several different formulations. Next, the plate is placed in the humidity chamber for 2 hrs and then its image is captured as shown in Figure 14.23. The image captured at this stage is processed and enhanced as described earlier. The enhanced image is an R-channel image as shown in Figure 14.24. This R-channel image is then converted to the gray-level image (Figure 14.25).

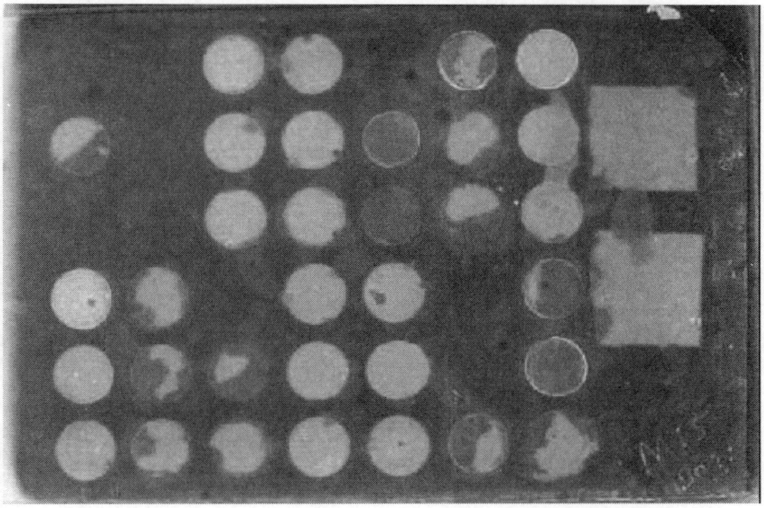

FIGURE 14.23. Original daughter plate after storage in the humidity chamber for 2 hr.

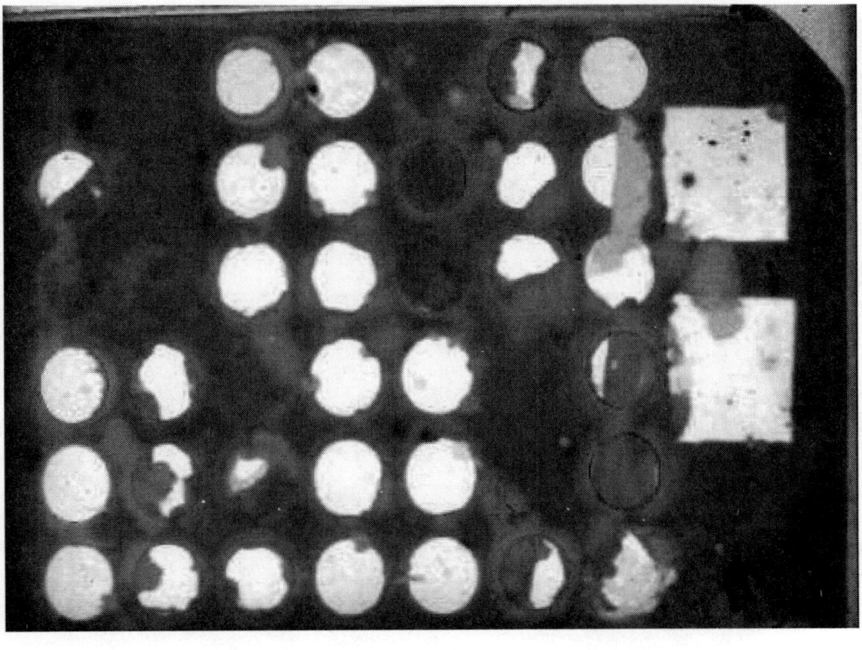

FIGURE 14.24. R-channel contrast-enhanced image.

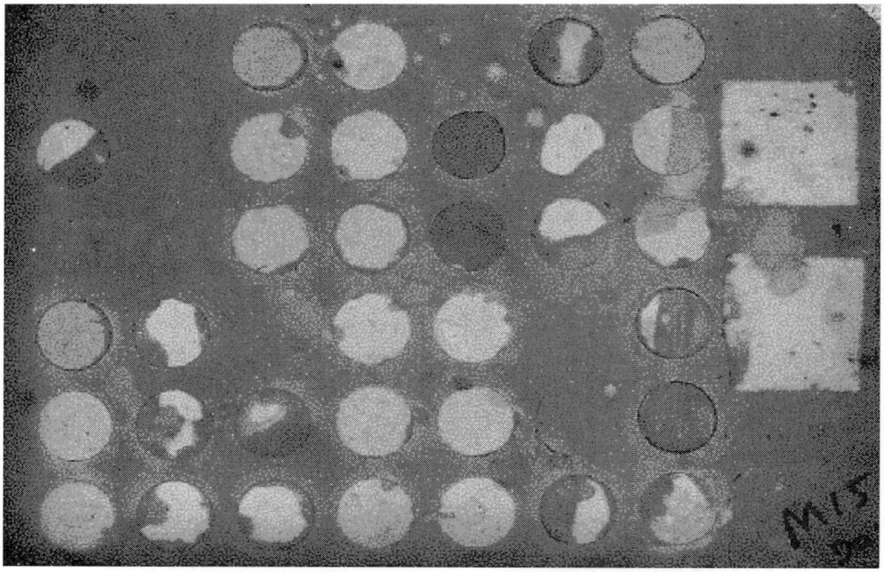

FIGURE 14.25. Gray-level image.

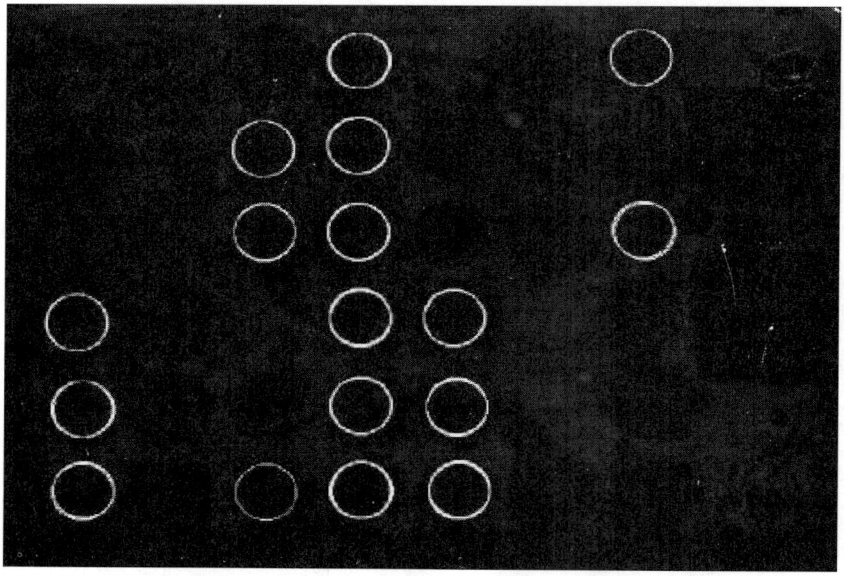

FIGURE 14.26. Good coatings are detected and marked by blue circles.

Figure 14.26 shows the final image with all good coatings represented by blue circles. This final image is translated into a binary matrix format, which contains information on good coatings. The binary matrix for this case is

$$\begin{matrix} \mathbf{0} & \mathbf{0} & \mathbf{0} & \mathbf{1} & \mathbf{0} & \mathbf{0} & \mathbf{1} \\ \mathbf{0} & \mathbf{0} & \mathbf{1} & \mathbf{1} & \mathbf{0} & \mathbf{0} & \mathbf{0} \\ \mathbf{0} & \mathbf{0} & \mathbf{1} & \mathbf{1} & \mathbf{0} & \mathbf{0} & \mathbf{1} \\ \mathbf{1} & \mathbf{0} & \mathbf{0} & \mathbf{1} & \mathbf{1} & \mathbf{0} & \mathbf{0} \\ \mathbf{1} & \mathbf{0} & \mathbf{0} & \mathbf{1} & \mathbf{1} & \mathbf{0} & \mathbf{0} \\ \mathbf{1} & \mathbf{0} & \mathbf{1} & \mathbf{1} & \mathbf{1} & \mathbf{0} & \mathbf{0} \end{matrix}$$

This binary matrix is used in future screening. It should also be noted that the coatings in daughter plates are prepared in triplicate in this particular case. Each three coatings in this plate column-wise are the same formulation. There is an obvious pattern in the final binary matrix showing that the coating failure occurred in the same formulation. This means that the defects appearing in the final coating materials are mainly due to their formulation. This is in agreement with the fact that some formulations usually fail upon dispensing (due to coagulation, gelation, and/or phase separation) and mixing in the first step. This is very useful information for material compatibility and applicability as the coating material.

CONCLUSIONS AND PATH FORWARD

In this chapter the results of novel high-throughput measurement methods for moisture and oxygen transmission rates and film haze have been presented. In addition, an image-processing algorithm was developed in order to screen the coating materials for the presence of defects and pinholes. The results of this study confirm the value of using combinatorial science for new materials discovery. The high-throughput test methods presented are currently being modified and optimized in an effort to improve accuracy while maintaining automation and speed.

ACKNOWLEDGMENTS

The authors would like to thank the National Institute of Standards and Technology (NIST) for providing the funding for this research through its Advanced Technology Program (ATP). The help and effort of the Materials Science & Characterization (MSC) Group at the Avery Research Center (ARC) must also be acknowledged.

REFERENCES AND NOTES

1. Devlin, J. P. *High Throughput Screening: The Discovery of Bioactive Substances*, **1997**, New York: Marcel Dekker.
2. Doyle, P. M. *J. Chem. Technol. Biotechnol.* **1995**, *64*, 317–324.
3. Gordon, E. M., Kerwin, J. F. Jr *Combinatorial Chemistry and Molecular Diversity in Drug Discovery*, **1998**, New York: John Wiley.
4. Aramendia, M. A., Borau, V., Jimenez, C., Marinas, J. M., Romero, F. J., Urbano, F. J. *J. Catal.* **2002**, *209*, 413–416.

5. Meredith, J. C., Karim, A., Amis, E. J. *MRS Bull.* **2002**, *27*, 330–335.
6. Sadagopan, D., Pitchumani, R. *J. Mech. Design* **1997**, *119*, 494–503.
7. Del Nobile, M. A., Fava, P., Piergiovanni, L. *J. Food Eng.* **2002**, *53*, 295–300.
8. Brennan, A. M. *TAPPI* **1992**, *75*, 145–148.
9. Yoon, D. S., Roh, J. S. *Adv. Func. Mater.* **2002**, *12*, 373–381.
10. Morgan, P. W. *Ind. Eng. Chem.* **1953**, *45*, 2296–2306.
11. Motegi, S. *Bull. Jpn. Soc. Sci. Fish.* **1979**, *45*, 205–209.
12. Park, J. H., Hong, D. H., Kim, Y. B., Choi, D.K. *J. Appl. Phys.* **2002**, *91*, 10022–10027.
13. Bird, R. B., Stewart, W. E., Lightfoot, E. N. *Transport Phenomena*, **1960**, New York: John Wiley.
14. Raimundo, I. M., Narayanaswamy, R. *Analyst* **1999**, *124*, 1623–1627.
15. Brook, T. E., Taib, R., Narayanasawamy, R. *Sens. Actuators B* **1997**, *38–39*, 272–276.
16. Otsuki, S., Adachi, K., Taguchi, T. *Sens. Actuators B* **1998**, *53*, 91–96.
17. Sadaoka, Y., Matsuguchi, M., Sakai, Y., Murata, Y. *Sens. and Actuators B* **1992**, *7*, 443–446.
18. Sadaoka, Y., Matsuguchi, M., Sakai, Y., Murata, Y. *J. Mater. Sci.* **1992**, *27*, 5095–5100.
19. Choi, M. M. F., Tse, O. L. *Anal. Chim. Acta* **1999**, *378*, 127–134.
20. Grunlan, J. C., Mehrabi, A. R., Chavira, A. T., Nugent, A. B., Saunders, D. L. *J. Comb. Chem.* **2003**, *5*, 362–368.
21. Saunders, D. L., Potyrailo, R. A. *High Throughput Screening for Moisture Barrier Characteristics of Materials*, US Patent Application 2001 0034063, **2001**.
22. Rharbi, Y., Yekta, A., Winnik, M. A. *Anal. Chem.* **1999**, *71*, 5045–5053.
23. Amao, Y., Asai, K., Okura, I., Shinohara, H., Nishide, H. *Analyst* **2000**, *125*, 1911–1914.
24. Amao, Y., Asai, K., Miyashita, T., Okura, I. Novel optical oxygen pressure sensing materials: platinum porphyrin-styrene-trifluoroethylmethacrylate copolymer film, *Chem. Lett.* **1999**, 1031–1032.
25. Barnikol, W. K. R., Gaertner, Th., Weiler, N. *Rev. Sci. Instrum.* **1988**, *59*, 1204–1208.
26. Carraway, E. R., Demas, J. N., DeGraff, B. A. *Langmuir* **1991**, *7*, 2991–2998.
27. Potyrailo, R. A. *Devices and Methods for Measurements of Barrier Properties of Coating Arrays*, US Patent 6,383,815, **2002**.
28. Russ, J. C. *The Image Processing Handbook*, **1992**, Boca Raton, FL: CRC Press.
29. Tong, X., Zhao, H. C., Tang, T., Feng, Z. L., Huang, B. T. *J. Polym. Sci. A Polym. Chem.* **2002**, *40*, 1706–1711.
30. Chang, J. H., Park, K. M., Cho, D. H., Yang, H. S., Ihn, K. J. *Polym. Eng. Sci.* **2001**, *41*, 1514–1520.
31. Li, Y. S., Weng, Z. X., Pan, Z. R. *Chin. J. Polym. Sci.* **1997**, *15*, 319–324.
32. Brown, R. A., Budd, P. M., Price, C., Satgurunathan, R. *Eur. Polym. J.* **1993**, *29*, 337–342.
33. Initial MHTS experiments used a simple linear extrapolation from only one reference film, rather than the multiple reference polynomial fit used more recently (see ref. 20). This linear approximation is only valid when the sample of interest has an MVTR within 10% of the reference film being used.
34. Bharadwaj, R. K., Mehrabi, A. R., Hamilton, C., Trujillo, C., Murga, M., Fan, R., Chavira, A., Thompson, A. K. *Polymer* **2002**, *43*, 3699–3705.

Chapter 15

Intrinsic Fiber-Optic Sensors for Spatially Resolved Combinatorial Screening

Peter Geissinger and Alan W. Schwabacher

INTRODUCTION

Computational methods for the calculation of the structure and properties of molecules have achieved a high degree of accuracy. However, structure-based drug design, while being a powerful approach, often does not afford the reliability needed for targeted synthesis of molecules that show a desired drug action. An alternative approach is the synthesis of large numbers of compounds and the systematic evaluation of these compounds for a desired effect. This approach led to the development of the field of combinatorial chemistry. For the combinatorial approach to be effective, three basic questions must be addressed.

1. *Library synthesis:* How can one create from a basic set of compounds and for a given number of reaction steps a library of all possible compounds that can result from the sequential combination of the members of the initial set?
2. *Library assay:* How can one identify *whether* a compound in this library has a desired property?
3. *Library decoding:* How can one determine *which* compound of the library has the desired property?

Various techniques have been developed to address the combinatorial synthesis of compound libraries. One such technique, the *split–mix* procedure [1], offers *fully parallel synthesis*, which means that each reactant has to be handled only once in any one reaction step. In this method, library members are synthesized on individual solid supports (usually *beads*); however, since these supports are mixed together after each reaction step, the *synthetic history* of the library is lost, meaning that at the completion of a library synthesis, the sequence of reactions that led to the formation of any specific member is not

Peter Geissinger and Alan W. Schwabacher • Department of Chemistry, University of Wisconsin-Milwaukee, 3210 N. Cramer Street, Milwaukee, WI 53211, USA
Contact author: Peter Geissinger, email: geissing@uwm.edu

High-Throughput Analysis, edited by Potyrailo and Amis
Kluwer Academic Publishers, New York, 2003.

known. For techniques based on a two-dimensional layout of the library members [2], the synthetic history at each position can be recorded, however, the synthesis is serial and, therefore, a much slower process.

Combinatorial libraries are frequently assayed with molecular probes that bind only to compounds with the desired properties. When *fluorescent molecular probes* are used, changes in the fluorescence properties of a library compound indicate that binding of the probe has occurred.

Having verified that one or more of the synthesized compounds have the desired properties, the library has to be decoded, that is, these compounds have to be identified and isolated. This problem is easily solved for two-dimensional arrays, where the identification of a compound on the two-dimensional support immediately provides the complete synthetic history. As described above, in the split–mix technique this is a more formidable problem: even if a bead that produces a positive assay can be identified and isolated, the synthetic history of the compound on the bead is usually unknown.

ONE-DIMENSIONAL COMBINATORIAL CHEMISTRY

Chapter 6 introduced a novel combinatorial chemistry method that allows for fully parallel synthesis while retaining the complete synthetic history. The key idea is to carry out the reaction sequence on a *linear* support. In short, the procedure works as follows (Figure 15.1) [3]. The linear support is wrapped around a cylinder of precisely determined circumference in a single spiral layer. The cylinder is then divided lengthwise into segments, each of which constitutes a separate reaction vessel. In the first synthesis step, one reactant is placed into each segment. The linear support then bears a set of repeating domains, the position of each encoding the identity of the compound at that position. In the next step, the linear solid support is removed from the initial cylinder, wrapped around another cylinder of a different appropriate circumference, and divided lengthwise again, so that adjacent regions in a newly formed segment bear unrelated substances. Then, a second set of reactants is applied, one to each of these segments. The linear support then contains all combinations of reactants in a repeating arrangement (an example of a linear

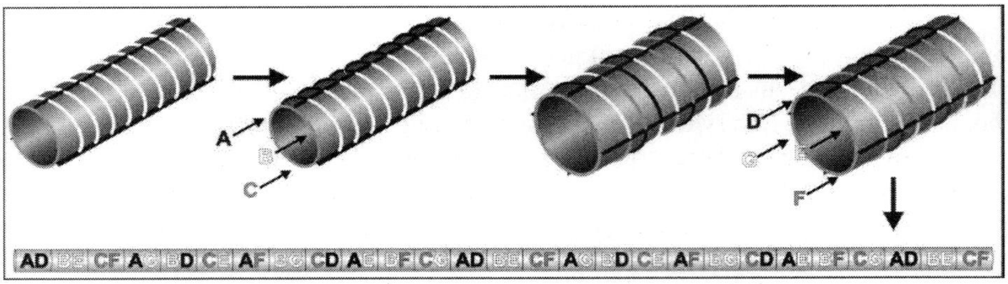

FIGURE 15.1. One-dimensional combinatorial chemistry scheme (for description see text). Also shown is the unwrapped linear support and the regions created on it after three synthesis steps. (Adapted from ref. 49. Copyright 2002 Association for Laboratory Automation).

support after two such synthesis steps is shown at the bottom of Figure 15.1). This process can be repeated with an arbitrary number of steps.

Decoding the combinatorial library is straightforward: the position of a compound along the thread encodes its complete synthetic history. Therefore the decoding problem is reduced to a measurement of the position along the thread of the library members that give positive assay responses. The scheme was first applied using a cotton thread as the linear support. The library was assayed using a fluorescent molecular probe. Pulling the thread through a fluorometer resulted in a fluorescence-versus-time trace, which provided the location of bound probe molecules along the thread.

The one-dimensional combinatorial chemistry method possesses advantageous features for the fabrication of compound libraries, since the method naturally creates a repeating pattern of compounds; however, the degeneracy of this pattern can be precisely controlled via the length of the one-dimensional support. Thus the library could consist of only a few different compounds, with the sequence repeating hundreds of times—an arrangement that lends itself to Fourier analysis (see Chapter 6)—or, conversely, the library could consist many thousands of compounds, each unique. Fourier analysis can be applied in this case as well, because of the periodicity of substructure identity.

The one-dimensional combinatorial chemistry scheme is not limited to cotton thread supports; optical fibers constitute a class of supports that are chemically more versatile than cotton. Library synthesis can be carried out in the cladding of an optical fiber or a suitable substance replacing the fiber cladding. When assayed with fluorophores, the compound libraries on the fiber constitute fluorescent optical-fiber sensor arrays, which can be evaluated using fiber-optic sensor technologies (Library assay). Using optical time-domain reflectometry (see below), the location of a fluorophore along the fiber can be determined (Library decoding).

Using optical fibers as linear supports expands the usage of the one-dimensional combinatorial chemistry method. In addition to the design of new drugs, sensors for a multitude of sensing tasks (e.g., immunosensors, gas and vapor sensors, sensors for toxins, sensors characterizing the chemical and physical environment including pH, metal ions, pCO, $p\mathrm{CO}_2$, $p\mathrm{O}_2$, solvent polarity, temperature, pressure, presence of explosives and other organics, etc.) can be fabricated directly onto the readout device, that is the optical fiber. Both libraries of fluorescent chemosensors and compound libraries assayed with fluorescent probes are evaluated optically using the same technique.

This chapter is structured as follows. First, the basic properties of optical fibers will be reviewed, with particular emphasis on the evanescent fields that extend from the fiber core into the fiber cladding. Experiments verifying the existence and the functional dependence of these evanescent fields will be discussed with the aim of illustrating their relevance to sensing applications.

As indicated above, for the purposes of optical evaluation, combinatorial libraries assayed with fluorophores constitute, in effect, fluorescent chemosensor arrays. It will be demonstrated how spatially resolved readout of these arrays can be achieved. Limitations on the spatial resolution will be discussed and a two-fiber scheme will be introduced that allows an increase in the spatial resolution by several orders of magnitude. The results of experiments to test the properties of the two-fiber scheme will be presented, with particular emphasis on the elucidation of the parameters required for the readout of large arrays manufactured with the one-dimensional combinatorial chemistry scheme.

FIBER-OPTIC BASICS

The reflection and refraction of light at the interface of media with different refractive indices is governed by Snell's law:

$$\frac{\sin \boldsymbol{\Theta}}{\sin \boldsymbol{\Gamma}} = \frac{n_{\mathrm{I}}}{n_{\mathrm{II}}} \tag{1}$$

The angles Θ and Γ are, respectively, the incident and refracted angles with respect to the normal of the interface, while n_{I} and n_{II} are the refractive indices of medium I and medium II (Figure 15.2). When light travels from the optically denser to the optically rarer medium, which is the situation pictured in Figure 15.2, there exists a range of incident angles for which refraction is no longer possible and all light is reflected in the optically denser medium. This phenomenon is called total internal reflection and it occurs for angles that are larger than a critical angle Θ_z, given by

$$\Theta_{\mathrm{Z}} = \arcsin(n_{\mathrm{I}}/n_{\mathrm{II}}) \tag{2}$$

Optical fibers can guide light by total internal reflection. The light propagates in the fiber core, which has a refractive index n_{co} that is larger than the refractive index of the surrounding fiber cladding n_{cl} (Figure 15.3). In the step-index fibers used in our work, the refractive index of the fiber core is constant with respect to the fiber cross section (inset to Figure 15.3).

From Eqn (1), one might conclude that, when greater than Θ_z, the range of Θ for guided modes is continuous. This is not the case. A discrete set of angles is obtained when demanding that the phase fronts, which the light rays represent, interfere constructively in

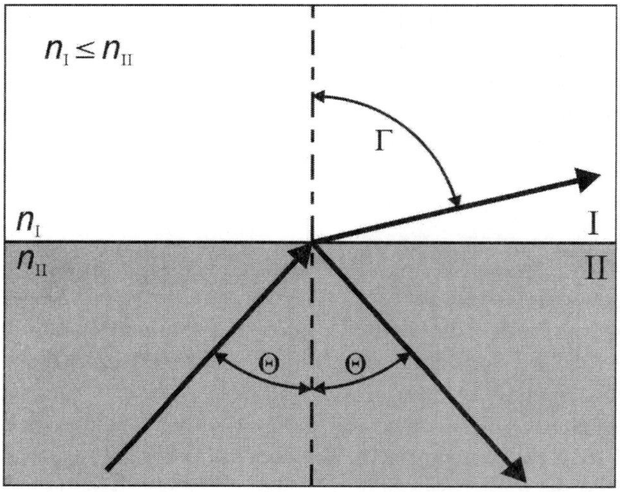

FIGURE 15.2. Reflection and refraction at the interface of two media with different refractive indices. Total internal reflection occurs for $\Theta > \Theta_{\mathrm{Z}}$.

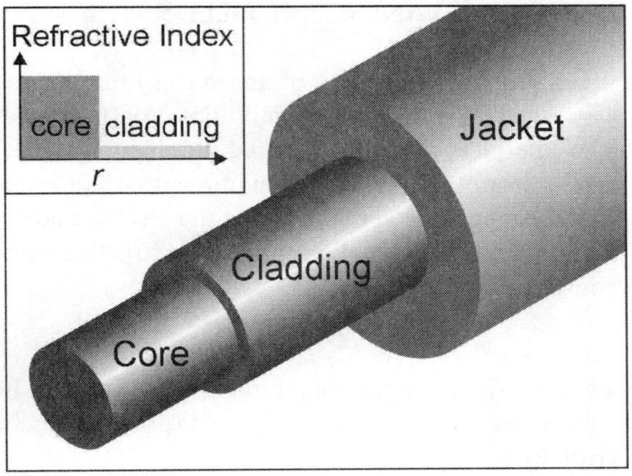

FIGURE 15.3. Layout of an optical fiber. The fiber core is surrounded by a cladding with a smaller refractive index. The fiber jacket surrounds the cladding for mechanical stability and protection. The inset shows the refractive index profile for the step-index fiber used in our work.

the fiber core after undergoing total internal reflections [4]. A full treatment of the problem is achieved by solving Maxwell's equations using the appropriate boundary conditions. However, light propagation in multimode fibers can be interpreted in terms of ray optics if the above restriction and the existence of evanescent fields (which will be described in detail below) are taken into account.

The guiding properties of an optical fiber can be described by the dimensionless fiber parameter [5]

$$V = (2\pi r/\lambda)\left(n_{\mathrm{co}}^2 - n_{\mathrm{cl}}^2\right)^{1/2} \tag{3}$$

where r is the radius of the fiber core and λ is the wavelength of the light in free space. In the case where $V \gg 1$ (as is true for the fibers studied in this work), the number of modes N that a fiber can support is given by [5]

$$N = V^2/2 \tag{4}$$

Clearly, for fibers to support many modes (each characterized by the respective incident angle Θ), the fiber radius has to be large with respect to the wavelength of the light, and the refractive-index difference between core and cladding has to be large. For example, the 200 μm core diameter fiber used in this work with $n_{\mathrm{co}} = 1.457$ and $n_{\mathrm{cl}} = 1.404$ supports approximately 2.64×10^5 modes at $\lambda = 337$ nm.

Finally, it has to be pointed out that Snell's law [Eqn (1)] is valid only for planar interfaces. For optical fibers it can only be applied to rays that cross the center axis of the fiber core (i.e. meridional rays); however, a class of guided rays exists that travel along a helical path through the fiber core without ever crossing the fiber axis. For these skew rays, Snell's law has to be modified (for more details see ref. 5). However, Snell's law as given in Eqn (1) remains a useful tool for interpreting the results presented below.

EVANESCENT FIELDS

Although light can propagate loss free in an optical fiber under guided conditions, the electric fields associated with the light rays do not drop to zero at the interface, but extend, normal to the interface, into the surrounding cladding. The existence of these evanescent fields results from the boundary conditions that are imposed by Maxwell's equations, namely that the tangential components of the electric and magnetic fields must be continuous at the interface of the two dielectric media [6]. The evanescent fields decay exponentially according to [7,8]

$$E(z) = E_0 \exp(-z/d_p) \tag{5}$$

where E_0 is the electric field amplitude at the core–cladding boundary, z is the distance normal to the core–cladding boundary, and d_p is the penetration depth of the evanescent fields, given by [7,8]

$$d_p = \frac{\lambda}{2\pi n_{co}(\sin^2\Theta - \sin^2\Theta_z)^{1/2}} \tag{6}$$

We can see that each mode has a different penetration depth. Furthermore, Eqn (6) shows that modes closer to Θ_z and longer wavelengths (λ/n_{co}) lead to increasing penetration depths. For a mode incident at Θ_z the penetration depth is infinity, but for $\Theta < \Theta_z$ it quickly decreases (Figure 15.4). Typical values of d_p are of the order of a wavelength; for example, a mode with $\lambda = 337$ nm incident at $\Theta = 1.1\ \Theta_z$ and propagating in a fiber with $n_{co} = 1.457$ and $n_{cl} = 1.404$, has $d_p = 162$ nm.

Over the past centuries, many elegant experiments have been performed to verify the existence of evanescent fields and their functional dependence [9–19]. In 1717 did Isaac Newton had already observed [9] that total internal reflection could be destroyed not only when a suitable object is brought into contact with the reflecting surface, but also when the object is brought sufficiently close to this surface. In his experiment, he brought a lens into contact with the hypotenuse of a right-angled prism. Light was coupled into the prism and experienced total internal reflection at the hypotenuse interface, except at the point of contact with the lens where it refracted out of the prism. However, he observed that total internal reflection was destroyed not only at the point of contact, but also where the prism

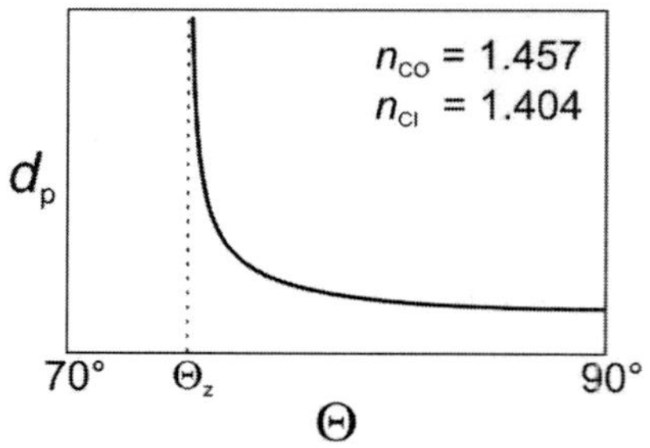

FIGURE 15.4. Dependence of penetration depth on incident angle for the step-index fibers used in this work. For incident angles equal to the critical angle, the penetration depth is infinity.

and lens were separated by a thin layer of air. In 1866, Quincke repeated the experiment and showed that total internal reflection is only restored when the thickness of the air layer reaches approximately four wavelengths [10].

Hall [12] placed a light-sensitive gel onto the totally reflecting surface of a prism (Figure 15.5). He found that the gel darkens only within a distance of a few wavelengths of light from the reflecting surface. In a modification of this experiment, Sélényi [13] replaced the gel layer with a solution containing the fluorescent dye fluorescein, and again found that only dye molecules within a narrow layer close to the reflecting surface emit fluorescence.

Since the existence of evanescent fields is a wave phenomenon, it applies to electromagnetic radiation of all wavelengths. Schaefer and Gross [14] used microwave radiation ($\lambda = 15$ cm) to verify the exponential dependence of the evanescent field [Eqn (5)]. In their experiment, the microwaves were coupled into a large right-angled paraffin prism, with total internal reflection taking place at the prism hypotenuse. The hypotenuse of a second movable prism was brought close to that of the first. If the gap between the prisms was sufficiently small, the microwaves traversed the gap to enter the second prism. The microwaves were detected with an antenna after leaving this prism. As expected, the reading at a voltmeter attached to the antenna changed exponentially with the width of the gap between the prisms. Drexhage [15] verified Eqns (5) and (6) for electromagnetic radiation in the visible spectral range in a beautiful experiment. Using Langmuir–Blodgett films to separate fluorophores, in nanometer-scale increments, from the surface of a glass slide, he measured the fluorescence signal as a function of this separation when propagating light through the glass slide under conditions of total internal reflection.

In all these experiments, the existence of evanescent fields was proven by "destroying" the total internal reflection (i.e. making it "less than total") [20], that is, by absorbing or redirecting light from the medium carrying it initially. It is precisely this feature of being able to couple out small amounts of light from an optical waveguide that is exploited in many applications, such as total internal reflection spectroscopy [7]. Before discussing the application of evanescent fields for sensing purposes, it is important to reiterate that if no absorbing species and no scattering centers are present within the range of the evanescent

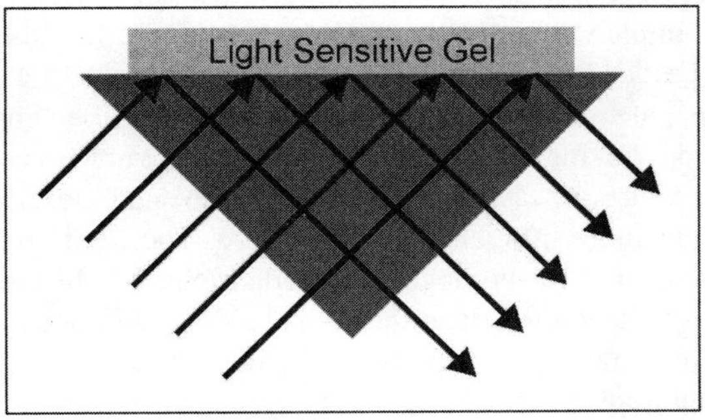

FIGURE 15.5. Total internal reflection in a prism. A light-sensitive gel is placed onto the totally reflecting surface of the prism. It is found that the gel darkens only within a distance of a few wavelengths of light from the reflecting surface.

fields, no light intensity will be lost from the optical waveguide (e.g., optical fiber). In this case the light can propagate over large distances, limited only by the optical clarity of the waveguide [21,22].

FIBER-OPTIC SENSING USING EVANESCENT FIELDS

Optical-fiber sensors can be defined [23] as a "means where light guided by an optical fiber can be modified in response to an external physical, chemical, biological, biomedical or similar influence." Sensing can take place along the entire length of the fiber (distributed sensing) or at discrete regions along the fiber (quasi-distributed sensing). As indicated above, we are interested in exploiting evanescent fields for sensing purposes. That is, the active sensor components (e.g., sensor molecules) are placed into the fiber cladding, or a suitable substance replacing the fiber cladding, within the range of the evanescent fields of the core modes. Since evanescent coupling is an inherently weak process, the attenuation of the light in the core is significantly less than for pure refractive coupling, making large arrays of chemosensors feasible. Moreover, since the evanescent field intensity drops off exponentially with distance from the core/cladding interface, only sensor molecules close to the interface can be excited.

Evanescent field absorption sensors monitor changes of the light intensity transmitted through the fiber, which occur when the absorbance within the range of the evanescent fields is modified as a response to a "triggering event." Such an event could be, for example, the diffusion of absorbing species into the range of the evanescent field, which results in a decrease of the overall transmission of the fiber [24,25]. Another example is an absorbance change of sensor molecules already present within the range of the evanescent fields as a response to changes in the chemical and physical environment (e.g., pH [26, 27], temperature [28], pressure [29, 30], etc.). Changes in the absorption energies that occur as a result of the specific binding of an analyte to a sensor molecule, for example, can be employed by spectrally resolving the light transmitted through the fiber with a dispersive element (e.g., monochromator) [31–33].

A second type of evanescent field sensor is based on the use of fluorescent sensor molecules. Following evanescent excitation, light emitted by fluorophores in the fiber cladding cannot be coupled refractively into guided modes of the fiber [34]. However, the fluorescence may be "captured" by the evanescent fields of the guided core modes enabling it to propagate under guided conditions to the fiber ends [35,36] where it can be detected. Modes having larger penetration depths are able to collect more light than modes with smaller values of d_p. Also, fluorophores located closer to the core/cladding boundary couple light more efficiently into the core. The light collection efficiency of a fiber is proportional to V for homogeneously distributed light sources in its cladding and to V^2 for sources located directly at the core/cladding interface [35]. Thus, in contrast to fiber-optic sensors that simply monitor cladding absorption, the fibers suited to fluorescent fiber-optic sensors are strongly guiding ($n_{co} \gg n_{cl}$) with large core diameters. A "triggering event" for a fluorescent sensor may result in changes of fluorescence intensity, fluorescence wavelength, and/or fluorescence lifetime.

SPATIALLY RESOLVED READOUT: OPTICAL TIME-DOMAIN REFLECTOMETRY

The use of a continuous light source to read out distributed or quasi-distributed sensors provides only the aggregate response of all the sensor molecules. For quasi-distributed optical-fiber sensors (or fiber-optic sensor arrays), the spatial location of sensors responding to triggering events can be determined using a method called optical time-domain reflectometry (OTDR), which was first used to locate faults in optical fibers [37–39]. The basic scheme is as follows. A laser pulse is coupled into the fiber, and upon encountering a defect in the fiber, the amount of Rayleigh-scattered light directed back along the fiber is altered, which may be detected at the same end of the fiber with a characteristic time delay (Figure 15.6). Thus the time delay encodes the spatial location of the scattering source along the fiber, and a defect that changes the backscattered intensity is easily located [40–42].

Evanescent excitation of fluorophores located in the fiber cladding and subsequent evanescent capture of the fluorescence by core modes was first combined with the OTDR method by Kvasnik and co-workers [43] and Chronister and co-workers [44]. Again, a laser pulse is coupled into the fiber core at the front end of the fiber (Figure 15.6). Following the absorption of the exciting light pulse by fluorophores in the cladding, a fluorescence pulse is coupled back into the fiber core and propagates to the fiber ends. When detected at the front end of the fiber, the location L_i of the ith emitting fluorophore can be calculated from the time delay t_i between exciting and returning pulses according to

$$L_i = (c/2n_{\mathrm{co}})t_i \tag{7}$$

where c is the speed of light in vacuum. This method is known as fluorescent OTDR [45].

At this point it is important to recall that the optical evaluation of combinatorial libraries built in the cladding of optical fibers using the one-dimensional combinatorial

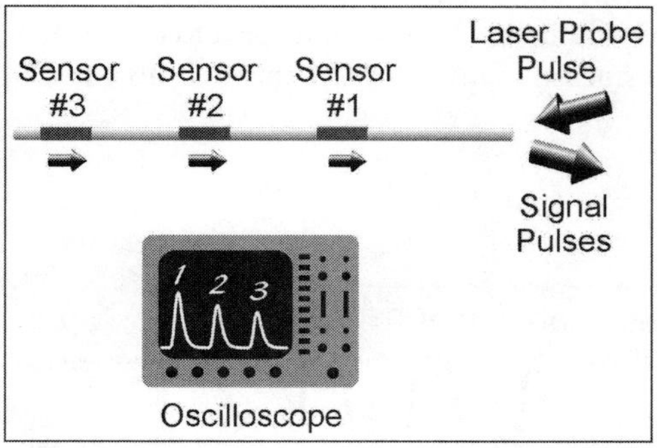

FIGURE 15.6. Principle of optical time domain reflectometry. A laser pulse is coupled into the fiber. Upon encountering a scattering center or a fluorophore, light is directed back to the front end of the fiber, where it is detected with a characteristic time delay. The time delay encodes the spatial location of the scattering source or fluorescence source along the fiber. (Adapted from ref. 51. Copyright 2002 Association for Laboratory Automation).

chemistry technique and assayed with fluorescent tags is identical to the readout of the quasi-distributed fluorescent chemosensor arrays as described above. It is an inherent virtue of our combinatorial chemistry method that the location of a fluorophore along the fiber encodes the complete synthetic history of assayed compound, meaning that the optical evaluation of combinatorial libraries is accomplished by simply measuring this location.

If there are many sensor regions on the fiber, the return signal consists of a train of pulses, each encoding in the time domain the spatial location of the emitters to which they pertain. The minimum separation ΔL_{i+1} between adjacent fluorescent regions required for the temporal resolution of the corresponding fluorescence pulses is determined by the fluorescence lifetimes $\tau_{\mathrm{fl}}(i)$ of the ith fluorophore and by the desired length w_{i+1} of the detection time window ($w_i > 1\ \forall i$):

$$\Delta L_{i+1} = L_{i+1} - L_i = (c/2n_{\mathrm{co}})w_{i+1}\tau_{\mathrm{fl}}(i) \tag{8}$$

where w_i is multiples of the fluorescence lifetime. For example, choosing $n_{\mathrm{co}} = 1.457$, $\tau_{\mathrm{fl}} = 10$ ns, and $w = 2.5$ yields a minimum separation of $\Delta L = 2.57$ m. For the large combinatorial libraries and the large arrays of fluorescent chemosensors envisioned here, the total length of fiber needed can be substantial.

HIGH SPATIAL RESOLUTION READOUT: TWO-FIBER SCHEME

We proposed to overcome this limitation by introducing the two-fiber readout scheme shown in Figure 15.7. A laser pulse is propagated through one fiber (the excitation fiber) and excites fluorophores in each of the sensor regions. A fraction of the emitted light from each region is then coupled evanescently into the second fiber (the detection fiber), which lies within the evanescent coupling range of the excitation fiber at each sensor region, and travels through this fiber to the detector. The distance between adjacent sensor regions along one of the fibers must be large enough to ensure that their fluorescence pulses are resolved at the detector (the detection fiber provides this delay in Figure 15.7); however, they may be spaced well within this limit with respect to the second fiber (as they are with respect to the excitation fiber in Figure 15.7). In this two-fiber readout scheme the location of a sensor is given by

$$L_i = (c/n_{\mathrm{co}})t_i - d_i \tag{9}$$

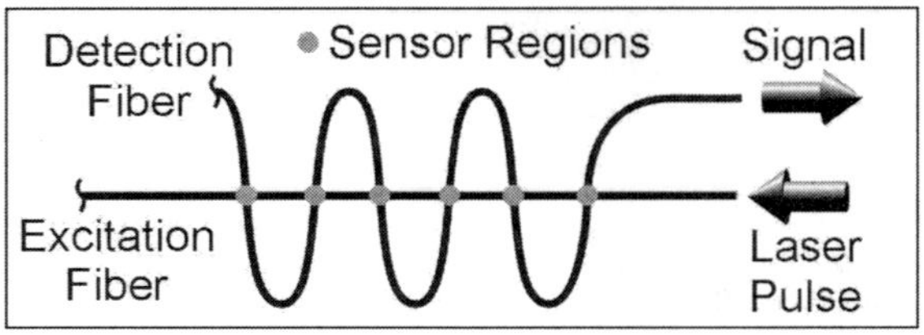

FIGURE 15.7. Basic idea of two-fiber scheme. As in the one-fiber scheme, laser pulses excite fluorophores in the cladding evanescently. However, the fluorescence pulses are now captured by a second fiber, which provides an optical delay between successive pulses to allow for temporal resolution.

where L_i is the position of the ith sensor region, with respect to a fluorescent reference region (see below), along the fiber that provides the optical delay between sensor regions [46], t_i is the time delay between the arrival of the signal from region i and that from the reference region, and d_i is the distance between sensor region i and the reference region along the fiber on which the regions are closely spaced. For arrays of sensors with $L_i \gg d_i$, as is the case for the studies presented here, this expression reduces to

$$L_i = (c/n_{\mathrm{co}})t_i \tag{10}$$

Note that the factor of 1/2 from Eqn (7) is absent from Eqns (9) and (10).

For both the one- and two-fiber schemes, a time reference is needed that is associated with a known value of L to which the value of L_i is measured. The front-end reflection of the input pulse provides this reference in the one-fiber scheme, while in the two-fiber scheme an additional fiber junction with a fluorescent reference region can be prepared for that purpose.

Many current fiber-sensor schemes consist of only one or a small number of sensor regions per fiber. The aim in many of these studies is to enhance the sensor response by enhancing the interaction of the evanescent fields with the sensor molecules, for example by tapering the fibers. This results in increased losses of the exciting light in the fiber core, which, given the small number of sensor regions, is not a critical factor in these applications. However, our one-dimensional combinatorial chemistry scheme allows the straightforward fabrication of very large fluorescent sensor arrays, which is why we have to be concerned with how the losses of the exciting light at one sensor region affect the signals that can be obtained from subsequent sensor regions. In general and in the ideal case, there should be equal excitation at each fluorescent region along the sensor array. However, even if there is only evanescent coupling, a gradual decrease in the amount of light in the excitation fiber, and therefore the extent of sensor-region excitation, is expected along this fiber. To minimize this decline, the percentage loss of light from the excitation fiber at each sensor region must be kept small. At the same time, however, for maximum detection sensitivity, the capture efficiency has to be maximized.

Compared with a one-fiber scheme, the two-fiber scheme offers considerably more flexibility for signal optimization since the parameters of the two fibers can be chosen separately and the fiber-junction geometry can be adjusted. Therefore one aim of the work described in the following was to determine the optimum parameters for the excitation and detection fibers in the two-fiber scheme. The results will be discussed in detail below. An additional benefit of the two-fiber scheme as an assay for combinatorial synthesis is that it allows the fiber providing the optical delay to be separated from the fiber that forms the solid support for library synthesis. The combinatorial library is created on one fiber, which is brought into contact with the second fiber for optical evaluation.

Another important issue concerns the properties of the fiber cladding. First, as described above, the cladding has to allow the diffusion of the analyte to the sensor molecule sites (or allow the diffusion of the sensor molecules themselves into the range of the evanescent fields). While commercially available fibers clad with silica or plastic might successfully perform these functions for some applications, generally a replacement cladding has to be used at the sensor regions. For example, fluorescent fiber-optic sensors for molecular oxygen were built by doping polysiloxane-based claddings with fluorescent molecules [34,47]. Secondly, a replacement cladding has to preserve the guiding condition of the fiber. The elucidation of the extent to which the latter condition restricts the permissible range of refractive indices of the replacement material was another focal point of our studies. The results will be discussed below.

PULSE DISPERSION AND SPATIAL RESOLUTION

Light pulses propagating through a fiber are found to broaden in time owing to the dispersive properties of the fiber. The pulse broadening increases with increasing length of the fiber. If the broadening becomes sufficiently large, initially well-separated adjacent pulses will overlap at some point. To avoid this problem, the temporal separation of the pulses has to be increased.

The pulse dispersion is due to intramodal and intermodal dispersion [5,48]. Intramodal dispersion, that is, the dispersion that a pulse experiences in one mode, results from the dependence of the propagation speed of the pulse on the wavelength of the light. Since a laser pulse has a finite frequency bandwidth, the different wavelengths contained in the pulse, while following the same path, propagate at different speeds, resulting in broadening of the pulse in time.

Intermodal dispersion results from the fact that the transit time of a ray propagating through a fiber depends on the path taken. Since path length is different for different modes, the components of a light pulse traveling on paths characterized by smaller incident angles will be delayed with respect to those propagating at larger incident angles, resulting in a broadening of the pulse. Intermodal dispersion is usually dominant for multimode fibers [4].

According to ray optics, the temporal intermodal dispersion per unit length ΔT_{rel} of a delta-shaped input pulse that populates all modes in a straight fiber with uniform intensity is given by [5]

$$\Delta T_{rel} = \frac{1}{c} \frac{n_{co}}{n_{cl}} (n_{co} - n_{cl}) \qquad (11)$$

where c is the speed of light in vacuum. This expression applies to bound rays irrespective of their skew angle (which is defined as the angle between the tangent to the core/cladding interface at a ray reflection point and the projection of the ray path into the core cross section [5]). Since Eqn (11) was derived by comparing the maximum and minimum path lengths for light in the fiber, it provides an upper limit for the expected pulse spreading. Equation (11) predicts that the intermodal dispersion broadens a delta-shaped input pulse after passage through 500 m of fiber with $n_{co} = 1.457$ and $n_{cl} = 1.404$ to 91.7 ns through 500 m of a fiber with $n_{co} = 1.457$ and $n_{cl} = 1.404$ is 91.7 ns. This estimate is based on the assumption that all modes are populated equally; the extent to which it applies when the light is coupled into the fiber from the cladding through the evanescent fields is discussed in more detail below.

In the one-fiber fluorescent OTDR scheme, dispersion affects both excitation and fluorescence pulses. The excitation laser pulses broaden as they propagate down the fiber so that fluorescent regions further from the light source are excited with increasingly broader pulses. The emitted fluorescence pulse is a convolution of the exciting pulse and the fluorescence decay curve. Therefore the fluorescence pulses emitted from regions far from the front end of the fiber (where the excitation pulse enters *and* detection take place) are increasingly broadened with respect to those from regions close to the front end of the fiber.

In the two-fiber scheme two cases have to be distinguished. If the detection fiber provides the optical delay between the signals from adjacent sensor regions, which are closely spaced with respect to the excitation fiber, then the fluorophores in all sensor regions are excited nearly simultaneously with laser pulses of the same width. The fluorescence pulses coupled into the detection fiber broaden as they propagate to the detector. Therefore, to ensure full temporal resolution, the length of the delaying fiber between adjacent fiber–fiber junctions has to increase the farther these junctions are from the detection end of the fiber. However, the required increase in the spacing is approximately half that needed in the one-fiber scheme.

If the excitation fiber provides the delay and the sensor regions are closely spaced with respect to the detection fiber, the fluorescent regions are excited with increasingly broadened pulses the farther the region is from the excitation source. This means that the fluorescence pulses coupled into the detection fiber are broadened correspondingly. However, on their way to the detector along the short detection fiber, these pulses experience essentially no dispersion effects. Again, to ensure full temporal resolution, the length of the delaying fiber between adjacent fiber–fiber junctions has to increase the farther the regions are from the source. Experimental data for the pulse dispersion are presented below.

EXPERIMENTAL SETUP

The results described below were obtained using three different experimental setups. Figure 15.8 shows a schematic representation illustrating the basic features of all three setups. In all cases, two fibers were used: the excitation fiber carried the 337 nm light pulses from a Photon Technology International PL2300 nitrogen laser (typically 0.6 ns pulse width and 1.4 mJ pulse energy with repetition rates of 3–10 Hz) to the sensor regions, and the detection fiber delivered the resulting fluorescence to a Burle C31034 (2.5 ns rise time) photomultiplier tube (PMT). The resulting signals were collected on either a Hewlett Packard HP54505B or LeCroy LC564DL digital storage oscilloscope (DSO) with bandwidths of 300 MHz and 1 GHz, respectively, and sampling rates of 500 MSa s^{-1} and 4 GSa s^{-1}, respectively. A trigger for the DSO was generated using a second PMT (RCA 1P28, 1.6 ns rise-time), which collected light scattered from the front of the excitation fiber. For setup 1 (described below), the excitation fiber passed through the housing of the 1P28 PMT (as depicted in Figure 15.8). In this case the PMT measured light escaping from the side of the fiber (largely cladding fluorescence, although there may have been contributions due to refractive losses of modes that were coupled into the fiber under nonguiding conditions, and scattered light from core inhomogeneities) and therefore provided a measure of the total intensity of light in the fiber, while still providing the trigger signal for the DSO. The spectral decomposition of the signal, if used, was obtained using colored glass filters or a Jarrell Ash 82-000 monochromator (0.5 m focal length, 1.6 nm/mm linear dispersion).

The sensor regions themselves were prepared by removing the fiber jacket and cladding at the junction(s) of the two fibers so that the two fiber cores, at an angle of 90° to

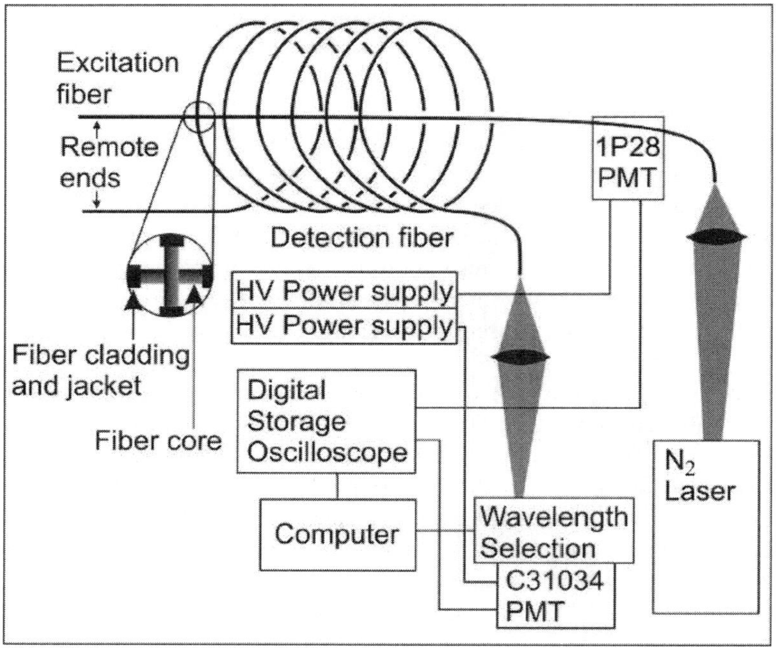

FIGURE 15.8. Basic experimental setup used for all experiments. See text for details. (Adapted from ref. 49. Copyright 2001 Society for Applied Spectroscopy.)

one another, were just touching. Therefore the contact pressure was minimal. The junctions were then immersed in a solution of a fluorescent dye in a solvent with a refractive index that preserved the guiding condition of the fiber. In all setups (as in Figure 15.8), the detection fiber provided the optical delay between the fluorescence from adjacent regions. In the two setups containing multiple fiber junctions, the separation between adjacent regions along the fiber providing the optical delay was typically two orders of magnitude greater than their separation with respect to the other fiber.

All setups used multimode silica fibers with a TECS™ (3M Corporation) cladding (Thorlabs/3M FT-200-UMT and FT-400-UMT with core diameters of 200 μm and 400 μm, respectively). All fibers had $n_{co} = 1.457$ and $n_{cl} = 1.404$. The silica/TECS fiber that was used in these studies satisfied two important criteria. First, for simple preparation of fiber–fiber junctions the TECS cladding could easily be removed with acetone, thereby preserving the core surface. This avoids the HF treatment often applied for the removal of silicone claddings, which may etch the core surface and lead to refractive losses. Second, the silica core enabled the use of a UV excitation source. However, when coupling 337 nm light pulses into these fibers, strong luminescence is observed from the cladding. Presumably this is fluorescence from the TECS cladding, which may be excited evanescently by the laser pulse (there may also be some excitation of the TECS cladding due to light coupled into the fiber under nonguiding conditions, but the population of these modes is expected to be small). As well as attenuating the laser pulse in the fiber, this luminescence can evanescently couple back into guiding modes of the fiber leading to a broad distribution of wavelengths within the fiber core. To minimize the transmission of light from the excitation fiber directly into the detection fiber at the point of contact, the fibers were crossed at right angles at the fiber junctions (see inset to Figure 15.8).

RESULTS

Sensitivity of Coupling

The initial experiments focused on the efficiency of evanescent fluorophore excitation and subsequent evanescent coupling into a second fiber, and the sensitivity of such a measurement. A single junction was located approximately 1 m along the excitation fiber (400 μm core diameter). The junction was immersed in an aqueous solution of the dye Rhodamine 590 (R6G). The captured R6G fluorescence signal traveled through 500 m of the detection fiber (200 μm core diameter) to the detector, which delayed the arrival time of the signal at the detector by approximately 2.5 μs. To determine the sensitivity of the measurements in the evanescent fiber–fiber coupling scheme, the R6G concentration was increased by factors of 10 starting from a solution of 10^{-10} mol L^{-1} until an R6G fluorescence signal could be discerned above the level of the background signal. The average of 1000 time traces obtained for each concentration was integrated and divided by the integrated reference time trace from the trigger PMT. The results are depicted in Figure 15.9. An R6G concentration of 10^{-7} mol L^{-1} can easily be detected above the background signal.

Intermodal Dispersion

Using the same setup we addressed the issue of intermodal fiber dispersion and its implications for the maximum spatial resolution that can be achieved. The single fiber–fiber junction was submerged in a 10^{-6} mol L^{-1} aqueous solution of R6G. Figure 15.10 compares the fluorescence decay curve of a solution of R6G (left) with the decay signal detected after passage through the detection fiber (delayed in time by

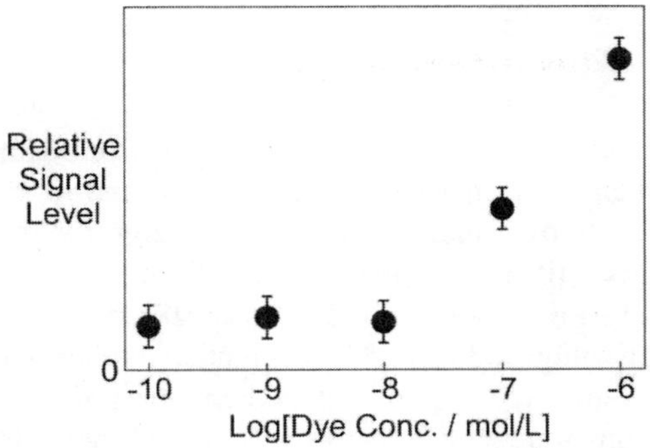

FIGURE 15.9. Signal intensity after passage through 500 m of fiber versus R6G concentration. (Adapted from ref. 50. Copyright 2002 Association for Laboratory Automation.)

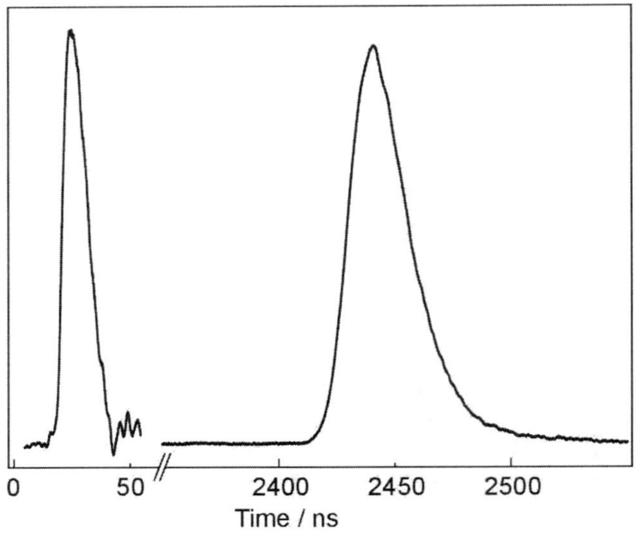

FIGURE 15.10. Fluorescence decay curve of R6G in water before entering the fiber (left) and after passage through 500 m of fiber (right). Adapted from ref. 49. Copyright 2001 Society for Applied Spectroscopy.

approximately 2.5 μs). Both pulses were recorded under the same experimental conditions (excitation pulse, detector response time, etc.), and so the broadening, which is clearly evident in the second pulse, is solely due to the dispersion of the fluorescence as it passed through the 500 m detection fiber. In order to quantify the extent of the dispersion, both signals were fitted with a convolution of a Gaussian function and an exponential decay. The fiber dispersion function was then deconvoluted from these two functions. Figure 15.11 shows the resulting function, which has a full width at half-maximum of approximately 25.5 ns. This broadening kernel, which represents the broadening of a delta-shaped input pulse, has a width significantly less than the value of 91 ns predicted using Eqn (11). This means that the loss of spatial resolution due to intermodal dispersion is not as severe as originally anticipated.

Attenuation of Light in the Fiber

A second setup was used to test the attenuation of the exciting and signal light pulses when multiple sensor regions are present along the fiber, and to investigate whether the evanescent fiber–fiber coupling scheme is suitable for large arrays of fiber sensors. For these experiments a 200 μm fiber was coiled up on a 20 cm diameter cylinder. Every nine windings (corresponding to a separation of 5.84 m or 28 ns), 1 cm of the fiber core was exposed. A 400 μm fiber with cladding and jacket removed touched each of these exposed regions. The crossing points were approximately 1 cm apart along the 400 μm fiber. Six fiber–fiber coupling regions were created in this way and submerged in a 2×10^{-5} mol L^{-1} aqueous solution of R6G. The refractive index of this solution, which is approximately that of water ($n_{wa} \approx 1.33$), ensured that the guiding conditions of both fibers were

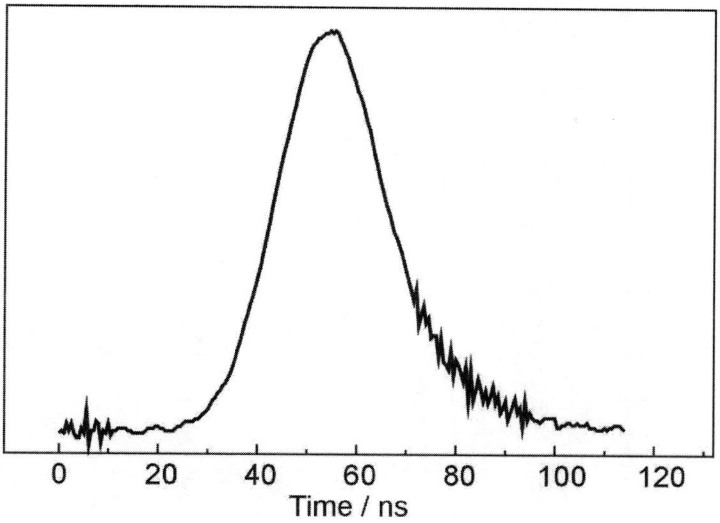

FIGURE 15.11. Fiber broadening function (deconvolution of the curve on the right of Figure 15.10, from that on the left). (Adapted from ref. 49. Copyright 2001 Society for Applied Spectroscopy.)

retained throughout the regions in which their claddings were removed. The 400 μm fiber was chosen to act as the excitation fiber because of its ability to support more modes [see Eqns (3) and (4)].

The four traces shown in Figure 15.12 were each obtained by averaging 64 time traces, and each trace shows six peaks corresponding to the six fiber–fiber junctions immersed in the R6G solution. Even though the R6G concentrations were the same in each junction, the resulting peak intensities were quite different. The reason for this observation lies in the fact that the evanescent fields decay exponentially [see Eqns (5) and (6)]. Therefore differences in the separation of the two fibers at a junction are reflected, "exponentially amplified," in the corresponding peak intensity differences. While we paid no particular attention to ensuring equal fiber separation in each junction, this issue will be of importance in the fabrication of large arrays.

The four traces shown in Figure 15.12 were recorded from each end of the detection fiber (labeled D_1 and D_2) for excitation from both ends, E_1 and E_2, of the excitation fiber. Therefore the labels E_i in Figure 15.12 indicate the fiber end that admits the laser pulses, while D_i specifies at which end of the detection fiber the fluorescence was collected. The sensor regions were numbered 1 to 6 from the E_1 (D_1) end of the fiber to the E_2 (D_2) end. In this setup, the detection fiber provided the optical delay for the fluorescence signals, so the subscript on D specifies the order in which the fluorescence signals from each region arrived at the detector; for E_1D_1 and E_2D_1, the data pulses arrived in order from 1 to 6, whereas for E_1D_2 and E_2D_2 the data pulse sequence is from 6 to 1.

The overall intensity of the four traces collected was expected to vary owing to slight differences in the efficiency of coupling light into or out of the ends of the fibers, but variations in the relative peak intensities of each trace provide information about the losses within the excitation and detection fibers. The differences in relative peak intensities

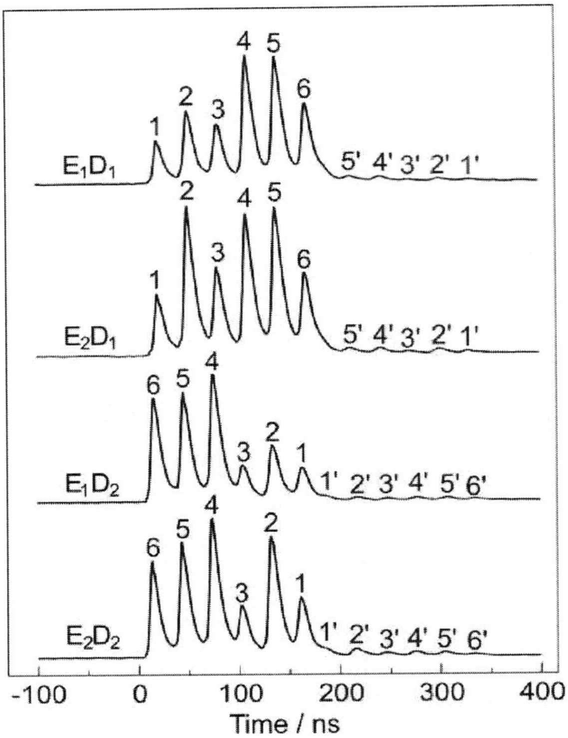

FIGURE 15.12. Signal traces obtained from a six-sensor array in a solution of R6G. The labels are explained in the text. All traces are plotted on the same scale. (Adapted from ref. 46. Copyright 2001 American Chemical Society.)

between the E_1D_1 and E_2D_1 traces, and between the E_1D_2 and E_2D_2 traces, can be attributed to attenuation of the exciting light in the excitation fiber between regions 1 and 6: with zero attenuation in this fiber these traces will be exact duplicates; with significant attenuation, the E_1D_i traces would be skewed in intensity towards region 1 relative to the E_2D_i traces. Similarly, the difference in relative peak intensities between E_1D_1 and E_2D_1, and between E_1D_2 and E_2D_2, can be attributed to attenuation of fluorescence in the detection fiber between regions 1 and 6: with zero attenuation in this fiber, these traces will exactly mirror one another; with attenuation, the E_iD_1 traces would be skewed in intensity towards region 1 relative to the E_iD_2 traces.

Although there are slight variations in the relative peak intensities of the four traces in Figure 15.12, the overall pattern remains, suggesting that there is little attenuation of light in the excitation and detection fibers. This observation is corroborated by the presence of a second set of peaks, which are labeled $1'$–$6'$ in Figure 15.12. These peaks arise from reflections of the sensor fluorescence from the "non-detection" end of the fiber. In fact, for D_1 detection, the $6'$ peak is unresolved from 6 since region 6 is too close to the end of the detection fiber. The ability to detect signals $1'$–$6'$, being reflected at the fiber end, traveling the entire length of the detection fiber, and passing all fiber–fiber junctions, demonstrates the low losses of the fluorescence signals at subsequent sensor regions and thus the potential of the two-fiber scheme for very large arrays of sensors.

Effects of Refractive Index Changes

Experiments using a sensor array consisting of six sensor regions containing dyes in solutions of agarose showed a significant change in absolute and relative signal strength with time [46]. Initially, when swollen with water, the agarose gel had a refractive index approximately equal to that of water. As the gel dried, however, the refractive index gradually increased to a final value of approximately 1.51—well in excess of the fiber core index—such that the coupling between the two fibers was refractive. The refractive index change had a profound influence on the measured signals.

For a quantitative study of the dependence of the coupling efficiency on the refractive index of the solution surrounding the fiber junctions, setup 2 was used, consisting of six fiber–fiber junctions immersed in glycerol–water mixtures of varying composition, maintaining a constant R6G concentration of $2.5 \times 10^{-5}\,\mathrm{mol\,L^{-1}}$. The refractive indices of these solvent mixtures ranged from 1.3797 to 1.471. Since the cladding of both fibers was removed at the sensor regions, the solution was in direct contact with the fiber cores. Thus the refractive index n_r of the solution determined the value of the critical angle and therefore the extent of guidance of the light modes in the fiber core at the sensor regions. The four signal traces shown in Figure 15.13 were obtained by averaging 32 individual time traces. Traces (b)–(d) were recorded at the same PMT voltage and plotted to the

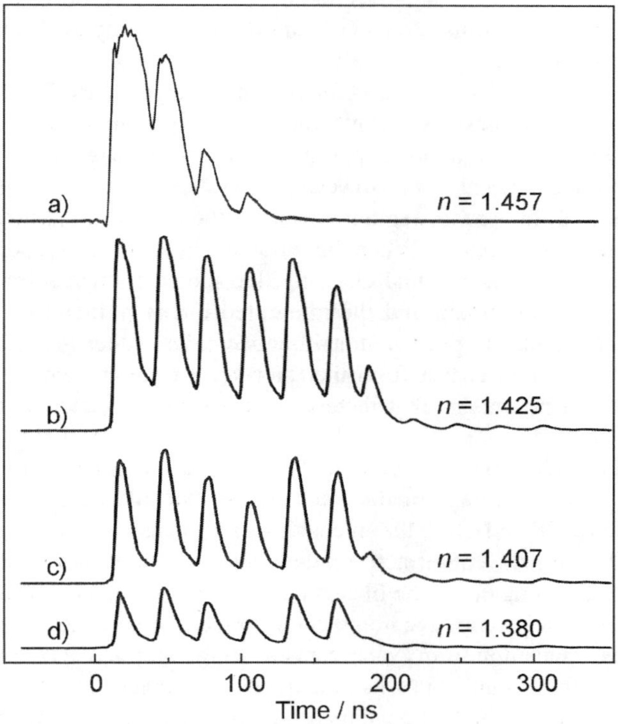

FIGURE 15.13. Refractive index effects: signal traces produced by six regions containing R6G in various glycerol–water mixtures. (Adapted from ref. 49. Copyright 2001 Society for Applied Spectroscopy.)

same scale, but the signal for trace (a) was so intense that the PMT signal saturated, necessitating a lower PMT voltage.

Trace (c) shows the case where the refractive index of the glycerol–water solvent ($n_r = 1.407$) was close to that of the original cladding of the excitation and detection fibers ($n_{cl,ex} = n_{cl,det} = 1.404$) so that only evanescent coupling between the fiber cores and sensor molecules was observed. Lowering the solvent refractive index to 1.380 [trace (d)] makes the fiber more strongly guiding at the sensor regions. While coupling is still evanescent, it is also weaker since those modes that lie close to the critical angle (and therefore have the largest evanescent penetration depth) when launched into the fiber, lie further from the critical angle of the fiber at the sensor regions and consequently have a smaller penetration depth. The result, as shown in trace (d), is that the relative intensities of the individual peaks are preserved, but the absolute intensities drop.

The solvent refractive index for trace (b) of $n_r = 1.425$ is above that of the original cladding, but still below the core refractive index of $n_{co} = 1.457$. In this case, modes that lie close to the critical angle when launched into the fiber propagate at angles greater than the critical angle at the sensor regions and therefore couple refractively out of the fiber. The improved efficiency of refractive coupling results in an increase in the absolute intensities of the peaks, while the change in the mechanism of coupling gives rise to a redistribution of relative intensity between the peaks. When the solvent refractive index is equal to, as in trace (a), or greater than the core refractive index, the guiding conditions no longer exist for any modes of the fiber at the sensor regions. This leads to a large redistribution in intensity towards the sensor regions that lie at the input end of the excitation fiber. The fluorescence signals from subsequent regions drop off rapidly as all intensity is quickly lost from the excitation fiber.

Figure 15.14 shows the refractive index dependence for a series of 12 solutions. The dashed curve represents the integrated intensity from all sensor regions. As the refractive index of the sensor regions approaches that of the original cladding, coupling is only via an evanescent mechanism. However, the evanescent penetration depth of populated modes increases at the sensor regions owing to the decreasing critical angle, so the integrated intensity also increases. When the refractive index of the sensor regions is between that of the fiber core and original cladding, the more efficient refractive coupling mechanism becomes predominant and the integrated intensity increases dramatically. Finally, the integrated intensity plateaus for refractive index values greater than that of the fiber core because all light is lost from the fiber via a refractive coupling mechanism at the first few sensor regions so that there can be no further increases in the excitation of sensor molecules.

The solid lines in Figure 15.14 represent the integrated intensities of the individual peaks as a percentage of the total intensity of all peaks. For solvent refractive indices between 1.38 and 1.41, the intensities of all peaks remain largely unchanged. The same holds true for solvent refractive indices above 1.46. In each of these ranges, the mechanism of light coupling out of the fiber is constant; it is completely evanescent in the former case, and completely refractive in the latter. In the intermediate region, where $1.41 < n_r < 1.46$, there is a transition from evanescent coupling to refractive coupling. As refractive coupling becomes the dominant process, the intensity changes vary dramatically for different peaks. The intensity of the first peak increases with increasing refractive index. The same behavior is observed for the second peak, although the increase is not as pronounced. All other

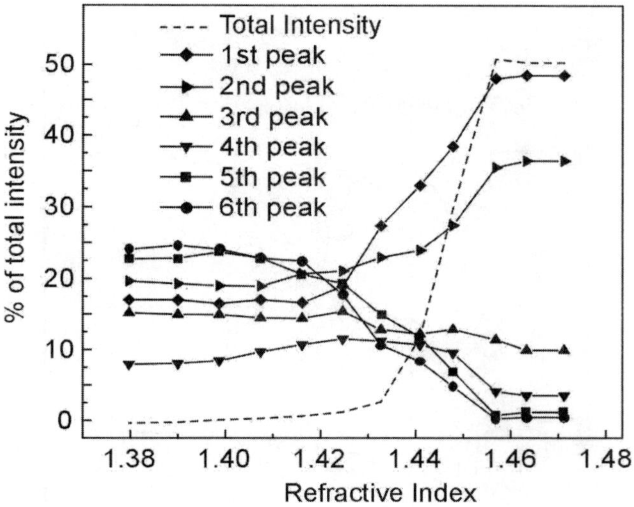

FIGURE 15.14. Integrated intensities for each peak in Figure 15.13 versus the solvent refractive index. The dashed curve represents the integrated intensity of all peaks. The *y*-axis scale applies to the individual peak intensities only. The core refractive index was $n_{co} = 1.457$, while the refractive index of the original cladding, which is retained outside the fiber junctions, was $n_{cl} = 1.404$. (Adapted from ref. 49. Copyright 2001 Society for Applied Spectroscopy.)

peaks decrease in intensity with increasing refractive index, although the magnitude of the decrease consistently rises from peak 3 to peak 6. Overall, there is a redistribution of relative intensity toward those sensor regions that lie closer to the input ends of the fibers.

Arrays Containing Multiple Dyes

While the first two setups were used to investigate the basic properties of the fiber–fiber coupling scheme, the third was prepared to test how a number of closely packed sensor regions containing different dyes [Stilbene 420 (S420), Coumarin 500 (C500), R6G, LDS698] can be spatially and spectrally resolved. First, five sensor regions were prepared, separated by 3.9 m (~19 ns) along the 200 nm diameter detection fiber and approximately 5 mm along the 400 mm diameter excitation fiber. These were immersed in a drop of agarose gel swollen with water, containing the dye molecules of interest which, from the remote end of the detection fiber, were LDS698, C500, LDS698, R6G, and S420.

Figure 15.15 shows a series of signals, each of which consists of 16 averaged time-traces. Trace (a) represents the fluorescence signals integrated over all wavelengths. The signal intensities depend on concentration, transition dipole moment, and fluorescence quantum efficiency of the fluorophore, as well as on the microscopic geometry of the fiber junction. Although the signals of the two LDS698-containing regions (3 and 5) are partially obscured by the stronger signals of the adjacent regions, the five sensor regions can be clearly discerned in time. By placing appropriate color filters before the PMT, spectral discrimination was possible. To record trace (b), a bandpass filter (transmission > 50% between 296 and 362 nm) was used. As expected, the S420 emission dominated. For

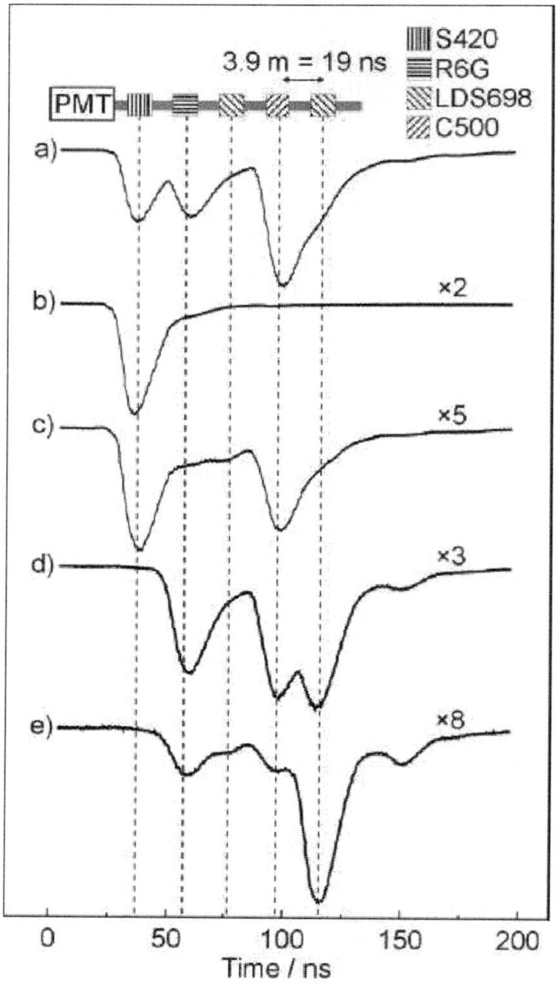

FIGURE 15.15. Response of five sensor regions containing different dyes in agarose gel. Trace (a) is the cumulative signal integrated over all wavelengths. For traces (b)–(e), color filters were used for wavelength discrimination. (Adapted from ref. 49. Copyright 2001 Society for Applied Spectroscopy.)

traces (c)–(e), the pass wavelengths of the filters (362–520 nm, > 584 nm, and > 664 nm ranges for transmission > 50%) moved toward longer wavelengths. In trace (c) the C500 peak is present in addition to the S420 peak. Trace (d) shows the S420 signal, effectively suppressed, with peaks due to R6G, C500, and LDS698 fluorescence clearly visible. Finally, in trace (e) the LDS698 peak from region 5 dominates the time trace, while the R6G and C500 fluorescence is further attenuated. Moreover, the weak LDS698 peak from region 3 is also visible. The intensity difference of the signals originating from the two LDS698-containing regions is again due to different geometries in the respective fiber–fiber junctions.

Again, additional peaks are present in the time traces detected at times after all fluorescence should have passed out of the ~25 m length of the detection fiber. These peaks, which are particularly evident in traces (d) and (e) of Figure 15.15, are due to reflections of the captured sensor fluorescence from the remote end of the detection fiber. The first reflected pulse to arrive at the detector consists of the LDS698 fluorescence of region 5. Upon changes of the optical filters, the intensity changes of the peak at 150 ns follow the intensity changes of region 5.

Subsequently, a new sensor array (also immersed in agarose–dye solution) was created with a total of 10 regions. These were separated by only three windings (~9.5 ns) of the detection fiber. The dyes in each sensor region, from the end of the detection fiber closest to the detector, were S420, LDS698, R6G, C500, S420, LDS698, C500, S420, LDS698, and C500. Since the spacing of the regions was of the order of the fluorescence lifetimes of these dyes, full temporal resolution of the fluorescence signals was not possible. However, by employing wavelength-selective devices, the signals from the individual regions could be distinguished. Using the 0.5 m focal length monochromator for the spectral decomposition of the light exiting the detection fiber and averaging 128 of the resulting traces yielded the surface plot shown in Figure 15.16, in which the emission of the fiber is plotted as a function of wavelength and time of detection. Although eight peaks are clearly discernible in this figure, in fact, all 10 sensor regions are represented.

These results show that even when the sensor regions are closely spaced along the fiber that provides the optical delay, they may still be spectrally resolved. It should be noted that the fluorescent signal from one region could be reabsorbed by a different dye in a subsequent region [e.g., the S420 signal from regions 5 and 8 passes two regions containing

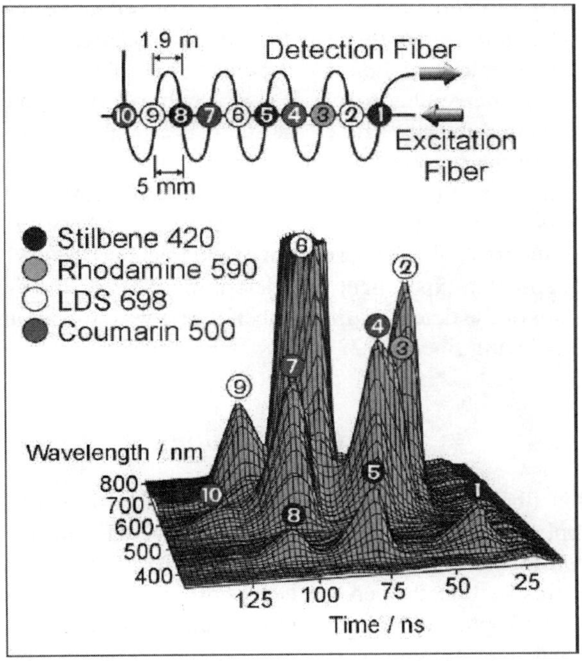

FIGURE 15.16. Spectrally and temporally resolved response of a 10-sensor array. (Adapted from ref. 50. Copyright 2002 Association for Laboratory Automation.)

LDS698 (2 and 6) which can be excited in this way]. The significance of reabsorption processes for the design of large fiber arrays requires further study.

DISCUSSION

Feasibility of the Two-Fiber Scheme

Our results demonstrate clearly that the two-fiber scheme can overcome the problem of limited spatial resolution found in applications involving fluorescent regions on optical fibers. Although in our studies these regions were separated by 5–10 mm, a further reduction is possible. A practical limit is given by the diameter of the fiber providing the optical delay; the penetration depth of the evanescent waves imposes a theoretical limit. As long as the fiber junctions are separated by approximately 5λ, the probability that the fluorescence from one such region is picked up by two regions on the detection fiber or that one sensor region is excited repeatedly from two regions on the excitation fiber is negligible.

In the one-fiber fluorescent OTDR scheme, two evanescent coupling steps are involved: transfer of exciting light and capture of emitted light. In contrast with the one-fiber scheme, however, the two-fiber scheme offers more flexibility since fiber parameters (like core diameter, refractive indices, numerical aperture) can be optimized separately for each of the excitation and detection fibers. This flexibility is particularly important for large arrays of fluorescent regions on fibers. The magnitude of the losses in the exciting fiber can be controlled to some extent by the choice of refractive indices. Likewise, the detection fiber can be optimized in this way for maximum capture efficiency. Furthermore, the values of E_0 [Eqn (5)] and d_p [Eqn (6)] depend on the fiber parameters and the particular core mode [7]. The range of core modes can be controlled to some extent by the cone angle of the light focused into the end of the fiber. The total amount of light that a fiber accepts increases with the number of guiding modes that it can support, which again is directly related to the basic fiber parameters [51]. For large arrays of fluorescent regions, the optical transmission characteristics of the delaying fiber have to be considered. The spectral position of the transmission maximum of the fiber and the excitation and emission wavelengths in a particular experiment will determine whether the delaying fiber is best used as the excitation or the detection fiber. Fiber clarity will impose an upper limit on the total length of the delaying fiber [21,22].

Attenuation Studies

The excitation fiber ideally has to provide equal intensities to each sensor region. Even if there is only evanescent coupling, a gradual decrease in the amount of light in the fiber, and therefore the extent of sensor region excitation, is expected along the fiber. To minimize this decline, the percentage loss of light from the excitation fiber at each sensor region must be kept small. Based on the results presented above, there are a number of initiatives that may be used to approach the characteristics of an ideal excitation fiber.

It has been shown [36,40] that the percentage of the total power within a fiber that is carried by the collective evanescent fields decreases as the fiber becomes more strongly guiding. That is, as the number of modes that a fiber can support increases, the percentage of light available for coupling through the evanescent fields decreases. Thus, by choosing very strongly guiding fibers for excitation, only a small proportion of the modes (those propagating close to the critical angle) would be available for evanescent coupling at each sensor region.

The efficiency of evanescent coupling from the excitation fiber could also be controlled by selectively populating a subset of its modes. For example, if only modes propagating at angles far from the critical angle are populated in a strongly guiding fiber [51,52], their evanescent penetration depths will be small and there will be small losses from the fiber at each sensor region. Alternatively, if the refractive index of the sensor regions (n_r) is smaller than that of the excitation fiber cladding ($n_{cl,ex}$), modes propagating close to the critical angle when launched into the fiber will be far from the critical angle in the sensor regions, and consequently these modes will experience smaller losses at each sensor region. In all cases, the more light that the excitation fiber can initially carry, the better, and so large-diameter fibers are preferred for the excitation fiber.

A different picture arises for the detection fiber. The light collection efficiency of a fiber is proportional to $N^{1/2}$ for homogeneously distributed light sources in the cladding and proportional to N for sources located directly at the core/cladding interface [35]. Therefore, for the detection fiber, increasing the fiber core radius r and/or $(n_{co}^2 - n_{cl,det}^2)^{1/2}$ {see Eqns (3) and (4) [5]} leads to increased collection efficiency and therefore to larger signals in the detection fiber. However, the consequence of increasing N by modifying the core and cladding refractive indices is increased dispersion [see Eqn (11)], which leads to a reduction of the spatial resolution.

Sensitivity

The sensitivity of the evanescent fiber–fiber coupling scheme is remarkable considering that the cladding of the fibers used in our experiments produced a background fluorescent signal. R6G concentrations of 10^{-7} mol L^{-1} could easily be detected against this background. By choosing a fiber with a silica cladding, more transparent to the UV excitation source, or by using a longer-wavelength light source for excitation, this background luminescence should be effectively eliminated, leading to a significant improvement in the signal-to-noise ratio (S/N) and therefore an enhancement of the sensitivity. It is also worthy of note that unlike the previous experiments of Kvasnik and co-workers [43] and Chronister and co-workers [44], which utilized a single fiber for both excitation and detection, and probed sensor regions which spanned ~25 cm and ~4 cm, respectively, of the fiber, the sensor regions probed here are orders of magnitude smaller. In this case the sensor regions encompass only the volume, which lies within ~5λ of both of the two fibers at the fiber junction, defined by the penetration depths of the modes of the two fibers. It is thus, all the more remarkable that this technique, in which such small volumes are probed, is as sensitive as we have determined here. For the data collected using setup 1, all detectable contributions to the noise were stochastic in nature. Thus, although the points plotted in Figure 15.9 were the result of averaging 10^3 traces, the LeCroy DSO has the capability of averaging 10^6 traces, providing the potential for a further 30-fold increase in S/N. At the present time such an experiment is not feasible because of the slow repetition rate of the laser (recording 10^6 traces at 10 Hz would take ~28 hr). Employing an excitation source with higher pulse repetition rates will further enhance the sensitivity of this readout scheme.

Refractive Index Dependence

Changing the refractive index of the substance replacing the cladding has a profound influence on the signals. Not only are the overall intensities affected, but also the relative intensities of the signals of the individual regions. Clearly, the losses that occur at the sensor regions under refractive conditions will prohibit the creation of large arrays of sensors. The case pictured in Figures 15.13 and 15.14 allows for a quantitative interpretation, since all junctions are immersed in the same dye solution. It shows that for a fiber in which the original fiber cladding is retained outside the fiber junctions, the refractive index of the substance replacing the cladding (e.g., a solution in the above case) has to be equal to or less than the refractive index of the original cladding for minimum attenuation. This is demonstrated by the results shown in Figure 15.12, which were obtained for such a case.

Fiber Dispersion

Dispersion of light pulses will reduce the attainable spatial resolution. We measured the dispersion of a pulse propagating through a 500 m fiber and found it to be significantly less than theoretically predicted. The theoretical expression [Eqn (11)] was calculated by comparing the time for light to travel along the maximum and minimum paths in a straight section of the fiber core. Therefore the calculated bound-mode dispersion is reached when all bound fiber modes are populated. In our case, however, the signal pulse enters the fiber from the cladding through evanescent coupling. The penetration depth of the evanescent field of a mode of the capture fiber depends on the mode itself (characterized by the incident angle Θ). Light emitted in the cladding (or in the sensor region) is preferentially captured by modes close to the critical angle Θ_z, since these evanescent fields extend farthest from the core into the cladding [see Eqn (6)]. This selective mode population could lead to reduced dispersion. Consequently, the extent of observed dispersion is not exclusively a property of the fiber itself, but also of the conditions under which the light is captured. This affords the opportunity to control pulse spreading, at least to some extent, by varying the geometry of the fiber–fiber junction. To minimize the pulse dispersion, the separation of the fibers may be increased, displacing the detection fiber from the fluorescent sensor molecules so that there is a greater prejudice towards the population of modes propagating close to the critical angle. However, an increased separation of the two fibers will lead to smaller signal intensities.

Of course, other factors could also have contributed to the lower than expected dispersion we have observed. Variations in the core diameter of the fiber will reduce the dispersion; however, in order to produce the observed reduction in the dispersion, these variations would have to be quite large. At this point, in the absence of data on the quality of the fiber, this contribution is hard to quantify, although it is anticipated that it will result only in a small dispersion reduction.

An additional effect could be due to the fact that majority of the 500 m long detection fiber was coiled on a cylinder with a radius of 10 cm. A bent fiber cannot sustain any truly bound modes [5], meaning that coiling of the fiber leads to mode-dependent losses of light intensity in the fiber which, in turn, could affect the measured pulse width. However, calculations have shown that the bending effect is neglible if the ratio of the bend radius to the radius of the fiber core (here 100 μm) is 10^3 or above [5], which is the case for our setup. We verified this prediction experimentally by coiling the fiber on a circular support with a radius of 60 cm. No change in the width of the signal peak was observed.

SUMMARY AND OUTLOOK

The aim of this work is to demonstrate a high spatial resolution readout scheme for combinatorial libraries built in the cladding of optical fibers. Assaying such a library with fluorescent compounds results in an array of fluorescent sensor regions which can be optically evaluated using fiber-optic detection methods. The fluorophores are probed using the evanescent fields of the light pulses propagating in the fiber core. The time delay between exciting laser pulses and fluorescence pulses returning through the fiber core to the detector provides the location of the emitter along the fiber. In the one-dimensional combinatorial chemistry scheme, the location of a compound on the fiber encodes its complete synthetic history. The work presented in this chapter demonstrated that the spatial resolution (and hence the packing density of compounds on the fiber) can be increased dramatically by using a second fiber as an optical delay. We have furnished the proof of principle for this scheme and determined the parameters that allow for the readout of large arrays. The future will bring the full implementation of the one-dimensional combinatorial chemistry scheme on optical fiber supports. For the implementation a material is needed that can act (1) as support for the chemical reactions leading to the combinatorial library and (2) as a fiber cladding, preserving the guiding conditions of the fiber. Efforts towards this goal and first results are described in Chapter 6.

ACKNOWLEDGMENTS

It is a pleasure to thank the members of our groups for their dedicated work. In the Geissinger group the main contributor to this work was Dr Barry Prince, with additional contributions from Dr Nadejda Kaltcheva and two undergraduate students, Janelle Scharon and Tami Zembrowski. In the Schwabacher group, this work was performed primarily by graduate students Christopher Johnson and Yixing Shen, and by Dr Maureen Prince. Contributions by undergraduates Michael Kordaszewicz, and Angela Eder are also gratefully acknowledged. This material is based upon work supported by the National Science Foundation under Grants CHE-9874241 (AWS) and CHE-0078895 (P.G.). P.G. would also like to acknowledge support from the University of Wisconsin-Milwaukee Campus Opportunity Fund.

REFERENCES

1. Lam, K. S., Lebl, M., Krchnák, V. The "one-bead-one-compound" combinatorial library method. *Chem. Rev.* **1997**, *97*, 411–448.
2. Pirrung, M. C. Spatially addressable combinatorial libraries. *Chem. Rev.* **1997**, *97*, 473–488.
3. Schwabacher, A. W., Shen, Y., Johnson, C. W. Fourier transform combinatorial chemistry. *J. Am. Chem. Soc.* **1999**, *121*, 8669–8670.
4. Marcuse, D. *Theory of Dielectric Optical Waveguides*, 2nd ed., **1999**, Boston, MA: Academic Press.
5. Snyder, A. W., Love, J. D. *Optical Waveguide Theory* **1983**, London: Chapman & Hall.
6. A second set of boundary conditions that is required by Maxwell's equations, namely the continuity of the normal components of the electric and magnetic field at the interface, is automatically satisfied for time-varying fields as found in an electromagnetic wave.
7. Harrick, N. J. *Internal Reflection Spectroscopy* **1967**, New York: Wiley–Interscience.
8. Harrick, N. J. Electric field strengths at totally reflecting interfaces. *J. Opt. Soc. Am.* **1965**, *55*, 851–857.
9. Newton, I. *Opticks*, vol. 3, 2nd ed., **1717**.
10. Quincke, G. Optische Experimental Untersuchungen. Über das Eindringen des total reflektierten Lichtes in das dünnere Medium. *Ann. Phys. (Leipzig)* **1866**, *127*, 1–29.
11. Quincke, G. Optische Experimental-Untersuchungen. II. Über die elliptische Polarisation des bei totaler Reflexion eingedrungenen oder zurück-geworfenen Lichtes. *Ann. Phys. (Leipzig)* **1866**, *127*, 199.
12. Hall, E. E. The penetration of totally reflected light into the rarer medium. *Phys. Rev.* **1902**, *15*, 73–106.
13. Sélényi, P. Sur l'existence et l'observation des ondes lumineuses sphériques inhomogènes *C.R. Acad. Sci. Paris* **1913**, *157*, 1408.
14. Schaefer, C., Gross, G. Untersuchungen über die Totalreflexion. *Ann. Phys. (Leipzig)* **1910**, *32*, 648–672.
15. Drexhage, K. H. Monomolecular layers and light. *Sci. Am.* **1970**, *222*, 108–119.
16. Goos, F., Hänchen, H. Über das Eindringen des total reflektierten Lichtes in das dünnere Medium. *Ann. Phys. (Leipzig)* **1943**, *43*, 383–392.
17. Goos, F., Hänchen, H. Ein neuer und fundamental Versuch zur Totalreflexion. *Ann. Phys. (Leipzig)* **1947**, *1*, 333.
18. Goos, F., Linberg-Hänchen, H. Neumessung des Strahlversetzungseffektes bei Totalreflexion. *Ann. Phys. (Leipzig)* **1949**, *5*, 251.
19. Renard, R. H. Total reflection: A new evaluation of the Goos–Hänchen shift. *J. Opt. Soc. Am.* **1964**, *54*, 1190–1197.
20. Goos and Hänchen [17, 18] furnished a proof for the existence of evanescent fields that does not suffer from this problem. They found that light rays undergoing total internal reflection are displaced, as if the rays were to enter the rarer medium "temporarily." It is interesting to note that Newton had already suggested that the "beam path is a parabola with the vertex in the rarer medium" [9].
21. MacChesney, J. B., DiGiovanni, D. J. Materials development of optical fiber. *J. Am. Ceram. Soc.* **1990**, *73*, 3537–3556.
22. Thomas, G. A., Shraiman, B. I., Glodis, P. F., Stephen, M. J. Towards the clarity limit in optical fibres. *Nature* **2000**, *404*, 262–264.
23. Culshaw, B. Basic concepts of optical fiber sensors. In B. Culshaw and J.P. Dakin (Eds.), *Optical Fiber Sensors: Systems and Components*, Vol. 1, 1988, Norwood, MA: Artech House, pp. 9–24.
24. Paul, P. H., Kychakoff, G. Fiber-optic evanescent field absorption sensor. *Appl. Phys. Lett.* **1987**, *51*, 12.
25. Blair, D. S., Burgess, L. W., Brodsky, A. M. Study of analyte diffusion into a silicone-clad fiber-optic chemical sensor by evanescent wave spectroscopy. *Appl. Spectrosc.* **1995**, *49*, 1636–1645.
26. Egami, C., Takeda, K., Isai, M., Ogita, M. Evanescent wave spectroscopic fiber optic pH sensor. *Opt. Commun.* **1996**, *122*, 122–126.
27. Gupta, B. D., Khijwania, S. K. Experimental studies on the response of the fiber optic evanescent field absorption sensor. *Fiber Integr. Opt.* **1998**, *17*, 63–73.
28. Dakin, J. P., Pratt, D. J. Fibre-optic distributed temperature measurement-a comparative study of techniques. *Proc. IEE Colloquium on Distributed Optical Fibre Sensors*, **1986**, *London*: The Institute of Electrical Engineers, pp. 10/11–16.
29. Culshaw, B., Davies, D. E. N., Kingsley, S. A. Acoustic sensitivity of optical-fibre waveguides. *Electron. Lett.* **1977**, *13*, 760–761.

30. Fields, J. N. Coupled waveguide acoustooptic hydrophone. *Appl. Opt.* **1979**, *18*, 3533–3534.
31. DeGrandpre, M. D., Burgess, L. W. Long path fiber-optic sensor for evanescent field absorbance measurements. *Anal. Chem.* **1988**, *60*, 2582–2586.
32. Vo-Dinh, T., Nolan, T., Cheng, Y. F., Alarie, J. P., Spaniak, M. J. A fiber optic antibody-based biosensor with time-resolved detection. Proc. *Conf. on Chemical, Biochemical, and Environmental Fiber Sensors*, **1989**, *Boston, MA*: The International Society for Optical Engineering (SPIE) pp. 266–272.
33. Hale, Z. M., Payne, F. P. Demonstration of an optimized evanescent field optical fibre sensor *Anal. Chim. Acta* **1994**, *293*, 49–54.
34. Lieberman, R. A., Blyler, L. L., Cohen, L. G. A Distributed fiber optic sensor based on cladding fluorescence. *J. Lightwave Technol.* **1990**, *8*, 212–220.
35. Marcuse, D. Launching light into fiber cores from sources located in the cladding. *J. Lightwave Technol.* **1988**, *6*, 1273–1279.
36. Christensen, D., Andrade, J., Wang, J., Ives, J., Yoshida, D. Evanescent-wave coupling of fluorescence into guided modes: FDTD analysis. *Proc. Conf. on Chemical, Biochemical, and Environmental Fiber Sensors*, **1989**, *Boston, MA*: The International Society for Optical Engineering (SPIE) pp. 70–74.
37. Ueno, Y., Shimizu, M. An optical fiber fault location method. *IEEE J. Quantum Electron.* **1975**, *QE-11*, 77D-78D.
38. Ueno, Y., Shimizu, M. Optical fiber fault location method. *Appl. Opt.* **1976**, *15*, 1385–1388.
39. Personick, S. D. Photon probe—An optical-fiber time-domain reflectometer. *Bell Syst. Tech. J.* **1977**, *56*, 355–366.
40. Barnoski, M. K., Jensen, S. M. Fiber Waveguides: A novel technique for investigating attenuation. *Appl. Opt.* **1976**, *15*, 2112–2115.
41. Barnoski, M. K., Rourke, M. D., Jensen, S. M., Melville, R. T. Optical time domain reflectometer. *Appl. Opt.* **1977**, *16*, 2375–2379.
42. Kharaz, A., Jones, B. E. A distributed fiber optic sensing system for humidity measurements. *Meas. Control* **1995**, *28*, 101–103.
43. Kvasnik, F., McGrath, A. D. Distributed chemical sensing utilising evanescent wave interactions. *Proc. Conf. on Chemical, Biochemical, and Environmental Fiber Sensors* 1989, *Boston, MA*: The International Society for Optical Engineering (SPIE), pp. 75–82.
44. Browne, C. A., Tarrant, D. H., Olteanu, M. S., Mullens, J. W., Chronister, E. L. Intrinsic sol–gel clad fiber-optic sensors with time-resolved detection. *Anal. Chem.* **1996**, *68*, 2289–2295.
45. Dakin, J. P. Distributed optical fiber sensors. *Proc. Conf. on Distributed and Multiplexed Fiber Optic Sensors II*, 1992, *Boston, MA*: The International Society for Optical Engineering (SPIE), pp. 76–108.
46. Prince, B. J., Schwabacher, A. W., Geissinger, P. A readout scheme for closely packed fluorescent chemosensors on optical fibers. *Anal. Chem.* **2001**, *73*, 1007–1015.
47. Blyler, L. L., Lieberman, R. A., Cohen, L. G., Ferrara, J. A., MacChesney, J. B. Optical fiber chemical sensors utilizing dye-doped silicone polymer claddings. *Polym. Eng. Sci.* **1989**, *29*, 1215–1218.
48. Gloge, D. Dispersion in weakly guiding fibers, *Appl. Opt.* **1971**, *10*, 2442–2445.
49. Prince, B. J., Schwabacher, A. W., Geissinger, P. Fluorescent fiber-optic sensor arrays probed utilizing evanescent fiber–fiber coupling. *Appl. Spectrosc.* **2001**, *55*, 1018–1024.
50. Prince, B. J., Schwabacher, A. W., Geissinger, P. An optical readout scheme providing high spatial resolution for the evaluation of combinatorial libraries on optical fibers. *J. Assoc. Lab. Autom.* **2002**, *7*, 66–73.
51. Gloge, D. Weakly guiding fibers. *Appl. Opt.* **1971**, *10*, 2252–2258.
52. Payne, F. P., Hale, Z. M. Deviation from Beer's law in multimode optical fibre evanescent field sensors. *Int. J. Optoelectron.* **1993**, *8*, 743–748.

Analysis of Mechanical, Electrical, and Thermal Properties

Chapter 16

Diffusion Multiples for High-Efficiency Alloy Design

J.-C. Zhao, L. A. Peluso, L. N. Brewer, and M. R. Jackson

INTRODUCTION

Unlike functional materials, which most often are used for one specific physical or chemical property, structural materials often need to satisfy simultaneously several different properties such as strength, ductility, modulus, fracture toughness, environmental resistance, etc. In addition, many structural materials undergo several different processes such as casting, forging, rolling, and the like before they are used in the final product. Thus it is not very practical to "screen" for structural materials directly. Instead, high-efficiency approaches can be used to map critical properties and data to accelerate the design of structural materials.

The strength of structural alloys is controlled by dislocation mobility and it comes predominately from three sources: solution strengthening, precipitation/dispersion strengthening, and microstructural strengthening. The first case involves atoms of different sizes distorting a crystal lattice, thus making dislocations more difficult to move. Alloy designers would like to know which elements are more effective than others in solution strengthening and by what amount. Currently, predictions from first principles are still not up to the challenge. Experimental mapping of the solution-strengthening effects would be very useful for alloy design, especially for new alloy systems. The second case involves in situ formation (precipitation) or intentional addition (dispersion) of second-phase particles in a crystalline matrix to slow/block dislocation movements. In this case, it is critical to know the stability and the nucleation and growth kinetics of the particles. To predict these properties requires phase diagrams and diffusion coefficients. Phase diagrams are also very useful to guide alloy designers to promote phases which have been identified as beneficial precipitates and to avoid detrimental ones. The third case involves microstructural imperfections such as dislocation tangles/pile-ups induced by alloy processing (such as cold rolling). The grain-size effect (the Hall–Petch effect) is an example of this type of strengthening. Microstructural strengthening is accomplished by

J.-C. Zhao, L. A. Peluso, L. N. Brewer, and M. R. Jackson • General Electric Company, GE Global Research Center, P.O. Box 8, Schenectady, NY 12301, USA
Contact author: J.-C. Zhao, email: zhaojc@crd.ge.com

High-Throughput Analysis, edited by Potyrailo and Amis
Kluwer Academic Publishers, New York, 2003.

processing and is not a necessary part of the alloy composition design. Therefore phase diagrams, solution-strengthening effects, and diffusion coefficients are among the very basic information required for alloy design.

This chapter describes the diffusion-multiple approach for efficient mapping of phase diagrams, solution-hardening effects, diffusion coefficients, and modulus variation with composition and phases. The emphasis is placed on explaining the detailed microanalytical probes that make efficient mapping possible and the detailed process of obtaining data from the diffusion multiples. The authors hope to show colleagues in other fields how to use tools such as electron probe microanalysis (EPMA), electron backscatter diffraction (EBSD), and nanoindentation to screen composition, crystal structure, and elastic modulus in other materials.

DIFFUSION MULTIPLE FABRICATION

The diffusion multiple shown in Figure 16.1 was made by cutting a slot of 1.8 mm width and 12.7 mm length from a 25 mm diameter pure chromium (Cr) disk of 3 mm thickness. Pure palladium (Pd), platinum (Pt), and rhodium (Rh) foils of 0.25 mm thickness are arranged in the bricklaying geometry shown in Figure 16.1 and put into the slot in the chromium disk along with a pure ruthenium (Ru) piece with two steps on it. The Ru piece had a thickness of 1 mm on one side and 0.5 mm on the other. Two pure Cr disks (without the slot) of 25.4 mm diameter and 3.2 mm thickness were placed on top and bottom of the slotted Cr disk containing all the precious metals. The assembly was then placed in a pure titanium (Ti) tube of 25.5 mm inner diameter and 32 mm outer diameter and 9.6 mm height. Two Ti disks of 32 mm diameter and 3.2 mm thickness were placed on the top and the bottom of the Ti tube filled with the three Cr disks. The assembly was welded using an electron beam in vacuum along the top and bottom outer circular edges of the Ti tube. The welded outer Ti tubes and caps served as a hot isostatic pressing (HIP) can. The whole assembly then underwent a HIP run of 1200°C at 200 MPa for 4 hr. The diffusion multiple was further annealed at 1200°C for 36 hr in a vacuum furnace, making the total diffusion exposure time 40 hr at 1200°C. The Cr disks and the Ti HIP can serve as protection for the precious metals from any interstitial contamination during heat treatment. The diffusion time was chosen to develop diffusion profiles measurable over lengths of $\sim$100 μm, so that neighboring diffusion sources would not interfere with the couple or triple being evaluated.

The annealed diffusion multiple was then cut into halves parallel to the two Ti caps and at the midpoint of the thickness direction. The sample was then ground and polished for EPMA, EBSD, and nanoindentation tests. Nanoindentation was performed using a Hysitron (Minnesota, USA) instrumented indenter. All results reported in this chapter were obtained from this single sample.

The bricklaying pattern allows the formation of 10 different diffusion triples in this diffusion multiple, as shown in Figure 16.1 highlighted with circles. The Cr–Pt–Ru tri-junction is shown in Figure 16.2. Based on the binary phase diagrams, there should be one intermetallic compound (Cr_3Pt with an A15 crystal structure) formed between Cr and Pt, and also one compound (σ with a tetragonal crystal structure) formed between Cr and Ru. There should be no compound between Pt and Ru, just the fcc–hcp equilibrium. The scanning electron microscopy (SEM) backscatter electron (BSE) image clearly shows

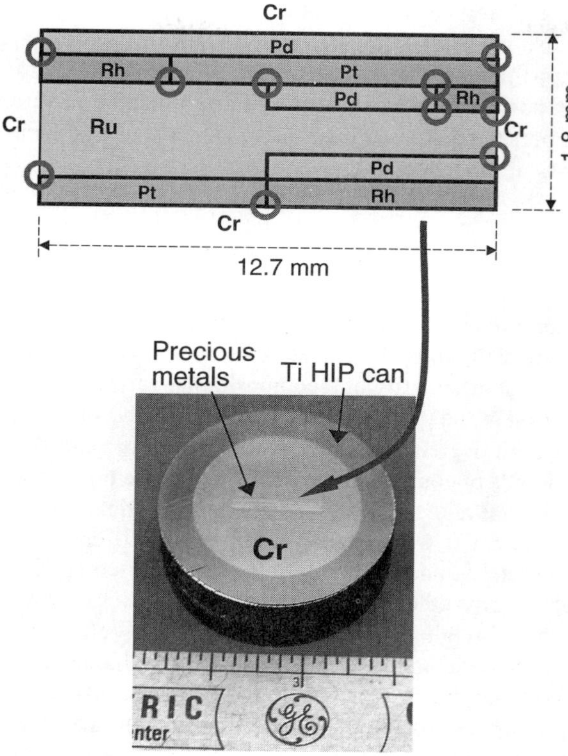

FIGURE 16.1. The size and geometry of the diffusion multiple. The thickness of the Pd, Pt, and Rh layers is 0.25 mm. The thickness of the Ru layer is 1 mm on one side and 0.5 mm on the other.

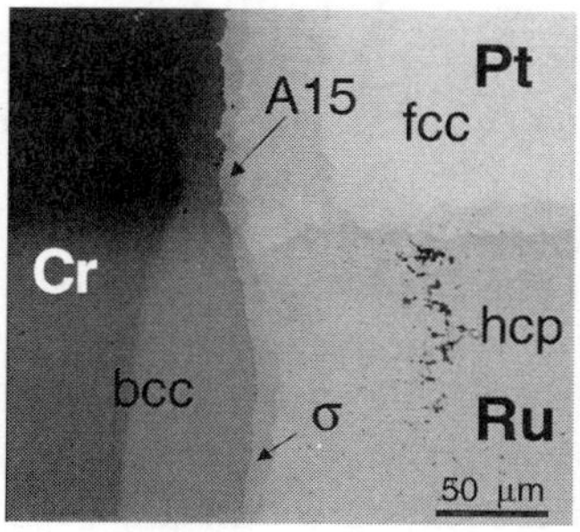

FIGURE 16.2. Backscatter electron image of the Cr–Pt–Ru tri-junction of the diffusion multiple showing the formation of the A15 and σ phases due to interdiffusion of Pt, Ru, and Cr.

the A15 (Cr_3Pt) and the σ phase and a clear fcc–hcp phase boundary. However, it is difficult to locate the phase boundary between the A15 and the σ phase using SEM BSE image alone. We will show how we can combine the EPMA and EBSD techniques to identify the phase boundaries and to obtain the tie-line information to construct the ternary phase diagram. In addition to the formation of all the intermetallic compounds, the diffusion multiple also creates a complete variation of all the single-phase compositions [1,2]. These compositions will allow us to map solution-hardening effects and modulus–phase-composition relationships. A brief introduction to various microanalytical techniques will be presented first to give the reader a sense of their spatial resolution, accuracy, and analytical procedures.

MICROANALYSIS

In choosing analytical techniques for high-throughput materials characterization, several factors are especially important: spatial resolution, automation, and spatial registry between techniques. The granularity of a compositional variation will be controlled either by the spatial resolution of the processing technique or by the spatial resolution of the analytical technique. A high degree of analytical automation is required for efficient characterization of the hundreds of compositional points and phase boundaries present in a single diffusion multiple. Spatial registry between techniques is equally important. To characterize a sample and connect it with physical properties, different analytical techniques must be able to accurately and reliably address the same compositional (*x*–*y*) point on a diffusion multiple or any other compositional library. Below we will discuss the use of EPMA, EBSD and imaging, and nanoindentation as characterization techniques capable of meeting these demands as high throughput characterization tools. Each of these techniques is already highly developed and commercially available. While we will review the basic principles of each technique, the emphasis will be on the use of the technique for diffusion-multiple studies with the precious metal diffusion multiple as a guiding example.

Microprobe

The electron probe microanalyzer is a powerful instrument for the microchemical analysis of inorganic materials. EPMA is a technique capable of chemically analyzing solid, vacuum-compatible materials with high spatial resolution and sensitivity. The volume excited under typical electron microprobe conditions is on the order of a cubic micron, which is equivalent to a sample size of a few picograms. EPMA can detect elements spanning most of the periodic table, from beryllium to uranium, with typical detection limits of 0.1 at % and down to a few parts per million in select cases. It can provide accuracy of less than 1% relative for most major sample components, and 3–5% relative for minor components, depending upon matrix effects. Most electron microprobe instruments operate in the 10^{-7}–10^{-6} torr range, which allows more rapid sample throughput than ultrahigh vacuum (UHV) instruments. Sample preparation requirements for EPMA are less stringent than those for many other instruments. EPMA can be performed on all conductive materials and on most nonconductive materials with the application of a thin conductive coating (e.g., carbon), which can be done with resistive evaporation or by common sputter coaters, to provide a pathway for electrons away from the electron beam–sample interaction area.

EPMA utilizes a finely focused high-energy electron beam which bombards a preselected location on a solid specimen. The resulting emission includes backscattered primary electrons, secondary electrons, low-energy photoelectrons, Auger electrons, and X-ray and optical photoemission (Figure 16.3). In EPMA, the radiation of interest is the characteristic X-ray emission. The specific X-ray energies (or wavelengths) are characteristic of the particular element which was initially excited. This phenomenon allows correlation of the measurement of characteristic X-rays to the composition of a probed region on a specimen.

Detection of X-rays and subsequent identification of elements present in a specimen can be carried out by solid state energy-dispersive spectrometry (EDS) or wavelength-dispersive spectrometry (WDS). WDS is preferred over EDS for data acquisition in EPMA for several reasons, including a much better energy resolution and higher peak X-ray count rates and peak-to-background ratios, which allow much lower minimum detection limits.

The basis for modern electron microprobe analysis was established in 1951 by Castaing in his Ph.D. thesis in which he set out its fundamental principles and its realization as a metallurgical tool [3]. He outlined a method for obtaining a quantitative X-ray analysis from a micrometer-sized region of a solid specimen. This was achieved by measuring the ratio K of characteristic X-ray emission intensity I of element A from the specimen to that of a pure elemental standard. This ratio was shown to be proportional, to a first approximation, to the composition C of element A in the analyzed area:

$$K_{\mathrm{A}} = \frac{I_{\mathrm{A,\ specimen}}}{I_{\mathrm{A,\ standard}}} = f(C_A) \tag{1}$$

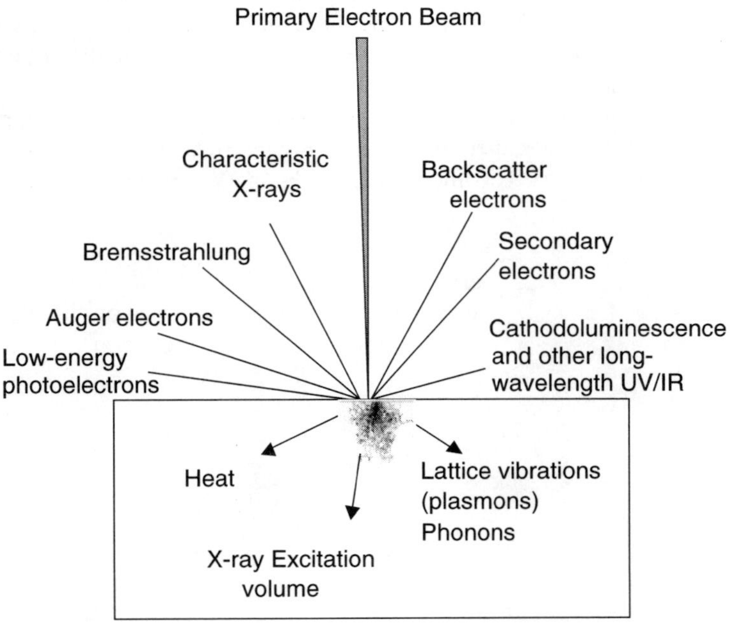

FIGURE 16.3. Schematic diagram showing the primary electron beam–sample interaction and generation of various signals from the sample.

Castaing's second-order approximations employed an empirical coefficient *ZAF* to account for certain matrix-dependent effects [4]. These physical phenomena, which significantly attenuate or amplify the characteristic X-rays generated within a specimen, include the electron scattering/atomic number effect *Z*, absorption *A* within the sample, and the fluorescence effect *F*.

Improvements in computer hardware and continued development of algorithms which model these effects have greatly increased the accuracy of quantitative EPMA. We can now simulate the electron trajectories in a single- or multielement sample to calculate the range of electrons along multiple-scattering events as a function of many variables such as incident beam energy and average atomic number. These Monte Carlo calculations [5] enable us to predict the energy distribution of the incident beam within the specimen, and the transfer of energy on the atomic level. From these range simulations, we can also model continuum and characteristic X-ray production. This ability is very helpful to the analyst in the preparation of an analysis traversing multiple varying composition regimes.

The resultant X-ray excitation volume exceeds that of the electron-scattering volume. We can see from Figure 16.4 that significant opportunity for X-ray production occurs for chromium Kα well below the surface, but less so for platinum. However, since only the emitted X-ray signal is measured, we must correct for the absorption of X-ray signal by the matrix (effectively reducing the volume sampled) as well as the "secondary" X-ray signal created by primary X-ray fluorescence (increasing the volume sampled). These and other factors result in an X-ray production density which is not constant with interaction volume [6].

Correction procedures for these phenomena have evolved for decades, and are still undergoing refinement [7]. Some of the most useful ones today are based upon $\phi\rho z$ distributions from various samples which carefully define the distribution of characteristic X-ray productions with depth.

The depth beneath the sample surface is often expressed as ρZ which represents mass thickness. From the simple depiction shown in Figure 16.5, we can see that the largest proportion of X-rays generated come from a fairly shallow region relative to the entire

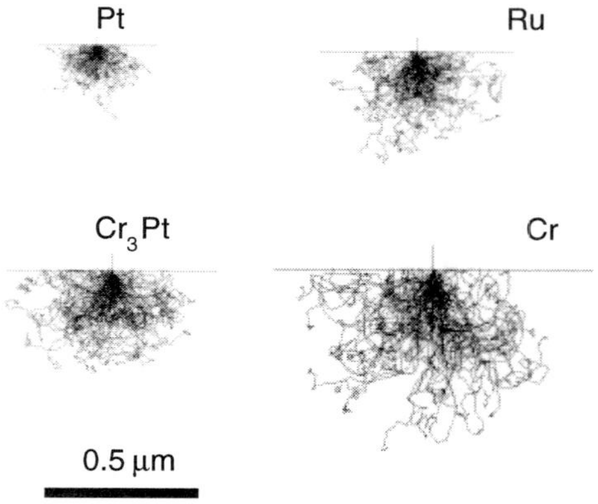

FIGURE 16.4. Monte Carlo simulation results of 15 keV electron trajectories at an 90° incident angle.

X-ray excitation volume. These $\phi(\rho Z)$ curves are widely used models which have been important in the calculation of X-ray intensities for multielement systems, (Figure 16.5).

The fundamentals of this analytical technique devised over a half-century ago remain intact. Great improvements in mechanical hardware have allowed precise multielement compositional profiling. The ability to move a specimen accurately in very exact and uniform increments while measuring characteristic X-ray intensities of many elements at each interval has advanced its value as a tool for materials science. The development of optically encoded WDS spectrometers allows multiple-element analysis of both peak and background X-ray signals on a single spectrometer with excellent reproducibility of positioning. The invention of synthetic multilayer "pseudocrystals" and anticontamination systems has improved the detection of light elements by orders of magnitude. Recently developed large-area crystals, such as lithium fluoride (LiF), pentaerythritol (PET), and thallium acid phthalate (TAP), offer microprobe operators huge count rate advantages, thus allowing productivity gains when applied to multielement composition profiling. These large count rates can indirectly help to improve spatial resolution in some cases by allowing the use of a lower beam current to achieve optimized X-ray counting statistics.

EPMA is naturally well suited to achieving efficient and accurate composition profiles with high spatial resolution. The marriage of high-speed computer automation and very precise sample stage and X-ray spectrometer hardware has provided a powerful analytical tool for high-throughput approaches to materials research. The microprobe work done in mapping of phase diagrams and in phase equilibria determination in the present study was performed on a Cameca SX50 with four vertical wavelength-dispersive spectrometers and a fully integrated PGT EDS system automated by a Sun Ultra1 Unix

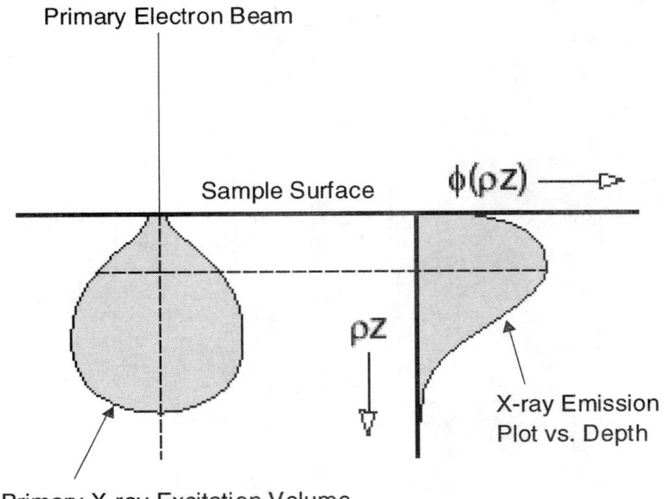

FIGURE 16.5. The $\phi(\rho Z)$ curves. Near the surface of the sample, X-ray emission is enhanced by scattered electrons from below. X-ray intensity continues to increase with depth due to scattering of the incident electrons which increases the mean path length per unit depth. As electron paths become random, the X-ray intensity per unit depth begins to decrease due to electron scattering and energy loss.

workstation (Figure 16.6). Modern EPMA systems are usually equipped with a motorized stage that has a positional precision of 0.5 μm in the *x*, *y*, and *z* directions, and can move over a range of several centimeters. Such a system allows automated data collection by logging the EPMA line scan positions ahead of time into an executable file and allows the instrument to collect data without the presence of the operator.

Data reduction (from X-ray intensity ratios to composition in weight percent) was typically performed using PAP, developed by Pichoir and Pouchou [8], a modified $\phi(\rho Z)$ program and good general model for a wide range of X-ray energies (100 eV to over 10 keV) and accelerating voltages (1–40 keV) which often accounts for the depth distribution of energies better than common matrix correction procedures (*ZAF*).

It is very important to achieve very high throughput EPMA while maintaining acceptable accuracy and precision in the resultant compositions measured. In measuring large numbers of phases and intermetallic compounds throughout this work with compositions varying only slightly among some of them, we need careful selection of the analysis parameters that would give the most productive results. A large effort is placed in both qualitative and quantitative examination of these diffusion multiples in advance of more thorough quantitative profiling in order to balance the accuracy needed with the total

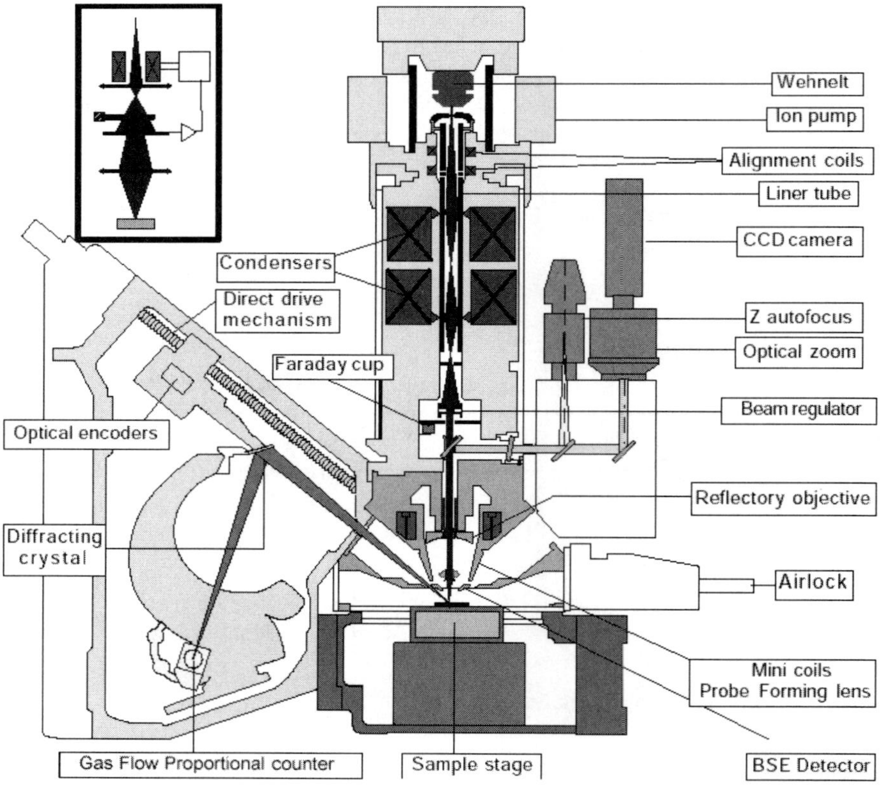

FIGURE 16.6. Schematic diagram of the Cameca SX50 electron microprobe. (Courtesy of Cameca Inc., France.)

acquisition time during analyses of on the order of 10^3–10^4 data points. Typical EPMA analyses were acquired in about 1 min per point.

The main procedures of automating the electron microprobe effort can be broken down into four activities: identify, screening, analysis design, and acquisition. The "identify" activity involves a combination of light optical and backscatter electron imaging coupled with some form of qualitative analysis to distinguish meaningful regions from artifact (topographic or crystallographic differences). Typically the multiple is given light optical inspection after the metallographic preparation. Next, secondary electron or backscattered electron micrographs are taken to identify the best sections within a given diffusion multiple for further evaluation from average atomic number contrast. In many cases, more than one section of a specific multiple needs to be analyzed because of porosity, cracking of intermetallic compounds during sample preparation, or other deleterious effects along a particular binary or ternary region in a given metallographic section.

The phases identified by backscattered electron contrast are compared with adjacent regions using semiquantitative surveys to determine the presence of unique phases or diffused regions—a critical step to avoid spending hours performing unnecessary quantitative analysis. This is often done using X-ray counting integration for specific elements present over a preset time, or by performing an EDS spectral acquisition on the areas for comparison. In some cases a quick WDS count rate meter comparison is used. The choice of comparison method and attention to detail of these inspections is also influenced by knowledge of the expected phases generated from the literature, especially the binary phase diagrams. This could help to predict and identify compositions generated as a function of the annealing parameters of the various diffusion multiples. This "identify" activity of analysis allows the locations of all significant phases in the multiple section to be logged into the computer.

The "screening" activity includes using EDS, WDS, or both for semiquantitative and quantitative analysis. Here we tried to obtain a matrix of count rates from identified phases for the elements present. Typically the screening was performed in automated mode and analysis was done over a very small subset of the actual analysis matrix to approximate the composition ranges to be measured. This gave vital information to allow selection of all analysis parameters.

In the "analysis design" activity, the goal is to optimize all the conditions for the impending analyses. The primary electron beam energy, beam current, choice of standard, spectrometer crystal, detector bias, and pulse height discrimination must be selected as results of the abbreviated "screening" trials. It is made certain that the statistical accuracy required in crucial regions is met and a predetermined number of composition profiles are tested to "map" the multiple most efficiently. Also important here are X-ray counting times, dead-time error, detection limits, and the accuracy required. We also determine the "step size" (distance between points along a profile) as a result of the screening. Since the acquisitions are performed in automated fashion, we can alter specific sections of the acquisition to accommodate needs of greater sensitivity, varying step size, counting time on peak and background positions, etc. as a function of the composition ranges we have found.

The Monte Carlo simulations such as those shown in Figure 16.4 provide us with critical information regarding the necessary thickness of the intermetallic compounds in the diffusion multiples and the "step size" for EPMA analysis. For instance, the X-ray excitation volume would be less than 1 μm^3 for the Cr_3Pt phase (Figure 16.4); thus theoretically the thickness of this phase only needs to be a few microns to allow reliable analysis of its composition. The thickness of the Cr_3Pt phase in the precious metal diffusion multiple is only about 10 μm (Figure 16.2), but this is more than enough to allow reliable EPMA analysis. In addition, since the thickness is small and the interaction volume is also small, it makes sense to use a step size of 1 μm for the analysis of the Cr–Pt–Ru region of the diffusion multiple.

The "acquisition" activity includes intermittent analyses of the standards as a measure of "drift" of beam current, stage, spectrometer position, and other instrument stability issues. This can help salvage data by restoring acquired values relative to known variation from the standard. It also helps identify long-term stability limits.

Owing to the enormous number of data points collected for a given EPMA acquisition, our ultimate challenge is to minimize the X-ray counting times at each data point. A few unnecessary seconds spent counting at each data point can represent nearly 24 hr of acquisition time for typical EPMA runs.

We can utilize established statistical relationships to help decide on the level of accuracy necessary in measured intensity data commensurate with our needs. The level of homogeneity (wt%) is given by

$$\pm\frac{W_{1-\alpha}}{C} = \frac{\pm(t_{n-1}^{1-\alpha})S_C(100)}{n^{1/2}\overline{N}} \tag{2}$$

and the analytical sensitivity (wt%) and its relationship to counting statistics is given by

$$\Delta C = C - C' \geq \frac{2^{1/2}C(t_{n-1}^{1-\alpha})S_C}{n^{1/2}(\overline{N} - \overline{N}_B)} \tag{3}$$

where C is the actual concentration in weight percent of the sample, S_C is the standard deviation of measured values, n is the number of repetitions, $\overline{N}$ and $\overline{N}_B$ are the average number of X-ray counts for the sample and continuum of the sample, respectively, and $t_{n-1}^{1-\alpha}$ (known as the "Student t value") represents a confidence level of $1-\alpha$ for $n-1$ degrees of freedom.

In EPMA, the confidence level is usually chosen between 0.95 and 0.99; that is, only between 5% and 1% of the repeated measurements of a composition will be outside of homogeneity. It is statistically more difficult to measure a given level of homogeneity as C decreases. To do so, the counting time per point must be increased.

Having some knowledge of the approximate levels of elemental concentration within a phase and along a gradient to be measured is extremely important in making decisions about the counting statistics and dwell times needed at each measurement for all elements and for both peak and background X-ray integrations. Having *a priori* knowledge also allows us to make predictions using the above equations to reduce the total number of steps along a composition gradient to avoid acquiring data below ΔC, thus reducing acquisition time and increasing efficiency.

Another important consideration is the amount of time spent measuring the background signal, and whether to measure it at all. It is common practice in EPMA to select only a "baseline" measurement for background counts associated with a given characteristic X-ray peak. In many cases this is acceptable, but as peak counts approach background counts, it becomes imperative to measure the background accurately in order to obtain the composition. In multielement systems, background measurements must be taken on both sides of a measured peak if there are interference peaks from the presence of elements which are variable in magnitude, or if the compositional variations present differences in effective absorption of the measured element.

In order to measure the peak and the continuum X-ray signals efficiently and minimize counting statistical uncertainties, we can try to appropriate the time spent according to the optimization equation below, which shows that the ratio of time T spent on the X-ray peak (p) relative to background (b) is equal to the square root of ratios of their respective count rates:

$$T_p/T_b = (R_p/R_b)^{1/2} \tag{4}$$

Of course, every multiple-element system offers unique challenges for characterization. One example outlined in detail in this chapter is the Ru–Pd–Rh–Pt–Cr diffusion multiple. The Lα X-rays for periodic table "neighbors" Pd, Rh, and Ru presented serious overlap issues. Using the PET crystal, a nearly direct peak overlap exists for the first-order Pd Lα_1 with first-order Ru Lβ_2, and first-order Rh Lβ_1. There are subsequent overlaps of second-order Cr Kα_1 and Ru Lβ_1 with first-order Rh Lα_1. In addition, a number of problem overlap peaks interfere with "off-peak" regions normally used to measure and fit background intensities.

In these severe conditions of overlapping peaks, we use interpolations from the standards in which we measure artifacts from overlapped peaks and extrapolate to the unknowns (specimens). This can introduce greater error into the analysis owing to nonlinearity of the model, dead-time errors, absorption edges causing nonlinear background effects, and others.

There are other approaches to deconvoluting overlapped peaks, such as the use of principle components analysis, which may generate better results for specific X-ray overlap cases. Other areas for possible improvement include X-ray collection using hybrid EDS–WDS methods. EDS can bring much higher X-ray collection efficiencies due to geometric advantage over WDS, and might be applied for simultaneous analysis of major elements or where the degree of accuracy or energy resolution required is less of a concern.

Electron Backscatter Diffraction

Electron backscatter diffraction (EBSD) and backscatter electron (BSE) imaging add crystallographic and compositional information to the quantitative compositional information obtained by electron microprobe. Backscattered electrons are high-energy electrons, (10–40 keV) which interact with the top layer ($< 0.5\ \mu$m) of the sample through high-angle virtually elastic scattering events. EBSD uses the diffraction of these high-energy electrons to identify the crystalline phase and its three-dimensional orientation at a given point on the surface. Backscatter imaging provides semiquantitative compositional information due to the dependence of the BSE scattering cross section on the average atomic number of the sample at a given point. A great deal of work has been done in this field, and good reviews can be found in texts by Randle [9], Schwartz et al. [10], and Goldstein et al. [11].

EBSD is performed by collecting diffracted BSEs from the surface of a tilted SEM sample (Figure 16.7). The sample is tilted to maximize the collection of diffracted BSEs. The diffracted BSEs strike a phosphor screen, generating light. This light is normally captured by either a CCD camera or a silicon-intensified target (SIT) camera. The output of this detector is then run through a digital signal processor and into computer-based acquisition software. The collected EBSD pattern resembles the Kikuchi electron diffraction patterns observed in transmission electron microscopy. The bands of intensity in the EBSD pattern are actually the intersection of diffraction cones with the phosphor screen. Each band corresponds to one type of $\{hkl\}$ crystallographic planar spacing. Most automated EBSD analysis packages use the Hough transformation to detect the diffracted $\{hkl\}$ bands. The angles between detected bands and their widths are compared with lists of calculated reflections for crystals input by the user. A solution criterion is used to identify the best fit between the experimental EBSD pattern and a given crystalline phase at a specific three-dimensional orientation. The final output of this calculation is the crystalline phase and a set of three Euler angles describing its orientation [Figure 16.7(b)].

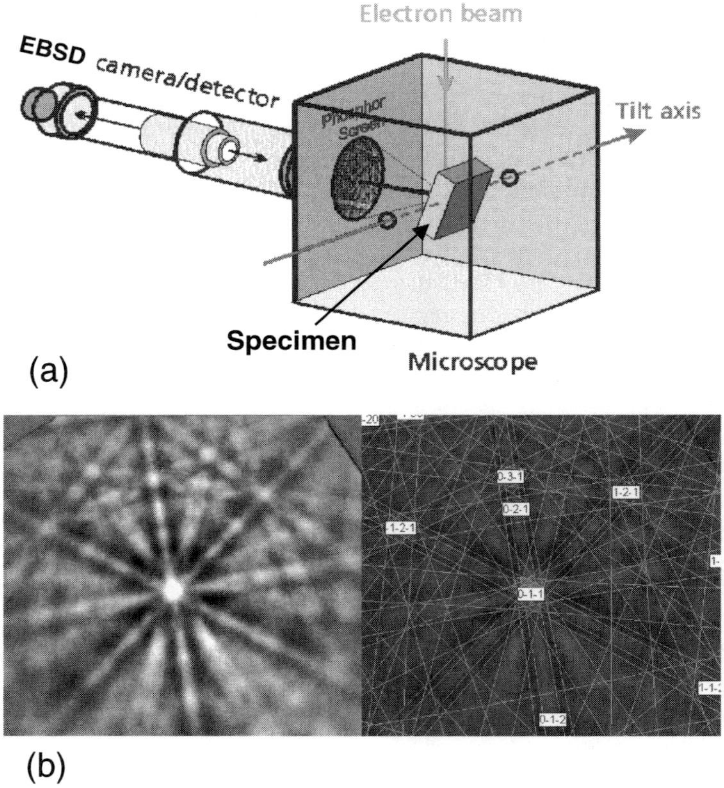

FIGURE 16.7. EBSD technique for crystal structure identification: (a) schematic of EBSD equipment and experiment; (b) EBSD pattern of Pt (fcc) and its indexed solution (1° mean angular deviation).

Both EBSD and BSE imaging are currently available as commercial techniques. Most commercial SEMs come equipped with some sort of BSE detector. EBSD systems are comprised of a special detector, acquisition hardware and software, and analysis software, and are available from a number of commercial vendors. The EBSD camera fits onto most standard SEMs. Commercial software can automatically index EBSD patterns of any crystal system at a rate of greater than 20 patterns per second at the fastest, although 10 patterns per second is more reasonable for a total experimental cycle time when scanning the beam. The measurements described in this section were performed on a CamScan CS44 SEM with a tungsten filament at a beam energy of 20 keV. The HKL Channel 5 EBSD acquisition and analysis system was used for collecting the EBSD data. This particular system used a Hammamatsu SIT camera for the detector.

Standard metallographic mounts can be used for EBSD samples with a few qualifications. Most metallographic procedures finish polishing with a 1 μm diamond medium. This finishing step leaves a damage layer on the surface on the order of 0.5–1 μm, particularly for metallic samples. This level of surface damage can severely degrade the quality of EBSD patterns due to the shallow depths (~100 nm) of beam interaction. To relieve this surface damage, vibratory polishing with a 0.05 μm silica suspension for many hours (usually more than 12 hr) is suggested. Mounting of samples in a conductive medium, when possible, is also beneficial. The use of standard mounting epoxies will cause sample charging, which will cause astigmatism to the optics of the SEM and cause sample drift, particularly at the 70° tilt used in EBSD experiments. Removing metallic samples from the mounting material or using conductive epoxies will help to alleviate sample charging. For insulating samples, such as ceramics, carbon coating may be necessary to alleviate charging even with a conductive mount.

The spatial resolution of EBSD is well suited to the compositional libraries created by diffusion multiples and is somewhat better than that of the electron microprobe. The physics governing the spatial resolution of EBSD is quite similar to that of the electron microprobe because both techniques are performed with focused electron beams. The principal difference between the two techniques is the signal collected. Diffracted BSEs emanate from a submicron volume around the probe's intersection with the sample. A minimum theoretical probe size of ~50 nm in the x direction and ~150 nm in the y direction (sample is tilted 70° about the x direction) has been suggested for tungsten filament instruments [12]. These estimates will vary strongly as a function of beam energy (larger energy gives a larger interaction volume) and as a function of sample atomic number (heavier elements have greater electron stopping power and consequently a smaller interaction volume). We have observed a practical resolution limit of 250–500 nm for standard metallurgical alloys.

The automation of EBSD is driven by the automated collection and analysis procedure and by the automation of the sample stage. The automation of pattern collection and analysis is the key point in the commercialization of the technique. As mentioned above, automated analysis routines are available from a number of vendors, which allow for rapid discrimination between a number of input phases (greater than 20 patterns per second). The automation of the sample stage is also important for rapid EBSD measurements. Most SEMs can collect grid patterns on the sample surface by deflecting the beam with the scan raster coils. Beam scanning can be done very rapidly with extremely fine step sizes (<0.1 μm). However, beam scanning cannot be done over large areas (>400 μm) without distorting the diffraction geometry. SEMs with motorized stages are increasingly the norm. A standard motorized stage has a positional precision of 0.5–1 μm and can move over a range of several centimeters. Using a motorized stage allows the point of interest always to be in the appropriate position for imaging and diffraction, but is much slower than beam scanning (300 ms per point), more expensive, and can suffer from mechanical hysteresis in the stage.

Microcrystallography is the primary use of EBSD. From a phase diagram mapping standpoint, EBSD's importance lies in its ability to detect phase boundaries and identify phases, and it is critical to efficient experimental planning in the "identify" and "screening" activities. If a list of expected phases in the sample can be generated, EBSD can be used for rapid detection of the spatial positions of the phase boundaries on the diffusion multiple. The spatial positions of these phase boundaries can be directly related to the quantitative compositions measured by electron microprobe, resulting in phase boundary positions on the phase diagram. BSE imaging can also be helpful in determining the position of phase boundaries on the sample. In addition to phase boundary detection, EBSD could be used to sample microtexture in samples with gradients in crystallographic texture or even gradients in plastic strain.

Another important use of EBSD is phase identification. Complete structure determination (point symmetry, point group, space group) for a crystallographic unknown is difficult for EBSD and requires detailed and relatively slow analysis of individual EBSD patterns [13]. The determination of lattice parameters with automated EBSD is quite limited but is an active area of research using individually analyzed EBSD patterns [14]. Phase matching is a much more rapid and readily available form of phase identification in EBSD. Phase matching describes the process of comparing an experimental EBSD pattern from an unknown phase with a list of possible crystallographic phases from either the literature or a crystallographic database. Commercial vendors now provide an automatic search procedure with standard crystallographic databases. The use of compositional information from either EDS (qualitative) or WDS (quantitative) is extremely valuable to phase identification or matching. Even a qualitative picture of the elements present in a given phase can reduce the number of possible phases significantly. BSE imaging is also helpful in identifying the phases in question because of its sensitivity to changes in composition. These procedures are not completely automated but can be performed on the order of minutes by a trained operator.

As an example of the utility of EBSD in the diffusion multiple approach, we will examine the Cr–Pt–Ru intersection. A BSE image of this region clearly shows five different phases as well as a region of diffusion of Ru and Pt into Cr (Figure 16.2). Indexing of individual EBSD patterns taken from these regions identifies the phases as Cr (bcc), Pt (fcc), Ru (hcp), Cr_3Pt (A15), and CrRu (σ phase) (Figure 16.8). An EBSD phase map of this same area locates the positions of the phase boundaries (especially the A15–σ phase boundary) which are used for intelligent placement of EPMA scan locations.

In brief, EBSD and BSE imaging are characterization tools capable of adding crystallographic and compositional information to the mapping of diffusion multiples. For diffusion-multiple studies, these techniques provide the position of phase boundaries for comparison with microprobe compositional data and assist in the identification of phases.

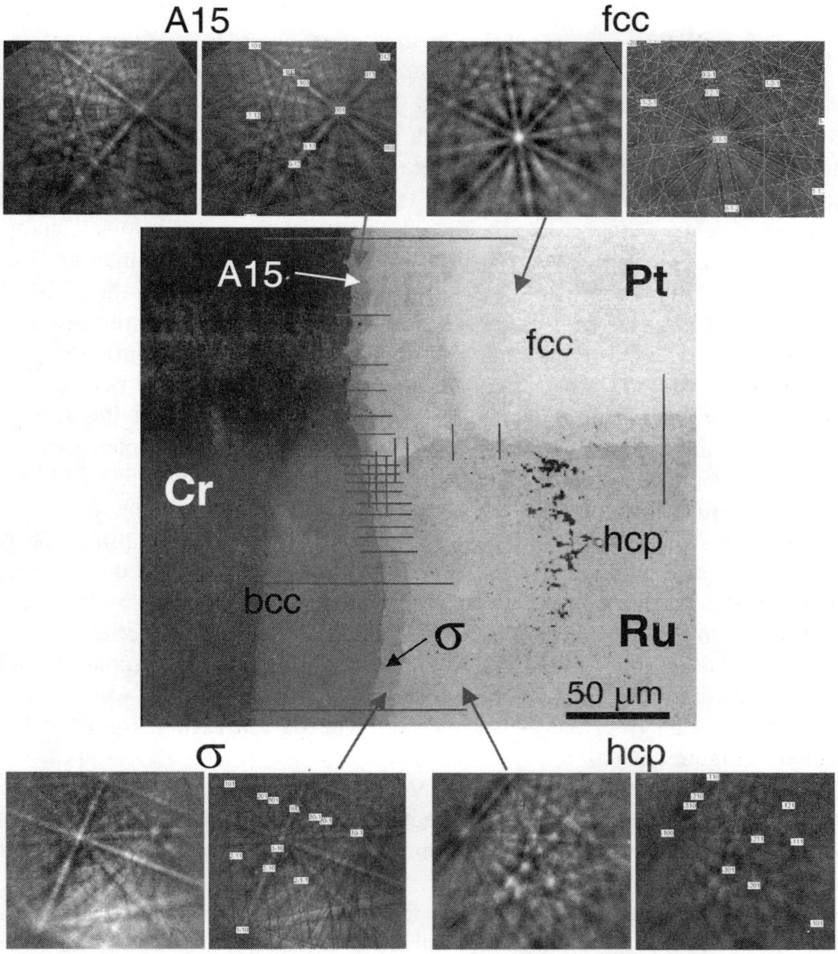

FIGURE 16.8. EBSD identification of phases in the Cr–Pt–Ru ternary system in the diffusion multiple. The lines on the SEM backscatter electron image (in the middle) represent the locations of the EPMA scans.

Example Analysis of the Precious Metal Diffusion Multiple

The novel diffusion-multiple approach has challenged the field of quantitative microanalysis to provide higher levels of efficiency, throughput, and mechanical and electron beam stability, while maintaining a high level of accuracy. Here we present the details of such a study.

The interdiffusion of elements at the tri-junction regions of the diffusion multiple allows the formation of all the intermetallic compounds and the generation of composition variations for all the single-phase regions. Detailed explanation of this can be found elsewhere [1,2]. An example of the analysis procedure for EPMA mapping of the Cr–Pt–Ru ternary system (tri-junction) is provided here (Figures 16.2 and 16.8). The interdiffusion of Cr and Pt formed the A15 phase, and that of Cr and Ru formed the σ phase. Close to the tri-junction region, ternary interdiffusion took place. In order to determine the tie-line compositions of the ternary system, EPMA analysis was performed using automatic stepping along the lines shown in Figure 16.8. These lines were chosen (1) to cover all the composition regions, (2) to provide enough tie-line and tie-triangle data to construct the phase diagram, (3) to produce more data in regions near the three-phase triangles (intersections) and less data elsewhere, and (4) to reduce the number of EPMA points for the reduction of cost and data-collection time. Longer scans were made to collect diffusion profiles for the three binary systems for diffusivity evaluation. Since the intermetallic compounds were very thin ($\leq 10\,\mu m$), $2\,\mu m$ steps were used in most of the EPMA scans shown in Figure 16.8. This decision was based on the very small electron beam–sample interaction volume, and thus very small X-ray excitation volume ($< 1\,\mu m$), for this particular ternary system (Figure 16.4). The $2\,\mu m$ steps ensured that we were not oversampling the region (not producing many unnecessary points while still producing enough data points in the thin intermetallic phases to allow the extrapolation to the interfaces to obtain reliable tie-line data). The entire Cr–Pt–Ru system was mapped with 558 EPMA points which were acquired in about 9 hr.

An example scan is shown in Figure 16.9(a) showing a $40\,\mu m$ scan from bcc(Cr) to A15 to fcc(Cr,Pt) phases. Six data points were collected in the thin A15 phase, which was sufficient to make reliable extrapolation to the interface to obtain the tie-line data. There was a data point (at the $24\,\mu m$ location) exactly at the A15/fcc(Cr,Pt) interface. Since it has sampled both phases, its composition was in the middle of the two-phase region. All the tie-line information obtained from the Cr–Pt–Ru tri-junction was collected in Figure 16.9(b) to construct the phase diagram. The tie-lines from Figure 16.9(a) are highlighted in Figure 16.9(b).

The local equilibrium at the phase interfaces was the basis for using diffusion multiples to map phase diagrams. The existence of such local equilibrium and its reliability in establishing equilibrium tie-line information has been demonstrated for many years in diffusion couples.

By performing EPMA and EBSD analysis of all the tri-junction regions in the diffusion multiple, we were able to map isothermal section phase diagrams of 10 ternary systems, as shown in Figure 16.10. The efficiency gain is tremendous—it would probably take approximately a thousand individual alloys to map these phase diagrams that we obtained from a single sample.

As we discussed earlier, the key balance in EPMA analyses is to obtain enough reliable data while reducing the total number of points and mapping time. This all relates back to our discussion about the counting time on peak and background, counting statistics, stepping size, scan length, scan locations, and selection of beam voltage and current. Based on our experience, the two key factors in achieving high-efficiency mapping are scan locations and stepping size. EBSD phase identification is extremely helpful in locating the phase boundaries, and thus facilitates greatly our ability to place scans at the proper locations (Figure 16.8) to obtain the most critical data. For instance, EBSD identification of the A15–σ phase boundary and the hcp–fcc boundary in the Cr–Pt–Ru system help us to choose appropriate line-scan positions to minimize the number of scans required for this system. The stepping size is very critical, since the analysis time can be reduced by half just by increasing the step size from 1 to $2\,\mu m$. The decision on stepping size should

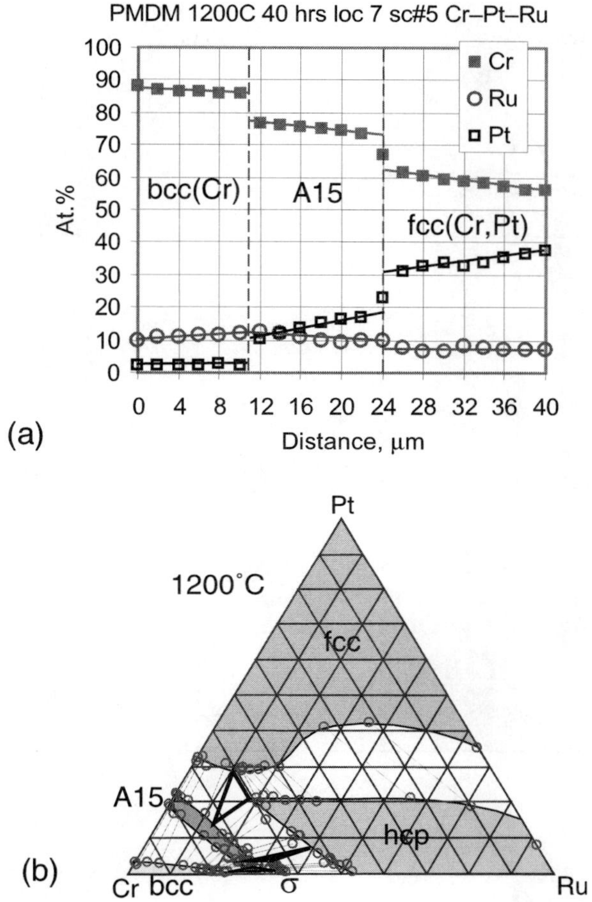

FIGURE 16.9. EPMA analysis of the Cr–Pt–Ru ternary system with the line scans shown in Figure 16.8. (a) A line scan showing a typical composition profile produced by EPMA. (b) All the tie-line information based on the EPMA scans for the construction of the Cr–Pt–Ru ternary phase diagram.

be based on the size of the phases and the interaction volume. Again using the Cr–Pt–Ru system as an example, the thickness of both the A15 and the σ phase was around 10 μm (Figure 16.8), and the beam interaction volumes were less than 1 μm. Thus it is reasonable to use 2 μm steps in EPMA analysis. On the other hand, it is important to generate enough data points to allow reliable extrapolation to the local equilibrium tie-line compositions. This has been demonstrated in Figure 16.9(a).

In addition to mapping phase diagrams, EPMA analysis was also required to evaluate the composition of the nanoindentation positions so that the hardness and modulus–location information, such as that shown in Figure 16.11(a), can be converted to hardness and modulus–composition information [Figure 16.11(b)]. This is usually done after the nanoindentation experiment.

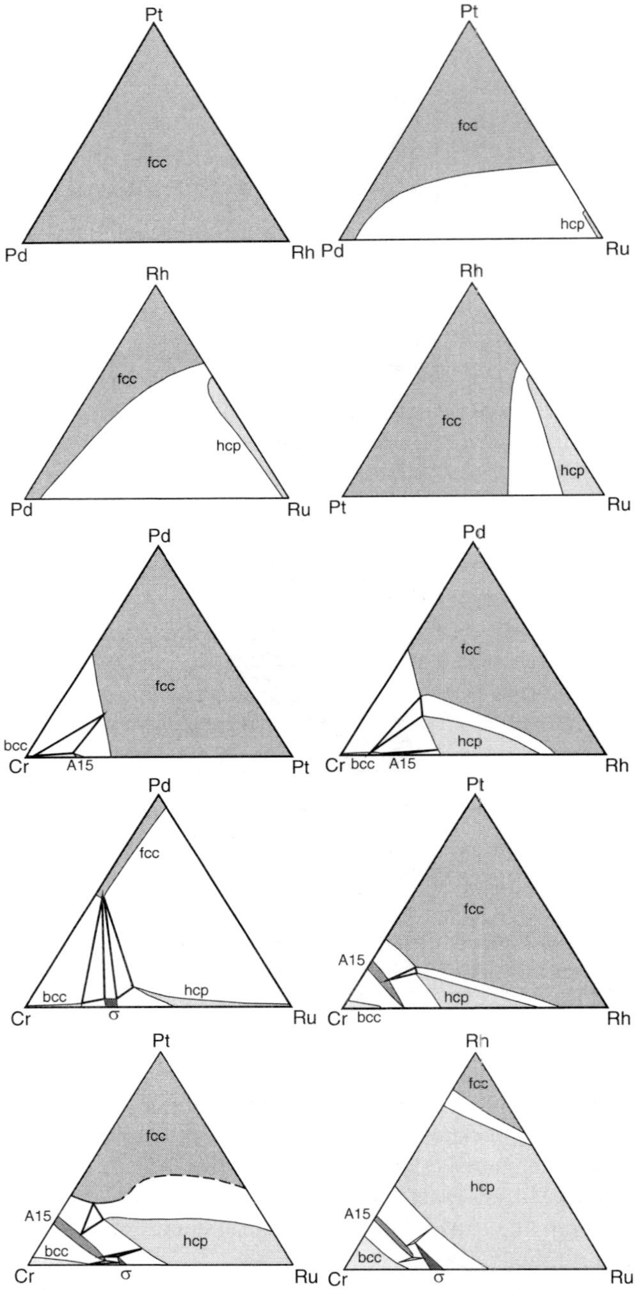

FIGURE 16.10. Ten ternary phase diagrams (isothermal sections at 1200 °C) obtained from the single diffusion multiple shown in Figure 16.1. They are plotted in atomic percent axes with the scale removed for simplicity.

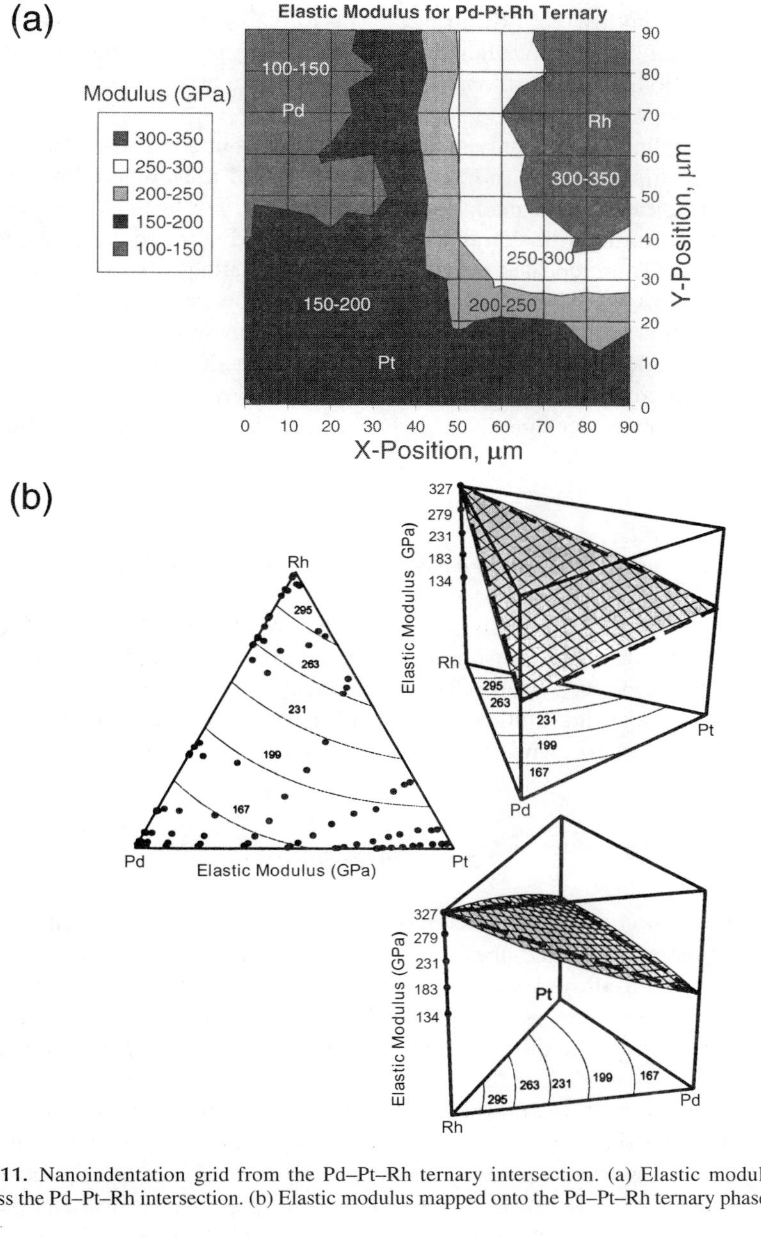

FIGURE 16.11. Nanoindentation grid from the Pd–Pt–Rh ternary intersection. (a) Elastic modulus versus position across the Pd–Pt–Rh intersection. (b) Elastic modulus mapped onto the Pd–Pt–Rh ternary phase diagram.

Since large amounts of EPMA data are generated during the analysis of the diffusion multiple, we have developed automatic plotting procedures based on the Microsoft Excel spreadsheet. During data reduction, we usually plot two figures for each scan to help in judging the tie-line compositions. The first plots compositions of individual points against location (distance in X or Y direction). This helps to extrapolate to the local equilibrium compositions at the phase interface [Figure 16.9(a)]. The second plots one element (e.g., Pt) against another (e.g., Ru in the Cr–Pt–Ru case). This plot basically shows the "diffusion path" of the scan location in the corresponding phase diagram format. The two plots define the tie-line compositions. The tie-triangles are usually obtained by extrapolating the three two-phase tie-lines.

For each tie-line, we can only obtain one set of data in one polished cross section of the diffusion multiple. This is different from analysis of individual alloy samples from which several repeats can be made for a single tie-line. Fortunately, the consistency of the tie-line trends in the diffusion multiple results [e.g., Figure 16.9(b)] give us as much confidence as that from repeated results from individual alloys.

The enormous increase in efficiency and reduction in analysis time of diffusion multiples compared with individual alloys is not just a mere reduction of the number of samples. It takes a long time for the EPMA to reach a good vacuum ($\sim 10^{-6}$ torr) and for the electron beam to stabilize. In the case of individual alloys, the time waiting for vacuum and beam stabilization is a significant part of the analysis time, whereas in the diffusion-multiple case it is negligible.

PHYSICAL AND MECHANICAL BEHAVIOR

The mapping of phase chemistry and behavior is the critical step for the alloy designer. Here we describe measurement techniques and the application of these techniques to complex phase systems. The mechanical behavior, represented by hardness, instructs us regarding solid solution effects, while the physical behavior, represented by the elastic modulus, instructs us regarding atomic interactions and bonding.

Nanoindentation

Nanoindentation provides the connection between the compositional and crystallographic information discussed above and the mechanical properties that are of ultimate interest in structural alloy development. In fact, it is the ability to obtain mechanical property information at the same length scale and locations as the microstructural information that transforms the diffusion multiple into a high-throughput approach for alloy development.

Nanoindentation, or instrumented indentation, is a fairly general technique for measuring the mechanical properties of solid materials at small length scales. Instrumented indentation at a basic level consists of inserting a shaft with a diamond tip into the surface of a sample and then withdrawing the tip. During this process, the load and the displacement of the shaft are continuously measured. The result is a load–displacement curve associated with an indentation ranging from 20 nm to 100 μm in depth and 100 nm to 700 μm in width, depending upon the load and the material (Figure 16.12).

A number of important mechanical properties can be calculated from the load–displacement curves. Most indenters use sharp tips with geometries that are self-similar (Vickers and Berkovich). Hardness, elastic modulus, and work of indentation can all be

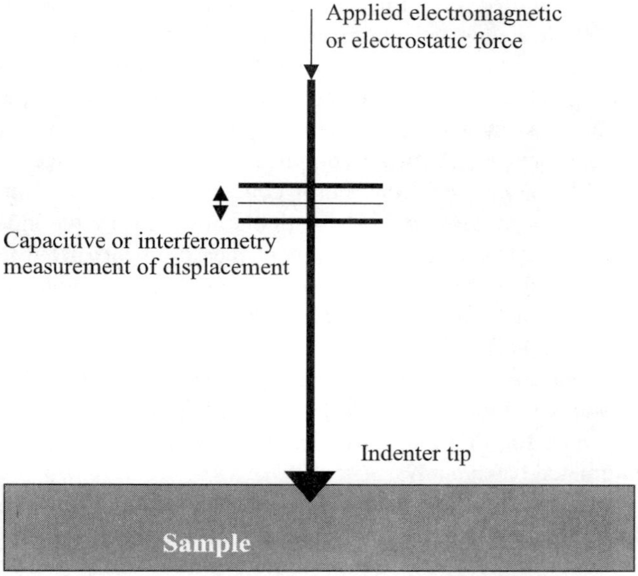

FIGURE 16.12. Schematic of instrumented indenter.

readily calculated from the load–displacement curve. Good reviews of the calculation of these mechanical properties are given by Pharr [15,16] and by Cheng and Cheng [17]. The self-similar nature of these tips prevents the direct measurement of yield strength from the indentation measurement using standard analysis techniques. The measurement of indentation yield strength using both spherical indenters and advanced analysis techniques is a topic of active research [18,19].

The spatial resolution of nanoindentation in the x–y plane is dictated completely by the depth of indentation (Z direction). Traditionally, nanoindentation has been focused on the depth or spatial resolution in the z (or sample thickness) direction. For work on coatings and thin films, the position on the x–y plane was typically of less importance. For characterizing compositional libraries, however, the x–y plane and z direction resolution are equally important. For a Berkovich indentation of 1 μm in width in the x–y plane, the depth of the indent is approximately 140 nm, which is certainly measurable by nanoindentation but is noisier than an indent, which is 500 nm or 1 μm deep. In general, as the indentation depth increases, the quality of the data increases because of decreasing sensitivity to surface roughness, geometric aberrations of the indenter tip, and noise in the displacement measurement. However, a deeper indent increases the lateral size of the indentation rapidly (width-to-depth ratio ~7:1). Indentation experimental practice suggests that indentations be spaced at a minimum of twice the width of the indentation in order to negate the overlapping stress fields associated with each indent [20]. For the measurements on the Pd–Pt–Rh–Ru–Cr diffusion multiple, a spacing of 10 μm between indentations was used with indentation depths of up to 225 nm.

Nanoindentation is now a commercial and highly automated technique. Standard instruments have sample stages with automated motion in the *x, y*, and *z* directions. These stages have an accuracy and precision of around 0.5–1 μm. Automated imaging using optical or topographic signals is also available, depending upon the manufacturer. The registry between the optic axis and the indentation axis is approximately 1 μm. The automation of the indentation experiment is somewhat complicated by concerns over mechanical drift and thermal drift. Mechanical drift can be of concern both in actual placement of the indentation (hysteresis of motors) and in vibrational noise during the indentation experiment (affecting the accuracy of the displacement measurement). Both these sources of mechanical drift are largely accounted for in commercial instruments. Thermal drift is caused by temperature fluctuations inside the indentation chamber during a single indentation experiment. Thermal drift between indentations is less of a concern because the load–displacement curve for each indent is analyzed separately. Most commercial systems use enclosures to reduce both mechanical and thermal drift and back-correction algorithms to account for thermal drift. Analysis of the indentation curves can be performed by automated batch analysis.

Measurable quantities on the load–displacement (*P–h*) curve, such as h_{max}, h_{final}, P_{max}, and $(dP/dh)/h_{max}$, can be related to useful mechanical quantities such as the stiffness, hardness, and reduced modulus of the material (Figure 16.13) [16,21]. The elastic modulus is calculated from the initial slope of the unloading curve. The loading portion of the indentation curve, or *P–h* curve, contains both elastic and plastic information, while the initial portion of the unloading curve contains elastic information only. The unloading portion of the *P–h* curve is fitted to the following power law relation:

$$P = B(h - h_{final})^m \tag{5}$$

The stiffness *S* is taken to be the initial derivative of the unloading curve, $(dP/dh)/h_{max}$. The contact depth h_C is then calculated by

$$h_C = h_{max} - 0.75\, P_{max}/S \tag{6}$$

The hardness is given by

$$H = P_{max}/A(h_C) \tag{7}$$

where P_{max} is the maximum applied load and $A(h_C)$ is the projected contact area of the indenter at the contact depth. The area function $A(h)$ is calculated from a polynomial fit to a calibration curve on an isotropic material of well-established modulus and hardness, such as a fused silica or a single-crystal aluminum standard [16]. The reduced modulus E_r is given by

$$E_r = \frac{\pi^{1/2}}{2[A(h_C)]^{1/2}} \tag{8}$$

and can be related to the actual modulus of the material by

$$E = (1 - \nu^2)\left(\frac{1}{E_r} + \frac{1-\nu_0^2}{E_0}\right)^{-1} \tag{9}$$

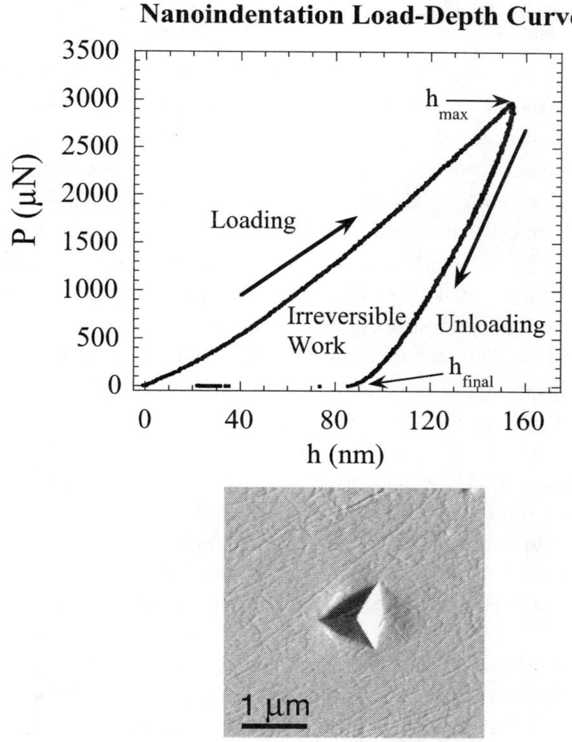

FIGURE 16.13. Load–displacement curve taken from indentation in Pt and an image of the indent.

where ν for the material is the Poisson's ratio, ν_o is the Poisson's ratio for the diamond indenter (0.07), and E_o is the modulus for the diamond indenter (1,141 GPa) [16].

Another quantity which can be calculated for characterizing the plasticity of a material is the irreversible work of indentation, which is the area inside the indentation curves [Eqn (6)] [17]:

$$W_{\text{irreversible}} = \int_0^{h_{\max}} P(h)_{\text{loading}} \mathrm{d}h - \int_{h_{\max}}^{h_{\text{final}}} P(h)_{\text{unloading}} \mathrm{d}h \qquad (10)$$

The first term in this expression calculates both elastic and plastic work of indentation, while the second term calculates only the elastic work upon unloading. This quantity is often expressed as the plasticity index P_i, which is given by

$$P_{\text{i}} = \frac{W_{\text{total}} - W_{\text{elastic}}}{W_{\text{total}}} \qquad (11)$$

From this, it is expected that materials which exhibit limited plastic deformation with a large elastic recovery should have a plasticity index approaching zero, while materials which exhibit large plasticity will have a plasticity index approaching unity.

Example of Analysis of Mapping Behavior With Chemistry

As an example of these methods on diffusion multiples, we will again consider the precious metal diffusion multiple. All of these measurements were performed on the Hysitron Triboindenter system using a Berkovich diamond indenter tip. Optical microscopy was used to position the sample at the Pd–Pt–Rh ternary intersection (the Pd–Pt–Rh system has completely mutual solubility during high-temperature heat treatment, and thus it formed a complete fcc solid solution for the entire ternary system compositions). A grid of 10 × 10 indentations with 10 μm spacing was made using a maximum load of 2,500 μN (Figure 16.14). Indentation line scans were also made across the binary interfaces under the same conditions. Multiple line scans were made across the Pd/Rh interface to investigate the reproducibility of the data. The Oliver–Pharr equations [16] were used to calculate the elastic modulus and hardness at each indentation point.

A clear transition in both elastic modulus and hardness can be seen for the line scans across the Pd/Rh interface [Figure 16.15(a)]. The reproducibility between the three line scans was quite high. EPMA was performed along these same lines allowing the replacement of the spatial position with a compositional axis [Figure 16.15(b)]. This result demonstrates the importance of spatial registry between techniques (nanoindentation and electron microprobe) and the efficiency of the diffusion multiple approach for alloy development (in contrast to creating several binary alloy compositions for modulus measurement).

The elastic modulus data from the ternary section also shows expected trends as a function of position and composition. A dramatic increase in modulus is observed in the upper right-hand portion of the grid, corresponding to increased Rh content [Figure 16.11(a)]. The moduli of Pd and Pt are similar, and this fact is demonstrated by the small measured differences in modulus from top left to bottom left. A much more useful representation of the mechanical property data comes from their combination with the compositional data from electron microprobe. The mechanical property information (more than 100 modulus measurements) can be plotted as a three-dimensional surface overlying the isothermal section of the Pd–Pt–Rh ternary phase diagram [Figure 16.11(b)].

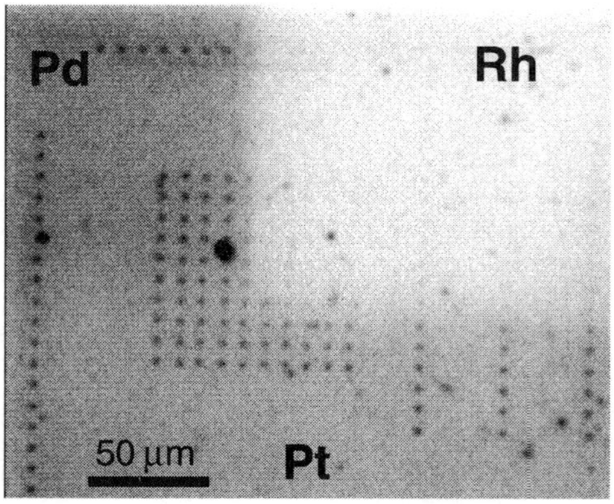

FIGURE 16.14. Optical image of an indentation grid made at the Pd–Pt–Rh ternary intersection.

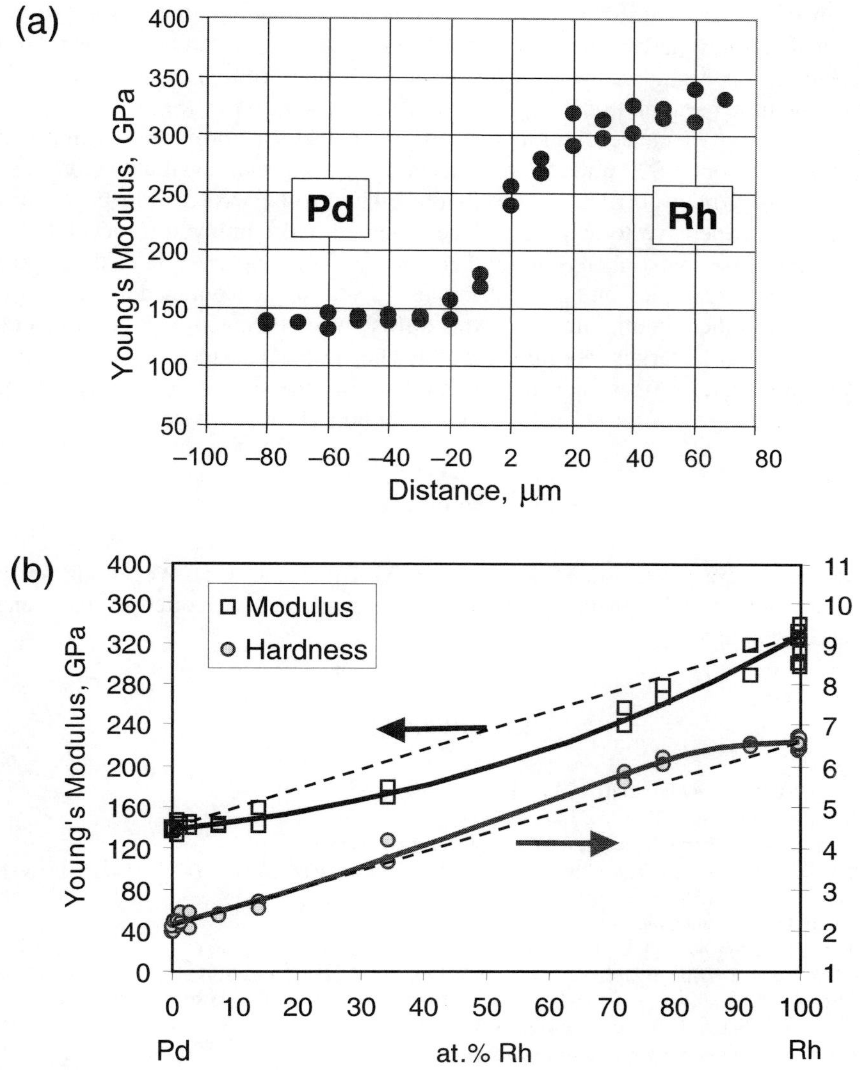

FIGURE 16.15. Nanoindentation line scans across the Pd/Rh interface. (a) Elastic modulus versus position across the Pd/Rh interface. (b) Elastic modulus and hardness as a function of Rh content.

This representation serves as an extremely efficient library for accessing the modulus or hardness at a given (Pd, Pt, Rh) composition.

CONCLUDING REMARKS

From a single diffusion-multiple assembly, 10 ternary isothermal phase diagrams have been determined in about 100 h of microprobe measurements, an extremely efficient generation of ternary equilibria. With a similar investment of time, information on hardness and modulus across this very extensive chemistry space can be mapped. For the alloy designer, this highly efficient diffusion-multiple approach can have a tremendous impact on the time required to define the most favored chemistry space.

This mapping of phase diagrams, resulting in material properties–composition relationships, is about 1,000 times more efficient than using individual alloys. With such a high efficiency, many critical materials data which otherwise would be too time-consuming and expensive to acquire can be obtained and employed to accelerate our understanding of the materials physics and chemistry. This chapter explains the approach and the mapping techniques, and gives examples. Readers are encouraged to read our earlier paper [22] on the potential impact of this diffusion-multiple approach for accelerated design of structural materials. We hope that this chapter has provided details concerning the various analytical probes such that researchers in other fields may be able to take advantage of these probes for efficient screening of high-throughput samples.

ACKNOWLEDGMENTS

The authors are grateful to M. F. Henry, A. M. Ritter, and J. H. Westbrook for their enthusiastic support and valuable discussions. This work was supported by the General Electric (GE) Company.

REFERENCES

1. Zhao, J.-C. *Adv. Eng. Mater.* **2001**, *3*, 143–147.
2. Zhao, J.-C. *J. Mater. Res.* **2001**, *16*, 1565–1578.
3. Castaing, R. Unpublished Ph.D. Thesis, University of Paris, **1951**.
4. Scot, V. D., Love, G. In V.D. Scot and G. Love (Eds.), *Quantitative Electron-Probe Microanalysis*, **1983** Chichester, UK: Ellis Horwood, p. 32.
5. Lifshin, E. Private communication.
6. Goldstein, J. I., Newbury, D. E., Echlin, P., Joy, D. C., Romig, A. D., Lyman, C. E., Fiori, C., Lifshin, E. (Eds.). *Scanning Electron Microscopy and X-ray Microanalysis*, **1981**, New York: Plenum Press, 110–111.
7. Armstrong, J. T. *Proc. Annual MSA/MAS Meeting*, **1999**, New York: Springer-Verlag pp. 560–561.
8. Pichoir, F., Pouchou, J. L. *Rech. Aerosp.* **1984**, *5*, p. 349.
9. Randle, V. *Microtexture Determination and Its Applications*, **1992**, London: Institute of Materials.
10. Schwartz, A. J., Kumar, M., Adams, B. L. *Electron Backscatter Diffraction in Materials Science*, **2000**, New York: Kluwer Academic/ Plenum.
11. Goldstein, J. I., Williams, D. B., Cliff, G. *Quantitative X-ray Analysis*, **1986**, New York: Plenum.
12. Dingley, D. J. The development of automated diffraction, In A.J. Schwartz, M. Kumar, and B.L. Adams (Eds.), *Electron Backscatter Diffraction in Materials Science*, **2000**, New York: Kluwer Academic/ Plenum, pp. 1–18.
13. Michael, J. R. Phase identification using electron backscatter diffraction in the scanning electron microscope. In A.J. Schwarz, M. Kumar, and B.L. Adams (Eds.), *Electron Backscatter Diffraction in Materials Science*, **2000**, New York: Kluwer Academic/ Plenum, pp. 75–89.
14. Michael, J. R., Eades, J. A. *Ultramicroscopy* **2000**, *81*, 67–81.
15. Pharr, G. M., Oliver, W. C. *MRS Bull.* **1992**, *July*, 28–33.
16. Pharr, G. M. *Mater. Sci. Eng.*, **1998**, *A253*, 151–159.
17. Cheng, Y. T., Cheng, C. M. *Appl. Phys. Lett.*, **1998**, *73* (5), 614–616.

18. Herbert, E. G., Pharr, G. M., Oliver, W. C., Lucas, B. N., Hay, J. L. *MRS Symp. Proc.* **2000**, *649*, Q3.4.1–3.4.6.
19. Dao, N., Chollacoop, N., VanVliet, K. J., Venkatesh, T. A., Suresh, S. *Acta Mater.* **2001,** *49*, 3899–3918.
20. Altan, T., Oh, S., Gegel, H. L. *Metal Forming—Fundamentals and Applications*, **1983**, Metals Park, OH: ASM.
21. Doerner, M. F., Nix, W. D. *J. Mater. Res.* **1986,** *1*, 601–610.
22. Zhao, J. -C., Jackson, M. R., Peluso, L. A., Brewer, L. *MRS Bull.* **2002**, *27*, 324–329.

Chapter 17

Combinatorial and Spread Techniques in the Fabrication of Organic-Based Photonic and Optoelectronic Devices

Ghassan E. Jabbour and Yuka Yoshioka

INTRODUCTION

In the last few years, organic light-emitting devices (OLEDs) [1] have moved from the laboratory to the market place [2,3]. It is now possible to buy devices such as car stereos and cell phones where the display is based on OLEDs. Efforts are also directed to larger displays, and prototypes of 13 and 17 in. displays have recently been demonstrated.

OLEDs are attractive for their wide viewing angle, ease of fabrication, color tunability, and ability to be made on any kind of substrate including flexible plastic and fabrics. Indeed, the ease of direct integration of OLEDs on plastic and/or fabric substrates opens a new window of opportunity for low-cost roll-to-roll fabrication of OLEDs. The flexible nature of OLEDs allows for new applications to emerge, such as wearable optoelectronics, roll-up displays, and luminescent lights.

Significant progress in the reliability of OLEDs, especially operational lifetime, has been made since the late 1980s. With starting brightness of 200 cd m^{-2}, OLEDs with an estimated operational lifetime of 35,000 hr have been reported [4]. In fact, reports of more than 40,000 hr of continuous OLED operation have appeared recently [5]. An acceptable lifetime for portable displays is currently around 10,000–15,000 hr at 100–150 cd m^{-2}. However, even with all the advances made so far, there is still much room for improvement, especially in the area of materials, device optimization, and lifetime.

There are two kinds of OLEDs: small molecule based and polymeric. Of course, one might consider hybrid OLEDs where both polymers and small molecules are used in device construction, as we shall show in one of the examples later in this chapter.

Ghassan E. Jabbour and Yuka Yoshioka • Optical Sciences Center, University of Arizona, Tucson, AZ 85721, USA
Contact author: Ghassan E. Jabbour, email: gej@optics.arizona.edu

High-Throughput Analysis, edited by Potyrailo and Amis
Kluwer Academic Publishers, New York, 2003.

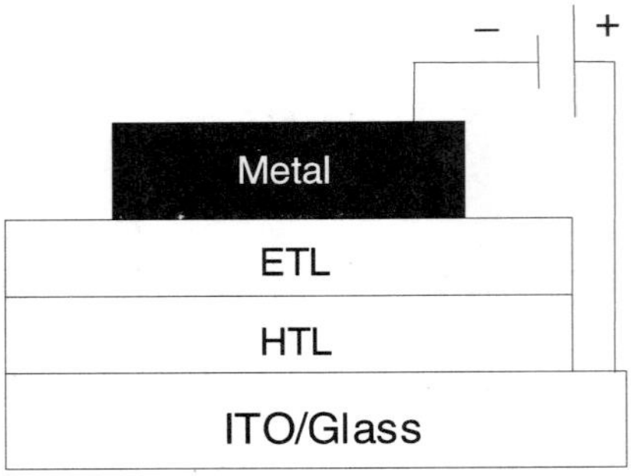

FIGURE 17.1. Schematic of a multilayer OLED.

Excluding the top and bottom contacts (used as electrodes), OLEDs can be made with single or multiple organic layers. The benefit of having multilayers is to increase device efficiency, stability, and lifetime. In its simplest forms, a typical multilayer OLED consists of a hole transport layer (HTL) and an electron transport layer (ETL) sandwiched between two electrodes, as shown in Figure 17.1. These layers are usually made of small molecules. One can simplify the design and reduce the number of layers to one by using polymers instead. This will result in a reduced applied voltage.

Although the exact operational mechanism of an OLED is rather more complex, the following simple description can be helpful in understanding how an OLED functions. Upon application of a voltage, holes are injected from the anode into the highest occupied molecular orbital (HOMO) of the HTL, and electrons are injected from the cathode into the lowest unoccupied molecular orbital (LUMO) of the ETL. The charges drift under the influence of the electric field and recombine on a molecular site in the emitting layer (or on a part of the polymer chain), which could be the HTL or the ETL, leading to generation of excited molecules. Some of the excited molecules decay radiatively, thus releasing light (Figure 17.2).

At least one of the electrodes must be optically transparent to allow light to exit the device towards the viewer. By far the most commonly used transparent electrode (anode) in OLEDs is indium tin oxide (ITO). ITO has a high transparency to visible light as well as acceptable electrical conductivity. It can be deposited on glass, silicon, and flexible plastic substrates. Many metals have been used as the cathode. The choice of the metal cathode and its composition has a significant effect on device performance and stability.

In general, organic materials have a higher hole than electron mobility. Among other factors, the high hole mobility, as well as the presence of a high barrier for electron injection at the cathode/organic layer interface, leads to unbalance between the hole charge density and the electron charge density near the interface of the two organic layers. This behavior has a negative effect on the device external quantum efficiency (which is defined here as the ratio of the number of photons collected in the forward direction to the number of charges injected in the device).

One way to enhance the external quantum efficiency is to increase the number of injected electrons. This can be achieved by decreasing the barrier height between the work function of the metal cathode and the LUMO of the ETL. OLEDs can have cathodes based on metals (or alloys) with relatively low work functions such as magnesium, calcium,

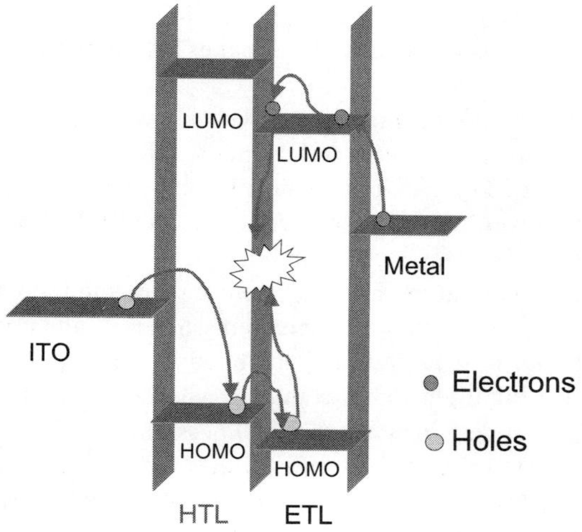

FIGURE 17.2. Simplified energy diagram showing electrons and holes in the process of generating light near the organic–organic interface.

an Mg–Ag alloy, and Li–Al alloy [2,3,6]. Other efficient electron-injecting cathode structures have been used, such as LiF/Al bilayers [7,8] and mixed layers of alkali halide and metal or alkali halide and organic material [9,10]. OLEDs using such cathode structures show higher external quantum efficiency and light output (at a given voltage) than similar devices with cathodes based on aluminum, copper, or silver.

The simplest form of efficient OLED based on small molecules is shown in Figure 17.1 [2]. Soon after, it was demonstrated that the performance of an OLED could be increased, significantly if luminescent organic dyes were used as dopants (guests) during OLED fabrication [11,12]. The dye dopant can be hosted by the HTL, the ETL, or both, and its concentration may vary from less than 1% to about 8%. More than two layers can also be used to optimize device stability further [12,13]. Although most OLEDs have been made with separate HTL and ETL, there are other device structures that showed enhanced stability by using a mixed layer of HTL and ETL materials [14]. Thus it is apparent that OLED performance depends on many factors including choice of electrodes, doping levels, placement of the doped layer (or layers) in the device structure, dopant concentration, host–dopant compatibility, thickness of device layers, electrode material, interfaces between layers, and device architecture, to mention a few. With so many variables, it would be impossible to devise a standard rule or recipe which, when used a priori in the fabrication process, would yield an optimized device with the desired characteristics. Moreover, photophysical and quantum calculations might guide in device optimization. However, given a set of materials and variables involved, it is impossible at this time for such methods to dictate and give complete insight on how to design the best OLED structure. Currently, for a given new material (e.g., a dopant) and a fixed HTL and ETL, the usual method of device optimization relies primarily on the fabrication and testing of many devices, each having a different dopant concentration, until the optimized performance is achieved. Similarly, for a new HTL or ETL material, the required thickness for maximum performance is found by fabricating numerous devices, each having different thickness of

ETL, HTL, or both. This is tedious and time consuming, especially if one has to vary not only the dopant material but also the host, the thickness of individual layers, and the device layer structure.

Combinatorial and spread techniques [15] have shown tremendous benefit in screening, optimizing, and discovering new materials that would have taken much longer to discover otherwise [16]. OLED development will benefit greatly from combinatorial and spread techniques. Moreover, not only the device aspects of OLEDs but also the materials synthesis efforts will benefit from such techniques.

Recently, the combinatorial method has been implemented in OLED fabrication and has been shown to be a helpful method for screening materials and optimizing device characteristics in much shorter time period than practiced currently [17–21].

In this chapter, we highlight three examples where combinatorial and spread techniques were used. The first demonstrates the thickness effects of the ETL on device properties, the second example relates to the effects the dopant concentration has on device characteristics, and the third demonstrates the use of inkjet printing as an alternative method of carrying out combinatorial and spread variation experiments. In this case, we will show how this can be used to modify the conductivity of polymeric anodes.

ETL LAYER THICKNESS EFFECTS ON OLED PERFORMANCE

The OLED structure for this study is shown in Figure 17.3. The HTL is deposited from solution using spin coating. In this case, the solute is a mixture of *N′*,*N′*-bis(3-methylphenyl)-*N′*,*N′* diphenyl benzidine (TPD), 5,6,11,12-tetraphenylnaphthacene (rubrene), and polycarbonate dissolved in chloroform at 10 mg ml^{-1} for a period of 24 hr at room temperature. Figure 17.4 shows the chemical structure of TPD and rubrene. The thickness of the HTL is about 55 nm as measured by a step profilometer. The thickness of the HTL varied about 6% over a given substrate. Moreover, all the HTLs among the various substrates had similar thicknesses.

The ETL is the green-light-emitting material 8-tris-hydroxyquinoline aluminum (Alq_3) (Figure 17.4) which was vacuum deposited on top of the HTL. A sliding-shutter mechanism driven by a computer-controlled robotic arm was used to deposit the various

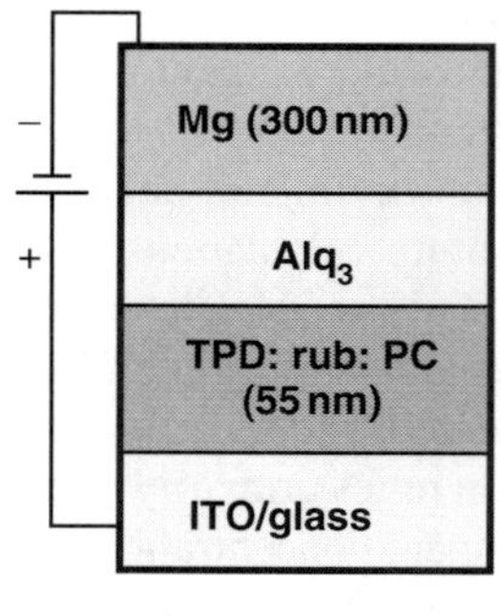

FIGURE 17.3. Cross section of OLEDs made with various thicknesses of ETL, which is Alq_3 in this case.

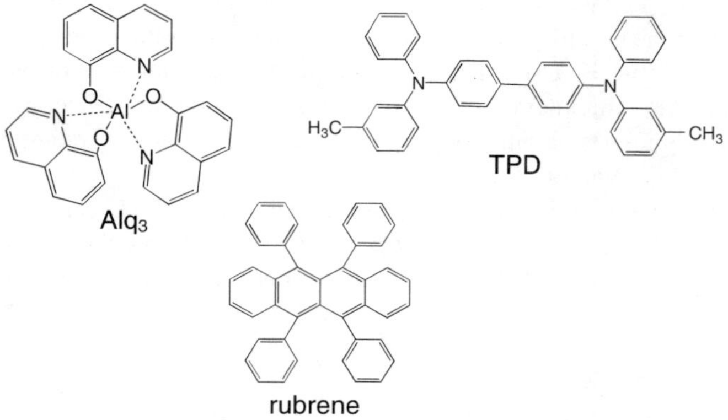

FIGURE 17.4. Chemical structure of the ETL (Alq_3), the HTL (TPD), and the emitter dopant rubrene.

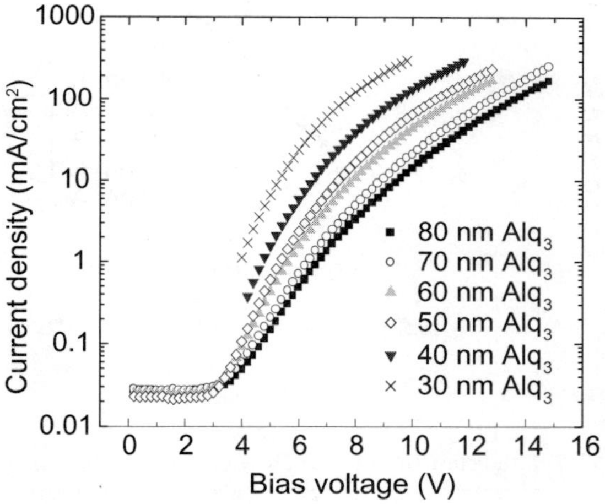

FIGURE 17.5. Current density versus bias voltage for various OLEDs having different ETL thicknesses ranging from 30 to 80 nm.

layers of Alq_3. The robotic arm controlling the movement of the shutter was programmed to move in a step fashion, thus resulting in decreasing ETL thickness as a new area of substrate is exposed to the Alq_3 deposition source. The throw distance between the Alq_3 source and the substrate is more than 1 m. This, along with the proper source geometry and loading of the Alq_3 material, allows for a uniform deposition over an area of 58 cm^2 with a variation in thickness uniformity of less than 3% over the substrate area. The Alq_3 was deposited at 0.8 Å s^{-1}.

The final step in device fabrication is the deposition of magnesium (Mg) cathode through a specially designed mask. The resulting devices are then transported in air and introduced into a nitrogen glovebox where they were tested and characterized. Although numerous Alq_3 thicknesses have been deposited in a single experiment, we shall limit our discussion to six of them (30, 40, 50, 60, 70, and 80 nm). Figure 17.5 shows the current density versus bias voltage. Owing to the increasing electric field, the current density increases with decreasing thickness, leading to an increase in the corresponding light output (Figure 17.6). Table 17.1 summarizes some of the characteristics of the devices made with the various Alq_3 thicknesses. The peak external quantum efficiency is reached for devices with Alq_3 thickness around 60 nm. In this case, the external quantum efficiency is 1.2% at an applied voltage of 10.8 V.

The changes in spectral emission with thickness of the ETL can also be monitored using combinatorial and spread techniques. Figure 17.7 shows the normalized electroluminescence spectra taken at the peak efficiency. For the devices in this study, the figure

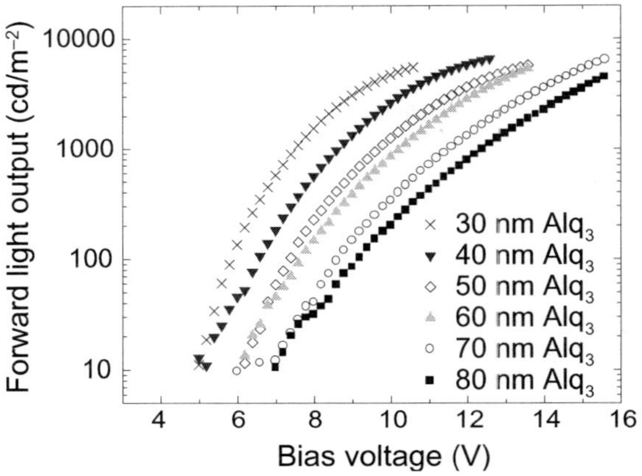

FIGURE 17.6. Forward light output versus bias voltage for OLEDs with ETL thicknesses ranging from 30 to 80 nm.

TABLE 17.1 Characteristics of OLEDs with Various Alq_3 Thicknesses

Thickness (nm)	Current density at 8 V (mA cm^{-2})	Forward light output at 8 V (cd m^{-2})	External quantum efficiency at 8 V (%)	Peak external quantum efficiency[a] (%)
30	124	3124	0.7	0.76 (7.6 V)
40	30	1303	1	1.05 (8.6 V)
50	16.5	556	1	1.1 (9.6 V)
60	10.8	365	1	1.2 (10.8 V)
70	4.9	138	0.8	1.07 (11.6 V)
80	3.3	85	0.76	0.97 (12.6 V)

[a]The bias voltage is given in parentheses.

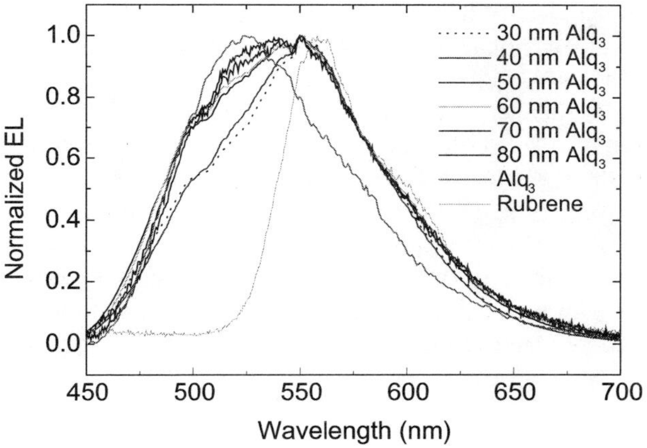

FIGURE 17.7. Electroluminescence spectra of various OLEDs with different thicknesses taken at peak efficiency. The emission spectra for rubrene and Alq_3 are also shown.

indicates that the recombination zone extends significantly into the Alq_3 layer for thicknesses between 50 and 80 nm. For thicknesses of 30 and 40 nm, the emission zone shifts towards the rubrene-doped HTL region. However, Alq_3 emission is still observed.

In displays, it is important that the light emission comes from the desired material (e.g., dopant) identifying the color of a given pixel, and not from the host or other layers in the device. Light emission from sources other than the dopant may lead to spectral contamination and in turn reduce image and display quality. Although this study was done by changing the thickness of the ETL only, it can be extended to the HTL, the cathode, or other layers in a given OLED structure.

DOPANT EFFECTS ON OLED PERFORMANCE

For given charge transport materials, the light output, efficiency, and lifetime of an OLED can be enhanced drastically by using dopants, usually with high luminescent efficiency. Depending on their chemical structure, dopants can be either fluorescent or phosphorescent. In either case, they require a host material when constructing an OLED. Although dopants can be highly luminescent, they can have several properties that preclude them from being used as pristine layers in the OLED architecture. Among the problematic features are the very poor charge mobility and aggregate formation (which leads to self-quenching and in turn lower device performance). The loading levels of the dopant can vary from 0.2 to 8% of the host material. Sampling the dopant concentration space, for a given host, can be tedious and inefficient. With traditional techniques it is also possible that the optimum concentration is missed. In this case, high-throughput techniques will be very helpful and allow a significant reduction in fabrication time.

Here we will give an example of the fabrication of OLEDs using combinatorial-based techniques that resulted in various dopant concentrations to be made in one run. The dopant is the red-light-emitting material 4-(dicyanomethylene)-2-*tert*-butyl-6(1,1,7,7-tetramethyljulolidyl-9-enyl)-4H-pyran (DCJTB) and an Alq_3 host material [22]. The chemicals used for this set of devices are shown in Figure 17.8.

A cross-sectional view of the device is shown in Figure 17.9. The copper pthalocyanine (CuPc), N,N′-di-(naphthalene-1-yl)-N,N′-diphenyl benzidine (NPB), and Alq_3 were deposited at 0.2–0.3 Å s^{-1}, 1 Å s^{-1}, and 0.3–2.5 Å s^{-1}, respectively. The deposition of the top Alq_3 layer was maintained at 0.8 Å s^{-1}. The Mg cathode was deposited at 10 Å s^{-1}. The CuPc was used to stabilize the hole injection contact between the ITO and the HTL.

Figure 17.10 shows the current density versus bias voltage for eight OLEDs having the structure shown in Figure 17.9 but with DCJTB concentrations of 0.1%, 0.3%, 0.6%, 1%, 1.5%, 2%, 2.5%, and 2.8%, respectively. The increase in voltage is mainly due to the trapping of holes at the dopant. Having the holes trapped at the dopant, for a given concentration of the dopant molecules, can lead to increase in quantum efficiency due to the direct recombination of electrons and holes on the DCJTB.

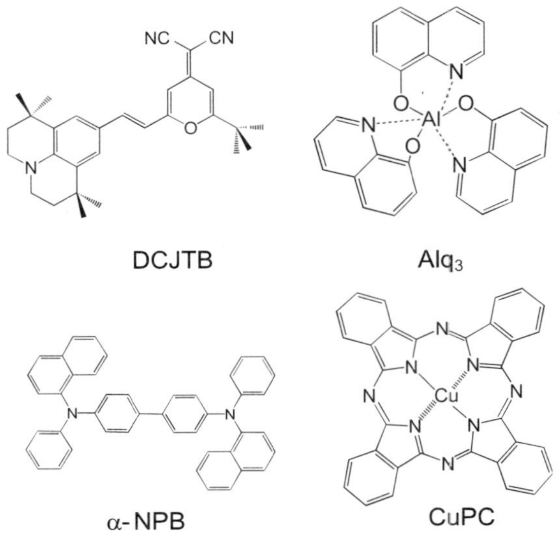

FIGURE 17.8. Organic chemicals used for the fabrication of red OLEDs.

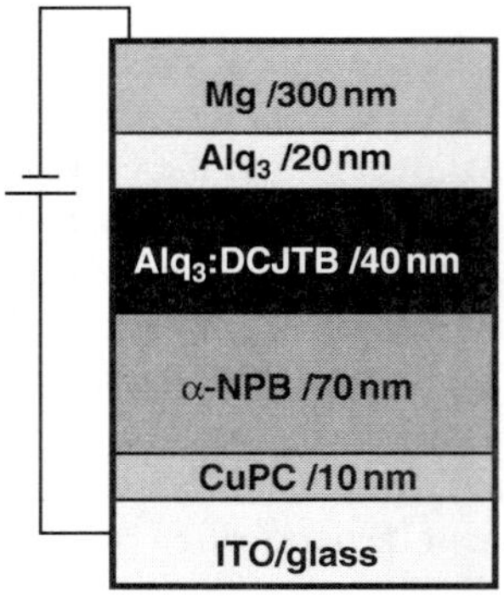

FIGURE 17.9. Schematic of red-light-emitting OLEDs using DCJTB as the light-emitting dopant. The thickness of each layer is shown in nanometers.

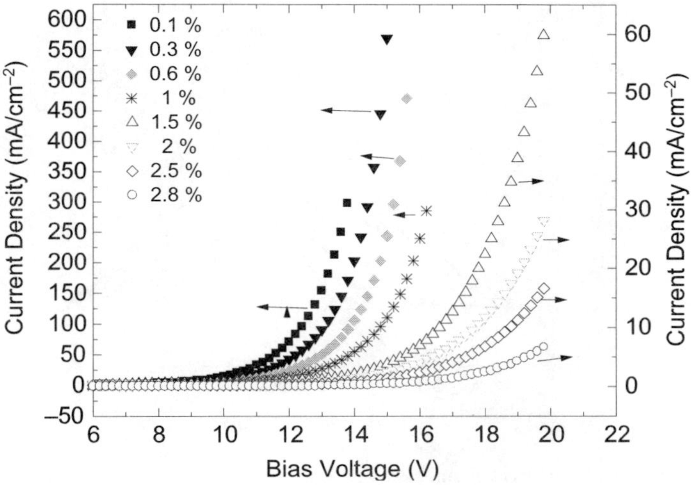

FIGURE 17.10. Current density versus bias voltage for OLEDs with different dopant concentrations. Note the increase in voltage with increasing dopant levels.

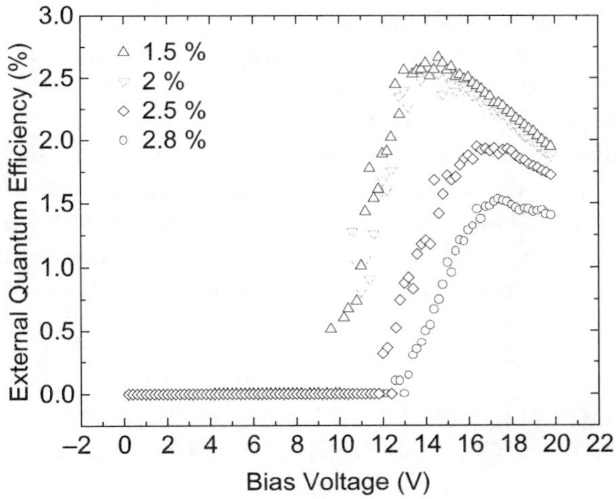

FIGURE 17.11. External quantum efficiency versus bias voltage for OLEDs with DCJTB concentrations of 1.5%, 2%, 2.5%, and 2.8%, respectively.

Although Förster energy transfer from the host to the dopant is the major channel for DCJTB excitation, direct charge trapping on the dopant site can also yield excited DCJTB states. A similar observation has been reported in OLEDs with 4-(dicyanomethylene)-2-methyl-6(p-dimethylaminostyryl)-4H-pyran (DCM) dye in an Alq_3 host [23]. However, increasing concentration may lead to dopant aggregation and self-quenching which in turn lower the quantum efficiency of the OLEDs, as indicated in Figure 17.11. The quantum

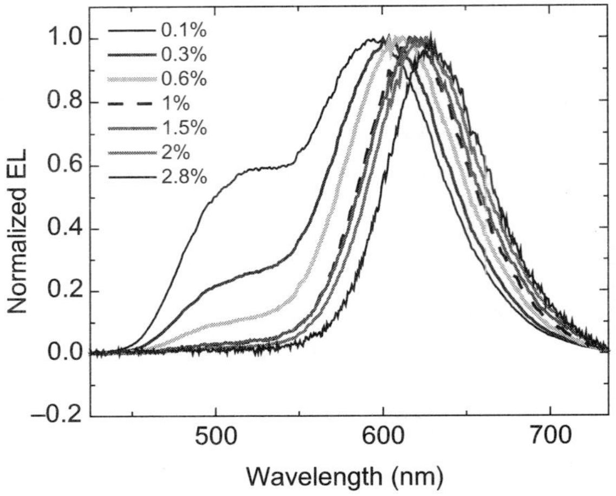

FIGURE 17.12. Dependence of spectral emission on dopant concentration. Notice the Alq_3 emission centered around 520 nm at low DCJTB concentrations.

efficiency reaches a maximum for DCJTB concentration in the range 1.5–2%, and then declines for higher dopant concentrations.

Figure 17.12 shows the normalized emission spectra for the DCJTB concentrations mentioned above. As the dopant concentration increases, the emitted spectra shift to resemble more that of DCJTB. If the Alq_3 emission is subtracted out, the resulting spectral shape and width does not change significantly as the concentration is increased, but rather moves deeper into the red region. The higher dipole moment of DCJTB compared with that of the host molecule might explain the observed red shift. Red shift in emission spectra can be due to an increase in the local polarization field near a given DCJTB molecule as the dopant concentration is increased. This behavior has been observed previously for laser dye molecules, namely DCM2 [23].

MODIFICATION OF SHEET RESISTIVITY VIA SIMPLE COMBINATORIAL INKJET TECHNIQUE

Inkjet printing is ideally suited to carry out high-throughput experiments. Here we will show an example of the use of commercially available inkjet printers to modify the sheet resistivity of a given conducting polymer.

The inkjet printer was slightly modified to accommodate our experiments. A special chemical (ink) with given viscosity was placed inside a precleaned cartridge. A computer is used to interface and command the printer to control the volume of ink disposed unto the substrate. PET substrates (11 in. × 8 in.) precoated with the conductive polymer poly(3,4-ethylenedioxy)-thiophene-poly(styrene sulfonate) (PEDOT-PSS), donated by Agfa, were used. They were placed in the paper compartment of the printer and the normal printing procedure was then carried out. Although more than one cartridge can be used, which results in much larger combinatorial library, we will focus here on the use of only the cartridge originally designed to dispose the black ink to carry out our experiments.

Although simple, this example demonstrates the power of inkjet technology in combinatorial science and device engineering.

The inkjet printing was controlled via the HSL (hue, saturation, luminosity or lightness) function using the existing values given in the PowerPoint software. The values for H, S, and L used in this experiment are shown in Figure 17.13(a). Only L was allowed to change, and nine different values are shown in Figure 17.13. The printed areas, which later were used in device fabrication, were 1 in. × 1 in. squares [Figure 17.13(b)]. After printing the ink, the substrates were processed and an HTL, ETL, and cathode were deposited on top of the treated PEDOT-PSS surface. The chemical structure of the conductive polymer is shown in Figure 17.14(a). Printing of the ink capable of altering the sheet resistivity of the PEDOT-PSS layer, along with the treated substrate, are illustrated in Figure 17.14(b).

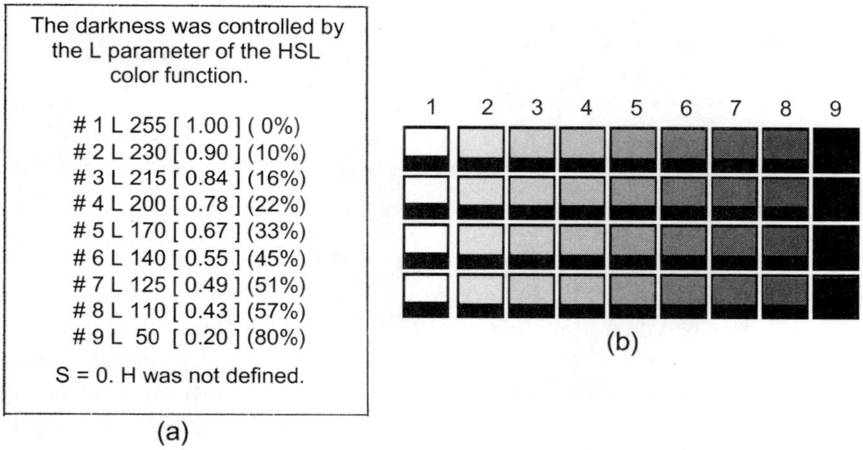

FIGURE 17.13. (a) HSL values used in the printing of the chemical upon placing it in the inkjet cartridge, and (b) schematic of 1 in. × 1 in. array of printed areas which were later used to fabricate devices.

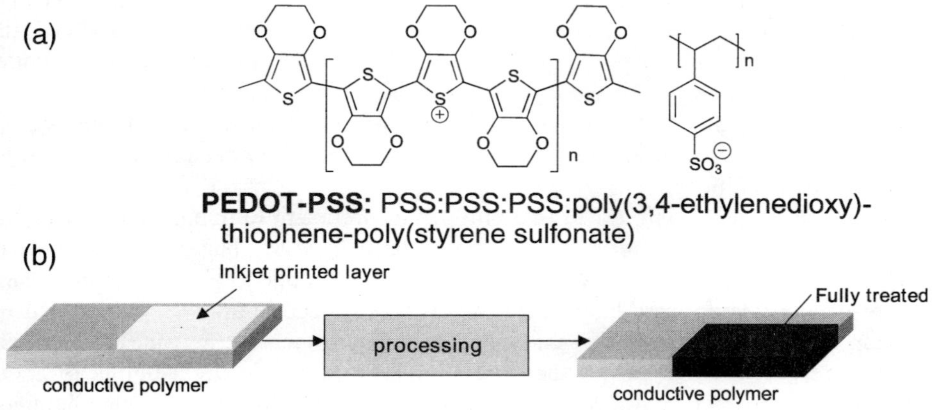

FIGURE 17.14. (a) Chemical structure of PEDOT-PSS; (b) illustration of inkjet printing of the chemical (left panel), and the resulting substrate upon processing (right panel).

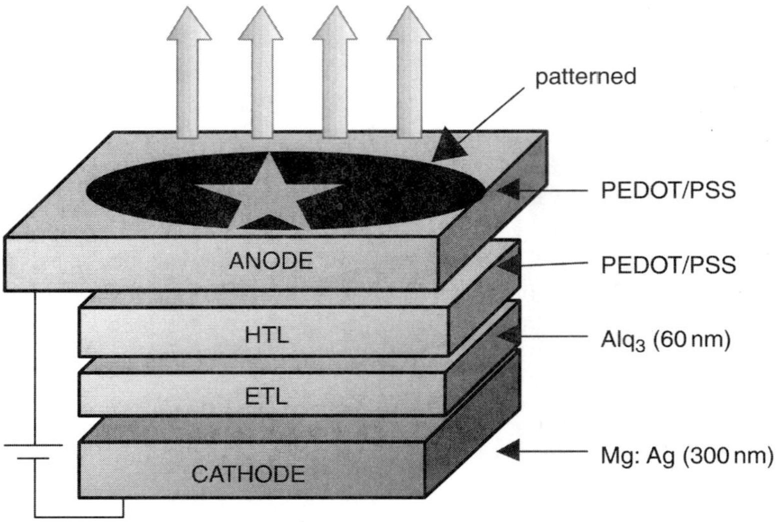

FIGURE 17.15. Schematic of OLEDs fabricated where the PEDOT-PSS was patterned using the inkjet. No patterning was carried on the HTL, ETL, or cathode.

The HTL was based on a solution of TPD, polycarbonate, and chloroform prepared at a solute mass ratio of 7:3 (TPD : polycarbonate) using 10 mg per 1 ml of solvent. The HTL was spun coated to yield a thickness of about 50–60 nm. Alq_3 was used as the ETL and was vacuum deposited at a rate of 1 Å s^{-1} to a thickness of 60 nm. Magnesium and Silver were co-deposited at a ratio of 10:1. The thickness of the metal cathode was 300 nm. Figure 17.15 shows the structure of the OLEDs. It is worth noting that the cathode was not patterned but rather deposited over the whole ETL. Only the PEDOT-PSS on the flexible PET substrate was patterned via the inkjet process described above. The resistivity for the various HSL values was measured using the four-point-probe method. The values obtained are shown in Figure 17.16. As the luminosity value decreases (increasing ink content), the sheet resistivity increases accordingly from $6 \times 10^2\ \Omega\ sq^{-1}$ for the untreated (white) region to $4 \times 10^4\ \Omega\ sq^{-1}$ for the fully treated black region (highest ink content). The values shown in Figure 17.16 represent some of the printed elements with various degrees of darkness as indicated in Figure 17.13(a). The sheet resistivity can also be affected by the processing condition.

The OLEDs made were tested in a nitrogen environment to minimize moisture and oxygen effects on device performance. The current density versus bias voltage is shown in Figure 17.17. As expected, the figure indicates a shift in applied voltage with increasing sheet resistivity. The forward light output versus bias voltage follows the current density behavior. Figure 17.18 shows that the voltage required to achieve a given level of light output increases with increasing sheet resistivity. For example, light output of about 110 cd m^{-2} requires 9.6 V, 10.2 V, 11.2 V, 12.8 V, and 14.8 V for devices having lightness values [Figure 17.13(a)] preset to 255, 230, 200, 140, and 110, respectively.

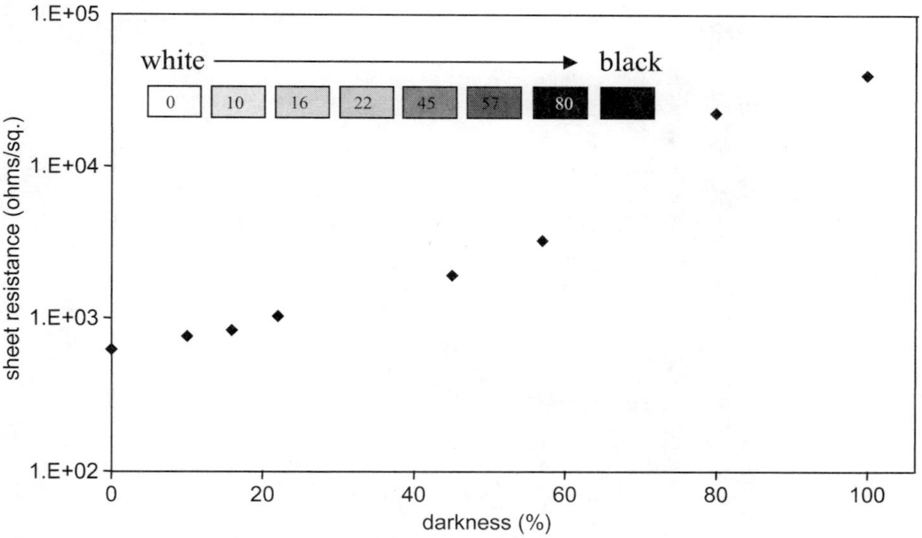

FIGURE 17.16. Measured sheet resistivity for various darkness values. The measurements were carried at room temperature in air using a four-point probe.

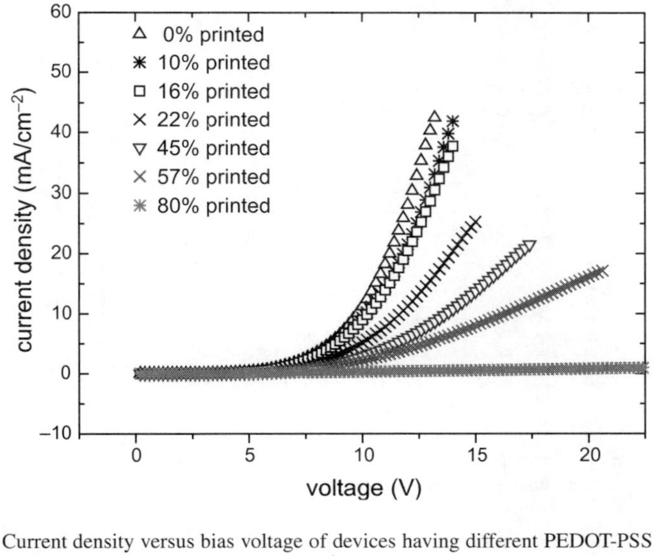

FIGURE 17.17. Current density versus bias voltage of devices having different PEDOT-PSS sheet resistivity.

Studies are currently under way to understand the mechanism behind the change in sheet resistivity upon the interaction of the printed ink with the polymeric anode. In this regard, preliminary studies using Fourier transform IR (FTIR) spectroscopy show no obvious changes in the chemical structure of the PEDOT-PSS upon interaction with the printed ink. Although Figure 17.19 indicates no detectable changes in the chemical structure of

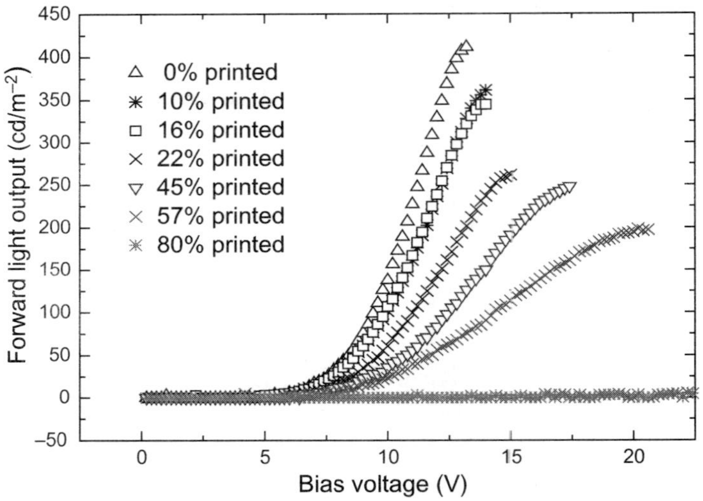

FIGURE 17.18. Forward light output versus bias voltage for OLEDs made with PEDOT-PSS anodes having various resistivities.

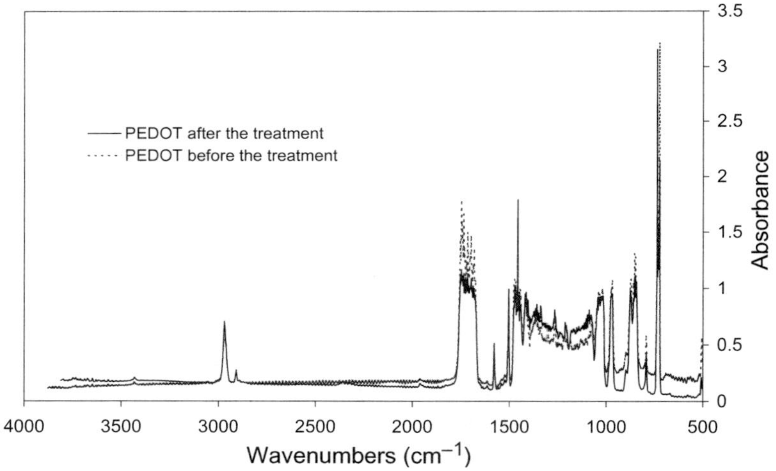

FIGURE 17.19. FTIR spectra of a PEDOT-PSS anode before (pink) and after (dark) treatment. Notice the overall similarity of the vibrational peaks in both spectra.

PEDOT-PSS before and after treatment with the deposited ink, these results do not necessarily exclude the possibility that some chemical changes might be occurring on a scale not easily resolved with the FTIR. However, the preliminary results indicate that the ink used in these experiments appears to have a direct effect on the surface morphology of the PEDOT-PSS. Figure 17.20 shows a decrease in mean surface roughness of the polymeric anode with increasing ink deposition (thus increased sheet resistivity).

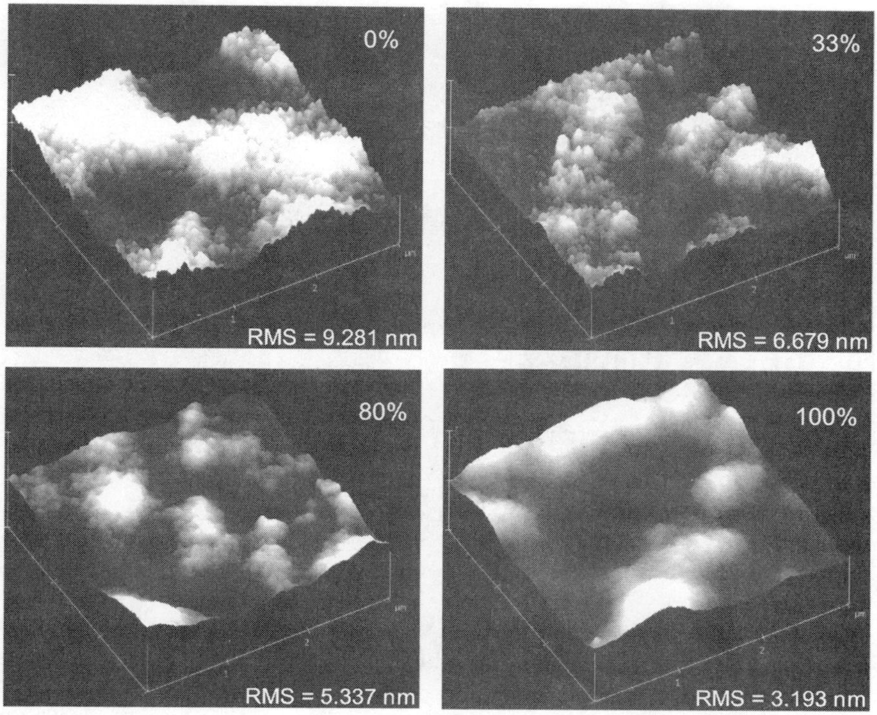

FIGURE 17.20. Atomic force microscope images of PEDOT-PSS layers with different sets of HSL values.

The above study of modifying the sheet resistivity of the polymer anode shows that the OLED fabrication process can be simplified significantly. In this case, fabricating an OLED of any shape or design simply requires inkjet printing of the image on the substrate and then the deposition of an HTL, an ETL, and a cathode. No specially designed deposition or lithography masks are needed, as is the case in current fabrication methods. For example, fabricating the light-emitting scorpion shown in Figure 17.21 required sending the image [Figure 17.21(a)] to the printer, then followed by the spin coating of the HTL and vacuum deposition of the ETL and cathode. No patterning is needed for the HTL, the ETL, or the cathode. On the other hand, fabricating the same image using traditional techniques cannot be achieved with such ease, especially if a gray region is to be shown in full detail.

In summary, these simple examples show the potential of combinatorial and high-throughput techniques in the rapid fabrication and optimization of OLEDs. It is anticipated that these tools will be instrumental in the rapid synthesis and screening of new materials, as well as in the discovery of new device structures. Moreover, rapid experimental verification of a given fundamental phenomenon can also benefit from such methods.

FIGURE 17.21. An image of a scorpion (left panel) that was printed on PEDOT-PSS surface and then made into an organic light-emitting device (right panel). The comet-shaped defects are due to dust particles present on the substrate during the spin coating of the HTL.

ACKNOWLEDGMENTS

We would like to acknowledge partial support of this project by NSF and DARPA. Also, we are grateful for the donation of the PEDOT-PSS substrates by Agfa, and the inkjet print heads by HP. We thank our colleagues P. Calvert and N. Peyghambarian for fruitful discussions.

REFERENCES AND NOTES

1. In this chapter OLEDs refer to either polymer-based or small-molecule-based devices.
2. Tang, C. W., Van Slyke, S. A. *Appl. Phys. Lett.* **1987**, *51*, 913.
3. Burroughes, J. H., Bradley, D. C., Brown, A. R., Marks, R. N., Mackay, R. N., Friend, R. H., Burns, P. L., Holmes, A.B. *Nature* **1990**, *347*, 539.
4. Sheats J. R. et al. *Science* **1996**, *273*, 884–888; Tang, C. W. *SID International Symposium. Digest of Technical Papers*, **1996**, pp. 181–184; for a review on the physics of OLEDs, see Campbell, I. H., Smith, D. L. *Solid State Physics* **2000**, *55*, 1.
5. Haskal, E. I., Büchel, M., Duineveld, P. C., van de Weijer, P. *MRS Bull.* **2002**, *27*, 864.
6. Wakimoto, T., Kawami, S., Nagayama, K., Yonemoto, Y., Murayama, R., Funaki, J., Sato, H., Nakada, H., Imai, K. *International Symposium on Inorganic and Organic Electroluminescence, Hamamatsu*, **1994**, p. 77.
7. Hung, L. S., Tang, C. W., Mason, M. G. *Appl. Phys. Lett.* **1997**, *70*, 152.
8. Jabbour, G. E., Kawabe, Y., Shaheen, S. E., Wang, J. F., Morrell, M. M., Kippelen, B., Peyghambarian, N. *Appl. Phys. Lett.* **1997**, *71*, 1762.
9. Jabbour, G. E., Morrell, M. M., Shaheen, S. E., Kippelen, B., Peyghambarian, N. *9th International Workshop on Inorganic and Organic Electroluminescence*, **1998**, p. 49.
10. Jabbour, G. E., Kippelen, B., Armstrong, N. R., Peyghambarian, N. *Appl. Phys. Lett.* **1998**, *73*, 1185.
11. Tang, C. W., Van Slyke, S. A., Chen, C. H., *J. Appl. Phys.* **1989**, *65*, 3610.
12. Van Slyke, S. A., Chen, C. H., Tang, C. W. *Appl. Phys. Lett.* **1996**, *69*, 2160.

13. Baldo, M. A., O'Brien, D. F., You, Y., Shoustikov, A., Sibley, S., Thompson, M. E., Forrest, S. *Nature*, **1998**, *395*, 151.
14. Choong, V.-E., Shi, S., Curless, J. *Appl. Phys. Lett.* **2000**, *76*, 958.
15. The term "combinatorial and spread techniques" has been used loosely in many recent publications, including this chapter, even when the experiments are not truly combinatorial. Today, the term is synonymous in many cases with high-throughput techniques.
16. Xiang, X.-D., Sun, X.-D., Briceño, G., Lou, Y., Wang, K.-A, Chang, H., Wallace-Freedman, W. G., Chen, S.-W., Schultz, P. G. *Science* **1995**, *268*, 1738.
17. Jabbour, G. E. Unpublished work, **1997**.
18. Jabbour, G. E. In Combinatorial methods fabricate libraries of organic light emitters: Newsbreaks. *Laser Focus World* **1999**, *13*.
19. Shmitz, C., Posch, P., Thelakkat, M., Schmidt, H. W. *Phys. Chem. Chem. Phys.* **1999**, *1*, 1777.
20. Schmitz, C., Thelakkat, M., Schmidt, H. W. *Adv. Mater.* **1999**, *11*, 821.
21. Zou, L., Savvate'ev, V., Booher, J., Kim, C.-H., Shinar, J. *Appl. Phys. Lett.* **2001**, *79*, 2282.
22. Chen, C. H., Tang, C. W., Shi, J., Klubek, P. *Thin Solid Films* **2000**, *363*, 327.
23. Bulović, V., Shoustikov, A., Baldo, M. A., Bose, E., Kozlov, V. G., Thompson, M. E., Forrest, S. R. *Chem. Phys. Lett.* **1998**, *287*, 455.

Chapter 18

High-Throughput Conductivity Measurements of Thin Films

Qi Wang

CONDUCTIVITY MEASUREMENT TECHNIQUES

Introduction

In studying the electrical properties of thin film materials, conductivity is a commonly used parameter which is relatively easy to measure. Almost everyone who knows how to use an ohmmeter can perform the measurement. This property often offers a quick screen or glimpse of the quality of the films, although other parameters, which are harder to measure, such as mobility and carrier lifetime are also important in characterizing the electrical properties of films. In a given electric field, higher-conductivity materials such as metals yield higher current densities and lower-conductivity materials such as insulators yield lower current densities. Conductivity is product of carrier mobility, carrier density, and electronic charge.

High-throughput methods deposit many different materials on a single substrate. Characterization of these very condensed material samples offers a great challenge. The most conventional conductivity measurement uses contacts and requires depositing electrodes to do the testing. This destructive measurement has some limits for high-throughput samples: the electrodes can be difficult to deposit on the condensed samples. A contactless conductivity measurement is preferred. Laser-induced microwave conductivity measurement [1] and the recently developed evanescent microwave microscope technique [2] look very promising. These nondestructive measurements and other scanning techniques fulfill the need for the high-throughput sample characterizations. In combination with high-throughput deposition, the likelihood of discovering new materials has been greatly improved.

In this chapter, many conductivity measurement techniques are presented—from conventional contact electrodes to modern evanescent microwave microscopy, from physical contact measurement to the contactless measurement, from a single measurement to the automatic-scanning two-dimensional sample. Finally, an example of measuring an

Qi Wang • National Renewable Energy Laboratory, Golden, CO, USA
Contact author: Qi Wang, email: qi_wang@nrel.gov

High-Throughput Analysis, edited by Potyrailo and Amis
Kluwer Academic Publishers, New York, 2003.

amorphous silicon-based high-throughput sample is used to demonstrate the feasibility of high-throughput electrical characterization.

Physical Contact

The most commonly used technique to measure the conductivity of thin films is measuring the current–voltage (I–V) dependence on a sample with patterned metal contact electrodes. To complete the testing circuit, the testing wire must make a solid physical contact to the electrode. Based on Ohm's law and other geometric parameters such as film thickness d, length l, and width w of the area under test, the conductivity σ can be easily calculated from

$$\sigma = \frac{I}{V}F \tag{1}$$

where F is a geometric factor of the material defined in terms of d, l, and w, which depends on the direction of current flow.

Two or four terminals are often used, depending on the conductivity of the material and the permitted measurement error. In a situation of high conductivity such as metal films, four terminals are preferred. For low-conductivity thin films such as semiconductors, two-terminal testing is sufficient. However, for thin films with very low conductivity, such as an insulators, care must be taken to prevent the surface conduction from dominating the measurement and a special electrode is required. In this section, various configurations of electrodes are presented to accommodate the need for the large variation of thin films with regard to conductivity.

Coplanar Structure

The simplest and most direct method of measuring conductivity is using a coplanar structure in which two-metal stripes are deposited on the samples with gaps between them. Figure 18.1 shows various configurations of contact patterns. By measuring the I–V characteristic, the conductivity of the film can be determined readily based on its thickness and $F = l/wd$. The conductivity is given by

$$\sigma = \frac{I}{V}\frac{l}{wd} \tag{2}$$

The conductivity of an a-Si:H intrinsic film that is 1 μm thick, is typically on the order of $10^{-10}\,\mathrm{S\,cm^{-1}}$. The configuration of $l/w = 0.1$ has been used with $w = 10\,\mathrm{mm}$ and $l = 1\,\mathrm{mm}$. If a voltage of 100 V is applied between the two electrodes, a current in the range of picoamperes will be expected. This configuration has another advantage in that it can be used to measure photoconductivity, in which the conductivity is enhanced under illumination. Light is shone onto the sample between the gap. This small ratio of l/w configuration is suitable for low-conductivity films. But consideration must be given as to whether the measurement is surface sensitive, and some unwanted surface conducting channel may cause uncertainty in the measurement. The metal contacts to the semiconductors often form a junction. This junction acts as a barrier to the carrier transport. However, the barrier is distant from the testing film region and this effect can be minimized.

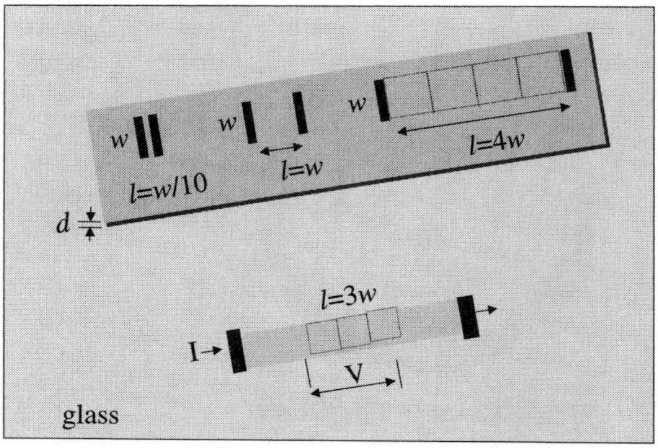

FIGURE 18.1. Metal-contact patterns of coplanar configurations for conductivity measurements: upper panel shows the various l/w ratio configurations of metal-contact patterns for low-conductivity and semiconducting films, and the lower panel shows four-terminal pattern for high-conductivity films. The current flow is in the plane of the film.

For semiconducting film, the configuration $l/w = 1$ can be used. What is measured is, by definition, the sheet resistance, and it is easy to convert this to conductivity based on film thickness. For conducting films, $l/w > 1$ is favored because increasing l results in increasing resistance and leads to a more reasonable reading for the current. Figure 18.1 shows $l/w = 4$, and the sheet resistance is a quarter of what was measured.

For highly conducting thin films such as metals or doped films, the contact resistance between the film and the end of electrode may cause a problem. To minimize or avoid this problem, four-terminal measurement is preferred. The configuration in the lower panel of Figure 18.1 shows the details. A constant current is passed through two outer electrodes, and the voltage that is measured across $l = 3w$ inside the two electrodes gives a more accurate value. The contact resistance at the current terminals is avoided. The sheet resistance is one third of what was measured.

Sandwich Structure

In some cases, a sandwich structure is needed to study transport in a direction normal to the film plane or for very low conductivity film. For example, microcrystalline film conductivity shows asymmetrical carrier transport behavior because of the grain size and the grain orientation. A sandwich structure is used to measure transport normal to film thickness and coplanar for in-plane transport. The geometric factor $F = d/S$, where S is the area in the direction of current flow.

The sandwich configuration is shown in Figure 18.2 [3]. Because the film is very thin, d is much less than the diameter of the top electrode; the edge effect can be neglected. However, metal electrodes often make a Schottky contact to the semiconductor or insulator and, even worse, the contact forms a barrier to block the flow of current. The barrier at the contact may cause a problem. To overcome this barrier, an ohmic contact made by inserting a very thin doped layer between the metal electrode and measured film is often used. An I–V curve will often be measured to check the contact. An ohmic contact shows a linear relation between I and V. At high voltage, nonohmic contact can be observed, for example, the space-charge-limiting current which depends on $V^{3/2}$.

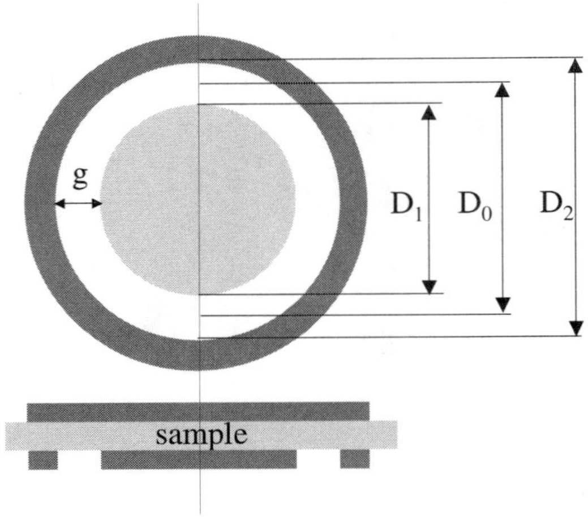

FIGURE 18.2. The definition and geometry of a circular sandwich electrode sample for measurement of very low conductivity.

For measurements of insulating thin films with very low conductivities, such as SiO_2, Si_3N_4, and SiC, dielectric films and some intrinsic a-Si:H films, the conductivities are in the range 10^{-11}–10^{-17} S cm^{-1}. The electrical measurement in this range of conductivity is more complicated in some ways: first, a good contact must be made between metal electrodes and insulator; secondly, surface leakage must be prevented; thirdly, the properties of the film depend on the growth conditions. For these special measurements, Wellard [4] developed a guard-ring technique using circular electrodes for very low conductivity films, which can prevent surface-state leakage. The sample configuration is shown in Figure 18.2. The circular sample configuration consists of a top circular electrode, a ring electrode, and a guarded electrode. The diameter D_1 of the guarded electrode and the inner diameter D_2 of the ring electrode are the most important factors that affect the measurement.

The conductivity σ is defined as the electrical conduction through the volume of an insulator material between top and bottom electrodes. It is measured in the sandwich or capacitance structure as shown in Figure 18.3. By applying a voltage across the top electrode and the bottom-guarded electrode of the sample and measuring the resultant current that passes through the volume of the sample, the conductivity σ can be calculated from:

$$\sigma = \frac{4d}{\pi D_1^2}\frac{I}{V} \quad (d \ll D_1). \tag{3}$$

When $d \ll D_1$, the edge effect can be neglected.

The structure of the film often depends on the growth conditions. Even with the same composition, the grain size and the orientation can be quite different in the plane of the film and in the direction of film thickness. Impurities or physical defects may cause a small leakage. All these behaviors are often observed in the I–V characterization of the insulator, in which Ohm's law is only obeyed at very low voltages and sub- or superlinear behavior is observed at high voltages.

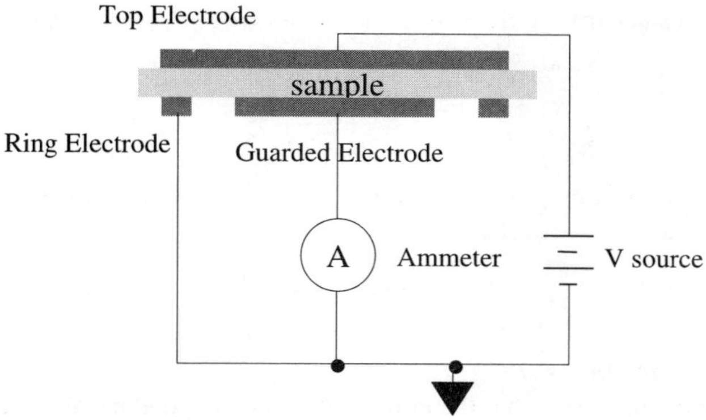

FIGURE 18.3. The circuit diagram of volume conductivity measurement. (Reproduced from ref. 3.)

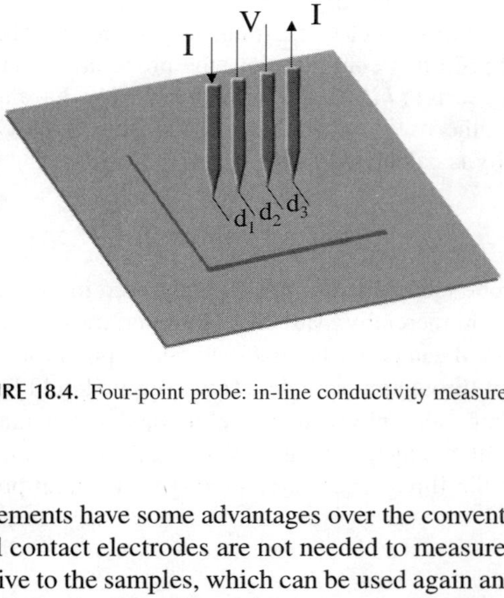

FIGURE 18.4. Four-point probe: in-line conductivity measurement.

Four-Point Probe

Probe-tip measurements have some advantages over the conventional contact-electrode technique. First, metal contact electrodes are not needed to measure the conductivity. This feature is less destructive to the samples, which can be used again and again for many other measurements. Second, the size of probes can be very small so that a variation of conductivity along the sample can be measured. Third, with the aid of computer-controlled automation, the probe technique can easily be modified to scan large two-dimensional high-throughput samples.

The most commonly used probe techniques are the four-point probe methods. An in-line four-point probe is illustrated in Figure 18.4. Four probes, which are separated by distances d_1, d_2, and d_3, make a physical contact to the film. The two outer probes pass current I and the two inner probes measure voltage V. A detailed description of in-line four-point probes has been given by Valdes [5]. The conductivity of the semi-infinite film is given by

$$\sigma = \frac{I}{V}\,\frac{1/d_1 + 1/d_3 - 1/(d_1 + d_2) - 1/(d_2 + d_3)}{2\pi} \tag{4}$$

When the distances separating the probes are equal ($d_1 = d_2 = d_3 = d_0$), a simple formula is obtained:

$$\sigma = \frac{I}{V}\frac{1}{2\pi d_0} \tag{5}$$

If the thin film is an infinitely thin slice resting on an insulating substrate, Eqn (5) can be further simplified to

$$\sigma = \frac{I}{V}\frac{\ln 2}{\pi d} = 0.221\frac{I}{V}\frac{1}{d} \tag{6}$$

where d is the film thickness.

The beauty of Eqn (6) is that the conductivity is independent of the probe spacing d_0 on the condition that the material is an infinitely thin film sitting on an insulator. However, for finite thickness, edge effects must be taken into account; interested readers can refer to the literature [6,7] for the exact corrections. For routine semiconductor wafer monitoring, a probe spacing of 25 μm is commonly used.

For high spatial resolution electrical testing, a square four-probe array [8] is preferred, as shown in Figure 18.5. In this configuration, the probe array makes a physical contact with the film, a constant current I is passed through any two adjacent probes, and voltage V is measured using the other two probes that are opposite to the two current conducting probes. The conductivity is calculated from

$$\sigma = 0.11\frac{1}{d}\frac{I}{V} \tag{7}$$

The four-point probe technique has been widely used in the semiconductor industry, and probe stations are commercially available. However, most stations are designated for wafer testing. This method can easily be modified and applied for characterizing the electrical properties of high-throughput samples. However, it should be noted that four-point probe techniques, like any other physical contact method, can damage the film, especially if the film is thin and soft and the probe tip is sharp, if the contact pressure is too high and the tip is dragged over the films. Therefore, although four-point probe techniques can be

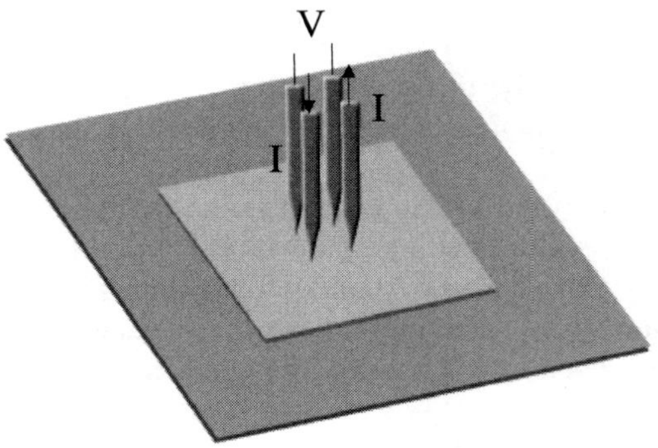

FIGURE 18.5. Four-point probe: square configuration conductivity measurement.

used to test various locations on the sample, they should not be used for x–y scanning across samples.

Gold and Mercury Probes

To enable probe techniques to be used for scanning a large sample area without damaging the film, a better method of physical contact must be found. Fortunately, two metals, gold and mercury, have provided the answer. Also, improved spring-load probe tips have reduced the uncertainty of contact between probes and samples. The pressure between the contacts can be controlled.

Gold has very high conductivity and is environmentally stable. It is the most commonly used metal in probes or probe tips, which are coated with a thin gold layer. A gold wire forms a spherical ball at the tip when liquefied using a torch. This very smooth spherical ball makes a better contact to the film without damaging it. In practice, one can freely move or drag the ball-shaped gold tip on the surface of the film for most thin-film measurements. The contact area between the ball-shaped gold tip and the thin film can be small ($\ll 1\,\text{mm}^2$) but ill defined. The actual area often depends on other conditions such as the contact pressure and the mechanical properties of the thin film. However, the area can be calibrated by measuring the capacitance of a film with known thickness and dielectric constant.

Despite safety concerns, the liquid metal mercury has been used for scanning conductivity measurement. At normal operational temperature mercury is in the liquid state and its conductivity is high. Mercury tip is almost an ideal tip for avoiding scratching the film and it can scan over the samples.

Figure 18.6 shows the technique of using a mercury-filled probe tip for electrical characterization [9]. In this testing setup, the film was deposited on a metal-coated substrate. This metal acts as the back-electrode to which one testing probe is attached. The mercury-filled probe tip ($< 1\,\text{mm}^2$) is used as the top electrode. Once the tip makes good contact with the film, the conductivity and dielectric constant of the film can be readily determined by measuring ac capacitance and conductance, the size of the mercury tip, and the thickness of the film. I–V testing can also be performed. To scan over the other locations of the sample, a computer-controlled or automated x–y position manipulator moves

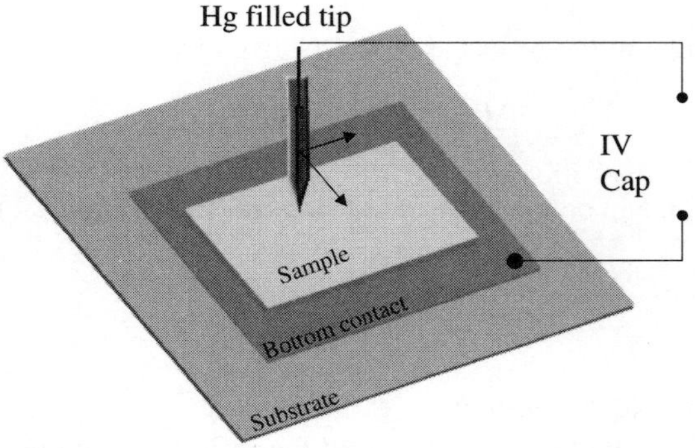

FIGURE 18.6. Schematic of scanning conductivity measurement with a mercury-filled probe tip.

the tip. This is an extremely useful low-cost technique for scanning high-throughput samples that require two-dimensional measurements. The drawback of the mercury tip technique is that the data, are sometimes, less reliable in I–V measurement because a contact barrier is formed between the mercury and the film. Therefore care must be taken and preliminary I–V testing is necessary before making the measurements. Gold or mercury probes are preferred in sandwich structure devices.

Noncontact Measurement

Noncontact (contactless) measurement techniques have advantages over the contact measurements described above in that the spatial resolution can be much higher and the measurement is nondestructive. Therefore samples can be used again for measuring other properties. This feature is especially useful for those high-throughput samples that have dense and different materials deposited on a single substrate—material library on chip. However, only a few techniques have been developed, such as laser-induced microwave conductivity measurement and the evanescent microwave microscope technique. Both are contactless measurements using the principles of microwave absorption in materials.

Scanning Laser-Induced Microwave Conductivity

The absorption and reflection of microwaves at the surface of a thin film depend on the geometric and dielectric properties of the film. As mentioned in an earlier section, conductivity increases when light of an energy greater than the bandgap illuminates a semiconductor. The change $\Delta\sigma$ in the conductivity is due to the photon-excited carrier increase in the semiconductor. Therefore the dielectric properties, or at least the imaginary part of the dielectric constant, change because of the increase in conductivity. The laser-induced microwave conductivity technique [1,10,11] measures the change in reflection of the microwave at the film according to the increase in conductivity with light.

If $P(\sigma)$ is the reflected microwave power at conductivity σ, $P(\sigma + \Delta\sigma)$ is the reflected microwave power when the light is shining on the sample. The change in reflected microwave power due to increased conductivity is

$$\Delta P = P(\sigma + \Delta\sigma) - P(\sigma) \tag{8}$$

and the relative change in the reflected power is given by

$$\frac{\Delta P}{P} = \frac{P(\sigma + \Delta\sigma) - P(\sigma)}{P(\sigma)} \tag{9}$$

For a small change in conductivity and using the Taylor expansion series, Eqn (9) can be simplified to

$$\frac{\Delta P}{P} = A\Delta\sigma \tag{10}$$

where A is a proportionality constant.

In the simple case, the relative change in the reflected microwave is proportional to the change in conductivity. In practice, Eqn (10) is not valid for materials that undergo

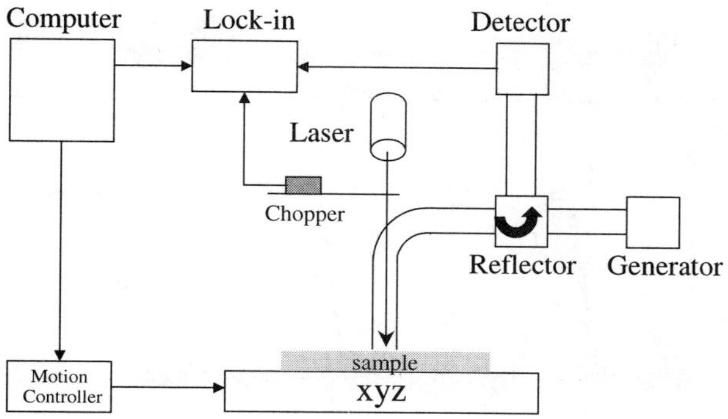

FIGURE 18.7. Experimental setup for a scanning contactless laser-induced microwave conductivity measurement. (Reproduced from ref. 1.)

a large change in conductivity after illumination. Unfortunately, for most semiconductors the change in photoconductivity is very large. Also, the analysis above assumes a uniform change of photoconductivity in the materials. In general, the photon absorption normal to the film direction is not uniform in thicker films. However, this assumption is valid for conditions in which the absorption length of the laser light is much greater than the film thickness.

The scanning light-induced microwave conductivity apparatus is shown schematically in Figure 18.7. The sample under test is mounted on an *x–y–z* stage that is controlled by a motion controller via a computer. A waveguide is used to reduce the geometrical problem that is related to the free electromagnetic field and also to reduce other background noise. The microwave power, which can change in frequency from 26.5 to 40 GHz (K_α band), is generated by a Gunn oscillator. This power (<1 mW) is guided to the sample through a broadband isolator and circulator. Then the power reflected by the samples is measured with a diode and the diode output is fed back to a lock-in amplifier. The sample, which is located at the center of the waveguide, is illuminated by a He–Ne laser (632.8 nm). The laser beam can be focused with an optical lens to a spot of 2 μm in diameter (not shown in the figure). The light intensity is adjusted using a neutral density filter and is always less than 3 mW. The chopped laser beam is used as the reference signal to the lock-in amplifier. The change in reflected microwave power due to the light is measured as the output from the lock-in amplifier. The scanning of the sample is done by the motor-drive *x–y* stage. The light-induced conductivity change is recorded and displayed on a computer.

Figure 18.8 shows a test result obtained using the scanning light-induced microwave apparatus. A p-type c-Si wafer with a dark conductivity of 20–30 S cm^{-1} was used for testing. The sample was masked with metal stripes that varied the gap. The sample cross section is shown in the upper panel of the figure. Position A has a gap of width 10 μm. The direction of the scan is perpendicular to the metal stripes. The microwave signal from the sample in the lower panel of the figure agrees well with the structure of the sample. There are no signals from the metal and the resolution is better than 5 μm (see position A in the figure).

The scanning light-induced microwave conductivity technique has its limits with regard to photosensitive semiconductor materials. However, this technique has proved the

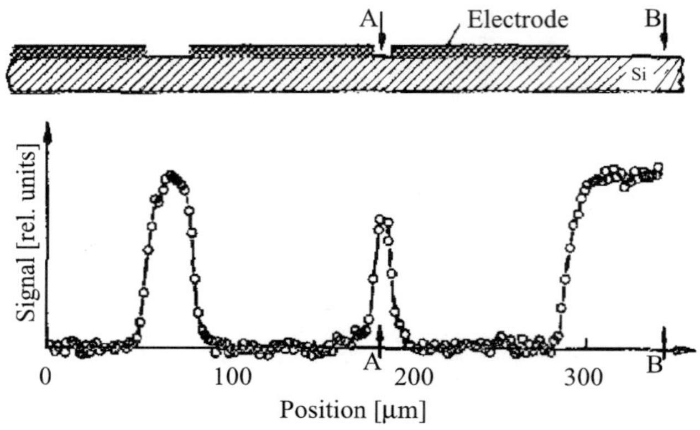

FIGURE 18.8. Experimental data for a microwave signal using the scanning setup shown in Figure 18.7. A p-type silicon wafer with metalized electrodes is shown in the upper panel. Position A is the center of 10 μm gap between metal electrodes, and position B is about 50 μm away from the electrode. An attenuated He–Ne laser ($\lambda = 632.8$ nm, power < 3 mW) was used and focused to about 2 μm in diameter and positioned at the center of the waveguide. (Reproduced from ref. 1.)

principle of contactless and scanning techniques for measuring conductivity. This provides a powerful tool for characterization of high-throughput samples.

High-Resolution Scanning Evanescent Microwave Microscope

The high-resolution scanning evanescent microwave microscope recently developed by Xiang and co-workers at the Berkeley National Laboratory [12] is a more generic testing tool for electrical property characterization. This microscope can be used for conductivity, surface resistance, superconductivity, dielectric constant, and magnetoresistance characterizations. This instrument has very high spatial resolution (~100 nm) [13], and the capability of scanning an area of up to 5 cm × 5 cm. It is a powerful tool for high-throughput samples with large materials libraries. In addition, temperature-dependent characterization can easily be conducted by implementing this microscope in a cryogenic dewar with a variable temperature from 4.2 to 300 K. In this section, the principle of evanescent microwave technique, the instrument apparatus, and the surface conductivity testing will be illustrated.

The evanescent microwave microscope utilizes the interaction between microwave and test material so that the properties of materials can be deduced by measuring and interpolating the interactions [2,14]. The sensitivity of the instrument has been greatly improved by using an evanescent microwave that enables high spatial resolution and the ability to measure a wide range of change in the properties.

The core of the instrument is the evanescent resonator probe shown in Figure 18.9. A $\lambda/4$ coaxial resonator is the basis for the probe. A needle-shaped metal tip is mounted in the center conductor and extended through a small aperture of the shield plate (see the enlargement on the left of Figure 18.9). The shield aperture plate is a sapphire disk with a small hole in the center. The coupling loops pass and receive the microwave as a source and a detector of the signals. The microscope operates at relatively low frequencies, from

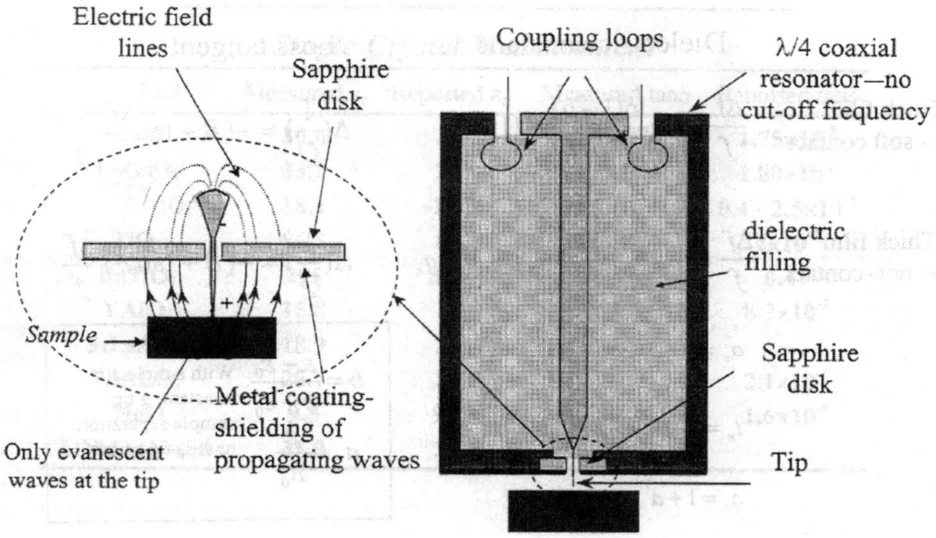

FIGURE 18.9. Details of the scanning evanescent microwave resonator probe. (Reproduced from ref. 12.)

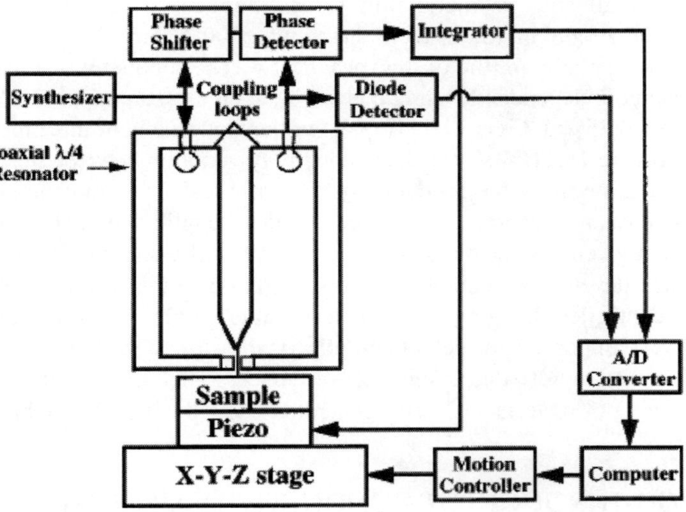

FIGURE 18.10. The experimental setup for a high-resolution scanning evanescent microwave measurement technique. (Reproduced from ref. 15.)

a few hundreds megahertz to 10 GHz. The electric field distribution is concentrated around the tip. The evanescent fields exist only when the tip is placed very close to the sample surface. All these designs significantly increase the quality factor Q and the sensitivity.

Figure 18.10 shows the schematic of the instrument setup. The evanescent microwave probe is placed on top of the testing sample. The sample is mounted on an x–y–z stage that is automatically controlled through a motion controller by a computer. The distance

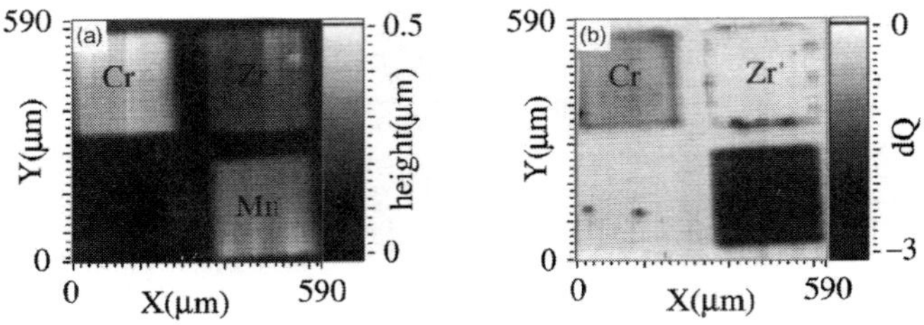

FIGURE 18.11. Experimental results of testing various thin-metal films using high-resolution scanning evanescent microwave measurement. (a) The topography and (b) The images of small chromium, manganese, and zirconium on thin platinum-coated film on a silicon substrate. The lighter regions correspond to higher regions on the sample in the topographic image, and lower losses correspond to higher conductivities in the loss images. (Reproduced from ref. 15.)

between the tip and the surface of sample is a crucial parameter [15] that affects the measurement. Therefore a feedback loop controlled by a piezo-stage is added to adjust the distance automatically. Other electronics, such as the diode detector, phase detector and shifter, and synthesizer, are used to detect the change in the microwave caused by various materials located close to the probe tip.

The experimental results of testing various thin-metal films using the high-resolution scanning evanescent microwave measurement are shown in Figure 18.11. The sample consists of three small squares metal thin films [manganese ($200\ \mathrm{S\,cm^{-1}}$), chromium ($555\ \mathrm{S\,cm^{-1}}$), and zirconium ($3{,}125\ \mathrm{S\,cm^{-1}}$)] about 2,500 Å thick deposited on a c-Si substrate coated with a platinum film of thickness 750 Å. The conductivity of each metal film was also measured independently using conventional dc measurements. The tip-to-sample distance is about 0.75 µm. Figure 18.11(a) shows the topography of three metal squares on the substrate. Figure 18.11(b) shows the conductivity variation between metals: the darker the imaging scale, the lower the conductivity. The gray scale is in good quantitative agreement with the measured conductivities. The detailed quantitative analysis of the surface conductivity at a given tip–sample distance is described elsewhere [16].

This newly developed technique possesses great potential for various applications. The ability of contactless imaging of many electrical properties within a submicron spatial resolution is very important for defect and electrical property studies of semiconductors, especially given the rapidly decreasing size of devices. This instrument can also be used for rapid screening of materials libraries generated using the high-throughput approach.

HIGH-THROUGHPUT CONDUCTIVITY MEASUREMENTS

High-Throughput Characterizations

High-throughput characterization deals with an array of samples (one-dimensional, two-dimensional, or continuous) on a single substrate. This type of approach allows the collection of more information in a shorter time. Therefore automation and computer assistance are needed. Fortunately, almost every conventional technique, such as four-probe conductivity measurement, can be modified for high-throughput characterization. In this section, high-throughput conductivity measurements are described.

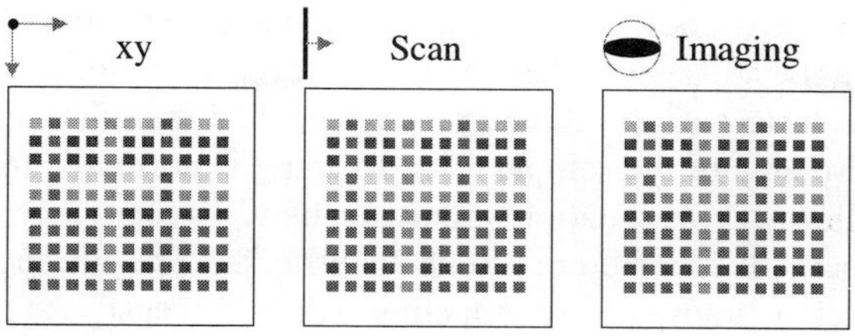

FIGURE 18.12. Three basic approaches to high-throughput characterizations.

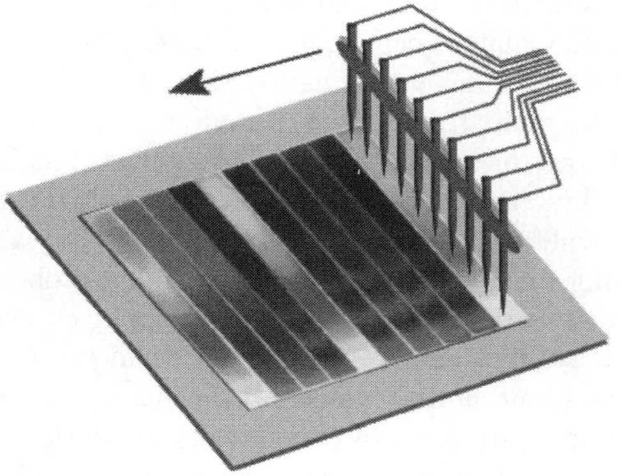

FIGURE 18.13. High-throughput conductivity measurement setup using an automated array of probe stations.

The conventional single-point measurement serves as the basis for high-throughput measurements. Without this knowledge, high-throughput measurements would be impossible. For example, to measure the conductivity of a film, one must know how to connect electrical wires, set up a testing circuit, use ammeters and voltmeters, and record and analyze the data. All these skills must be mastered for this technique to be modified and to become a high-throughput measurement.

Figure 18.12 shows the three basic ways of performing high-throughput measurements: x–y scanning, linear scanning, and imaging. In x–y scanning, the measurement is made at one position and then moves to another position in the x–y plane, and so on. This process can be either manual or automated, but automation speeds up the characterization. The four-point probe technique, gold and mercury probes, laser-induced microwave conductivity, evanescent microwave microscope, and even contact electrode measurements can all be modified for x–y scanning. In linear scanning, the measurement is made at one linear array at a time, and then moves linearly to other positions. The best example is the photoscanner. A linear array of photodetectors measures the light intensity in one direction and scans in the other direction to finish the two-dimensional (2D) mapping. In the conductivity measurement, a similar approach can be achieved by grouping an array of probes or detectors, as shown in Figure 18.13. This type of measurement is faster than the x–y scan because measurements need be taken in only one direction. However, the characterization system can be complicated and costly because of the array of probes and detectors. The spatial resolution can be limited by the size of the probes and the detectors. In the case of four-probe measurements, the distance between the adjacent probes can be as small as 10 μm.

Imaging is the fastest of the three approaches. The measurement is done simultaneously for the entire area of the sample. This imaging is normally done with an array of detectors which collect 2D information simultaneously. The best example is a camera, in which a digital camera takes a picture as one snapshot. At present, the dc conductivity of conducting films such as transparent conducting oxides can easily be measured, in principle, by taking a snapshot of reflected or transmitted 2D imaging at a critical or various wavelengths because of the absorption of light by free carriers [17]. A high-resolution IR imaging 256×256 pixel detector array is available commercially but it is expensive. Also, a more powerful light source is needed to image a large area of sample (greater than $2.5\ \text{cm} \times 2.5\ \text{cm}$). A video camera can be used for more accurate results in which the timescale will be related to various wavelengths. More powerful computers and software are needed to process the multiarray of data. Based on similar physics but operated at a lower frequency, a 2D array of microwave detector cameras, could probably be used for one-shot imaging of conductivity. Optical and microwave imaging would allow conductivity characterization of all materials.

Measurements using linear or 2D arrays of four-point probes, or gold and mercury probe tips, are practically useful for the high-throughput conductivity measurement. This technique is well studied and has been applied in the semiconductor industry. This instrument can be home-built because all the parts are commercially available. Optical and microwave measurements of conductivity are currently well suited for measuring relative changes in conductivity, but more work is needed for high-throughput characterization.

High-Throughput Samples

Figure 18.14 shows a picture of a high-throughput sample that consists of ten stripes of various amorphous-silicon-based thin films. Each stripe is 50.8 mm long and 4.6 mm wide. This sample was deposited using the high-throughput hot-wire chemical vapor deposition (CVD) system developed at NREL. The details of the deposition system are described elsewhere [18]. This system uses a set of physical masks of varying size, and an *x–y–z* sample manipulator that can position the substrate in any position inside the vacuum chamber. A motorized shutter is used to create a thickness-graded "wedge" stripe.

The a-Si:H film [19, 20] is deposited using SiH_4 gas that is decomposed through a hot-wire filament (2,000°C). By mixing H_2 gas with SiH_4, μc-Si film can be grown [21], by mixing SiH_4 with PH_3 gas, n-type film can be grown, and by mixing SiH_4 with trimethylboron (TMB) or B_2H_6, p-type film can be deposited. The high-throughput HWCVD system is able to deposit all types of film on a single glass substrate. Also, each type of film has a thickness grading. The thickness variation can easily be seen from the color fringes in the picture.

To test the electrical properties of the material library sample, a conventional metal contact method has been used for demonstration. Many parallel palladium metal pads have been evaporated on each stripe as shown in Figure 18.14(b). By measuring the current

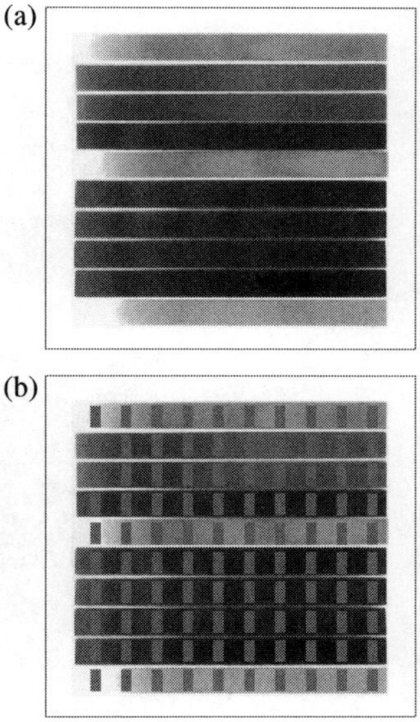

FIGURE 18.14. High-throughput a-Si-based thin-film sample using a commercial white light scanner. (a) Ten stripes of thin a-Si:H-based film on a 63.5 × 63.5 mm glass substrate. Each stripe is 4.6 mm wide × 50.8 mm long. Ten different types of thin film were deposited on each stripe as shown. Along each stripe, the film thickness is graded as indicated by the gray change. (b) Configuration of the metal bar contacts for conductivity measurement.

between the parallel bars in response to the applied voltage, the conductivity can readily be deduced. To accelerate the testing, an array of automated probes be added to the conventional approach, as shown in Figure 18.13. The 10 probes contact at 10 metal bars simultaneously, and move to the next stripe after performing the *I–V* measurement. Measurements which would take 5 hr if performed manually now take less than 20 min with the automated probe array approach. With the aid of a computer-controlled sample stage, multichannel data acquisition, and data analysis software, a task that was once tedious now looks easy.

Figure 18.15 shows the thickness mapping of the sample in Figure 18.14(a). The thickness is measured using optical reflection spectra and fitting the refractive index *n*, the extinction coefficient *k*, and thickness using the n & k Model 1280 instrument (n & k Technology Inc.). Nine positions at the center of every two metal bar pairs [Figure 18.14(b)] across 50.8 mm were measured on each stripe. The total thickness varied from 150 to 4,500 Å as indicated in the color scale plate. These thicknesses were later used to calculate the conductivity at each position. In Figure 18.14(a), the fringes of each stripe that come from optical reflection under white light agree well with the thickness plotted in Figure 18.15. The μc-Si stripes range in thickness from 200 to 600 Å, and the a-Si-based stripes vary from 1,500 to 4,500 Å. The thicknesses of all the films are intentionally selected to cover the range of device applications, such as n–i–p or the p–i–n solar cells.

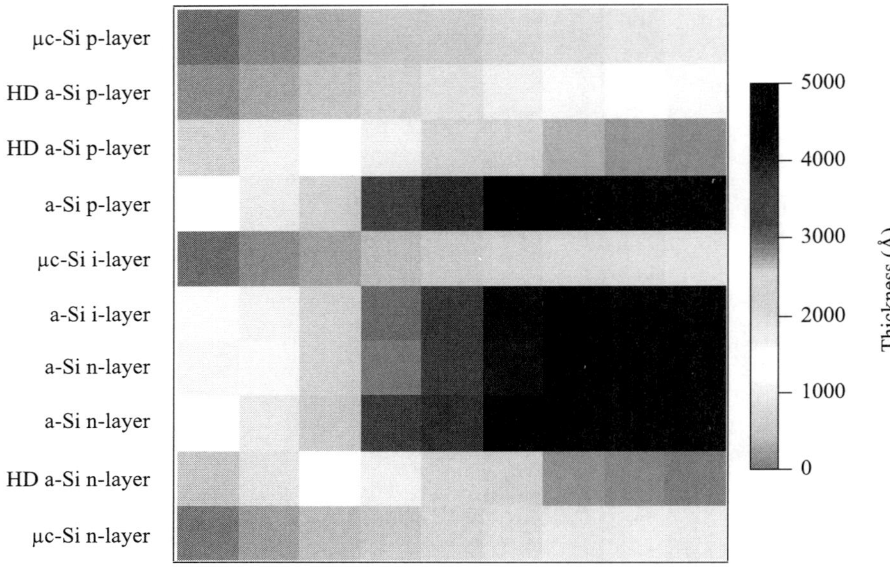

FIGURE 18.15. Thickness imaging of the high-throughput sample in Figure 18.14(a).

Dark and One-Sun Conductivity

Figure 18.16 shows conductivity mapping of the sample in Figure 18.14(a). At each position, a dark and one-sun (AM1.5) conductivity was plotted, as indicated in the inset. Note that the logarithmic color scale presents a change in conductivity of more than 10 orders of magnitude for all the films that were deposited on a single substrate. The first stripe, μc-Si (n-type), has the highest conductivity with close to $1\,\mathrm{S\,cm^{-1}}$, and the one-sun photoconductivity is the same as dark conductivity. The second stripe, H-diluted n-type a-Si:H, has moderate conductivity with a small one-sun photoconductivity. Stripe 5, a-Si:H, has the lowest conductivity of $10^{-10}\,\mathrm{S\,cm^{-1}}$, but also has the largest one-sun photoconductivity of $10^{5}\,\mathrm{S\,cm^{-1}}$. The ratio of photoconductivity to dark conductivity has reached 10^5. This material is widely used as an absorber in a-Si:H-based solar cells. Stripe 6, intrinsic μc-Si, has moderate conductivity and photoconductivity better by a factor of 10. Currently, it is a most promising material for μc-Si solar cells or high-mobility thin-film transistor applications. Normally the doping efficiency of p-type a-Si:H is low and results in a low conductivity. For stripe 7, a-Si p-layer, the conductivity is $10^{-6}\,\mathrm{S\,cm^{-1}}$. However, by adding H_2 while mixing with TMB and SiH_4, a higher conductivity film has been achieved in stripes 8 and 9. Stripe 10, μc-Si p-type Si, has a much higher conductivity of $10^{-2}\,\mathrm{S\,cm^{-1}}$, which is still less than μc-Si n-type films, but much higher than a-Si p-type films.

The speed of measurement depends on the materials. For the example shown above, a single two-electrode *I–V* measurement and computer analysis takes less than 10 ms. However, the a-Si:H has localized states that trap carriers. In the dark conductivity measurement, it takes a few seconds or more to stabilize the reading when a voltage is applied.

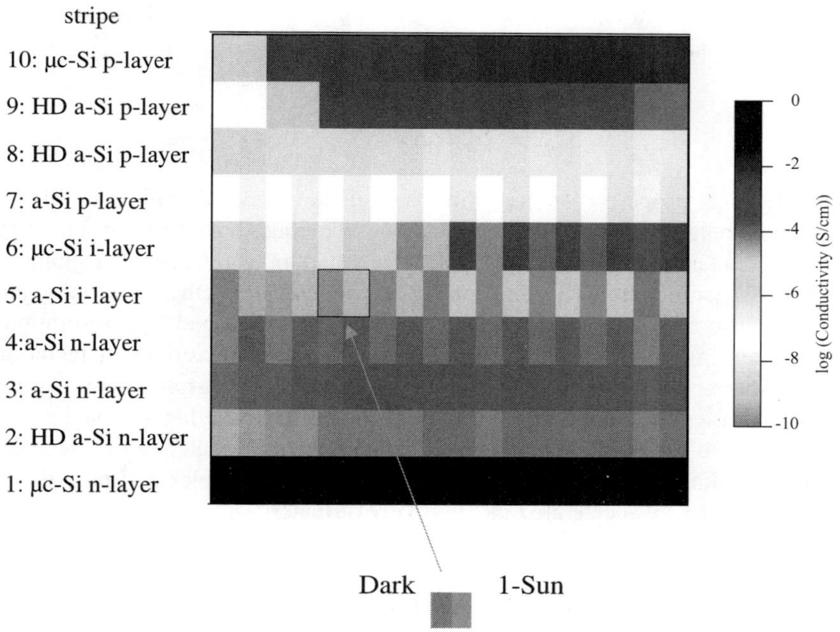

FIGURE 18.16. Dark conductivity and photoconductivity of the high-throughput sample in Figure 18.14. Note the logarithmic scale in the gray plate. The conductivity spans close to 10 orders of magnitude ranging from 10^{-10} to $1\,\mathrm{S\,cm^{-1}}$ on the sample. Each small square, as shown in the inset, presents the conductivity at one position of the film. The dark conductivity is on the left and the one-sun photoconductivity is on the right.

For the photoconductivity measurement, the setting time is much faster. The single measurement can be done in less than 1 s. Fortunately, the time to move the array of probes mechanically to the other position is not a limiting factor for the measurement. It can be positioned within 10 ms. However, care must be taken when the probe tips contact the electrodes or thin films. Underpressure will result in no or poor contact, and overpressure may damage the contact or film. An automatic feedback is desired to have a reliable measurement, although it will take a longer time. Finally, the data analysis and especially presentation are challenging (these topics are beyond the scope of this chapter) but the final results are rewarding.

Conductivity Thickness Dependence

The thickness dependence of conductivity is a very interesting subject. Very thin films may not be continuous, and so the conductivity may vary. For example, μc-Si films do not form a uniform film because of the discontinuity of the connection between the grains. Figure 18.16 shows the effects—the dark conductivity in the thinner film (about 200 Å) is lower than that in the thicker film. This study has a significant impact on solar cell devices because the doped layers in these devices are very thin (about 100 Å). The results from testing thicker films may not be applicable to the very thin films that are used in these devices. The results from Figure 18.16 clearly demonstrate the effect, especially for doped μc-Si films.

SUMMARY

The electrical properties of thin-film materials have been described in this chapter, especially with regard to the conductivity. Various techniques have been presented, such as conventional coplanar and sandwich (capacitance structure) electrodes, four-point probes, gold and mercury probe tips, physical contact and noncontact methods, scanning photoinduced microwave conductivity techniques, and recently developed high-resolution scanning evanescent microwave techniques. The conductivity measurement technique for high-throughput applications has been discussed. Finally, the results of measuring conductivity that varies by more than 10 orders of magnitude on a high-throughput sample have demonstrated the feasibility of various techniques to characterize the electrical properties of thin films. The combination of high-throughput samples and high-throughput characterization greatly accelerated the discovery of materials.

ACKNOWLEDGMENTS

The author gratefully acknowledges Dr X.–D. Xiang for helpful discussions, Yueqin Xu for metalization, Richard S. Crandall for reviewing the manuscript, and Susan Moon for editing. This work is supported by the US Department of Energy under Contract No. DE-AC36–99-GO10337.

REFERENCES

1. Beck, G., Kunst, M. Contactless scanner for photoactive materials using laser-induced microwave absorption. *Rev. Sci. Instrum.* **1986**, *57* (2), 197–201.
2. Wei, T., Xiang, X.-D. *Scanning tip microwave near-field microscope. Appl. Phys. Lett.* **1996**, *68*, 3506.
3. Keithley Instruments, *Model 6517a Electrometer User's Manual*, **1996**.
4. Wellard, C. L. *Resistance and Resistors*, **1960**, New York: McGraw-Hill.
5. Valdes, T. B. *Proc. IRE* **1954**, *42*, 420.
6. Logan, M. A. *Bell Syst. Tech. J.* **1961**, *40*, 885.
7. Kane, P. F., Larrabee, G. B. *Charact. Semicond. Mater.* **1970**, *68*, 3506.
8. Zrudsky, D. R., Bush, H. D., Fassett, J. R. *Rev. Sci. Instrum.* **1966**, *37*, 885.
9. Schneemeyer, L., Dover, B. V. The application of continuous compositional phase spread approaches to materials discovery, In *Combi 2000, 2nd Combinatorial Approaches for New Materials Discovery*, **2000**, San Diego, CA: Knowledge Foundation.
10. Kunst, M., Werner, A. Material characteristics by contactless photoconductivity measurements. *In Materials Characterization Symposium, 1986, Palo Alto, CA*, **1986**, Pittsburgh, PA: Materials Research Society, pp. 361–366.
11. Kunst, M., Beck, G. The study of charge carrier kinetics in semiconductors by microwave conductivity measurements. II. *J. Appl. Phys.* **1988**, *63* (4), 1093–1098.
12. Xiang, X.-D. Identication of phase boundaries and singular phases using continuous phase-diagram mapping. In *Combi 2000, 2nd Combinatorial Approaches for New Materials Discovery*, **2000**, San Diego, CA: Knowledge Foundation.

13. Gao, C., Wei, T., Duewar, F., Lu, Y., Xiang, X.-D. High spatial resolution quantitative microwave impendance microscopy by a scanning tip microwave near-field microscope. *Appl. Phys. Lett.* **1997**, *71*, 1872.
14. Gao, C., Duewar, F., Xiang, X.-D. Quantitative microwave evanescent microscopy. *Appl. Phys. Lett.* **1999**, *75*, 3005.
15. Duewar, F., Gao, C., Takeuchi, I., Xiang, X.-D. Tip–sample distance feedback control in a scanning evanescent microwave microscope. *Appl. Phys. Lett.* **1999**, *74*, 2696.
16. Takeuchi, I., Wei, T., Duewar, F., Yoo, Y. K., Xiang, X.-D. Low temperature scanning-tip microwave near-field microscopy of $YBa_2Cu_3O_{7-x}$ *films*. *Appl. Phys. Lett.* **1997**, *71*, 2026.
17. Madelung, O. *Introduction to Solid-State Theory*, **1978**, Berlin: Springer-Verlag, p. 284.
18. Wang, Q., Yue, G. Z., Li, J., Han, D. X. A combinatorial study of materials in transition from amorphous to microcrystalline silicon. *Solid State Commun.* **2000**, *113*, 175–178.
19. Mahan, A. H., Nelson, B. P., Salamon, S., Crandall, R. S. Deposition of device quality, low H content a-Si:H by the hot wire technique. *J. Non-Crystalline Solids*, **1991**, vol. 137 & 138, 657.
20. Nelson, B. P., Wang, Q., Iwaniczko, E., Mahan, A. H., Crandall, R. S. The influence of electrons from the filament on the material properties of hydrogenated amorphous silicon growth by the hot-wire chemical vapor deposition technique. *Mater. Res. Soc. Proc.* **1998**, *507*, 927–932.
21. Han, D. X., Yue, G. Z., Lorentzen, J. D., Lin, J., Habuchi, H., Wang, Q. Optical and electronic properties of microcrystalline silicon as a function of microcrystallinity. *J. Appl. Phys.* **2000**, *87*, 1882.

Chapter 19

Development of High-Throughput Methods for Polymer Nanocomposite Research

J. W. Gilman, R. Davis, M. Nyden, T. Kashiwagi, J. Shields, and W. Demory

INTRODUCTION

This chapter will present an overview of the development of two high-throughput (HT) methods: (1) preparation of formulated polymer libraries using extrusion; (2) screening of flammability properties using flame-spread measurements.

Nature and evolution are possibly the best evidence of the effectiveness of combinatorial and HT processes. As a research tool the HT concept is not only itself a *disruptive technology*, but it is likely to be the major source of new *disruptive technologies* in any field where it is applied [1]. The application of these types of concepts to the development of materials began with Hanak and co-workers at the RCA Corporation in the early 1970s. Specific to the point of these first papers were the rapid processing methods for sample preparation (cosputtering of continuum compositions), the characterization methods, and the inherent greater efficiency of Hanak's "multiple-sample concept" [2]. Today the sample architecture that Hanak described in his original papers is referred to as "composition-spread" or "gradient" type.

For the purposes of our research we are utilizing the gradient approach and developing experimental methods which simply offer "higher throughput" or faster data generation. Of course, more data is not the final goal; we strive for the opportunity to explore more completely the nearly unlimited number of ideas one has, with the ultimate output being the rapid generation of useful knowledge.

The focus of this research is the development of fundamental structure–property relations for polymer nanocomposites. Of primary interest are the governing mechanisms

J. W. Gilman, R. Davis, M. Nyden, T. Kashiwagi, J. Shields, and W. Demory • National Institute of Standards and Technology, 100 Bureau Drive, Gaithersburg, MD 20899, USA
Contact author: Jeffrey W. Gilman, email: jeffrey.gilman@nist.gov

High-Throughput Analysis, edited by Potyrailo and Amis
Kluwer Academic Publishers, New York, 2003.

behind the enhanced mechanical properties and the improved flammability properties of nanocomposites.

Polymer nanocomposites are prepared by mixing a polymer (or monomer) with some dissimilar material, or additive, that has one or more dimensions on the nanometer scale. Over the last few decades a wide variety of materials and synthesis approaches have been developed that allow molecular-level control over the design and structure of nanocomposite materials. Nanocomposites have been prepared by sol–gel methods [3], by in situ polymerization routes, and by using simple compounding methods [4]. All these approaches share a common theme—the intermingling, on the nanometer scale, of dissimilar materials for the purpose of creating new materials with properties not available from either of the component pure materials. For example, simple organic polymers modified with layered silicates [5, 6] have been prepared with improved heat distortion temperatures, twice the modulus, a factor of 10 lower permeability to gases and solvents, improved thermal stability, a fourfold lower flammability [7–11], enhanced ablative performance [12], and reduced rates of degradation in space [13]. All these attributes derive from incorporation of only 5–10% (by mass) of the layered silicate, and only occur if the surface area between the two phases is very high, that is, the particle size of the additive is on the nanometer scale and the degree of phase mixing is homogeneous on the nanometer scale. In other words, the fraction of material in an "interphase" must be high. Other types of nanocomposites show similar enhanced properties as long as the same conditions are met. Polyoligosilsesquioxane (POSS) materials blended or copolymerized with various polymers [14], sol–gel hybrid materials [15], nano-silica composites [16], and polymer nanocomposites based on graphite [17,18] and carbon nanotubes [19] are other types of nanostructured materials with unique properties.

One might ask why the study of nanocomposites requires the development of HT methods. The answer is simply that there are a large number of parameters which influence polymer nanocomposite performance, and to develop a detailed understanding of these materials a large volume of the associated multidimensional property space should be investigated. The multidimensional parameter space for polymer nanocomposites, shown in Table 19.1, consists of the obvious list of different material types under consideration, such as "polymer" and "nano-additive." All these materials must be miscible with one another so that a nanocomposite can form, and so that there is a strong stabilizing interaction at the interphase between the two dissimilar materials. This is essential both for the phase stability of the nanocomposite and for optimal physical properties. These requirements introduce the next parameter— "surface chemistry." Control of surface chemistry is most often accomplished in layered silicate nanocomposites by modification of the inorganic surface with an organophillic reagent, such as an alkyl ammonium, or a chelating agent. In POSS materials many different functionalities can be incorporated directly into the structure. While this means that no additional "surface" treatment is required, there are still a large number of possible POSS materials which need to be evaluated to obtain the optimal property improvement for a specific application [20].

Like layered silicates, both carbon nanotubes and silica nanoparticles require secondary functionalization to render them miscible with polymers. This brings us to the very critical parameter "processing conditions." The optimal processing conditions for a given nanocomposite system will depend on complex interactions with the previous three parameters. We have found that careful control of processing conditions is critical both for

TABLE 19.1. Multidimensional Parameter Space for Polymer Nanocomposites

Polymer	Nano-additive	Surface chemistry	Processing conditions	Conventional additives	Flame retardant
PE	Layered	Alkylammonium	Temperature	Processing	Phosphorus
PP	silicate	Imidazolium	Shear	stabilizers	Halogen
PS	POSS	Chelates	Residence time	UV-stabilizers	Silicon-based
PA6	Carbon	Silated	.	Antioxidant	.
PU	nanotubes	alkyl	.	Fillers	.
PVC	Silica	carboxylate	.	Pigments	.
PC	.	.	.	.	.
PEO	.	.	.	.	.
Epoxy	.	.	.	.	.
.	.	.	.	.	.
.	.	.	.	.	.
~10	~10	~10	~10	~10	~10

PE, polyethylene; PP, polypropylene; PS, polystyrene; PA6, polyamide-6; PU, polyurethane; PVC, polyvinyl chloride; PC, polycarbonate; PEO, polyethyleneoxide.

the preparation of a nanocomposite for the prevention of degradation of the nanocomposite during processing [21]. For the purpose of considering combinations of nanocom posites with other "conventional additives" one also needs to include the investigation of possible synergistic and antagonistic interactions with processing and UV stabilizers, pigments, dyes, fillers, and, for our particular interests, flame-retardant additives. This list of parameters generates on the order of 10^6 combinations or formulations worthy of investigation. To try to study this property space, HT methods must be developed for nanocomposites.

EXPERIMENTAL

Homogeneous and compositionally graded samples containing ammonium polyphosphate (APP), pentaerythritol (PER), and/or organic-modified layered silicate (OLS) (dimethyl dehydrogenated tallow ammonium montmorillonite, 15A, from Southern Clay Products, www.nanoclay.com) in polystyrene (PS) (Shell, Styron 663) were produced in our twin screw extruder (B&P 18 mm, 25 : 1 $L:D$; feed rates, 2–3 kg/hr^{-1}). The samples were strips approximately 1.5 m long, 7 mm wide, and 2 mm thick (Figure 19.1) consisting of PS blended with varying amounts of additives. In the gradient samples we attempted to create a linear concentration gradient (from $C = 0\%$ to $C = 30\%$ additive by mass) by increasing the rate of feed from the hopper containing additive linearly with time. The evaluation of the flame-spread properties of the samples was carried out in our modified flooring radiant panel device (Figure 19.2). The homogeneous composition samples were evaluated in a *gradient* flux environment (see Figures 19.3 and 19.4). The gradient composition samples were evaluated for flame spread under a *constant* flux of $16.8 \pm 0.4\,\mathrm{kW\,m^{-2}}$ in the radiant panel device. This was accomplished by fixing the methane burner parallel to the sample holder (see Figure 19.3).

FIGURE 19.1. Extruded strips of PS with various concentrations (30%, 5%, 40%) of a 3 : 1 mixture of APP and PER.

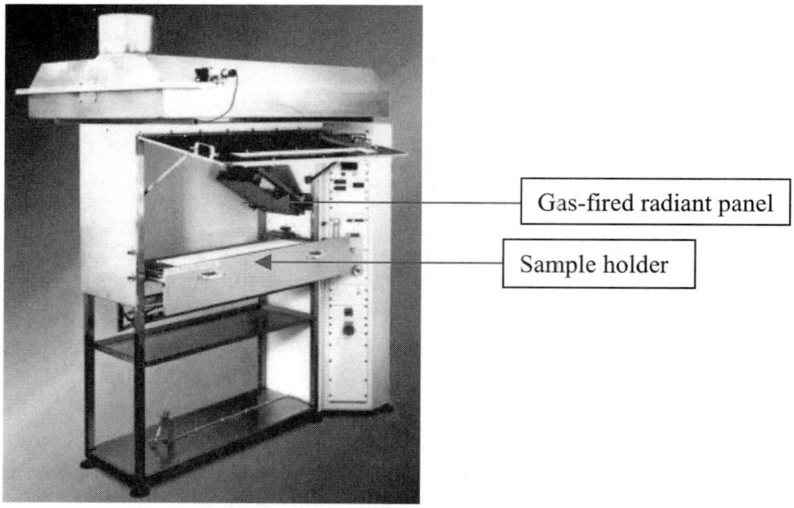

FIGURE 19.2. Flooring radiant panel apparatus.

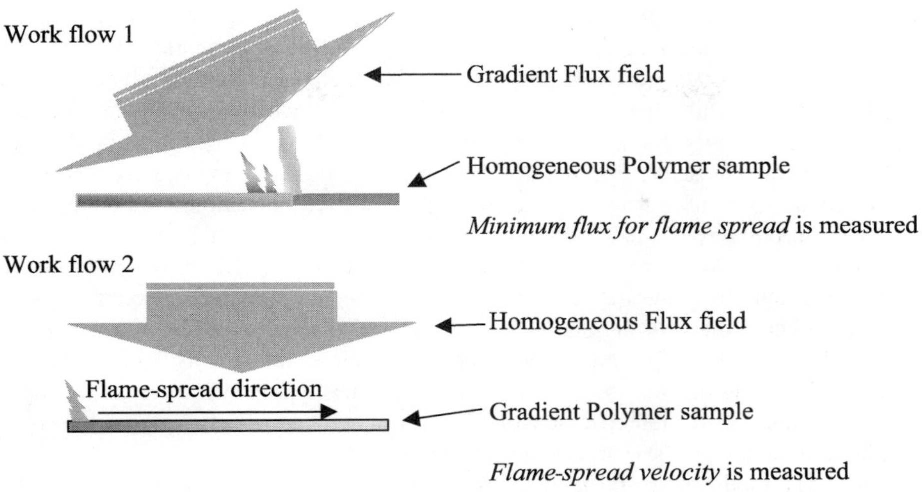

FIGURE 19.3. Schematic drawings of the two different flame-spread approaches: one where the sample is homogenous and the flux field has a gradient in intensity (work flow 1), and the other utilizing a compositional gradient within the sample and a homogenous flux field (work flow 2). A color version of this figure appears between pages 424–425.

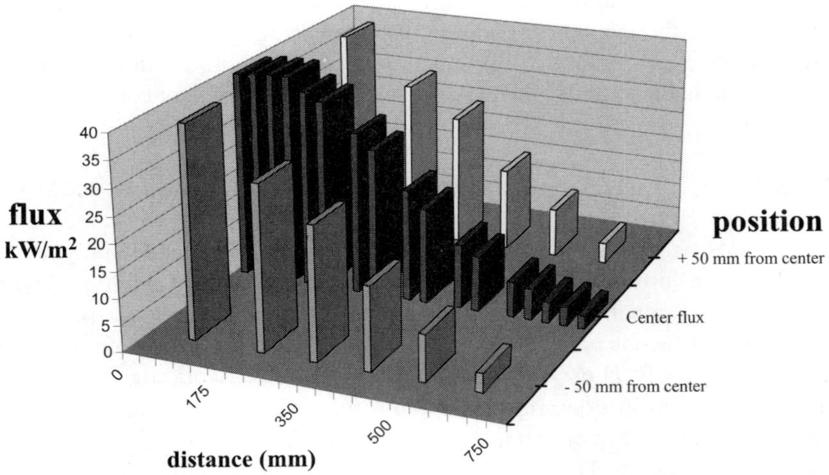

FIGURE 19.4. Flux gradient map in the sample holder region of the radiant panel apparatus.

RESULTS AND DISCUSSION

The approach we are pursuing focuses on the inherent HT capabilities of twin-screw extruders for rapid preparation of compounded samples (or libraries). The inherent HT nature of an extruder is derived from three important capabilities: (1) the high mass flow rates, (2) the ability to automate changes in the feeders used to deliver polymer and additive, and (3) the ability to change the processing conditions, such as residence time and shear, easily. While these attributes offer HT preparation of compounded nanocomposite samples, they also introduce a bottleneck in our work flow: the characterization of the library. Nanocomposites present an especially difficult, albeit interesting, challenge in this regard. In contrast with conventional fillers and additives, where simple measurement of their concentration might suffice, characterization of nanocomposites must be done with resolution at the nanoscale. Specifically, one needs to determine the degree of mixing of the individual nanoscale particles. In addition, the effect of this nanomixing on the overall order and morphology of the system must be determined. Traditionally this is done using transmission electron microscopy (TEM), X-ray diffraction (XRD), and other methods (small-angle neutron scattering (SANS), solid state nuclear magnetic resonance (NMR), rheometry). These methods are not usually considered HT, although some have been converted to HT systems [22]. Our recent efforts focus on removing this "characterization" bottleneck by using *in-line* sensor methods. We are developing two tools: (1) an optical sensor [23] and (2) a dielectric sensor [24]. Both sensors are directly in-line on the extruder. The details of these sensors will be published separately.

Once HT methods are in hand for preparing nanocomposite samples and for their in-line characterization, the next task in the HT work flow is property characterization. Our interests in nanocomposites are focused on flammability and mechanical properties. The traditional approach to evaluating these properties often involves injection molding of test coupons and property testing using standard procedures [tensile testing, dynamic mechanical analysis, flammability testing (UL 94 V, Cone Calorimeter, UL 910)]. Instead, we utilize the inherent HT nature of nanoindentation for mechanical properties measurements and new flammability characterization techniques developed involving measurement of flame spread, which take advantage of the *gradient* concept.

Work Flow 1: Extrusion and Flame Spread with Gradient Heat Flux

The extruder is an excellent tool for mass producing compounded polymer. Typical compounding of a formulated system involves setting the ratio of polymer feedstock to additive at a fixed number, and then optimizing the processing conditions so as to extrude a consistent homogeneous product out of the extruder's die, in the form of a strand or strip, which is subsequently chopped into pellets. In the first work flow described here the extruder is used as just described to manufacture many constant composition (homogeneous) strips rapidly. Typical samples are shown in Figure 19.1.

When developing an HT method, standard samples with well-characterized behavior are used first, and the results from the HT method are compared with those of the standard test method. A set of standard samples was made using polystyrene (PS) and a combination of additives (3 : 1 mixture of APP and PER) that impart reduced flammability to the polymer. Bourbigot and co-workers [25] reports that PS formulated with a mass fraction of 29% or greater of (3 : 1) APP–PER exhibits self-extinguishing properties. The standard flammability test used for comparison is the Underwriters Laboratories 94 vertical test for electronic components (UL94) [26]. This test is widely accepted, but because of the nature of the test it is not HT. In an attempt to develop a faster and more reliable method of characterizing the flammability properties of polymers we have used the *gradient* concept, but in somewhat of a different manner. Instead of building in a compositional gradient in the sample, a gradient was used in the radiant flux environment to which the homogeneous samples were exposed (Figure 19.3).

The samples were burned under a gradient flux using a device similar to the standard flooring radiant panel apparatus [27], which is shown in Figure 19.2. This approach was designed in an attempt to develop an HT method which could be used to predict UL94 test performance. The UL94 test is a pass–fail type of test. The burning configuration is essentially upward flame spread. For the most common rating, V-0, the sample must self-extinguish in less than 10 s following each of two applications of a Bunsen burner flame (10 s application) to the sample. In upward flame spread, radiative heat transfer from the flame to the sample is the dominant heating mechanism which drives the burning; however, for small flames, as in the UL94 test, convective heating dominates. The imposed flux in the radiant panel apparatus approximates the heat transfer in the UL94 V test, and both tests measure self-extinguishing behavior. However, by using a *flux gradient* this new approach can determine under what combination of conditions (imposed flux, polymer type, and flame-retardant concentration) the sample will exhibit self-extinguishing behavior, but with only *one set of experiments*. A flux map in the region where the sample strips sit in the test apparatus is shown in Figure 19.4. Two to three samples were run at a time. The samples were ignited at the end, in the high-flux region (Figure 19.5), and were allowed to burn until they self-extinguished. The burn length was converted to a minimum flux for flame spread, using the flux map. Typical data showed that samples with a higher concentration of flame-retardant additive self-extinguished sooner and had a shorter burn length and higher flux than samples with lower flame-retardant concentration. The repeatability of this HT flammability test is excellent ($2\sigma = 0.5$ kW). The exact position where the sample self-extinguishes is determined visually during the experiment, or afterward by

FIGURE 19.5. NIST radiant panel apparatus with three samples just prior to ignition.

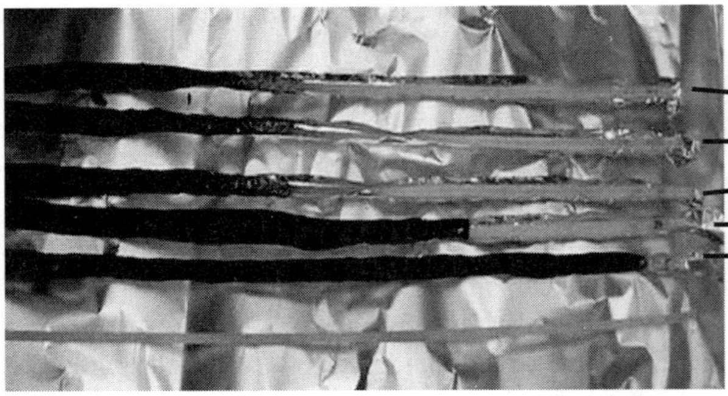

FIGURE 19.6. Post-burn samples of extruded strips of PS with various concentrations of a 3 : 1 mixture of APP and PER.

viewing the video of the experiment [28]. A video of a typical experiment, run on a single sample strip exposed to a flux gradient, is shown in the digital video clip in the CD. Typical post-burn samples are shown in Figure 19.6.

To evaluate the effectiveness of these two HT methods (sample extrusion and flame-spread flammability screening) a library was designed which encompassed a wide range of combinations of an OLS additive (15A) and APP–PER flame retardant. Since both these additives are known individually to reduce the flammability of PS, we were interested in evaluating combinations for synergistic or antagonistic interactions.

The 38 formulations (four replicates of each sample) were extruded in 2–3 days, and the flammability was also characterized in only 2–3 days. The data are shown in Figure 19.7. Compared with the data generation rate for the UL94 V tests, this HT approach (work flow 1) is 50–100 times faster. The same enhancement in data generation rate is found when comparing the efficiency of this HT method with characterizing the flammability of materials using the cone calorimeter. This faster method of evaluating many combinations of polymer and additives reveals some interesting behavior for the systems studied. In the samples which contained only low mass fractions of either APP–PER in PS (0%, 5%, and 10% 3 : 1 (APP–PER in PS) or low mass fractions of OLS (2%, 4%, 6%, 8%, and 10%) we observed no self-extinguishing behavior. This was also the case for pure PS. However, introduction of 2% OLS into the 10% (APP–PER)–PS system enabled the sample to self-extinguish (sample 10-2 in Figure 19.7). Synergistic behavior is observed for samples containing 2% OLS with 15% and 20% mass fractions of APP–PER (i.e., samples 15-2 and 20-2 in Figure 19.7). We define synergistic behavior as self-extinguishing behavior which lies above that for the pure APP–PER–PS, that is, above the red curve which connects the pure APP–PER–PS sample data.

For another subset of the library we observed "rule of mixtures" behavior, that is, the data for combinations of APP–PER and OLS falls very near the red curve. These combinations are samples 10-4, 10-6, 15-4, 15-6, 25-2, and 30-2 in Figure 19.7. One might expect an overall better performance of both these sets of samples (synergistic and rule of mixtures) if the effect of the OLS on the mechanical properties was included in the evaluation [29]. For all the other members of the library we observed antagonistic self-extinguishing behavior, that is, the minimum flux for flame spread was well below the red curve. These samples included 10-8, 10-10, 15-8, 15-10, 20-4, 20-6, 20-8, 20-10, 25-4, 25-6, 25-8, 25-10 30-4, 30-6, 30-8, and 30-10. It appears (for the 20%, 25%, and 30%

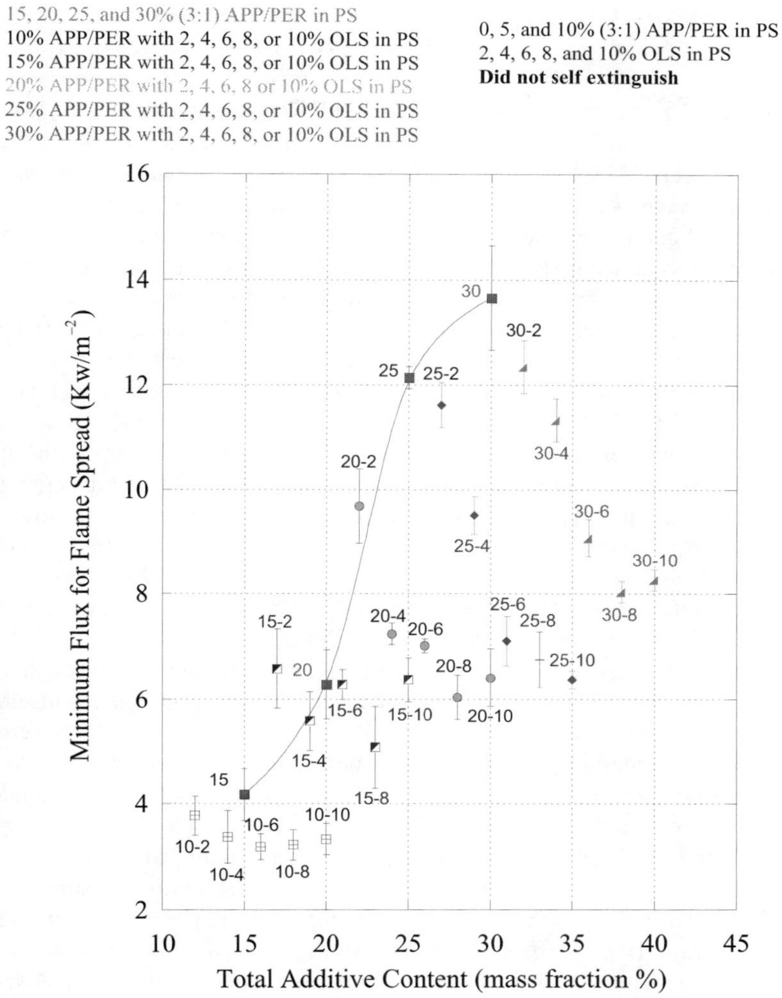

FIGURE 19.7. Plot of minimum flux for flame spread versus total additive content in PS samples. A color version of this figure appears between pages 424–425.

APP–PER combinations) that as the mass fraction of OLS increased, the minimum flux for flame spread decreased linearly. We propose three possible mechanisms for this behavior.

1. OLSs often raise melt viscosity and the APP–PER system is an intumescent foaming–charring system which is very sensitive to melt viscosity: too high and the char does not foam; too low and the bubbles do not stay in the char.
2. Upon thermal decomposition (burning) the APP–PER produces phosphoric acid, which may undergo a reaction with the basic OLS, essentially destroying the layered structure and reducing the effectiveness of the OLS [30].
3. We have observed in other flammability experiments that OLS nanocomposites shorten the ignition time (e.g., in the cone calorimeter). We shall return to this issue in our discussion of the flame-spread test of gradient samples.

Work Flow 2: Extrusion and Flame Spread with Gradient Composition

As discussed above, the other new flammability characterization technique developed in our laboratories at NIST involves measurement of flame spread and takes advantage of the *gradient* concept [31]. In this approach we again take advantage of the inherent HT capabilities of the twin-screw extruder: (1) the high mass flow rates, and (2) the ability to automate gradual changes in the feeders used to deliver polymer and additive. We evaluated two gradient systems: APP–PER in PS and OLS in PS. In this case we varied the ratio of additive to polymer over the extrusion experiment to prepare a compositional gradient within the extruded sample. A compositionally graded sample containing APP–PER in PS was produced in the extruder. In contrast with the flame-spread test in work flow 1 we did not use a gradient flux. Instead the samples are burned in our flooring radiant panel device under a *constant* flux of $16.8 \pm 0.4\,\text{kW}\,\text{m}^{-2}$ (Figure 19.3). We chose this flux because, as we see from the data from work flow 1, it is above the minimum flux for flame spread for all the loadings of additives in Figure 19.7. That is, this incident flux is high enough to prevent self-extinguishing behavior, even for materials with very high concentrations. The sample was a strip (approximately 1.5 m long, 7 mm wide, and 2 mm thick) consisting of PS blended with varying amounts of the 3 : 1 APP—PER mixture. We attempted to create a linear concentration gradient (from $C = 0\%$ to $C = 30\%$ additive by mass) by increasing the rate of feed from the feeder containing the additive linearly with time.

The flame velocities were measured by pulling the extruded strip through a 300 mm heating zone such that the flame front remained at a fixed position. The results are plotted in Figure 19.8. The data (time t) were collected at 0.050 m intervals, which were marked off on the sample holder strip, with the initial point (at $x = 0$) corresponding to the pure polymer ($C = 0\%$ at $x = 0$). The solid line was obtained by fitting the experimental data to a hypothetical function $\Delta x \approx (v_0/b)[\exp(bt) - 1]$ that was derived by assuming a linear dependence between flame velocity v and additive concentration, which was also assumed to decrease linearly with distance. The validity of these assumptions is supported by the fact that this function provides a good representation of the experimental data. The derivative of the function $[v = v_0 \exp(bt) = bx + v_0]$ in Figure 19.8 is a linear function of x (or of C, since $C = ax$) with a *negative* slope b, indicating a reduction in the flame-spread rate with increasing concentration of additive as expected.

Comparable data was collected for an OLS–PS gradient with an approximately linear concentration gradient of OLS varying from 0% to about 14% by mass. In this case, the flame-spread rate was *positive* ($0.95 \pm 0.22 \times 10^{-3}\,\text{ms}^{-1}$), that is, the flame-spread rate was observed to *accelerate* with increasing clay concentration (Figure 19.9). We propose the same explanation for this enhanced flame-spread as proposed above to explain the antagonistic behavior in some of the samples in Figure 19.7, that is, that the shorter ignition times observed for OLS polymer nanocomposites in cone calorimetry experiments may have a larger effect on the flame spread behavior than the effect of the reduced peak heat release rate (HRR). Tewarson and co-workers [32] have shown that the ignition time is indeed a more important parameter in upward flame-spread models than the peak HRR. Eventually, we would like to make simultaneous measurements of both flame-spread rate and HRR to determine the optimal composition(s) which results in the best compromise between flame-spread and HRR.

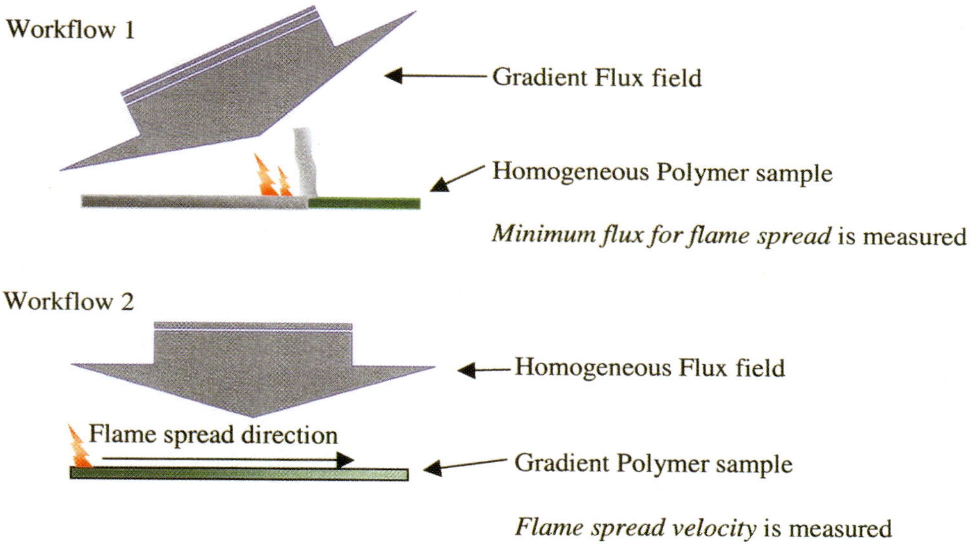

FIGURE 19.3. Schematic drawings of the two different flame-spread approaches: one where the sample is homogenous and the flux field has a gradient in intensity (work flow 1), and the other utilizing a compositional gradient within the sample and a homogenous flux field (work flow 2).

FIGURE 19.7. Plot of minimum flux for flame spread versus total additive content in PS samples.

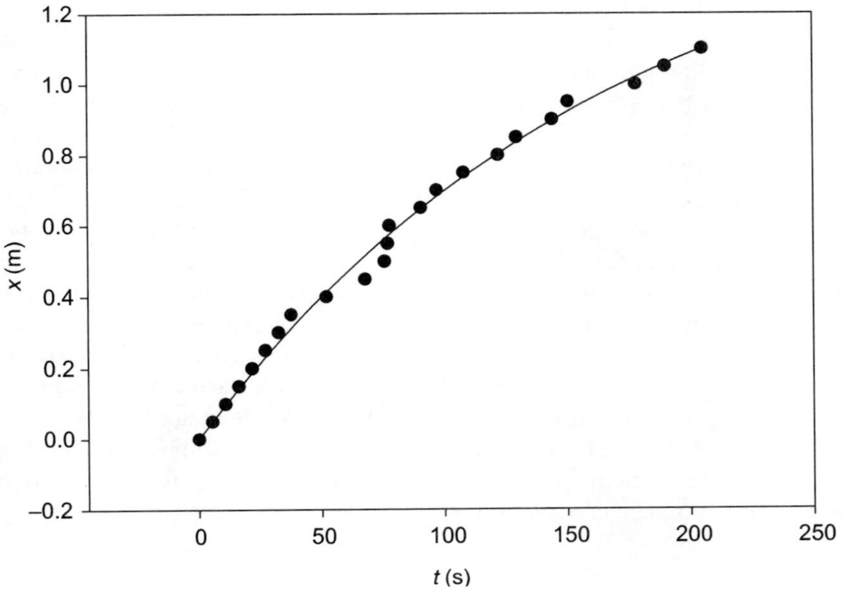

FIGURE 19.8. The progression of the flame front as a function of time measured for the APP–PER in PS gradient. The solid line was obtained by fitting the experimental data (circles) to a hypothetical function derived on the basis of assumptions stated in the text.

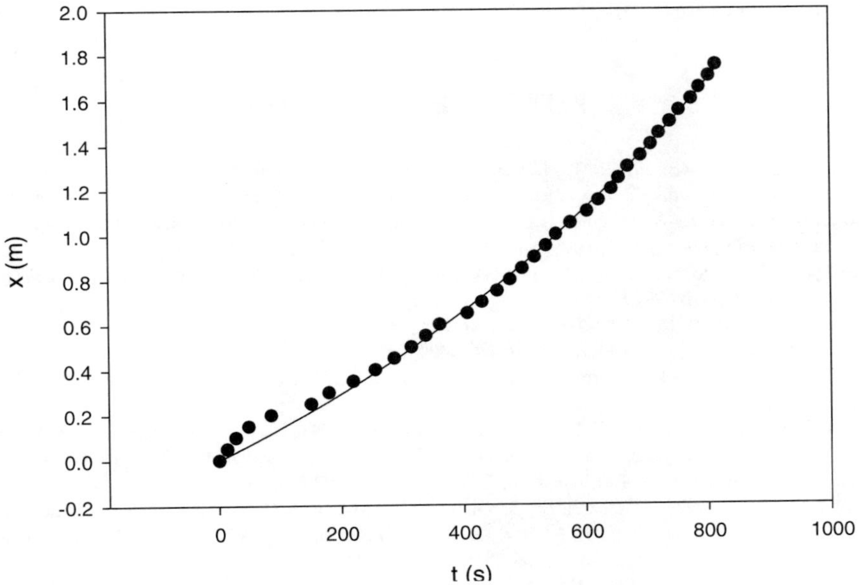

FIGURE 19.9. The progression of the flame front as a function of time measured for OLS–PS gradient. The solid line was obtained by fitting the experimental data (circles) to a hypothetical function derived on the basis of assumptions stated in the text.

CONCLUSIONS

An HT method has been developed which offers rapid preparation of nanocomposite and conventional-filler polymer libraries using twin-screw extrusion. This method allows preparation of homogeneous and gradient-type samples. An HT method has also been developed for determining the flammability properties of the above polymer libraries. The flame-spread measurements that are used to evaluate the flammability properties take advantage of the gradient approach in two different ways. First, the homogeneous polymer samples are screened using a *flux gradient*. This determines under what combination of conditions (imposed flux, polymer type, and flame-retardant concentration) the sample will exhibit self-extinguishing behavior. Secondly, *gradient polymer samples* are screened using a constant flux. By combining these HT methods (extrusion and flame spread), the rate of data generation has been accelerated by at least a factor of 50! Furthermore, the structure–property relationships revealed in this work would not have been so easily established without these HT approaches. Work is planned to combine the *gradient flux* and a series of different homogeneous samples in a larger flame-spread device so that we can have a compositional variation orthogonal to the flux gradient.

ACKNOWLEDGMENTS

The authors would like to thank Southern Clay Products for the donation of OLS (15A) samples. We also wish to thank the Federal Aviation Administration (DTFA 03-99-X-9009) and the Air Force Office of Scientific Research (AFOSR-ISSA-01-0001) for partial funding of this work.

REFERENCES AND NOTES

1. According to Christensen, C. M. *MRS Bull. a disruptive technology* **2002**, *27*, 278, is "technology that initially under performs the incumbent technology, but typically offers more flexibility, convenience, or a lower price. A *disruptive technology* can serve to create new markets, but it can also upset the equilibrium of companies with good reputations and good products."
2. Hanak, J. J., *J. Mater. Sci.* **1970**, *5*, 964; Dagani, R. A faster route to new materials. *Chem. Eng. News*, March 8, **1999**, pp. 51–60.
3. Carrado, K.A., *Appl. Clay Sci.* **2000**, *17*, 1.
4. Alexandre, M., Dubois, P. *Mater. Sci. Eng.(R)* **2000**, *28*, 1.
5. Kojima, Y., Usuki, A., Kawasumi, M., Okada, A., Fukushima, Y., Kurauchi, T., Kamigaito, O. *J. Mater. Res.* **1993**, *8*, 1185.
6. Messersmith, P. B., Giannelis, E. P. *J. Polym. Sci. A.* **1995**, *33*, 1047; Okada, A., Fukushima, Y., Kawasumi, M., Inagaki, S., Usuki, A., Sugiyama, S., Kurauchi, T., Kamigaito, O. US Patent 4,739,007, **1988**.
7. Gilman, J. W., Kashiwagi, T., Lichtenhan, J. D. *SAMPE J.* **1997**, *33*, 40.
8. Gilman, J. W., Kashiwagi, T., Lomakin, S., Giannelis, E., Manias, E., Lichtenhan, J., Jones, P. In M. Le Bras, G. Camino, S. Bourbigot, and R. Delobel (Eds.), *Fire Retardancy of Polymers: The Use of Intumescence*, **1998**, Cambridge, UK: Royal Society of Chemistry, pp. 203–221.
9. Gilman, J. W. *Appl. Clay. Sci.* **1999**, *15*, 31.
10. Gilman, J. W., Jackson, C. L., Morgan, A. B., Harris, R. H., Manias, E., Giannelis, E. P., Wuthenow, M., Hilton, D., Phillips, S. *Chem. Mater.* **2000**, *12*, 1866.
11. Giannelis, E. *Adv. Mater.* **1996**, *8*, 29.

12. Vaia, R. A., Price, G., Ruth, P. N., Nguyen, H. T., Lichtenhan, J. D. *J. Appl. Clay Sci.* **1999**, *15*, 67–92.
13. Fong, H., Vaia, R. A., Sanders, J. H., Lincoln, D., Vreugdenhil, A. J., Liu, W. D., Bultman, J., Chen, C. G. *Chem. Mater.* **2001**, *13*, 4123–4129.
14. Lichtenhan, J. D., Vu, N. Q., Carter, J. A., Gilman, J. W., Feher, F. J. *Macromolecules* **1993**, *26*, 2141–2142; Lichtenhan, J. D., Gilman, J. W., Ismail, I. M. K., Burmeister, M. J. *Chem. Mater.* **1996**, *8*, 1250–1259.
15. Ahmad, Z., Mark, J. E. *Chem. Mater.* **2001**, *13*, 3320.
16. Frisch, H. L., Mark, J. E. *Chem. Mater.* **1996**, *8*, 1735–1738.
17. Uhl, F. M., Lamelas, F. J., Wilkie, C. A. ABSTR PAP Am. Chem. Soc. 220: 66-PMSE Part 2 AUG 20 **2000**.
18. Liu, Z.-H., Wang, Z.-M., Yang, X., Ooi, K. *Langmuir* **2002**, *18*, 4926–4932.
19. Ajayan, P. M., Schadler, L., Giannaris, C., Rubio, A. *Adv. Mater.* **2000**, *12*, 750.
20. For a description of the many POSS materials available see www.hybridplastics.com
21. VanderHart, D. L., Asano, A., Gilman, J. W. *Chem. Mater.* **2000**, *13*, 3796.
22. For example, commercially available equipment developed by Bruker for high-throughput screening of combinatorial libraries using X-rays is discussed at www.bruker-axs.de.
23. Thomas, C. L., Bur, A. J. *Polym. Eng. Sci.*, **1999**, *3*, 1619.
24. Bur, A. J., Roth, S. C., McBrearty, M. *Rev. Sci. Instrum.* **2002**, *73*, 2097.
25. Bourbigot, S., Le Bras, M. "Synergy in intumescence: Overview of the use of zeolites" In M. Le Bras, G. Camino, S. Bourbigot, and R. Delobel (Eds.), *Fire Retardancy of Polymers: The Use of Intumescence*, Cambridge, UK: Royal Society of Chemistry, pp. 223–235.
26. *UL 94. The Standard for Flammability of Plastic Materials for Parts in Devices and Appliances.* Available at www.ul.com/services/flame.html
27. See http://www.fire-testing.com/frp.htm for additional details of this apparatus.
28. Attempting to measure the burn length after the test introduces significant error. This is because tar and char spread over the molten polymer surface, thus making determination of the exact burn length difficult.
29. The nanoindentation data on the library were not complete at the time of writing.
30. Bourbigot, S., Le Bras, M., Dabrowski, F., Gilman, J. W., Kashiwagi, T. *Fire Mater.* **2000**, *24*, 201.
31. Nyden, M., Gilman, J. W., Davis, R. D., Shields, J. R. *Proc. 47th International Society for the Advancement of Materials and Process Engineering (SAMPE) Symposium and Exposition, SAMPE* May **2002**, p. 728.
32. Tewarson, A., Khan, M., Wu, P. K., Bill, R. G. *Fire Mater.* **2001**, *25*, 31.

Electrochemical Analysis

Chapter 20

Combinatorial Electropolymerization: Concept, Equipment, and Applications

Vladimir M. Mirsky and Valentin Kulikov

INTRODUCTION

Most combinatorial chemistry technologies are based on solid phase synthesis. They can be realized using different formats and addressation techniques. Typically, reagents are dispensed into micro- or nanotiter plates [1]. Further investigation can be performed using a variety of techniques [2] including binding and inhibitor assays, screening of biological activity, mass spectroscopy, chromatography, and optical [1,3] or electrochemical [4] methods.

Other methods described in literature include a light-directed immobilization [5] and an addressation using microfluidic systems [6,7]. Only few authors have described electrical addressation [8–10]. We focus our efforts on application of the combinatorial approach with electrical addressation to the synthesis of electrochemically polymerized materials and multilayer structures. The synthesis is performed by combining parallel and serial electropolymerization. In the next step, the electrical properties of the combinatorial library obtained are screened. Then the data are analyzed according to a customer-defined program.

Electrochemical polymerization of organic molecules on electrodes was developed more than 50 years ago, possible mechanisms of this process have been reviewed in many publications [11–19]. Electropolymerization in the presence of several monomers can lead to the formation of copolymers and block copolymers [20–27]. In many cases, a synthesized polymer (or a substantial amount of it) is adsorbed on the electrode surface; this phenomenon led to a simple technology for the deposition of thin polymer layers on conductive surfaces. Since the 1960s this technology has been used in industry for corrosion

Vladimir M. Mirsky and Valentin Kulikov • University of Regensburg, Institute of Analytical Chemistry, Chemo- and Biosensors, 93040 Regensburg, Germany
Contact author: Vladimir M. Mirsky, email: vladimir.mirsky@chemie.uni-regensburg.de

High-Throughput Analysis, edited by Potyrailo and Amis
Kluwer Academic Publishers, New York, 2003.

protection [28–39]. The use of polyaniline or polyphenol corrosion protection for steel is reviewed in refs 28 and 37, respectively. Polyoxyphenylenes obtaned by the polymerization of phenol [33,36], phenol–2-allylpheno mixtures [34], or phenolate–allylamine mixtures [35] have been used to protect carbon steels and magnesium alloys. Electrodeposition of polymethylthiophene [29] or poly-(5-amino-1-naphtol) [30] reduces the corrosion rate of mild steel.

An intensive investigation of conductive polymers performed in the 1970s and 1980s was an additional motivation for further development of electropolymerization. It was shown that most of the conductive polymers, including substituted and nonsubstituted polythiophene [40], polypyrrole, polyaniline, polyazulene, polycarbazole, polyindole, polytriphenilene, poly(3,4-ethylenedioxythiophene)-poly(styrene sulfonate) (PEDOT) [41], polyazine [42–44], and many others, can be very simply synthesized by electrochemical oxidation of the corresponding monomers [45]. The oxidized conductive polymers can be electrochemically (or chemically) reduced; this corresponds by the loss of counter-ions and by dramatic changes in the optical and electrical properties. If the polymer is conductive, electrochemical synthesis can lead to the formation of very thick polymer layers (one can obtain layers about 1 mm thick). If the polymer is a poor conductor, the gradual increase in the electrical resistance of the organic layer suppresses electropolymerization, and this is the principal limitation on layer thickness.

The potential for application of conductive polymers is very impressive [17,46]. Applications include such different fields as corrosion protection, antistatic coatings [47], or biosensors [48–50] where conductive polymers are used for coupling electron transfer [51], immobilization of biomolecules [52–53] or as selective filters [54]. In the field of amperometric Electrochemical polymerization became a very popular technology [42,55–58]. Conductive polymers are also used for the preparation of pH electrodes [59,60] or reference electrodes [61,62], the development of ion exchangers and catalysts [63,64], the fabrication of electrochromic windows [65], and the development of gas sensors [66–70]. An array consisting of 342 impedometric chemosensors, based on conductive polymers, has been described [71]. The selectivity of conductive polymers to different odorants can easily be modified by variation of plasticizers [72]. However, the main field of future applications of conductive polymers is probably in organic electronics—the application of these materials to the fabrication of individual electronic devices such as solar cells [73], transistors [74,75], diodes [75,76], capacitors [77], light-emitting diodes and displays [78], and complete integrated circuits [79]. The requirements of these application fields are very different or even conflicting. To satisfy these requirements, a technology for the fast and cheap synthesis and screening of novel polymers should be established. We describe one possible solution here.

A CONCEPT OF COMBINATORIAL ELECTROPOLYMERIZATION

The possibility of controlling polymerization on solid conductive surfaces by applying appropriate electrochemical potentials suggests a natural realization of this process as a computer-controlled technology. Organization of the electrodes as an array removes the necessity for complicated and expensive dispensing systems. However, such geometry alone does not guarantee that the polymerization occurs only on the electrode connected to a potential source with a corresponding potential value. The problem is that if the distance between electrodes is too small (which minimizes reagent consumption), the electrical potential of nonconnected electrodes can be influenced by the potential of connected electrodes, thus leading to undesirable electropolymerization on uncoated electrodes (Figure.20.1). This problem can be overcome by fixing the potentials of all electrodes [Figure 20.1(a)]. A special investigation demonstrated that this approach indeed leads to electrically addressable immobilization.

Electrical addressation allows both serial and parallel electrochemical synthesis of polymers to be realized (Figure 20.2). The serial synthesis is carried out stepwise on different spots of the electrode array; each step is independent and can be performed with a specific mixture of reagents and under specific conditions. Serial synthesis is performed on several electrodes simultaneously. A combination of serial and parallel synthesis can be used to vary the thickness of a polymer layer or to form multilayer electrode structures.

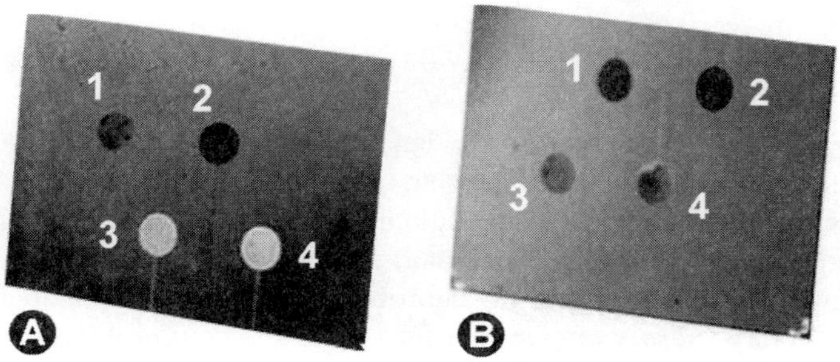

FIGURE 20.1. Application of an appropriate electrical potential allows one to control an electrochemical reaction. For example, the electropolymerization of aniline can be switched on by application of the electrode potential +900 mV/vs. SCE [electrodes 1 and 2 in (A) and (B)] and completely blocked by application of zero potential [electrodes 3 and 4 in (a)] during electropolymerization on other two electrodes. However, if the electrodes are disconnected, electropolymerization on the adjacent electrodes (electrodes 1 and 2) can also result in polymerization on the disconnected electrodes [electrodes 3 and 4 in (B)].

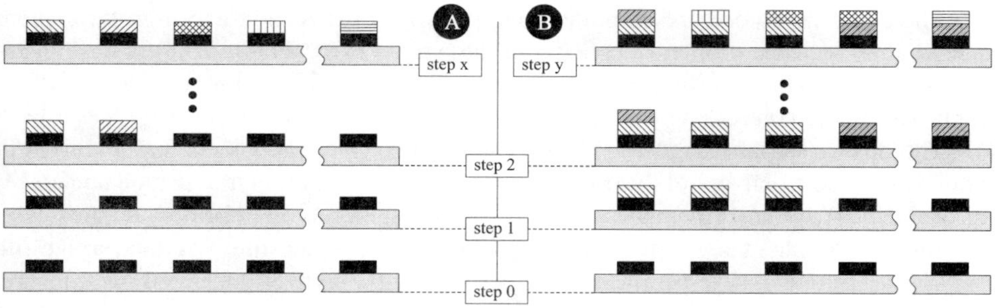

FIGURE 20.2. (A) Serial synthesis of different polymer layers and (B) parallel synthesis of multilayer structures can be combined giving up to N^n variations for n-layer structures. Variation of the layer includes any quantitative or qualitative modification of its chemical nature (reagent, ratio of reagents in the mixture, additives, etc.) or modification of any physical parameter (temperature, voltage, deposition time, etc). The adjacent layers in the scheme can be formed from the same substance, thus providing a variation of thickness.

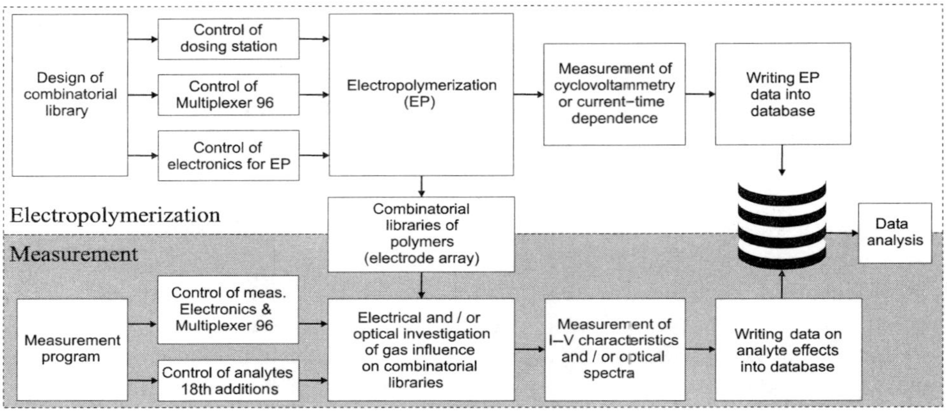

FIGURE 20.3. Data flow during high-throughput synthesis of combinatorial libraries of polymers and multilayer polymeric structures.

Combinatorial electrochemical synthesis of polymers on an electrode array is combined with an investigation of their electrical properties. This allowed us to design and develop a complete system for high-throughput synthesis and characterization of combinatorial libraries of polymeric structures on the same chip. A data flow in this system is shown in Figure 20.3, and the setup is described below. Information about a suggested combinatorial library (*library preparation* file) is used to control mixing of reagents and conditions of electropolymerization. The electrical properties of the obtained combinatorial library are investigated in situ according to settings defined in the *library screening* file. Then the measured and settings data are saved and used for further analysis.

THE SETUP FOR COMBINATORIAL ELECTROPOLYMERIZATION AND SCREENING

The setup developed for combinatorial electropolymerization and high-throughput screening of electrical properties consists of three main parts (Figure 20.4): a dosing station, an electronic block with a computer, and an electrode array. All parts of the setup are synchronized and controlled by a computer program.

The dosing station, (Figure 20.5) provides a mixing of monomers and additives in defined ratio and a delivery of this mixture to the cell for electropolymerization and cleaning of the flow system. All processes in the dosing station are performed automatically according to the parameters (mixture composition, timings, cleaning) defined in the file *library preparation*. These parameters are defined by the user or generated by software. The *library preparation* file describes electropolymerization conditions on each of 96 interdigital electrodes.

Up to four different reagents (monomers and additives) are injected by dosing pump syringes (DP, Schott Titronic Universal) from reservoirs A–D into the mixer bin (MB) which is equipped with a magnetic stirrer. Then the mixture is transported into the reaction

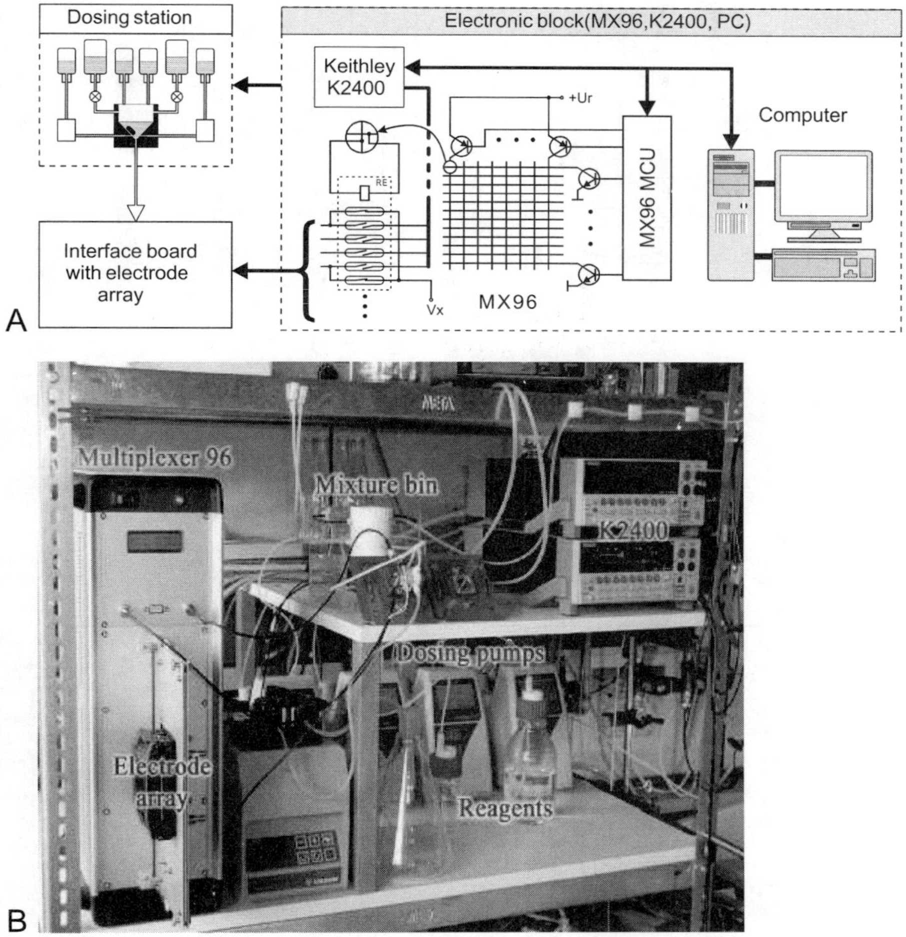

FIGURE 20.4. (A) Block diagram and (B) photograph of the setup for electropolymerization and screening of synthesized polymers.

cell (RC) by a peristaltic pump (P1, Gilson Minipuls 3), and the electropolymerization process starts. The electropolymerization is performed step by step on each interdigital electrode of the electrode array (EA), so that every electrode is coated by one or more polymer layers of specific chemical content defined in the *library preparation* file. If required, the electrodes can be coated simultaneously. After each polymerization step, the dosing station cleans the reaction cell, mixer bin, and transporting tubes. Two reservoirs (CL1, CL2) are used for cleaning solvents. Transport of cleaning solvents is driven by hydrostatic pressure and switched on and off by two-way valves (2W, NR Research).

The second part of the setup is an electronic block consisting of a home-made high-impedance electronic multiplexer MX96, a source monitor unit (SMU) (Keithley K2400), an interface board with a test clamp socket (Yamaichi), and a miniature thermostat for the polymerization cell (Figure 20.6).

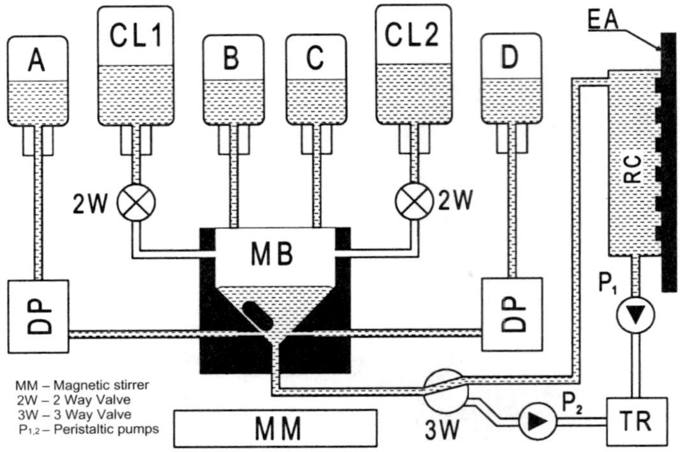

FIGURE 20.5. Dosing station and reaction cell.

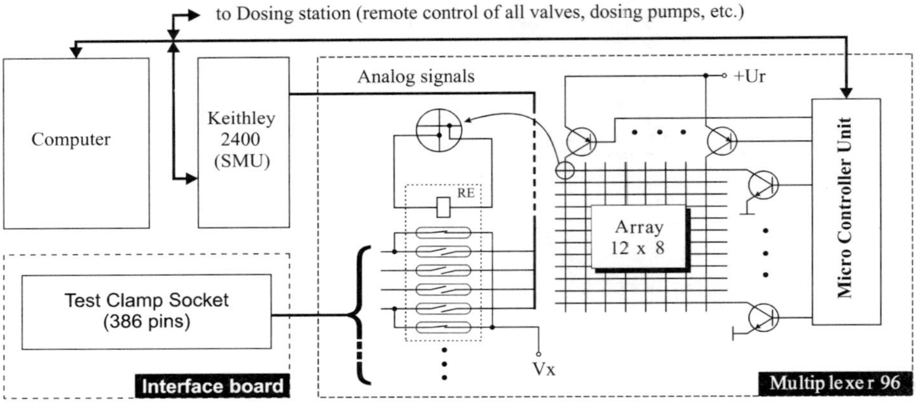

FIGURE 20.6. Electronic block.

The electronic multiplexer includes 96 channels of a 4A (four independent contacts—normally open) plus 2B (two contacts—normally closed) switch configuration. This configuration provides four-point measurements of electrical properties of synthesized polymeric combinatorial libraries. An application of the four-point technique allows potential drops at the electrode/polymer interface to be taken into account and bulk and contact resistance of polymers to be distinguisted. An additional 2B switch configuration is used to maintain the potential of the unused (already coated or still uncoated) electrodes outside the polymerization range, thus protecting them from undesirable electropolymerization. The SMU provides electropolymerization, which can be done galvanostatically, potentiostatically, or by potential cycling, and measurements of the electrical properties of the

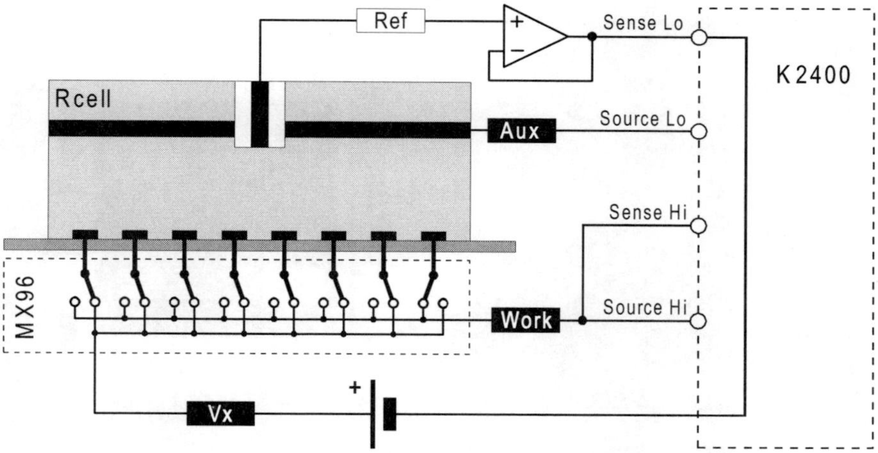

FIGURE 20.7. Connection of the SMU for electropolymerization. Ref, Aux, and Work refer to signals from the reference, auxiliary, and working electrodes, respectively.

polymer structures obtained. The detailed circuit for connecting the SMU for electropolymerization is shown in Figure 20.7. An Ag/AgCl electrode with a salt bridge filled with saturated KCl solution is used as a reference electrode and the auxiliary electrode is a platinum wire.

The third part of the setup includes an interface board with a test socket consisting of 436 contacts (of which 384 are used) for connection to an electrode array. The array is based on an oxidized Silicon wafer, with dimensions 61 mm × 61 mm, on which a metallic layer is deposited (platinum or gold and corresponding adhesive layers). The metallic layer was treated by lithography to form array of 96 electrode groups [Figure 20.8(a)]. Every group consists of four specially designed [80] interdigital electrodes [Figure 20.8(b)], allowing two- and four-point measurements of electrical properties. To protect connections between contact pads and interdigital electrodes from undesirable electropolymerization, the electrode array is coated with an insulating layer with 96 windows located in the centers of the interdigital electrodes.

The software to control the setup consists of three parts: a program for electropolymerization, a program for screening, and a program for data analysis. The first program provides electropolymerization on each electrode of the electrode array according to procedures defined in the *library preparation* file. This text file can be generated by an external database for combinatorial analysis or manually using a specially developed utility. The program controls each electropolymerization step and all components of the setup including the dosing station and electronic block. During polymerization, it generates files containing information on settings and the electrical measurements. Then the electrical properties of the prepared combinatorial library and changes of these properties on addition of different reagents, are screened. This screening is performed with the same setup and is controlled by the second program according to the *library screening* file. The general structure of this program is similar to that of the first one, but it allows control of a gas

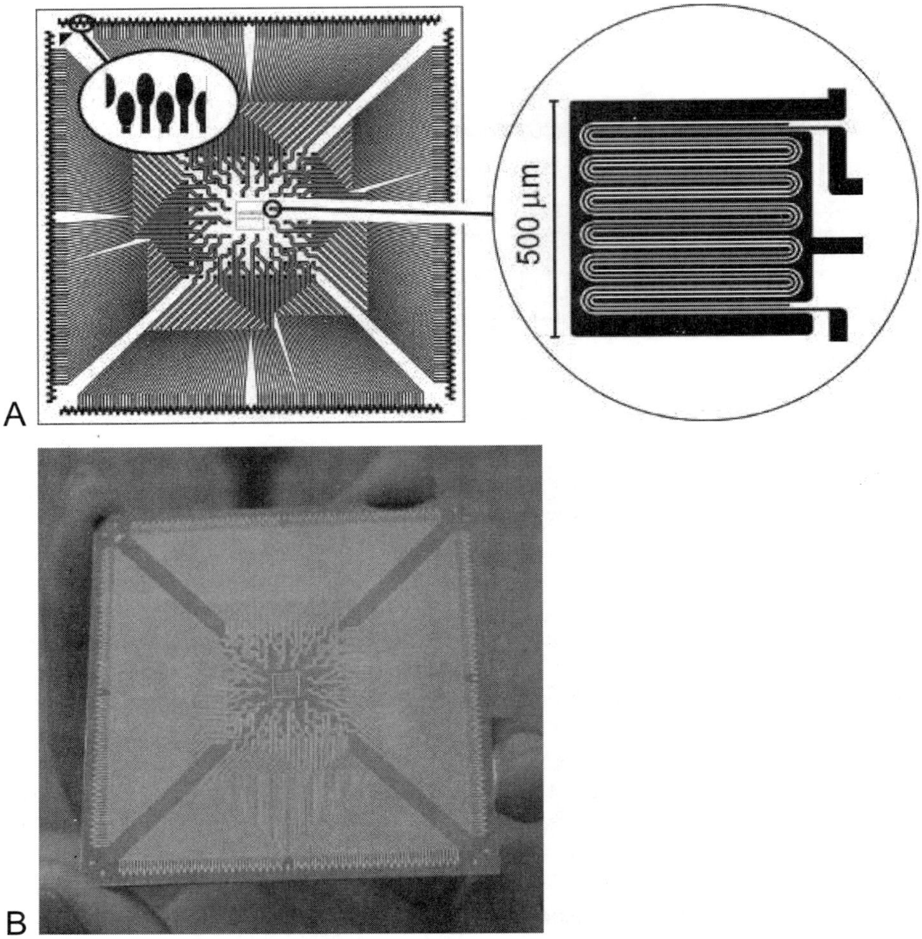

FIGURE 20.8. (A) design and (B) photograph of the electrode array containing 96 interdigital structures for four-point measurement.

flow system making serial additions of different gases in specified concentrations. The measured data and settings are written into files for further analysis which is performed by the third part of the software.

EXAMPLE OF EXPERIMENTS

The method has been tested on the electropolymerization of aniline. The relation between the electrode position and polymerization order was randomized. It was shown that the electrical resistance of polymers obtained in different positions under the same electrochemical conditions shows only statistical deviation and does not depend on either the order of polymerization or the electrode position on the array.

The electrical properties of polymers formed by the electropolymerization of a mixture of 4-aminobenzoic acid and aniline were investigated. The results are shown in Figure 20.9. As expected, incorporation of 4-aminobenzoic acid perturbs the conductive polymer chains and leads to a large decrease in the polymer conductance. Exposure to stepwise increased concentrations of gaseous hydrogen chloride leads to a dramatic decrease in polymer resistance [Figure 20.9(b), (c)]. The 35 curves describing the dependence of the relative resistivity of polymers on HCl concentration show almost no deviation (except for three curves, probably because of impurities on the polymerization sites) [Figure 20.9(b)]. This confirms the high reproducibility of the technology. Because the initial resistance is much higher than the resistance after the addition of HCl, the curves can be considered as normalized to maximal values. This allows us to estimate a binding constant for the gas molecules. The value is about 0.1 ppm for all polymers used. The dependence of relative conductivity on HCl concentration is strongly influenced by the polymer content [Figure 20.9(c)]. This dependence is evaluated quantitatively in [Figure 20.9(d)].

The results illustrate the high potential of the approach for developing polymeric gas sensors. The results were obtained on one chip and required about 20 "person mintues"

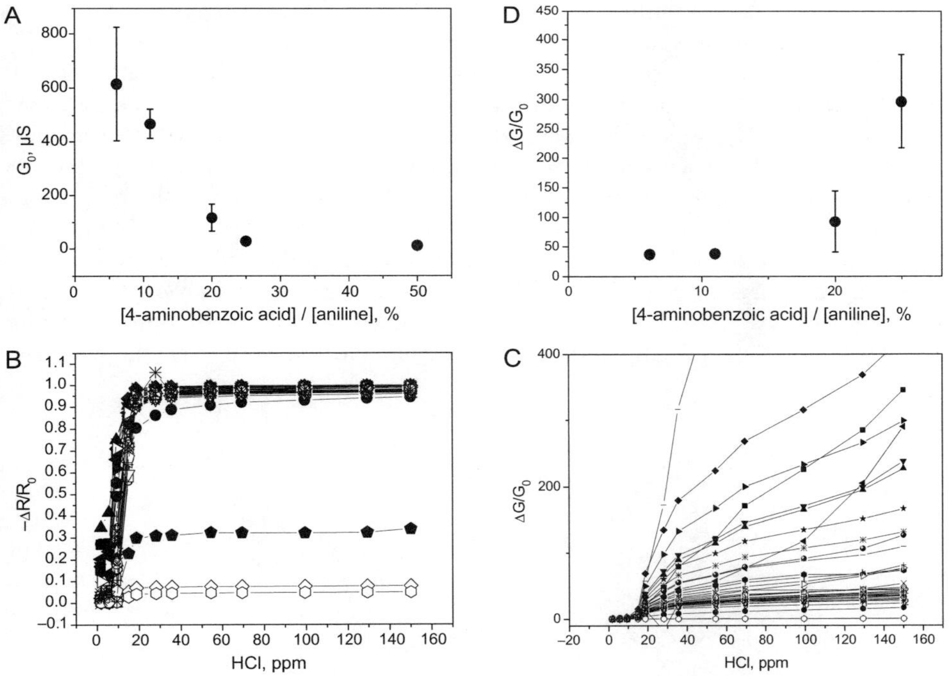

FIGURE 20.9. Influence of the ratio of aminobenzoic acid to aniline on (A) the conductance of polymers synthesized from these monomers and (B, C) on the sensitivity of these polymers to gaseous HCl. The data were obtained at 38 spots. The concentration dependences of the relative resistivity ($\Delta R/R_0$) and relative conductivity on gas concentration are presented in (B) and (C), respectively. (D) Curve obtained by taking a cross-section of all curves in (C) at an HCl concentration of 150 ppm. Electropolymerization conditions: serial electropolymerization at constant potential +900 mV vs. Ag/AgCl(sat); 600 s per spot. The molar ratio of 4-aminobenzoic acid to aniline is indicated. The total concentration of these reagents was 100 mM in 0.5 M H_2SO_4.

and about 14 hr of computer-controlled experiment. Performance of this work by the usual methods would require about 2 weeks of laboratory work.

COMBINATORIAL SYNTHESIS OF BIOLOGICAL SENSORS

For more than 20 years conductive polymers produced by electropolymerization have been widely used for the preparation of enzymatic biosensors (Figure 20.10) [48–50, 81–88]. This procedure involves polymerization of appropriate monomers (typically, pyrrole) in the presence of enzyme [89]; the enzyme can be preliminarily modified by pyrrole or by another electrochemically polymerizable group [90]. To accelerate the electron transfer between the polymer and the redox site within the enzyme, redox mediators can be added or copolymerized [91,92].

Optimization of enzymatic biosensors requires a variation of many parameters including the type of enzyme or a combination of enzymes, redox mediators, and polymerization conditions. By using high-throughput methods which are already established in molecular biology—a fast combinatorial synthesis of redox mediators and a wide range of several physical parameters (concentrations of all reagents, polymerization time, and conditions)—one can design enormous combinatorial libraries for screening. Development of multienzymatic biosensors results in a further increase in the size of these libraries.

As has been proven in preliminary experiments, the developed device can be used to screen all reagent concentrations and physical parameters without any modifications; screening of different preformed compounds (libraries of enzymes and mediators) requires combination of this device with an automated probe sampler. Further screening of the libraries can be performed on the same device; in this case, instead of the reagents for synthesis, the reservoirs are filled with analytes (enzyme substrate) and possible interferents. The device dilutes the substances, delivers them into the cell with the library of sensors, and measures the dependence of the electrode currents on the analyte (or interferent) concentration; these data can be used further for programmed analysis of biosensor parameters. Depending on the polymerization time and screening program, the estimated

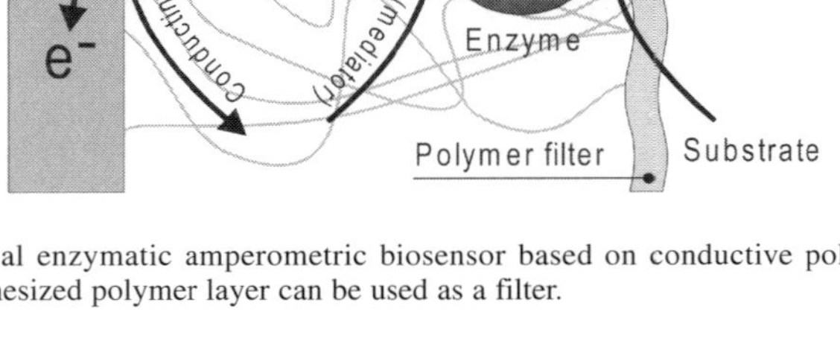

FIGURE 20.10. Typical enzymatic amperometric biosensor based on conductive polymers. An additional electrochemically synthesized polymer layer can be used as a filter.

throughput of such experiments in one device is about 100–300 spots (i.e., one to three libraries) per day.

As well as optimization of biosensors, the setup can be used for the synthesis and screening of an outer filter layer. The layer can be formed by electropolymerization as an additional polymer layer after biosensor synthesis [54]. The sensor based on electrically polymerized polymers can also investigate nonelectrical detection. For example, sensors for DNA analysis, based on electrochemical polymerization with subsequent fluorescent detection, have been described [83,93]. In this case, the polymer matrix has a mechanical function only; therefore nonconducting polymers can also be used. Possible uses of this approach for synthesis of corresponding combinatorial libraries do not require any further explanation. Other possible applications of combinatorial electropolymerization to the development of chemical and biological sensors include combinatorial electrochemical synthesis of molecularly imprinted polymers (this has been described elsewhere using a noncombinatorial approach [87, 94–97]), optimization of embedding of biomolecules into a polymer matrix, development of impedometric biosensors based on electrochemically polymerized films [98,99], and many others.

COMBINATORIAL SYNTHESIS OF ORGANIC ELECTRONIC DEVICES

The unique mechanical (flexibility and low weight), electrical, and optical properties of electrochemically synthesized polymers and the inexpensive fabrication technology required make them very attractive for applications in electronics. The use of organic materials in display technology (Organic light-emitting diodes or TFT organic displays) and in the fabrication of Schottky contacts, diodes, field-effect transistors, and ultrahigh capacitors [79] is particularly important.

Organic light-emitting Diodes (OLEDs) [78] are typically formed on glass or plastic substrate with indium tin oxide (ITO) electrodes [78,100–102]. These electrodes are transparent and provide good hole emission. The electrodes are covered by a thin protective layer (e.g., copper phthalocyanine), and then by polymer layers with p- and n-type conductivity. Finally, a cathode alloy is added. Other types of OLEDs use an organic emitter. OLEDs based on fluorene polymers [100,103,104] have as good electrical and light-emitting properties as traditional (inorganic) light-emitting diodes; however their temporal stability is still rather poor. Some types of OLEDs are already produced commercially by Cambridge Display Technology [105], DuPont Displays [106], IBM [107], and other companies. A fast search for novel polymeric materials and their combinations for OLED development can be performed by combinatorial electropolymerization. The preparation technique is outlined in Figure 20.11. Two possible structures for the preparation of OLED structures are shown. The first involves deposition of a common metallic anode formed on the OLED structure (by electroplating in situ, sputtering, or evaporation). In the second approach, both cathode and anode are placed on the electrode array. A combination of serial and parallel combinatorial electropolymerization leads to the formation of combinatorial library of OLEDs. The current configuration of the setup allows preparation of libraries containing up to 96 basic OLED structures but this can easily be increased up to 384.

Schottky diodes formed by a metal–semiconducting polymer structure can easily be prepared by electropolymerization [108]. The preparation approach is similar to that of

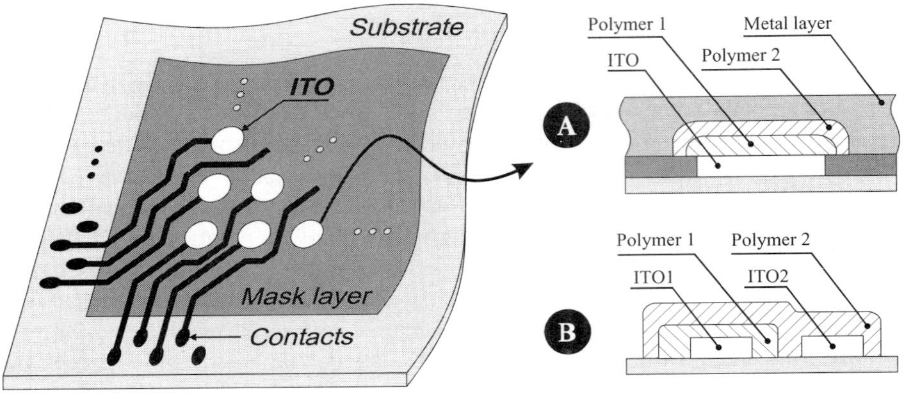

FIGURE 20.11. Suggested approaches for the preparation of combinatorial libraries of OLEDs and possible structures for these devices.

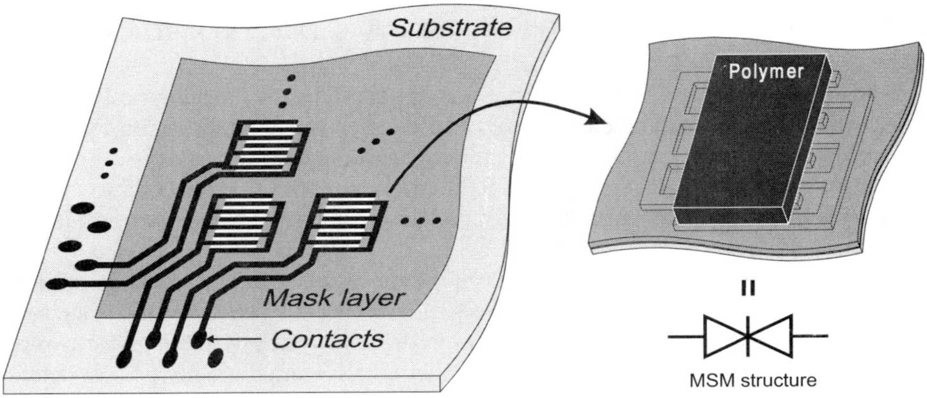

FIGURE 20.12. Electrode array for preparation of MSM structures (left) and a single interdigital electrode with a polymer layer (right).

OLEDs (Figure 20.11), but the electrode material can be different [108–111]. The same principles can also be used to prepare classic p–n junction diodes [109]. The combinatorial synthesis of metal–semiconductor–metal (MSM) structures consisting of two Schottky contacts is particularly simple (Figure 20.12). These structures are often used in high-speed photodetectors. Many light-sensitive polymers can be used in these devices [112]. The MSM structure can also be used for simple transducers in chemical sensors.

Organic field-effect transistors are described in the literature [74,108,113–118]. Although several operating devices have been described, the search for new materials and structures is still in its infancy and a combinatorial approach could certainly accelerate technological development in this field. A typical organic field effect transistor structure [75,76] can be prepared by electropolymerization of the source and drain electrodes

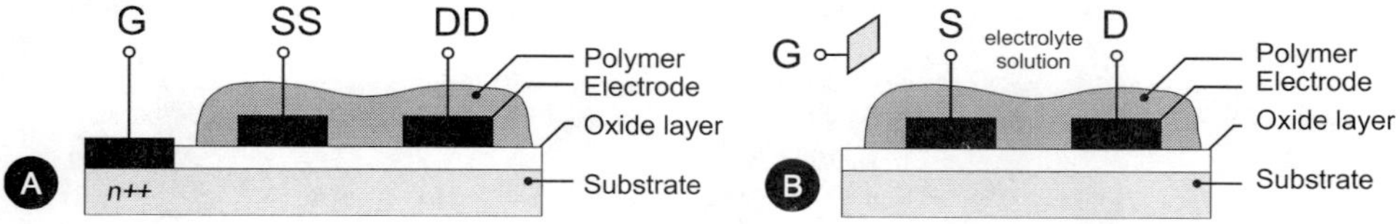

FIGURE 20.13. Structures of organic field-effect transistors (A) for applications in electronics and (B) with a gate contacting a solution for applications in transducers in chemical and biological sensors.

formed on highly doped silicon substrate coated with an oxide layer [Figure 20.13(a)]. It is known that electrochemically synthesized polymers grow much faster in the lateral direction than in the transverse direction; this allows the formation of a polymeric gap operating as a conductive channel. Another topology [Figure 20.13(b)] can be used to prepare organic field-effect transistors for applications in chemical sensors and biosensors [117] instead of traditional inorganic ion-sensitive field-effect transistor (ISFETs) [119–121]. Other applications of the combinatorial approach in organic electronics include the development of capacitors [77], high-efficiency solar cells [73], batteries, printed circuit boards [122], and many others.

CONCLUSION

We have described a concept and setup for the combinatorial electrochemical synthesis of polymers and multilayer polymeric structures. Of course, possible applications are not limited to the few examples described here. The approach can be used for almost every technology where the electrical properties of polymers or polymeric structures are important. It can be extended by combination with photochemical polymerization techniques or with optical investigation methods. Even in the configuration presented here, it provides about 100-fold increase in throughput for polymer synthesis and investigation. The aim of creating a common platform for different applications and the desire not to make the first prototype too complicated resulted in some restrictions on electrical parameters and array order. Development of equipment specialized for particular applications will lead to a further increase in its throughput and screening facilities.

ACKNOWLEDGMENTS

This work was supported by the project KOMBISENS, German Ministry for Science and Technology. The authors are grateful to Professor O. S. Wolfbeis for fruitful discussions and to Mrs Qingli Hao for help with the experimental work.

REFERENCES

1. Beck-Sickinger, A., Weber, P. *Kombinatorische Methoden in Chemie und Biologie*, **1999**, Heidelberg, Germany: Spektrum Akademie Verlag.
2. Croston, G. E. *Trends Biotechnol.* **2002**, *20*, 110–115.
3. Schiedel, M. S., Briehn, C. A., Bauerle, P. *Angew. Chem. Int. Ed.* **2001**, *40*, 4677–4680.

4. Briehn, C. A., Schiedel, M. S., Bonsen, E. M., Schuhmann, W., Bauerle, P. *Angew. Chem. Int. Ed.* **2001**, *40*, 4680–4683.
5. Gallop, M. A. *Solid-Phase Synthesis of Compound Libraries and Their Applications in Drug Discovery*, **2000**, Geel, The Netherlands: Acros Organics.
6. De Mello, A. *J. Analyt. Bioanalt. Chem.* **2002**, *372*, 12–13.
7. Salimi-Moosavi, H., Tang, T., Harrison, D. J. *J. Am. Chem. Soc.* **1997**, *119*, 8716–8717.
8. Yudin, A. K., Siu, T. *Curr. Opin. Chem. Biol.* **2001**, *5*, 269–272.
9. Pilard, J. F., Marchand, G., Simonet, J. *Tetrahedron* **1998**, *54*, 9401–9414.
10. Niyazymbetov, M. Use of electrochemical methods in synthesis of organic compounds for biological screening. Presented at 10th International Forum on Electrolysis in Chemical Industry, Electrosynthesis, Lancaster, New York, **1996**.
11. Breitenbach, J. W., Olaj, O. F., Sommer, F. *Fortschr. Hochpolym.-Forsch.* **1972**, *9*, 47–227.
12. Subramanian, R. V. *Adv. Polym. Sci.* **1979**, *33*, 33–58.
13. Wolf, M. O., Zhu, Y. *Adv. Mater.* **2000**, *12*, 599–601.
14. Mengoli, G., Musiani, M. M. *Prog. Org. Coat.* **1994**, *24*, 237–251.
15. Yuan, W., Iroh, J. O. *Trends Polym. Sci.* **1993**, *1*, 388–394.
16. Mengoli, G., Musiani, M. M. *J. Electrochem. Soc.* **1987**, *134*, 643C–652C.
17. Waltman, R. J., Bargon, J. *Can. J. Chem.* **1986**, *64*, 76–95.
18. Sabouraud, G., Sadki, S., Brodie, N. *Chem. Soc. Rev.* **2000**, *29*, 283–293.
19. Guerra, G. D. Electroinitiated polymerization. In G. C. Eastmond, A. Ledwith, S. Russo, P. Sigwalt (Eds.), *Chain Polymerization*, Part 2, Vol. 4, **2002**, Oxford: Pergamon Press, pp. 453–458.
20. Rifi, M. R., Covitz, F. H. *Introduction to Organic Electrochemistry* **1974**, New York: Marcel Dekker.
21. Ye, S. Y., Do, N. T., Dao, L. H., Vijh, A. K. *Synth. Met.* **1997**, *88*, 65–72.
22. Chen, J., Burrell, A. K., Collis, G. E., Officer, D. L., Swiegers, G. F., Too, C. O., Wallace, G. G. *Electrochim. Acta* **2002**, *47*, 2715–2724.
23. Hosseini, S. H., Entezami, A. A. *Polym. Adv. Technol.* **2001**, *12*, 524–534.
24. Glidle, A., Swann, M. J., Hadyoon, C. S., Cui, L., Davis, J., Ryder, K. S., Cooper, J. M. *J. Electron. Spectrosc. Relat. Phenom.* **2001**, *121*, 131–148.
25. Aubert, P. H., Neudeck, A., Dunsch, L., Audebert, P., Capdevielle, P., Maumy, M. *J. Electroanal. Chem.* **1999**, *470*, 77–88.
26. Lee, J. Y., Su, X. H., Cui, C. Q. *J. Electroanal. Chem.* **1994**, *367*, 71–78.
27. Noftle, R. E., Pletcher, D. *J. Electroanal. Chem.* **1987**, *227*, 229–235.
28. Spinks, G. M., Dominis, A. J., Wallace, G. G., Tallman, D. E. *J. Solid State Electrochem.* **2002**, *6*, 85–100.
29. Rammelt, U., Nguyen, P. T., Plieth, W. *Electrochim. Acta* **2001**, *46*, 4251–4257.
30. Meneguzzi, A., Ferreira, C. A., Pham, M. C., Delamar, M., Lacaze, P. C. *Electrochim. Acta* **1999**, *44*, 2149–2156.
31. Kinlen, P. J., Menon, V., Ding, Y. W. *J. Electrochem. Soc.* **1999**, *146*, 3690–3695.
32. Bernard, M. C., Goff, A. H.-L., Joiret, S., Dinh, N. N., Toan, N. N. *J. Electrochem. Soc.* **1999**, *146* (3), 995–998.
33. Musiani, M. M., Furlanetto, F., Guerriero, P., Heitbaum, J. *J. Appl. Electrochem.* **1993**, *23* (10), 1069–1075.
34. Fabrizio, M., Mengoli, G., Musiani, M. M., Paolucci, F. *Corros. Sci.* **1993**, *35*, 1527–1533.
35. Kendig, M., Jeanjaquet, S. *Proc. Electrochem. Soc.* **1989**, *89*, 186–197.
36. Mengoli, G., Musiani, M. M. *Polym. Mater. Sci. Eng.* **1988**, *58*, 163–167.
37. Mengoli, G., Musiani, M. M. *Makromol. Chem., Macromol. Symp.* **1987**, *8*, 265–283.
38. Kendig, M., Anderson, D., Jeanjaquet, S. *Proc. Electrochem. Soc.* **1986**, *86*, 149–162.
39. Pham, M. C., Mourcel, P., Lacaze, P. C., Dubois, J. E. *Bull. Soc. Chim. Fr.* **1985**, *6*, 1169–1173.
40. McCullough, R. D. *Adv. Mater.* **1998**, *10* (2), 93–116.
41. Groenendaal, L., Jonas, F., Freitag, D., Pielartzik, H., Reynolds, J. R. *Adv. Mater.* **2000**, *12*, 481–494.
42. Karyakin, A. A., Karyakina, E. E., Schuhmann, W., Schmidt, H. L. *Electroanalysis* **1999**, *11*, 553–557.
43. Karyakin, A. A., Karyakina, E. E., Schmidt, H. L. *Electroanalysis* **1999**, *11*, 149–155.
44. Schlereth, D. D., Karyakin, A. A. *J. Electroanalt. Chem.* **1995**, *395*, 221–232.
45. Bloor, D. Electrical conductivity. In C. Booth and C. Price (Eds.), *Polymer properties*, Vol. 2, **1989**, Oxford: Pergamon Press, pp. 687–705.
46. Inzelt, G., Pineri, M., Schultze, J. W., Vorotyntsev, M. A. *Electrochim. Acta* **2000**, *45*, 2403–2421.
47. Khan, M. A., Armes, S. P. *Adv. Mater.* **2000**, *12*, 671–674.

48. Gerard, M., Chaubey, A., Malhotra, B. D. *Biosens. Bioelectron.* **2002**, *17*, 345–359.
49. Palmisano, F., Zambonin, P. G., Centonze, D. *Fresenius J. Anal. Chem.* **2000**, *366*, 586–601.
50. Wallace, G.G. Smyth, M., Zhao, H. *Trends Anal. Chem.* **1999**, *18*, 245–251.
51. Habermuller, K., Mosbach, M., Schuhmann, W. *Fresenius J. Anal. Chem.* **2000**, *366*, 560–568.
52. Cosnier, S. *Appl. Biochem. Biotechnol.* **2000**, *89*, 127–138.
53. Cosnier, S. *Biosens. Bioelectron.* **1999**, *14*, 443–456.
54. Christie, I. M., Rigby, G. P., Treloar, P., Warriner, K., Maines, A., Ashworth, D., Eddy, S., Reddy, S. M., Vadgama, P. *NATO ASI Ser., Ser. 2* **1997**, *38*, 41–56.
55. Karyakin, A. A., Vuki, M., Lukachova, L. V., Karyakina, E. E., Orlov, A. V., Karpachova, G. P., Wang, J. *Anal. Chem.* **1999**, *71*, 2534–2540.
56. Ouerghi, O., Senillou, A., Jaffrezic-Renault, N., Martelet, C., Ben Ouada, H., Cosnier, S. *J. Electroanal. Chem.* **2001**, *501*, 62–69.
57. Trojanowicz, M., Geschke, O., Krawczynski vel Krawczyk, T., Cammann, K. *Sens. Actuators B* **1995**, *B28*, 191–199.
58. Gambhir, A., Gerard, M., Mulchandani, A. K., Malhotra, B. D. *Appl. Biochem. Biotechnol.* **2001**, *96*, 249–257.
59. Kaden, H., Juttner, K., Jahn, H., Berthold, M., Mangold, K. M., Schafer, S. *Chem. Inge. Tech.* **2000**, *72*, 1534–1538.
60. Kaden, H., Jahn, H., Berthold, M., Juttner, K., Mangold, K. M., Schafer, S. *Chem. Eng. Technol.* **2001**, *24*, 1120–1124.
61. Mangold, K. M., Schafer, S., Juttner, K. *Synth. Met.* **2001**, *119*, 345–346.
62. Mangold, K. M., Schafer, S., Juttner, K. *Fresenius J. Anal. Chem.* **2000**, *367*, 340–342.
63. Bouzek, K., Mangold, K. M., Juttner, K. *J. Appl. Electrochem.* **2001**, *31*, 501–507.
64. Weidlich, C., Mangold, K. M., Juttner, K. *Synth. Met.* **2001**, *119*, 263–264.
65. Heuer, H. W., Wehrmann, R., Kirchmeyer, S. *Adv. Funct. Mater.* **2002**, *12*, 89–94.
66. Schaller, E., Bosset, J. O., Escher, F. *Lebensm.-Wiss. Technol.* **1998**, *31*, 305–316.
67. Barisci, J. N., Conn, C., Wallace, G. G. *Trends Polym. Sci.* **1996**, *4*, 307–311.
68. Amrani, M. E. H., Payne, P. A., Persaud, K. C. *Sens. Actuators B* **1996**, *B33*, 137–141.
69. Gardner, J. W., Bartlett, P. N. *Synth. Met.* **1993**, *57*, 2665–2670.
70. Miasik, J. J., Hooper, A., Tofield, B. C. *J. Chem. Soc., Faraday Trans.1* **1986**, *82*, 1117–1126.
71. Cyr, J. A. *Am. Environ. Lab.* **1998**, *10*, 14–15.
72. Freund, M. S., Lewis, N. S. *Proc. Natl Acad. Sci. USA* **1995**, *92*, 2652–2656.
73. Fahlman, M., Salaneck, W. R. *Surf. Sci.* **2002**, *500*, 904–922.
74. Garnier, F. *Chem. Phys.* **1998**, *227*, 253–262.
75. Kagan, C. R., Mitzi, D. B., Dimitrakopoulos, C. D. *Science* **1999**, *286*, 945–947.
76. Mitzi, D. B., Chondroudis, K., Kagan, C. R. *IBM J. Res. Dev.* **2001**, *45*, 29–45.
77. Andrieu, X. *New Trends Electrochem. Technol.* **2000**, *1*, 521–547.
78. Bernius, M. T., Inbasekaran, M., O'Brien, J., Wu, W. *Adv. Mater.* **2000**, *12*, 1737–1750.
79. Plastic Logic, http://www.plasticlogic.com
80. Scheying, G., Schulte, T., Brinz, T., Kulikov, V., Mirsky, V. German patent application DE 10131581, **2001**.
81. Trojanowicz, M., Lewenstam, A., Krawczyk, T. K. V., Lahdesmaki, I., Szczepek, W. *Electroanalysis* **1996**, *8*, 233–243.
82. Krawczynski vel Drawczyk, T., Trojanowicz, M., Lewenstam, A., Moszczynska, A. *Biosens. Bioelectron.* **1996**, *11*, 1155–1165.
83. Bidan, G., Billon, M., Galasso, K., Livache, T., Mathis, C., Roget, A., Torres-Rodriguez, L. M., Vieil, E. *Appl. Biochem. Biotechnol.* **2000**, *89* (2–3), 183–193.
84. Gerard, M., Chaubey, A., Malhotra, B. D. *Biosens. Bioelectron.* **2002**, *17*, 345–359.
85. Dequaire, M., Heller, A. *Anal. Chem.* **2002**, *74*, 4370–4377.
86. Emr, S. A., Yacynych, A. M. *Electroanalysis* **1995**, *7*, 913–923.
87. Panasyuk, T. L., Mirsky, V. M., Piletsky, S. A., Wolfbeis, O. S. *Anal. Chem.* **1999**, *71*, 4609–4613.
88. Situmorang, M., Gooding, J. J., Hibbert, D. B., Barnett, D. *Biosens. Bioelectron.* **1998**, *13*, 953–962.
89. Quinto, M., Losito, I., Palmisano, F., Zambonin, C. G. *Anal. Chim. Acta* **2000**, *420*, 9–17.
90. Yon-Hin, B. F. Y., Smolander, M., Crompton, T., Lowe, C. R. *Anal. Chem.* **1993**, *65*, 2067–2071.
91. Reiter, S., Habermuller, K., Schuhmann, W. *Sens. Actuators B* **2001**, *B79*, 150–156.
92. Schuhmann, W. *Rev. Molc. Biotechnol.* **2002**, *82*, 425–441.

93. Bidan, G., Billon, M., Livache, T., Mathis, G., Roget, A., Torres-Rodriguez, L. M. *Synth. Met.* **1999**, *102*, 1363–1365.
94. Panasyuk, T., Dall'Orto, V. C., Marrazza, G., El'skaya, A., Piletsky, S., Rezzano, I., Mascini, M. *Anal. Lett.* **1998**, *31*, 1809–1824.
95. Cheng, Z., Wang, E., Yang, X. *Biosensors Bioelectron.* **2001**, *16*, 179–185.
96. Peng, H., Zhang, Y., Zhang, J., Xie, Q., Nie, L., Yao, S. *Analyst* **2001**, *126*, 189–194.
97. Malitesta, C., Losito, I., Zambonin, P. G. *Anal. Chem.* **1999**, *71*, 1366–1370.
98. Farace, G., Lillie, G., Hianik, T., Payne, P., Vadgama, P. *Bioelectrochemistry* **2002**, *55*, 1–3.
99. Lillie, G., Payne, P., Vadgama, P. *Sens. Actuators B* **2001**, *B78*, 249–256.
100. Bernius, M., Inbasekaran, M., Woo, E., Wu, W., Wujkowski, L. *Thin Solid Films* **2000**, *363*, 55–57.
101. Bernius, M. T. Patent application WO A2 0265558, **2002**.
102. Mitschke, U., Bauerle, P. J. *Mater. Chem.* **2000**, *10*, 1471–1507.
103. Herguth, P., Jiang, X., Liu, M. S., Jen, A. K. Y. *Macromolecules* **2002**, *35*, 6094–6100.
104. Jung, B. J., Lee, J. I., Chu, H. Y., Do, L. M., Shim, H. K. *Macromolecules* **2002**, *35*, 2282–2287.
105. Cambridge Display Technology, http://www.cdtltd.co.uk
106. DuPont Displays, http://www.dupont.com/displays/oled
107. IBM, http://www.almaden.ibm.com/st/projects/oleds/fabrication/index.html
108. Torsi, L., Malitesta, C., Sabbatini, L., Zambonin, P. G., Dodabalapur, A., Katz, H. E. *Mater. Sci. Eng. C* **1998**, *C5*, 233–236.
109. Torsi, L., Malitesta, C., Morea, G., Sabbatini, L., Zambonin, P. G. *Ann. Chim. (Rome)* **1992**, *82*, 259–274.
110. Singh, R., Srivastava, D. N., Singh, R. A. *Synth. Met.* **2001**, *121*, 1439–1440.
111. Narasimhan, M., Hagler, M., Cammarata, V., Thakur, M. *Appl. Phys. Lett.* **1998**, *72*, 1063–1065.
112. Kudo, T., Kimura, M., Hanabusa, K., Shirai, H. J. *Porphyrins Phthalocyanines* **1998**, *2*, 231–235.
113. Horowitz, G., Fichou, D., Peng, X., Xu, Z., Garnier, F. *Solid State Commun.* **1989**, *72*, 381–384.
114. Lovinger, A. J., Rothberg, L. J. *J. Mater. Res.* **1996**, *11*, 1581–1592.
115. Tsumura, A., Koezuka, H., Ando, T. *Synth. Met.* **1988**, *25*, 11–23.
116. Koezuka, H., Tsumura, A., Fuchigami, H., Kuramoto, K. *Appl. Phys. Lett.* **1993**, *62*, 1794–1796.
117. McCoy, C. H., Wrighton, M. S. *Chem. Mater.* **1993**, *5*, 914–916.
118. Jones, E. T. T., Chyan, O. M., Wrighton, M. S. *J. Am. Chem. Soc.* **1987**, *109*, 5526–5528.
119. Sandifer, J. R., Voychek, J. J. *Mikrochim. Acta* **1999**, *131*, 91–98.
120. Janata, J. *Analyst* **1994**, *119*, 2275–2278.
121. Van der Schoot, B. H., Bergveld, P. *Biosensors* **1988**, *3*, 161–186.
122. Goosey, M. *Plastics for Electronics* (2nd ed.) **1999**, Dordrecht, The Netherlands: Kluwer.

Chapter 21

Principles and Applications of Combinatorial Electrochemistry

Rongzhong Jiang and Deryn Chu

INTRODUCTION

Over the past decade, the combinatorial approach has received much attention as a powerful tool for simultaneously studying large numbers of parallel experiments, and its applications have been widely extended to DNA recognition [1], antibody research [2], drug discovery [3], chemical syntheses [4–6], and material discovery [7–13]. Combinatorial methods have been applied to almost every part of science and technology, and greatly accelerated research and development. Combinatorial methods were first used in electrochemistry research in 1998 when Reddington et al. [14] created a 645-member electrocatalyst array on a conductive carbon substrate and electrochemically studied the samples under constant potential. The catalyst activity was identified at each spot of the array by an optical screening method with a fluorescent acid–base indicator.

Terminologically, electrochemistry is a part of the science that studies interfaces and reactions between electrode and electrolyte. So far electrochemistry has branched into studies of batteries, fuel cells, electrolysis, electroplating, electrosynthesis, ultracapacitors, sensors, electrochromic displays, and metal corrosion. Electrochemistry is also closely related to other fields of science and technology, including spectroscopic electrochemistry, chromatographic electrochemistry, mass electrochemistry, photoelectrochemistry and bioelectrochemistry. Generally, only a few electrochemical experiments are carried out at the same time in traditional electrochemistry. In order to discover an advanced electrochemical material such as an electrode catalyst hundreds or even thousands of experiments are needed to identify the best from numerous catalyst combinations. Complex electrochemical parameters (e.g., current, voltage, impedance, temperature, pressure, concentration, mass transfer, thermodynamics, and electrode kinetic factors), as well as unexpected variations in experimental conditions, may cause difficulties for fast and efficient completion of the research. However, a combinatorial method could provide an opportunity of

Rongzhong Jiang and Deryn Chu • U.S. Army Research Laboratory, Adelphi, MD 20783-1197, USA
Contact author: Rongzhong Jiang, email: rjiang@arl.army.mil

High-Throughput Analysis, edited by Potyrailo and Amis
Kluwer Academic Publishers, New York, 2003.

solving this problem and realizing high-throughput analysis of the electrochemical parameters simultaneously. This chapter emphasizes the understanding of the principles and applications of combinatorial electrochemistry.

TRADITIONAL ELECTROCHEMICAL CELL AND ELECTRODES

A single cell (Figure 21.1), containing two or three electrodes (working, counter, and reference electrodes) soaked in or in contact with homogeneous electrolyte, is generally used in electrochemical research. The three electrodes are usually located in one, two, or three compartments depending upon the purpose of the electrochemical research. As long as the electrode contacts the electrolyte, a potential difference is generated at the interface between the electrode and electrolyte, and the magnitude of the potential is dependent on the type and concentration of electrochemical reactants in the electrolyte and at the electrode surface. There is a potential difference between the working and counter electrode. The electrochemical reaction occurs when the working and counter electrodes are linked by a conducting wire, and the electric current passes along the wire. Conversely, the electrochemical reaction can also be forced to occur by applying an external potential difference between the working and counter electrodes. The reference electrode, as a standard point of potential, is used to compensate the *iR* drop between the working and counter electrodes caused by the electrolyte. Negligible or zero current flows between the working

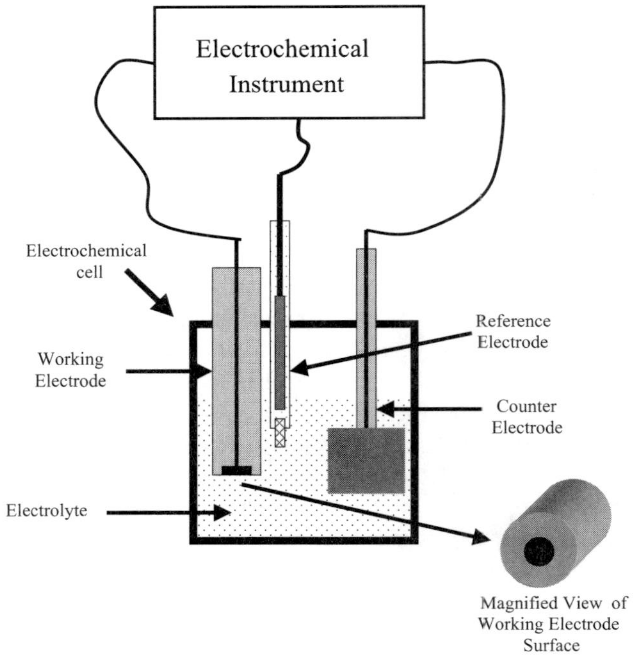

FIGURE 21.1. A typical electrochemical cell linked to an electrochemical instrument.

and the reference electrodes. This kind of single cell is commonly used for half electrochemical cell research, that is, for studying the electron transfer and chemical reactions at the working electrode rather than the counter electrode.

Practical applications of two-electrode cells are seen in most electrochemical systems, such as batteries, fuel cells, and ultracapacitors. The two-electrode cell is commonly formed like a sandwich where the electrodes, electrolyte, and separator are layered in parallel, and the cathode and anode electrodes are positioned in a face-to-face configuration. In this simplified version of the cell configuration (Figure 21.1), the counter and reference electrodes are combined in a single electrode.

These traditional electrochemical systems are unsuitable for combinatorial studies. In order to realize simultaneous parallel electrochemistry studies, the primary step is to generate an electrochemical array.

ELECTROCHEMICAL ARRAYS AND SCREENING METHODS FOR COMBINATORIAL MEASUREMENTS

Direct Electrode Array and an Automated Screening Method

Generally, we focus on the working electrode for electrochemical studies. The lower right-hand panel in Figure 21.1 shows a magnified view of a working electrode surface, the center of which is the electrode made of carbon substrates or metallic materials. The electrode is sealed with an electrically insulating material such as plastic, glass, or ceramic. An electrochemical array of working electrodes can be generated directly by integration of a number of single electrodes on the same geometrical plane. Figure 21.2 shows a working electrode array containing 256 members. Each electrode in the array is sealed separately. The electric leads from these electrodes are led into a cable with multiple channels. The direct electrode array can be placed in a single cell where all the working electrodes in the array share the same counter electrode and reference electrode. Therefore this kind of electrochemical system is similar to a single electrochemical cell except that the working electrode is replaced by a multimember array. This kind of working electrode array requires a special electrochemical instrument with multiple channels for data input. Unfortunately commercial electrochemical instruments containing 256 channels are rarely available, which causes a difficulty in parallel electrochemical screening. In order to solve this problem a computer-controlled device with automatic electrode connection and disconnection functions has been designed. The working electrodes in the electrode array can be connected individually, in groups, or all together. Figure 21.3 shows a schematic view of an automated screening system for combinatorial electrochemical measurements. In 1999 Sullivan et al. [15] described a similar automated screening device for studying organosulfur monolayers on gold electrode surfaces. The primary advantage of the direct electrode array is the ability to distinguish small differences in current and voltage with high accuracy. Secondly, combinatorial electrochemistry research with the direct electrode array can also be used with a variety of traditional electrochemical methods, such as cyclic voltammetry, chronoamperometry, chronopotentiometry, potential step, and ac impedance. Thirdly, all the electrodes in the array can be polished to refresh the surfaces so that reproducible experimental results are obtained.

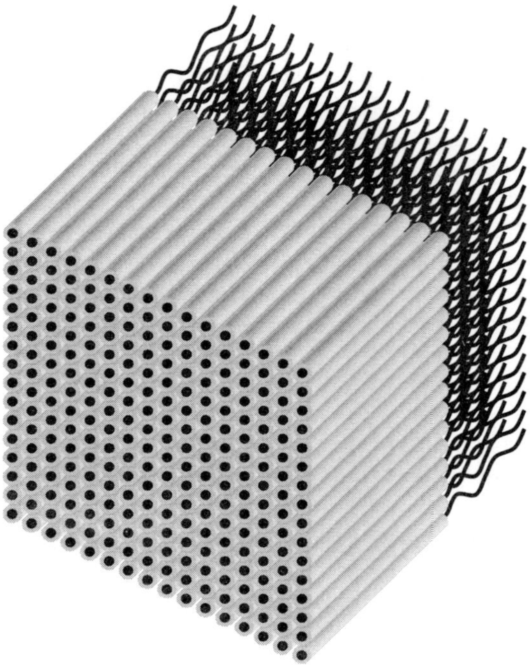

FIGURE 21.2. A direct electrode array containing 256 working electrodes.

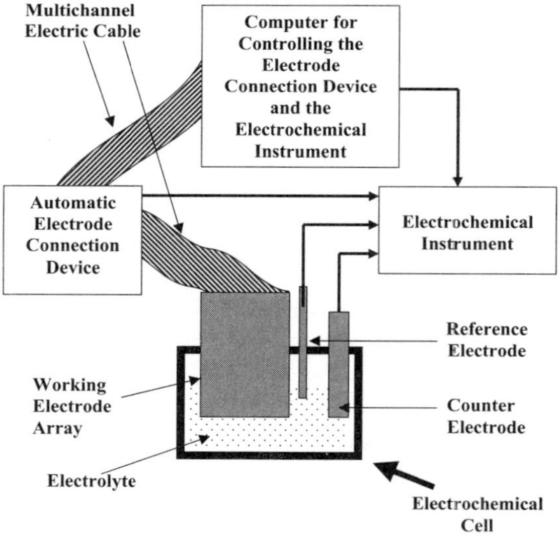

FIGURE 21.3. Operating principle of an automatic electrode-connection device and a direct electrode array containing 256 working electrodes in a single cell.

The disadvantages of the direct electrode array are the significantly increasing complexity and cost associated with electrode processing and equipment design as the number of members in the array increases. Some multiple-channel devices and instruments for combinatorial research have been developed [15–17] and improved in recent years.

Material Spot Array on a Single Electrode and an Indirect Screening Method

Formation of a Material Spot Array for a Single Cell

A number of target electrochemical materials can be studied simultaneously by forming a material spot array on a single electrode. Figure 21.4 shows a schematic view of a material spot array containing 400 spots on a planar electrode surface. Each of the spots is used for depositing different electrochemical materials or combinations. For example, if one wants to study electrode catalysts, one can coat the catalysts as spots on an electrode and conduct parallel electrochemical measurements. The electrode substrates are generally graphite, carbon, or metals. This system is very similar to the traditional single electrochemical cell shown in Figure 21.1, except that the working electrode is elaborately tailored into many small regions or spots to adopt the targeting electrochemical materials. In the material spot array all the spots are electrically shorted, which is very different from

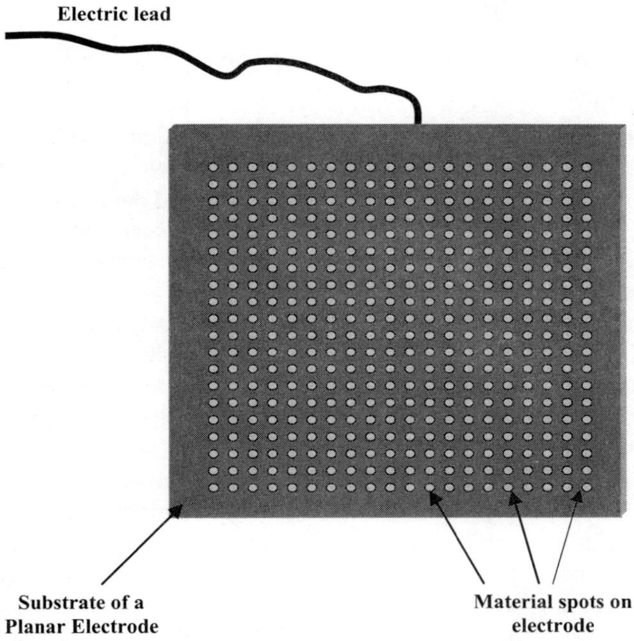

FIGURE 21.4. A material spot array containing 400 spots on a planar electrode surface. Each of the spots is used for depositing different electrochemical materials or combinations.

that of the direct electrode array where all the electrodes are electrically connected but separated. Reddington et al. [14] and Choi et al. [18] described a computer-controlled printing technique to deposit hundreds of combinations of binary, ternary, and quaternary catalysts on a planar carbon substrate and form material spot arrays on a single electrode surface. Traditional electrochemical methods, such as cyclic voltammetry, chronoamperometry, chronopotentiometry, potential step, and ac impedance, can also adopt combinatorial electrochemistry techniques with the material spot array and indirect screening methods. As long as a potential is applied between the working and counter electrodes, all the spots in the array will have the same potential but the current will be different because they have different electrochemical activities. The most obvious advantage of this kind of array is that only one electrochemical measurement is needed for analysis of all the spots in the array and no special electrochemical instrument or device, like that shown in Figure 21.3, is required. Furthermore, it is much easier to generate a spot array with a large number of members than a direct electrode array. After creation of the electrochemical array, the second step is to find an appropriate screening method for high-speed electrochemical analysis of all members of the array.

An Optical Screening Method with Fluorescent Proton Indicators

Reddington et al. [14] used fluorescent proton indicators as an indirect screening method for the electrochemical investigation of methanol oxidation on hundreds of catalyst combinations that were coated on a carbon electrode surface to form spot arrays. The electrochemical and optical reactions are

$$CH_3OH + H_2O \rightarrow CO_2 + 6H^+ + 6e^- \qquad (1)$$

$$\text{indicator} + H^+ \rightarrow \text{fluorescent compound} + \text{fluorescence} \qquad (2)$$

Quinine and the Ni^{2+} complex of 3-pyridin-2yl-(4,5,6)triazolo-[1,5-a] pyridine (PTP) are used as indicators at neutral and low pH, respectively. The more active electrocatalyst will generate more current and produce more protons as a potential difference is applied between the working electrode (i.e., material spot array) and the counter electrode, and, in turn, the protons stimulate the indicator to generate fluorescence on the spots. The higher concentration of protons results in brighter spots because the spots in the array are spatially isolated on the electrode surface.

The advantages of the optical screening method are its fast response, that no special electrochemical equipment is required, and that all these samples can be examined by a single electrochemical measurement. The disadvantage is the dependence on proton generation or consumption in the electrochemical reaction. Because many electrochemical reactions are studied in concentrated bases or acids, or do not involve significant proton change, or take place in nonaqueous media where most acid–base indicators cannot work, the application of the optical screening method is greatly limited. The accuracy of the optical method is also an issue, as it is dependent on visually comparing the strength of the fluorescence which is more likely to be affected by errors and irreproducibility than are direct electrochemical measurements [15].

Most recently, a number of reports [15,18–20] described an optical screening method with some improvements in accuracy using the same or different fluorescent indicators. Sullivan et al. [15] reported that a thin-layer cell with an array of 28 spots was used to

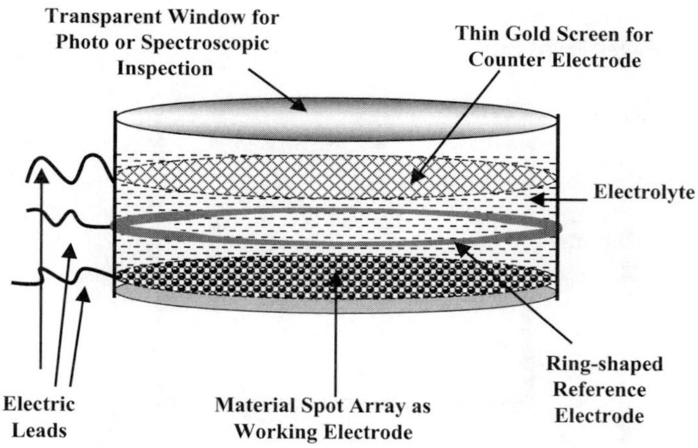

FIGURE 21.5. A schematic view of a thin-layer electrochemical cell with a material spot array on a planar conductive substrate for working electrodes.

study the electrochemical reactions of 1,4-benzoquinone and alkanethiols using fluorescein as an indicator. However, the accuracy of the optical screening method is only slightly improved because of the minimization of the fluorescent contribution from the background by such thin-layer electrochemical cell.

Figure 21.5 shows a thin-layer configuration of an electrochemical cell for combinatorial studies. Here, three electrodes are used in the cell. The working electrode is replaced by a material spot array on the electrode surface. The counter electrode is a thin gold screen which is able to pass a photobeam for inspection of fluorescence from the spots in the array. The reference electrode is ring-shaped to ensure optimum compensation of the *iR* drop in the electrolyte between the array and the counter electrode. Here, the fluorescent contribution from the background is greatly reduced due to the short distance from optical inspection window to the array compared with the conventional cell design in Figure 21.3.

Indirect Electrode Array on a Single Conductive Substrate and a Screening Method with an Electrolyte Probe

Indirect Electrode Array on a Single Conductive Substrate

Jiang et al. [21] reported an electrolyte probe screening method for combinatorial electrochemical analysis. They designed a working electrode array on a single conductive substrate, which is shown in Figure 21.6 as a collection of electrode array containing 144 electrodes arranged on a single graphite substrate. The surface of the graphite is covered with a plastic coating except for certain areas reserved for electrode loading. The plastic material is used for insulation between electrodes. The electrodes in the array are coated with different combinations of catalysts for electrochemical reactions, and microporous separators are used for loading electrolyte that contains electrochemical reactants. The electrodes in the working electrode array are electrically shorted to each other because they

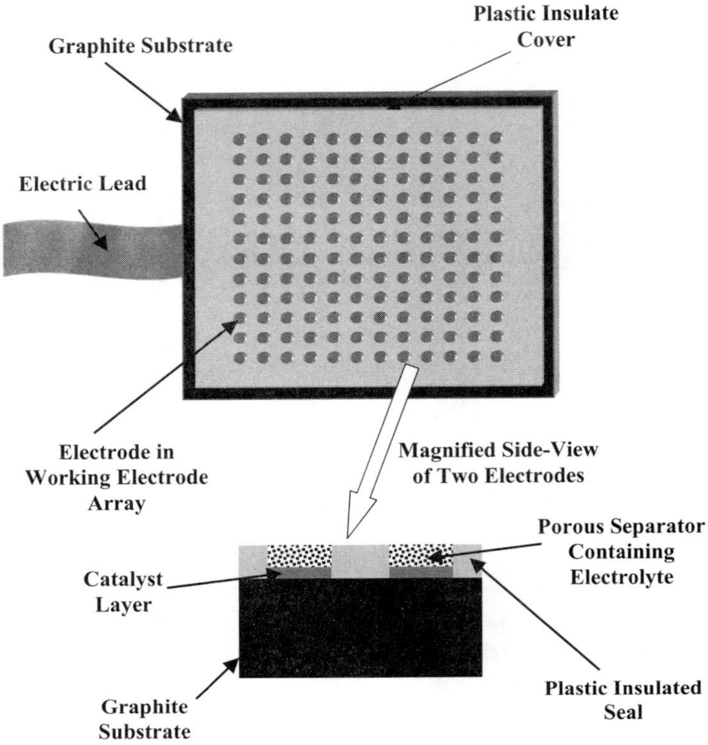

FIGURE 21.6. An indirect electrode array containing 144 electrodes arranged on a single graphite substrate and the magnified side-view of two adjacent electrodes.

share the same electrically conductive substrate. This kind of electrode array is different from that of the direct electrode array shown in Figure 21.2 where the working electrodes are insulated from each other. Therefore we define this kind of electrode array as an indirect electrode array in order to indicate its difference from the direct electrode array. The microporous separators in the indirect electrode array are isolated from each other, that is, these electrodes are ionically insulated. Therefore this kind of array is also different from the material spot array shown in Figure 21.4 where the material spots in the array are ionically shorted when the array is used in a single cell.

Electrolyte Probe Screening Method

The upper panel in Figure 21.7 shows a design for an electrolyte probe that is formed by filling a flexible tube, whose terminals are sealed by microporous material, with concentrated electrolyte. The ionic conductance can be adjusted by the section area and length of the electrolyte probe, and by the concentration of the electrolyte. A tiny reference electrode can be inserted in the electrolyte tube where it is closest to the terminal facing toward working electrodes. In studies of two-electrode cells, such as batteries or fuel cells, the reference electrode is not necessary. All the working electrodes in the electrode array

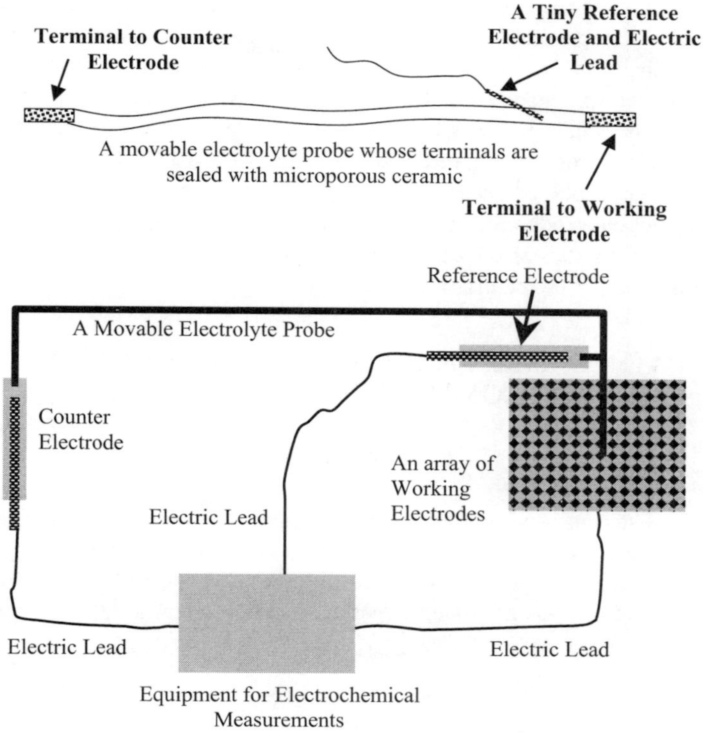

FIGURE 21.7. Combinatorial measurement with an indirect electrode array for working electrodes and a corresponding movable electrolyte probe for fast screening. (Reproduced from ref. 21.)

have a common counter electrode that is located separately at a specified distance from the indirect electrode array.

The lower panel in Figure 21.7 shows the principle of combinatorial measurement with the indirect electrode array and the movable electrolyte probe. During combinatorial analysis, an individual electrochemical cell is formed at the instant when one terminal of the electrolyte probe is linked with the counter electrode and another terminal is linked with one of the working electrodes in the electrode array. The electrochemical responses (e.g., current, voltage, impedance, conductivity, etc) recorded by the electrochemical instrument relate only to the working electrode spot that the electrode probe touches, and other working electrode spots, which are ionically insulated, do not make any contribution. After completing analysis of one working electrode analysis, the terminal of the electrolyte probe is moved on to the next working electrode for the next analysis, and so on, until all the working electrodes have been screened. The major difference between the configuration of this electrochemical cell and that of the traditional cell is that in the traditional cell the working electrode always faces toward the counter electrode at a very short distance, whereas the counter electrode for the combinatorial analysis is physically separated from the working electrode array.

The screening method with the electrolyte probe has many advantages. First, it does not need special electrochemical instrumentation, and most of the tradition electrochemical methods, such as cyclic voltammetry, chronoamperometry, chronopotentiometry, potential step, and ac impedance, can be used directly. Secondly, it uses electrochemical cells as array members because each individual electrode in the electrode array can form a different electrochemical cell. Therefore it can be used for studying a wide range of electrochemical materials, such as catalysts, electrolytes, and reactants, by simply arranging them in the indirect electrode array for independent cells. For example, it can be used for studying a group of parallel electrochemical reactions that occur in different electrolytes or have different reactant concentrations.

TWO COMMON ELECTROCHEMICAL METHODS FOR COMBINATORIAL RESEARCH

Potential Step

The potential-step method is used for measuring the current variations from one constant potential, where the electrochemical reaction does not occur, to another potential where the electrochemical reaction does occur. Figure 21.8 shows the potential–time and current–time curves for a potential-step experiment. The current at time t is given by [22]

$$i(t) = \frac{nFAD_O{}^{1/2}C_O^*}{\pi^{1/2}t^{1/2}(1 + \xi\theta)} \tag{3}$$

$$\xi = \left(\frac{D_O}{D_R}\right)^{1/2} \tag{4}$$

$$\theta = \frac{C_O(0, t)}{C_R(0, t)} = \exp\left[\frac{nF}{RT}(E - E^{0'})\right] \tag{5}$$

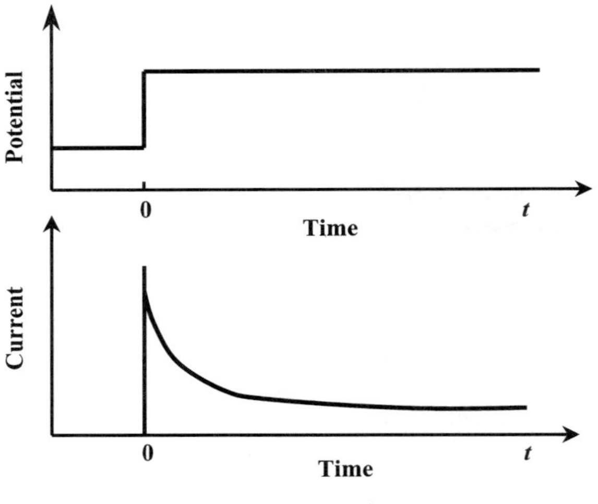

FIGURE 21.8. Potential–time curve (upper plot) and the corresponding current response (lower plot) during a potential-step experiment.

where $i(t)$ Ampere is the current at time t, n is the electron number per molecule oxidized or reduced, F is the Faraday constant, A is the electrode area (cm^2), D_O and D_R ($cm^2\ s^{-1}$) are the diffusion coefficients of the oxidative and reductive species, respectively, C_O^* ($mol\ cm^{-3}$) is the concentration of oxidative species in the bulk solution, and t(s) is time. $C_O(0, t)$ and $C_R(0, t)$ are the concentrations of oxidative and reductive species, respectively, at the electrode surface. E and $E^{0\prime}$ (V) are the potential versus a reference electrode and the formal potential of an electrode, respectively.

For a diffusion-controlled electrode reaction, the Eqn (3) becomes

$$i_d(t) = \frac{nFAD_O{}^{1/2}C_O^*}{\pi^{1/2}t^{1/2}} \tag{6}$$

where i_d is the current for a diffusion-controlled process.

Cyclic Voltammetry

The second widely used electrochemical method for combinatorial measurement is cyclic voltammetry. When a triangular potential (upper panel in Figure 21.9) is applied to the working electrode, the typical current response when plotted versus potential is as shown in the lower panel. For a reversible electrochemical reaction the peak current i_p is given by [22]

$$i_p = (2.69 \times 10^5)n^{3/2} AD_O{}^{1/2}\, \nu^{1/2}\, C_O^* \tag{7}$$

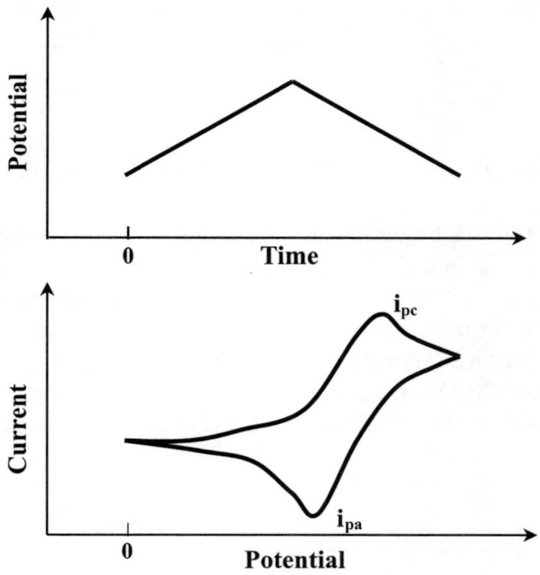

FIGURE 21.9. Triangular potential (upper curve) and the correspondent current–potential curve (lower curve) for a typical electrochemical reaction.

where ν (Vs^{-1}) is the linear potential scan rate. The other parameters are as defined above.

For an irreversible electrochemical reaction, the peak current is [22]

$$i_p = (2.99 \times 10^5) n (\alpha n_\alpha)^{1/2} A C_O^* D_O{}^{1/2} \nu^{1/2} \quad (8)$$

where α is the electron transfer coefficient and n_α is the number of electrons involved in the rate-determining step. The other parameters are as defined above.

APPLICATIONS OF COMBINATORIAL ELECTROCHEMISTRY

Discovery of Better Electrocatalysts

Reddington et al. [14] first reported the use of combinatorial methods for the discovery of electrocatalysts. They used a computer-controlled printing technique to map 645 members of a catalyst array, containing five elements (platinum, ruthenium, osmium, iridium, and rhodium) and 80 binary, 280 ternary, and 280 quaternary combinations, on a carbon substrate. The compounds initially printed on the carbon substrate were metal salts. After chemical reduction by sodium boron hydride, alloys formed from the metal salts. The catalyst array was used as a single electrode in a cell for electrochemical experiments under constant potential. The better catalysts were more electrochemically active and hence generated higher currents at the spots where they were located and, in turn, more protons were produced. Because a fluorescent indicator was dissolved in the electrolyte solution, the greater number of protons produced resulted in brighter fluorescence at the spot. Therefore the most active catalyst was identified by the brightest spot in the catalyst array. Later, Choi et al. [18] used the same combinatorial method to identify a better quaternary platinum-based catalyst containing platinum, ruthenium, molybdenum, and tungsten elements.

Applications to Electrochemical Analysis

Jiang et al. [21] applied the combinatorial method to the search for a better methanol oxidation catalyst by using an indirect electrode array. Many electrochemical factors were investigated, such as catalyst loading, reagent concentration, potential response, and ac impedance. Working electrodes coated with different catalysts were arrayed on a single graphite plate, which had a common counter electrode. The accuracy of this method was examined by measuring 10 electrodes containing the same catalyst. The mean value for current from 10 analyses of the 10 electrodes is 0.317 mA, and the standard deviation is 0.0108 mA.

Effect of Catalyst Loading at Electrode

The effect of catalyst loading for a methanol–air cell was studied by depositing different amounts of catalyst on the anode electrodes in the indirect electrode array. Each electrode had an area of 0.03 cm^2. Figure 21.10(a) shows discharge voltage–current curves for various cells with different anode catalyst loadings. When the anode catalyst loading increases from 0.02 to 0.40 mg, the open-circuit voltage increases from 0.55 to 0.80 V and the discharge current increases significantly. The plot of open-circuit voltage versus anode catalyst loading is shown in Figure 21.10(b). For catalyst loading in the range 0–0.1 mg the open-circuit voltage increases rapidly. If the catalyst loading is more than 0.1 mg the open circuit voltage levels off, and eventually reaches an almost constant value at 0.80 V.

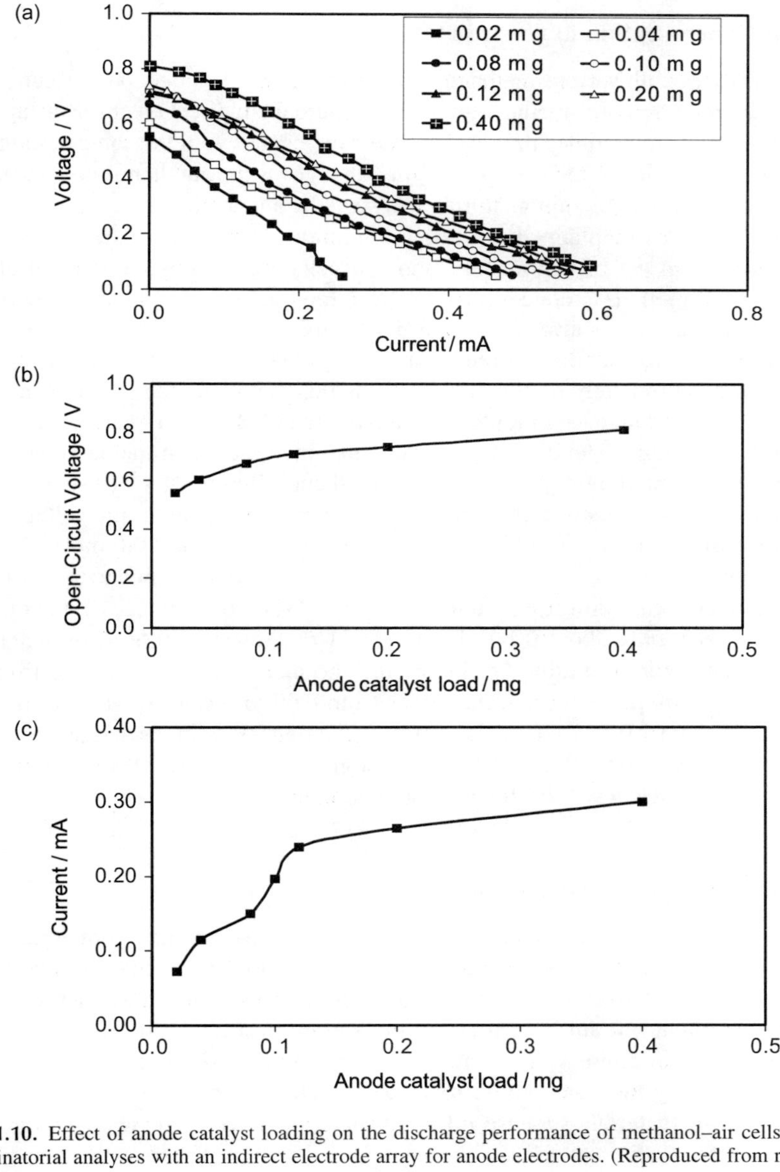

FIGURE 21.10. Effect of anode catalyst loading on the discharge performance of methanol–air cells obtained from combinatorial analyses with an indirect electrode array for anode electrodes. (Reproduced from ref. 21.)

The variation of discharge current with anode catalyst loading seems to be similar to that of the open-circuit voltage with anode catalyst loading. The plot of discharge current versus anode catalyst loading is shown in Figure 21.10(c). When the catalyst loading increases from zero to 0.12 mg, the discharge current changes rapidly from zero to 0.25 mA at 0.4 V and then levels off. For 0.4 mg of anode catalyst loading the discharge current is about 0.3 mA at 0.4 V. The variation of discharge current in this electrochemical system is not very sensitive when the catalyst loading is more than 0.12 mg.

Reagent Concentration in Electrolyte

Electrolytes with various methanol concentrations were added to the electrodes in the indirect electrode array for methanol–air cells, where the indirect electrode array was used for anodes. All the electrodes in the array had been coated with the same amount of PtRu catalyst (0.4 mg). In order to minimize the change in the methanol concentration due to evaporation during the combinatorial analyses, the anode electrode array was placed in a chamber that was maintained at saturated humidity. Figure 21.11(a) shows discharge voltage–current curves of various methanol–air cells obtained by combinatorial analyses with different methanol concentrations. A large range of methanol concentrations (0.05–10.0 mol L^{-1}) was investigated. At 0.05 mol L^{-1} methanol the open-circuit voltage is only 0.24 V, the discharge current is also very low, and the voltage–current curve is shown at the bottom left of the figure. With increasing methanol concentration the open-circuit voltage increases rapidly to a maximum at 0.8 V for a methanol concentration of 3 mol L^{-1} or more. The discharge current also increases with methanol concentration until it reaches a maximum at 10.0 mol L^{-1} methanol. Figure 21.11(b) shows the plot of open-circuit voltage versus methanol concentration. The open-circuit voltage increases much faster from zero to 1.0 mol L^{-1} methanol than from 1.0 to 10.0 mol L^{-1} methanol. The open-circuit voltage is 0.72 V for 1.0 mol L^{-1} methanol; however, on increasing the methanol concentration from 1.0 to 10.0 mol L^{-1} (i.e., by 9 mol L^{-1}), the open-circuit voltage increased only about 0.1 V. Figure 21.11(c) shows the plot of discharge current versus methanol concentration. On increasing the methanol concentration from 0.05 to 1.0 mol L^{-1}, the discharge current increases from 0.09 to 0.27 mA, but a further increase of 9 mol L^{-1}, from 1.0 to 10.0 mol L^{-1}, results in an increase in discharge current of only 0.03 mA (from 0.27 to 0.30 mA). This result suggests that methanol concentrations above 1 mol L^{-1} are not necessary for the methanol–air cells.

Potential Response at Material Surface

Table 21.1 shows the potential responses for various conductive materials obtained by combinatorial analyses. An electrolyte probe was used with one terminal connected to the reference electrode (reversible hydrogen electrode) and the other was used for screening the material surfaces arrayed on a graphite plate. All these measurements were carried out in ambient air. Different materials have significantly different potential responses varying from −0.02 to +1.05 V. This method can be used for the fast identification of materials because different materials have their own specific potential responses.

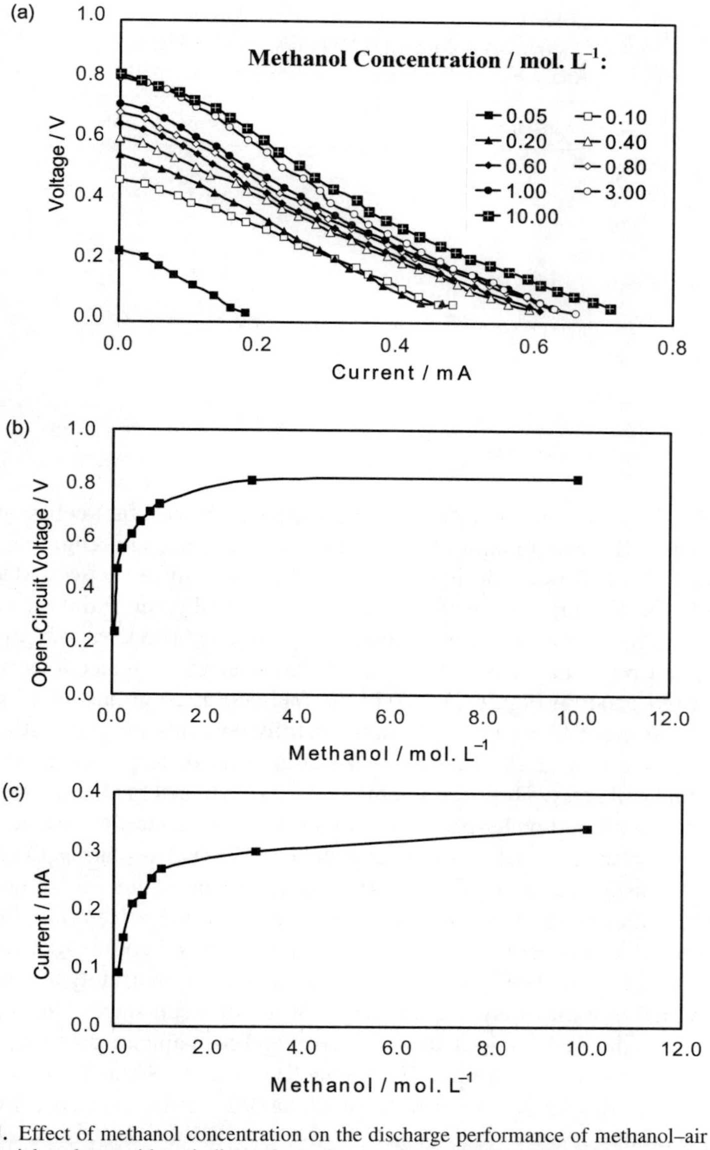

FIGURE 21.11. Effect of methanol concentration on the discharge performance of methanol–air cells obtained from combinatorial analyses with an indirect electrode array for anode electrodes. (Reproduced from ref. 21.)

Fuel Cell Array for Portable Power Source

For practical applications, a number of single fuel-cells are assembled together to form a fuel-cell stack. Generally, fuel-cell stack design is a complex and expensive engineering problem. The most common methods for engineering a number of fuel cells into a power source are bipolar and monopolar designs [23]. In order to use expensive

TABLE 21.1. Potential Responses for Various Conductive Solid Materials Obtained by Combinatorial Analyses with an Electrolyte Screening Method

Conductive materials	Potential response (V/RHE)
Mn	−0.022
Al	+0.011
W	+0.395
Cu	+0.397
Ag	+0.530
Mo	+0.542
Graphite	+0.641
Rh	+0.809
Pd	+0.895
Pt	+1.049

RHE, reversible hydrogen electrode.

materials effectively and to maximize the performance of the fuel-cell stack Jiang and Chu [23] proposed a combinatorial method for configuring and engineering fuel cell stacks. Figure 21.12 shows a design for the configuration of a fuel-cell stack where an 80-member fuel-cell array is arranged on the same material plane. All these cells are electrically linked in series to provide maximum stack voltage. The front side of the fuel-cell array (upper left panel in Figure 21.12) has different electric connections with the back side (upper right panel in Figure 21.12). In the fuel-cell array groups of 10 cathodes and 10 anodes are arranged linearly on a Nafion electrolyte membrane. The cathode lines and the anode lines are placed alternately to form four pairs of anode–cathode lines for the 80-member fuel cell array. There are sealing materials between the boundaries of the electrode lines, but no sealing materials within each line of 10 electrodes. The lower panels in Figure 21.12 show the air- and fuel-flow channels for the fuel-cell array. On the front side of the fuel-cell array there are four air-flow channels and four fuel-flow channels to provide air and fuel to the four lines of cathode electrodes and the four lines of anode electrodes; the same channel design is repeated on the back side of the fuel-cell array. This design provides all the fuel cells in the array with the same opportunity to receive fuel and air in order to achieve maximum performance for the fuel-cell stack. The stack endplates are used to carry the fuel- and air-flow channels and to maintain pressures between the current collectors and the electrodes. The fuel cells in the array can be screened by measuring the potential difference between the anode and the cathode of each fuel cell under constant current discharge.

Applications for High-Energy Battery Development

In traditional battery development each cell is built and investigated individually. Jiang and co-workers from the U.S. Army Research Laboratory proposed a combinatorial method for simultaneous high-throughput construction and investigation of a number of cells. For example, in the development of lithium ion batteries we need to find higher-capacity and more reversible carbon materials for the anode electrode. Figure 21.13 shows

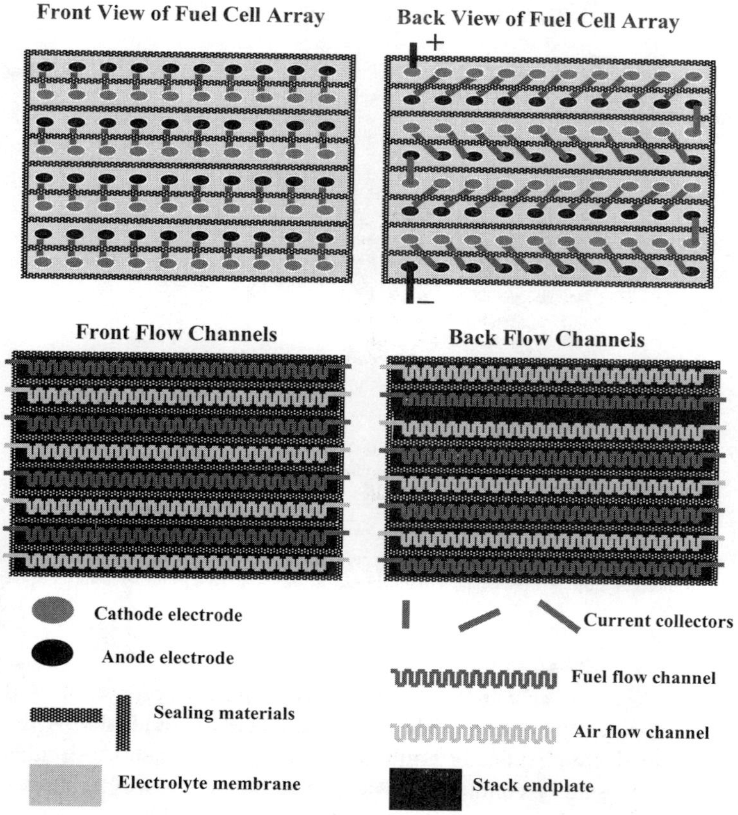

FIGURE 21.12. Combinatorial method for the configuration of fuel-cell stack containing 80 single fuel cells that are arrayed in the same plane.

a direct electrode array containing 56 carbon electrodes on a copper substrate for studying lithium ion intercalation. The various carbon materials or combinations can easily be processed onto the copper surfaces. All the carbon electrodes in the array share a common lithium counter electrode. A common separator is placed between the array and the counter electrodes, where the electrolyte is filled. During lithium ion intercalation the carbon array works as a cathode array. Conversely, during the de-intercalation the carbon array works as an anode array. A common lead is used for the counter electrode. A cable with multiple channels that can be controlled automatically, as described in Figure 21.3, is used for the carbon electrodes. Because a battery test instrument containing multiple channels is commercially available, the approach described in Figure 21.13 is convenient to use.

Applications in Electrochemical Syntheses

Many chemical compounds can be synthesized by electrochemical methods as long as the chemical reactions involve electron transfer, such as radical cationic vinyl polymerization

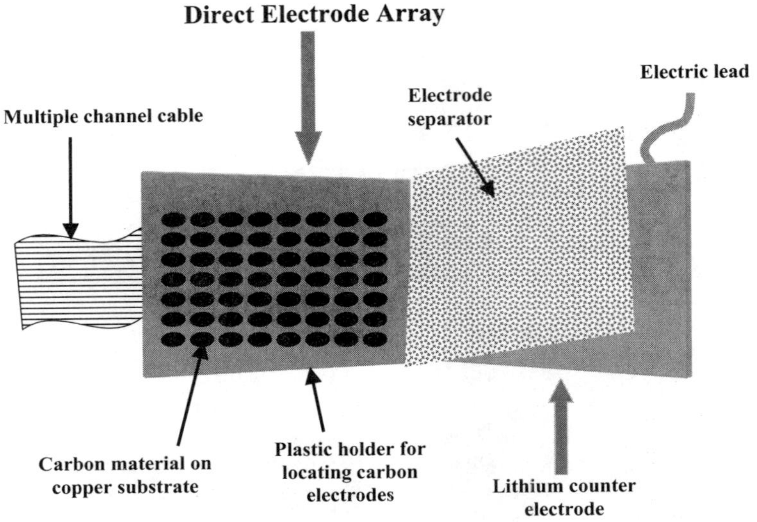

FIGURE 21.13. A direct electrode array containing 56 members of carbon material electrodes on a copper surface for studying Li ion intercalation. All the carbon material electrodes share a common lithium counter electrode.

of proto-porphyrins [24] and electrosynthesis of polythiophene, polypyrrole, and polyaniline [25]. Traditionally, electrochemical syntheses are conducted individually. With combinatorial methods we can electrochemically synthesize a group of compounds simultaneously. Siu et al. [26] have successfully conducted parallel electrosyntheses of α-alkoxycarbamates, α-alkoxyamides, and α-alkoxysulfonamides with a spatially addressable electrolysis platform containing a 16-member cell array. The advantages of this method are not only that high-throughput electrochemical syntheses is possible, but also that it minimizes the effects of other experimental factors that are changing with time, such as temperature, humidity, pressure, and reactant concentrations. Brandli et al. [27] have described a combinatorial method of fabricating porous alumina membranes by electrochemical synthesis. They generated a 256-member array of porous aluminum oxides with different pore sizes and pore densities. These pore sizes are strongly dependent on the anodization potential, and the uniformity of the pores depends on anodization time.

SUMMARY

This chapter has described the principles and applications of combinatorial electrochemistry. Several types of arrays and screening methods for electrochemical research have been summarized. High-throughput electrochemical analysis is strongly dependent on how large an electrochemical array can be generated and how fast the corresponding screening method can be performed. The direct electrode array and automated screening method give more accurate electrochemical measurements, but are more difficult to use as the number of electrodes in the array increases, and this method has special requirements for electrochemical instrumentation. The material spot array on a single electrode and the corresponding indirect screening methods using such an array are much easier to use for a large size array but are less accurate for electrochemical measurements. Indirect electrode arrays on a single conductive substrate and the corresponding electrolyte probe screening method are relatively accurate and easy to use, they have diverse features for studying catalysts, electrolytes, and reactants in either half or full electrochemical cells, and they do not need special instrumental requirements. The disadvantage of the electrolyte probe screening method is that it is not easy to perform automatic measurements.

The traditional electrochemical methods most commonly used for combinatorial electrochemistry are potential-step and cyclic voltammetry. Because combinatorial electrochemistry has a wide variety of applications in many fields of science we could not cite all of them in this chapter. Here, we have summarized the applications of combinatorial electrochemistry in catalyst research, electrochemical analyses, fuel-cell developments, high-energy battery research, and electrochemical syntheses. The most frequently reported applications of combinatorial electrochemistry are in the search for electrode catalysts. In short, in order to realize high-throughput electrochemical measurements by combinatorial electrochemistry we need to generate large-size electrochemical arrays and find an appropriate screening method that is fast, accurate, and easy to use.

ACKNOWLEDGMENTS

The authors would like to acknowledge support from the US Army Research Laboratory Director's Research Initiative Program, and to thank Dr Sol Gilman, Dr Kang Xu, and Mr Charles Walker for helpful discussions.

REFERENCES

1. Pei, D., Ulrich, H. D., Schultz, P. G. A combinatorial approach toward DNA recognition. *Science* **1991**, *253*, 1408–1411.
2. Janda, K. D., Lo, L. C., Lo, C. H., Sim, M. M., Wang, R., Wong, C. H., Lerner, A. Chemical selection for catalysis in combinatorial antibody libraries. *Science* **1997**, *275*, 945–948.
3. Lam, K. S., Salmon, S. E., Hersh, E. M., Hruby, V. J., Kazmierski, W. M., Knapp, R. J. A new type of synthetic peptide library for identifying ligand-binding activity. *Nature* **1991**, *354*, 82–84.
4. Yoshida, J. I., Suga, S., Suzuki, S., Kinomura N., Yamamoto, A., Fujiwara, K. Direct oxidative carbon—carbon bond formation using the "cation pool" method. 1. Generation of iminium cation pool and their reaction with carbon nucleophiles. *J. Am. Chem. Soc.* **1999**, *121*, 9546–9549.
5. Schiedel, M. S., Briehn, C. A., Bauerle, P. C–C cross-coupling reactions for the combinatorial synthesis of novel organic materials. *J. Organometall. Chem.* **2002**, *653*, 200–208.
6. Gray, N. S., Wodicka, L., Thunnissen, A. M. W. H., Norman, T. C., Kwon, S., Espinoza, F. H., Morgan, D. O., Barnes, G., LeClerc, S., Meijer, L., Kim, S. H., Lockhart, D. J., Schultz, P. G. Exploiting chemical libraries, structure, and genomics in the search for kinase inhibitors. *Science* **1998**, *281*, 533–538.
7. Xiang, X. D., Su, X., Briceno, G., Lou, Y., Wang, K. A., Chang, H., Wallace-Freedman, W. G., Chen, S. W., Schultz, P. G. A combinatorial approach to materials discovery. *Science* **1995**, *268*, 1738–1740.
8. Briceno, G., Chang, H., Sun, X., Schultz, P. G., Xiang, X. D. A class of cobalt oxide mangnetoresistance materials discovered with combinatorial synthesis. *Science* **1995**, *270*, 273–275.
9. Wang, J., Yoo, Y., Gao, C., Takeuchi, I., Sun, X., Chang, H., Xiang, X. D., Schultz, P. G. Identification of a blue photoluminescent composite material from a combinatorial library. *Science* **1998**, *279*, 1712–1714.

10. Siu, T., Yekta, S., Yudin, A. K. New approach to rapid generation and screening of diverse catalytic materials on electrode surfaces. *J. Am Chem. Soc.* **2000**, *122*, 11787–11790.
11. Wang, Q., Perkins, J., Branz, H. M., Alleman, J., Duncan, C., Ginley, D. Combinatorial synthesis of solid state electronic materials for renewable energy applications. *Appl. Surf. Sci.* **2002**, *189*, 271–276.
12. Sohn, K. S., Seo, S. Y., Park, H. D. Search for long phosphorescence materials by combinatorial chemistry method. *Electrochem. Solid-State Lett.*, **2001**, *4*, H26–H29.
13. Seo, S. Y., Sohn, K. S., Park, H. D., Lee, S. Optimization of Gd_2O_3-based red phosphors using combinatorial chemistry method. *J. Electrochem. Soc.* **2002**, *149*, H12–H18.
14. Reddington, E., Sapienza, A., Gurau, B., Viswanathan, R., Sarangapani, S., Smotkin, E. S., Mallouk, T. E. Combinatorial electrochemistry: A highly parallel, optical screening method for discovery of better electrocatalysts. *Science* **1998**, *280*, 1735–1737.
15. Sullivan, M. G., Utomo, H., Fagan, P. J., Ward, M. D. Automated electrochemical analysis with combinatorial electrode arrays. *Anal. Chem.* **1999**, *71*, 4369–4375.
16. Zhou, L., Savvatelev, V., Booher, J., Kim. C. H., Shinar, Combinatorial fabrication and studies of intense efficient ultraviolet-violet organic light-emitting device arrays. *J. Appl. Phys. Lett.* **2001**, *79*, 2282–2284.
17. Liu, H., Felten, C., Xue, Q., Zhang, B., Jeddrzejewski, P; Karger, Foret, F. Development of multichannel devices with an array of electrospray tips for high-throughput mass spectrometry. *Anal. Chem.* **2000**, *72*, 3303–3310.
18. Choi, W. C., Kim, J. D., Woo, S. I. Quaternary Pt-based electrocatalyst for methanol oxidation by combinatorial electrochemistry. *Catal. Today* **2002**, *74*, 235–240.
19. Chen, G., Delafuente, D. A., Sarangapani, S., Mallouk, T. E. Combinatorial discovery of bifunctional oxygen reduction—water oxidation electrocatalysts for regenerative fuel cells. *Catal. Today* **2000**, *67*, 341–355.
20. Gurau, B., Visvanathan, R., Liu, R., Lafrenz, T. J., Ley, K. L., Smotkin, E. S., Reddington, E., Sapienza, A., Chan, B. C., Mallouk, T. E., Sarangapani, S. Structural and electrochemical characterization of binary, ternary, and quaternary platinum alloy catalysts for methanol electro-oxidation. *J. Phys. Chem. B* **1998**, *102*, 9997–10003.
21. Jiang, R., Chu, D. A combinatorial approach toward electrochemical analysis. *J. Electroanal. Chem.* **2002**, *527*, 137–142.
22. Bard, A. J., Faulkner, L. R. *Electrochemical Methods, Fundamentals and Applications*, **1980**, New York: John Wiley, pp 143–159, 218–222.
23. Jiang, R., Chu, D. Stack design and performance of polymer electrolyte membrane fuel cells. *J. Power Sources* **2001**, *93*, 25–31.
24. Dong, S., Jiang, R. Research on chemically modified electrodes. XIX. preparation of iron protoporphyrin film electrode by electrochemical polymerization and its catalysis. *J. Inorg. Biochem.* **1987**, *30*, 189–201.
25. Jiang, R., Dong, S. J. Chromatic reaction of polyaniline film and its characterization. *Chem. Soc., Faraday Trans. I*, **1989**, *85*, 1585–1597.
26. Siu, T., Li, W., Yudin, A. K. Parallel electrosynthesis of α-alkoxycarbamates, α-alkoxyamides and α-alkoxysulfonamide using the spatial addressable electrolysis platform (SAEP). *J. Comb. Chem.* **2000**, *2*, 545–549.
27. Brandli, C., Jaramillo, T. F., Ivanovskara, A., McFarland, E. W. Automatic synthesis and characterization of diverse libraries of macroporous alumina. *Electrochim. Acta* **2000**, *47*, 553–557.

Combinatorial Libraries, Data Management, and Scale-Up of Discovered Materials

Chapter 22

Ink-Jet Methods in Combinatorial Materials Synthesis

David B. Wallace and Michael E. Grove

INTRODUCTION

Ink-jet printing incorporates a family of different technologies with a common end result to form and direct small fluid droplets to a specific location with high accuracy. These technologies fit into two broad classes. In continuous-mode ink-jet printing technology, a jet of liquid is formed by forcing a fluid under pressure through an orifice. Surface tension forces create instabilities in the cylindrical jet, causing the jet to break up into drops [1,2]. By providing a controlled single-frequency disturbance, the jet can be forced to break up into uniform diameter and velocity droplets. The droplets can be charged during their breakup and deflected from a straight trajectory by an electrostatic field. Charging the drops to different levels allows the drops to be placed at one of several locations on a substrate, or to be directed in to a catcher or gutter for recycling or disposal [3]. Figure 22.1 shows a schematic of a continuous-mode ink-jet printing system, and Figure 22.2 shows a photomicrograph of an array of 70 μm diameter jets of water issuing from a droplet generator device and breaking up into ~140 μm diameter droplets at 40 kHz.

Continuous-mode ink-jet printing systems produce droplets that are approximately twice the orifice diameter of the droplet generator. Droplet-generation rates for commercially available continuous-mode ink-jet systems are usually in the 80–100 kHz range, but systems with operating frequencies up to 1 MHz are in use. Droplet sizes can be as small as 20 μm in a continuous system, but 150 μm is typical. Droplets as large as 1 mm (~0.5 μL) have been observed.

Continuous-mode ink-jet systems are currently in widespread use for industrial marking, principally for product labeling of food and medicines. They have high-throughput capabilities, especially array continuous-mode systems, and are best suited for very high duty cycle applications. Few continuous-mode ink-jet systems are multicolor

David B. Wallace and Michael E. Grove • MicroFab Technologies Inc., 1104 Summit Avenue, Plano, TX 75074, USA
Contact author: David B. Wallace, email: dwallace@microfab.com

High-Throughput Analysis, edited by Potyrailo and Amis
Kluwer Academic Publishers, New York, 2003.

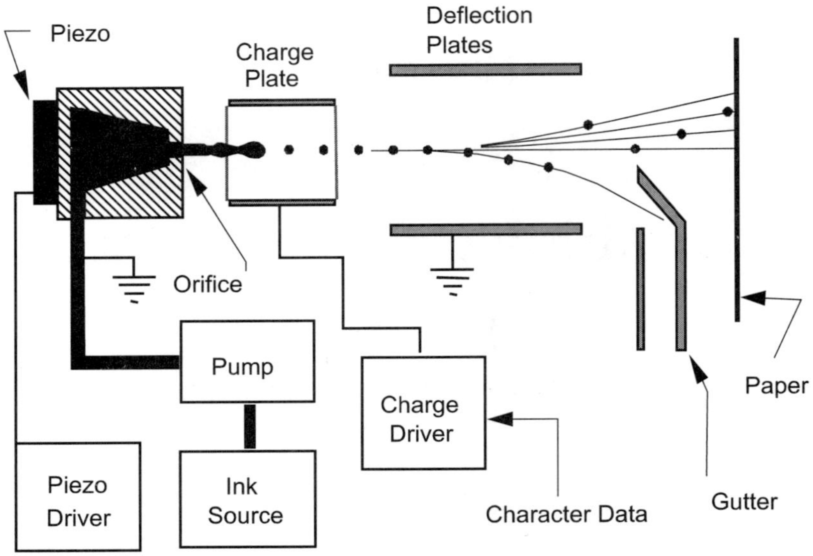

FIGURE 22.1. Schematic of a continuous mode ink-jet printing system.

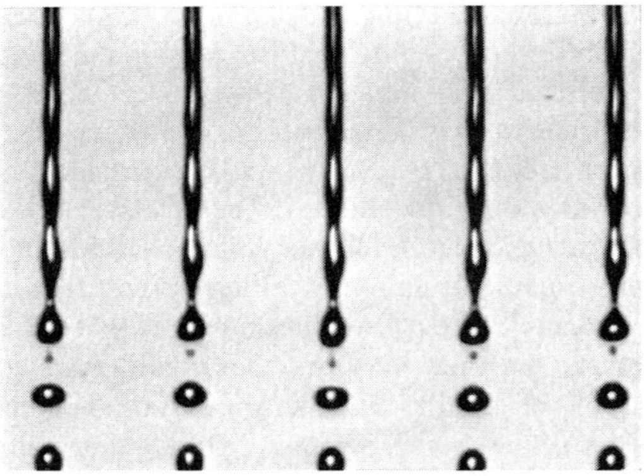

FIGURE 22.2. An array of 70 μm jets of water breaking up into ~140 μm droplets at 40 kHz.

(i.e., multifluid). Continuous-mode systems have been employed in novel applications such as rapid prototyping [4], electronics manufacturing [5], uniform sphere or balloon generation [6], and medical diagnostics manufacturing [7].

Although it is fairly easy to create a continuous jetting system in the laboratory, there are a number of practical limitations to their use in combinatorial synthesis. Because they generate drops continuously, as the name implies, continuous systems require unused drops to be recirculated or wasted. Recirculation causes a change in fluid properties over time, owing to solvent loss, and can change the properties of the jetted material in other ways owing to the stress of repeated pumping and filtration. Drop velocities and diameters are generally higher in continuous systems than in demand-mode systems (discussed below), both of which can lead to difficulties in depositing materials for synthesis applications.

Demand-mode ink-jet technology is employed in all small office/home office (SOHO) ink-jet printers. In demand-mode systems, a transducer is used to create a pressure–volume disturbance in the ink, and the resulting pressure wave is directed so as to produce a drop that issues from an orifice [8,9]. The physics of the drop formation are complex [10], but, in brief, a pressure is created at the orifice for a few microseconds and this pressure is high enough for a small amount of fluid to overcome viscous and surface tension forces, and thus "escape" from the orifice. A single drop may be generated, or a group of drops at arbitrary intervals of time. Thus the drops are created "on demand." Demand-mode drops are typically the same diameter as the orifice diameter. Drop diameters of 15–100 μm (2–500 pL) can be achieved with demand-mode systems at maximum droplet generation rates of 10–25 kHz. Demand-mode ink-jet systems have no fluid recirculation requirement and, as Figure 22.3 indicates, they are conceptually less complex than continuous-mode systems. On the other hand, demand-mode droplet generation requires the transducer to deliver three or more orders of magnitude greater energy to produce a droplet, compared with continuous-mode generation, and there are many "elegant" (i.e., complex) array demand-mode printhead designs [11]. Figure 22.4 shows a demand-mode ink-jet device generating 60 μm diameter drops of butyl carbitol from a device with a 50 μm orifice at 4 kHz.

The transducer in demand-mode systems was piezoelectric in early systems, but in many modern systems a thin-film resistor is substituted for the piezoelectric transducer. When a high current is passed through this resistor, the ink in contact with it is vaporized, forming a bubble over the resistor [12]. This vapor bubble serves the same functional

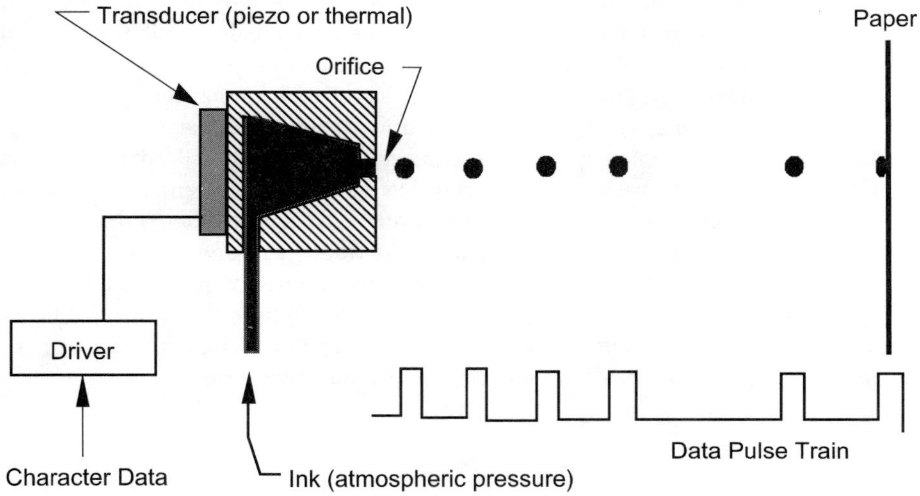

FIGURE 22.3. Schematic of a drop-on-demand ink-jet printing system.

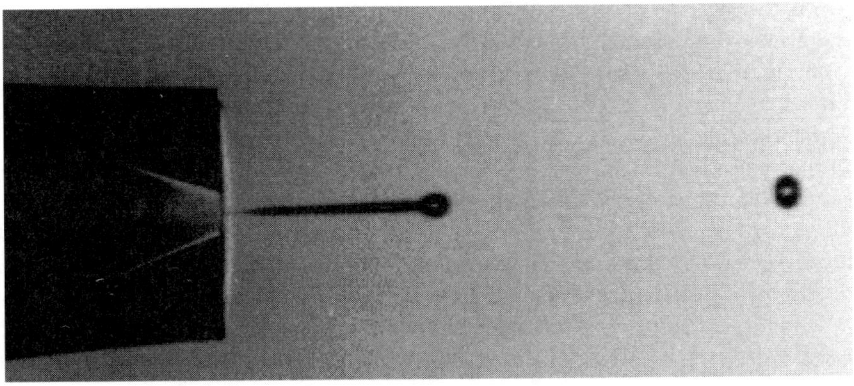

FIGURE 22.4. Demand-mode ink-jet device generating 60 μm diameter drops at 4 kHz.

purpose as the piezoelectric transducer. This type of printer is usually referred to as a thermal ink-jet printer. Making the fluid an integral part of the energy input process places additional constraints on the fluid properties. In practice, this means that thermal demand-mode ink-jet technology is rarely used for novel (i.e., nonink) applications, although some researchers have employed it [13].

Since piezoelectric demand-mode ink-jet technology does not require recirculation of the working fluid, does not create thermal stress on the fluid, and does not depend on a thermal process to impart acoustic energy to the working fluid, it is the most adaptable of the ink-jet printing technologies to fluid microdispensing in general, and combinatorial synthesis in particular. As a noncontact printing process, the volumetric accuracy of ink-jet dispensing is not affected by how the fluid wets a substrate, as is the case when positive displacement or pin transfer systems "touch off" the fluid onto the substrate during the dispensing event. In addition, the fluid source cannot be contaminated by the substrate, or contamination on the substrate, in a noncontact dispensing process. Finally, the ability to free-fly the droplets of fluid over a millimeter or more allows fluids to be dispensed into wells or other substrate features (e.g., features that are created to control wetting and spreading).

Valve-jet methods are sometimes considered as a third group, along with demand mode and continuous-mode. In a valve-jet system, a high-speed solenoid valve [14] is used to rapidly turn on and off a pressure source connected to an orifice. Generally, larger volumes (50–100 nL minimum) are generated per "event" and repeatability is dependent on the electromechanical behavior of the solenoid. Unlike demand-mode and continuous-mode ink-jet systems, valve-jet methods do not usually produce highly repeatable drops. The dispensing event produces a filament (jet) of fluid issuing from the orifice that may or may not break up into random drop sizes before impacting the substrate. If the substrate is stationary or moving slowly, the fluid will impact at a single location.

MATERIALS

Generally, the fluid property requirements for demand-mode ink-jet dispensing are a viscosity less than 40 cP and surface tension greater than 20 dyns cm^{-1}. Very low viscosities can lead to difficulties with satellite formation and lack of acoustic damping, but organic solvents such as methanol, with viscosities less than that of water (1 cP), can be jetted. Very high surface tensions present unique difficulties, but molten solders with surface tensions greater than 400 dyn cm^{-1}, or roughly six times that of distilled water, have been jetted, as shown in Figure 22.5. If the fluid is heated or cooled, the above properties are required *at the orifice*. Higher viscosities can usually be tolerated in the fluid delivery system.

Newtonian behavior is not strictly required, but the fluid properties at the orifice flow conditions must be less than 40 cP. Thus a shear thinning fluid could have a low shear rate viscosity much higher than 40 cP. Viscoelastic behavior causes significant performance problems by increasing the amount of deformation that the fluid can withstand without breaking off from the orifice. Viscoelastic fluids have been observed to for filaments 1–2 mm long (20–40 orifice diameters) without breaking off into a drop! Figure 22.6 shows drop formation for a non-Newtonian fluid.

FIGURE 22.5. Posts of tin–lead solder (~80 μm at the base) printed onto thermoelectric elements using Solder Jet™ technology.

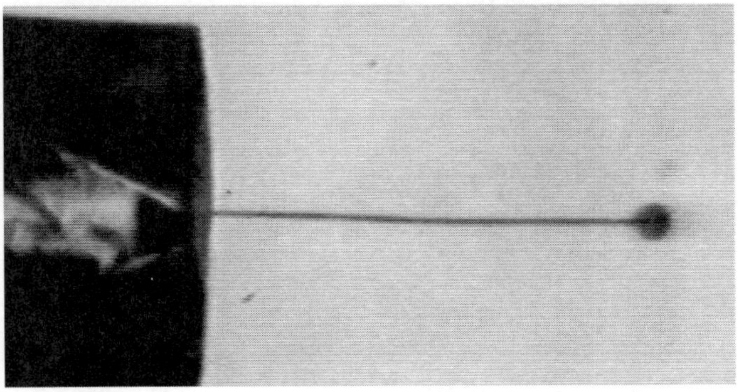

FIGURE 22.6. Non-Newtonian fluid forming a long filament during the drop-formation process.

Particle suspensions, such as pigmented inks, are acceptable as long as the particle/agglomerate size and density do not cause the suspension to depart from the fluid properties range given above. Particles or agglomerates that are $>5\%$ of the orifice diameter will cause at least some instability in drop-generation behavior, but may still be acceptable in low concentrations. An example of display phosphor particles printed using ink-jet technology is given in Figure 22.7.

The "window" of fluids and suspensions that can be dispensed using ink-jet technology has been stretched by heating, cooling, stirring, wiping, purging, pre-oscillating, diluting, and other methods. However, this window is unavoidably narrowed as orifice diameter decreases, frequency increases, and number of jets in an array increases.

The diversity of fluids that have been dispensed using ink-jet technology is impressive, given the fluid property restrictions described above. Inks alone represent a broad class of materials. Dye-based aqueous solutions are the most commonly used, but aqueous–pigment, volatile solvent (e.g., methyl ethyl ketone)–dye, volatile solvent–pigment, and low-volatility solvent–dye formulations are all in common use.

Table 22.1 lists both materials classes and specific materials that have been dispensed using ink-jet technology. Some of the materials listed have a wide latitude in their formulation, and thus represent a broad class of materials. Particle suspensions, including inks, inherently fall into this category. Polymers do also, but, in practice, having the knowledge, intellectual property, equipment, or budget required to tailor polymeric formulations to ink-jet dispensing can be problematic. Metals, oligonucleotides, and precursor formulations used for chemical synthesis (e.g., peptides, DNA) represent fluids for which there is less latitude in modifying the fluid formulation to adapt to ink-jet dispensing method. A number of examples are given in Figures 22.8–22.15.

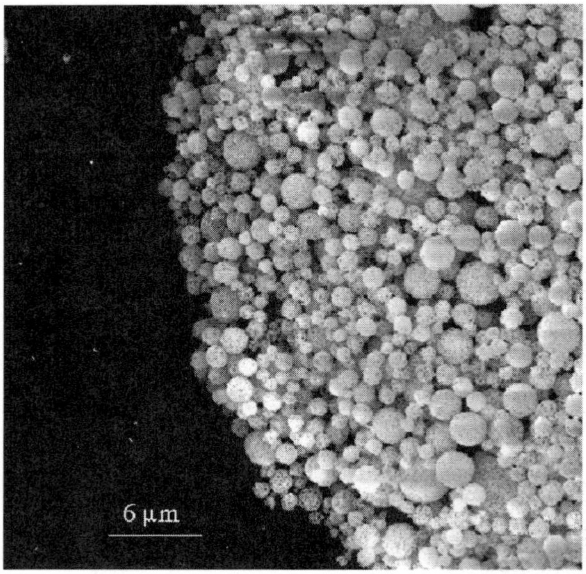

FIGURE 22.7. Phosphor particles (SMP Corporation) deposited using ink-jet technology.

TABLE 22.1 Materials That Have Been Dispensed Using Ink-Jet Printing Technology

Electronic/ optical materials	Biological fluids	Organic solvents	Other materials	Particle suspensions
Organometallics	DNA	Alcohols	Sol–gels	Pigments
Cyanoacrylates	Nucleic acids	Ketones	Thermoplastics	Cells
Photoresists	Amino acids	Aliphatics	Thermosets	Teflon
Solders	Proteins	Aromatics (xylene, toluene)	Acrylics	Phosphors
Polyimides	Lipids	Dipolar solvents (NMP, DMF)	> 1 M salt solutions	Ferrites
Electroactive materials	Biodegradable polymers		Photographic developer	Zeolites
Fluxes			Fuels	Latex spheres
Epoxies			Aqueous adhesives	Metal particles (Ag, W, Cu)

NMP, N-methyl pyrrolidone.
DMF, dimethyl formamide.

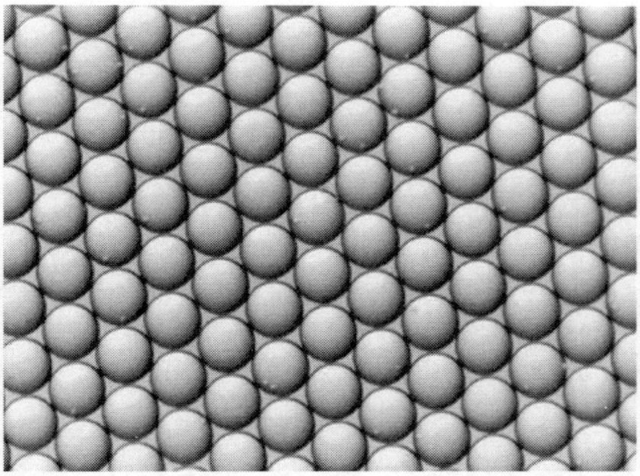

FIGURE 22.8. Spheres (80 μm) of biodegradable polymer created using continuous ink-jet technology.

INK-JET PROCESSING AND CREATION OF COMBINATORIAL LIBRARIES

During the past 10 years, biomedical and pharmaceutical industries have broadly adopted the idea of synthesizing many compounds and then testing them by simple high-throughput procedures. This collective process of drug discovery has been termed combinatorial chemistry. It is also applied to arrays of compounds that are individually prepared and to collections of existing reagents prepared for purposes unrelated to the study in which they are used. The term combinatorial library refers to the various compounds that are formed in the synthesis. The high-throughput combinatorial procedure for studying small molecules as

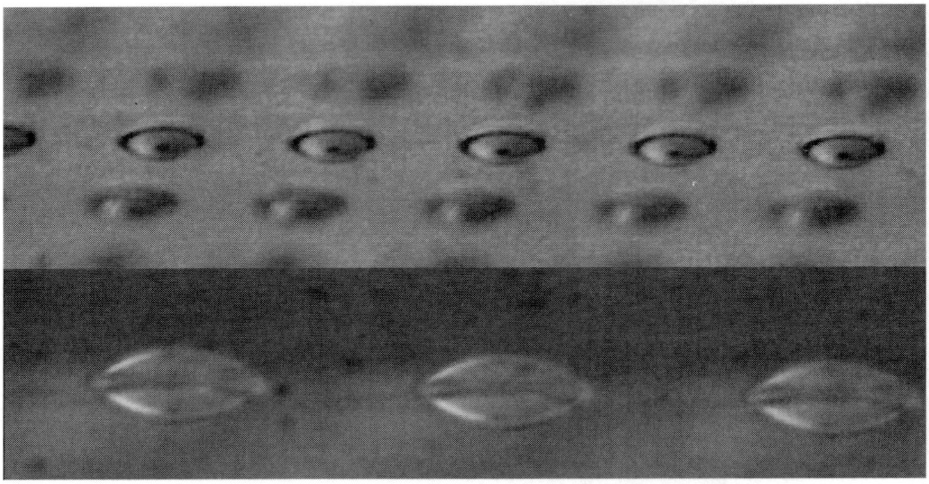

FIGURE 22.9. Spots (100 μm) of cholesterol printed using ink-jet technology.

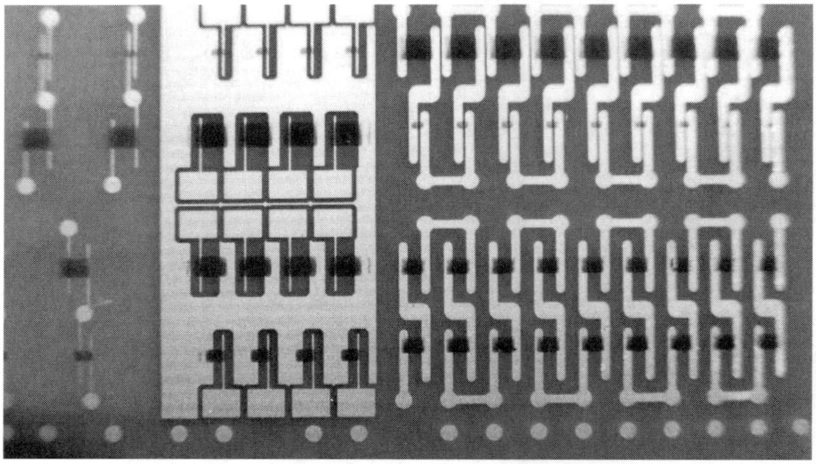

FIGURE 22.10. Conductive polymer ink-jet printed as resistive elements onto circuit board. The minimum printed feature size is 0.2 mm.

potential drugs has been attractive to the pharmaceutical industry because the structural characteristics required for a particular desired biological activity are often uncertain.

Materials scientists can use, and are using, the advantages of combinatorial screening procedures to help in their search for improvements in nonbiological materials systems. Application of the precision dispensing abilities of ink-jet technology to the synthesis side of combinatorial chemistry can rapidly expand the number of samples that can be made with a given expenditure of person-hours.

FIGURE 22.11. Silver nanoparticle lines (150 μm) ink-jet printed onto a ferrite nanoparticle layer, also ink-jet printed.

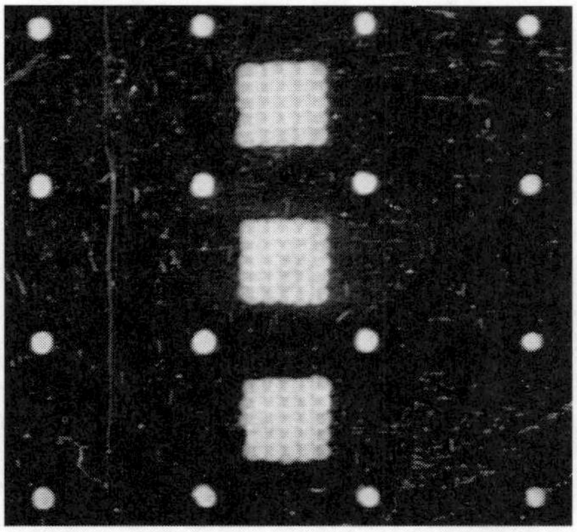

FIGURE 22.12. Patterns etched into polystyrene by ink-jet printing acetone.

Detection

While the small volumes of reactants which ink-jet devices are capable of dispensing may have only a minor impact on devising chemical reaction strategies, whichever pooling strategies are used, they can certainly have an impact on detection strategies. The potential cycle-time advantages of ink-jet synthesis can only be realized if a detection strategy with comparable throughput is utilized.

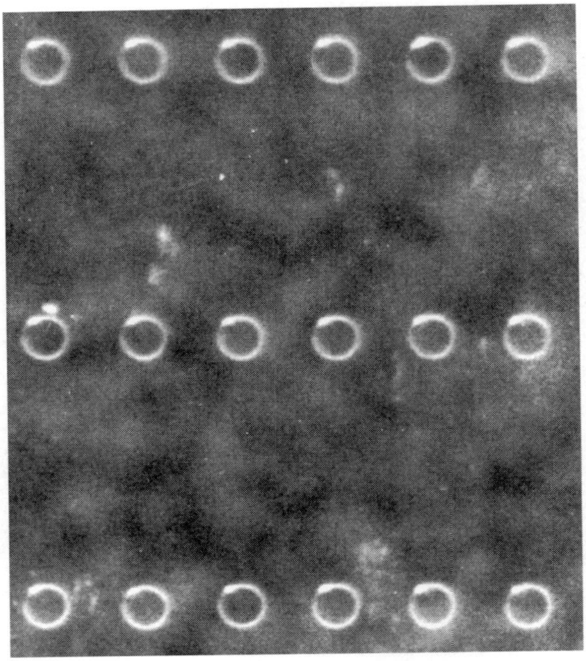

FIGURE 22.13. Spots (150 μm) of cyanoacrylate ink-jet printed onto glass.

FIGURE 22.14. Polymer lenses (100 μm) ink-jet printed onto 1 mm spherical glass lens.

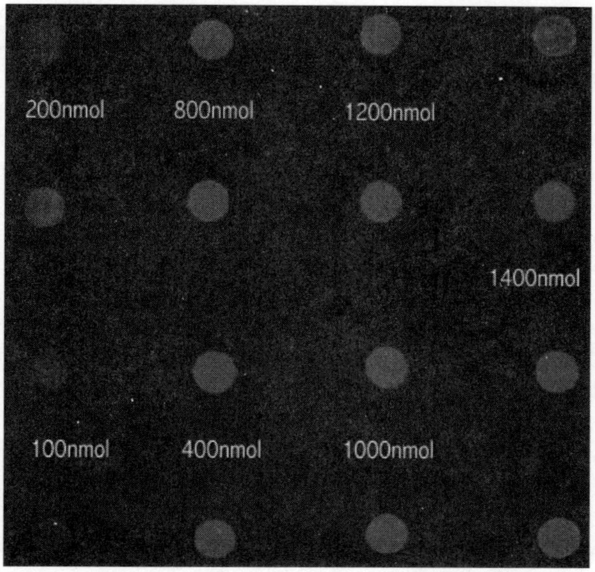

FIGURE 22.15. Seven concentrations of cytochrome-*c* protein ink-jet printed into 80 μm spots.

Mass spectrometry is a preferred detection method due to its high sensitivity (<pmol) and ability to define structure. It can be used for direct identification of library compounds after they are recovered from the substrate. Matrix-assisted laser description time-of-flight (MALDI-TOF) spectrometry has emerged as a powerful tool for sequencing DNA [15] and peptides formed on supports. Indirect mass spectrometry is also being used to identify library compounds by identification of surrogate tags. In addition to MALDI-TOF, electrospray ionization, flow injection analysis mass spectrometry, Raman microspectroscopy, high-performance liquid chromatography–mass spectrometry (HPLC–MS), and tandem mass spectrometry are useful analytical tools in nonlabel detection strategies. Other analytical techniques used in library characterizations include optical evanescent wave methods such as surface plasmon resonance (SPR), X-ray diffraction, especially X-ray microprobe analysis, fluorescence spectroscopy, UV–Vis spectrophotometry, scanning/transmission electron microscopy, and liquid and gas chromatography. These detection methods are used in serial analysis of the samples in a library. The time required to perform serial analyses has to be a consideration in developing the overall strategy. Progress in the development of combinatorial libraries using ink-jet technology can be so rapid as to overwhelm the characterization methodology. One problem is that many of the classical analytical tools require bulk samples rather the ultra-low-mass solids which are the results of ink-jet combinatorial syntheses.

Characterization of combinatorial libraries by parallel analytical strategies is attractive where compound syntheses are being rapidly implemented by ink-jet techniques since a single substrate can easily contain hundreds to thousands of compounds. One example of parallel analysis is IR thermography, which has been used for the evaluation of thousands of heterogenous hydrogenation catalyst samples by imaging them on their substrate while it is functioning in a reactor with hydrogen and an alkene [16]. The most popular parallel detection strategies have been those using a label, which can be an isotope tag, a fluorescent tag, surrogate tags, or, for biological libraries, enzyme assays, and sandwich assays.

A factor in any ink-jet combinatorial approach will be the quantity of the anticipated compounds that are needed for reliable detection of a significant characteristic. In the microbiological or biochemical arena, mass spectroscopic detection of subpicomolar amounts matches very well with the ink-jet's single-drop volume capabilities, as does the use of fluorescent tags. Raman microspectroscopy and SPR with their 2 μm and 100 μm minimum fields of view also match well with the ink-jet's drop deposition capabilities. Where larger amounts of a compound are required to detect an important characteristic that will not be identified by mass spectroscopic analysis, ink-jet-assisted synthesis can be programmed to deliver multiple drops of a fluid from a jetting device onto one location.

Potential Library-Generation Applications

In the next section, we will briefly review several combinatorial chemistry applications as reported in the literature, and discuss the advantages that ink-jet technology could bring to library generation in these applications.

Example 1

An example may help identify the advantages of ink-jet assisted library generation. The sol–gel reaction for materials synthesis is very adaptable to ink-jet dispensing techniques. It will be illustrated using the V_2O_5–polyphenol nanocomposite library reported by Wang and Gonsalves [17]. The goal of the work is the optimization of optical properties due to the delocalized electrons in the polyphenol over the nascent microcrystalline V_2O_5. Vanadium pentoxide deposited gels have been examined for commercial applications as multilayer large-area coatings on glass or plastic for their antistatic properties, for switching and electrochromic properties, and as reversible cathodes [18]. The V_2O_5 coatings show mixed electronic conduction properties due to a electron hopping between V^{4+} and V^{5+} ions. This semiconductor has recently been found to have a large third-order nonlinear optical susceptibility [19].

The combinatorial work was divided into two parts with the first devoted to developing the procedure required to obtain nanocrystalline V_2O_5 by sol–gel synthesis since other forms of vanadium oxide such as VO_2, V_2O_3, V_2O_4, V_3O_5, and V_4O_7 can also be produced by this process. An array of 20 mL vials was used to prepare solutions of acetone and water at various molar ratios. To these were added dropwise vanadium oxytripropoxide to achieve various molar ratios of $VO(OC_3H_7)_3$–acetone–water. A matrix of 90 compositions was planned and 38 of them were made. Rapid analysis of this library was performed by Raman microspectroscopy, with attention being given to bands at 995, 700 cm^{-1} and at 285, 147 cm^{-1}, to screen for the best component ratio to produce crystalline V_2O_5. One of the 38 compositions was selected for use in the nanocomposite preparation.

In the second part, previously prepared, purified, and dried poly(*p*-ethylphenol) (PEP) was dissolved in acetone. This solution became the acetone component in a repetition of the sol–gel V_2O_5 synthesis employing the molar ratios of the three components found to produce the best results. The PEP was used in two different weight ratios (5 wt.% and 30 wt.%) to the V_2O_5 in the preparation of the nanocomposite samples. Detailed characterization of the materials was performed by Raman microspectroscopy, Fourier transform IR microspectroscopy, and transmission electron microscopy. This paper Wang and Gonsalves' [17] reports on a nice piece of work utilizing the flexible synthetic advantages of the sol–gel process and instrumental methods of analysis that work well on small samples.

The use of ink-jet-deposition-assisted combinatorial library preparation could certainly have shortened the time it took to generate the samples reported and allowed more samples to be made within the scope of this project. PEP could have been dissolved in the acetone at the outset. The acetone–PEP solution, water, and $VO(OC_3H_7)_3$ could have all been dispensed onto a substrate, or substrates, with cavities to hold the solutions during the reaction period. The substrates could have been of a size that matched the requirements of the analytical requirements and the automated alignment capabilities of an ink-jet printing platform. Rather than examining 38 V_2O_5 samples and two polymer-to-V_2O_5 ratios, there is a real possibility of examining hundreds of nanocomposite samples within a shorter time period.

Example 2

A second example of a combinatorial experiment where ink-jet deposition could be used to easily expand the size of a library can be found in the generation of 128 combinatorial libraries of thin crystalline three-color display phosphor films by Isaacs et al. [20]. The phosphors made in the study are rare-earth-activated refractory metal oxides similar to those used in projection television cathode-ray tubes. Oxide-based phosphors are preferred for this application over the normal television tube sulfide phosphors because of the higher voltages and operating temperatures in this application.

The combinatorial libraries reported in this study were prepared principally on single-crystal silicon wafers by sputtering La_2O, GaF_3, Al_2O_3, EuF_3, and $SrCO_3$ targets in a chamber using masks and annealing steps to create phosphor films approximately 1 μm thick with area dimensions of 1 mm × 2 mm and 1 mm intervals between films. Similar films were prepared on $LaAlO_3$, but are not part of this study. A requirement in this process of depositing each precursor in a layer-by-layer sequence is adequate post-deposition annealing to assure uniform diffusion of the precursors and phase formation. Obviously, it required some time to work through this process to create the combinations of the five elements represented in the 128 film libraries. The letter reports the results of X-ray fluorescence analysis of three films in particular which had the elemental compositions $Gd_{0.64}Sr_{0.26}Al_{1.03}O_3{:}Eu_{0.07}$, $Sr_{0.38}Al_{1.52}O_3{:}Eu_{0.098}$, and $Gd_{0.63}Sr_{0.25}Al_{1.01}O_3{:}Eu_{0.11}$.

Owing to the large number of sputtering targets, masks, and annealing steps, structural characterization by X-ray diffraction is crucial to evaluating the libraries resulting from this combinatorial procedure as a large number of phases are possible in addition to the desired perovskite and sapphire. Evaluation of well-crystallized granular films by X-ray diffraction for phase identification is a prerequisite to selecting materials combinations for synthesis on larger scale. Near Edge X-ray Absorption Fine Structure (NEXAFS) is used to determine the valence state of the europium atom, either Eu^{3+} or Eu^{2+}, which along with the crystal structure of the cubic perovskite $(GdSr)AlO_3$ or $SrAl_2O_3$ determines the color that the phosphor will emit. These three X-ray based characterizations form a rational basis for selecting libraries to move up to film sizes suitable for photoluminescent and cathodoluminescent examinations.

One feature of the X-ray microprobe techniques in this work [20] is the small size of the beam spot on the sample, an area of only 3 μm by 20 μm. Given these dimensions, the area of a single sample could be smaller than the 2 mm^2 that was made. While the X-ray microprobe instrumentation offers superior advantages in many respects, access to beamlines and beamtime at Argonne National Laboratory can be complicated and protracted [21]. Brookhaven National Laboratory offers a similar service. Conventional X-ray techniques, including X-ray imaging, diffraction, fluorescence, and spectroscopy, can generally work with a spatial resolution of 0.1–1 mm, depending on the instrument. Consideration should be given to availability and timeliness in determining detection strategies for combinatorial synthesis. In the case of ink-jet-assisted synthesis, design of a substrate that is acceptable to an accessible instrumental technique may require a longer time to dispense all the fluids, but could return meaningful results in a shorter time. The promise of ink-jet synthesis is that a lot of samples can be created in a short time period, but determining their value needs to be as rapid.

Example 3

Display phosphors for projection televisions such as the europium-activated $Gd(La,Sr)AlO_3$ example above are used in the form of powders, not thin films. A patent issued to Phillips [22] describes the fabrication of a powder oxide phosphor of similar composition. The oxide phosphors of this invention, like those described above, are not as high in luminescent efficiency as sulfide phosphors, but offer other advantages in projection television or field emission display service. The invention in part makes use of a hydrothermal synthesis method, the elemental combinations being the other part, which is stated to demonstrate a photoluminescent efficiency advantage over conventional solid state synthesis for the same phosphor example, YVO_4:Eu. Examples of several other rare-earth-activated refractory metal oxides, which are known display phosphors and are usually prepared by high-temperature solid state processing, prepared by the hydrothermal process described are also given.

Hydrothermal synthesis is said to produce well-crystallized oxide powders with good control over particle-size distribution. The patent notes that lanthanum and gadolinium can replace yttrium in the oxides. The elemental compositional claims of the patent are very broad, encompassing the rare earths, including all the lanthanides, the transition metal vanadium, and the nonmetal phosphorus in the form of mixed oxides. The oxide phosphors prepared in the patent's examples seem less ambitious in their use of the metals and rare earths than the patent's claims. This may be due to the multigram quantities of reagents used in the preparations given in the examples and the length of time taken for processing. It has not been unusual to find that in phosphor research the use of metals has been limited to three or less. Employment of combinatorial processes should enable the preparation of libraries built with four or five metals in a relatively short time period that would be suitable for X-ray detection strategies. Because the hydrothermal method is based on fluids, ink-jet techniques can advantageously be employed to dispense the reactants as solutions onto locations of suitable substrate plates to begin the processing.

The last example of phosphor preparation given in the patent will be described to show applicability of ink-jet-assisted processing. $La_{0.95}Ce_{0.05}PO_4$ is prepared by dissolving La_2O_3 and $Eu(NO_3)$. $6H_2O$ in proportions of 19:1 in 6N HCl so that the lanthanide concentration is 1.00 M. The mixture is heated and stirred to effect dissolution. This solution is combined with an equal volume of 2M H_3PO_4 in water, which produces a precipitate. The precipitate is removed, dried, added to an equal weight of water, and heated to 600°C at 2,500 bar in an autoclave for 24 hr. The product consisted of 2 μm crystallites of pure $La_{0.95}Ce_{0.05}PO_4$ which were annealed at 1000°C for 2 hr. In the patent's first example, the preparation of the phosphor $Y_{0.95}Eu_{0.05}VO_4$ is given. The reagents used are $Y(NO_3) \cdot 5H_2O$, $Eu(NO_3) \cdot 6H_2O$, and NH_4VO_3. They are dissolved in deionized water, and a given amount of guanidine carbonate is added for pH adjustment. The suspension is placed in a pressure vessel which is closed and heated to 200°C for 5 days. Using silica substrate plates with 1 mm depressions, and using ink-jet technology to deposit all the reactants as solutions in the proper order, the hydrothermal method as described and post-reaction annealing can be employed to produce many more combinations of metals in oxide phosphor libraries, which can then be characterized by X-ray analysis, as above, for their composition, crystalline phases, and valence states of the rare earths, than a normal experimentalist would care to contemplate if he or she had to do it using the procedures given in this patent.

Sulfide phosphors could similarly be prepared as combinatorial libraries since many preparations begin with soluble cations, which are precipitated, fired in air to convert to oxides, and then annealed in an atmosphere of H_2S or CS_4 to produce the sulfide phosphor [23]. Metals of interest in this patent include Na, K, Rb, Cs, Gd, La, Sc, Y, Ag, Cu, Eu, Ce, Mn, Sn, Sb, Pb, Pr, Sm, Tb, Dy, Ho, Er, Tm, and Bi. Even though the inventors were interested in a number of elements, only 23 phosphors are shown as being made and analyzed, and none had more than three metals. And this is work of a large outstanding industrial laboratory!

Ink-jet-assisted combinatorial library generation can be used in nonbiological materials research to multiply the developmental power of the experimental chemist where components in fluids or fluid reactants can be used, and where realistic detection strategies can be devised. Advantages being enjoyed by the biomedical and pharmaceutical industries are available to other researchers in materials science.

INK-JET-BASED COMBINATORIAL SYNTHESIS

Oligomeric Molecules

Oligomeric molecules represent an important class of materials in biomedical applications. From a combinatorial synthesis view, they represent the simplified case of a two-variable system: the length of the molecule, in terms of the number of discrete subunits, and selection of each subunit from a small set of subunits with similar structure.

Ink-jet printing methods have been used to synthesize arrays of both DNA and peptide oligomers. These arrays have a number of applications, including clinical diagnostics, single nucleotide polymorphism (SNP) detection, point mutation detection, drug screening, and expression studies. Ink-jet methods have been used to dispense large numbers of different oligonucleotide solutions that have been fabricated off-line. These systems employ a large number of ink-jet devices (up to 96) that can each be loaded with a different fluid [24]. By repeatedly loading, printing, cleaning, and reloading, these systems can produce arrays of more than 10,000 different oligonucleotides. The chief difficulty in the deposition of oligonucleotides is the number of fluids that must be dispensed.

Synthesis of DNA arrays using ink-jet technology greatly decreases the number of different fluids required. Only the four constituent bases (A, G, C, T) of DNA, plus an activator, are jetted, as illustrated in Figure 22.16. The complexity of multistep chemical synthesis in an anhydrous environment is an added problem, but a number of investigators have overcome this difficulty [25,26], and DNA arrays manufactured in this way are available from Agilent.

Peptide arrays for drug and expression screening studies [27] can also be synthesized using ink-jet printing technology, in a manner similar to DNA arrays, except that there are 20 naturally occurring amino acids, making the dispensing system more complex, as illustrated in Figure 22.17. However, an anhydrous environment is not required for peptide synthesis. A number of companies are pursuing this approach, but have not yet reported on their results.

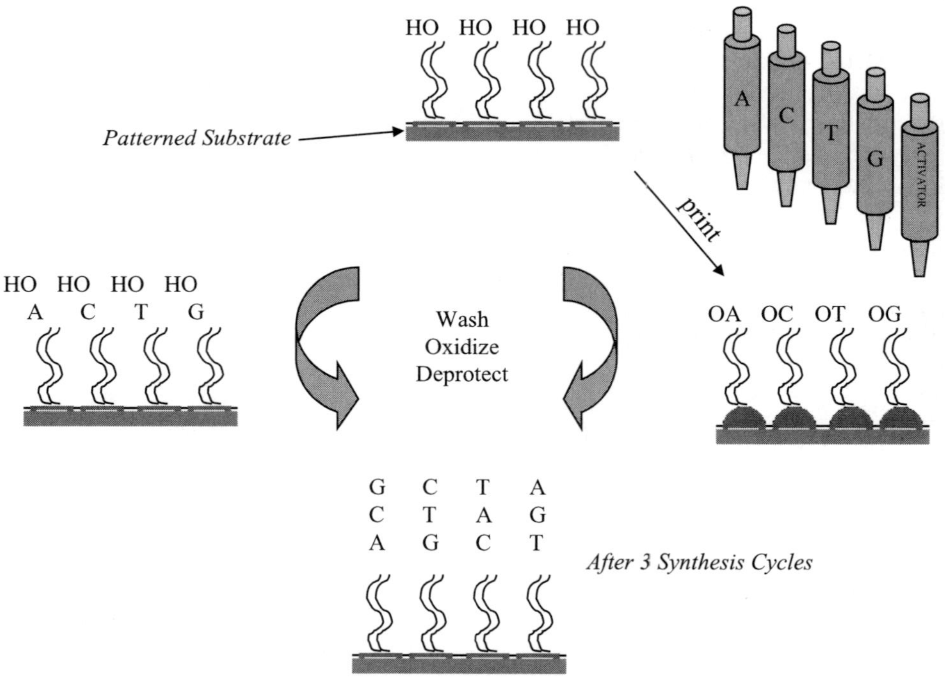

FIGURE 22.16. Schematic of ink-jet-based DNA array synthesis.

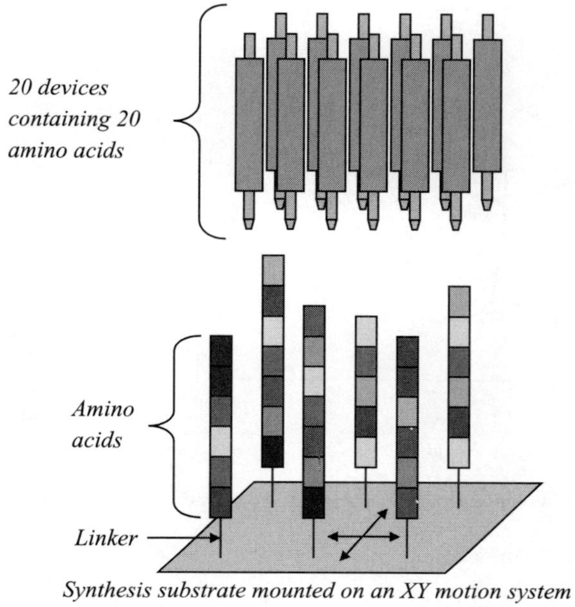

FIGURE 22.17. Schematic of peptide synthesis using ink-jet technology.

Electronic Materials

Transparent conducting oxides (TCOs), such as indium tin oxide (ITO, $InSnO_x$), SnO_2, and ZnO, are used in a number of applications including flat-panel displays employing light-emitting polymers, solar cells, low-emissivity windows, electrochromic devices, and antistatic coatings [28]. Increased transparency (at specific wavelengths, or broadband) and conductivity of TCO materials would result in improved performance in these applications. Matching the work function of the TCO to the adjacent material would result in improved polymer display performance. Improved stability (chemical, thermal, plasma, interfacial) and decreased deposition cost/temperature would result in lower-cost products employing TCO materials. Finally, creation of p-type TCO materials paired with current or novel materials would allow the creation of metal-oxide-based logic that would be competitive with the polymer-based electronics that is under development by many of the world's electronics manufacturing giants [29].

Currently used TCO materials are a small set of the possible phase space of indium, tin, and zinc oxides. In addition, cadmium and gallium oxides can be used in conjunction with the other three to create a five-dimensional phase space for n-type TCO materials, as illustrated in Figure 22.18, that one could modestly describe as "target rich"! Adding the use of MgO, dopants, and other processing variations, the possible space for discovery of new TCO materials becomes extremely large.

Thus because of both the high economic impact and the large space available for new materials discovery, TCOs are an excellent application for high-throughput combinatorial

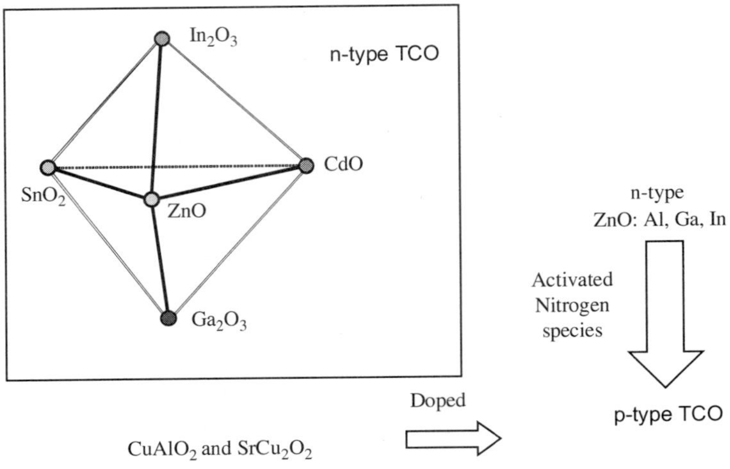

FIGURE 22.18. Graphical illustration of the large phase space available for TCO optimization. (Courtesy of David Ginley, NREL.)

methods. Researchers at the National Renewable Energy Laboratory (NREL) in Golden have applied combinatorial methods to TCO discovery using a multigun sputtering system with either radiofrequency or dc capabilities to generate libraries with continuous composition gradients [30]. Evaporative, pulsed laser deposition, and chemical vapor deposition (CVD) methods [31] can also be employed to generate TCO libraries.

Use of ink-jet methods to create combinatorial libraries would have a number of advantages over current methods. Deposition rates in sputtered libraries limit the dynamic range deposition to approximately 4:1, and even less for CVD systems. With ink-jet deposition, the amount of material dispensed and thus the stochiometry can be controlled digitally by the number of drops of each fluid dispensed onto each site. For an even wider range of concentration ratios, multiple concentrations of the same material can be dispensed from different ink-jet devices. This is analogous to the high-quality ink-jet photoprinters that employ six colors (plus black), where three of the colors have a lower dye concentration for printing highlight areas more accurately [32]. Sputtering of more than three materials becomes increasingly complex to the point of becoming impractical. Jetting four color inks to produce over a million colors is possible in all desktop ink-jet printers; jetting five fluids to synthesize DNA has been demonstrated by a number of researchers, both in the laboratory and commercially; and peptide arrays using 20 or more fluids are under development, as discussed previously. Finally, CVD methods require masking, which implies multiple tooling or configuration changes to produce a library, whereas ink-jet is inherently digital and thus maskless.

Combinatorial synthesis of TCO libraries has been proposed by NREL and MicroFab, but not yet demonstrated. Here, we will outline two approaches involving different precursor materials and synthesis processes. One approach is organometallic decomposition [33]. The other is a sol–gel approach. Sol–gel-based ITO synthesis has been reported for both nonaqueous and aqueous procedures [34, 35], but the requirements of ink-jet deposition necessitate some changes from conventional processing.

For the organometallic decomposition approach, commercially available organometallic compounds are selected for the desired metals and formulated with nonaqueous solvents to make the precursors jettable. The precursors are printed onto the substrate in a manner to produce the desired stochiometry, as discussed above. Complete mixing is critical to formation of the desired compound and is achieved by a combination of solvent selection, printing process (drops can be deposited in an arbitrary order), and/or external driving forces such as vibration or ultrasonics. Unlike the ordered libraries that gradient sputtering creates, compounds can be created in an arbitrary geometric and temporal arrangement, allowing systematic variation in either synthesis or measurement to be removed statistically. Also, many repetitions of the same compound can be included, allow for doping variations, localized treatment variations (e.g., laser), and/or higher experimental confidence.

After printing the organometallic precursors, the substrate is heated to evaporate the remaining solvents. Then it is processed in an oven or furnace to convert the organometallics into oxides. Annealing is accomplished in a tube furnace in air, an inert atmosphere, or under other conditions.

Conventional sol–gel processing methods have been used to create TCOs in the form of zinc oxide electrodes [36] and can produce ultrafine films and very pure oxides through the use of metal alkoxides [37]. As with organometallic decomposition, solvents, which will include water and precursor formulations, are selected to to make the sol–gel precursors jettable. Printing methods are similar to those for metal–organic decomposition.

After printing, the precursors are hydrolyzed and transformed into a metaloxo-based macromolecular gel through a polycondensation process. Drying the gel, which is a mixture of liquid and solid phases, removes the liquid and produces a porous xerogel. Using heat treatments, the xerogel is densified and transformed into a solid material, either an amorphous glass or a crystalline ceramic. Densification and annealing is done in a tube furnace in air, nitrogen, or an atmosphere with a controlled partial pressure of oxygen.

Catalysts

Fuel cells are a major focus in energy and power research because they are quiet and emmision free [38]. Recently, government and industry have decided to shift from emphasis of battery-powered electric vehicles to fuel-cell-powered vehicles as part of the Clean Vehicle Initiative [39]. A key to the economic widespread use of fuel cells as a power source is high-performance low-cost manufacturable electrocatalysts.

Ink-jet technology has been used for library preparation for methanol fuel cell catalysts discovery by researchers at Penn State University and Illinois Institute of Technology [40]. Using a modified Epson ink-jet printer, they printed metal salt solutions of platinum, rhodium, osmium, ruthenium, and iridium to form 220 compositions in the phase space of ternary and quaternary alloys of these metals. After the metal salt solutions were printed, they were reduced to metallic form and screened for performance using a pH-sensitive fluorescence indicator. This study resulted in identification of new fuel cell catalysts, $Pt_{62}Rh_{25}Os_{13}$ and $Pt_{44}Ru_{41}Os_{10}Ir_5$, which provided 40% higher current densities.

In recent results reported by Reichenbach and McGinn at Notre Dame University [41], a modified Pechini powder synthesis process was using together with a valve-jet deposition process to produce combinatorial libraries of the mixed oxide system $Cu_{1-x}Ce_xO_3$.

After printing the precursor materials, mixed-oxide crystalline powders were formed at 500°C and below. The resulting powders had particle sizes in the 20.0–85.5 nm range. The library was evaluated for catalytic activity in the CO oxidation reaction at temperatures between 150 and 350°C. All elements in the library were found to have a measurable catalytic activity.

In addition to the $Cu_{1-x}Ce_xO_3$ system, Reichenbach and McGinn used their ink-jet-based deposition protocol to generate a library of oxide powders of the perovskite-like $La_{1-x}Sr_xCoO_{3-*}$, where $0 \leq X \leq 1$ system. The La–Sr–Co–O system exhibited lower average surface areas due to higher decomposition temperatures, but acted as an active catalyst for the CO oxidation reaction.

LOCALIZED POST-DEPOSITION PROCESSING, REACTION SITE CREATION, AND ACTIVATION

Localized postdeposition processing, and in particular laser annealing, of ink-jetted sites or patterns has been proposed by a number of researchers [42] but not reported in the literature. Localized annealing would allow for different conditions to be utilized for different sites and allow the use of substrates that cannot withstand higher temperatures for times that are typical of ovens or hot plates. If all the material in a single site is deposited first and then laser annealed, there is no coupling between the two processes. However, a key feature of ink-jet printing is precise temporal (digital) control, and so it is possible to synchronize the dispensing of multiple drops of one material and/or the dispensing of multiple materials onto the same location to produce a layered microstructure. Although this concept has not been demonstrated, it is similar to the dye-assisted laser ablation process that has been employed in several biomedical procedures [43].

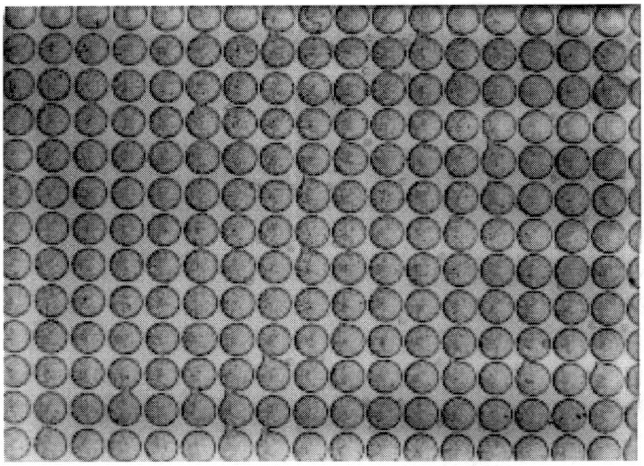

FIGURE 22.19. Spots (120 μm) of nitrocellulose printed on glass using ink-jet printing.

In addition to printing the precursor/component materials used in combinatorial synthesis, ink-jet technology can be used to to create or modify solid support structures to be used as attachment or synthesis sites. This can include creation of a site for local control of wetting or reactivity. Solid phase materials such as nitrocellulose (Figure 22.19), methyl cellulose, sol–gels, and biotinylated Poly(lactic-co-glycolic acid) (PLGA) have been dispensed onto substrates using ink-jet technology. Chromic acid has been used to modify polypropylene, and acetone has been used to modify polystyrene (Figure 22.12). Finally, cleavable linkers such as succinate and amidate have been dispensed.

SUMMARY

Ink-jet printing technology has been shown to be capable of dispensing a wide range of materials of interest in combinatorial library generation. This can be accomplished at high deposition rates and with precision in the picoliter scale with digital control. The use of ink-jet technology for synthesis of oligomeric molecules, such as DNA and peptides, and catalysts was described. The potential use for synthesis of TCOs, emissive materials for displays, and nonlinear optical composites was also discussed. Using the temporal precision of ink-jet technology for interactive dispensation and post-processing was discussed, along with modification of the physical and chemical characteristics of the synthesis surface using ink-jet methods. Finally, the interaction of ink-jet library generation and detection strategy were discussed.

ACKNOWLEDGMENTS

The authors would like to thank David Ginley and his associates at the National Renewable Laboratory for many valuable written and oral communications on the use ink-jet technology in combinatorial library generation.

REFERENCES AND NOTES

1. Chaudhary, K. C., Redekopp, I. G., Maxworthy, T. The non-linear capillary instability of a liquid jet. *J. Fluid Mech.* **1979**, *96*, 257–312.
2. Pimbley, W. T. Drop formation from a liquid jet: a linear one-dimensional analysis considered as a boundary value problem. *IBM J. Res. Dev.* **1984**, *29*, 148–156.
3. FIlmore, G. L., Buehnet, W. L., West, D. L. Drop charging and deflection in an electrostatic ink jet printer. *IBM J. Res. Dev.* **1984**, *21*, 37–47.
4. Yoo, J., Cima, M., Sachs, E., Suresh, S. Fabrication and microstructural control of advanced ceramic components by three dimensional printing. *Ceram. Eng. Sci. Proc.* **1995**, *16* (5), 755–762.
5. Muntz, E. P., Orme, M., Pham-Van-Diep, G., Godin, R. An analysis of precision, fly-through solder jet printing for DCA components. *Proc. ISHM '97, Philadelphia*, **1997**, pp. 671–680.
6. Hendricks, C. Inertial confinement fusion targets. *Proc. 2nd International Colloquium on Drops and Bubbles*, **1982**, NASA Conf. Rep. NASA-CR-168848, JPL 82-7, pp. 88–93.
7. Hayes, D. J., Wallace, D. B., VerLee, D., Houseman, K. *Apparatus and Process for Reagent Fluid Dispensing and Printing*, U.S. Patent 4,877,745, **1989**.
8. Dijksman, J. F. Hydrodynamics of small tubular pumps. *J. Fluid Mech.* **1984**, *139*, 173–191.

9. Adams, R. L., Roy, J. A. One dimensional numerical model of a drop-on-demand ink jet. *J. Appl. Mech.* **1986**, *53*, 193–197.
10. Wallace, D. B. *A Method of Characteristics Model of a Drop-On-Demand Ink-Jet Device Using an Integral Method Drop Formation Model*, **1989**, ASME Publ. 89-WA/FE-4.
11. Pies, J. R., Wallace, D. B., Hayes, D. J., *High Density Ink Jet Printhead*, U.S. Patent 5,235,352, **1993**.
12. Aden, J. S., Bohorquez, J. H., Collins, D. M., Crook, M. D., Garcia, A., Hess, U. E. The third generation HP thermal inkjet printhead. *Hewlett-Packard J.* **1994**, *45* (1), 41–45.
13. Okamoto, T., Suzuki, T., Yamamoto, N. Microarray fabrication with covalent attachment of DNA using bubble jet technology. *Nat. Biotechnol.* **2000**, *18* (4), 438–441.
14. For example, valves from The Lee Company: http://www.microhydraulics.com/EFSWEB/INKJET.htm
15. Wu, K. J., Shaler, T. A., Becker, C. H. *Anal. Chem.* **1994**, *66*, 1637–1645.
16. Holzwarth, A., Schmidt, H.-W., Maier, W. *Angew. Chem. Int. Ed.* **1998**, *37* (19), 2644–2647.
17. Wang, J., Gonsalves, K. E. *J. Comb. Chem.* **1999**, *1*, 216–222.
18. Barboux, P., Baffer, N., Morineau, R., Livage, J. *Solid State Protonic Conductors*, Vol. 3, **1985**, Paris: Odense University Press, p. 173.
19. Ando, M., Kadono, K., Haruta, M., Sakaguchi, T., Miya, M. *Nature* **1995**, *374*, 625.
20. Isaacs, E. D., et al. *Appl. Phys. Lett.* **1998**, *73*(13) 1820–1822.
21. See www.aps.anl.gov/xfd/communicator/useroffice or www.aps.anl.gov
22. Phillips, M. L. F. Method for synthesizing fine-grained, phosphor powders US Patent 5,743,955, **1998**.
23. Yocom, N., Dismukes, J. Luminescent sulfides of monovalent and trivalent cot ions US Patent 3,981,819, **1976**.
24. James, P., Papen, R. A new innovation in robotic liquid handling. *Drug Discov. Today* **1998**, *3* (9), 429–430.
25. Blanchard, A. P., Kaiser, R. J., Hood, L. E. High density oligonucleotide arrays. *Biosens. Bioelectron.* **1996**, *11*, 687–690.
26. Brennan, T. M. *Apparatus for Diverse Chemical Synthesis Using Two-Dimensional Array*, U.S. Patent 6,00,1311, **1999**.
27. Frank, R., Overwin, H. Spot synthesis: Epitope analysis with arrays of synthetic peptides prepared on cellulose membranes. *Methods Mol. Biol.* **1996**, *66*, 149–169.
28. Ginley, D. S., Bright, C. Transparent conducting oxides, *MRS Bull.* **2000**, *25*, 15–18.
29. International Microelectronics and Packaging Society (IMAPS) 1st Advanced Technology Workshop (ATW) on Printing an Intelligent Future: Printed Organic Electronic Technologies, March 8–10, 2002, Incline Village, NV.
30. Perkins, J., Readey, D., Alleman, J. del Cueto, J., Li, X., Coutts, T., Stauber, R., Duncan, C., Young, D., Parilla, P., Keyes, B., Gedvilas, L., Balzar, D., Wang, Q., Ginley, D. A combinatorial approach to TCO synthesis and characterization. *Mater. Res. Soc. Symp. Proc.* **2001**, *623*, 211–213.
31. Freeman, A. J., Poeppelmeier, K. R., Mason, T. O., Chang, R. P. H., Marks, T. J. Chemical and thin-film strategies for new transparent conducting oxides. *MRS Bull.* **2000**, *25* (15), 45–51.
32. For example, Epson Stylus Photo 890: www.epson.com
33. Gallagher, D., et al. *J. Mater. Res.* **1993**, *8*, 3135.
34. Mattox, D. M. *Thin Solid Films* **1991**, *204*, 25–32.
35. Perry, Carole C., et al. *Proc. SPIE* **2000**, *3934*, 270–279.
36. Hoyer, R., Weller, H. *J. Phys. Chem.* **1995**, *99*, 14096.
37. Klein, Lisa *SOL-GEL Technology for Thin Films, Fibers, Preforms, Electronics and Specialty Shapes*, Part II, **1988**, Park Ridge, NJ: Noyes Publications.
38. Vanderborgh, N. E., Inbody, M. A. *The Future of Fuel Cell Vehicles*, **1998**, Los Alamos, NM: Los Alamos National Laboratory.
39. *Beyond the Internal Combustion Engine: The Promise of Methanol Fuel Cell Vehicles*: retrieved from The American Methanol Institute, http://www.methanol.org/fuelcell/special/ami.pdf
40. Reddington, E., Sapienza, A., Gurau, B., Viswanathan, R., Srangapani, S., Smotkin, E. S., Mallouk, T. E. Combinatorial electrochemistry: A highly parallel, optical screening method for discovery of better electrocatalysts, *Science* **1988**, *280*, 1735–1737.
41. Reichenbach, H. M., McGinn, P. J. Combinatorial synthesis of oxide powders. *J. Mater. Res.* **2001**, *16* (4), 967–974.
42. Ginley, D. S., private communication; Poulikakos, D., private communication; Hayes, D., private communication.
43. Arcoria, C. J., Frederickson, C. J., Hayes, D. J., Wallace, D. B., Judy, M. M. Dye microdrop assisted laser for dentistry. *ASLM Abstracts, Laser Surg. Med.* **1993**, Supplement *5*, 17.

Chapter 23

Miniaturized Reactor Concepts and Advanced Analytics for Primary Screening in High-Throughput Experimentation

Torsten Zech, Jens Klein, Stephan A. Schunk, Thorsten Johann, Ferdi Schüth, Stefan Kleditzsch, and Olaf Deutschmann

INTRODUCTION

The development of novel and better catalysts for chemical processes is mainly an empirical process utilizing existing technical expertise and experience. The aim of reducing time and costs has proved a powerful incentive for the development of high-throughput technologies for applications in heterogeneous and homogeneous catalysis.

Obviously, there are strong analogies between the development of combinatorial chemistry for drug discovery and combinatorial materials science, of which high throughput experimentation in catalysis is an important part. It is worth reading the excellent paper by Lebl [1] about the development of combinatorial chemistry and comparing it with today's situation in combinatorial material science and high-throughput experimentation. The first papers in combinatorial chemistry were published in the mid-eighties. Since then, activities in this field have grown exponentially, fundamentally changing old research paradigms. The first companies operating in this field were founded during the late eighties

Torsten Zech, Jens Klein, and Stephan A. Schunk • hte Aktiengesellschaft, Kurpfalzring 104, 69123 Heidelberg, Germany
Thorsten Johann and Ferdi Schüth • Max-Planck-Institut für Kohlenforschung, Kaiser-Wilhelm-Platz 1, 45470 Mülheim an der Ruhr, Germany
Stefan Kleditzsch • Universität Heidelberg, Interdisziplinäres Zentrum für Wissenschaftliches Rechnen, Im Neuenheimer Feld 368, 69120 Heidelberg, Germany
Olaf Deutschmann • Institut für Chemische Technik, Universität Karlsruhe, Kaiserstr. 12, D-76131 Karlsruhe, Germany
Contact author: Torsten Zech, email: torsten.zech@hte-company.de

High-Throughput Analysis, edited by Potyrailo and Amis
Kluwer Academic Publishers, New York, 2003.

and at the beginning of the nineties. Nowadays, combinatorial chemistry and high-throughput screening are core technologies in all major pharmaceutical companies.

In addition, parallel testing technologies have been used for a long time in materials science [2] and heterogeneous catalysis [3, 4]. Nevertheless, the landmark publication by Xiang et al. in 1995 [5] attracted a great deal of attention. Since then, numerous applications of library screening techniques for the different stages in the development of heterogeneous catalysts have been reported [6–15]. Approaches to high-throughput experimentation in catalysis are generally divided into two stages: catalyst discovery (primary screening) as the first stage, and catalyst optimization (secondary screening) as the second stage.

The main target in primary screening is the identification of hits out of large catalyst libraries. The number of catalysts tested in parallel should be a few hundred on a small reactor footprint. Therefore, specific library preparation methods resulting in very small amounts of the individual materials are required. The depth of information of the analytical methods employed is often reduced in favor of a high analytical speed. Accordingly, only the most promising materials from a library are characterized in order to gain additional insights into their properties. Primary screening approaches are usually employed in ambitious catalyst discovery programs with little or no previous knowledge about the target reaction and the class of materials to be studied.

In contrast, the secondary screening of catalytic materials aims at optimization of known catalyst formulations and reaction conditions at a more sophisticated level. Results and relations discovered in primary screening are explored in more detail in all phases of the work flow, encompassing (1) increasing the quantities of catalysts prepared by conventional synthesis procedures, (2) more detailed screening processes involving wide variation and precise control of reaction parameters, (3) complete and accurate product analysis, and (4) careful characterization of the tested materials. In all phases of a secondary screening work flow, the conditions are "close to conventional" compared with standard "one catalyst at a time" testing procedures.

Catalyst preparation is carried out predominantly by well-known conventional methods (e.g., impregnation, precipitation, ion exchange). In most cases, a large number of parallel fixed-bed reactors, each filled with 20–200 mg of catalyst, is operated in parallel under a specific set of steady state operating conditions and continuous flow. Product analysis is generally carried out in a sequential manner by well-established analytical methods (e.g., gas chromatography, gas chromatography–mass spectrometry, or Fourier transform IR spectroscopy) using multiposition valves for channel selection. If carefully designed and operated, the performance of secondary screening reactors is similiar to or better than that of conventional laboratory reactors.

Because of the successful application of secondary screening tools, the demand for primary screening devices and methodologies has increased, with miniaturization being one of the key issues. However, only a few primary screening techniques are described in the literature, with a strong focus on the applied analysis technique. In the first publication in the field, Moates et al. [6] used a simple setup comprising a substrate-based library and an IR camera to detect catalytic activity. Holzwarth et al. [8] extended this principle to emission-corrected IR thermography with high-temperature resolution to resolve reaction heats spatially over catalyst libraries composed of transition metals in amorphous microporous mixed oxides on a flat substrate over which the reactants flowed. Obviously, this technique cannot be used to identify reaction products and thus estimate selectivities, but it is particularly useful for detecting catalyst activity sensitively and effectively during primary screening. Flat substrates with catalysts produced as thin films were also used by Cong et al. [9]. Two concentric capillaries were used to supply and withdraw reactants, and the catalysts were tested sequentially in a transient mode with a special mass spectrometer. Later, this methodology was simplified and extended by Orschel et al. [12], who used small amounts of catalyst particles deposited in wells on the substrate. The reactor characteristics were also improved, allowing a conventional mass spectrometer to be used. However, the catalysts were still tested in a sequential and transient mode. Thus it was not possible to follow activation and deactivation phenomena over long periods of time. Senkan [7] developed the resonance-enhanced multiphoton ionization method (REMPI) to detect a selected molecule in the effluent of a continuously operated arrangement of fixed-bed catalysts and a multichamber reactor system [16]. The products are detected by an array of spatially addressable microelectrodes directly exposed to the gas stream after the absorption and ionization features of the target molecule have been determined. However, this can cause problems in measurements over a long period because of drift of the electrode response. Furthermore, the system has not been miniaturized to allow larger catalyst libraries to be tested simultaneously. Another interesting technique used for primary screening is photothermal deflection [17], which allows fast sequential analysis based on two crossed laser beams.

Techniques based on optical readout of catalyst activity are very suitable for real-time parallel deflection since first analysis is possible. The first use of this approach was by Reddington et al. [18], who used fluorescence detection of protons formed at the anode catalyst of a direct methanol fuel cell by a fluorescent dye. This work was carried out in solution where the lack of mixing allowed the fluorescence signal to be correlated with the catalyst, but direct parallel analysis of products is also possible in a gas stream from a parallel reactor. A selective reaction of the product or an educt of the reaction with a selective color reagent is used to do this [19]. Such a system, which can be parallelized to a rather high degree, has recently been used to detect NO remaining in the offstream of potential $DeNO_x$ catalysts arranged in a 49-channel parallel reactor [20].

As can be seen, one of the crucial elements of a primary screening setup is an analysis technique which is able to deal with very low sample volumes while providing good spatial resolution and sufficient chemical information about the nature and amount of the products. Generally, fast sequential, quasiparallel and parallel analysis techniques can be distinguished. Usually, simple parallel detection methods such as IR thermography [6,8] provide a lower depth of information than sequential methods like mass spectrometry [21], but are lot faster.

Primary screening approaches also require an integration of catalyst preparation, reactor, and analysis methods. This may result in unconventional preparation methods or reaction conditions that complicate the transfer of results from primary to secondary screening. In addition, the question of miniaturization so that several hundred catalysts can be tested simultaneously has not yet been addressed sufficiently, although it is known from combinatorial and analytical chemistry that miniaturized reactors and microfluidics can be applied successfully. One reason for this is related to the challenging requirements that have to be met by primary screening reactors: the reactor design has to accomodate catalysts made by a simple, fast, and scalable catalyst preparation method; many catalysts should be tested simultaneously under steady state conditions and continuous flow; and the analysis method should have a high spatial resolution and allow a good estimate of conversion degree and selectivity, while still being very fast and reproducible.

Although microstructured reactors have attracted a great deal of interest in recent years, little is known about continuously operated miniaturized reactors for applications in high-throughput discovery programs in catalysis. Jensen and co-workers recently presented a novel differential packed-bed microreactor designed for the investigation of heterogeneously catalyzed gas phase reactions [22,23]. Utilization of the reactor for the determination of the reaction kinetics of CO oxidation on a Pd–Al_2O_3 catalyst was discussed. The proposed cross-flow design led to a uniform flow distribution over the catalyst bed, which was confirmed by a two-dimensional computational fluid dynamics (CFD) simulation based on a finite-element approach. The authors state that this type of reactor could also be applied to high-throughput testing of heterogeneous catalysts, although no example is given.

Another approach was explored by Bergh et al. [24], who used an arrangement of reaction chambers with thin- or thick-film catalysts on a flat substrate for parallel catalyst testing. These workers also described a method for continuously supplying reactants to the reaction chambers [25]. However, the description of the actual design of the reactor is not unambiguous, which makes a technical evaluation of this approach difficult.

From our own experience with microchannel-based approaches for catalyst discovery [26,27] we concluded that microstructured reactors present a number of interesting properties which are worth investigating with respect to primary screening. Microstructured reactors provide excellent heat and mass transfer properties and can be operated at isothermal conditions. The response times to changes of the reaction conditions are very low and a number of additional static or active elements (mixers, valves, etc.) can be integrated into the device besides the reactor itself. Owing to their potential mass fabrication, miniaturized reactors may prove cost efficient. However, much development work is necessary to set up a microstructured reactor system, as many microfabrication technologies are still be considered “experimental” and the behavior of miniaturized reactors is still not well understood.

Therefore the application of CFD calculations should be considered to facilitate system design. A variety of commercial and academic computational tools which can assist in the development and design of reactors for heterogeneously catalyzed gas phase reactions have been developed in the last few years. CFD codes which can numerically predict the flow field, the temperature, and species concentration distributions in three-dimensional configurations, sometimes even including detailed models for the heterogeneous reactions on the catalytic surface and potential homogeneous reactions in the gas phase, are also available [28]. Provided that the assumptions of continuum mechanics remain valid, these tools can be used to simulate processes in microreactors. A frequent advantage of small reactors is that the flow field remains laminar; thus no turbulence model needs to be applied. For instance, CFD simulations have led to a better understanding of the reactive flows in the small channels (radius 125–500 μm) of catalytic monoliths [29,30] and around catalytic wires [31–33] for selective oxidation reactions and microburners [34–36].

"SINGLE-BEAD REACTORS" FOR PRIMARY CATALYST SCREENING

Catalytic Materials and Reactor Architecture

The basis of the single-bead concept is the use of single shaped bodies as the catalytic material of interest. In principle, these particles may be of any shape, but are usually spherical. In accordance with approaches known from combinatorial chemistry, such spherical particles are called "beads", although they fulfil very different functions in comparison with their application in combinatorial setups in organic and biochemistry.

Each bead represents one catalyst as a member of a library of solid catalysts. It consist of an nonporous material (e.g., α-Al_2O_3 or Steatit) or a typical porous support material (e.g., γ-Al_2O_3, SiO_2, or TiO_2). These beads can be subjected to different synthesis procedures and sequences such as impregnation, coating, etc. In addition, full mixed metal oxide catalysts can also be formed to spherical particles.

Using single beads for high-throughput experimentation has a number of advantages. First, such beads are comparable to well-known fixed bed catalysts and the synthesis pathways can be the same as for conventional materials, which may facilitate scale-up procedures. A number of common preparation procedures, which can be carried out in standard laboratory environments, are available for these beads.

The second main advantage is that each bead represents a single entity that can be handled independently from other beads or the final reactor configuration. Starting from master batches, a large diversity of materials can easily be prepared. Furthermore, different beads can be treated individually, for instance, subjected to different preparation steps or pretreatments like calcination or steaming, rather than handling the complete library as is necessary for substrate-bound thin- or thick-film catalysts. This property allows the use of synthesis procedures different from the parallel approach. Furthermore, the potential use of *ex situ* synthesis procedures may present a significant advantage, especially in microchemical systems, as in situ preparation of the catalysts in the reactor could cause problems with contamination or the thermal and chemical stability of the reactor material.

The reactor suitable for testing single-bead catalysts in a highly parallel fashion, the so-called "single-bead reactor," is designed as a two-dimensional arrangement of microreaction chambers which can each hold one catalytic bead. The reactor can be divided into two parts, a base part and a top part. The base part is filled with beads, and the top part is then pressed, sealed, or bonded against the base part to encapsulate each bead in a single independent microreaction chamber. The beads sit loosely in the reaction chambers, allowing continuous fluid flow from openings in the top part of the reactor, around the bead, and through openings in the base part. To avoid mass transfer problems while reacting the fluid on the bead, eggshell-type catalytic beads can be used.

Using the top-down direction as fluid flow path and the small footprint of a single microreaction chamber allows extension of the reactor in two dimensions. This is not possible using microchannel-based approaches, as the microchannels extend as straight long channels on a substrate and therefore block one dimension [26]. In most cases parallel microchannel reactors on a substrate are one-dimensional arrangements of channels, having a much larger footprint than a two-dimensional arrangement of reaction chambers such as the single-bead reactor. The resulting array density is much higher for the single-bead reactor, reaching up to 60 or more catalytic reaction chambers per square centimeter [37].

Reactant Distribution Options for Miniaturized Reactors

In order to carry out comparable catalytic experiments in a high-throughput parallel mode, equal reaction conditions for each library member have to be ensured. This is especially true for reactant distribution, as variations in residence time and space velocity can have a dramatic impact on the performance.

For conventional laboratory reactors, reactant flow is usually controlled by mass flow controllers that use valves to adjust the pressure drop for the fluid flow through the controller, thus regulating mass flow. Such an active control of fluid flow for each reaction chamber would probably not be possible, for reactors with several hundred parallel reaction chambers, or at least would not be economic as long as there are no dramatic developments and improvements in micro electro-mechanical systems (MEMS).

Adjusting the pressure drop along the fluid flow path is the key to controlling the fluid flow rate [38]. This also holds true for passive control of pressure drop by geometric design of the flow path. In contrast with active control elements, passive flow restrictors like capillaries or orifices can already be miniaturized and tightly integrated into a microreactor design.

An approach using capillary manifolds or microchannel manifolds, known as "binary trees," has been described by Bergh and Guan [25], among others. A channel network is built onto a flat wafer substrate that splits a common feed channel recursively into two new feed channels. As a result, a tree of splitting channels is generated which connects 256 reaction chambers with the common feed channel (Figure 23.1). The channel length and channel geometry to each of the chambers is the same, resulting in an equal pressure drop and therefore an equal reactant distribution. However, the channel network is very complex, and all reaction chambers are connected by the network. These interconnections may present a considerable limitation. For instance, a single particle in the feed stream may clog and block a whole subtree of the channel network. Furthermore, the tree becomes increasingly complex with increasing number of reaction chambers, therefore limiting *scalability*. Finally, the channel network occupies a considerable amount of space on the wafer, limiting the density of the reaction chambers.

As presented, there may be a number of drawbacks resulting from the single fact that the channels, that is, the flow restrictors causing pressure drop, are connected to each other. However, if we consider only the required pressure drop, there is no need to connect all the channels in a tree format.

An obvious solution to this problem is to use single independent flow restrictors for each reaction chamber. In an early design study, the pressure drop of fluid flow through short narrow channels or pores was calculated using the Hagen–Poiseuille law:

$$\Delta p = \frac{8}{\pi} l_p \dot{V}_B \eta \frac{1}{r_P^4 N} \tag{1}$$

with l_P is the pore length, $\dot{V}_B$ is the flow rate, η is the viscosity, r_P is the pore radius, and N is the pore number.

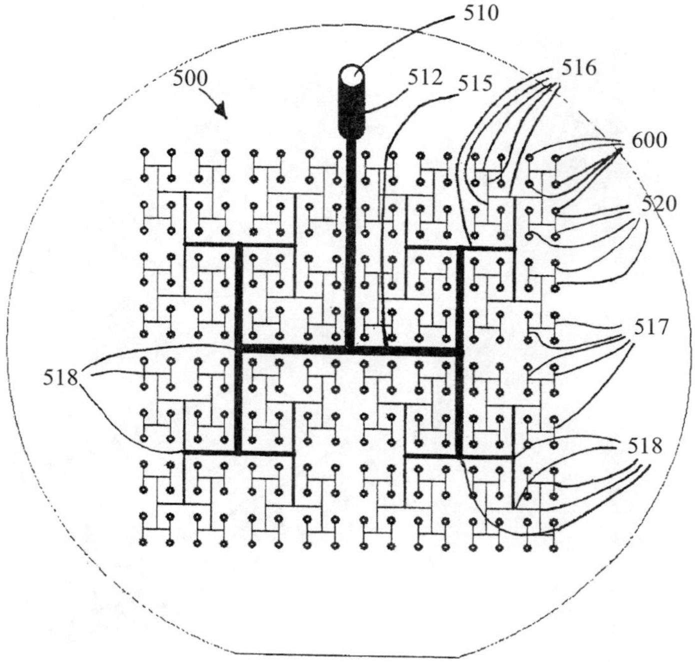

FIGURE 23.1. Reactant distribution using a binary-tree channel network [25].

As can be seen in Figure 23.2, a considerable pressure drop can be generated depending on the pore number and the pore radius. In particular, small pore radii result in a strong increase in the pressure drop. From this rough calculation, it is clear that such arrangements of short narrow pores can be sufficient to generate an equal flow distribution.

Therefore the fluid access ports of the microreaction chambers have been designed as flow restrictors in our "single-bead approach." These flow restrictors are microstructured membranes comprising a collection of a defined number of short straight pores with a very small diameter, as described below.

Computational Fluid Dynamics Calculations

The development of miniaturized reactors is supported by CFD calculations. We used CFD to study several crucial items of the design of the high-throughput reactor: (1) the realization of uniform inlet flow conditions in individual microreaction chambers, (2) the spatial profiles of the pressure, velocity, temperature, and species concentrations inside these chambers, and (3) potential interference of the microreaction chambers, mainly by thermal conductivity. The first two items will be discussed in more detail in this section.

The CFD simulation of our reactor is based on the solution of the steady state three-dimensional Navier–Stokes equations for gaseous flows. The flow fields are always laminar due to the small reactor sizes. A mass source term, gravitational and external body

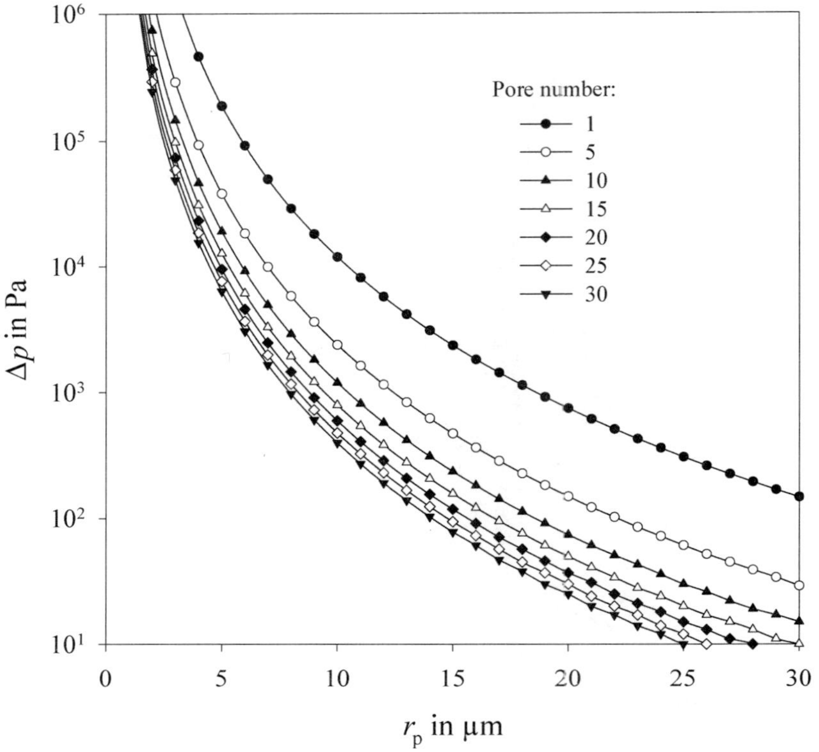

FIGURE 23.2. Pressure drop for fluid flow through short narrow pores as function of pore radius and pore number: $l_P = 100$ μm, $\dot{V}_B = 1$ ml min^{-1}, $\eta = 28$ μPa s.

forces, thermal diffusion, viscous heating, and homogeneous chemical reactions in the gas phase are not significant in the reactor. We do not consider the thermal interference between different reactor chambers, which will be described in a future publication, but assume an isothermal reactor. This results in the following system of conservation equations.

Continuity:

$$\frac{\partial(\rho v_j)}{\partial x_j} = 0 \tag{2}$$

Momentum:

$$\frac{\partial(\rho v_j v_i)}{\partial x_j} + \frac{\partial p}{\partial x_i} - \frac{\partial \tau_{ij}}{\partial x_j} = 0 \tag{3}$$

with stress given by

$$\tau_{ij} = \mu\left(\frac{\partial v_i}{\partial x_j} + \frac{\partial v_j}{\partial x_i} - \frac{2}{3}\delta_{i,j}\frac{\partial v_k}{\partial x_k}\right)$$

Species mass:

$$\frac{\partial(\rho v_j Y_i)}{\partial x_j} + \frac{\partial j_{i,j}}{\partial x_j} = 0 \quad (i = 1, \ldots, N_g) \tag{4}$$

with the diffusion mass flux in the x_j direction given by

$$j_{i,j} = -\rho D_i^M \frac{\partial Y_i}{\partial x_j}$$

In the above equations, the x_j ($j = 1,2,3$) are cartesian coordinates, v_j is the cartesian component of the velocity vector, ρ is the mass density, p is the pressure, Y_i is the mass fraction of species i in the mixture, N_g is the number of gas phase species, and D_i^M is the effective diffusion coefficient of species i into the mixture. The transport properties, viscosity μ and diffusion coefficients D_i^M, depend on temperature and composition [39,40]. The Einstein convention is used; that is, whenever the same index appears twice in any term, summation over that index is implied, except if the index refers to a chemical species. The system is closed by the equation of state

$$p = \frac{\rho RT}{\bar{M}} \tag{5}$$

where T is temperature, R is the universal gas constant, and $\bar{M}$ is the mean molar mass of the mixture.

On the reactive surface of the catalytic bead the following boundary conditions are used for the species continuity equation:

$$\mathbf{n}\mathbf{j}_i = \dot{s}_i M_i F_{\mathrm{cat/geo}} \quad (i = 1, \ldots, N_g) \tag{6}$$

where $\mathbf{n}$ is the outward-pointing unit vector normal to the surface, $\dot{s}_i$ are the chemical reaction rates due to heterogeneous reactions at the catalytic surface, and $F_{\mathrm{cat/geo}}$ is the ratio of catalytic active surface area to geometric surface area. No-slip boundary conditions are chosen for the momentum equation at all solid surfaces; that is, all velocity components vanish at the gas–surface interface.

The molar net production rate $\dot{s}_i$ of a gas phase species or an adsorbed species due to heterogeneous reactions on the solid surface is then given by

$$\dot{s}_i = \sum_{k=1}^{K_s} \nu_{ik} k_{fk} \prod_{i=1}^{N_g + N_s} c_i^{\nu'_{ik}} \quad (i = 1, \ldots, N_g + N_s) \tag{7}$$

where K_s is the number of surface reactions, ν_{ik} and ν'_{ik} are stoichiometric coefficients, k_{fk} is the rate coefficient of reaction k, c_i are species concentrations, and N_s is the number of adsorbed species. For complex heterogeneous reactions the expression for the rate coefficient k_{fk} can become rather complex:

$$k_{fk} = A_k T^{\beta_k} \exp\left(\frac{-E_{a_k}}{RT}\right) \prod_{i=1}^{N_s} \theta_i^{\mu_{ik}} \exp\left(\frac{\varepsilon_{ik}\theta_i}{RT}\right) \tag{8}$$

where A_k, β_k, and E_{ak} are the pre-exponential factor, the temperature exponent, and the activation energy, respectively, for reaction k. The parameters μ_{ik} and ε_{ik} describe the dependence of the rate coefficients on the surface coverages θ_i of species i [29,41,42].

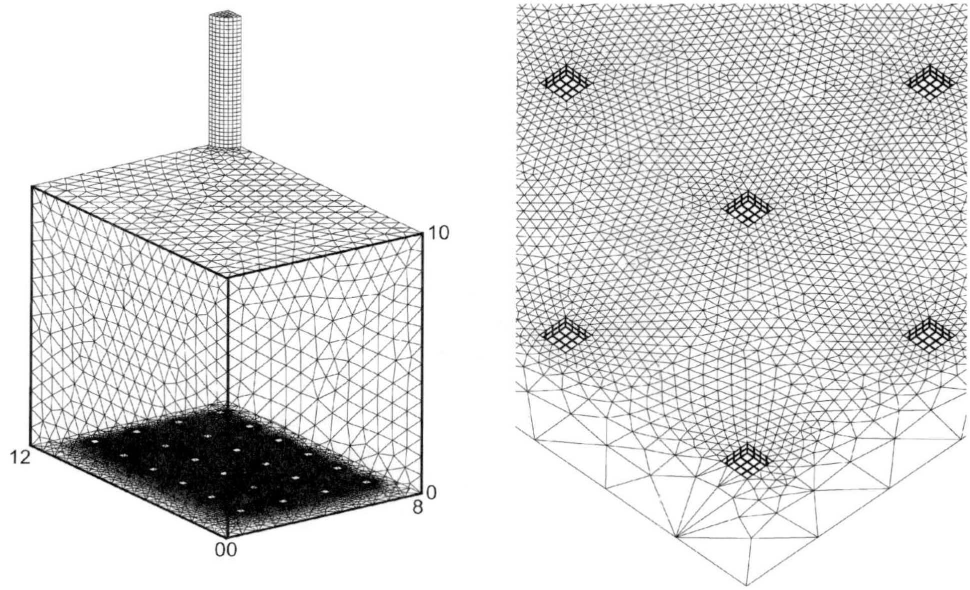

FIGURE 23.3. Computational grid at the walls of the reactor (left) and at the outlets (right) (scales in millimeters).

The numerical simulation is based on the commercial finite-volume CFD code Fluent [43], which was coupled with our software package DETCHEM [42] for simulation of the chemical reactions.

First, CFD was used to ensure an even flow distribution over the microreaction chambers; that is, each reactor is loaded with the same mass flow. Therefore the flow field distribution in the reactor head and its inlet and outlets was calculated. The reactor geometry is as follows. The premixed reactants flow though a tube 1.6 mm in diameter and 5 mm long that enters the reactor head in the center. Owing to symmetry, only a quarter of the reactor head needed to be analyzed (Figure 23.3). The transition from the reactor head to a single-bead reactor with 96 microreaction chambers is achieved by membranes at the entrance to the reactor.

The grid generator GAMBIT [43] was used to construct an unstructured grid. Since the sizes of the reactor head and the membranes differ tremendously, approximately 200,000 vertices were needed to produce a computational grid that leads to a converged solution. This grid is shown in Figure 23.3.

Since the reactants are premixed and no reactions occur in the reactor head, the species mass equations are not solved. The membranes are modeled by the porous media model, which is provided by the CFD code Fluent [43]. The membranes consist of sets of thin vertical channels. The pressure drop over these membranes can be estimated by applying the Hagen–Poiseulle law and then used to derive the parameters internal resistency and permeability for Fluent's porous media model [43]. Since the pressure drop is also predicted by the CFD code, this value can be used as an additional empirical convergence check.

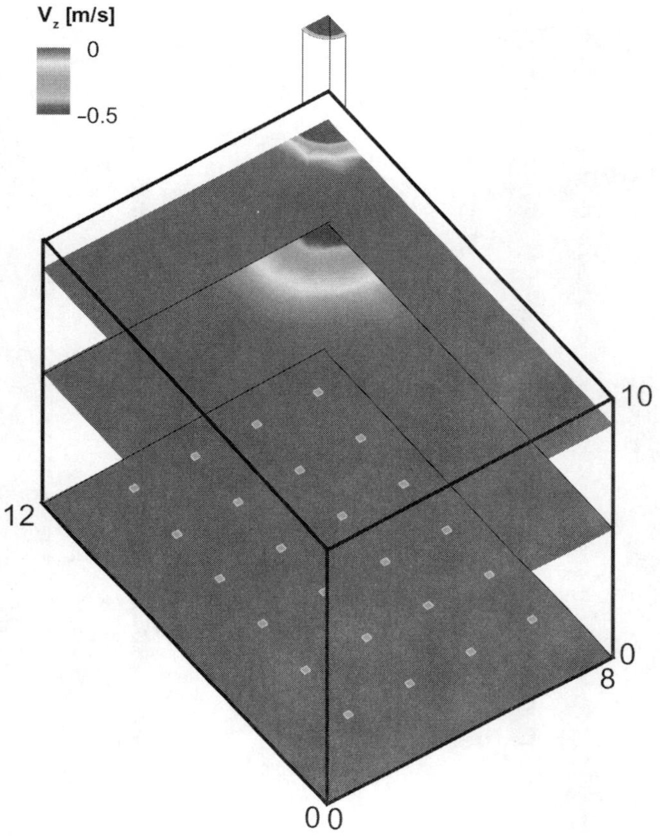

FIGURE 23.4. Vertical velocity at different horizontal planes in the reactor head (scales in millimeters). The actual minimum velocity is $-1.4\ \mathrm{m\,s^{-1}}$.

As expected, the simulation revealed that under the given conditions, and also within a wide range around those conditions (e.g., reactor geometry, membrane parameters, flow rate), the mass flow through all the membranes is the same. The maximum deviation from the mean value is less than 0.01%, which is within the numerical error. Figure 23.4 illustrates this fact by showing the vertical flow velocity for different levels inside the reactor head. It should be noted that the levels are shown in reverse order to give a better view of the interesting lowest level, where the fluid inlets to the reaction chambers are located.

The second point of interest was the description of the reactive flow field inside the microreaction chamber shown in Figure 23.9. The simulation is carried out for the region between wafers 2 and 4, that is, from the top of the inlet membrane to the bottom of the outlet membrane. The membranes are again treated as porous media. The heterogeneous reactions occur on the spherical bead inside the reaction chamber consisting of rectangular pyramid sections below the inlet and above the outlet membrane and in the cylindrical middle part. The bead diameter was varied in the simulation. Because of symmetry, only an eighth of the reactor needs to be simulated. Nevertheless 20,000 vertices forming

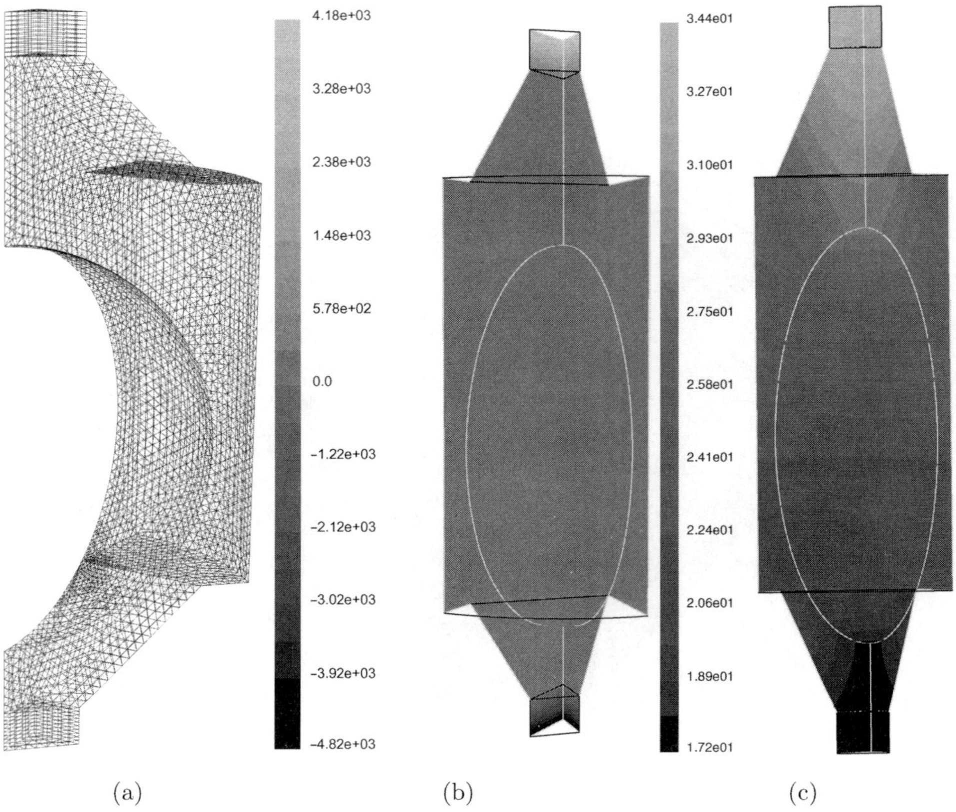

FIGURE 23.5. (a) Computational grid of the simulated section (one eighth) of the microreaction chamber. Only knots on the bead, reactor, and section walls are shown. (b) The calculated pressure drop (Pa) and (c) the CO mass fraction inside the microreaction chamber.

80,000 cells were generated using GAMBIT [43]. Lower spatial resolution did not lead to successfully converged solutions. As an example, the computational grid generated for a reactor with a bead of diameter 1 mm is shown in Figure 23.5(a).

The catalytic formation of butane from synthesis gas (CO and hydrogen) serves as an example reaction. The kinetics was arbitrarily set to first order in hydrogen and CO.

In the example discussed, the rate coefficient was set to a value of $10^{-2}\,\mathrm{m}^6\,\mathrm{mol}^{-1}\,\mathrm{s}^{-1}$ which ensures almost complete conversion. Under these conditions, a potential transport limitation would be most significant. The initial composition (mass fractions) is given as 0.344, 0.056, and 0.60 for CO, hydrogen, and argon, respectively. An isothermal reactor of 400 K was assumed, even though the reaction will release significant heat. The heat distribution in microreactors for high-throughput screening has also been studied and will be discussed in a future publication.

Figures 23.5(b) and 23.5(c) show the numerically predicted pressure drop and CO mass fraction for a bead with a diameter of 1 mm. The pressure drop in the reactor occurs almost exclusively over the two membranes; this also holds for larger beads. Strong vertical gradients of the CO concentration occur under the chosen conditions (high conversion). The single reaction chamber is far from being ideally mixed. Rather, it resembles a plug-flow reactor.

We have shown that CFD calculations can assist in the design of miniaturized reactors for high-throughput techniques. Different reactor concepts and varying reactor parameters and conditions can be tested much faster by using computational tools than by constructing a new reactor. For the concept discussed here, CFD simulations confirmed the even flow distribution over the single reaction chambers and revealed the processes inside the single-bead reactor. Since these were three-dimensional configurations and varying length scales, a complex computational grid had to be constructed, necessitating a very large computing effort when using complex kinetics.

Layout and Manufacture

The cross-section of a simple single-bead reactor layout is depicted in Figure 23.6. Wafers 1 and 2 are bonded together and represent the base part of the reactor; wafer 3 is the top part. The silicon wafers are structured by wet etching procedures such that microreaction chambers are formed the top and bottom parts are pressed together. These microreaction chambers consist of a cuboid and two frustums of a pyramid. Microstructured membranes are fabricated at the top and the bottom of the reaction chambers by deep reactive ion etching. One reaction chamber has an internal volume of appoximately 0.8 mm^3 and can hold a single catalytic bead. Beads of different sizes can be used; however, in this simple setup, no measures were taken to fix the position of the bead in the center of the reaction chamber.

Figure 23.7 shows the corresponding top view of the base part of such a reactor: A total of 105 reaction chambers were fabricated as a 15×7 arrangement on a very small reactor footprint. Excluding the outer margins used for sealing, the array density is approximately 60 reaction chambers per square centimeter.

A number of scanning electron microscope (SEM) of the microreaction chambers of this reactor type are shown in Figure 23.8. The dense arrangement of the chambers is obvious [Figure 23.8(a)]. The structure of the mircoreaction chambers of the base part can

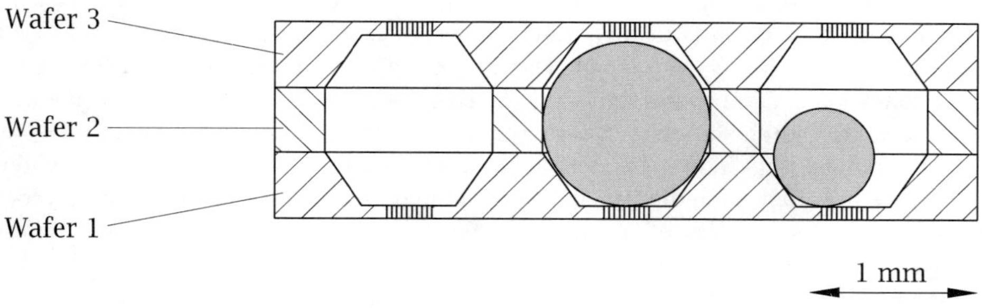

FIGURE 23.6. Partial cross-section of a simple single-bead reactor, showing three microreaction chambers without beads and with beads of different sizes in the reaction chambers.

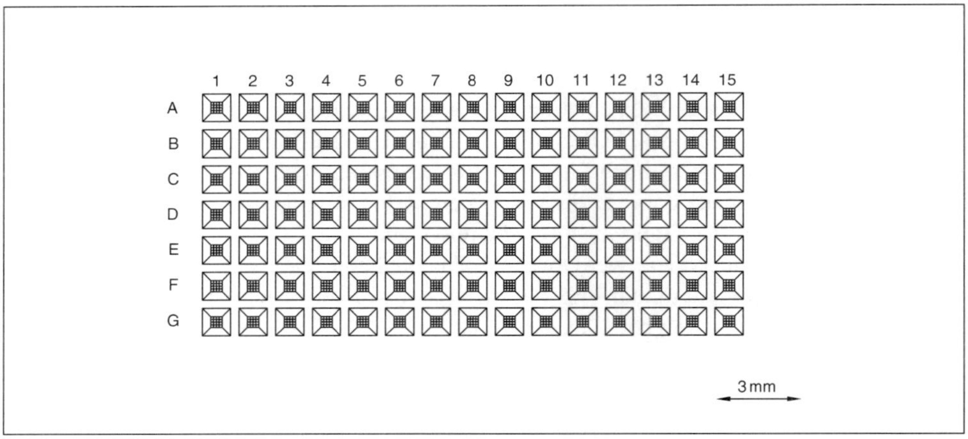

FIGURE 23.7. Top view of a simple 105-parallel single-bead reactor array. The positions of the reaction chambers are indicated.

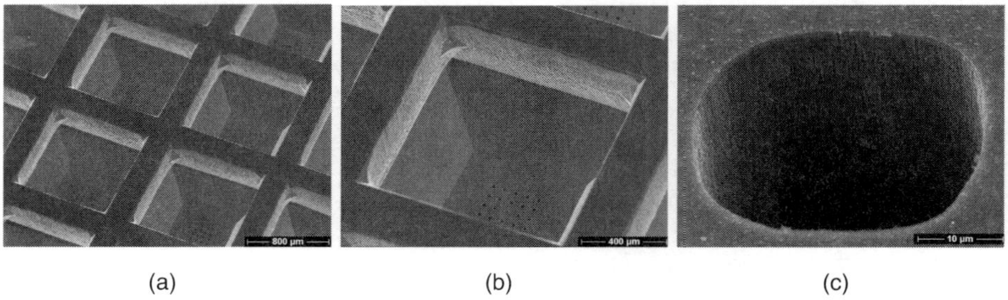

FIGURE 23.8. Scanning electron microscope (SEM) pictures illustrating the structure of a single-bead reactor consisting of a base part and a top part: (a), (b) reaction chambers in the base part; (c) single pore of a pore membrane.

be seen in Figure 23.8(b) as a stack of a cuboid and a frustum of a pyramid. The entrance to a single pore of a membrane is shown in Figure 23.8(c).

The partial cross-section of an advanced single-bead reactor is shown in Figure 23.9. The complete setup consists of four wafers, where wafers 1–3 form the base part of the reactor and wafer 4 forms the top part. The microreaction chambers are larger than in the simple setup and contain means for fixing the position of the bead. A number of possible designs can be envisaged; however, a simple frustum of a pyramid was chosen for compatibility with the well-known wet etching processes. Another advantage of the pyramid is its tolerance to variations in the size and form of the beads. A spherical bead cannot block the quadratic cross-section of the frustum. Furthermore, different bead sizes can be employed; however, the bead size is usually kept constant during an experiment. In addition, the beads are not in contact with the pore membranes.

Wafers 2 and 4 are processed identically to reduce fabrication costs. The additional wafer 1 contains a product withdrawal zone, which can be used for sampling with a narrow

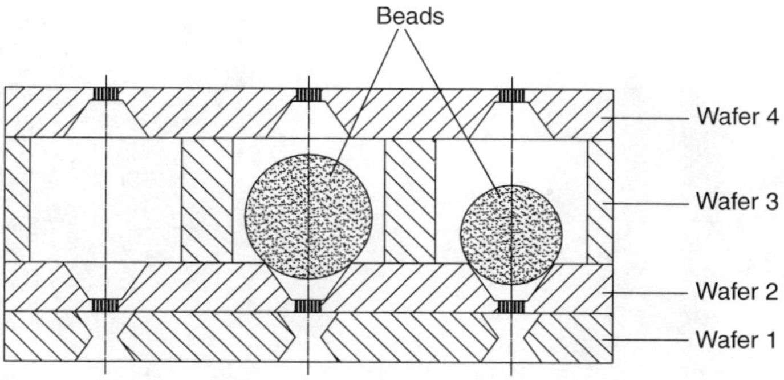

FIGURE 23.9. Advanced configuration of a single-bead reactor shown in partial cross-section with and without beads in the reaction chambers.

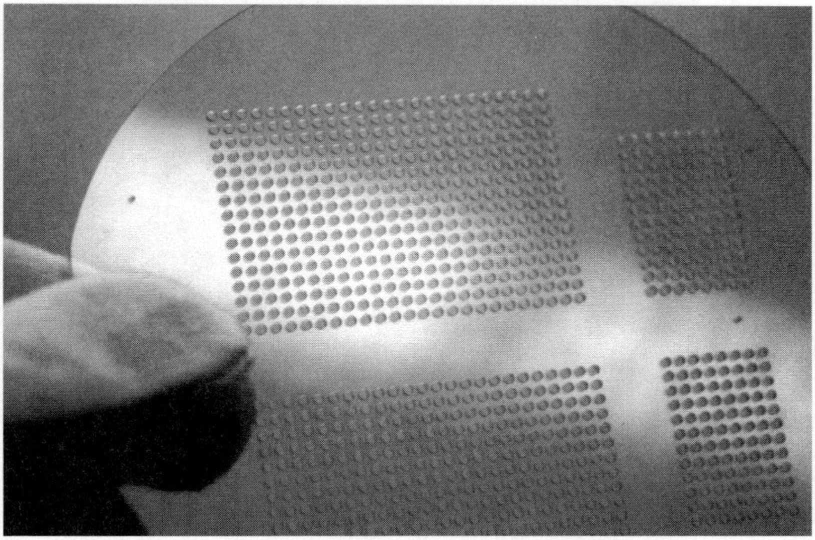

FIGURE 23.10. An ultrasonically micromachined Pyrex wafer forming part of the single-bead reactor. Two 384-parallel arrays and two 96-parallel arrays can be prepared on a single 100 mm Pyrex substrate.

capillary in order to allow a fast sequential analysis. Wafer 1 can be bonded to wafer 2 by suitable bonding technologies.

While standard silicon wafers with a thickness of 200–800 μm can be used for wafers 1, 2 and 4, wafer 3 is usually thicker and has cylindrical holes to increase the height of the reaction chambers. Therefore wafer 3 was constructed from Pyrex and the holes were fabricated by parallel ultrasonic drilling. A structured Pyrex wafer is shown in Figure 23.10. Two 384-parallel arrays as well as two 96-parallel arrays can be fabricated on a 100 mm substrate, illustrating once again the high array density of the single-bead approach. Photographs of the fabricated arrays of 105 and 384 parallel reactors are shown in Figure 23.11.

FIGURE 23.11. View of different 105-parallel and 384-parallel single-bead reactors.

Application and Results

Several sets of experiments were carried out to evaluate the performance of the different single-bead reactors in continuous flow catalytic experiments. Beads of 2% Pd–γ-Al_2O_3 with a diameter of 1 mm were prepared from γ-Al_2O_3 by wet impregnation and subsequent calcination at 450°C. The mass of one catalytic bead was approximately 700 μg.

Such Palladium catalysts are known to be active in hydrogenation reactions. Together with inactive reference beads, a 105-parallel single-bead reactor was charged as shown in Figure 23.12 and tested in the partial hydrogenation of 1,3-butadiene at 60°C. The base and top were pressed together using a suitable heatable flange system and mounted into a test rig. The reactants were continuously supplied by mass flow controllers and preheated to the reaction temperature before entering the reactor. The product analysis was carried out in a fast sequential mode by spatially resolved mass spectrometry, as described in detail later in this chapter. Figure 23.13 shows a section of the reactor base together with the sampling capillary in front of a specific product outlet.

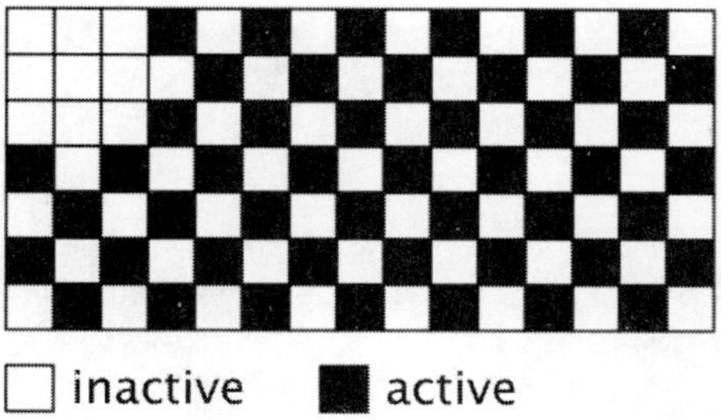

FIGURE 23.12. Arrangement of catalysts in a single-bead reactor for the validation of the reactor system: inactive catalyst, pure γ-Al_2O_3 beads; active catalyst, 2% Pd–γ–Al_2O_3 beads.

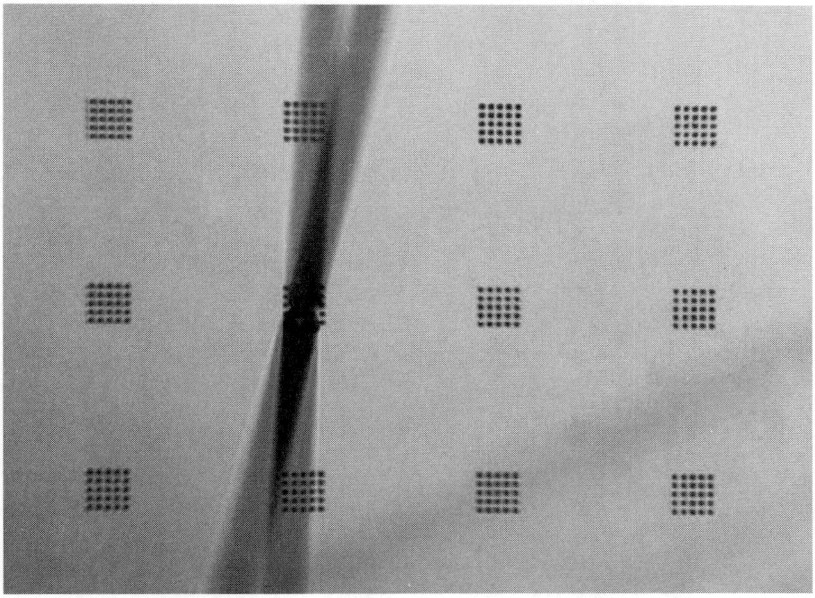

FIGURE 23.13. Product outlets in the base of the 105-parallel single-bead reactor and the capillary probe in front of a product outlet.

An example of the results of a screening are shown in Figure 23.14. At a residence time of about 10 ms, conversion degrees of approximately 50% are achieved on the active catalysts [Figure 23.14(a)]. In contrast, only very low conversions are measured for the inactive samples. This indicates that the cross talk between adjacent members of the library is almost negligible. Furthermore, only slight differences in the cumulative selectivities to the *n*-butenes are measured [Figure 23.14(b)].

The results for a residence time τ of 32.5 ms are shown in Figure 23.14(c). Compared with $\tau = 10.8$ ms, the resulting conversion degrees for the active catalysts are higher, but the conversion for the inactive reference catalysts is still very low. Nevertheless, a slight trend in the conversion degree from position B15 to F1 can be observed for both residence times. As identical Palladium catalysts from the same preparation batch were used, this trend is believed to arise from the reactor configuration. To exclude a potential reactant

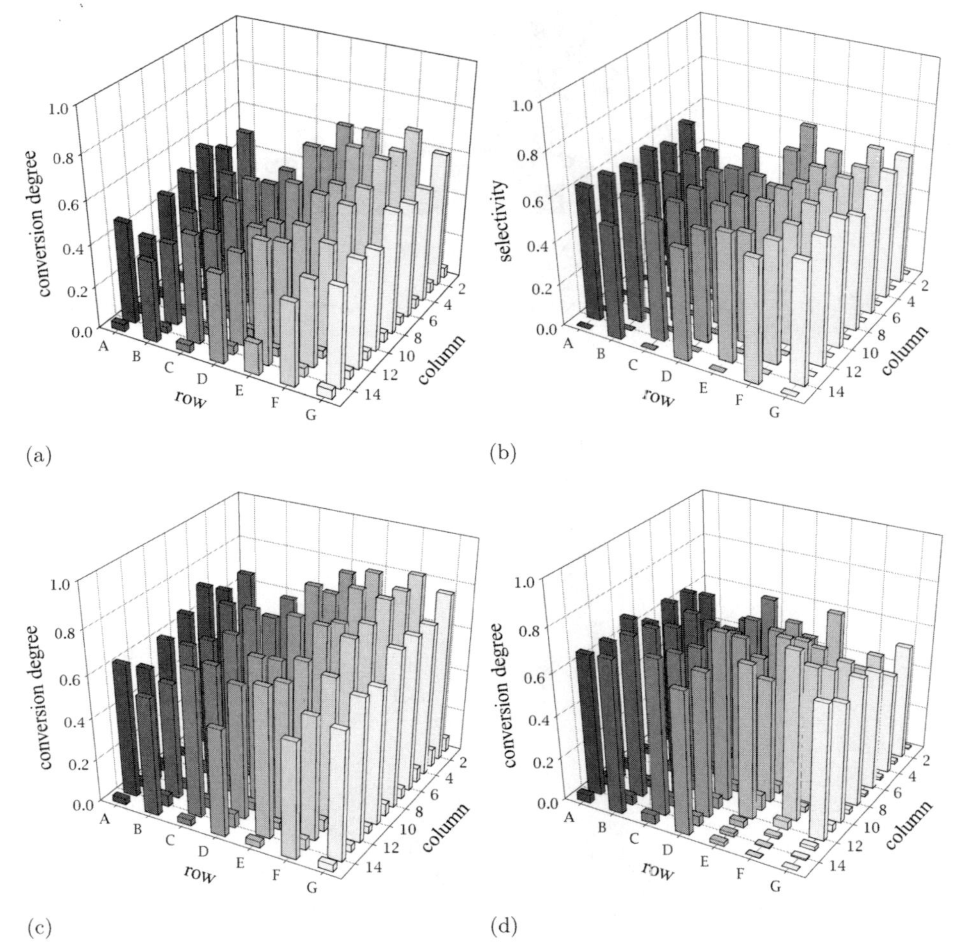

FIGURE 23.14. (a) Results from the high-throughput screening of identically prepared 2% Pd–γ-Al_2O_3 beads and inactive γ-Al_2O_3 beads in the partial hydrogenation of 1,3-butadiene. Reaction conditions: $T_R = 60°C$, 3 vol.% C_4H_6; 3 vol.% hydrogen; balance, argon. (a) Conversion degree, $\tau = 10.8$ ms. (b) Cumulative selectivity to *n*-butenes, $\tau = 10.8$ ms. (c) Conversion degree, $\tau = 32.5$ ms. (d) Conversion degree, $\tau = 10.8$ ms, reactor rotated by 180°.

maldistribution induced by the flange or the heating system, the library was rotated by 180° and the experiment was repeated under the same reaction conditions ($\tau = 10.8$ ms). The results are shown in Figure 23.14(d). Again, a trend in the conversion is observed, but now in the opposite direction.

The reason for this trend was identified by measuring the pore sizes of the single pores of the membranes across the reactor (Table 23.1). As can be seen, there are rather large deviations in the pore size and the resulting pore area, which determines the resistance to fluid flow. The pores in column 15 of the reactor are considerably larger than the pores in column 1. This means that the resistance to flow is higher in column 1, which leads to a maldistribution of reactants with lower flow rates and thus higher residence times per

TABLE 23.1 Variation in Pore Size in the 105-Parallel Single-Bead Reactor

	Position	a_{min}	a_{max}	b_{min}	b_{max}	A_{pore}	ΔA_{pore}
Base	A 1	0.92	0.96	1.00	1.08	0.98	−0.02
	G 1	1.02	1.08	0.96	1.12	1.09	0.09
	A 15	1.08	1.15	1.08	1.19	1.27	0.27
	G 15	1.05	1.15	1.00	1.15	1.19	0.19
Top	A 1	0.89	0.92	0.85	0.96	0.82	−0.18
	F 1	0.86	0.92	0.85	0.92	0.79	−0.21
	G 1	0.86	0.92	0.85	0.85	0.76	−0.24
	A 15	1.02	1.04	0.96	1.12	1.07	0.07
	B 15	1.02	1.08	1.04	1.12	1.13	0.13
	G 15	0.95	1.08	0.96	0.96	0.98	−0.02

reaction chamber in column 1. As indicated by the results shown in Figures 23.14(a) and 23.14(c), longer residence times lead to an increase in conversion. Thus the observed trend in the conversion degree from one region of the reactor to another can be explained by the differences in the pore sizes of the pore membranes. Finally, we determined that the pore size differences resulted from the use of an inexpensive emulsion mask process for fabricating the pore membranes.

A major improvement in terms of flow distribution could be achieved by the more accurate fabrication process that was used to develop 384-parallel single-bead reactors with a 24×16 format. Before using this reactor in catalytic reactions, the pore sizes were determined by optical microscopy. The pore size distribution (Figure 23.15) indicates excellent pore uniformity.

As an example, the 384-parallel single-bead reactor was applied in a partial oxidation reaction. In an initial experiment, the reactor was filled with inactive and active catalysts according to the pattern shown in Figure 23.16(a). A continuous flow screening experiment was carried out at 400°C and a reactant flow V of 1 ml min^{-1} per bead, again using spatially resolved mass spectrometry for sequential product analysis. The normalized results for the conversion degree are shown in Figure 23.16(b). According to the mass spectrometric intensities, white colors represent low conversions of the hydrocarbon and dark colors represent high conversions. The maximum conversion in this experiment is approximately 60%. As can be seen, the results for the conversion degree are in good agreement with the filling pattern of the reactor. Cross talk between adjacent reaction chambers can be observed to some minor extent, although it does not limit the applicability of the single-bead reactor for primary screening purposes. Furthermore, it is expected that cross talk can be reduced by bonding the base and the top of the reactor, instead of just pressing them together as presented here.

It can be concluded that the single-bead reactor concept provides a novel system architecture for the high-throughput primary screening of solid catalysts. By using single beads as catalytic materials and a scalable method for the fluid distribution, array densities of up to 60 catalysts per square centimeter have successfully been applied. This array density vastly exceeds those used by other groups. Thus it is possible to realize parallel reactors for heterogeneous catalysis with a much higher degree of parallelization as shown successfully with the 384-parallel single-bead reactor.

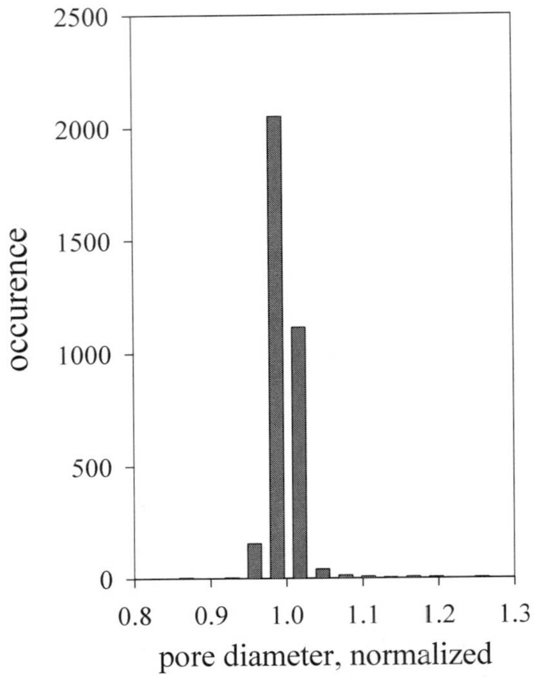

FIGURE 23.15. Pore size distribution of pore membranes prepared by a deep reactive ion etching process on a silicon wafer.

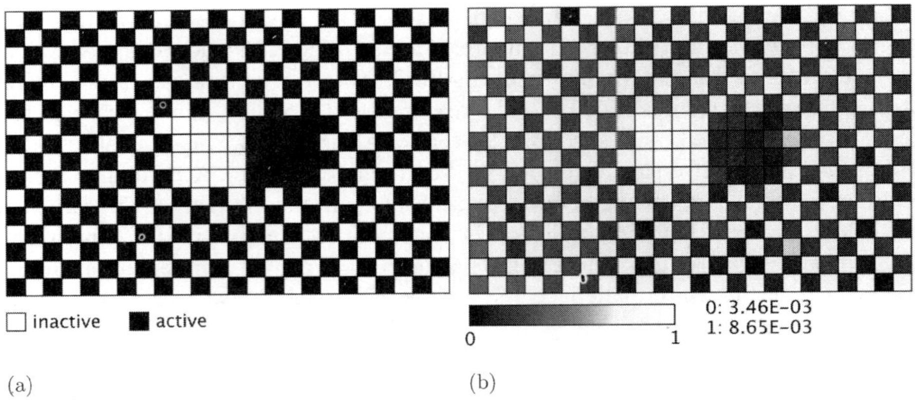

FIGURE 23.16. Results of high-throughput screening of catalysts in a 384-parallel single-bead reactor in a partial oxidation reaction: (a) arrangement of inactive and total oxidation catalysts in the reactor; (b) screening results for the conversion of a hydrocarbon at 400°C and a reactant flow of 1 ml min^{-1} per bead.

The multifunctional reactor architecture developed enables handling, identification, and storage of the catalysts, and the automation and integration of the complete high-throughput workflow. This includes the possibility of adapting different sequential or parallel analysis techniques for the detection of interesting products. The versatility of the system facilitates its application to different classes of reaction. Operation at conditions relatively close to conventional and/or secondary catalyst test systems can be ensured.

Novel Preparation Techniques for the Single-Bead Reactor Concept

As a consequence of the increasing parallelization and integration of reactor systems, requirements for synthesis methodologies are becoming more extreme. The synthesis of minute amounts of material (i.e., different multicomponent mixed oxides in the microgram or milligram range) must be reproducible and fully automated. Meeting these challenges demands a combination of different expert skills in inorganic synthesis, robotics, automation, and related fields.

Related to the previously unmet degree of parallelization possible with the single-bead reactor concept, new methods are needed to avoid bottlenecks in the preparation of the catalyst libraries. Of course, it may be possible to apply highly parallel preparation methods for single beads (e.g., by suitable micropipetting robots) where each bead is placed in a separate well on a synthesis platform. However, the resources needed to install such a work flow scale at least linearly with the number of beads to be prepared in a given timeframe. Furthermore, only simple catalyst preparation methods like wet impregnation seem to be manageable on such a small scale.

Several strategies for the preparation of large diverse molecular libraries are well known in combinatorial approaches in organic and bio-chemistry. One of the most powerful concepts for the generation of very large libraries is the so-called "Split & Pool" technique [44,45] (Figure 23.17). This technique provides an effcient strategy for assembling

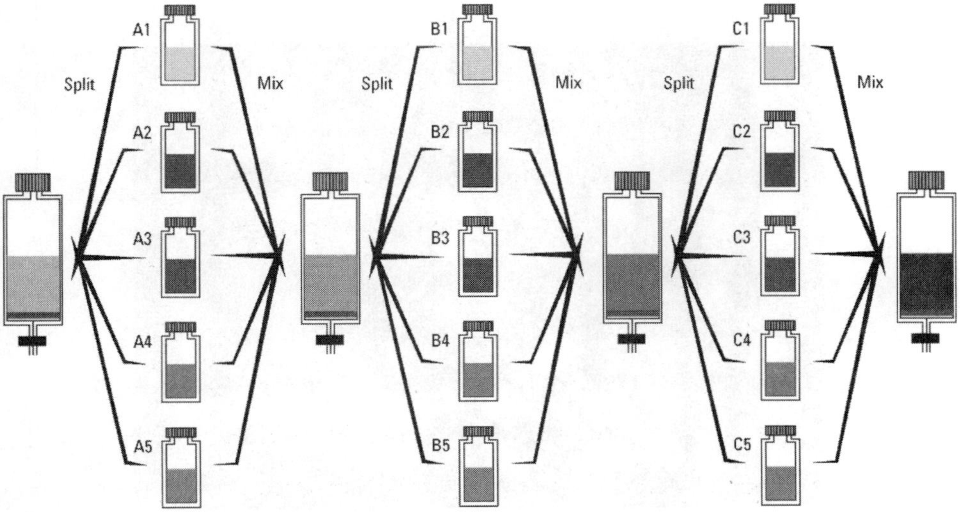

FIGURE 23.17. Principle of the Split&Pool method [52].

combinatorial libraries. The solid support is divided into portions, each of which undergoes a reaction with a single building block. Pooling of these portions results in a single batch of solid support bearing a mixture of components. Repetition of the divide, couple, and recombine processes results in a library where each discrete particle of solid support carries a single library member, and the number of members is equal to the product of the number of building blocks incorporated at each step.

Furthermore, libraries prepared by Split & Pool are fully combinatorial libraries containing all possible combinations of building blocks. Therefore it is possible to produce highly diverse libraries containing 10^4–10^8 samples with a very simple work flow.

Using single beads as basis of the single-bead reactor concept enables us to adopt equivalent strategies for the preparation of inorganic material libraries, even if the preparation of such libraries presents different challenges than those encountered in the organic realm so that a self-contained independent methodology is required. For instance, if wet impregnation is used as preparation method, the percentage loading of a support has to be controlled precisely, the resulting elemental distributions have to be controlled, and other factors involving the solid state materials have to be taken into consideration. Furthermore, the application of such pooled synthesis techniques in inorganic chemistry reaches far beyond the elemental/compositional diversity known from Split & Pool for molecular entities. In inorganic chemistry, the *diversity* parameters include not only compositions but also precursors, supports, pre- and post-treatment, etc.

Split & Pool, perhaps carried out via special routes such as "directed sorting" [47,48] or "direct divide" [49], provides a very powerful method for generating diverse libraries of inorganic materials. An example of part of a library is shown in Figure 23.18.

Further prospects, implications, and results from the application of this method for combinatorial material science are published elsewhere [50].

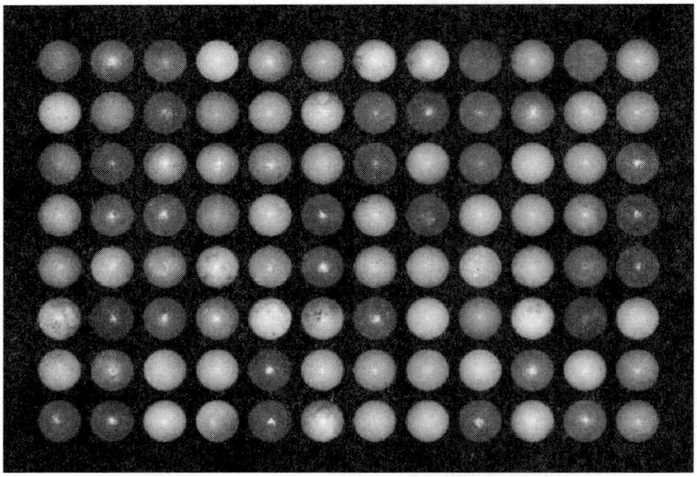

FIGURE 23.18. A 96-parallel single-bead reactor filled with beads prepared by inorganic Split & Pool.

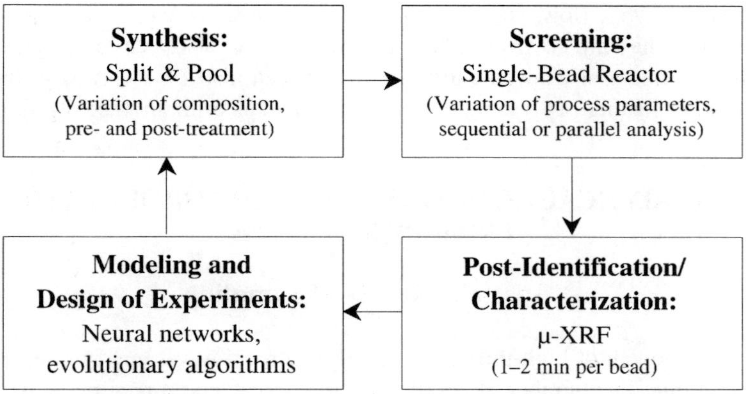

FIGURE 23.19. Simplified work flow for primary screening in heterogeneous catalysis applying the "single-bead" concept.

Work Flow for the Screening of Bead Catalysts

To avoid bottlenecks in the screening work flow, the different elements of a high-throughput research program, especially in primary screening, need to be tightly integrated. Therefore the single-bead reactor concept is embedded in the closed-loop work flow shown in Figure 23.19. Starting with the Split & Pool-method, suitable libraries of single beads can be prepared very effciently. This preparation includes the variation of compositional parameters as well as process parameters in potential pre- and post-treatment steps, such as calcination or steaming. The resulting library is a batch of different materials, where the identity of a single bead is usually not known unless tagging is carried out during the sequential preparation steps.

In the next step, the pooled library is transferred to the single-bead reactor, where each microreaction chamber is filled with one bead. Thus the elements of the former pool are now individually spatially addressable in the reactor. This combination of pooled and parallel strategy allows the acquisition of performance data for each bead individually in the next processing steps. First, the catalysts are tested under continuous flow and steady state conditions in the single-bead reactor, where a set of different process conditions (reactant concentrations, residence time, temperature, etc.) is usually applied. At this stage, spatially resolved fast sequential or parallel analysis methods can be used to characterize the product composition and to identify hits.

Then the composition of a selection of or all the individual beads is determined by X-ray fluorescence analysis with a focused beam (μ-XRF), which allows the identification of the elemental composition according to the synthesis procedure. The elemental composition of the beads can be used for tagging and also contains information about the synthesis sequence that was applied to an individual bead. Depending on the time needed for this analysis, only a subset of the library is analyzed to avoid bottlenecks.

The data generated and acquired in the previous steps is submitted to data analysis and modeling in order to generate recommendations and predictions for the preparation of the next library generation. Here, neural networks and evolutionary algorithms are used.

Beyond this work flow, single-bead reactor architecture can be used as "integrated materials chip." The chip serves not only as a reactor, but also as a sample holder for the μ-XRF characterization. Conveniently, the library can also be archived in the chip and reused later after further processing or regeneration steps or in another reaction.

ANALYTICAL CONCEPTS FOR HIGH-THROUGHPUT PRIMARY SCREENING

Spatially Resolved Sampling

A key component of high-throughput catalyst screening systems is a suitable product analysis technique. The number of parallel reactors in catalyst screening is, in most cases, limited by the speed of the product analysis.

Therefore a sequential sampling device was developed [21], shown schematically in Figure 23.20. The device consists of a capillary probe, which can be micropositioned in xyz direction within the parallel reactor configuration. The capillary probe transfers the samples to an appropriate analytical instrument [e.g., a mass spectrometer (MS), a gas chromatograph (GC), or a GC-MS]. In the present study, a conventional commercially available MS (Hiden Analytical, Warrington, UK) was used. The gas samples were continuously transferred to the MS through the capillary, allowing fast on-line analysis.

The capillary can be positioned with high accuracy under hazardous chemical conditions and high thermal demands. The parameters and the position of the capillary, which can be adjusted and monitored by a CCD camera equipped with magnifying optics, are

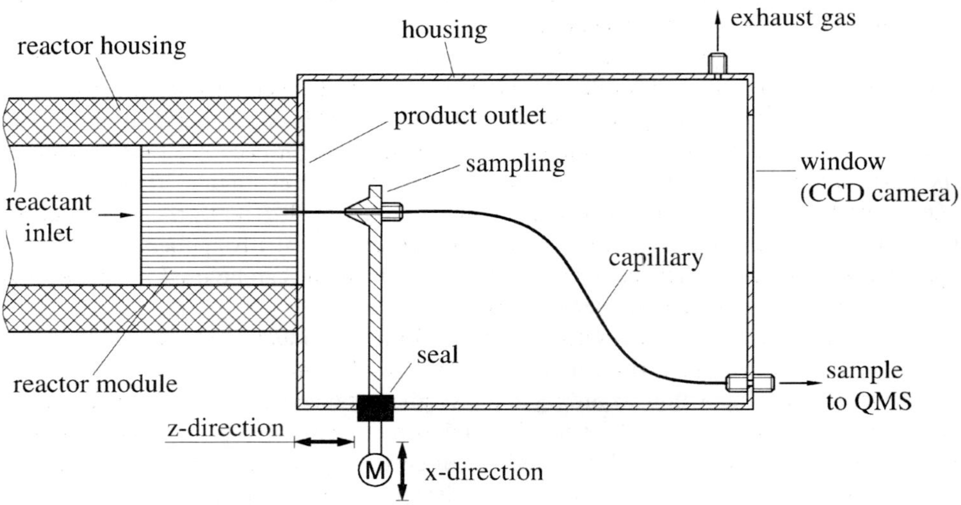

FIGURE 23.20. Sketch of the sampling device developed for spatially resolved gas analysis. The gas samples are taken sequentially by the *xyz*-positioned capillary probe and transferred to the analytical instrument (e.g., a mass spectrometer).

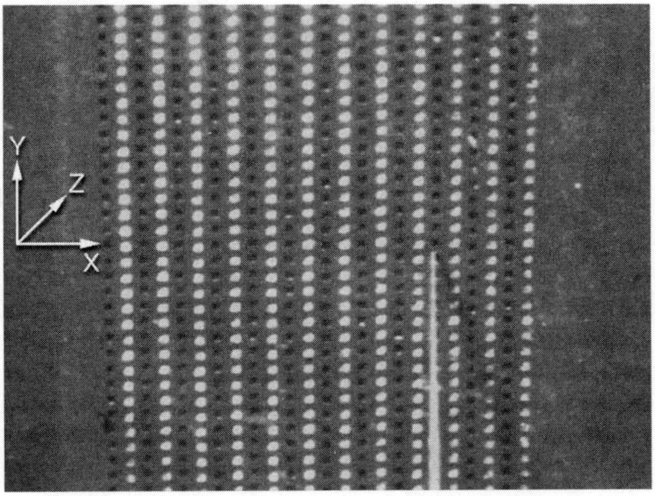

FIGURE 23.21. The microchannel array and the capillary probe of the sampling device.

freely programmable. The materials of the housing and the seal were carefully chosen to withstand temperatures up to 450°C. The speed of analysis can be determined by choosing the right MS conditions and the application of short narrow transfer capillaries. Depending on the nature of the products to be investigated, analysis times of less than 60 s can easily be achieved.

In primary screening, miniaturization is essential to increase the number of catalysts tested. The question was how far the catalytic reactors could be miniaturized while maintaining a good analysis speed and sample purity with the sampling technique described. An arrangement of microchannels was used to investigate the spatial and analytical resolution of the instrument. The outlet of the microchannel array consisted of a very large number of parallel microchannels with a cross-section of 100 μm × 100 μm on a pitch of only 200 μm. Each of these microchannels delivered a separate gas stream. The composition the substreams was analyzed using this spatially resolved analysis technique. A CCD picture of the channel outlets is shown in Figure 23.21 where lines of light and dark channel outlets can be observed resulting from light reflection due to different outlet angles. During an experiment, microchannels with light channel outlets deliver argon while those with black channel outlets deliver nitrogen. The capillary scanning the microchannels at a distance of 50 μm can be seen on the right-hand side of Figure 23.21.

A sampling capillary with an outer diameter of 30 μm and an inner diameter of 20 μm is used. In a typical experiment, the capillary scanned the fluid composition at different positions and distances from the channel outlets. The MS continuously analyzed the samples at a rate of about three samples per second, resulting in about 60,000 data sets per experiment. The fluid temperature was kept constant at 25°C.

A typical result obtained by scanning of the microchannel array in the x direction at a defined distance of 50 μm is shown in Figure 23.22. The diagram shows the molar concentration of argon in the fluid versus the capillary position. Strong variations in the argon

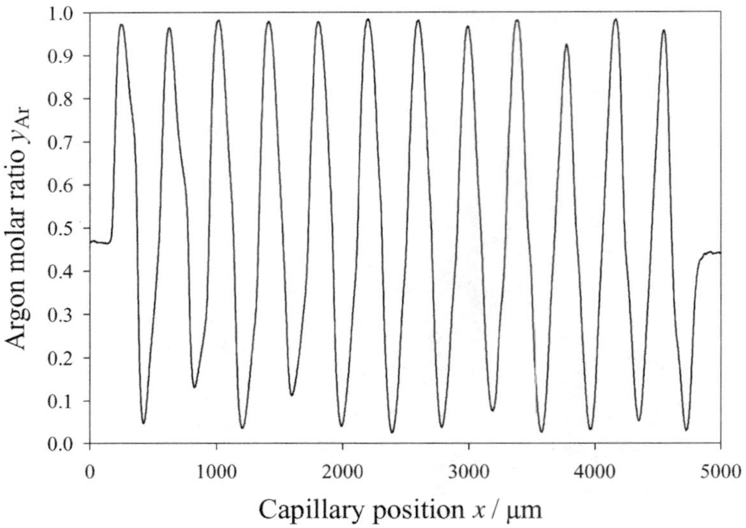

FIGURE 23.22. Screening result at $\dot{V} = 5{,}000$ ml min^{-1} per fluid, a capillary distance of 50 µm from the channel outlets, a capillary velocity of 2.5 µm s^{-1}, and a temperature of 25°C.

concentration can be observed. The variations are very regular, representing the structure of the microchannel array at the outlet. When the capillary reaches the front of a microchannel delivering argon, the argon concentration increases to a value of almost unity. Similarly, the argon concentration decreases to almost zero when the capillary moves to a microchannel with a nitrogen substream. This illustrates that even an extremely miniaturized reactor array on a pitch of 200 µm can be screened by means of the sampling capillary while still yielding almost pure samples.

It is obvious that parallel gas streams may mix very quickly when flowing into a common fluid chamber as an array of such microchannels acts as static micromixer. This is illustrated in Figure 23.23. When the microchannel array is scanned at a distance of 550 µm from the outlets, the gas streams are almost completely mixed so that separate analysis of the streams is impossible. Consequently, the probe should be positioned as close as possible to the outlet of the corresponding reaction chamber. Furthermore, it may prove invaluable to take additional precautions to avoid mixing the product stream with adjacent streams at the sampling position by carefully designing the reactor outlets.

Based on our experience, a combination of the sampling device with an appropriate analysis technique can be regarded as an almost universal sequential analysis tool for gas phase reactions in combinatorial catalysis. The device can be used in combination with conventional commercially available analytical equipment (e.g., *MS, GC*, and *GC-MS*) and with almost every reactor configuration and application requiring temporally and spatially resolved sampling and the analysis of complex gaseous mixtures. In several applications, such an *xyz*-sampling device can present a significant advantage compared to channel selection via multi-position valves.

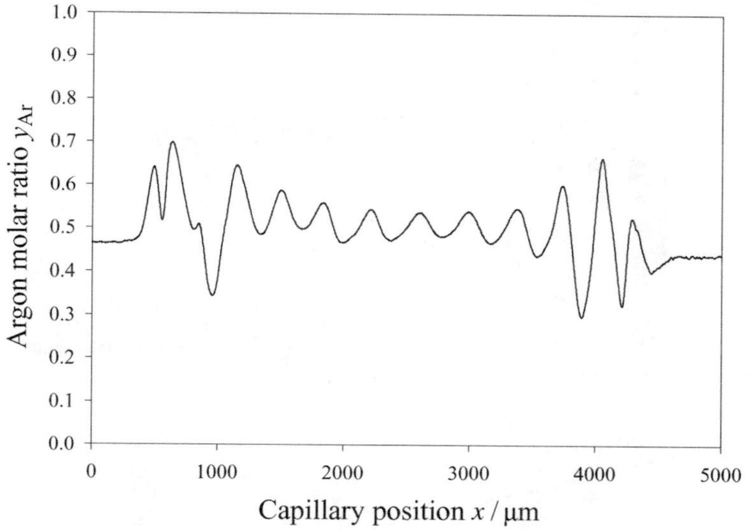

FIGURE 23.23. Screening result at $\dot{V} = 5{,}000$ ml min^{-1} per fluid, a capillary distance of 550 μm from the channel outlets, a capillary velocity of 2.5 μm s^{-1}, and a temperature of 25°C.

Parallel and Quasiparallel Product Analysis by Photoacoustics

Sequential analysis may be too slow for the reliable comparison of different catalysts in a primary screening involving several hundred catalysts operated in parallel. Therefore an effective test method has to be real-time parallel or quasiparallel to accumulate information about activity of all catalysts at the same time. We have developed a parallel detection system which is based on the photoacoustic effect [51]. It is based on the detection of pressure pulses, which are generated by exciting a selected type of molecule with a laser pulse. The wavelength of the laser determines which molecules are excited, because they must have an absorption band at this frequency. The molecule absorbs energy and produces a pressure pulse (if the laser is pulsed or modulated) by internal radiationless conversion of the energy into molecular translation. The intensity of the pressure pulse is correlated with the concentration of the product.

In order to demonstrate the versatility of this system, we evaluated this method in two different setups, the real parallel "free-field setup" and the quasiparallel "resonance setup."

Free-Field Setup

For molecules with a high extinction coeffcient and therefore strong light–matter interactions we developed a cell arrangement for parallel on-line detection of up to eight gas flows, which can in principle be expanded to more channels (Figure 23.24). Information on the concentration is obtained from the intensity of the pressure pulse, and information on the position of the channel is obtained by measuring the time delay

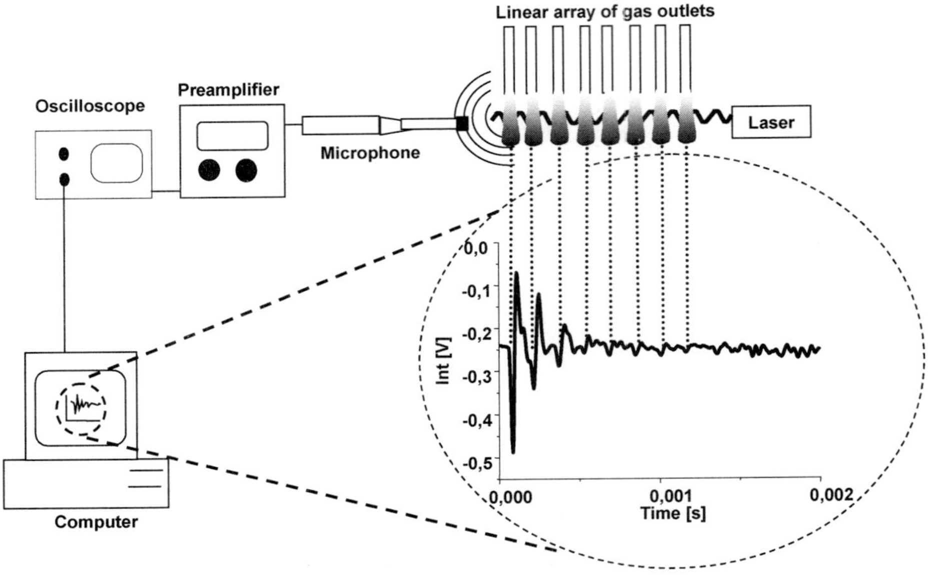

FIGURE 23.24. Free-field setup for the parallel detection of products from parallel experiments by photacoustics.

between the laser pulse and the pressure pulse detected at the microphone. The spatial resolution of the setup depends on the damping of the microphone and the length of the excitation pulse. A spatial resolution in the centimeter range can be achieved with this setup. In our model reaction system, the oxidative dehydrogenation of ethane, we could easily detect ethene in an ethane–air–water background because of the strong nonoverlapping band of ethene at around $950\,cm^{-1}$. The gas is excited by a modulated DEOS 25 W CO_2 laser which was operated at 10 or 100 Hz and a pulse length of 35 μs. The pressure wave generated, which contains the information about the position and the concentration of ethene in all gas flows at the same time, is detected by a commercial condenser microphone. The signal from the microphone was recorded on an oscilloscope and then read out by a computer. For the free-field setup, a linear correlation of product concentration and signal intensity was proven experimentally.

To improve the signal-to-noise ratio, the oscilloscope was operated in an averaging mode, adding the response of 256 single laser pulses. An automatic peak-detection and integration routine produced a direct illustration of the conversion to ethane on-line on screen. We attached this detection cell to a 16-channel parallel reactor and were able to accumulate the complete data set of concentrations of ethene in eight gas flows within 3 s. The long measurement time is due to the averaging of 256 spectra for one measurement and the communication between the amplifier and the computer. Physically, one measurement takes only a few milliseconds.

The time difference between the signals gives a maximum spatial resolution of less than 3 cm, which could be improved by further reducing the pulse length of the laser to

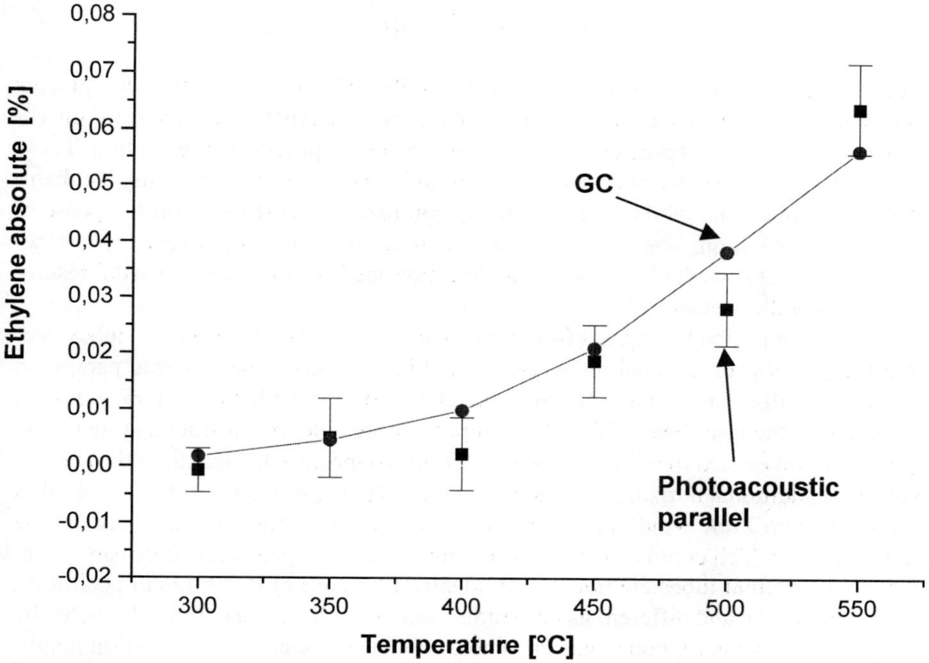

FIGURE 23.25. Analysis results from a photoacoustic parallel detection compared with a single-channel reactor with GC analysis.

reduce the propagating pressure pulse to a minimum and increasing the sensitivity of the microphone and its damping characteristics. However, this resolution was sufficient for our purpose of evaluating the applicability of the free-field measurement, since the outlets were arranged in a line with equal spacing of 5.3 cm between neighboring gas channels. For the best spatial resolution, the damping of the microphone has to be strong enough to reduce the oscillation to the offset level within the time that the pressure pulse of the next channel is detected, which is especially important for the channels close to the microphone where high pressure intensities are typically recorded. For the channels which are further away from the microphone, the damping should be low enough to detect the pulse which is reduced by $1/r^2$. We investigated different filters and found that a filter with low level of 20 Hz and an upper level of 30 or 100 kHz was best for the detection of all pulses.

When operating the system in an averaging mode, the lower detection limit is below 0.015% for the channels close to the microphone, and 0.3% ethane can be detected even at the remote channels where the intensity is decreased by $1/r^2$. Several small libraries of mixed oxide catalysts were analyzed using the photoacoustic detection method and the results were compared with those obtained in a single-channel reactor equipped with GC analytics [51]. The agreement was excellent (Figure 23.25), which proves that the method can be used for the reliable analysis of strongly absorbing molecules in a truly parallelized fashion in the free-field setup. No special requirements with respect to noise reduction in the environment are necessary.

Resonance Setup

The free-field setup cannot be used for molecules with low extinction coefficients because of the low intensity of the pressure pulses which cannot be detected by microphones with the damping characteristics necessary for the required time resolution. These very fast microphones usually have low detection limits. However, by increasing the diameter of the microphone membrane one can easily optimize the sensitivity at the expense of decreased time resolution. Therefore for CO oxidation processes with a very low extinction coeffcient of CO_2 we had to change the detection method and give up spatial resolution and truly parallel detection.

Instead we amplified the signals from each channel by means of resonance tubes integrated at the end of each channel in the gas outlet [52]. These can be arranged parallel in one line and therefore are suitable for quasiparallel operation without complex guidance and alignment of the laser beam. With this setup, we were able to construct a system containing up to 16 tubes next to each other within 50 cm. To increase the length of the absorption path short additional horizontal tubes were attached to the resonance tube. These tubes were equipped with ZnSe windows on one side to avoid cross-contamination between the different channels which could result from diffusion out of the open tubes. Resonance conditions for the different tubes are not exactly identical because of the different conditions (i.e., beam divergence and different gas mixtures) which needed to be corrected. Each tube was equipped with a microphone, and the whole array was connected to a lock-in amplifier which was sequentially switched from channel to channel. By operating the amplifier in a phase-angle-dependent mode, we could suppress the phase of a second superimposing constant wave generated by the absorption of the tubes and detect only the signal generated by the product molecules.

With this system attached to a 16-parallel reactor, we were able to monitor 16 CO_2 concentrations in less than 1 min. The rate-limiting factor for these measurements is the dead-time of the switching process due to the lock-in amplifier, and not the analysis itself which is completed in less than 1 s. The reliability of this resonance detection method was checked against results from a single-channel reactor equipped with non-dispersive IR analysis and found to be in excellent agreement [51].

If there is one product which provides a characteristic absorption band in a frequency range which is accessible by laser sources, the photoacoustic method described here is a fast alternative to other methods like GC or MS. Of course, photoacoustic detection only allows information to be obtained about the absorbing species, whereas the whole product spectrum can be analyzed with GC or MS. For routine applications one has to choose either fast real-time detection (typically in the discovery mode) or a method delivering information about the complete product distribution which is more typical of the optimization mode. The shortcomings of the photoacoustic detection are possible overlap of absorption bands of different molecules and the fact that in the IR, which is the most convenient frequency range to use since many specific relatively narrow absorption bands are present, high-power laser sources are limited. However, the whole frequency range in the mid-IR can be covered with solid state lasers, albeit at reduced intensity and a high cost.

SUMMARY

As a result of an interdisciplinary research effort, we have developed a novel concept for high-throughput primary screening in heterogeneous catalysis. Based on the idea of using individual shaped bodies or "beads," as the catalytic materials of interest, a new parallel reactor type was developed. This microfabricated "single-bead reactor" can accomodate large numbers of catalytic beads in individual microreaction chambers. By using CFD calculations it was possible to achieve the design goal of a very compact arrangement of reaction chambers together with an intrinsic scalable method for equal reactant distribution. Therefore a 384-parallel single-bead reactor could be realized as a chip-based device with an overall size less than half that of a credit card. Several experiments proved the applicability of the new reactor concept in partial hydrogenation and partial oxidation reactions.

The single-bead concept allows invaluable flexibility in terms of library preparation. An outstanding feature is the possibility of applying the "Split & Pool" strategy for the synthesis of large diverse catalyst libraries in high-throughput discovery programs in heterogeneous catalysis. By applying combination of a pooled synthesis strategy and a parallel screening methodology, a very effcient automated work flow was established.

Furthermore, suitable product analysis techniques for primary screening were developed. Among the fast sequential methods, spatially resolved sampling combined with mass spectrometry proved to be a versatile tool for fast and selective analysis in very compact reactor configurations. An even higher speed of analysis can be reached by applying photoacoustic detection in a real-time parallel or quasiparallel mode.

These technologies have a number of potential applications in high-throughput experimentation in general, and primary screening in particular.

REFERENCES

1. Lebl, M. Parallel personal comments on classical papers in combinatorial chemistry. *J. Comb. Chem.* **1999**, *1* (1), 3–24.
2. Hanak, J. J. The "multi-sample concept" in materials research: Synthesis, compositional analysis and testing of entire multicomponent systems. *J. Mater. Sci.* **1970**, *5*, 964–971.
3. Milberger, E. C. *Automatic Catalyst Screening Unit*, US Patent 4,099,923, to The Standard Oil Company, **1977**.
4. Perez-Ramirez, J., Berger, R. J., Mul, G., Kapteijn, F., Moulijn, J. A. The six-flow reactor technology—A review on fast catalyst screening and kinetic studies. *Catal. Today* **2000**, *60*, 93–109.
5. Xiang, X.-D., Sun, X., Briceño, G., Lou, Y., Wang, K.-A., Chang, H., Freedman, W. G. W., Chen, S.-W., Schultz, P. G. A combinatorial approach to materials discovery. *Science* **1995**, *268*, 1738–1740.
6. Moates, F. C., Somani, M., Annamalai, J., Richardson, J. T., Luss, D., Willson, R. C. Infrared thermographic screening of combinatorial libraries of heterogeneous catalysts. *Ind. Eng. Chem. Res.* **1996**, *35*, 4801–4803.
7. Senkan, S. M. High-throughput screening of solid-state catalyst libraries. *Nature* **1998**, *394*, 350–353.
8. Holzwarth, A., Schmidt, H.-W., Maier, W. F. Detection of catalytic activity in combinatorial libraries of heterogeneous catalysts by IR thermography. *Angew. Chem. Int. Ed.* **1998**, *37* (19), 2644–2647.
9. Cong, P., Doolen, R. D., Fan, Q., Giaquinta, D. M., Guan, S., McFarland, E. W., Poojary, D. M., Self, K., Turner, H. W., Weinberg, W. H. High-throughput synthesis and screening of combinatorial heterogeneous catalyst libraries. *Angew. Chem. Int. Ed.* **1999**, *38* (4), 483–488.

10. Jandeleit, B., Schaefer, D. J., Powers, T. S., Turner, H. W., Weinberg, W. H. Combinatorial materials science and catalysis. *Angew. Chem. Int. Ed.* **1999**, *38* (17), 2494–2532.
11. Newsam, J. M., Schüth, F. Combinatorial approaches as a component of high-throughput experimentation (HTE) in catalysis research. *Biotechnol. Bioeng. (Comb. Chem.)* **1998/1999**, *61* (4), 203–216.
12. Orschel, M., Klein, J., Schmidt, H.-W., Maier, W. F. Detection of reaction selectivity on catalyst libraries by spatially resolved mass spectrometry. *Angew. Chem. Int. Ed.* **1999**, *38* (18), 2791–2794.
13. Senkan, S., Krantz, K., Ozturk, S., Zengin, V., Onal, I. High-throughput testing of heterogeneous catalyst libraries using array microreactors and mass spectrometry. *Angew. Chem. Int. Ed.* **1999**, *38* (18), 2794–2799.
14. Hoffmann, C., Wolf, A., Schüth, F. Parallel synthesis and testing of catalysts under nearly conventional testing conditions. *Angew. Chem. Int. Ed.* **1999**, *38* (18), 2800–2803.
15. Hoffmann, C., Schmidt, H.-W., Schüth, F. A multipurpose parallelized 49-channel reactor for the screening of catalysts: Methane oxidation as the example reaction. *J. Catal.* **2001**, *198*, 348–354.
16. Senkan, S. M., Ozturk, S. Discovery and optimization of heterogeneous catalysts by using combinatorial chemistry. *Angew. Chem. Int. Ed.* **1999**, *38* (6), 791–795.
17. Liu, Y., Cong, P., Doolen, R. D., Turner, H. W., Weinberg, W. H. High-throughput synthesis and screening of V–Al–Nb and Cr–Al–Nb oxide libraries for ethane oxidative dehydrogenation to ethylene. *Catal. Today* **2000**, *61*, 87–92.
18. Reddington, E., Sapienza, A., Gurau, B., Viswanathan, R., Sarangapani, S., Smotkin, E. S., Mallouk, T. E. Combinatorial electrochemistry: A highly parallel, optical screening method for discovery of better electrocatalysts. *Science* **1998**, *280*, 1735–1737.
19. Brenner, A., de Oliveira, A. L., Schüth, F., Schunk, S. A., Stichert, W., Unger, K. *Verfahren zum Nachweis eines Produktes im Abstrom eines katalytischen Materials einer Vielzahl von katalytischen Materialien*, German Patent DE 19,830,607 to hte Aktiengesellschaft, July **1998**.
20. Busch, O., Hoffmann, C., Johann, T., Schmidt, H.-W., Strehlau, W., Schüth, F. Application of a new color detection based method for the fast parallel screening of DeNOx catalysts. *J. Am. Chem. Soc.*, 2002, vol. 124, 13527–13532.
21. Zech, T., Lohf, A., Golbig, K., Richter, T., Hönicke, D. Simultaneous screening of catalysts in microchannels: Methodology and experimental setup. In W. Ehrfeld (Ed.), *Microreaction Technology: Industrial Prospects, IMRET3: Proceedings of the Third International Conference on Microreaction Technology*, **2000**, Berlin: Springer-Verlag, pp. 260–266.
22. Ajmera, S., Delattre, C., Schmidt, M., Jensen, K. Microfabricated differential reactor for heterogeneous gas phase catalyst testing. *J. Catal.* **2002**, *209*, 401–412.
23. Ajmera, S. K., Delattre, C., Schmidt, M. A., Jensen, K. F. Microfabricated cross-flow chemical reactor for catalyst testing. *Sens. Actuators B* **2002**, *82*, 297–306.
24. Bergh, H. S., Engstrom, J. R., Hagemeyer, A., Lugmair, C., Self, K., Erden, L. V., Cong, P., Guan, S., Liu, Y., Markov, V., Turner, H., Weinberg, W. H. High-throughput screening of combinatorial heterogeneous catalyst libraries. Presented at IMRET4: 4th International Conference on Microreaction Technology, Topical Conference Proceedings, Atlanta, GA, AIChE National Spring Meeting, **2000**.
25. Bergh, S. H., Guan, S. *Chemical Processing Microsystems, Diffusion-Mixed Microreactors and Methods for Preparing and Using the Same*, World Patent WO 00/51720 to Symyx Technologies Inc., **1999**.
26. Zech, T. *Miniaturisierte Screening-Systeme für die kombinatorische heterogene Katalyse*, Dissertation, Technische Universität Chemnitz, VDI-Verlag Düsseldorf, Reihe 3, Nr. 732, **2002**.
27. Claus, P., Hönicke, D., Zech, T., Miniaturization of screening devices for the combinatorial development of heterogeneous catalysts. *Catal. Today* **2001**, *67* (4), 319–339.
28. Deutschmann, O., Schwiedernoch, R., Maier, L., Chatterjee, D. Natural gas conversion in monolithic catalysts: Interaction of chemical reactions and transport phenomena. In *Natural Gas Conversion VI, Studies in Surface Science and Catalysis 136*, **2001**, Amsterdom: Elsevier, pp. 215–258.
29. Deutschmann, O., Schmidt, L. D. Modeling the partial oxidation of methane in a short-contact-time reactor. *AIChE J.* **1998**, *44*, 2465–2477.
30. Zerkle, D., Allendorf, M., Wolf, M., Deutschmann, O. Understanding homogeneous and heterogeneous contributions to the partial oxidation of ethane in a short contact time reactor. *J. Catal.* **2000**, *196*, 18–39.
31. Rinnemo, M., Deutschmann, O., Behrendt, F., Kasemo, B. Experimental and numerical investigation of the catalytic ignition of mixtures of hydrogen and oxygen on platinum. *Combust. Flame* **1997**, *111*, 312–326.
32. de Smet, C., de Croon, M., Berger, R., Marin, G., Schouten, J. C. An experimental reactor to study the intrinsic kinetics of the partial oxidation of methane to synthesis gas in the presence of heat-transport limitations. *Appl. Catal.* A **1999**, *187*, 33.

33. O'Connor, R., Schmidt, L., Deutschmann, O. Simulating cyclohexane millisecond oxidation: Coupled chemistry and fluid dynamics. *AIChE J.* **2002**, *48*, 1241–1256.
34. Veser, G. Experimental and theoretical investigation of H_2 oxidation in a high-temperature catalytic microreactor. *Chem. Eng. Sci.* **2001**, *56*, 1265–1273.
35. Raimondeau, S., Norton, D., Vlachos, D. G., Masel, R. I. Modeling of high-temperature microburners. *Proc. Combust. Inst.* **2003**, *29*, 901–907.
36. Maruta, K., Takeda, K., Ahn, J., Borer, K., Sitzki, L., Ronney, P. D., Deutschmann, O. Extinction limits of catalytic combustion in microchannels. *Proc. Combust. Inst.* **2003**, *29*, 957–963.
37. Zech, T., Hönicke, D., Klein, J., Schunk, S. A., Demuth, D. A novel system architecture for high-throughput primary screening of heterogeneous catalysts. Presented at IMRET5: 5th International Conference on Microreaction Technology, Strasbourg, France, **2001**.
38. Idelchik, I. E. *Handbook of Hydraulic Resistance* (3rd ed.) **1994**, Boca Raton, FL: CRC Press.
39. Bird, R., Stewart, W., Lightfoot, E. *Transport Phenomena*, **1960**, New York: John Wiley.
40. Warnatz, J., Dibble, R., Maas, U. *Combustion: Physical and Chemical Fundamentals, Modeling and Simulation, Experiments, Pollutant Formation*, **1996**, New York: Springer-Verlag.
41. Coltrin, M., Kee, R., Rupley, F. *SURFACE CHEMKIN (Version 4.0): A Fortran Package for Analyzing Heterogeneous Chemical Kinetics at a Solid-Surface–Gas-Phase Interface*, Rep. SAND90-8003B, **1990**, Albuquerque, NM: Sandia National Laboratories.
42. Deutschmann, O., Kleditzsch, S., Correa, C., Tischer, S., Chatterjee, D., Warnatz, J. *DETCHEM, User Manual, Version 1.4.2*, **2002**. Available at http://reaflow.iwr.uniheidelberg. de/~dmann/DETCHEM.html.
43. *Fluent, Version* 5, **2000**, Lebanon, NH: Fluent Inc.
44. Furka, A., Bennett, W. D. Combinatorial libraries by portioning and mixing, *Comb. Chem. High-Throughput Screen.* **1999**, 105–122.
45. Lam, K. S., Lebl, M., Krchnak, V. The "one-bead–one-compound" combinatorial library method. *Chem. Rev.* **1997**, *97*, 411–448.
46. Krchnak, V., Dalton, C. Grabbing the golden bead. *Modern Drug Discov.* **2002**, *8*, 22–28.
47. Xiao, X.-Y., Zhao, C., Potash, H., Nova, M. P. Combinatorial chemistry with laser optical encoding. *Angew. Chem. Int. Ed.* **1997**, *36*, 780–781.
48. Guiles, J. W., Lanter, C. L., Rivero, R. A. A visual tagging process for mix and sort combinatorial chemistry, *Angew. Chem. Int. Ed.* **1998**, *37*, 926–992.
49. Baldwin, J. J., Horlbeck, E. G. *Synthesis of Combinatorial Libraries*, U.S. Patent 5,663,046 to Pharmacopeia Inc., **1994**.
50. Klein, J., Zech, T., Schunk, S. A. Split & Pool synthesis of functional inorganic materials, *Appl. Catal. A, Special Issue* **2003**, in press.
51. Johann, T., Brenner, A., Schwickardi, M., Busch, O., Marlow, F., Schunk, S., Schüth, F. Real-time photoacoustic parallel detection of products from catalyst libraries, *Angew. Chem. Int. Ed.* **2002**, *41* (16), 2966–2968.
52. Johann, T., Brenner, A., Schwickardi, M., Busch, O., Marlow, F., Schunk, S., Schüth, F. Listening to catalysis—A real time parallel method for high throughput product analysis. *Catal. Today* **2003**, *8* (3), 449–455.

Chapter 24

A High-Throughput Cross-Linker Evaluation for Polyurethane Dispersions

Hermann Bach, Charles A. Gambino,
Phillip D. Lunney, and Douglas A. Wicks

INTRODUCTION

High-Throughput Experimentation

Materials science is on the verge of a fundamental change of its research and development (R&D) methodology. High-throughput experimentation, combinatorial chemistry, and parallel synthesis have entered laboratories and promise a dramatic acceleration in the process of discovering and developing new products. Although Hanak had already devised the basic concepts in the 1970s [1], it took the leading example of the pharmaceutical industry and the pioneering work of Schultz and co-workers and Symyx Technologies to spark activities in almost all branches of materials science [2–13]. The combinatorial or high throughput screening approach to materials development involves four key steps.

1. Design of experiment: A large set of materials—referred to as a library—rather than a single compound is conceived, whereby the library members ideally cover the spectrum of one or more properties of the materials class. Descriptors for materials and material components support the selection of a diverse set of library members.
2. Fabrication and measurement: Samples are prepared on a small scale, usually in form of arrays, and characterized using parallel or fast sequential automated screening techniques. Laboratory automation and a high degree of parallelism are a prerequisite for high sample throughput.

Hermann Bach, Charles A. Gambino, and Philip D. Lunney • Bayer Corporation, Coatings & Colorants Division, 100 Bayer Road, Pittsburgh, PA 15205, USA
Douglas A. Wicks • University of Southern Mississippi, Box 10076, Hattiesburg, MS 39406, USA
Contact author: Hermann Bach, email: hermann.bach.b@bayer.com

High-Throughput Analysis, edited by Potyrailo and Amis
Kluwer Academic Publishers, New York, 2003.

3. Data analysis: In high-throughput screening experiments, large multidimensional data sets are often generated within days. Advanced visualization and statistical software packages are employed for their analysis.
4. System integration: In order to achieve high throughput, all steps of a work flow need to be linked by an IT infrastructure. Ideally, the results of a screening experiment are fed into the experimental design of the next run.

As a result of these efforts, a dramatic expansion in sample throughput is achieved, which it is hoped will be translated into a substantial productivity increase of the R&D process. Two mechanisms contribute to this productivity increase. First, provided that experiments are carefully chosen, increasing the laboratory capacity by two or more orders of magnitude enhances the odds of discovery. Maybe even more important, by being able to explore the experimental parameter space systematically and more comprehensively, quantitative structure-property relationships may be readily accessible, facilitating in turn the rational design of new materials.

The concepts of high-throughput experimentation seem to be particularly suited for tackling formulation problems [14–21]. Many of today's industrial products are based on multicomponent formulations. Examples include pharmaceutical or agrochemical dispersions, polymer blends, heterogeneous catalysts, and paint formulations, to mention just a few. The plethora of choices for components, mixing ratios, and process parameters, and the limited knowledge on quantitative structure-property relationships, result in an expensive, tedious, and time-consuming optimization of formulations.

In this chapter, we introduce a fully integrated, parallel, and automated high-throughput screening work flow for studying polymeric coating formulations. The heart of the work flow is a screening test to assess the relative degree of cross-linking of polymer films. Using a polyaziridine cross-linker evaluation for waterborne dispersions as an example we will illustrate how the work flow can be used to support the R&D process.

Polyaziridine Cross-Linkers

Figure 24.1 depicts the chemical structures of six major polyaziridine cross-linkers, which are differentiated by the residues R′ and R. These products can be grouped in two categories, namely the cross-linkers XAMA7 and XAMA2 based on ethylene-imine, where R′ is a hydrogen atom, and the cross-linkers PFAZ322, XAMA220, XAMA720, and NEOCRYL CX-100 based on propylene-imine, where R′ is a methyl group.

Polyaziridines have been known and used as cross-linkers for water-based resins for more than 40 years. The most important areas of usage have been photography, coatings, and adhesives. Polyaziridines react with carboxylic acid groups, which are present in many acrylic emulsions and polyurethane dispersions (PUDs) as internal solubilizers. The cross-linking reaction takes place via protonation of the basic aziridine nitrogen atom followed by a nucleophilic attack of the carboxylate anion at the neighboring carbon atom and aziridine ring-opening. A mechanistic study revealed that the presence of an active hydrogen to protonate the aziridine ring is crucial [22]. Alternative cross-linker chemistries for water-based resins include polycarbodiimides, water-dispersible polyisocyanates, epoxy silanes, and water-dispersible epoxies. Like polyaziridines, polycarbodiimides and epoxy systems are designed to react with carboxylic acid groups. Isocyanate groups react with

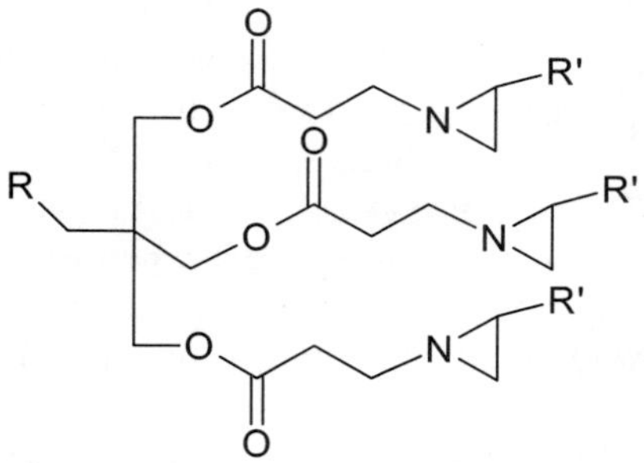

	R'	R	
XAMA 7	H	OH	
XAMA 2	H	CH_3	
PFAZ 322	CH_3	CH_3	} Different process
XAMA 220	CH_3	CH_3	
XAMA 720	CH_3	OH	
NEOCRYL CX-100	CH_3	CH_3	

FIGURE 24.1. Polyaziridine cross-linkers.

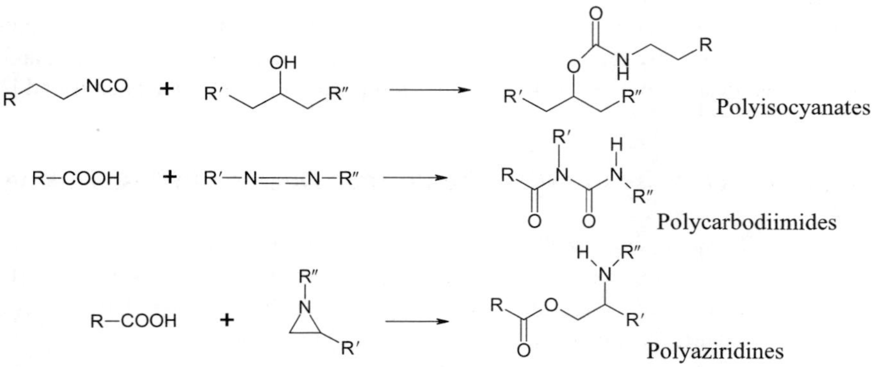

FIGURE 24.2. Cross-linker chemistries for water-based acrylics and polyurethane dispersions.

hydroxyl functionalities of the resin dispersion. The most common cross-linker chemistries for water-based resins are shown in Figure 24.2.

A number of studies comparing cross-linker chemistries for water-borne resins have been published [23,24]. In brief, it was found that polyaziridines performed better than polycarbodiimides and epoxy systems as cross-linkers for PUDs containing carboxylic acid groups. Among the polyaziridines, the ethylene-imine-based cross-linkers XAMA7 and XAMA2 performed slightly better with respect to modulus and solvent swell than the propylene-imine-based cross-linker NEOCRYL CX-100, which is equivalent to PFAZ322.

In this chapter, we present a cross-linker evaluation for water-based dispersions by means of high-throughput experimentation that extends beyond the scope of studies previously published. The goal of this work was twofold. First, we aimed at elucidating the effect of various formulation and process parameters—factors—on the cross-linker performance. In particular, we included the independent factors *resin, cross-linker, cross-linker level, resin pH*, and *time differential* between mixing and film formation, which relates to the pot-life of a formulation. The effect of the independent factors *base for pH adjustment* and *cosolvent* was also examined. Having a comprehensive knowledge of formulation and process parameters affecting product performance is crucial. Secondly, we explored whether we would be able to establish structure–property relationships with respect to the resin makeup, which would in turn enable us to tailor resins for polyaziridine cross-linker formulations.

With respect to the first goal we were looking not only for main effects (i.e., how *resin pH* affects cross-linker performance) but also for higher-order interactions between factors (i.e., how *resin pH* in conjunction with *cross-linker level* affects the cross-linker performance). This is where high-throughput experimentation becomes an extremely valuable tool. A fairly large number of combinations of factors (e.g., factor *resin*, factor *cross-linker*) and levels (e.g., the factor *resin* has the *levels* BAYHYDROL 110, BAYHYDROL 121, etc.) need to be screened in order to reveal higher-order interactions. In very general terms, higher-order interactions between factors—either antagonistic or synergistic—might play an increasingly important role in materials research. There is some reason to believe that "the potential groundbreaking material innovations of the next generation will be found in high level synergies" between known materials and material components [25]. Evidence for this statement has already been found in high-throughput experiments in the field of heterogeneous catalysis, where "by screening combinations of five elements (Pt, Ru, Os, Ir, and Rh) several good catalysts in unexpected regions of composition space" have been found [8].

Fluorescence Assay to Assess the Relative Degree of Cross-Linking

In our experiment, we screened the cross-linker performance by means of a very versatile fluorescence extraction assay that has already been successfully employed in other high-throughput screening studies of solvent-based formulations [26,27]. Figure 24.3 schematically illustrates the steps involved in this assay.

First, a dye (e.g., a fluorescent dye) is dissolved in the coating formulation. The dye should be chemically inert under the reaction conditions and soluble in common organic solvents. Polyaromatic hydrocarbons meet these requirements well. Secondly, a film is prepared by casting the formulation into a container followed by curing (e.g., thermal or UV). In the third step, an organic solvent is added to the container. The solvent swells the polymer film and extracts the dye, whereby the dye concentration extracted in a certain time interval is related to the swelling characteristics of the film. In the fourth step, the dye concentration in the supernatant is analyzed by means of an absorption or fluorescent intensity measurement.

For a given polymer system, the swelling characteristics are expected to correlate with the relative degree of cross-linking. However, one needs to exercise caution when comparing different polymer systems since the swelling characteristics depend, among other things, on an interaction parameter χ which describes the specific interaction energy between a polymer segment and a solvent molecule [28]. Also, cross-linking needs to be understood in a very broad sense, encompassing both chemical cross-linking and

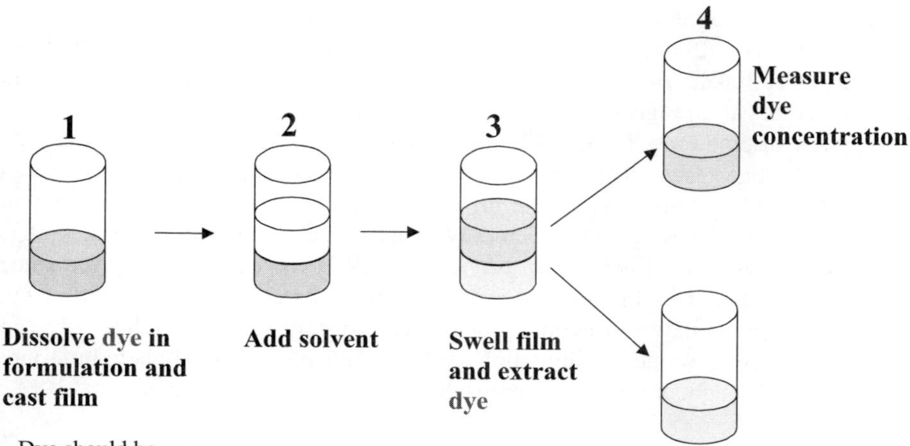

Dye should be

- chemically inert under reaction conditions
- soluble in organic solvents

FIGURE 24.3. Dye assay to assess the relative degree of curing of polymer films.

other factors affecting the solvent resistance, like particle coalescence. If a cross-linker facilitated particle coalescence, but completely failed to form chemical bonds between the particles, the assay would still yield a favorable result. This is particularly important for waterborne systems. In that sense the assay actually measures a composite effect. Usually, a relative degree of cross-linking is derived from the extraction data by including reference points in the screening. In this cross-linker evaluation, the same formulation with and without cross-linker was always tested for this purpose.

Formulation Stability and Pot-Life

In addition to the relative degree of cross-linking, we determined the formulation stability qualitatively (pass or fail). Thereby we considered a formulation unstable if it gelled, which was the dominant mode of failure. Gelation due to cross-linking is a well-known failure mechanism limiting the pot-life of polyaziridine formulations. The other main mechanism limiting the pot-life of polyaziridine formulations is hydrolysis of the polyaziridine. Polyaziridine hydrolysis leads to a degradation of film properties, but does not manifest itself in an increase of formulation viscosity. It is even suggested that the pot-life can be extended by adding more polyaziridine later on to make up for hydrolyzed cross-linker [22]

Findings

In this comprehensive survey involving the preparation and testing of several thousand films it was found that the factors *resin type* and *cross-linker level* dominate cross-linker performance whereby the carboxylic acid content of the resin plays an important, but not the only, role. The type of polyaziridine cross-linker accounts for smaller performance differences only. Similar behavior was observed for the propylene-imine-based cross-linkers on the one hand, and the ethylene-imine-based cross-linkers on the other hand. On average, the ethylene-imine-based products exhibit shorter pot-life than the other aziridine cross-linkers. All polyaziridine cross-linkers perform better than the polycarbodiimide tested. The effects of *pH, cross-linker, time differential* between blending and film formation, and *base for pH adjustment* significantly vary depending on the resin. They need to be evaluated on a case-by-case basis. One major second-order interaction, namely *cross-linker level * resin pH* was unveiled. The analysis has given some indication on how the cross-linking performance is related to dispersion properties but failed to reveal a quantitative structure–property relationship. This study has provided valuable information on scope and limitations of the methodology and will help in further refining the high-throughput screening approach to formulation development.

EXPERIMENTAL DETAILS

Design of Experiment

Materials, Factors, and Levels

Tables 24.1 and 24.2 list the resins and cross-linkers included in this study. The commercial products were used as supplied without further purification. Lab products 1–3 [29], 4 and 5 [30], and 6 were synthesized using standard procedures. On the resin side, we aimed at assembling a diverse set of resins including aliphatic PUDs, aromatic PUDs, dispersions having different soft blocks in their backbone, polyacrylic dispersions, polyacrylic emulsions, and PUD–acrylic blends. Owing to the lack of descriptors, the selection of dispersions was based on the aforementioned qualitative criteria. The resins were diluted to a solid level of 30% in de-ionized water and used at two different pH values. The lower resin pH was the pH of the resin as supplied (usually between pH 6 and pH 8); the higher resin pH was chosen about two pH units higher (pH range 8–10). NH_4OH, triethylamine (TEA), or *n*-methylmorpholine were used for pH adjustments. All Bayer Corporation polyaziridines were included in the study along with a competitive propylene-imine-based polyaziridine (NEOCRYL CX-100) and a polycarbodiimide (UCARLINK XL 29 SE) for comparison purposes. The cross-linkers were used at 4% and 8% on solid resin, respectively. In doing so, the amount of cross-linker was not based on the carboxylic acid content of the dispersion, which needs to be taken into account in the analysis. For a dispersion having a carboxylic acid (COOH) content of roughly 1% on solid resin, which is a typical value for many dispersions, 4% of PFAZ 322 on solid resin corresponds to an aziridine-carboxylic acid index of about 1. The six time differentials between blending and film formation were 1 hr, 5 hrs, 1 day, 2 days, 3 days, and 6 (7) days. Either toluene (HPLC grade) or butyl-cellosolve (*n*-butyl-glycol-ether, 99%) were used as a cosolvent at 5% on total formulation. The fluorescence dye perylene (99+%) was purchased from ACROS and used without further purification. Toluene (HPLC grade) was used for the dye extraction assay.

TABLE 24.1. Resins Included in this Study

Resin[a]	Type	Supplier
BAYHYDROL 110	Anionic dispersion of an aliphatic polyester urethane resin in water–*n*-methyl-2-pyrrolidone, 35% solids	Bayer Corporation
BAYHYDROL 121	Anionic dispersion of an aliphatic polycarbonate urethane resin in water–*n*-methyl-2-pyrrolidone, 35% solids	Bayer Corporation
BAYHYDROL 123	Anionic dispersion of an aliphatic polycarbonate urethane resin in water–*n*-methyl-2-pyrrolidone, 35% solids	Bayer Corporation
BAYHYDROL VP LS 2273	Polyester PUD in water–*n*-methyl-2-pyrrolidone, 55% solids	Bayer AG
BAYHYDROL VP LS 2917	Water—thinnable fatty acid—modified PUD in water–*n*-methyl-2-pyrrolidone–2-butoxyethanol, 45% solids	Bayer AG
BAYHYDROL VP LS 2952	PUD in water, 40% solids	Bayer AG
BAYHYDROL PR 435	Aqueous aliphatic PUD, 35% solids	Bayer Corporation
BAYHYDROL XP 7110	PUD in water	Bayer Corporation
CARBOSET CR-720	Self-cross-linking acrylic emulsion in water, 42% solids	B.F. Goodrich
CARBOSET CR-785	Acrylic copolymer emulsion in water, 42.5% solids	B.F. Goodrich
DISPERCOLL U KA 8713	Linear aliphatic polyester urethane dispersion in water, 40% solids	Bayer Corporation
DISPERCOLL U KA 8758	Anionic PUD in water, 40% solids	Bayer Corporation
DISPERCOLL U 53	Linear aliphatic polyester urethane dispersion in water, 40% solids	Bayer Corporation
EASY STREET	PUD–acrylic blend in water	Basic Coatings
IMPRANIL DLV	Polyester PUD in water	Bayer Corporation
Lab product 1	Anionic polyacrylic dispersion in water	
Lab product 2	Anionic polyacrylic dispersion in water	
Lab product 3	Anionic polyacrylic dispersion in water	
MINWAX POLYCRYLIC	Acrylic in water	Minwax
Lab product 4	Anionic polyester urethane dispersion in water–*n*-methyl-2-pyrrolidone	
Lab product 5	Anionic polyester urethane dispersion in water–*n*-methyl-2-pyrrolidone	
NEOCRYL A-639	Styrenated acrylic copolymer in water, 45% solids	Avecia BV
NEOCRYL A-640	Styrenated acrylic copolymer in water, 40% solids	Avecia BV
NEOREZ R-940	Aromatic PUD in water, 31% solids (for use with NEOCRYL CX-100)	Avecia BV
NEOREZ R-9637	Aliphatic PUD in water, 36% solids	Avecia BV
SANCURE 1514	Aromatic PUD in water, 35% solids	B.F. Goodrich
SANCURE 825	Aliphatic PUD in water	B.F. Goodrich
SANCURE 850	Aliphatic PUD in water	B.F. Goodrich
STREET SHINE	PUD–acrylic blend	Basic Coatings
Lab product 6	Polyester dispersion in water	

[a]BAYHYDROL® is a trademark of Bayer AG, CARBOSET® is a trademark of B.F. Goodrich Company, DISPERCOLL® is a trademark of Bayer AG, EASY STREET® and STREET SHINE® are trademarks of Basic Coatings, IMPRANIL® is a trademark of Bayer AG, MINWAX® POLYCRYLIC® is a trademark of MINWAX Company, NEOCRYL® and NEOREZ™ are trademarks of Avecia BV, and SANCURE® is a trademark of Sanncor Industries Inc.

TABLE 24.2. Cross-Linkers Included in this Study

Cross-linker[a]	Type	Supplier
NEOCRYL CX-100	Propylene-imine-based polyaziridine	Avecia BV
DESMODUR XP 7063	Water dispersible isocyanate	Bayer AG
PFAZ 322	Propylene-imine-based polyaziridine	Bayer Corporation
XAMA 2	Ethylene-imine-based polyaziridine	Bayer Corporation
XAMA 220	Propylene-imine-based polyaziridine	Bayer Corporation
XAMA 7	Ethylene-imine-based polyaziridine	Bayer Corporation
XAMA 720	Propylene-imine-based polyaziridine	Bayer Corporation
UCARLINK XL 29 SE	Multifunctional carbodiimide	Union Carbide

[a]DESMODUR® is a trademark of Bayer AG, PFAZ™ is a trademark of Bayer Corporation, XAMA® is a trademark of EIT Inc., and UCARLINK® is a trademark of Union Carbide Corporation.

High-Throughput Screening Design Strategy

Owing to the lack of descriptors for the formulation components, in particular the dispersions, all factors were treated as nominal variables. As already stated in the Introduction, we were looking for higher-order interactions between factors in this evaluation. The combinations of factors and levels to be used for screening is dictated by the order of the interaction of interest. In our example, where we deal with seven independent factors, a full factorial design would need to be run in order to obtain all interactions up to seventh order. A full factorial design would require the testing of 38,880 combinations (30 *resins* × 2 *resin pHs* × 9 *cross-linkers* × 2 *cross-linker levels* × 6 *time differentials* between blending and film formation × 3 *bases for pH adjustment* × 2 *cosolvents*), which is a daunting task even with the high-throughput experimentation methods at hand. We decided to restrict the design (and the analysis) to second-order interactions since higher-order effects are often obscured if the random noise level in the data is rather high, which was the case in this experiment. If interactions up to second order are of interest, the minimum number of combinations required to determine all second-order interactions according to a custom design calculation using JMP V4.0.2 [31] would be 684 and the recommended number of combinations would be 10,080. Screening the minimum of 684 combinations would be sufficient to reveal all second-order interactions. Based on a fit of a linear model function to these 684 data points, JMP also enables predictions of values for combinations that have not been run. However, the quality of the fit might be low if just the minimum number of experiments is run and it seems prudent to check some of these predictions.

In our screening, we tested 10,656 combinations. Thereby we did include some combinations suggested by the experimental design calculation mentioned above but we did not strictly adhere to the design proposed by JMP. The following circumstances and reasons led us to deviate from JMP's suggested path.

1. The number of factors and levels changed throughout the course of this screening study. This requires recalculating the experimental design whenever a factor or a level is added, which complicates matters significantly. In particular, we added the factors *base for pH adjustment* and *cosolvent* after part of the study had already been completed. Also, new resin levels were frequently added as we proceeded.

2. We found that some combinations suggested by the experimental design calculation could not be run since the formulations turned out to be unstable. Thus films could not be cast and no data on the relative degree of cross-linking could be obtained. In principle, it seems to be feasible to design around such points without jeopardizing the fit of a statistical model function, but we did not bother to do that given the relative ease with which we could run a large number of combinations of factors resulting in stable formulations instead.

3. Our main interest was in the five factors *resin, resin pH, cross-linker, cross-linker level*, and *time differential* between blending and film formation. Considering the work flow of our high-throughput screening experiment, it turned out to be much easier to run a full factorial design of these factors than just a minimum number of selected combinations. The added benefit of this experimental strategy is that the quality of the statistical model becomes better as the data set becomes larger (provided that systematic errors are insignificant). The factors *base for pH adjustment* and *cosolvent* were of secondary importance, and we decided to screen full factorial designs of these factors for selected resins and cross-linkers whereby both cross-linker levels, both resin pHs, and all six time differentials were always included. The result of this design strategy is various subsets of full factorials comprising different numbers of factors and levels each. This approach still enables us to determine the most important factors controlling the cross-linker performance based on the analysis of the different subsets. What is no longer possible are predictions for combinations outside the subset of full factorials chosen for the model calculation.

Fabrication and Measurement

Fluorescence Extraction Assay for Waterborne Dispersions

Applying the fluorescence extraction assays to waterborne formulations requires some experimental adjustments. The concept of this assay calls for (1) fluorescence dyes that are inert under the reaction conditions and (2) dyes that can be dissolved in a polymer film. This is why polyaromatic hydrocarbons like perylene, which lack functional groups and are highly nonpolar, were selected for this type of assay. When dealing with water-based resins the question arises as to how the dispersions can be doped with dyes that are not water soluble. In this experiment, we used either toluene or butyl-cellosolve as a transfer agent. First, dye stock solutions having a dye concentration of 3×10^{-3} M were prepared. These stock solutions were then used as cosolvents at 5% on total formulation. Since the resins were used at 30% solids level, the amount of cosolvent on solid resin is not much more than 1%. Since the cosolvent might have an effect on resin performance (e.g., by swelling the particles), even if only used at 1% on solid resin, we included it as a factor in our screening experiment.

In order to assess the robustness of the fluorescence extraction assay adjusted for waterborne resins, we ran the assay on several films originating from the same formulation. We found a larger random noise level in these data than in similar data based on solvent-based formulations. Therefore we prepared several films from each formulation to minimize this effect by averaging.

Preparation of Stock Solutions

All resin dispersions were diluted to 30% solids in the appropriate amount of de-ionized water to give a total weight of 200 g. The samples were then mixed for 10 min. After mixing, 10 ml of the perylene dye stock solution (dye concentration 3×10^{-3} M) in either toluene or butyl-cellosolve was added to each stock solution and mixed by shaking for another 30 min. Upon addition of the cosolvent, two phases were clearly visible. However, after shaking for 30 min the solvent phase had usually vanished and originally white resin dispersions had turned to a greenish color due to the uptake of perylene and the cosolvent by the dispersed particles. Some resins (e.g., the acrylics NEOCRYL A-640 and NEOCRYL A-639) did not pick up the cosolvent and the dye completely when toluene was used. Other resin–cosolvent combinations gelled upon addition of the cosolvent and could not be included in the study. The solutions were then split into two equal parts. One set of samples maintained the pH as supplied; the pH of the other set was adjusted to a level between 8.5 and 10 by adding one of the three bases. These stock solutions were made 1 day prior to the start of the experiment. All polyaziridines, which come at 100% solids level, were diluted to 80% solids using an appropriate amount of de-ionized water and shaking for 10 min to reduce the viscosity immediately before the experiment was started.

Formulation and Film Preparation: Work Flow

Figure 24.4 depicts the work flow of the experiment. A hand pipette was used to dispense 6 ml of resin stock solution into each of 24 glass vials arranged on a rack in four rows and six columns. In the first two rows, the resin stock solution having a pH as delivered was used. In the third and fourth rows, the pH-adjusted resin stock solution was dispensed. On a similar rack, five different cross-linker stock solutions were provided so that each column (1–5) contained a different cross-linker. The racks containing the resins and the cross-linkers were then put onto the working table of a four-arm liquid handler equipped with disposable pipetting tips (TECAN Genesis; TECAN® is a trademark of TECAN AG Analytische Instrumente) and cross-linker was added to the resin stock solutions at 4% on solids and 8% on resin solids columnwise in parallel such that all combinations of resin pH, cross-linker, and cross-linker level were formulated. The tips were exchanged after each addition cycle in order to prevent cross-contamination.

No cross-linker was added to the sixth column of the formulation rack in order to obtain a film consisting of the straight resin for reference purposes. After addition of the cross-linker, the resin rack was taken to an overhead shaker and mixed for 30 min. After letting the formulations sit for another 5–10 min, films were cast by transferring 80 μl from each vial of the formulation rack to a corresponding vial on a similar rack using the liquid handler.

The films were cured at 90°C for 10 min after they had sat for 10–20 min to allow for the release of bubbles originating from the liquid-handling step. Curing at 90°C for 10 min yields equivalent results to curing at room temperature for 1 week. Three copies (repeats) were generated each time for statistical analysis. We found that this significantly improved the quality of our data set since it enabled us to account for both random noise and systematic errors (e.g., pipetting errors due to a loose tip). This procedure was repeated at various time differentials after blending resin and cross-linker. It should be noted that

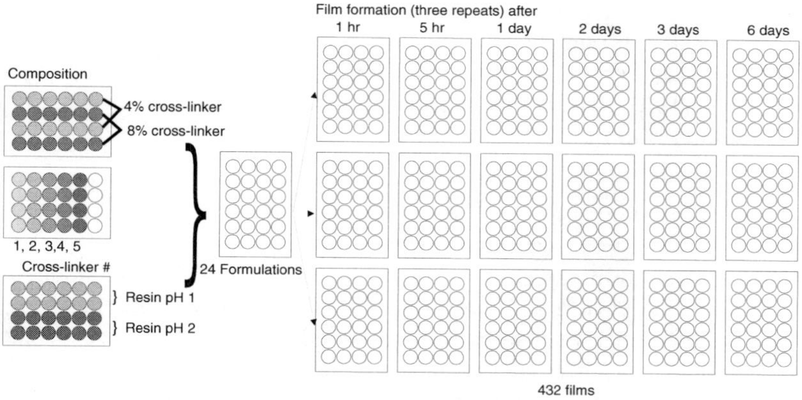

FIGURE 24.4. Schematic representation of the experimental work flow. The water-based resins were diluted to 30% solids with water followed by mixing in a dye stock solution (perylene in toluene or butyl-cellosolve) at 5% on total formulation by shaking for at least 30 min. The resulting formulation was split into half and the pH of one part was raised to about pH 9 (pH 2) while the other part was used as delivered (pH 1). The resin stock solutions were than partitioned into 8 ml glass vials as indicated above. Five cross-linkers were added at two different levels and mixed by shaking for 30 min. After letting the formulation sit for another 5–10 min, films were cast in glass vials by transferring 80 μL using a TECAN liquid handler. The formulations were recapped and the films were cured at 90°C for 10 min. This procedure was repeated at various times after blending resin and cross-linker. Three copies were generated every time for statistical analysis. The films were conditioned in a constant temperature and humidity room and tested all at once (usually the day after the last copies were made). The formulations were also kept in a controlled atmosphere. Six to eight resins were tested within a week, resulting in a test load of up to 3,500 films per week.

a film with cross-linker and a film with no cross-linker were prepared at all time differentials. The formulations were always properly recapped and kept in a controlled temperature and humidity environment between film preparation sessions.

Whenever films were cast, we assessed the formulation stability by monitoring viscosity changes qualitatively (1 = unstable, 0 = stable). Detection of viscosity increase was facilitated by our automated work flow since the liquid handler only works properly at viscosities below about 150 cP (mPa s). If it failed to transfer the formulation, it was a clear sign of instability due to gelation. Formulation instability could thus be detected almost automatically since obviously we did not extract any fluorescence dye from the corresponding vials.

The films were conditioned in a constant temperature and humidity room for at least 24 hr and then subjected to the fluorescence extraction assay concurrently (usually on the day after the final copies were made). The fluorescence extraction was performed by dispensing 900 μL of toluene onto the films. After extraction for 10 min time the supernatant was mixed to remove any concentration gradient and 200 μL of the supernatant were transferred to a microtiter plate cuvette, which was read out employing a fluorescence reader (TECAN Spectra Fluor). Since polymer film swelling sensitively depends on temperature and humidity, we monitored both parameters during the film-casting step as well as the extraction step. It turned out that conditioning the formulations and films in a constant temperature and humidity environment until shortly before their use was sufficient to minimize environment-related effects. Six to eight resins were usually tested within a week, resulting in a test load of up to 3,500 films per week. The work flow shown in

Figure 24.4 is effectively a blocked experimental design, that is, formulations are not run in random order, but all formulations of one dispersion are run at the same time. Therefore some of the resins were tested twice to check for systematic errors. We kept track of the experiment using an electronic notebook developed in-house. The complete high-throughput experimentation workflow is shown in Figure 24.5.

Analysis

A formulation stability reading (0 = stable, 1 = unstable) was obtained for all 10,656 combinations of factors and levels tested in this experiment. For the stable formulations, three fluorescence intensities were obtained from the fluorescence extraction assay. The mean and the standard deviation were calculated and obvious outliers (e.g., due to pipetting errors) were excluded. From the average intensities, we computed a relative intensity by dividing the fluorescence signal from a formulation with cross-linker by the corresponding fluorescence signal of the same formulation with no cross-linker. As described above, a formulation with no cross-linker was always included in the screening for reference purposes. Analyzing the data based on relative intensities enables easy comparison of cross-linker performance between different resins. Furthermore, we account for effects that might arise from an incomplete uptake of perylene by the dispersion. The results presented in the remainder of this chapter are for the most part derived from relative intensities.

All data generated throughout the screening experiment were combined in a single column-oriented file comprising 10,656 rows, which is amenable to an analysis using

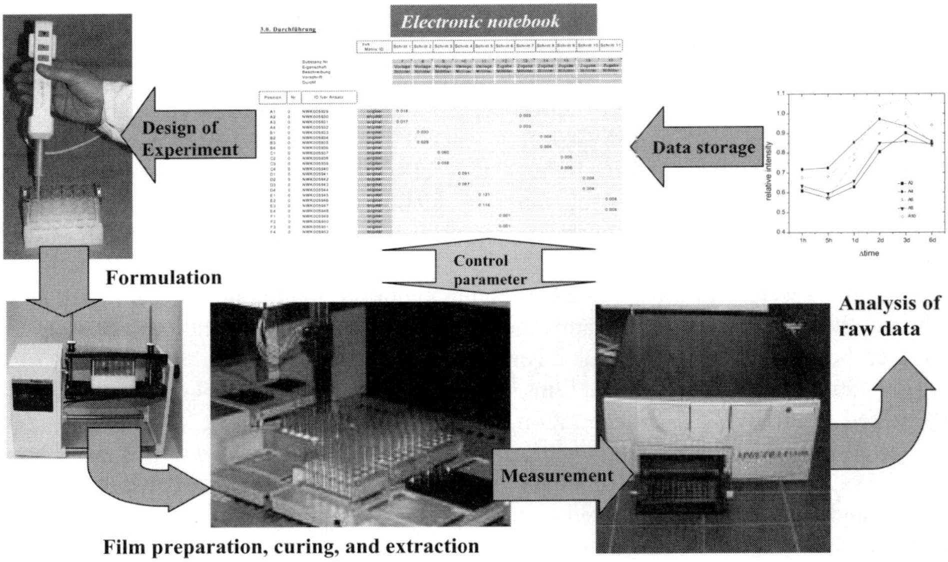

FIGURE 24.5. High-throughput experimentation work flow comprising design of experiment using an electronic laboratory journal, formulation, film preparation, film curing, dye extraction, and dye concentration measurement.

graphical and statistical software packages. Spotfire [32] was employed for the visualization of the multidimensional data set. JMP V4.0.2 was used for the statistical analysis.

Validation by Traditional Paint Testing

Since it is difficult to assess film appearance, and in particular compatibility, in glass vials, drawdowns were made of some interesting resin formulations. In addition, methylethylketone (MEK) double rubs were performed for some formulations for comparison with results obtained using the fluorophor extraction assay. In particular, we employed MEK double rubs to validate that the fluorescence extraction assay can be applied to measure the degree of cross-linking in waterborne formulations.

MEK double rubs are a common test in the coatings industry to assess solvent resistance of films. They were performed as follows. A drawdown bar was used to coat a steel panel with 180 μm of formulation (wet-film thickness). The curing and conditioning were done as described earlier. For the MEK double-rub experiment, four layers of felt cloth were wrapped around the ball-peen part of a 2 lb hammer. The cloth on the hammer was then soaked in MEK and rubbed back and forth on the coating. Care was taken not to exert vertical forces other than those originating from the mass of the hammer when rubbing. One stroke up and down is considered one MEK double rub and the cloth was kept wet with MEK after every 20 double rubs. The number of double rubs until breakthrough of the film were recorded.

Qualitatively, high fluorescence intensities were observed for systems failing at a small number of double rubs and vice versa. We concluded from this that the fluorophor extraction assay can indeed be applied to waterborne systems as well.

RESULTS AND DISCUSSION

Graphical Analysis: Trends

Two responses, namely formulation stability and relative degree of cross-linking, were measured in this study. We will mainly restrict our discussion here to the response relative degree of cross-linking and touch on formulation stability only where needed.

There are several ways of analyzing this multidimensional data set. We have chosen a hierarchical approach to the analysis. That is, we will first be concerned with general trends, and then delve further into detail to unveil subtle effects for selected formulations. Graphical analysis using software packages like Spotfire is a powerful tool to visualize and comprehend general trends in such data sets. Figure 24.6 is a scatter plot of all relative fluorescence intensities sorted by the factor *resin*. The arrow size represents the standard deviation of the signal, and the orientation encodes the factor *time differential* between blending and film formation. A vertical arrow symbolizes a data point obtained for a small time differential. The arrows rotate clockwise with increasing time differential. Blue arrows are associated with formulations containing toluene; red color indicates formulations containing butyl-cellosolve. A relative intensity of one means that the solvent resistance of a formulation with cross-linker is not improved compared with the same

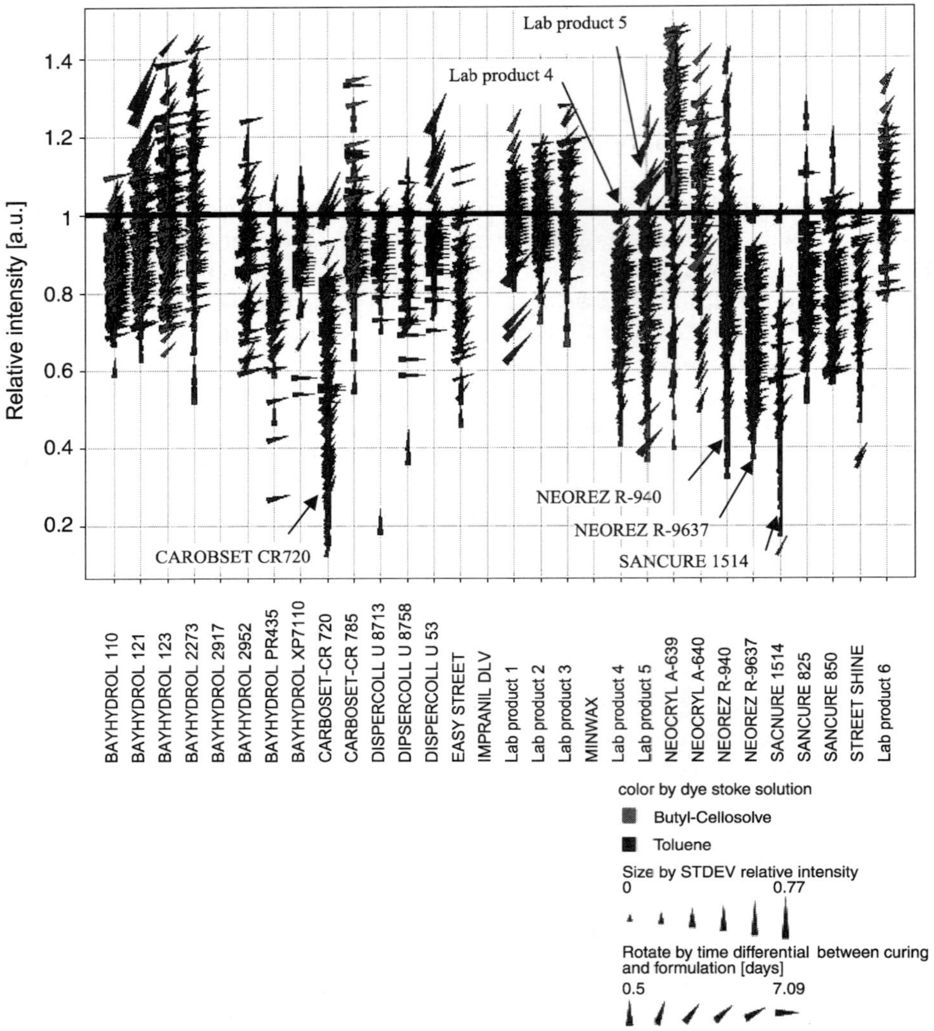

FIGURE 24.6. Scatter plot of relative fluorescence intensity (intensity/intensity control) versus resin type; 8,251 out of 10,656 records are visible. Relative intensities below unity are observed when a film exhibited a lower fluorescence signal with cross-linker than without cross-linker. See the legend on the right-hand side for details of the plot.

formulation with no cross-linker. Relative intensities <1 are obtained if the cross-linker improves the solvent resistance, and relative intensities >1 are obtained if the solvent resistance is adversely affected by the cross-linker.

From the scatter plot in Figure 24.6, the following conclusions regarding the cross-linker performance can be drawn:

1. Not surprisingly, the factor *resin* plays a crucial role in the cross-linker performance and is obviously much more important than the cross-linker itself. The lowest

relative intensities are obtained for the acrylic CARBOSET CR-720, the PUDs SANCURE 1514, NEOREZ R-940, NEOREZ R-9636, and the PUDs Lab product 4 and Lab product 5. In order to confirm the hits, drawdowns were made and the film appearance was assessed visually. Thus we noticed that a drawdown of CARBOSET CR-720 with toluene exhibits cracks, whereas a smooth film surface was obtained in the presence of a polyaziridine. Here, the polyaziridine seems to work as a coalescing agent and it could well be that the observed improvement in solvent resistance is due to this process rather than to a chemical cross-linking.

2. Fluorescence intensities $\ll 1$ are observed for small time differentials between blending and curing (vertical arrows). For long time differentials there is only a slight or no improvement of the solvent resistance (horizontal arrows), which can be explained by degradation of polyaziridine through hydrolysis.

3. For most resins, the choice of cosolvent (toluene or butyl-cellosolve) has no clearly visible affect on the cross-linker performance. The distributions of red and blue arrows along the vertical axis overlap. This is actually very reassuring from a methodological point of view, since the reason for using a cosolvent was to dope the resins with a fluorescence dye. If the cosolvent had a major effect on the cross-linker performance, the results of the whole study would have been in jeopardy. Exceptions are the acrylics NEOCRYL A-639 and NEOCRYL A-640. Toluene as a cosolvent yields better results than butyl-cellosolve. As already pointed out in the experimental section, these two resins were unusual since toluene could not be completely incorporated. Results from the statistical analysis, to be discussed later, led us to exclude some of these data from the analysis.

Statistical Analysis: Main Effects and Second Order Interactions

A more detailed and quantitative understanding of the factors controlling cross-linker performance, in particular with respect to higher-order interactions, can be obtained using statistical analysis [33,34]. Though crucial in terms of the validity of the results, a thorough discussion of methodological aspects of this analysis would clearly go far beyond what is helpful in explaining cross-linker performance. In the interest of the readability from the chemist's perspective, we chose to restrict the discussion here to the required minimum.

As a first step an outlier analysis was performed on the raw data, which enabled us to discern data points likely to be caused by systematic experimental error. Interestingly, part of the extraction data associated with the acrylics NEOCRYL A-640, NEOCRYL A-639, CARBOSET CR-785, and CARBOSET CR-720 were among those, and we excluded them from the analysis. Thus 918 of the 10,656 data points were excluded from the data set.

Second, we excluded all combinations of factors and levels from the raw data containing the cross-linker "none" since the relative intensities for this "cross-linker" are of course one and useless for the analysis of cross-linker performance. Thus another 1,776 data points were excluded from the analysis.

Third, we selected subsets of data amenable to a statistical analysis of main effects and higher-order interactions. As explained earlier, our aim was to obtain information on main effects and second-order interactions from the data. The main effects can be obtained from an analysis of all data with outliers removed [35]. To obtain second-order interactions, screening results for certain combinations of factors and levels need to be available. Because of the high-throughput experimentation work flow employed, the data set comprises various subsets of full factorials, which can be utilized for this purpose (full factorial designs are well balanced). In particular, we will present the results obtained for eight different sets of data.

1. *All data*: This data set comprises 7,962 combinations of factors and levels (10,656 − 918 − 1,776).
2. *Subset 1* comprising 12 resins and five polyaziridine cross-linker. Toluene was used as a cosolvent and NH_4OH as a base.
3. *Subset 2* comprising six Bayer resins, five polyaziridines, and both cosolvents.
4. *Subset 3* comprising three resins, five polyaziridine cross-linkers, and all three bases.
5. *Subset 4* comprising 15 Bayer resins and three polyaziridine cross-linkers. This subset only contains the time differentials 1 and 5 hr.

6–8. Subsets containing the resins NEOREZ R-9637, NEOREZ R-940, and Lab product 4, respectively. All polyaziridines and the carbodiimide UCARLINK XL 25 SE were included in these subsets.

If not specified otherwise, both cross-linker levels, both resin pHs, and all time differentials were contained in all subsets.

Fourth, we chose to treat all factors as nominal factors even if they happen to be continuous factors like the cross-linker level or the resin pH. This enables us to perform a classical ANOVA analysis, which has certain advantages in terms of the comparability of the results of effect tests. In particular, it enables us to consistently compare the statistical significance of factors controlling either the formulation stability or the relative intensity. This is done using F tests, which result in F ratios. The F ratio compares variation in the data due to controllable factors (e.g., the factor resin) with variation due to random noise. A high F ratio usually means that a factor makes a significant contribution to the response of an experiment; the responses here are formulation stability or relative intensity. For the resin pH, we created a new variable called nominal pH, which can assume the values pH1 (as delivered, usually around pH 7–8) and pH2 (as delivered +2). This is a simplification, since we actually dealt with a distribution of resin pHs. However, we found that the result of our analysis is qualitatively the same no matter whether we treated *resin pH* as a nominal or a continuous factor.

Figure 24.7 shows the F ratios as a function of all main effect and second-order interaction terms for the response relative intensity for four different subsets (*All data, Subset 1, Subset 2, Subset 3*). Let us first focus on the main effects obtained from the analysis of *All data*. The factor *cross-linker level* accounts for most variations in the data, followed by the factor *resin* [36]. *Base for pH adjustment, time differential, nominal pH, cross-linker,* and *dye-stock-solution* are statistically significant, but compared with the other two factors are likely to be of less practical importance. With respect to the factors *resin* and *dye-stock-solution* this confirms what has already been concluded from the graphical analysis. Using either toluene or butyl-cellosolve as transfer agents for the fluorescence dye obviously has only a slight effect on the cross-linker performance.

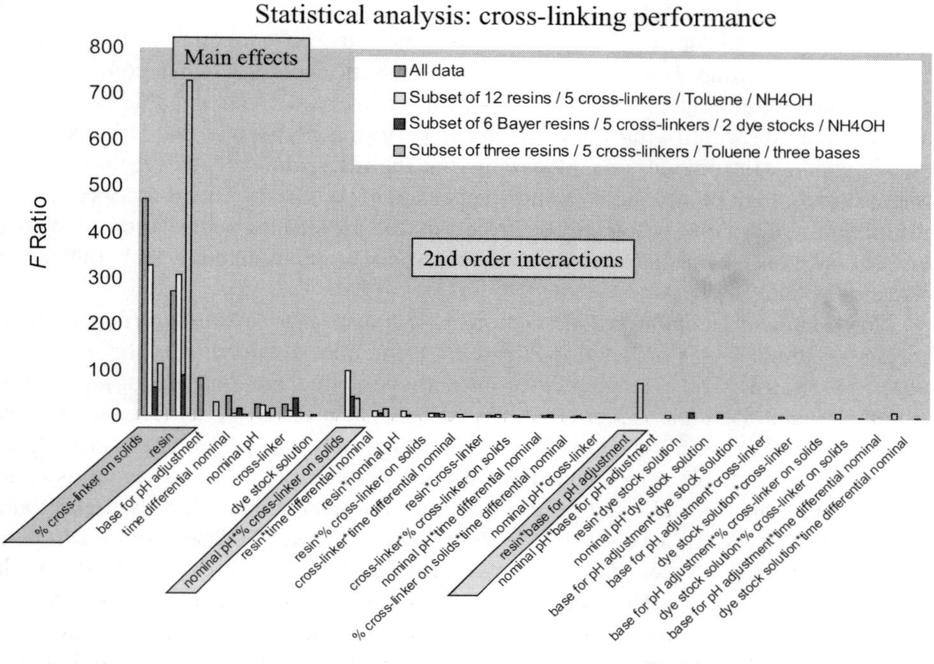

FIGURE 24.7. Main and second-order effects controlling the cross-linking performance for different data sets.

In general, it is not very difficult to find main effects. More interesting are second-order interactions that are not as easily recognized. For this purpose, we need to resort to data sets comprising the required combinations of factors and levels. An analysis of the three subsets of full factorials including second-order interaction terms is required to reveal second-order interactions. The F ratios of these analyses are also plotted in Figure 24.7. It should be noted that the absolute size and the relative size of the F ratios obtained for main effect terms and interaction terms depend on the subset chosen, which is expected for subsets of nominal variables. However, for most subsets we find that three terms have the highest F ratios. The three terms are the two main effect terms *cross-linker level* and *resin*, which have already shown up in the main effect analysis of *All data*, and one second-order interaction term, namely *cross-linker level * resin pH*. Since these terms stand out across all subsets, we assume them to be of general importance. We view the variability of the result for various subsets as a good indication of the validity of a conclusion based on a statistical analysis.

What are the effects and how can they be explained? The main effect *cross-linker level* is easily rationalized. Simply put, the more cross-linker these is, the better is the cross-linking, which is certainly not a very surprising revelation. The effect of the factor *resin* cannot be as easily described and a detailed comparison of resin properties with screening results is required to rationalize the properties of a "good" *resin*. For this purpose we analyzed *Subset 4* comprising 14 Bayer resins and three polyaziridine cross-linkers since detailed information on the makeup of these resins is readily available. Given the cross-linking mechanism, we evaluated the correlation between carboxylic acid content and average relative intensity first. This analysis showed that, for dispersions having a similar makeup, the carboxylic acid content indeed correlates with the relative degree of cross-linking. However, this did not necessarily hold true when comparing dispersions having different structural features. Unfortunately, we were not able to reveal the structural features controlling the cross-linker performance beyond the carboxylic acid content, which is the obvious one. More experiments with specifically designed resins covering a broad range of structural features appear to be needed to obtain more conclusive information. One of the problems might be that we are dealing with various responses here, but have only 14 resins made of a number of different building blocks with which to correlate the data.

The dominant second-order interaction *cross-linker level * nominal pH* found in the analysis of all subsets is depicted in Figure 24.8 (the plot obtained from the analysis of *Subset 1* is shown). The average relative intensity is plotted as a function of *nominal pH* for both 4% and 8% cross-linker on solid resin based on linear model fitted to the data. At 4% cross-linker on solid resin, pH $\approx$ 9 yields slightly better results; at 8% cross-linker on solid resin, resin pH $\approx$ 7 is favorable. We can only speculate at this point as to the rational behind this second-order interaction. It is conceivable that self-polymerization of polyaziridine occurs at a level of 8% on resin solids at a low pH. The result could be an interpenetrating network of self-polymerized polyaziridine and cross-linked PUD with a better solvent resistance. However, we do not have experimental evidence for this.

The goal of this study was to perform a cross-linker evaluation for waterborne resins. To this end, however, we have not been concerned with the various cross-linkers at all. Both the graphical analysis and the statistical analysis presented above suggest that all factors except the cross-linker are important in controlling cross-linker performance. This is not actually true, or is only true in a relative sense. Compared with the factors *resin*

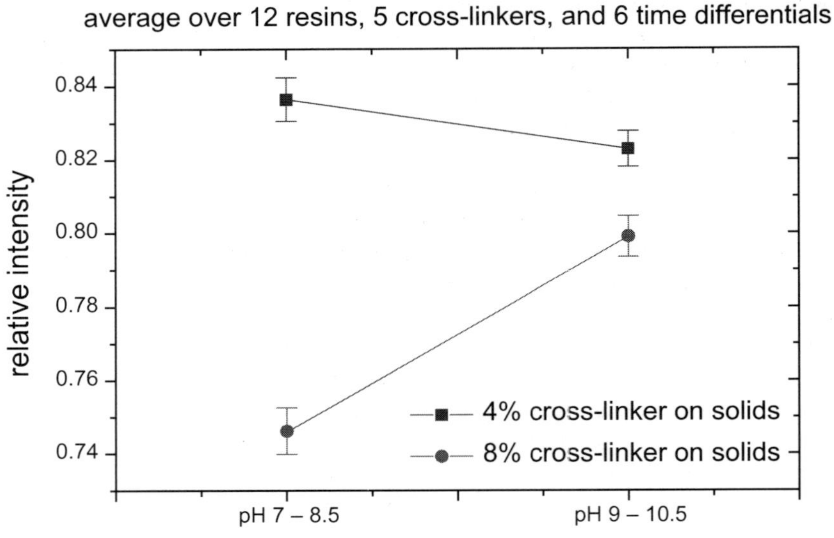

FIGURE 24.8. Interaction term *nominal pH * cross-linker level* obtained by fitting a linear model to *Subset 1*.

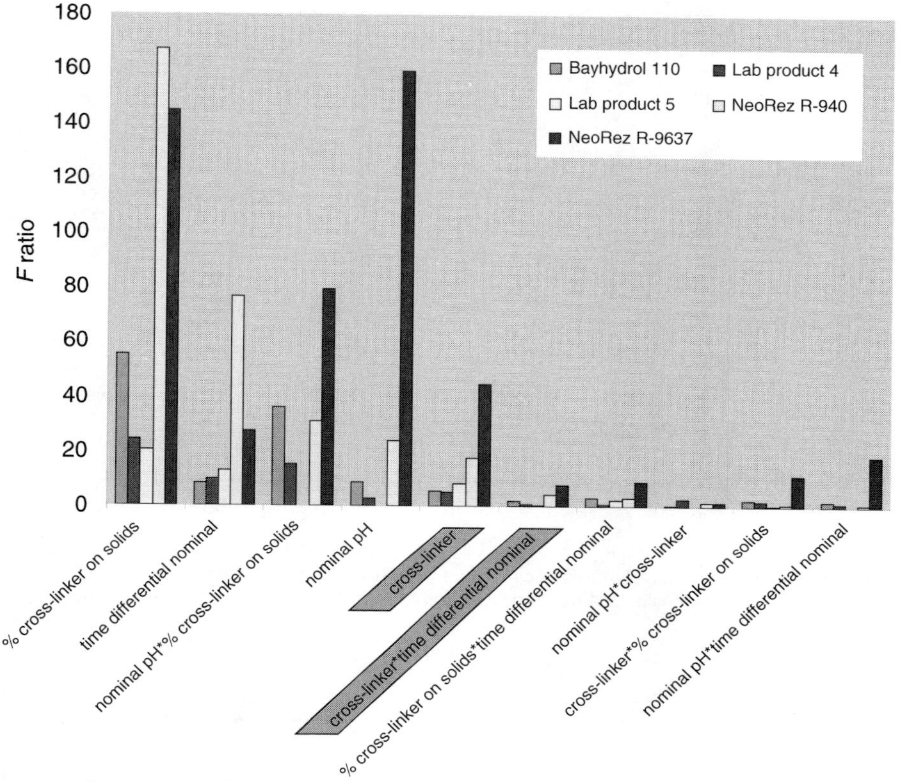

FIGURE 24.9. Ratios for various main and second-order terms obtained from a statistical analysis of resin subsets. The effects *cross-linker* and *cross-linker * time differential* nominal are highlighted.

and *cross-linker level*, the cross-linker type is relatively unimportant. However, there are differences between the various cross-linkers that were obscured by more important factors in the global analysis presented so far. In this respect, statistical analysis based on regression analysis on a global level can be quite misleading since it disregards hierarchical aspects. The differences between cross-linkers can be seen if the statistical analysis is performed by resin, which removes the overwhelming power of this important factor. Figure 24.9 shows F ratios obtained from the analysis of five resin subsets. The factor *cross-linker level* is still highly significant and dominates for almost all resins, but other factors that were previously unnoticed gain in importance. Note that there are major differences between the various resins in terms of the factors controlling cross-linker performance. For example, while the factor *nominal pH* is relatively unimportant compared with *cross-linker level* for NEOREZ R-940, these two factors are of almost equal importance for NEOREZ R-9637.

We will now focus on the effects associated with the terms *cross-linker* and *cross-linker level * time differential* that are highlighted in Figure 24.9. Figures 24.10, 24.11, and 24.12 show predictions of relative intensities as a function of the time differential for various cross-linkers for the resins NEOREZ R-9637, NEOREZ R-940, and

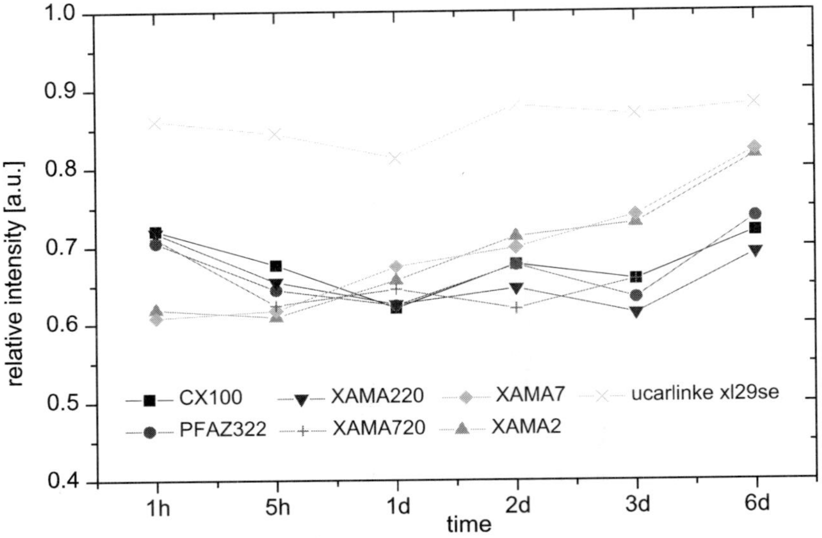

FIGURE 24.10. Statistical analysis of subset containing NEOREZ R-9637. Relative intensity as a function of the time differential between blending and film formation for NEOREZ R-9637 based on a linear model.

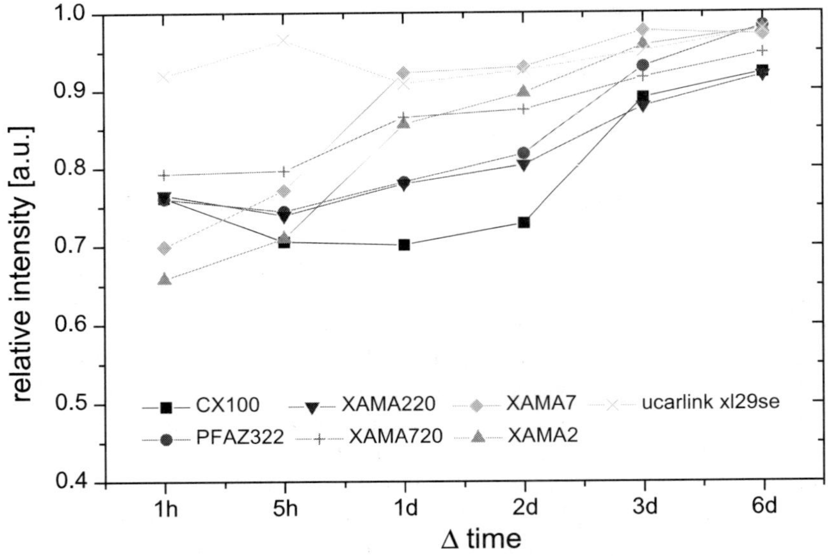

FIGURE 24.11. Statistical analysis of *Subset 1* by resin. Relative intensity as a function of the time differential between blending and film formation for NEOREZ R-940 based on a linear model.

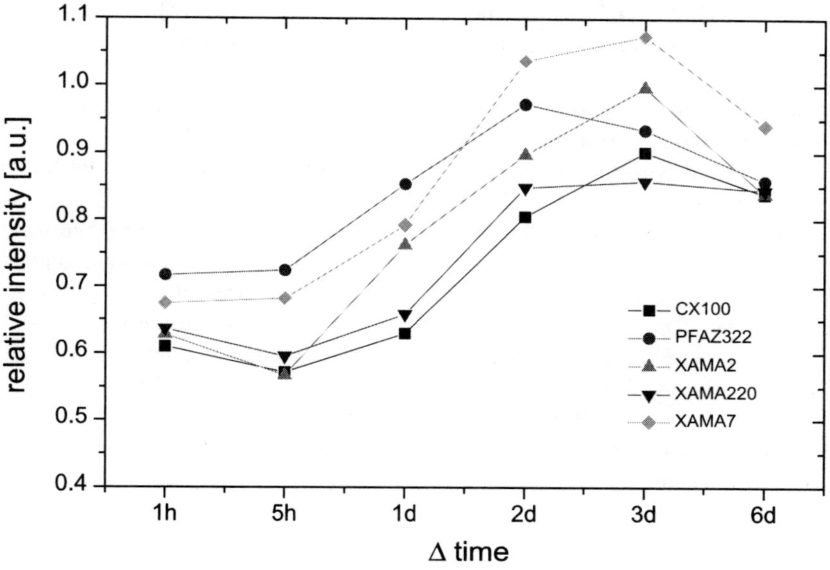

FIGURE 24.12. Statistical analysis of *Subset 1* by resin. Relative intensity as a function of the time differential between blending and film formation for Lab product 4 based on a linear model.

Lab product 4 respectively. The predictions are based on a linear model fitted to the respective subsets [37].

As can clearly be seen there are significant differences between cross-linkers. For NEOREZ R-9637 (Figure 24.10) we observe similar behavior for the ethylene-imine-based cross-linkers XAMA2 and XAMA7 on the one side and the propylene-imine-based cross-linkers PFAZ322, XAMA220, XAMA720, and CX100 on the other side. Compared with the propylene-imine-based system, the ethylene-imine-based cross-linkers exhibit better initial performance, but shorter pot-lives due to faster hydrolysis, which is expected and well known. It seems as if the pot-life of the formulations containing propylene-imine-based systems would extend well beyond 6 days without any performance loss. It even looks as if their performance improves with time. The carbodiimide cross-linker UCARLINK XL 29 SE does not perform nearly as well as the polyaziridines, which confirms earlier reports [23].

When comparing these results with those obtained for NEOREZ R-940, some similarities and some striking differences are observed. As with NEOREZ R-9636, the propylene-imine-based systems and the ethylene-imine-based systems exhibit similar behavior although there seem to be some differences within these categories. Again, the ethylene-imine-based systems perform better initially, and the propylene-imine-based systems perform better later. In contrast with NEOREZ R-9637 the pot-lives for both categories are much shorter. For formulations containing XAMA7 and XAMA2 the pot-life seems to be well below 1 day.

For Lab product 4 (Figure 24.12) the situation changes again. Here, the relative intensities obtained at small time differentials are similar for the propylene-imine-based cross-linkers NEOCRYL CX-100 (NeoResins) and XAMA220 and the ethylene-imine-based cross-linker XAMA2. As in the other cases discussed, the pot-lives of formulations containing ethylene-imine cross-linkers are shorter than the pot-lives of those containing propylene-imine.

When comparing all three graphs, XAMA2 seems to always perform slightly better than XAMA7. From the variety of behaviors observed for the three resins it also becomes clear why the factor *resin* turned out to be such an important factor in the statistical analysis.

It should be emphasized that the interaction term *cross-linker*time differential* discussed above was statistically significant but virtually unimportant when analyzing the data set on a global level. It is unlikely that any one would argue that this interaction term is still important. The three examples presented here demonstrate the variety of different effects that can be revealed when analyzing this multidimensional data set. Analyzing it in a hierarchical way enabled us to observe both the general trends and main effects, as well as more subtle details and higher-order interactions. As one delves into the details, the wealth of information contained in such a data set becomes overwhelming. The question arises of how to report such an experiment.

CONCLUSIONS

We have performed an evaluation of polyaziridine cross-linkers by means of high-throughput experimentation. The main findings were as follows:

1. The main factors controlling cross-linker performance are *cross-linker level* and *resin*. Five PUDs (NEOREZ R-940, NEOREZ R9637, SANCURE 1514, Lab product 4, and Lab product 5) have been found to be the most potent in conjunction with polyaziridines with respect to relative degree of crosslinking. Although the impact of *cross-linker level* is straightforward (the more, the better), the effect of *resin* could not be traced back to a single structural feature. A careful analysis of results obtained for Bayer AG/ Bayer Corporation resins revealed that the carboxylic acid content is obviously important but is not the only parameter controlling cross-linker performance.
2. Compared with the factors *resin* and *cross-linker level*, the type of polyaziridine cross-linker accounts for only small differences in the cross-linking performance. Similar behavior is observed for the propylene-imine-based PFAZ322, NEOCRYL CX-100, XAMA720, and XAMA220, and the ethylene-imine-based cross-linkers XAMA2 and XAMA7. On average, XAMA2 and XAMA7 exhibit a shorter pot-life than the other aziridine cross-linkers. All polyaziridine cross-linkers perform better than the carbodiimide (UCARLINK XL 29 SE) tested. These results are in agreement with data published previously [23].
3. The effects of *pH, cross-linker, time differential* between blending and film formation, and *base for pH adjustment* significantly vary depending on the resin. They need to be evaluated case by case.
4. We found one previously unpublished major second-order interaction, namely *cross-linker level * resin pH*. At 4% cross-linker on solid resin a resin pH $\approx$ 9 yields slightly better results; at 8% cross-linker on solid resin a resin pH $\approx$ 7 is favorable. At this point we can only speculate as to the rationale behind this effect.

From a methodological point of view this study touched on two important aspects of high-throughput experimentation that deserve further discussion and will need to be addressed in future experiments.

Experimental Design

In retrospect, the question arises as to whether the experimental design chosen was appropriate. This question concerns the factors, the levels, the number of levels, and the design strategy. The answer probably depends strongly on the type of information to be obtained from the experiment.

1. The design enabled us to reveal main effects and second-order interactions controlling cross-linker performance, which was the primary aim of this study. However, it was probably not the most efficient design for this purpose. We have been able to reveal these effects by analyzing subsets of full factorials. In doing so we have discarded a considerable number of data points. A more rigorous application of statistical tools in designing the experiment along with careful prior consideration of which factors and levels to include would have helped to speed up this evaluation.

2. However, we were not able to obtain detailed information relating formulation performance and structural features of the dispersions, which might be due to inadequacies in the selected resin set and/or a lack of descriptors for the dispersions. What are the criteria for selecting a meaningful set of resins and cross-linkers for establishing such quantitative structure–property relationships? A medicinal chemist skilled in combinatorial synthesis techniques would advise selecting a diverse library of compounds covering a particular parameter space (e.g., log *P* values [38] or structural descriptors). What would a diverse library of dispersions be? What would descriptors for dispersions that would enable us to select a diverse set of resins be? One can even go one step further and ask what a diverse library of material components and process parameters would be in the light of synergistic effects that might finally lead to a superior formulation? Once descriptors are available, the selection of a set of dispersions could be based on principal components analysis (PCA) [39].

Analysis

What level of detail is needed when dealing with huge multidimensional data sets? How can the enormous amount of information be turned into knowledge? These are the basic questions regarding the analysis of our screening results. In this study, we have employed statistical analysis based on a hierarchical approach and regression analysis to reveal the factors controlling cross-linker performance. The path we chose and the examples we selected are just one of many ways of approaching the analysis. Even the statistical method to be used can be debated. If a descriptor-based selection of a set of dispersions is available a partial least-squares (PLS) analysis of the various responses might be an interesting alternative, in particular for establishing quantitative structure–property relationships [40–42].

High-throughput experimentation with all its methodological advancements in statistically guided experimental design, sample fabrication, rapid material property screening, and analysis will play an increasingly important role in formulation development. This study has provided valuable information on the scope and limitations of the methodology and will help in further refining the high-throughput screening approach to the development of coatings formulation.

ACKNOWLEDGMENTS

We thank the Bayer Corporation for supporting this work, Lyuba Gindin, Peg Kendi, and Stephanie Strazisar for valuable discussions and experimental assistance, and many colleagues at Bayer AG Central Research and LS Research for their contributions to the development of high-throughput screening methodologies.

REFERENCES AND NOTES

1. Hanak, J. J. The "multiple-sample concept" in materials research: Synthesis, compositional analysis and testing of entire multicomponent systems. *J. Mater. Sci.* **1970**, *5*, 964.
2. Xiang, X. D., Sun, X. D., Briceno, G., Lou, Y., Wang, K. A., Chang, H., Wallace-Freeman, W. G., Chen, S. W., Schultz, P. G. *Science* **1995**, *268*, 1738.
3. McFarland, E. W., Weinberg, W. H. *Trends Biotechnol.* **1999**, *17* (3) 107.
4. Wang, J., Yoo, Y., Gao, C., Takeuchi, I., Sun, X., Chang, H., Xiang, X.-D., Schultz, P. G. *Science* **1998**, *279*, 1712.
5. Van Dover, R. B., Schneemeyer, L. F., Fleming, R. M. *Nature* **1998**, *392*, 162.
6. Danielson, E., Devenney, M., Giaquinta, D. M., Golden, J. H., Haushalter, R. C., McFarland, E. W., Poojary, D. M., Reaves, C. M., Weinberg, W. H., Wu, X. D. *Science* **1998**, *279*, 837.
7. Jandeleit, B., Schaefer, D. J., Powers, T. S., Turner, H. W., Weinberg, W. H. *Angew. Chem. Int. Ed.* **1999**, *38*, 2494.
8. Reddington, E., Sapienza, A., Gurau, B., Viswanathan, R., Sarangapani, S., Smotkin, E. S., Mallouk, T. E. *Science* **1998**, *280*, 1735.
9. Meredith, J. C., Karim, A., Amis, E. J. *Macromolecules* **2000**, *33*, 5760.
10. Meredith, J. C., Smith, A. P., Karim, A., Amis, E. J. *Macromolecules* **2000**, *33*, 9747.
11. Crosby, A. J., Karim, A., Amis, E. J. *Polym. Prepr.* **2001**, *42* (2), 645.
12. Senkan, S. M. *Nature* **1998**, *394*, 350.
13. Brocchini, S., James, K., Tangpasuthadol, V., Kohn, J. *J. Am. Chem. Soc.* **1997**, *119*, 4553.
14. Wicks, D. A., Bach, H. The coming revolution for coatings science: High throughput screening for formulations. *Proc. Water-Borne and Higher Solids and Powder Coatings Symposium, New Orleans, LA*, Pittsburg, PA: Bayer Corporation, **2002**.
15. Chisholm, B., Potyrailo, R., Cawse, J., Brennan, M., Molaison, C., Shaffer, R., Whisenhunt, D., Olson, D. Combinatorial chemistry methods for coating development III. An illustration of an experiment conducted with the combinatorial factory. *Proc. Water-Borne and Higher Solids and Powder Coatings Symposium, New Orleans, LA*, Pittsburg, PA: Bayer Corporation, **2002**.
16. Schrof, W., Lehmann, S., Hadeler, J., Oetter, G., Dralle-Voss, G., Beck, E., Paulus, W., Bentz, S. Combinatorial materials research to speed up the development of coatings. *Proc. 27th Conference on Coating Science and Technology, Athens, Greece, July* **2001**, pp. 283–296.
17. *An Improved Combinatorial Testing Method and Apparatus for Coat Material Formulations and Methods*, PCT Int. Appl. WO 01/32320, **2001**.
18. *Systems and Methods for Fabrication of Coating Libraries*, PCT Int. Appl. WO 02/34381, **2002**.

19. Alakhov, V., Pietrzynski, G., Kabanov, A. Combinatorial approaches to formulation development. *Curr. Opin. Drug Discov. Dev.* **2001**, *4* (4), 493.
20. Web site www.tranformpharma.com
21. *Formulation Arrays and Use Thereof.* PCT Int. App. WO 00/59627, **2000**.
22. Pollano, G. *Polym. Mater. Sci. Eng.* **1997**, *77*, 383–384.
23. Coogan, R. G. Post crosslinking of water-borne urethanes. *Prog. Org. Coat.* **1997**, *32*, 51–63.
24. Pajerski, A. D., The effectiveness of poly-aziridine used in high performance waterborne urethane adhesives. *Adhes. Age* **2001**, *June*, 43.
25. Cawse, J. N. Experimental strategies for combinatorial and high-throughput materials development. *Acc. Chem. Res.* **2001**, *34* (3), 213–221.
26. Bach, H., Guertler, C., Nowak, S. Effective and efficient: High throughput screening of catalysts for 1K PUR coatings. *Farbe Lack* **2002**, *108* (4), 30.
27. *Test Method for Optimizing Film Forming Polymers.* PCT Int. Appl. WO 02/42745 A2 20020530, **2002**.
28. Eisele, U. *Introduction to Polymer Physics*, **1990**, Berlin: Springer-Verlag.
29. EP 1024 184.
30. Rosthauser, J. W., Nachtkamp, K. Water-borne polyurethanes. *Adv. Urethane Sci. Technol.* **1987**, *10*, 121–162; Dieterich, D. Aqueous emulsions, dispersions and solutions of polyurethanes: synthesis and properties, *Prog. Org. Coat.* **1981**, *9*, 281–340.
31. Web site www.sas.com
32. Web site www.spotfire.com
33. Box, G. E. P., Hunter, W. G., Hunter, S. J., *Statistics for Experimenters*, **1978**, New York: John Wiley.
34. Cochran, W. G., Cox, G. M. *Experimental Designs*, **1957**, New York: John Wiley.
35. When interpreting the result of this analysis one needs to be very careful since this data set is not well balanced. For example, we tested some resins with five cross-linkers and others with seven. Some resins were tested with more than one cosolvent, other resins with just one, and so on. In a design based on nominal variables balancing is crucial since otherwise one might overinterpret the importance of factors otherwise.
36. Again, this analysis is based on an unbalanced design and so the ordering of the *F* ratios might be misleading.
37. It should be noted that JMP predicts relative intensities for all combinations of factors and levels in the subset. This can be misleading since we already know from the formulation stability analysis that some combinations result in unstable formulations (e.g., at high time differentials) and consequently no relative intensity can be obtained. We were not able to find a way to suppress the calculation of the statistical model for combinations that we know are unstable. Thus it is prudent to compare the predictions with the formulation stability.
38. Log *P* is a partitioning coefficient describing the partitioning of a substance between a polar (aqueous) and a nonpolar phase.
39. Wold, S., Esbensen, K., and Geladi, P. Principal component analysis. *Chemom. Intell. Lab. Syst.* **1987**, *2*, 37.
40. Geladi, P., Kowalski, B. R. *Anal. Chim. Acta* **1986**, *185*, 1.
41. Gabrielsson, J., Nyström, Å., Lundstedt, T. Multivariate methods in developing and evolutionary strategy for tablet formulation. *Drug Dev. Ind. Pharm.* **2000**, *26* (3), 275.
42. Andersson, P. M., Linusson, A., Wold, S., Sjöström, M., Lundstedt, T., Nordén, S. Design of small libraries for lead exploration. In P. M. Dean and R. A. Lewis (Eds.), *Molecular Diversity in Drug Design*, **1999**, Boxton, MA: Kluwer Academic, pp. 197–220.

Chapter 25

Data Management for Combinatorial Heterogeneous Catalysis: Methodology and Development of Advanced Tools

D. Farrusseng, L. Baumes, and C. Mirodatos

INTRODUCTION

As catalytic chemical processes become increasingly mature, innovation in this field is less and less likely. The industry is facing such a slow rate of discovery because the empirical development of new active solids by a trial and error process from a few selected samples remains highly speculative. As well as the low rate of investigation, this research strategy based on exhaustive studies and complete understanding is very time consuming. After decades of very intensive efforts, new and major breakthroughs based on experience and common knowledge cannot be expected in most research areas dealing with bulk chemistry. Therefore new research strategies must be developed to produce breakthroughs and a paradigm shift in chemical research to revitalize innovation.

Drug development underwent a drastic and successful change in the 1990s by means of the fast synthesis and screening of large libraries of diverse formulations using fast fully automated workstations and analytical equipment. This combinatorial approach, which rapidly spread to other research domains such as materials science and catalysis, relies on the systematic screening of a population surface which combines all the parameters of the study [1–4]. Thus the investigation strategy shifts from essentially qualitative studies to highly quantitative studies with data work flow increased by several orders of magnitude. Indeed, the unprecedented rate of exploration involved in this screening approach directly

D. Farrusseng, L. Baumes, and C. Mirodatos • Institut de Recherches sur la Catalyse—CNRS, 2 avenue Albert Einstein, F-69626 Villeurbanne, France
L. Baumes • Equipe de Recherche en Ingénierie des Connaissances, Université Lumière Lyon 2, Bâtiment L, 5 avenue P. Mendès-France, F-69676 Bron, France
Contact author: David Farrusseng, email: farrusse@catalyse.univ-lyon1.fr

High-Throughput Analysis, edited by Potyrailo and Amis
Kluwer Academic Publishers, New York, 2003.

increases the probability of success. In addition, instead of generating a single material with improved properties, the combinatorial strategy produces a series of leads with similar overall catalytic properties. As a logical result, the diversity of these leads increases the probability that at least one of the candidates could be marketed successfully.

Rapid automated screening of large libraries of solids and materials is now possible because of the fast-growing technologies for automation and miniaturization. Fully automated robots specially designed for fast synthesis and testing of solids are now commercially available. In contrast with the technological breakthroughs outlined above, information tools capable of managing the high-throughput (HT) work flow of data for heterogeneous catalysis are not commercially available [5]. The extremely complex nature of solid and heterogeneous catalysis makes data analysis a challenge [6–8]. In addition, data management is a large and complex process including instrument and software integration, databasing, statistical studies, and data-mining functions. New information tools have to be designed and developed for every single application, which represents a new and highly valuable market for the computer science industry linked to science.

This chapter aims at pointing out the major and often underestimated issues involved in data management for heterogeneous catalysis. All management steps ranging from raw data acquisition to data feedback will be addressed, including data storage, identification, pretreatment, and coding, together with the algorithms related to catalyst discovery and optimization. A methodology for facilitating data management for combinatorial heterogeneous catalysis is also provided. In addition, a database of general interest for heterogeneous catalysis and a powerful algorithm for the iterative discovery and optimization of catalysts are presented.

DATABASING

Motivation

A simple file consisting of arrays of rows and columns can be used as a database (DB) for *a single target application* in such a way that catalyst performance (conversion, selectivity, yield, etc.) can be linked to the main features of the catalysts (synthesis recipe and characteristics) and the test conditions. However, this two-dimensional file structure cannot accommodate various categories of solids and different catalytic reactions. Thus a completely new file has to be created when the context changes even slightly, making cross-analysis of data difficult. Therefore it is necessary to design robust relational DBs that are sufficiently flexible to accomodate changes. A database management system (DBMS) provides users with access to their data and helps them to transform the data into information. Such DBMSs include dBase, Paradox, IMS, and Oracle. These systems allow users to create, update, and extract information from their databases. Compared with a manual filing system, the main advantages of a computerized database system are speed, accuracy, reliability, accessibility, and integrity.

In combinatorial heterogeneous catalysis, the information is not concentrated in a single place but spread over various spots and functionalities, more or less automated, such as synthesis robots for preparation, parallel reactors for fast testing, characterization tools, and analytical equipment. Thus the main role of the database is to store in a single location all information available concerning the catalytic systems generated which include both their performances and their history. As a case study, the Symyx Company claims to have stored more than 10 T bytes of data regarding catalyst development for various processes over the last decade. These data, accumulated in large DBs, are believed to accelerate discoveries in new projects by means of data mining.

What is a Relational Database?

A database is a structured collection of data referring to any type of system such as items (catalysts) and/or events (catalytic performance). Relational DBMSs store data in tables and enable users to define relationships between the tables. Each table, easily visualized as a tabular arrangement of data, consists of fields and rows of records. The field names (column names) are fixed during the construction of the DB, although the field values may differ. A field has little meaning unless it is seen within the context of other fields. For example, the selectivity code field is informative only if associated with other related data (e.g., reaction name). The fields related to a particular system are bundled together to form a single and complete set of data. The link between tables is based on one or more field values common to the related tables. In each column of a table, a specific category of information has to be related to the table subject. Every table has a field or a combination of fields that uniquely identifies each row in the table. This unique identifier is called the primary key (Pk). The primary key distinguishes each set of records from all the others in a given table. It allows the user and the DB system to identify, locate, and refer to one particular row in the table.

Example

As an example, a small part of the StoCat™ DB scheme we have developed for data management of HT catalyst screening is presented in Figure 25.1. This part is related to the identification of catalyst precursors; it includes the name, formula, CAS number, type of precursor, purity grade, provider, and commercial reference number. Precursors with a given chemical formula, but with different purity grade or purchased at different companies, have the same CAS number. Therefore the CAS number cannot be used as a unique identification (ID) number for a given precursor or, generally speaking, for any chemical. The relationship between precursor formula, CAS number, commercial chemicals, provider, and purity grade can be seen in Figure 25.1 as linked tables where all the information is split according to a rational procedure. Each chemical is assigned an ID number (Id_commercial_product), which is linked to a unique precursor formula and name via the identification number of the precursor (Id_precursor). Table 25.1, which represents the conventional way of storing data, encompasses all the relevant information one needs about precursors. When applying the DB scheme to this table, data are split into different smaller tables (Table 25.2), which are linked by the key Id values. This example clearly shows how relational DBs can optimize data storage in that that the same *n*-uplet is written just once, and each time these values are used the corresponding primary key value has to be given. Thus, barium acetate is written three times in the two-dimensional array but just once in the DB. In addition, a well-designed DB strongly minimizes empty records.

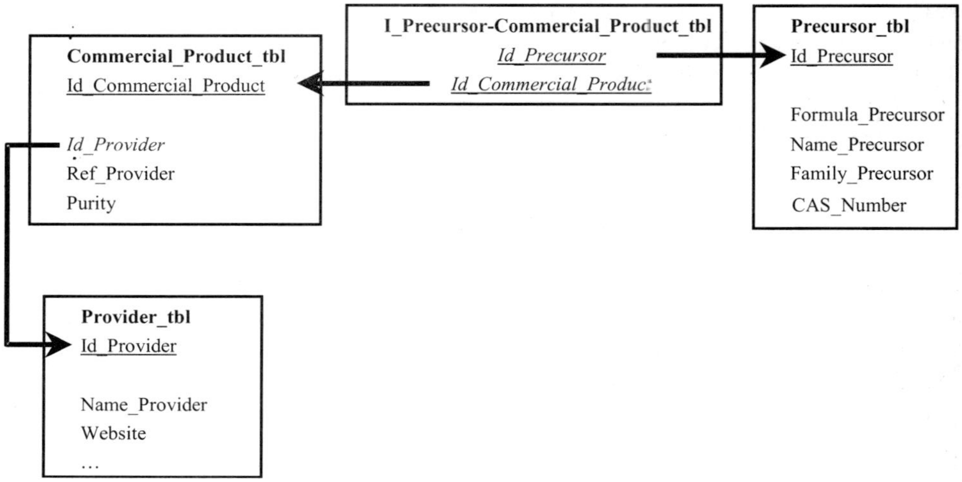

FIGURE 25.1. Example of a relational scheme. All the fields correspond to the name of the column in a common spreadsheet (Excel type). The primary key (underlined) is the unique identifier for the set of all the other fields.

How to Retrieve Data from the Database (Query)

SQL stands for Structured Query Language According to ANSI (American National Standards Institute) and is the standard language for relational DBMSs used to communicate with a DB. SQL statements are used to perform tasks such as updating or retrieving data from a DB. When one queries the DB, the rows corresponding to matching values can be retrieved from different tables. Depending on the DB structure, these queries, which join tables, may be rather complex and generally require an advanced knowledge of the whole DB scheme to be formulated (Fields, Primary keys, Foreign keys). An interface for users' queries has been specially designed for the present StoCat™ DB which consists of 120 tables and 450 fields.

StoCat™ Database

The challenge was to create a DB that was robust and flexible enough to accommodate most of the reactions using solids as catalysts for both biphase and triphase systems. The StoCat™relational DB has been designed and developed using Oracle and, to our knowledge, is the first extensive DB for heterogeneous catalysis. The adaptative skills of the StoCat™ DB rely on the choice of relevant descriptors (field names) and on a complex structure linking these descriptors. As examples of the flexibility, one can create and store in tables new chemicals, precursors, or products by a combination of elements from the periodic table. It is also possible to store systems as complex as fluid catalytic cracking (FCC) or three-way automotive catalysts. This means that investigations are not restricted to a set of reactions or catalysts, but can be matched to most studies in academic and industrial laboratories.

TABLE 25.1. Data Stored in a Spreadsheet Format

Id_Prec	Formula_Prec	Name_Prec	Family_	Id_Com_Product	Id_Provider	Name_Provider	Web_site_Provider	Ref_Provider	Purity	CAS
1	H6C4O4Ba	Barium acetate	Acetate	1	1	Aldrich	sigma-aldrich.com	24,367-1	99	543-80-6
1	H6C4O4Ba	Barium acetate	Acetate	2	1	Aldrich	sigma-aldrich.com	25,591-2	99,999	543-80-6
1	H6C4O4Ba	Barium acetate	Acetate	3	2	ABCR	abcr.de	11724	99	543-80-6
30	Bi2O3	Bismuth (III) oxide	Oxide	4	2	ABCR	abcr.de	15103	99	1304-76-3
30	Bi2O3	Bismuth (III) oxide	Oxide	5	1	Aldrich	sigma-aldrich.com	33,595-9	99,99	1304-76-3

TABLE 25.2. Data Stored in the Relational DB Shown in Figure 25.1

Id_Precursor	1	...	30
Formula_Precursor	H6C4O4Ba	...	Bi2O3
Name_Precursor	Barium acetate	...	Bismuth (III) oxide
Family_Precursor	Acetate	...	Oxide
CAS_Number	543-80-6	...	1304-76-3

Id_Precursor	*1*	*1*	*1*	*30*	*30*
Id_Commercial_Product	*1*	*2*	*3*	*4*	*5*

Id_Commercial_Product	1	2	3	4	5
Id_Provider	1	1	2	2	1
Ref_Provider	24,367-1	25,591-2	11724	15103	33,595-9
Purity	99%	99,999	99	99	99,99

Id_Provider	1	2
Name_Provider	Aldrich	ABCR
Website	www.sigma-aldrich.com	www. abcr.de

Basically, StoCat™ permits the storage and control of any information linked to (1) catalyst synthesis (precursor name, origin, volume, concentration, synthesis method with parameters, heat treatments, etc.) and (2) operating conditions for testing (carrier gas, temperature, total and partial pressure, gas flow rates, time on stream, etc.). Additional information, such as the name of the chemist or precursor batch numbers, can also be stored. Data on relevant characteristics, such as crystallinity, surface area, porosity of porous materials, etc., can be stored as well. The performance (e.g., rate of conversion, selectivity, yield, etc.) of a single catalyst under given operating conditions are automatically calculated and stored in appropriate tables. All data dealing with catalyst synthesis and testing can be retrieved automatically from synthesis robots (or optimization algorithms) and from analytical equipment currently used in laboratories, respectively, and stored. This is achieved via interfaces developed in Delphi. Their design reflects DB flexibility and has to be convenient for use by chemists. For example, vial location in a synthesis robot is displayed through the interface in a way which corresponds to the actual physical location on the *xy* surface, thus ensuring data integration by allowing export of files to robots.

A Key Tool for Advanced Combinatorial Methodology

On the laboratory scale, a single DB can connect several robots and on-line users, which permits a real networking strategy for data management. Thus StoCat™ is used as a single database by a consortium of 10 European organizations including academia and industries working in the framework of the European Union project "Combicat." This DB allows the consortium partners to collect and manage data produced at several locations in Europe. The aim of this strategy is to acquire a sufficiently large amount of data to be processed by means of statistics or artificial intelligence tools in order to discover key features of catalysis, such as structure–activity relationships.

An unexpected feature has emerged from this work—the compulsory data standardization which can be seen as a new attempt for establishing agreed rules of normalization in the rather confused area of heterogeneous catalysis.

FROM DATA TO ALGORITHMS

What Kind of Data are we Dealing with?

In combinatorial catalysis, catalysts are usually poorly characterized or uncharacterized, partly because of the lack of fast and inexpensive in situ techniques. As a consequence HT catalysts under reacting conditions can be seen essentially as black boxes. As shown schematically in Figure 25.2, the learning process aims at finding relationships between catalyst features, process variables, and catalytic performances. The input data are called endogenous—they describe the catalyst systems (descriptors)—whereas the output data are called exogenous—they describe the result of the descriptors. Establishment of all the links between endogenous and exogenous data requires the discovery of mathematical functions which model the whole surface. However, in most cases HT studies aim to discover and optimize formulations, and hence only the optimum properties of the surface are of interest.

Prior to the data analysis, the type of data to be manipulated must be determined. In general, data can be categorized into five distinct classes: continuous, quantitative ordered, qualitative ordered, binary, and non-ordered discrete (Figure 25.3). As an example, different types of data are described in Figure 25.4 according to synthesis parameters, reaction conditions, catalyst characteristics, and catalytic performances. One of the most relevant variables for a solid is the overall composition. In many processes catalysts are mixed oxides, prepared by mixing appropriate precursors in a well-defined way.

Let us consider V–Nb–Mo catalysts, recently screened for the reaction of oxidative dehydrogenation of ethane (ODHE) [9]. These elements were selected a priori because of their individual redox properties. The combinatorial approach systematically combines the elements by predicting promotion, cooperation, or synergy effects. In the corresponding ternary diagram, every single point represents an elemental composition, though different phases may coexist at the atomic level (Figure 25.5). Phase characterization is not carried out systematically, because it would take too much time and the results would still be questionable. Therefore, in order to discover a possible relationship between catalyst composition and ODHE performance, every single point of the ternary is associated with

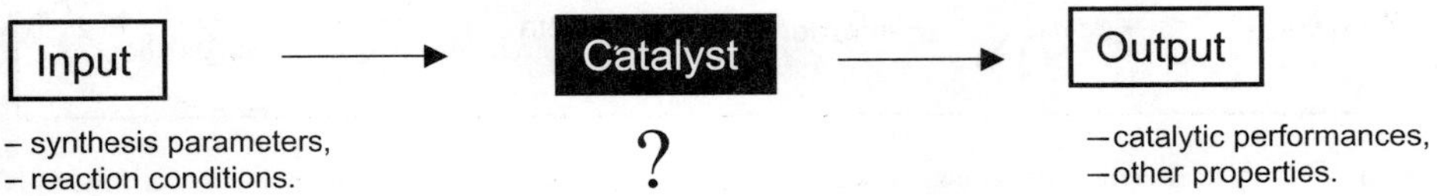

FIGURE 25.2. Catalyst as a black box function f: output $= f$(input).

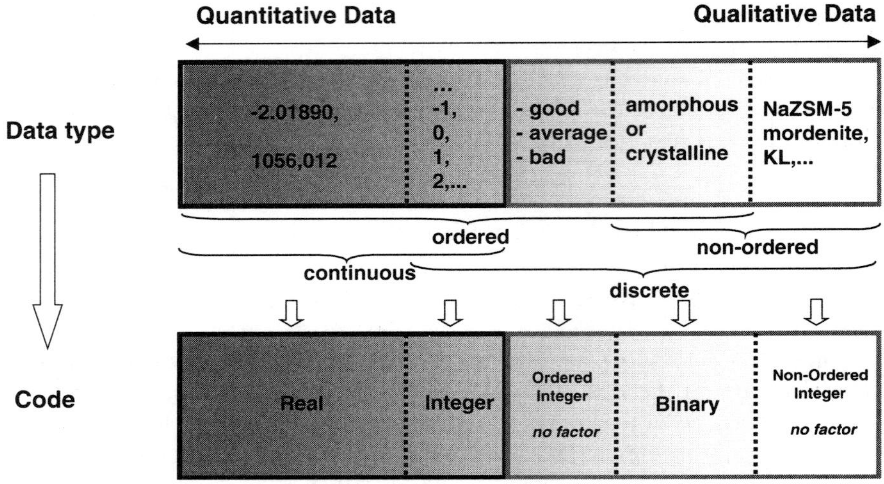

FIGURE 25.3. Encoding of different types of data related to heterogeneous catalysis.

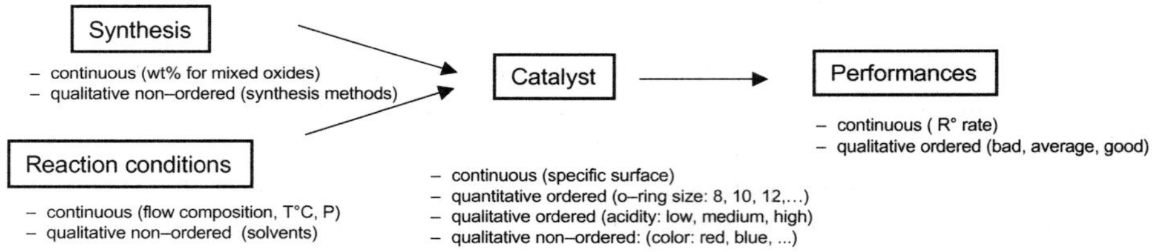

FIGURE 25.4. Examples of different types of data in heterogeneous catalysis.

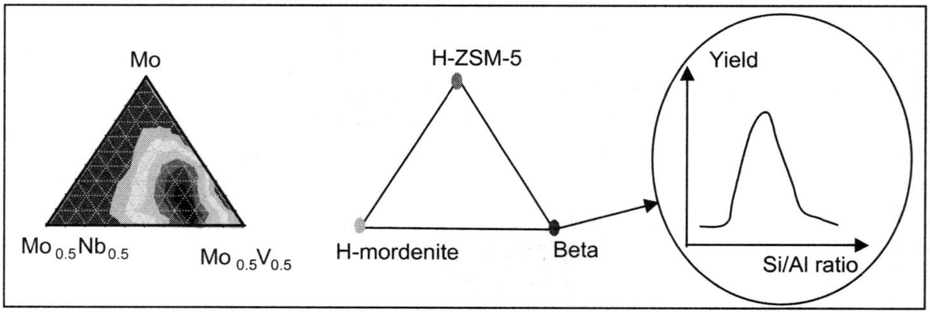

FIGURE 25.5. Visualization of parameter spaces and respective performance. Continuous case: Mo–V–Nb ternary where the darkest areas represent the highest ethylene yield for ODHE [9]. Mixed discrete and continuous case: three separate zeolites with a continuous Si/Al ratio.

catalytic data, thus creating an [X|Y] matrix. The function relating [X] and [Y] represents the catalyst–property relationship.

As a second example, let us consider a way of finding a new zeolite for hydrocracking. Based on experience and reports in the open literature, three zeolites are selected (e.g., HZSM-5, H-Mordenite, H-Y). Obviously, we could expect improved performance from zeolites exhibiting close characteristics. However, there is no continuous space between one zeolite and another. A structural combination of two zeolites would create a third zeolite with its own properties rather than a hybrid which would have the average properties of its two constituents. On the other hand, the same zeolite structure can have different Si/Al ratios, thus creating a continuous space around a given zeolite structure. In terms of surface visualization, the search space can be described as separated peaks with a local continuous space around the peaks. This simple example show how complex a search space can be for heterogeneous catalysis.

The method for catalyst synthesis also provides a good example of non-ordered discrete data. Let us consider a series of VMgO catalysts that have been prepared by varying the vanadium content, following three different methods: (1) the Pechini method, (2) vanadium impregnation onto MgO, and (3) the sol–gel process. These preparation techniques are distinct from each other and one cannot predict a priori any relationship between these methods. This feature generates a space which consists of three orthogonal areas of VMgO catalysts. These catalysts were tested for the oxidative dehydrogenation of ethylbenzene under unsteady state conditions [10]. From the plot of conversion versus vanadium content presented in Figure 25.6, it can be seen that the optimal vanadium content depends on the preparation method: the sol–gel method appears to be more appropriate for low vanadium contents (below 14 wt.%), while the Pechini method is most suitable at high contents (up to 30 wt.%). These apparently distinct areas generated by a nonordered parameter like the preparation method could be merged into a continuous domain by considering another more relevant parameter. In the present case, the degree of crystallization of the top vanadium phase (continuous parameter) could directly monitor the catalyst performance, as suggested in several advanced characterization studies of redox reactions carried out on vanadium-based catalysts. Therefore nonordered parameters

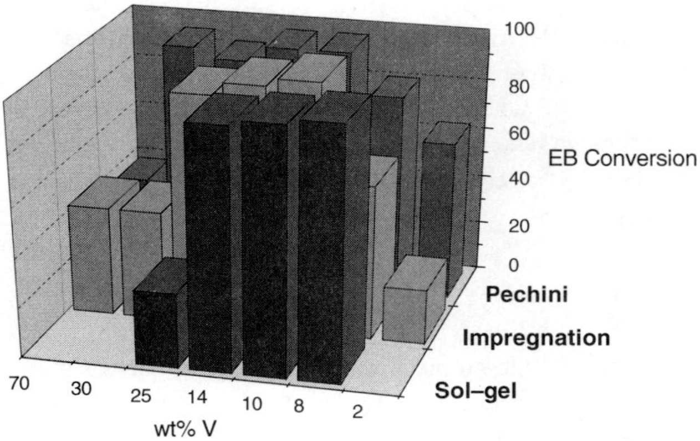

FIGURE 25.6. Conversion rate for the oxidative dehydrogenation of ethylbenzene on V–Mg–O catalysts as function of the vanadium content and the synthesis method [10].

generate a distinct search area a priori, even if a posteriori relationships could be empirically found between these parameters.

Data of a different type can also be found for reaction conditions. For example, temperature and flow rates are continuous variables, whereas for bi- or triphasic reactions the nature of carrier/solvent can be considered as a nonordered discrete variable. The catalytic performances can be evaluated either with continuous ordered data, such as the degree of conversion, yield, etc., or with discrete ordered data, such as a qualitative ranking into "bad," "average," and "good" categories. The latter can be considered as a preliminary data treatment, aimed at grouping data into classes. This type of data can be quite relevant for subsequent data mining because it has already been categorized by experts (see next section).

Thus data identification is essential before any data treatment because, depending of the nature of the data, access to treatment methods can be denied. This issue is underestimated because the search spaces that have been investigated in the literature are generally very simple. Nevertheless, it is likely that in the near future combinatorial approaches will encounter catalytic problems exhibiting complex search spaces. In addition, when advanced characterization tools are adapted to the HT combinatorial approach, the diversity and complexity of data will increase substantially.

Data Pretreatment

When data are coded, numerous operations aiming at preparing the data still have to be performed before starting the proper data analysis. This preliminary step, which accounts for about 80% of the computing time in data analysis, will determine the quality and efficiency of sub data mining.

Database Sampling

The first question that arises is whether to sample the whole DB or only part of it. Analyzing the whole DB will ensure that all possible relationships are discovered, but depending on the DB size this may require excessive computing time. In the case of heterogeneous catalysis, where typically data related to 500–10,000 catalysts have to be considered for a given reaction, the amount of data produced can be analysed exhaustively. However, when hyper-HT screening is achieved, data sampling might become an issue.

Missing data

Missing data correspond to empty fields and may originate from distinct events such as robotics errors, human error, or parameters that have no meaning in the particular context. These events have to be identified and corrected before running data analysis techniques because most of them will not proceed properly. For instance, missing data can be estimated by using a predicting analysis from the data existing in the other DB fields.

Such a strategy may consume computing time and will not necessarily be relevant. The best approach is to enter data which will not have much effect on the data analysis process (e.g., median or average for continuous data) in the missing fields.

Atypical Data

In a screened library of 500 samples, the occurrence of erroneous data due to artifacts or technical problems may be quite likely. These atypical data have to be detected before data analysis. For continuous data, values are checked with respect to a distribution generally assumed to be normal type and the domain of possible values is extended to x times the standard deviation. Treatment of these data is similar to that described above for missing data.

Feature Selection for Data Analysis

One of the main issues in data analysis is to find a space of description that matches with the problem addressed. Too many irrelevant variables (called features here) will create noise in the search algorithms and decrease their performance. In the case of heterogeneous catalysis, the choice of relevant data for describing the catalysts to be optimized is typically fixed by the chemist for screening purposes and is in general restricted to the nominal composition for primary screening. In contrast, many properties can be found in the *Handbook of Chemistry* or in the literature as raw numerical data. These can also be used as descriptors to facilitate the discovery of structure–property relationships. However, it is necessary to select relevant parameters in order to avoid data analysis "pollution." In addition, the lower the number of catalysts studied, the more restricted is the selection of the parameters.

Data Encoding: Discretization

Most algorithms and data-mining techniques can only operate on either continuous or discrete data. Therefore the nature of the data will impose a constraint and restrict the choice of data-treatment techniques. The appropriate algorithm classes that can be used for a certain type of data and for a specific goal are summarized schematically in Figure 25.7.

As already emphasized, catalysis encompasses data of different nature, which have to be pretreated in order to make them uniform. Most data-mining techniques developed in the past operate for discrete data either as classes or as binary systems. Therefore data discretization, which consists in converting continuous data to discrete data by defining intervals, is often used, although it may lead to some information loss (Figure 25.8). The best way to minimize this loss of data quality while maximizing data quantity is to design an algorithm with its own encoding system, as discussed in this chapter.

Conclusion

Preparing data is a key step in data mining. For treating parameters, automated and interactive processes have to be closely linked: the former to unravel highly intricate situations and the latter to transform real data to computer-readable data.

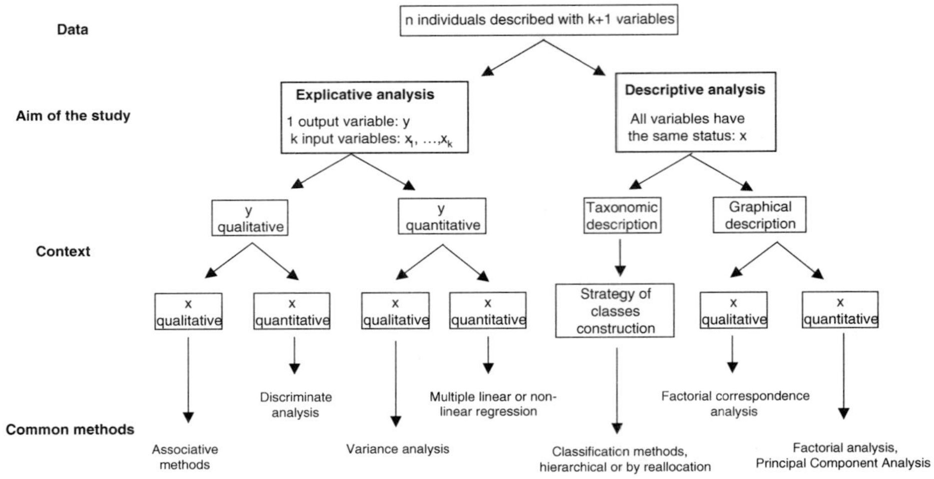

FIGURE 25.7. Appropriate common analysis methods for different types of data and study aims.

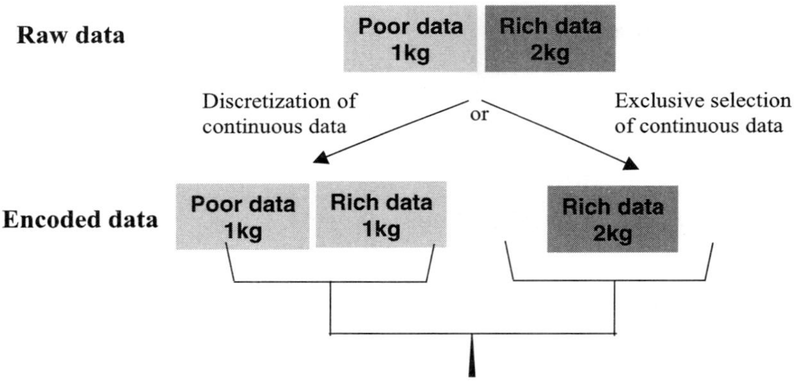

FIGURE 25.8. Choice of coding system.

INVESTIGATION STRATEGIES FOR CATALYST DISCOVERY

The space of all existing solutions is called the search space. The quality of each solution can be evaluated by its performance with respect to the addressed problem. Looking for a solution is then looking for a minimum or a maximum, or "hits," in the search space. For catalyst discovery, data analysis is used to reduce the number of samples to be tested while maintaining a high probability of discovering new solids. In terms of informatics, the ratio of exploitation to browsing has to be optimized.

Typically, when one starts a new study using the combinatorial approach, only a few selected items of information deriving from common knowledge will be considered, although a large literature exists. On the other hand, data work flow produced by hardware systems (synthesis robots, parallel reactors) will be discontinuous in time, since, for practical reasons, catalysts screening is performed by sequential analysis populations of samples, called libraries. Investigation will be conducted by an iterative optimization process, starting from a very limited number of rules but able to "learn" while proceeding. Feedback of data treatments to the design of libraries will prevent investigation of unproductive areas while speeding up the rate of convergence toward "hits."

Two-Stage Screening Strategy Versus Single Iterative Process

Based on the pharmaceutical experience, a specific two-stage strategy was successfully applied for material discovery [2,3]. It consists of a broad or primary screening followed by a more restricted or secondary screening. Basically, the primary stage deals with the ab initio identification of new materials called "hits" via the systematic combination of elements for the preparation of multicomponent materials. A high throughput of about 10,000 formulations per month is generally considered for such an initial screening. Secondary screening seeks to optimize the primary "hits" in order to identify "leads." This second step of the combinatorial process combines HT preparation and evaluation of materials under conditions that are as realistic as possible with a precise strategy following predetermined pathways. The number of experiments is reduced to about 500 per month.

Recently a combination of primary and secondary screening has been successfully demonstrated. Such "integrated" screening aims at speeding up the investigation by reducing the number of experiments while still using stochastically based screening. This is achieved with a smart strategy of guidance toward active catalysts by using evolutionary concepts [11] (see later in this chapter). Well-known optimized catalysts have been "rediscovered" during the proving testing reactions, while novel compositions have been highlighted for real case studies ("Combicat" program). This newly developed approach shows that alternative screening strategies assisted by advanced software can be still be developed.

Design of the First Library

Diversity of Catalysts

To date the initial choices regarding the spectrum of periodic table elements, chemical precursors, and concentration ranges have been made by the chemists themselves. For special cases, the intuition of chemists has been computerized in Expert Systems in order to assist catalyst design for targeted reactions [12–18]. However, one can argue that an a priori selection of synthesis parameters including a choice of elements based on common knowledge would not favor the probability of discovering new materials. In contrast, the screening of very diverse catalysts with synthesis parameters very different from the data in the literature is more likely to increase the chance of finding new catalysts. The concept of diversity is usually used for the combinatorial discovery of drugs. In organic chemistry, descriptors exist in terms of well-identified functional groups, which are structural molecular patterns, so that the main features of a molecule can be described by an exhaustive listing of its functional groups. Using this concept, molecules can be classified into families, sorted, and even prescreened via computer simulation, which makes the evaluation of library diversity feasible. In the case of heterogeneous catalysis, for which these exhaustive descriptors do not exist, the development of tools for assessing diversity is a challenging task. If such tools existed, algorithms would be able to select the most relevant synthesis parameters and then generate many initial libraries exhibiting maximized diversity [19].

Experimental Design

The parameter space is defined by all the samples which could be generated by varying each parameter stepwise. Typically, for mixed oxides based on 10 selected elements and several supports and promoters, more than 50,000 formulations could be designed. This number is increased even further if one takes into account preparation parameters and activation and testing conditions. In an investigation of a basic reaction catalyzed by zeolites, considering a restricted number of synthesis parameters (i.e., nature of zeolites and cations, precursor types, preparation methods, and cation contents), a parameter space of more than 300,000 samples can easily be generated. It should be emphasized that the size of these parameter spaces that reflect catalyst complexity is directly related to the probability of discovering new hits or leads. However, a *high complexity* obtained by varying a large number of parameters does not, necessarily mean that the *catalyst diversity of a library* will be maximized with respect to the total number of catalysts. Considering the presence or the absence of elements rather than variations in catalyst composition can also enhance the diversity of a library. A methodology for partitioning the search space which takes into account synthesis methods, supports, and active phase synthesis was developed in our laboratory. Such a methodology facilitates data analysis because it can be coded directly in a binary file without loss of information.

ALGORITHMS FOR CATALYST DISCOVERY IN ITERATIVE PROCESSES

There are many methods of finding a satisfactory solution in a complex parameter space, including hill climbing, tabu search, simulated annealing, and evolutionary algorithms. A solution is often considered as good solution, since it is generally impossible to prove that it represents the effective global optimum. Evolutionary algorithms (EAs) are ideally adapted for the iterative screening of catalysts because they proceed by creating an ensemble of solutions instead of a unique solution, unlike the conventional stochastic optimization method (e.g., simulated annealing). In addition, EAs do not involve gradient calculation and are insensitive to nonlinearity and complexity of the solution space

Basis of Evolutionary Algorithms

Introduction to Evolutionary Algorithms

Many optimization methods inspired by Darwin's theory of evolution (genetic algorithms, evolutionary strategies, etc.) have been developed for a variety of purposes.

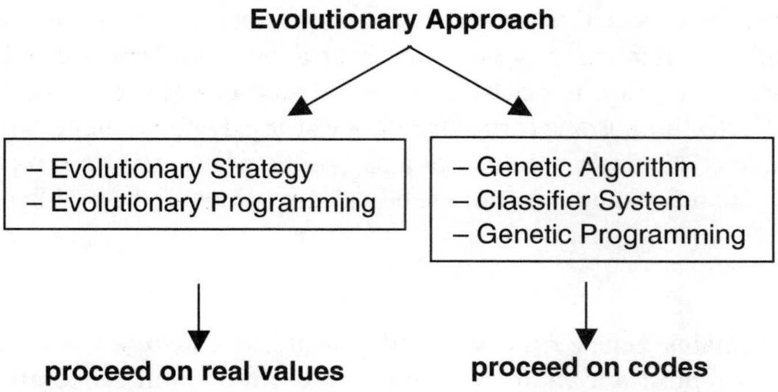

FIGURE 25.9. Classification of EAs.

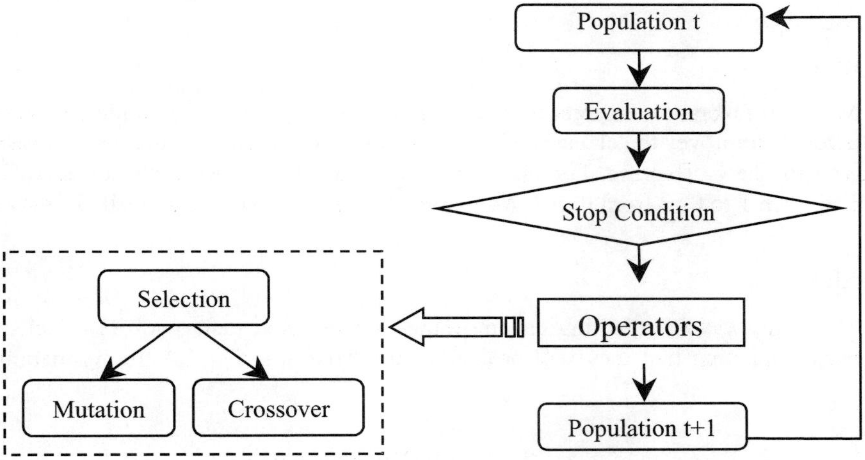

FIGURE 25.10. Overall principles of EAs

They are commonly grouped under the name of evolutionary algorithms (Figure 25.9). Their common features are the evolutionary principles of the natural laws based on population mutation, crossover, and selection (Figure 25.10). Genetic algorithms (GAs) generally show a high rate of convergence, even in complex solution spaces, but they run with binary codes. Evolutionary strategies (ESs) can proceed using continuous variables although they generally converge toward accurate values in a slower rate.

Basic Elements of Evolutionary Algorithms

Population

The algorithm starts with a set of solutions (represented by chromosomes) called a population. Solutions from the last population are taken and used to form a new population which is expected to be better than the former one. Solutions are selected to form new solutions (offspring) based on their their fitness (performance). This is repeated until a satisfaction criterion is reached. The usual way of encoding a chromosome is a binary string. Each chromosome forms one binary string. Each bit in this string can represent some characteristic of the solution, or a part of the whole string can represent a number. Note there are many other ways of encoding, depending on the problem to be solved.

Crossover

Crossover mates genes from selected parent chromosomes and creates a new offspring. The simplest way is to choose one crossover point randomly and to copy everything before this point from the first parent and everything after this point from the second parent (Figure 25.11). Much more complex crossover methods can be chosen (e.g., with several crossover points), depending on the chromosome encoding.

Mutation

Mutation favors the emergence of new genetic configurations by widening the pool and therefore improves the chances of finding optimal solutions. It proceeds by randomly changing the new offspring. For binary encoding, a few randomly chosen bits can be switched from 1 to 0 or from 0 to 1. Mutation depends on encoding as well as crossover.

Selection

The main way of selecting chromosomes for the next generation is to select the best ones, expecting that they will produce better offspring. For EA, the probability of

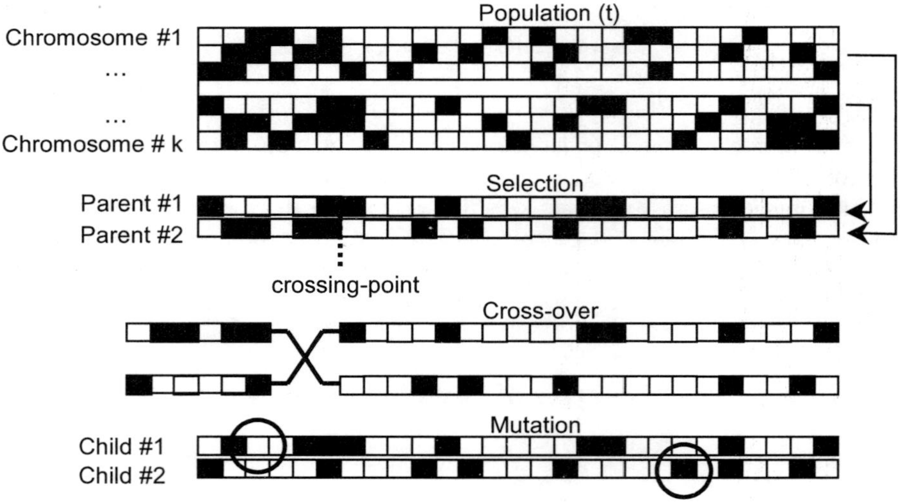

FIGURE 25.11. Examples of EA operators: selection, crossover, and mutation

Individual #	Individual	Fitness	Relative Fitness
1	10111	2.23	0.11
2	10001	3.45	0.22
...	...	...	...
i	i	f(i)	$\frac{f_i}{\sum_{i=1}^{K} f_i}$

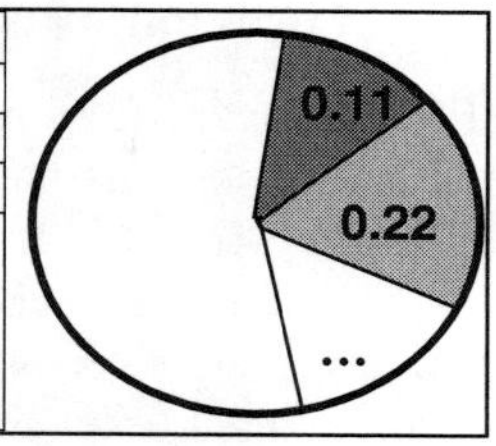

FIGURE 25.12. Roulette type selection.

reproduction directly depends on the fitness of each subject. Parents are selected according to their fitness (Figure 25.12). One selection method is the roulette wheel containing all the chromosomes of a current population, each occupying a sector according to its fitness. Though quite easy to implement, this selection method faces a major problem. When a "super-chromosome" occupies too large a sector of the wheel at the very beginning of the study, it will very often be selected as the parent for the next generation. This will result to a drastic loss in diversity and the algorithm will be trapped in a local maximum. There are various methods of solving this problem.

1. Ranking selection: First the population is ranked and then every chromosome is assigned a fitness from this ranking. The worst will have fitness 1, second worst fitness 2, etc., and the best will have fitness N (i.e., number of chromosomes in the population). Thus, whatever the difference between the subjects, strong or weak, the difference between probabilities of reproduction only depends on the ranking of the subjects.
2. Windowing: The fitness of the worst subject is subtracted from the fitness of each subject. This strengthens the strongest subject and leads to a zero-based distribution.
3. Exponential: This consists in taking the square roots of the fitness plus one. This reduces the influence of the strongest subjects.

Thus the principles of EA can be summarized as follows:

1. encoding the problem;
2. random generation of a population;
3. assignment of a fitness value to each subject;
4. selection of the subjects;
5. genome crossover and mutation;
6. start again from point 3.

Application of Evolutionary Strategy to Heterogeneous Catalyst Optimization

Key features of heterogeneous catalysts with respect to a given objective function can be improved through natural laws of selection, and crossover. In addition, the stochastic mechanism and the mutation enables new features to be discovered.

Case Study of Continuous Search Spaces

An ES was successfully used to optimize catalysts for the oxidative dehydrogenation of propane in two separate studies [11,20]. Two sets of elements were chosen. The first comprised eight elements of the periodic table selected a priori according to common knowledge, whereas the second consisted of 13 elements exhibiting medium metal–oxygen bonding energy. As example, the first library consists only of ternary compounds with the same element frequency (Figure 25.13) (presence/absence of vizualisation). After evaluation in a test setup, the catalysts are ranked with respect to an objective function which, in this case, was chosen as minimizing the difference between the experimental yield of propene and 100% (Figure 25.14). At the selection stage, three of the four best catalysts in the first library contain vanadium. Two other elements (manganese and Iron) also appear frequently in the selected materials. The next generation is designed according to the following algorithm features and operators (Figure 25.15):

1. selection rate (the best four formulas were selected);
2. mating (one of the selected samples + one stochastic sample);
3. qualitative crossover rate (add or remove elements);
4. qualitative mutation rate;
5. quantitative mutation (used to vary the composition).

Optimization occurs through generations of populations of trial solutions with increasing average fitness. The best fourth-generation catalyst composition indicates a convergence toward a single formulation containing V, Fe, Mn, and Ga (Figure 25.16). In the first study, the conversion–selectivity plot as a function of the number of generations clearly shows the convergence of the algorithm after four generations toward a zone at 15% conversion and 55% selectivity (Figure 25.17). Again, the lines formed by the stacking of catalyst genes at the fourth generation, which represent the key features, clearly indicate a convergence toward a well-defined composition consisting of V, Mg, Mo, and Ga.

The balance between convergence rate and the probability of obtaining the best leads can be adjusted by altering the crossover and mutation rates. In the early stages of the study, exploitation dominates by using a relatively high crossover rate, whereas at the end of the study mutation rates have to be increased to avoid fast conversion toward local maxima. The rates were tuned automatically by analysing the difference between the mean and average yield, which is used to evaluate the population "diversity."

Limitations of Evolutionary Strategy

The previous case study demonstrates that an appropriate strategy for designing generations of libraries can greatly speed up the optimization process. However, this type of ES suffers from several limitations in the context of catalyst discovery.

1. ESs apply essentially to systems which can be described/encoded by continuous variables, like multiple oxides (mixed oxides) or metal (alloys). ESs are not applicable to other systems generating discrete and nonordered data, such as libraries of exchanged zeolites.
2. Additional information (ionic radius, oxidation state, etc.) or characteristics such as surface area and pore size cannot be integrated with ES, although this knowledge would be relevant for selection of parents.

	V	Mo	Mn	Fe	Zn	Ga	Ge	Nb	W	Co	Ni	Cd	In
1				0.79		0.02		0.19					
2		0.44									0.23		0.33
3					0.7		0.08			0.22			
4	0.33			0.43								0.24	
5						0.009		0.33			0.66		
6		0.33			0.42								0.25
7							0.11		0.33			0.56	
8	0.26		0.33			0.41							
9								0.16		0.33			0.5
10		0.42		0.53			0.44						
11	0.47								0.19		0.33		
12			0.41			0.51		0.08					
13		0.45								0.22			0.33
14				0.75			0.05		0.2				
15	0.33		0.43								0.23		
16					0.67			0.1		0.23			
17		0.33		0.42								0.25	
18						0.065			0.33		0.6		
19			0.33		0.41								0.25
20							0.14			0.33		0.53	

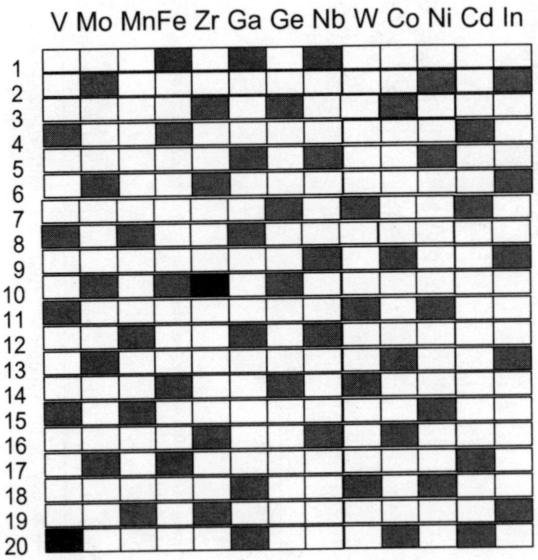

FIGURE 25.13. Composition of a first library consisting 20 mixed-oxide ternaries. The chromosome-like visualization shows a good distribution of the elements over the whole library (each element is used four times) [20].

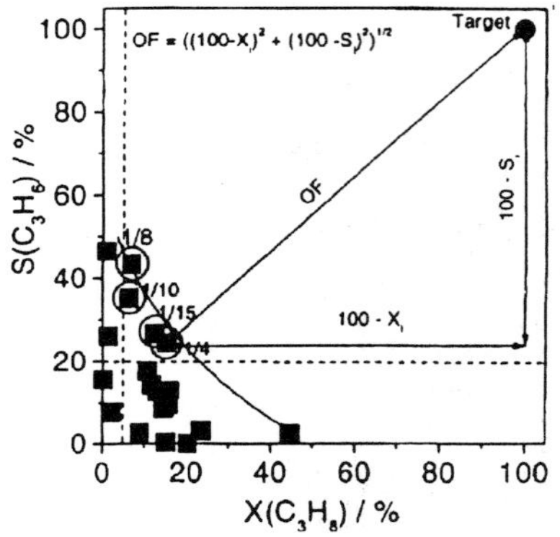

Selection of the four best ranked catalysts

1 - #4- $V_{0.33}$ $Fe_{0.43}$ $Cd_{0.24}$
2 - #8- $V_{0.26}$ $Mn_{0.33}$ $Ga_{0.41}$
3 -#15- $V_{0.33}$ $Mn_{0.43}$ $Ni_{0.23}$
4 - #10- $Mo_{0.42}$ $Fe_{0.53}$ $Ge_{0.44}$

FIGURE 25.14. Selectivity–conversion plot of the first-generation catalysts and catalyst ranking according to the objective function and additional threshold requirements.

Operators	Qualitative Crossover	Qualitative mutation	Quantitative mutation
Catalyst selection	1 selected + 1 randomly	random selection	fitness-proportional selection
		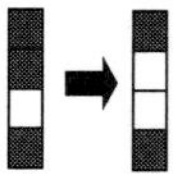	$x_i = x_i + (-1)^t x_i/2$

FIGURE 25.15. Operators derived for the design of subsequent libraries. The presence or absence of elements proceeds through qualitative crossovers and mutations, while the composition varies with quantitative mutations.

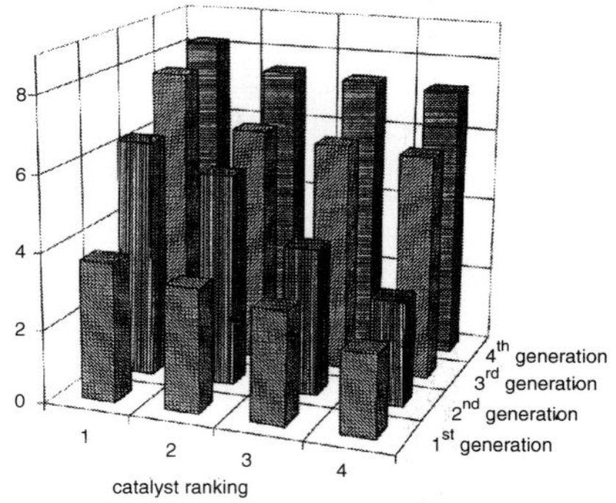

1 - $V_{0.19}$ $Mn_{0.24}$ $Fe_{0.32}$ $Ga_{0.32}$
2 - $V_{0.06}$ $Mn_{0.02}$ $Ga_{0.92}$
3 - $V_{0.49}$ $Mn_{0.13}$ $Ga_{0.38}$
4 - $V_{0.47}$ $Fe_{0.19}$ $Ga_{0.33}$

FIGURE 25.16. Visualization of the yield improvement of the four best catalysts of each generation of the evolutionary process (left). The composition of the four best catalysts of the fourth generation is shown on the right.

3. It is not possible to integrate results of previous screenings for a target reaction. Let us consider, for example, a strategy which would consist initially of a broad primary screening. An iterative optimization of the initial hits could be then started by EA. However, this approach implies a selection of catalysts for the first generation, and the question arises as to how to perform this selection by integrating the knowledge acquired during the primary screening.

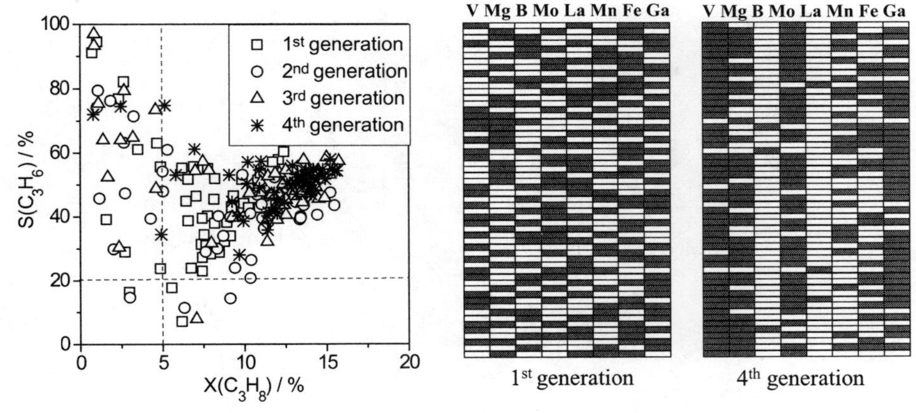

FIGURE 25.17. Selectivity–conversion plot of four generations of catalysts (left). Chromosome-like visualization of the presence (full square) or absence (empty square) of elements for each catalyst of the first and fourth generation which shows a convergence toward V–Mg–Mo–Ga catalysts (right).

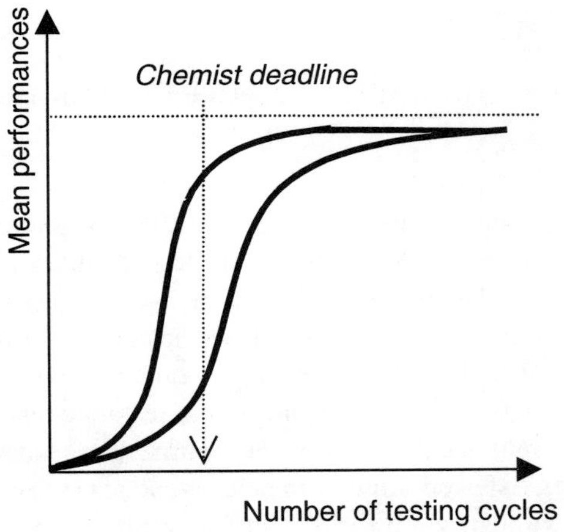

FIGURE 25.18. Scheme of typical convergence rates for EAs.

4. The browsing/exploitation ratio can be tuned by changing the EA parameters. However, there are no general rules for the a priori optimization of these parameters because it depends strongly on the context.
5. The iteration number for EA is usually not limited. However, a rapid convergence toward catalysts satisfying the chosen criteria is expected within a reasonable period of time. In this case, convergence of ESs may not be reached within a given time constraint, especially for complex systems (Figure 25.18).
6. Algorithms able to converge not only toward a global maximum but also toward several local maxima would greatly improve the chances of discovering very different catalysts exhibiting similar catalytic activities (Figure 25.19).

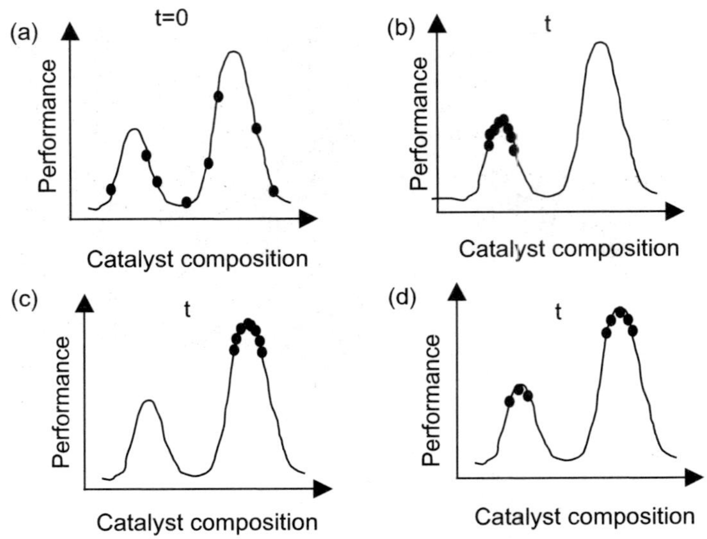

FIGURE 25.19. Scheme of typical convergence behavior. (a) Location of the individuals of the first generation over the search space at time $t = 0$. (b) Premature convergence toward a local optimum. (c) Direct convergence toward the optimum. (d) Convergence with sharing.

Original Hybrid Algorithm for Iterative Optimization of Catalysts

Motivation

It has recently been shown that combining scientific understanding and iterative screening of generations using an evolutionary strategy resulted in a breakthrough in the optimization of eniantoselective enzymes [21]. In this study, after three loops of optimization applying the "pressure" of natural laws, three leads were fully characterized and a common feature believed to be a key feature for outstanding performance was found (Figure 25.20). Next, the iterative optimization was restarted using catalysts possessing this key feature, and a rapid acceleration of the optimization was observed. Though this last example cannot be transferred directly to heterogeneous catalysis, we strongly believe that efficiency will be gained when the evolutionary process is directed by knowledge obtained from powerful data analysis techniques performed on all the catalysts already tested. In the following, we describe a hybrid algorithm in which a genetic algorithm used to generate parents in a stochastic manner is mixed with a knowledge-based system (data-mining technique) for directing the selection of parents (Figure 25.21). This learning method is based on the extraction of equivalence classes, which accumulates strategic information on the whole search space previously explored to gain dynamic control of the browsing/exploitation ratio.

Search Spaces

Because GAs cannot proceed on intrinsic characteristics or properties of solids, two distinct search spaces are defined: the GA search space, which covers all the parameters of

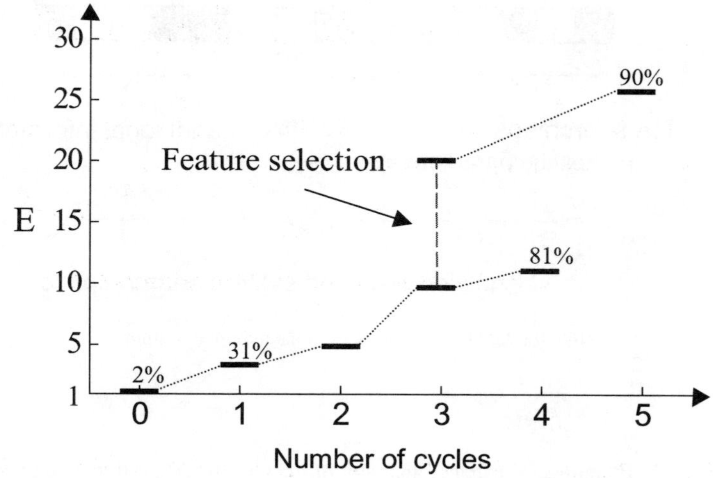

FIGURE 25.20. Iterative optimization of enantioselective enzymes [21]. Characterization of selected samples at the third generation allowed the design of libraries with well-identified structural key features which led to a breakthrough in the optimization process.

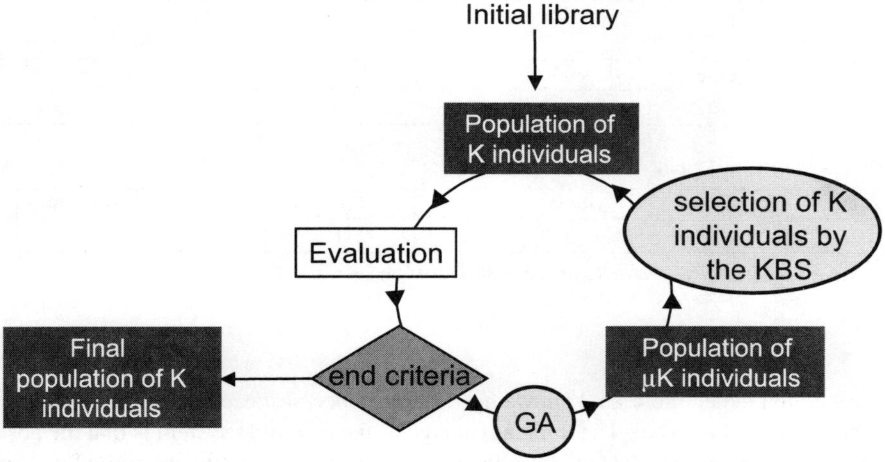

FIGURE 25.21. Mechanism of the hybrid algorithm. The KBS system directs evolution by selecting the individuals.

the synthesis and the testing, and the knowledge-based system (KBS) search space, which includes the GA search space, the results, and all the additional information (Figure 25.22).

GA Search Space

The hybrid genetic algorithm can run on mixtures of discrete ordered and nonordered variables as well as continuous data coded in a binary system. As a simplified example of

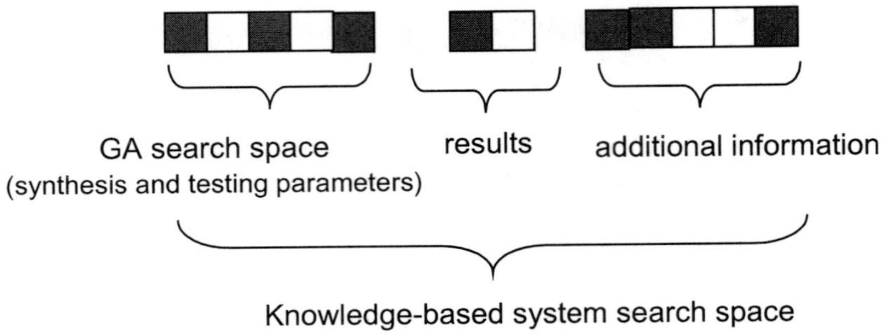

FIGURE 25.22. Search spaces in the hybrid algorithm.

TABLE 25.3. Examples of Binary Encoding of the Genetic Algorithm Search Space[a]

	V1	V2	V3	V4
Variable Name	Fe	Cu	Concentration range	Precursor type
presence/absence		presence/absence	[0,1]]1,2[[2,5]	nitrate/acetate/hydroxide

Ex #1: $[Fe(C_2H_5O_2)] = 0.25$ mol/l

Value	presence	absense		0.25 (mol/l)			acetate	
Code	1	0	1	0	0	0	1	0

Ex #2: $[Cu(NO_3)_2] = 1.5$ mol/l

Value	presence	absense		1.5 (mol/l)			hydroxide	
Code	0	1	0	1	0	0	0	1

[a]Continuous variables are discretized and coded by a set of binary values.

a binary coded gene, Table 25.3 shows how the presence/absence of continuous and nominal variables can be coded. The main advantage of the encoding system is that the absence of elements can be identified and classification can be carried out by enlarging knowledge.

Knowledge-Based System Search

KBS space is only binary and combines GA search space and external data which can be integrated. Table 25.4 shows how, on a set of zeolites used as supports, additional knowledge (pore diameter, Si/Al ratio) can be inserted in catalyst descriptors. The classification process is preferred because it makes the search of relationships easier.

Class Extraction as a Learning Process

The goal of the learning process in this hybrid algorithm is to identify classes of catalysts which could correspond to "empirical rules" deduced from conventional chemical

TABLE 25.4. Example of Information Entries in the Different Search Spaces

Sample	AG search space					Additional information							
	Zeolite Presence/absence					Ring		Pore Diameter			Si/Al		
	KL	NaZSM5	NaY	NaX	NaB	10	12	[5..6,9]	[7..7,1]	[7,2..7,4]	<5	]5..20[	>20
1	0	1	0	0	0	1	0	1	0	0	0	0	1
2	0	0	0	1	0	0	1	0	0	1	1	0	0
3	0	0	1	0	0	0	1	0	0	1	0	1	0
4	0	0	0	0	1	1	1	1	1	0	0	0	1
5	1	0	0	0	0	0	1	0	1	0	0	1	0
6	0	0	1	1	0	0	1	0	0	1	1	1	0
7	1	0	0	0	1	1	1	1	1	0	0	1	1

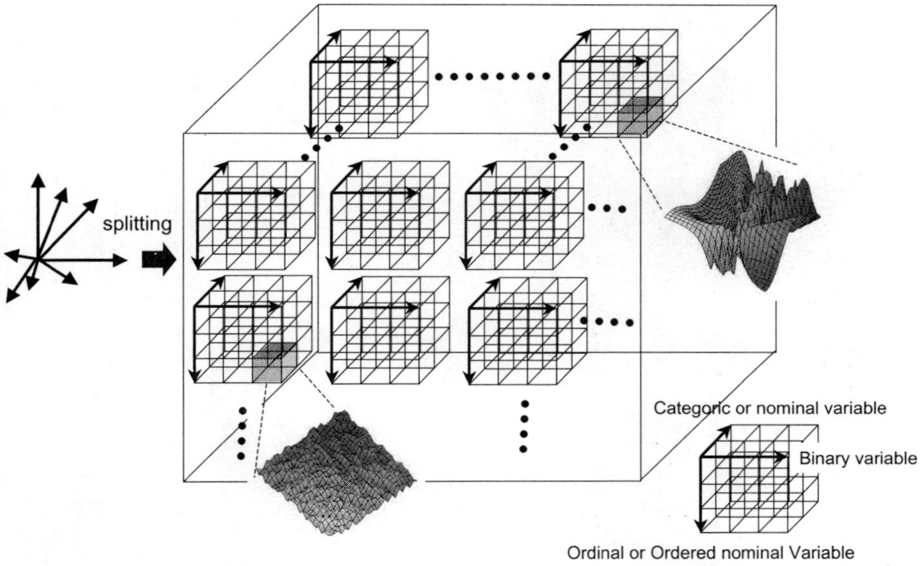

FIGURE 25.23. Splitting of parameter spaces into smaller search spaces consisting of continuous data.

expertise. In addition, the class system enables the difference between catalysts and how diverse they are to be evaluated. The identification of classes proceeds as follow.

1. The KBS search space first has to be split into hypercubes such that each hypercube corresponds to a single value of the discrete variables and contains a continuous surface (Figure 25.23). As an example, we consider the oxidation of ethylbenzene to styrene discussed earlier. Let us assume that the total search space and in turn the widest hypercube is constituted by the relationship between styrene yield, vanadium content, temperature, and preparation method. It can be divided into many smaller cubes by fixing discrete variables in which a continuous space can be found $Y = f(\%V)$ (Figure 25.24). In this simplified example, additional binary encoded information is not mentioned, but it is considered as well. For example, we could envisage splitting the space as a function of the size of the zeolite lattice opening rings.

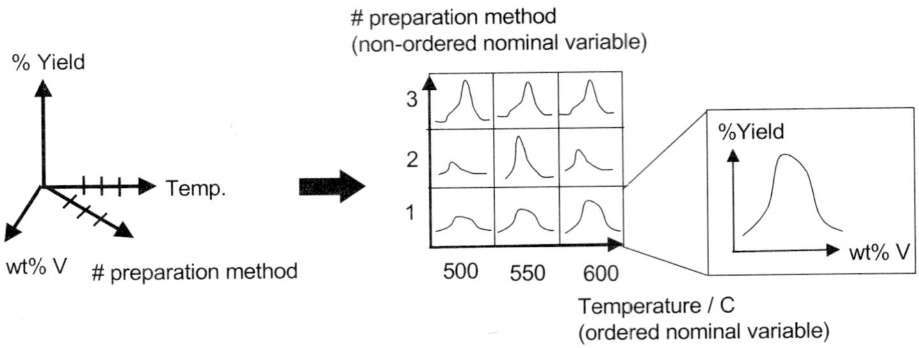

FIGURE 25.24. Example of splitting of parameter spaces for VMgO catalysts.

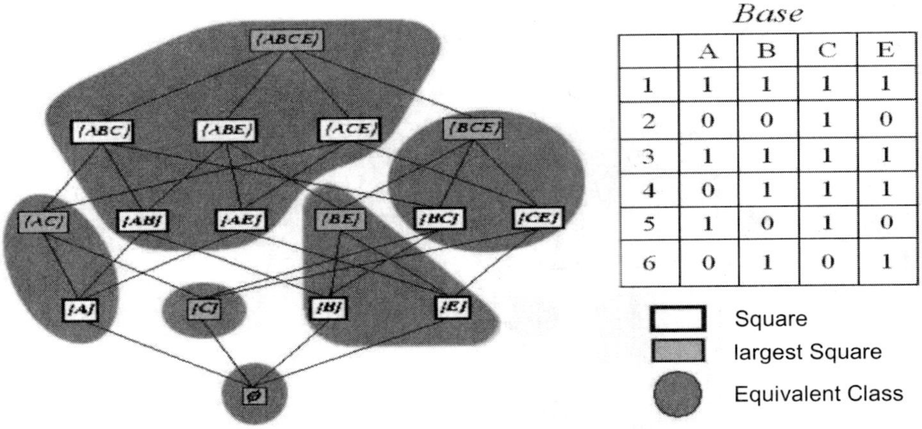

Base

	A	B	C	E
1	1	1	1	1
2	0	0	1	0
3	1	1	1	1
4	0	1	1	1
5	1	0	1	0
6	0	1	0	1

	C	A	B	E
1	1	1	1	1
2	1	0	0	0
3	1	1	1	1
5	1	1	0	0
4	1	0	1	1
6	0	0	1	1

	B	E	A	C
1	1	1	1	1
3	1	1	1	1
4	1	1	0	1
6	1	1	0	0
2	0	0	0	1
5	0	0	1	1

	A	C	B	E
1	1	1	1	1
3	1	1	1	1
5	1	1	0	0
2	0	1	0	0
4	0	1	1	1
6	0	0	1	1

	B	E	C	A
1	1	1	1	1
3	1	1	1	1
4	1	1	1	0
6	1	1	0	0
2	0	0	1	0
5	0	0	1	1

	B	E	C	A
1	1	1	1	1
3	1	1	1	1
4	1	1	1	0
6	1	1	0	0
2	0	0	1	0
5	0	0	1	1

FIGURE 25.25. Example of the mechanism of class formation starting with six individuals described by four binary variables (A, B, C, E). The five equivalent classes identified are shown as largest squares in the tables and are represented in the lattice scheme.

2. The equivalent classes correspond to the largest rectangles that are obtained by systematic permutations of the discrete variables and individuals. Figure 25.25 represents a simplified extraction of equivalent classes by looking at the largest rectangle of the research area (base) after having split the variables into smaller sets of variables.In this example, five distinct classes (or zones) can be determined which can be linked to other classes, thus making a lattice of classified variables. The more individuals (catalysts) that

are tested, the more defined will be the zones and the empirical rules which can be associated with them.

Hybridization GA: KBS coupling

Any GA can work in this hybrid algorithm provided that it generates individuals through its operators. Here, the GA is run n times starting from the same set of data in order to create n times more individuals than would generally be the case. The key concept of hybridization is that classification of individuals allows the most relevant individuals to be selected for the next generation. The question of the criteria of relevance arises. Two distinct evaluation functions of the relevance of each individual x of the search space are defined. The first function f_1 (x) provides an evaluation of space browsing; a high value is expected if x is located in a zone which has not been screened before and a low value if x is in a zone where many individuals have been already tested. In turn, this function enables the "real" diversity of an individual with respect to the occupancy rates of all the classified zone to be estimated and quantifies its contribution to the extent of browsing of the total search space. The second function f_2 (x), which evaluates the performances of x, takes into account the results of all individuals in the same classification zones as x (e.g., the mean value). This second function quantifies the relevance of an individual for improving exploitation. In order to obtain a unique selection criterion, a third function $R[f_1(x), f_2(x), t]$ which combines the two previous functions and the number t of individuals still to be tested, is defined. This enables the global relevance of an individual with respect to its interest for both the browsing and exploitation to be evaluated. In this manner, the relevance of individuals constituting each new generation can be increased and controlled as a function of the time. The choice of the functions f_1, f_2, and R is of major importance and completely defines the control of the optimization process. For example, if good catalysts have been found early in the same zone, the algorithm will be forced to perform the search in unexplored zones rather than in the same "hot" zone.

An additional advantage of GA–KBS hybridization is that the largest-rectangle analysis is independent of the order of variables whereas the GA are not. Indeed for GA, the greater the separation between two variables in a chromosome, the higher is the probability that there will be a crossover between. However, it is important not to cut blindness variables groups (called elementary units) for which values are correlated with high performances. Hybridization allows a dynamic reallocation of the positions of variables which increases the probability of efficient crossover.

CONCLUSIONS

The importance of robotics with respect to scientific creativity is probably overestimated in the HT approach. Most breakthroughs speeding up the discovery of new materials are unlikely to come from faster or highly parallel techniques, but probably from smart concepts allowing synthesis, screening, and further optimization via data mining. In addition, the combinatorial approach of heterogeneous catalysis will soon have to deal with even wider parameter spaces, taking into account both HT characterization data (development of

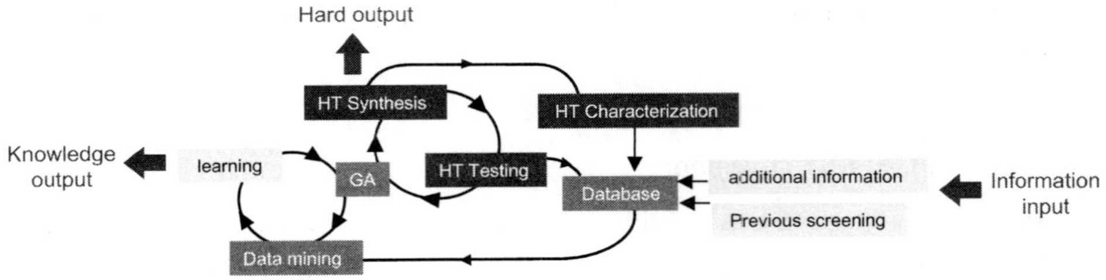

FIGURE 25.26. Scheme of advanced data management systems.

new tools in progress) and process operations and kinetics. Therefore the development of new and adapted data management tools is obviously and urgently required. The hybrid algorithm presented here, which combines data-mining and GA techniques, matches that demand quite well. Indeed, it will accelerate and adjust the convergence rate while generating understanding of both continuous and discrete systems. As a prospective methodology, we propose a search mechanism based on iterative optimization directed by data-mining techniques where a DB stores all the information and links together high-throughput screening by the hardware and the generation of understanding by the software (Figure 25.26).

ACKNOWLEDGMENT

The EU "Combicat" program is fully acknowledged for supporting part of the work reported here.

REFERENCES

1. Senkan, S. Combinatorial heterogeneous catalysis—a new path in an old field. *Angew. Chem. Int. Ed.* **2001**, *40*, 312–329.
2. Jandeleit, B., Schaefer, D. J., Powers, T. S., Turner, H. W., Weinberg, W. H. Combinatorial materials science and catalysis. *Angew. Chem., Int. Ed.* **1999**, *38*, 2494–2532.
3. Newsam, J. M., Schuth, F. Combinatorial approaches as a component of high-throughput experimentation (HTE) in catalysis research. *Biotechnol. Bioeng.* **1999**, *61*, 203–216.
4. Harold, M. P., Mills, P. L., Nicole, J. F. *Stud. Surf. Sci. Catal.* **2001**, *133*, 87–98.
5. Cohan, P. Results and commercialization—progress in the practice of combinatorial materials science. *Abstracts, 221st American Chemical Society Meeting*, **2001**, Paper BTEC-056.
6. Yoneda Y. Prospect, rather than retrospect, on the impact of computers in catalytic research and development. *Catal. Today* **1995**, *23*, 305–310.
7. Harmon, L. A., Vayda, A. J., Schlosser, S. G. Informatics challenges in combinatorial materials discovery. *Abstracts, 221st American Chemical Society Meeting*, **2001**, Paper BTEC-067.
8. Dorsett, D. R., Jr. Capturing the combinatorial workflow. *Abstracts, 221st American Chemical Society Meeting*, **2001**, Paper BTEC-064.
9. Cong, P., Dehestani, A., Doolen, R., Giaquinta, D. M., Guan, S., Markov, V., Poojary, D., Self, K., Turner, H., Weinberg, W. H. Combinatorial discovery of oxidative dehydrogenation catalysts within the Mo–V–Nb–O system. *Proc. Natl Acad. Sci. USA* **1999** *96*, 11077–11080.

10. Geisler, S., Vauthey, I., Zanthoff, H., Muhler, M. Presented at European Workshop on Combinatorial Catalysis (EuroCombiCat), Ischia, Italy, **2002**.
11. Wolf, D., Buyevskaya, O. V., Baerns, M. An evolutionary approach in the combinatorial selection and optimization of catalytic materials. *Appl. Catal. A* **2000**, *200*, 63–77.
12. Banares-Alcantara, R., Westerberg, A. W., Ko, E. I., Rychener, M. D. DECADE-A hybrid expert system for catalyst selection. I. Expert system consideration. *Comput. Chem. Eng.* **1987**, *11*, 265–277.
13. Banares-Alcantara, R., Ko, E. I., Westerberg, A. W., Rychener, M. D. DECADE—A hybrid expert system for catalyst selection. II. Final architecture and results. *Comput. Chem. Eng.* **1988**, *12*, 923–938
14. Kito, S., Hattori, T., Murakami Y. Expert systems approach to computer-aided design of catalysts. *Appl. Catal.* **1989**, *48*, 107–121.
15. Koerting, E., Baerns, M. Use of expert systems in catalyst development. Chemie Ingenieur Technik **1990**, *62* (5), 365–372.
16. Sun, Y. H., Li, Y. W. Expert system approach to the preparation of supported catalyst. *Chem. Eng. Sci.* **1992**, *47*, 2799–2804.
17. Prevoo, H., Derouane, E. G., Vercauteren, D. P. Development of a prototype expert system for catalysis by zeolites. *AIP Conf. Proc.* **1995**, *330*, 775–781.
18. Selvam, T., Iyer, D. N., Deka, R. C., Chatterjee, A., Vetrivel, R. A computational "expert system" approach to design synthesis routes for zeolite catalysts. *Stud. Surf. Sci. Catal.* **1997**, *105A*, 133–140.
19. Klanner, C., Baumes, L., Farrusseng, D., Mirodatos, C., Schueth, F. QASAR, in press.
20. Buyevskaya, O. V., Wolf, D., Baerns, M. Ethylene and propene by oxidative dehydrogenation of ethane and propane—Performance of rare-earth oxide-based catalysts and development of redox-type catalytic materials by combinatorial methods. *Catal. Today* **2000**, *62*, 91–99.
21. Reetz, M. T., Jaeger, K. E. Enantioselective enzymes for organic synthesis created by directed evolution. *Chem. Eur. J.* **2000** *6*, 407–412.

Chapter 26

Scaling Up of Catalysts Discovered from Small-Scale Experiments

Maureen L. Bricker, Ralph D. Gillespie, Jennifer S. Holmgren, J. W. Adriaan Sachtler, and Richard R. Willis

INTRODUCTION

The chemical industry faces a challenging business climate due to difficult economic conditions in much of the world, strong international competition, and worldwide environmental concern. In addition, innovation in this industry has slowed as catalyst and process technology has matured. The need for a methodology that can increase catalyst innovation while continuing to decrease cycle times has been recognized by the Council for Chemical Research (Catalysis Roadmap, Vision 2020) [1]. In the 1980s, the pharmaceutical industry faced similar circumstances. Downward pressure on drug prices became incompatible with the high cost of drug discovery. Combinatorial chemistry, based on advances in laboratory automation, high-throughput synthesis, and activity screening, allowed the pharmaceutical companies to break the innovation impasse [2,3].

We believe that the creation of tools and methods which enable the adaptation of combinatorial science in the chemical industry will result in a similar step change in laboratory productivity. This belief is shared by others and has fueled academic and industrial efforts to create a combinatorial toolbox for use in the invention of materials, catalysts, and adsorbents [4–29].

SINTEF was the first company in the world [20] to perform combinatorial hydrothermal synthesis of molecular sieves, materials used extensively as basis for catalysts and adsorbents in the chemical industry. The partnership between SINTEF and UOP LLC began in 1997 [30–32]. This alliance has developed a range of combinatorial test platforms and tools for synthesis of materials and catalysts. These tools include the following:

1. material synthesis, including parallel inorganic synthesis;

Maureen L. Bricker, Ralph D. Gillespie, Jennifer S. Holmgren, J.W. Adriaan Sachtler, and Richard R. Willis • UOP LLC, 25 East Algonquin Road, Des Plaines, IL 60017, USA
Contact author: Maureen L. Bricker, email: MLBricke@uop.com

High-Throughput Analysis, edited by Potyrailo and Amis
Kluwer Academic Publishers, New York, 2003.

2. catalyst preparation, including metal loading, oxidation/oxichlorination, and reduction;
3. catalyst evaluation, including probe reaction assays for catalyst acidity, basicity, metal function (dehydrogenation), and bifunctional (acid/metal) activity and commercial assays for propane dehydrogenation;
4. informatics, including flexible data management, analysis and visualization, mining and predictive performance modeling.

The approach was to develop the preparation and testing components on a modular basis. Automation of the synthesis and characterization systems has been reported [33]. To date, about a number of UOP LLC projects have been addressed successfully using the combinatorial approach [34–36].

In this chapter, an overview of the tools that are used as well as examples of their applicability to the commercialization work process will be presented.

CAPABILITIES

Library Design

Once a catalyst project has been initiated and the preparation methods identified, the experimenter must begin to design the experimental space that will be examined. The variables can include the major catalyst components, their concentration ranges, the reagents that will be used, the order in which experimental steps will be carried out, and which of the many possible treatments will be chosen for study.

An example of the choice between possible experimental protocols is given in Figure 26.1. This example compares two options for adding two reagents to a solid. In

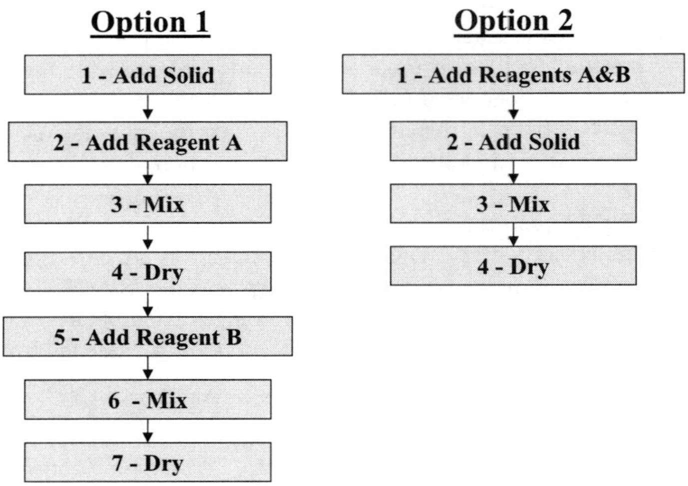

FIGURE 26.1. Experimental protocol examples. In Option 1, catalyst support solid is added to liquids, while the opposite protocol is utilized in Option 2.

option 1, reagent A is added followed by two treatment steps and then reagent B is added. Alternatively, the order of A and B could be reversed. The second option involves adding reagents A and B together and then adding the solid to the mixture followed by post-treatment. The choice between these and other protocols represent the first step in preparing the experimental design.

Figure 26.2 demonstrates the next step in developing an experimental design. Here a single protocol has been chosen. Of course, it is possible for the experimenter to use two or more protocols in a design—the protocol choice becomes a design variable. In this case with one protocol, the possible process variables available for variation in the design have been identified. Step 1 has two variables, step 2 has four, step 3 has three, and so on. At this point it is generally preferable to limit the number of process variables that will be studied. For example, one experimental protocol of seven steps had 22 variables. If all 22 variables were studied at only two levels, a full factorial would give over 4 million experiments, while studying four variables at three levels produces only 81 experiments. While 81 experiments once required a major commitment of time and resources with the conventional approach, it is a relatively small number when using combinatorial tools.

One of the major factors that will help guide the definition of the experimental space is whether the study will be a screening study or an optimization study. Generally, when first approaching a new project, it is desirable to start with screening studies. For a new exploratory system, the design that has 81 diverse samples may be the best for the exploratory stage. On the other hand, if a catalyst prototype has been defined, a week or two of testing the sensitivity of the preparation variables may help direct the commercialization steps. Screening studies can be as simple as examining a number of different supports with the same active component or something as complex as examining a ternary or higher

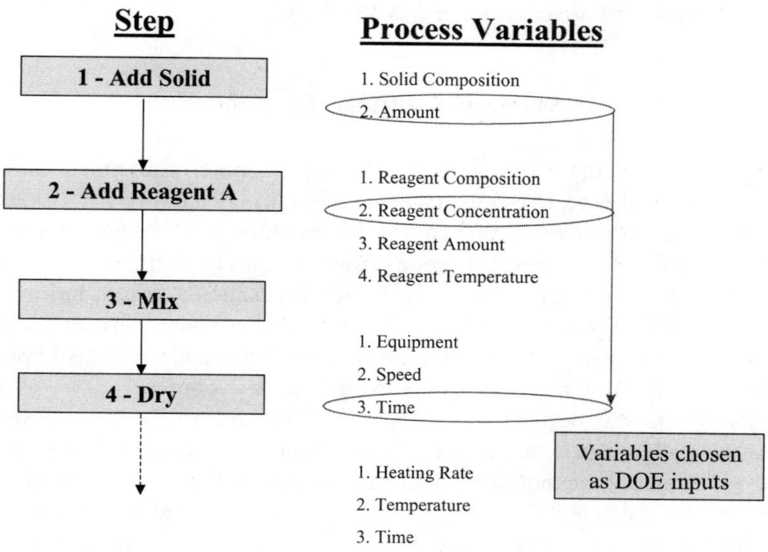

FIGURE 26.2. Process variable selection. Each step of an experimental protocol has process variables that can be selected for evaluation.

system in a true combinatorial study. As the studies shift toward optimization, the designs can shift to factorial designs, optimal coverage designs or other response surface designs. In all of these cases, the large number of samples that can be studied in a short amount of time greatly enhances the experimenter's ability to apply statistical design tools. Complex studies that took 3–6 months or more using traditional tools can now be done in a week or less.

After the design of experiment (DOE) has been generated, it must be translated into specific experimental operations. These operations include solution preparation, powder weighing, solution pipetting, and heat treatment. Because most of the combinatorial synthesis tools are automated, this requires the production of control files for each type of apparatus. Efficient translation tools are needed to convert the design to the instrument control files. These tools must also send data to a central database that will track each preparation step. If manual steps are needed, these also must be tracked by the database system.

Although a single control program could operate all instruments, this limits flexibility. In our system, each synthesis instrument is operated independently. There is no overall system that enforces a particular order to the steps. The experimenter uses the control file that is made for the current step to run the laboratory protocol. This is a very useful way of conducting the combinatorial synthesis. It allows complete flexibility in the order in which the experimental steps are conducted.

It is critical to have meaningful descriptors for a particular sample. These descriptors must then be combined with the assay data. Catalyst composition is a simple starting point for describing a sample. However, additional data related to the specifics of the preparation steps are also required. For example, if samples were treated at a range of temperatures, the treatment temperature, time, and gas composition would be used as descriptors. In another case, the order of reagent addition may be varied. Thus, reagent addition order would become a desired catalyst descriptor. The descriptors will change for each set of experiments as the process variables being examined change. The use of a database with powerful query tools that are easy to use is a necessity.

Material Synthesis Module

In order for combinatorial chemistry to impact material synthesis methods in a positive, meaningful, and relevant manner, tools must be continually developed and validated versus standard scale-up procedures. For example, most combinatorial materials synthesis is carried out in quiescent mode. However, in commercial materials synthesis plants, reactors are typically stirred and/or reactor contents mixed during synthesis. Therefore evaluation of various internally designed and commercially available mixing devices has been an ongoing effort. Some commercially available systems have provided meaningful stirred reaction results down to about 5 mL gel volume.

Another system was designed and constructed in-house to allow for mixing during synthesis at the 1 mL scale. A variety of reaction mixtures that should give different results depending on the mixing method were selected for study. All the reaction mixtures had to have common synthesis conditions (temperature and time), and all the reaction mixtures had to be homogeneous for ease of addition to the combinatorial mixed reactor system. The combinatorial mixing systems were heated in quiescent and mixing mode. Meanwhile, typical laboratory-scale reactors were also operated quiescent and mixed

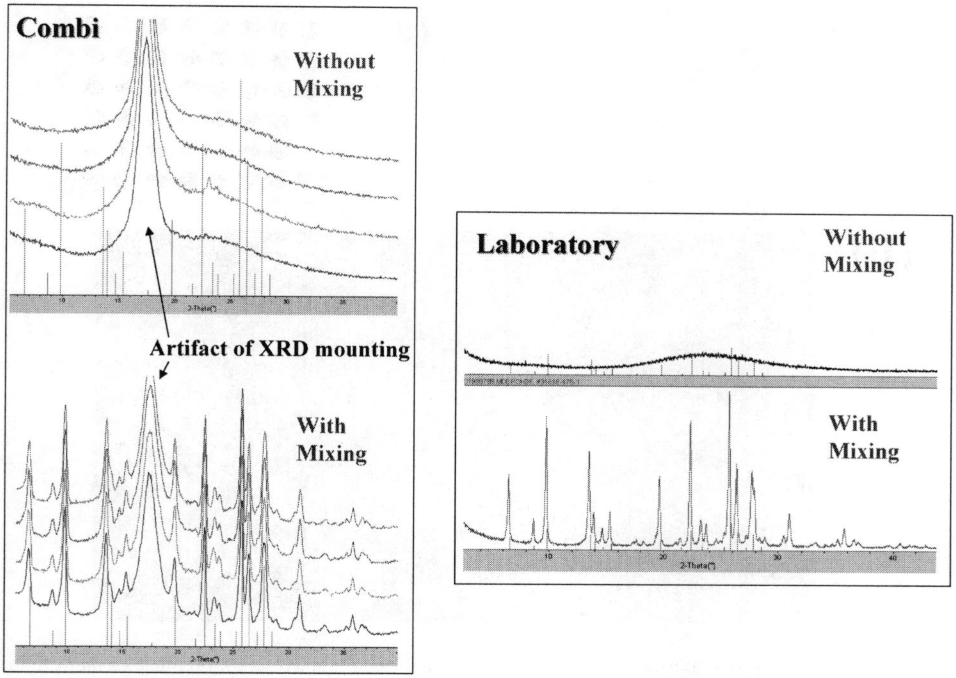

FIGURE 26.3. Production of the zeolite MOR: mixing versus quiescent conditions. Combinatorial and typical laboratory-scale MOR synthesis products derived from quiescent and homogenized reaction environments are shown. MOR only forms in a homogenized reaction environment.

in order to serve as reference points. All reaction products were evaluated by X-ray diffraction (XRD). Among the materials chosen for evaluation in this study was the zeolite mordenite (MOR).

In the MOR recipe example, it is known that quiescent reaction conditions produce only amorphous solids. Mixing provides good quality MOR (Figure 26.3). Thus, the MOR preparation showed that combinatorial 1 mL scale mixed and quiescent conditions provide the same results as typical laboratory mixed and quiescent conditions, respectively.

Characterization

The most common combinatorial characterization technique is XRD. XRD data collection can be accomplished in true high-throughput fashion. Nearly all inorganic materials, including many catalysts, can be characterized via XRD. However, there are instances where XRD data are inappropriate or insufficient to differentiate samples. For example, there is a general situation where high-throughput qualitative XRD data collection, coupled with high-throughput scanning electron microscopic (SEM) imaging, can be utilized very effectively. That is, it is possible that XRD-identical products can be differentiated by

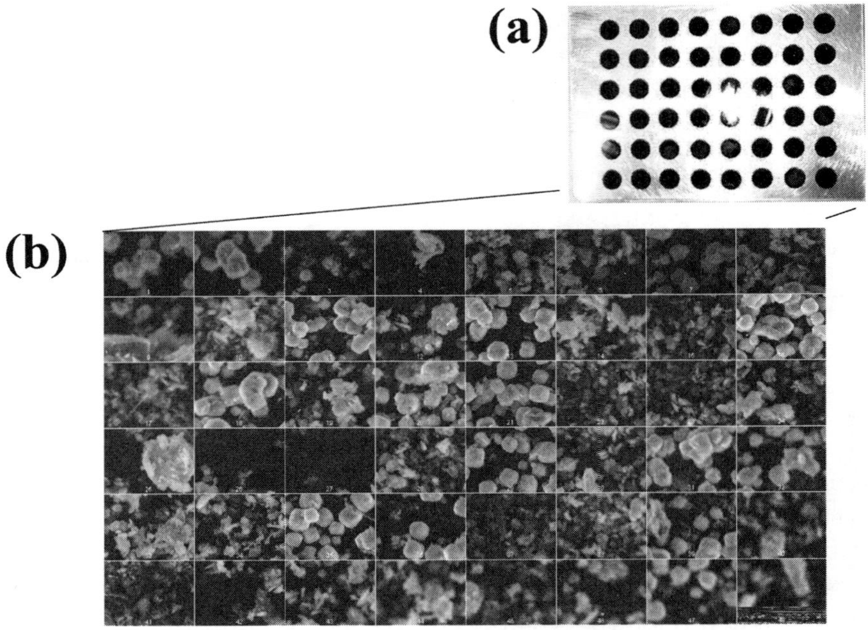

FIGURE 26.4. Characterization example: low-vacuum scanning electron microscopy. (a) Samples on plate for combinatorial high-throughput SEM analysis. (b) Montage of 48 images collected at 1,000× magnification.

changes in product morphology, particle size, aspect ratio, or some other property readily characterized by high-throughput SEM.

In order to demonstrate this point, a set of sodium A-type (NaA) zeolite [37] samples were prepared in the combinatorial material synthesis system. Among the variables examined were silica to alumina framework oxide ratios—from 1.48 to 2.69. This set of NaA syntheses was selected because varying framework oxide ratios are known to provide products with different morphologies. The expected morphologies are beveled cubic when $X = 1.48$, and cubic with sharp edges when $X = 2.69$. A thumbnail montage of NaA SEM images showed that there was good dispersion of material on the disks of the plate at a magnification of 1,000× (Figure 26.4), and that both beveled cubes and sharper-edged cubes are observable. Image analysis software can also be utilized to obtain more quantitative information from high-throughput SEM images.

Catalyst Preparation Module

The combinatorial preparation and evaluation of catalysts has some unique challenges that set it apart from other applications of combinatorial chemistry. Catalysts are extremely sensitive to the preparation scheme used for their production. Catalyst preparation frequently involves five to ten steps or more. Each step can have a number of key input variables. The net result is millions of different possibilities for even a simple catalyst problem. Even with the higher preparation and testing rates brought about by applying high-throughput methods, experiments must be carefully designed for efficient evaluation of a subset of the possible experimental space. The catalytic expertise of the experiment designer is an important factor in determining the success of the combinatorial catalysis effort.

Catalyst preparations are often scale and equipment dependent. A catalyst that works well when prepared in small-scale equipment may perform poorly when prepared in commercial-scale equipment. It is important to have high-throughput preparation methods that will successfully translate to the methods used in large-scale preparations. Care must be taken during the development of high-throughput equipment to account for the scale effect. Likewise, scaling-down of difficult preparations to small preparations must be done in a relevant manner to make this practice of small-scale preparations acceptable. The translation between laboratory preparations and commercial productions are important to UOP LLC where we invent, develop, and commercialize materials and catalysts.

An example of how combinational-scale equipment can be utilized to provide relevant data is the generation of zeolite ion exchange isotherms. Such isotherms are essential in the design and operation of commercial zeolite manufacturing plants. As an example, a detailed potassium for hydronium ion exchange isotherm was generated readily in combinational-scale equipment for the zeolite ferrierite (FER). A plot of mole fraction of potassium in the solids versus mole fraction of potassium in the solution at equilibrium is shown in Figure 26.5. These data suggest that FER is selective toward potassium at low ratios of potassium ion containing solution to ion exchange sites in the solid, but that it is not selective for potassium at high ratios of potassium ion containing solution to ion exchange sites in the solid. These results are consistent with reference data at low ratios of potassium ion containing solution to ion exchange sites in the solid. Results at higher ratios are new.

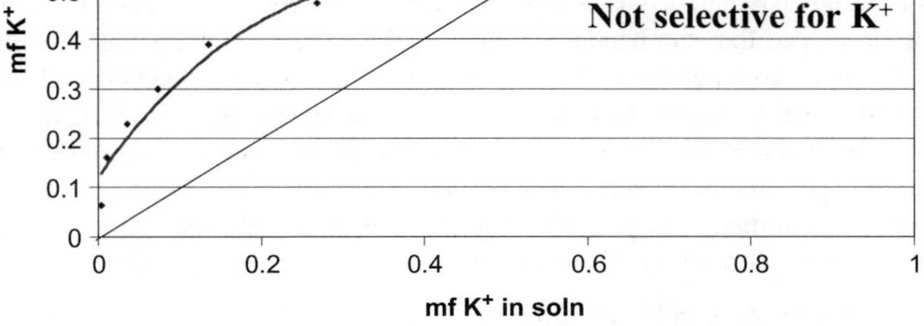

FIGURE 26.5. Full potassium for hydronium ion-exchange isotherm for FER zeolite. FER selectivity for potassium cation decreases as solution concentration increases.

Heat Treatment

Heat treatment under various gases is another critical unit operation for many inorganic materials. At UOP LLC we have a heat treatment unit capable of combinatorial time, temperature, gas composition, and flow studies. This system has been validated for drying, calcination, oxidation, and steaming unit operations. In one example, an objective was to determine what conditions are required to return a commercial catalyst to reference catalyst performance. Our response was the combinatorial methylcyclohexane (MCH) reactor assay described in detail below.

Variables in this study were chlorine level, time, and temperature. Even though the experimental design (three-variable three-level Box-Behnken [38]) required only 20 samples, 60 samples were prepared at about the 0.5 mL scale. This enabled built-in replication in order to provide more trustworthy and relevant results. Combinatorial MCH test assay results are summarized in Figure 26.6. It is clear from these results that chlorine level was the only variable to contribute significantly to the catalytic response. In other words, the midpoint and highest chlorine levels provided reference catalyst performance at all times and temperatures, while the lowest chlorine level provided poorer performing catalysts at all times and temperatures. Therefore the key variable was identified as chlorine concentration. This information can be used to assist the development of the commercial process.

Reactor Assay

Testing Hardware

The reactor system that is used is a complex system comprised of feed system, system hardware, reactor module, analytical system, hardware control, automation, and data acquisition. Because the system is modular, the required flexibility can be designed into a part of the system without having to redevelop the entire test train. Development can be focused on the part of the system that needs to be adapted for the next application. For example, in one experiment, the feed system was modified and the rest of the test system was left intact. In another case, only the sampling system and analytical assay were changed.

Feed Systems

The feed system is the starting point of the test system. The current generation of test systems is designed to handle gas blends or single-component liquid feeds. However, systems are being developed that can handle complex liquids, which can vary from a simple model feed to a complex commercial feed from a refinery. The choice of feed depends on the goal of the experimental program. A model feed may be acceptable for screening new materials. A commercial feed may be necessary as the last step in deciding on the prototype system for optimization. Typically the choice of feed is application dependent, and the experimenter must decide what compromises can be made. Whatever the choice, the feed must be delivered in a uniform manner to all reactors. There are several options for feed systems:

- a gas blend that can be fed directly
- a saturator in which as a gas stream passes through a liquid it picks up vapor from the liquid
- a gas blending system to combine multiple gas streams.

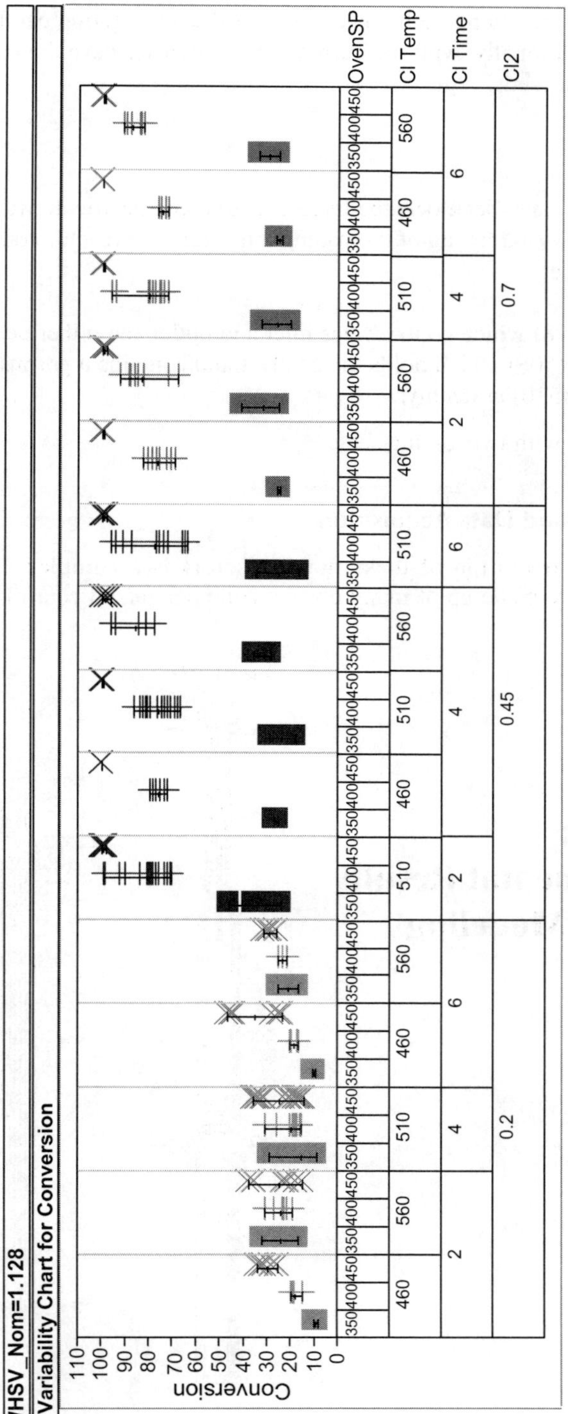

FIGURE 26.6. Oxichlorination study results. Combinatorial MCH reactor assay data suggest that chlorine is the only significant input variable.

With the saturator or gas blending hardware, the feed concentration can be changed during a single test. Additionally, with the current unit design we have the option of evaluating up to six different feeds.

Reactor Hardware

Special reactors have been designed and constructed specifically for parallel operation. Figure 26.7 shows a drawing of the patented reactor. The reactor contains three components which include:

- reactor top (518) which contains gas inlet and outlet and a thermowell assembly
- reactor insert (508) which holds the catalyst and contains a porous bottom
- reactor outer (510) assembly.

The reactor is sealed with two O-ring seals.

Hardware Control and Data Acquisition

Supporting the operation of these novel reactors is a complex infrastructure of hardware and software made up of many diverse components, all controlled by dedicated

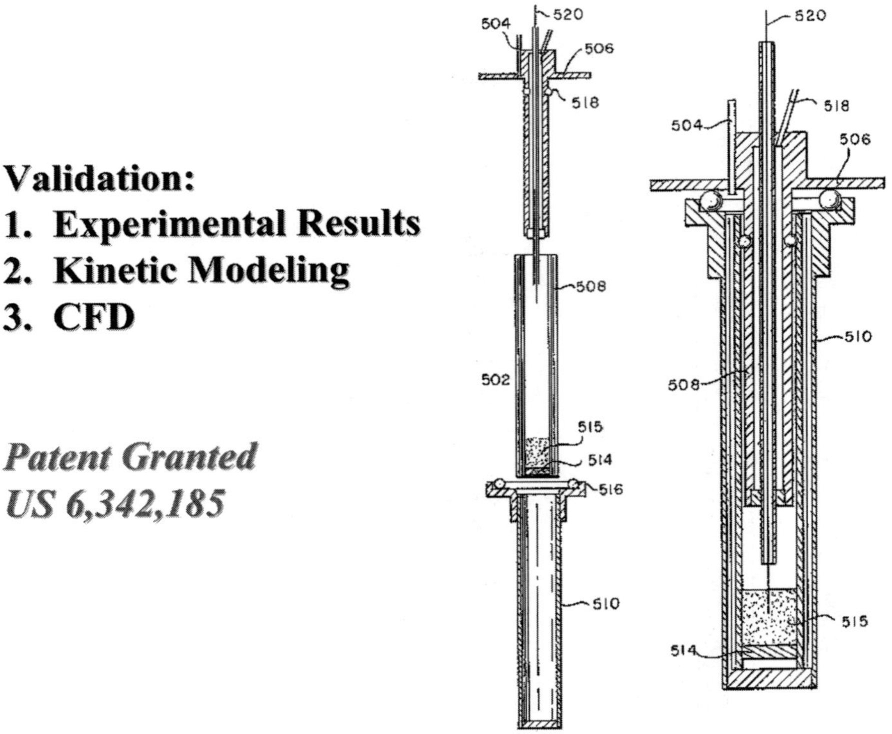

FIGURE 26.7. Combinatorial catalyst reactor. A combinatorial reactor design example [39].

software. The modularity of the system was an advantage when building large complex prototypes. The final number and types of Input–Output signals were not in place until the end the project. Having this flexibility was helpful for adding or removing modules as needed. The hardware was as follows:

- multiposition valves for gas stream selection
- mass flow controllers for individual control of gas streams
- multiloop temperature regulators for a variety of heated zones in the system
- modules for registering temperatures inside the reactors and other essential parts
- modules for recording pressure on the reactor lines and elsewhere
- modules for general analog input and for operating valves
- dedicated setups with specially designed multiposition valves for the sampling system
- analytical instrumentation, gas chromatographs (GCs)
- dedicated tools for automated calibration of pressure, flow and temperature sensors.

Automation

Lab VIEW Software, available from National Instruments Corporation, has been used to write a program named TestRoom™ to control the reactor. This program has evolved over time. It is used to control the oven, the analytical train, the mass flow controllers, and the sampling system. Typically, the tests are set up during the day and run unattended at night, and the data are summarized and uploaded to databases during the next day. The target time for a typical test, including pretreatment and analytical time, is 10–20 hr. Usually, the plants run every day.

The main automation system features the following attributes.

1. Complete control over the analysis software and hardware via ActiveX communication.
2. Reporting and logging of data in either ASCII or Microsoft Excel format (Microsoft Excel is available from the Microsoft Corporation) is selected by the operator.
3. Advanced setting up of tests via a dedicated editor (MethodBuilder™ software).
4. Dynamic change of run parameters during a test. There are options for pausing or editing remaining parts of a test during a run.
5. Four levels of alarm/error handling causing various actions. The highest level aborts the tests and starts an emergency shutdown method that flushes the system with inert gas and cools the oven.
6. Error/alarm messages sent via e-mail to operator and system administrator for notification.
7. Dedicated editor for the sampling/analysis module.
8. Subroutines for automated calibration of the large sets of sensors and controllers.
9. Dynamic controls for setting 48 flow set-points, individually, by column, by row, or using a single value for all reactors.

Key process parameters can be viewed on screen and the process conditions followed in real time. Typically the most important of these are the pressure, temperature, and gas flow. Two monitors are used. The controls and indicators that represent the hardware are displayed on the left monitor, while all charts and graphs are displayed on the right monitor. Because much of the test is done in an unattended mode, the ability to scroll back the plant history for the previous 24 hr was made possible. This review allows any unusual plant parameters to be evaluated. For instance, it may be observed that the reactor flow did not reach the set-point. The reasons for this would be investigated before the next run. In this case, this may be due to the pressure buildup in the reactor, a valve not turning, or a flowmeter failure.

A critical requirement of the system was ease of planning and setting up an experiment. To aid this task, a dedicated editor was developed. Through easy to use step-by-step operations, this tool provides the operator extensive flexibility in setting up all parameters, including the sequencing and timing of various actions within the test. Because the whole test procedure can be saved in a method file, complex test procedures can be preserved for documentation and reused.

Data Analysis

Data Flow: Step 1 Analysis Quality Control

After the run is completed, the first step is to review the GC data. A dedicated application, DataPainter™ software, is used for this task. This program imports the GC ASCII report files (typically several thousand files per run) and renders the data range with a color code that is similar to how a weather map shows temperatures. Low values are rendered in blue, intermediate values in green, and high values in red. Within a color, lower values are shown in darker shades and higher values in lighter shades. The data are arranged in a grid of reactor ID versus injection number. Using this color visualization, a block of thousands of numbers can be reviewed in only a few seconds (Figure 26.8).

One GC item is visualized at a time. Drop-down selection boxes allow easy selection of summary items such as total area counts, unidentified peak area counts, number of peaks, or any individual GC peak. For any individual peak, the retention time, area counts, or concentration can be displayed. For each peak, the minimum and maximum limit can be specified and any points outside the acceptable range are automatically colored differently. This feature allows the viewer to refine the visual range of key samples for a look at a further distribution. Additionally, summary values such as the sum of cracked products, the sum of selected isomers, or heavies can be calculated and viewed. This tool provides the first quality screen of the run by revealing whether any chromatograms have problems. If a problem is found, the cell can be selected and, by clicking a button, the entire GC trace is imported and graphed. The operator decides if the GC injection can be "rescued" by reanalysis with the GC software (e.g., a small retention time shift caused a peak to be missed), or whether this analysis needs to be discarded (e.g., if a sample valve failed to inject and the chromatogram contains no peaks). In the latter case, the GC is marked as invalid and this quality flag is carried forward through the subsequent calculations and uploaded to the database.

Although the DataPainter™ program was developed as a GC quality screening tool, it can easily be used to obtain a first look at catalyst performance. Activity can be evaluated by visualizing the concentration of unconverted feed and selectivity by rendering sums of desired or undesired components.

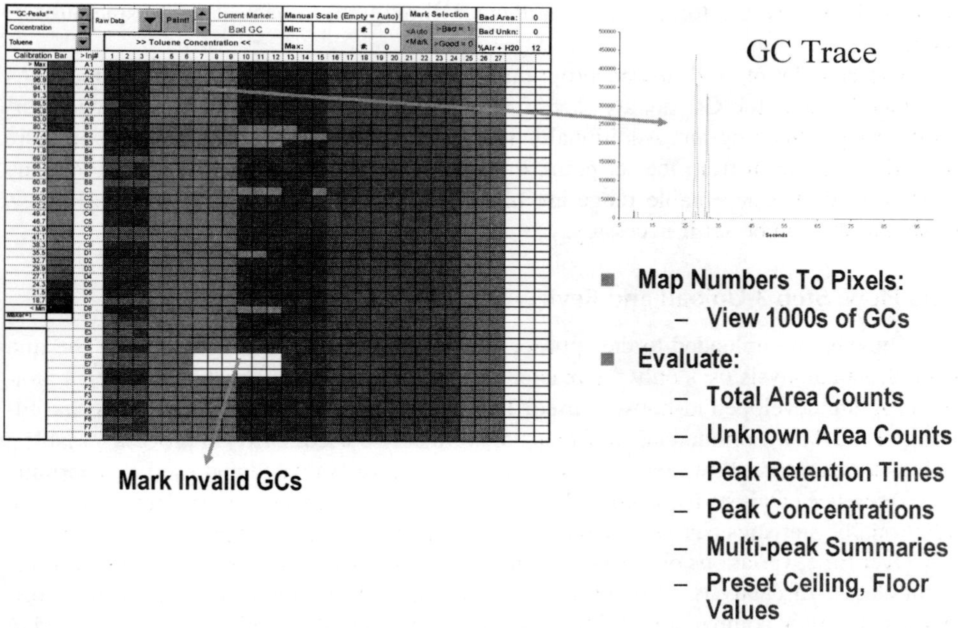

FIGURE 26.8. Reactor assay data visualization (DataPainter™). A snapshot view of the entire data set is possible on one screen. Single catalyst reactor data can also be screened.

Data Flow: Step 2 Calculations and Process Quality Control

The calculation and upload program takes the GC data from the DataPainter™ software, imports the process ASCII file that matches the GC injections, the zero-time-on-stream file, and the catalyst loading information file. Other information, such as the feed analysis and component information, is retrieved from the database. Then, several groups of calculations are performed.

1. Calculations on process variables, such as the vapor pressure of feed hydrocarbons in the saturator based on measured saturator temperature and pressure, weight hourly space velocity (WHSV), liquid hourly space velocity (LHSV), etc. These calculations take into account information on the process flow scheme, which for the current systems primarily consists of the feed system configuration. Thus the calculations can easily be toggled between using premixed feed gas blends, a hydrocarbon saturator, or a gas blending system.
2. Generalized conversion, yield, and selectivity calculations from the GC results and the feed analysis. These calculations are done for each component using a process-independent generalized formula.
3. Process-specific calculations, such as selectivities or yields for groups of components, product isomer ratios, approach to thermodynamic equilibrium, etc.

The calculation and upload program provides a number of graphs to allow the operator to review the calculated data and look for any data outliers. At this time, one can easily switch back and forth to the DataPainter™ program to inspect any suspicious GC results.

The calculation and upload program provides additional quality control tools. As mentioned above, the GC quality flags from the DataPainter™ software are carried forward through this program. Additionally, specifications can be set for maximum pressure drop, flow deviation from the set-point, or temperature deviation from the set-point. Any points outside the acceptable range are highlighted and can be marked as bad. Finally, a manual flag can be set if necessary.

Data Flow: Step 3 Upload and Review

The data are uploaded to the appropriate database tables (see Figure 26.9). The first level of data analysis uses only the reactor data described above. For this analysis, a plotting program developed in-house is used. This program allows extensive grouping and filtering of the data. Commercial statistical software is also used. Drifts in process variables, such as temperature, flow, and reactor pressure drop, are evaluated and used to determine if the reference catalysts included in the run match the performance of prior reference runs. Additionally, statistics can be calculated on data quality and reactor productivity.

Over time, variations on a given test method develop that may be better suited for different classes of catalysts, or be preferred by a customer to provide a match with historical data. To allow comparison of different runs the method information is also uploaded to the database where any differences can be quickly identified using a special algorithm.

Once the reactor test has been reviewed, the reactor and catalyst preparation databases are combined (Figure 26.10). Various queries have been written to combine and sort particular data from the database for further analysis. The merged catalyst descriptors and assay results are used for analysis and visualization. The resulting data set can be used to determine performance–property relationships, lead identification, and catalyst optimization. The analyses can be relatively simple, for example, finding the best results from the sample set using color-coded performance data. The data analyses can be more complex, generating sophisticated response surfaces and model generation. The former is generally

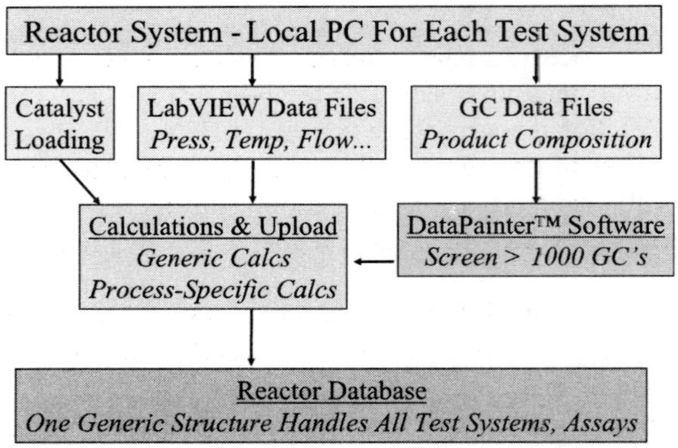

FIGURE 26.9. Reactor assay data flow to database schematic. One generic structure handles all reactor assay systems.

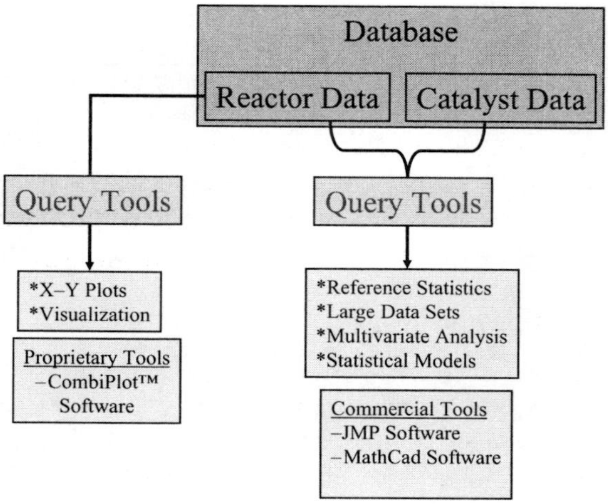

FIGURE 26.10. Catalyst preparation and reactor assay module data flow from database schematic. Query tools enable combination of catalyst preparation variables with reactor assay results.

used in the early discovery phase of the process and the latter is used once the optimization phase has begun. This analysis relies heavily on statistical tools. An internal software application, CombiPlot^{TM}, or the commercial JMP5 statistical software (SAS Institute Inc.) can be used to evaluate the data. With JMP, the data analysis tools are used to compare the mean performance of duplicate catalysts and dissect the experimental variables with the variability plot. Models can be generated and trends can be predicted for optimization. Typically a single run is not evaluated at this level. It is more common to generate a large data set for all the tests in the DOE.

However, as the dimensionality of the system increases, more sophisticated visualizations are needed that require more mathematically oriented software. Mathcad Professional 2001i (Mathsoft Engineering and Education Inc.) was used to generate the visualization shown in Figure 26.11. The tetrahedron representing the quaternary compositional information has "used up" all the degrees of freedom for making a three-dimensional graph. Since it is also desirable to show performance in the same diagram, a "fourth dimension" is introduced by coding the symbols based on performance, in this example the toluene yield from the MCH dehydrogenation reaction.

Test Assay

In one application, micro-GCs are used for analysis of feed and product. This type of micro-GC has thermal conductivity (TC) detectors. Micro-GCs are used for systems that analyze gas because they can give fast analysis. A proprietary sampling module stores effluent samples from all reactors simultaneously. This has the advantage that all reactors are sampled at the same time-on-stream. A mixed parallel–serial approach was used. Three GCs with two modules each provide six analytical channels. In systems where the feed or products may contain heavies, this sample storage system is not used. Instead, the serial injections result in offsets in the time-on-stream values of the analyses. The analytical

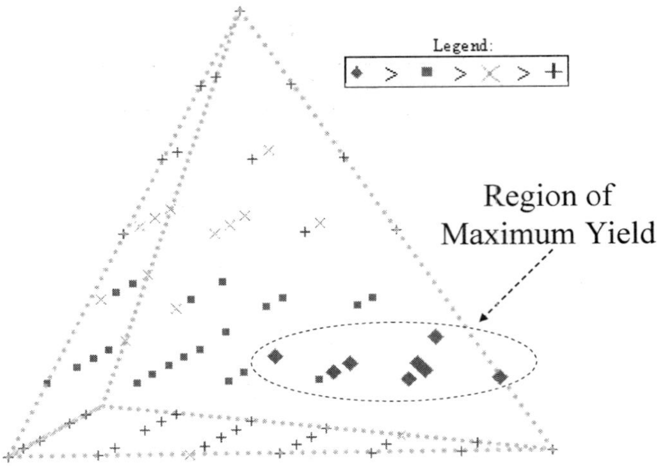

FIGURE 26.11. "Five-dimensional" visualization. An example of how to evaluate five variables is shown. The tetrahedron figure represents four metals that add up to a constant level. The space within the tetrahedron is colored by the results of the testing. Red represents more active catalyst composition regions.

methods are optimized for each type of test. The goal is to have all samples analyzed in about 30 min. A compromise on data analysis for screening is acceptable. Separation is not needed for every single peak, but only for those that can be meaningful for drawing a conclusion about the activity or selectivity of a catalyst.

Strategy of Reference Testing

Reference catalysts are included in every test. Multiple reference testing allows for the consistency of catalyst comparisons. Typical screening runs are done with a minimum of six reference catalysts, one in each row. This allows the six analyzer channels to be checked in every run. In this scenario, the positions of the references within the row change in every test. Periodically, all reactors are evaluated with reference catalysts. This strategy allows catalysts to be compared easily within a single run and across multiple runs. This also allows tracking of the optimization of each reactor's performance. The reactor, or components of the reactor, can be changed as necessary.

Reference Testing/Reactor Consistency

Every test is evaluated initially in terms of the reference performance. The reference data are compared with past references. An example of reference performance over time is shown in Figure 26.12. This analysis shows an offset in conversion at run 59, which returned to the prior baseline conversion after run 61. It is possible to determine that this offset occurred after a group of catalysts containing a new component, which may not have been thermally stable, was evaluated. One possible conclusion was that the decomposition product of those specific catalysts adsorbed on the reactor metals and poisoned the hydrogenation activity of the subsequent reference catalyst. After several runs, the poison was

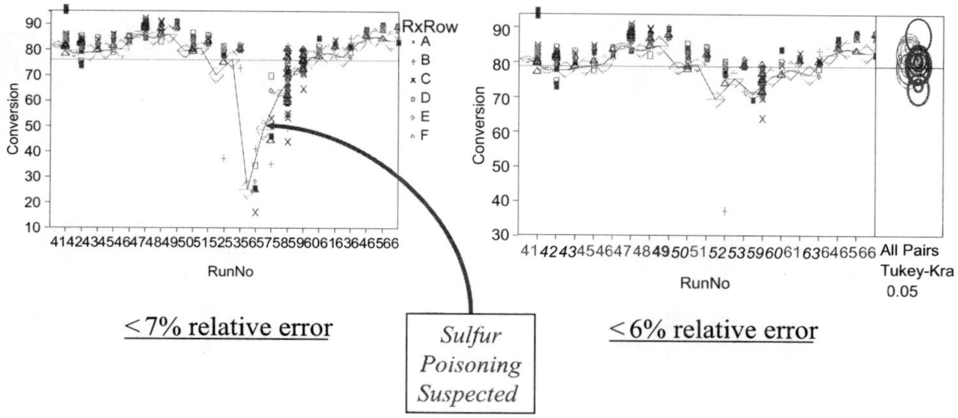

FIGURE 26.12. Reactor assay module system III reproducibility. These data represent MCH test reference catalysts by run at 275 and 300°C.

flushed from the system and the activity returned to baseline. In a typical reactor system, there is less than 7% relative error over more than 25 runs. Over a shorter span of tests, the relative error is reduced to less than 4%. The precision of this test is not sufficient for fine resolution of subtle differences between catalysts. Instead, this test is meant to evaluate catalysts with widely different activities and compositions.

Experimental Catalyst Evaluation

For testing, a group of catalysts are assembled. This group can come from one block design, or be assembled from various catalyst designs, or come from reference catalyst lots. The catalyst identifications, loaded volumes, and weights are recorded in a file as described above.

For optimal performance the catalysts are pretreated in situ. Air, nitrogen, hydrogen, or dilute hydrogen sulfide in hydrogen can be used to prepare the proper catalyst function. The test parameters to be investigated are set up in method file. The typical tests are designed to evaluate the catalyst performance over a wide range. In a typical test, both temperature and space velocity are varied, covering nine or ten different conditions. Ideally, the test method is set up for the duration of an entire program so that all the data can be evaluated in the same way.

The catalysts are placed in the reactor inserts and then dropped into the outer reactor assembly. The array is placed in the reactor holder and then all 48 reactors are sealed in one operation. Before every test, the system is pressure tested with nitrogen to check that the system will hold pressure. If any leaks are observed, they can often be fixed by opening and closing the block, thereby re-engaging the O-ring seals. If the system does not seal after several attempts, some component of the reactor is changed.

The performance of the system was first validated by showing the correct ranking of catalysts with known performance, that is, performance measured in an established non-combinatorial test. Four samples of different generations of Platforming™ catalysts were

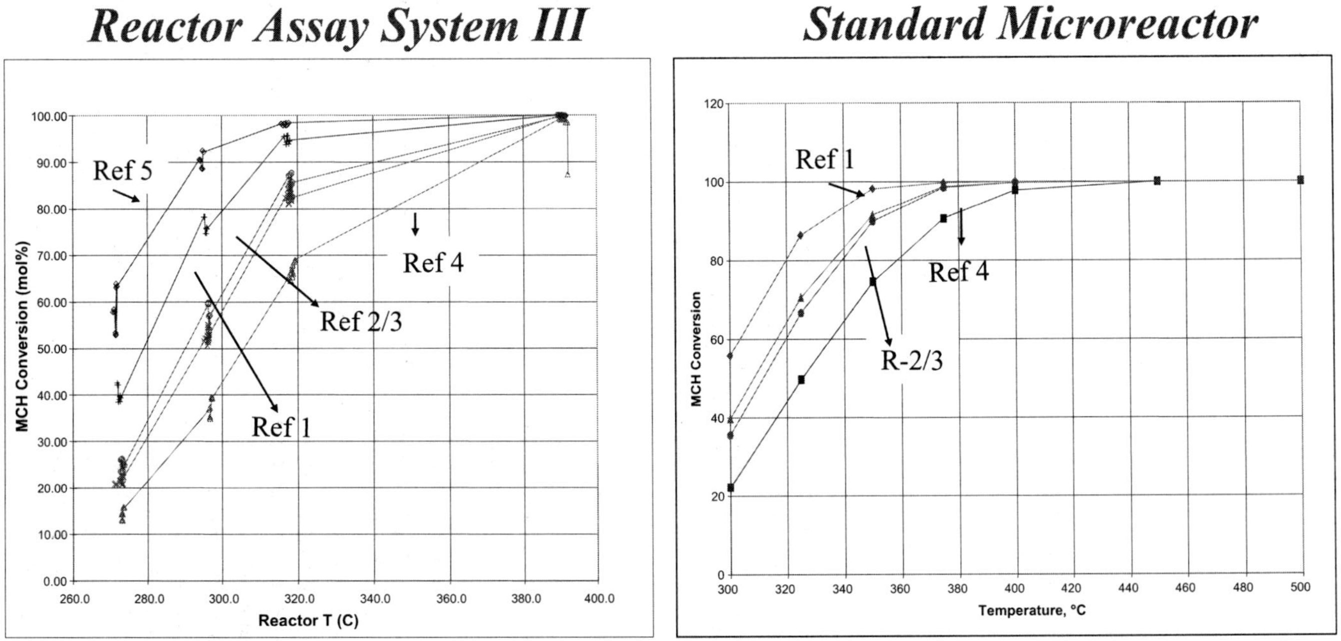

FIGURE 26.13. System III ranking of Platforming catalysts. Reference catalyst testing in the combinatorial reactor assay module system III and in a typical microreactor reveals identical trends.

evaluated. These samples were chosen for the diversity that was expected in the MCH dehydrogenation tests. These catalysts are designated here as Ref 1, Ref 2, Ref 3 and Ref 4. They differ in the amount of active metal and the degree of attentuation by a second modifier. These catalysts were treated with the same pretreatment conditions. As the benchmark, test results were chosen from the standard microreactor test using similar conditions. A comparison of the combinatorial test and the standard microreactor test is shown in Figure 26.13. The combinatorial test and the microreactor test correctly rank the activity of the catalysts showing Ref 1 > Ref 2~Ref 3 > Ref 4. The data further support our conclusion that this test can differentiate catalysts that are 5% different.

CASE STUDIES

Materials Synthesis

In this section, a material synthesis example will be used to illustrate the utilization of combinatorial chemistry tools for developing cheaper and more robust product manufacturing recipes. In the example, high-quality meaningful data were generated to help guide more expensive laboratory- and pilot-scale experimentation and development. Additionally, this material synthesis example will demonstrate how the results from such experimentation can be used to develop a predictive model suitable for more typical laboratory- and pilot-scale product development sensitivity studies.

For the researcher operating in an exploratory mode [30–32,34,35], qualitative XRD information is required. Such data can be acquired in true high-throughput fashion. However, often slightly different material synthesis recipes can result in high-purity zeolites and molecular sieves. For these materials, the best measure of product quality is crystallinity as determined by quantitative XRD. It is important to realize that, when a researcher operates in such an optimization mode, the quality of such quantitative XRD data is of paramount importance. To obtain such data, better instrumentation and/or more time-consuming analyses may be required.

The following material synthesis recipe example shows the application of combinatorial synthesis to generate sensitivity data around a proposed Material C synthesis problem. The primary objective of this work was to provide a more robust and less expensive recipe to scale-up and commercial operations. In addition, it was a goal to provide a mathematical model with material synthesis conditions and formulation information as input variables and product crystallinity as the response to the input variables.

UOP LLC's laboratory and commercial experience with Material C is extensive. The overall experimental plan was to eliminate the use of one expensive reagent (R1), which was postulated to be unnecessary, and to reduce the amount of, or substitute for, another reagent (R2). The designs also included a lower crystallization temperature and less expensive sources of framework oxides.

A concurrent set of material synthesis experiments was conducted at combinatorial and development scales. The scale-up group focused on the potential removal of R1 and reduction and/or replacement of R2 from the synthesis recipe, while the combinatorial group also examined lowering crystallization temperature and utilization of less expensive framework oxide sources. The combinatorial experimental objectives were to replicate and

TABLE 26.1. Recipes from Material C Optimization Screening Design That Provided Pure Material C and Did Not Contain R1 or R2[a]

Recipe	R2-R (mole)	NaOH (mole)	Nucleating agent (level)
1	6	3	2
2	6	3	1
3	6	1.5	2
4	6	1.5	1
5	6	1.5	0
6	3	1.5	2
7	3	1.5	1

[a]Mole ratios are referenced to 1 mol of alumina in each recipe.

augment the larger-scale work and to explore further Material C synthesis space. This approach was to have two outcomes: (1) the validation of combinatorial scale results with pilot-scale results, and (2) a more detailed definition of the relationship(s) between synthesis variables in this system. Combinatorial recipes that produced pure Material C but contain no R1 and R2 are shown in Table 26.1.

A second set of combinatorial experiments was carried out to refine further the synthesis composition range in the less expensive oxide system. In Figure 26.14, the compositional experimental design is outlined in scatter-box format. XRD spectral results, summarized in Figure 26.15, show the strong effects of crystallization temperature (higher is better) and nucleating agent component (presence is good) on product crystallinity and phase composition. The results were used to generate a simple model that confirmed that crystallinity increases with increasing R2 replacement and decreasing NaOH content.

Scale-up experiments were performed based on the newly derived cost-effective Material C recipes. Pure Material C properties were further examined for acidity as measured by pyridine IR (pyIR) adsorption, particle size and morphology as measured by scanning SEM, and catalytic activity as measured in a microreactor screening test. All desired product characteristics matched or exceeded those of the reference material (Figure 26.16). A commercial-scale cost reduction of about 40% for Material C was achieved as a result of this work.

The objective of providing valid and relevant data to scale up hydrothermal synthesis operations was met. In fact, some pilot work being carried out in parallel to the work described above was halted early owing to the rapidly completed and successful combinatorial work. It is clear that combinatorial chemistry provided a deeper understanding of material synthesis variables versus product formation in a short period of time.

Catalyst Preparation

Paraffin isomerization is an important refinery process that is used to increase the octane of light paraffin refinery streams. UOP LLC is active in this market with several available catalyst formulations including I-8/I-80, HS-10, and LPI-100. All the catalysts have their place in today's refinery, but there is a trade-off between catalyst activity and

Multivariate

Correlations

	R1	R2-R	NaOH	Temperature
R1	1.0000	−0.0147	0.0562	0.0000
R2-R	−0.0147	1.0000	0.1608	−0.0000
NaOH	0.0562	0.1608	1.0000	−0.0000
Temperature	0.0000	−0.0000	−0.0000	1.0000

Scatter-plot Matrix

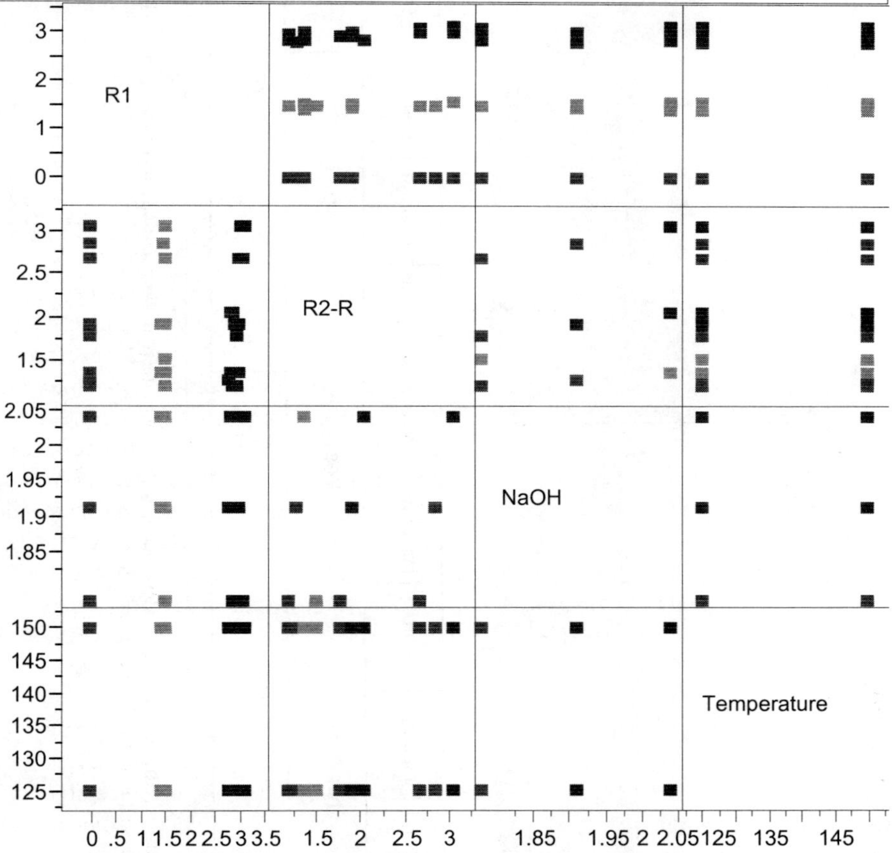

FIGURE 26.14. Material C synthesis experimental design. The scatter-plot matrix shows the range of experimental parameters. The data points shown reflect the actual contents of each well. The minor correlation of the R1 and NaOH variables is the unintended result of experimental error in volumes of liquid reagents dispensed.

operating conditions. It was considered desirable to find a new catalyst composition that would have high activity at conditions dictated by existing isomerization units.

The generation of a new catalyst lead was conducted in several stages. The first stage consisted of demonstrating that the performance of known catalysts could be equaled by

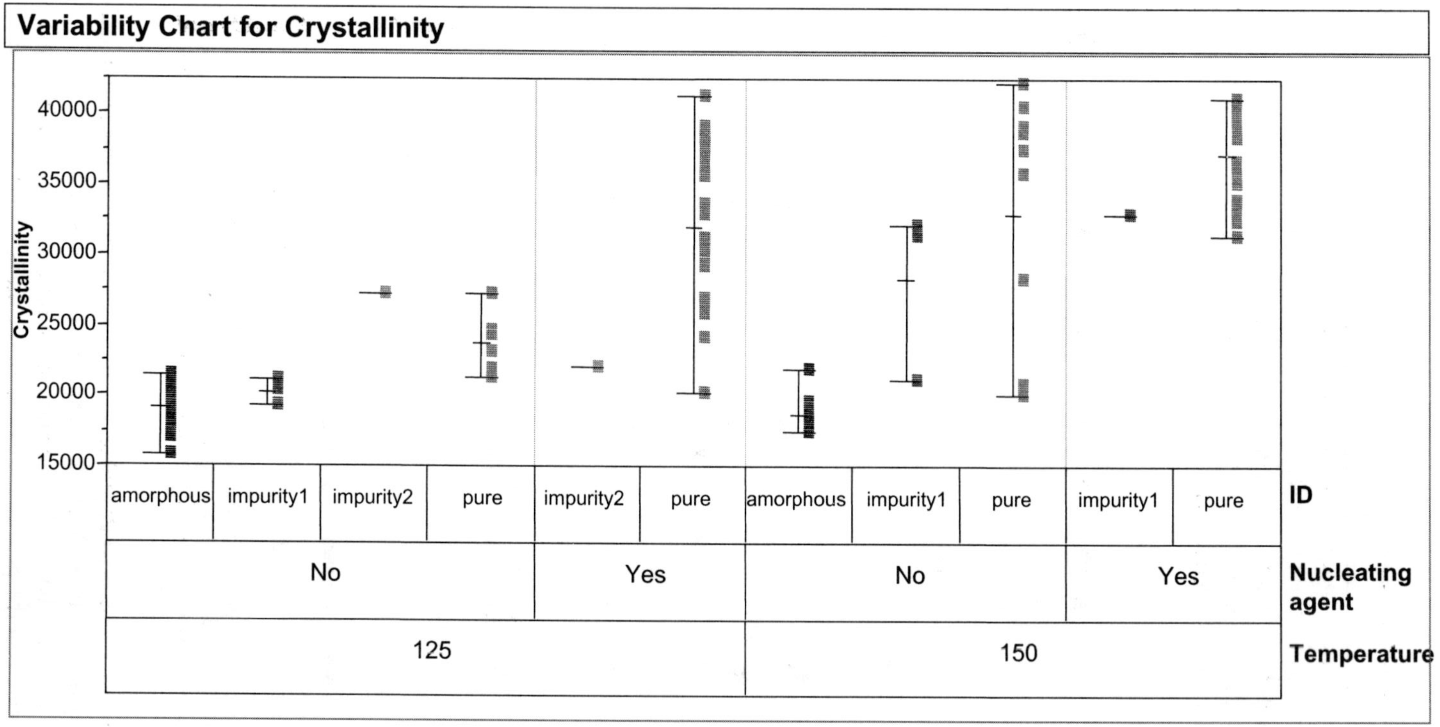

FIGURE 26.15. Material C synthesis XRD spectral results. XRD data show the strong effects of crystallization temperature (higher is better) and nucleating agent component (presence is good) on product crystallinity and phase composition.

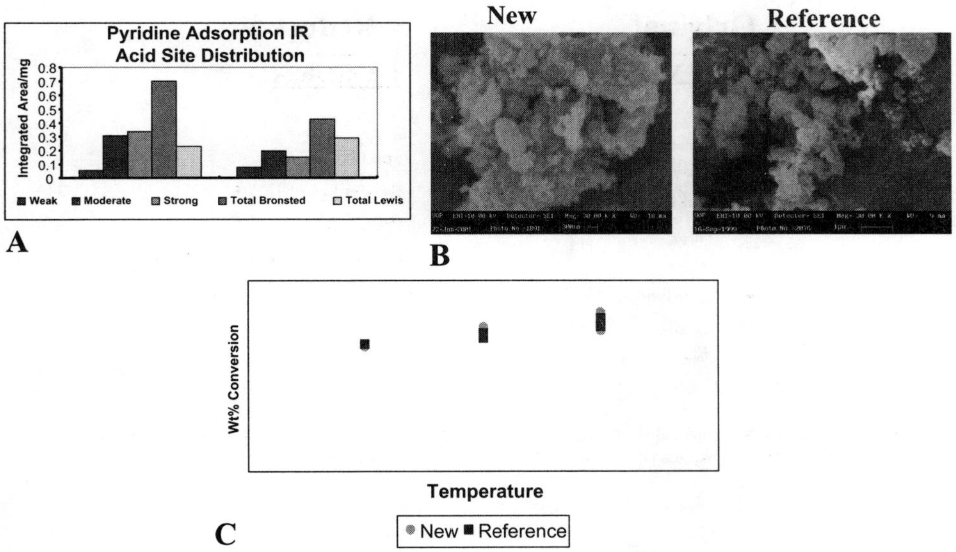

FIGURE 26.16. Scaled-up Material C properties. Material C scaled up from the combinatorial recipe is as good as or better than reference Material C. Among properties determined were acidity as measured by (A) pyIR adsorption, (B) particle size and morphology as measured by SEM, and (C) catalytic activity as measured by a microreactor screening test.

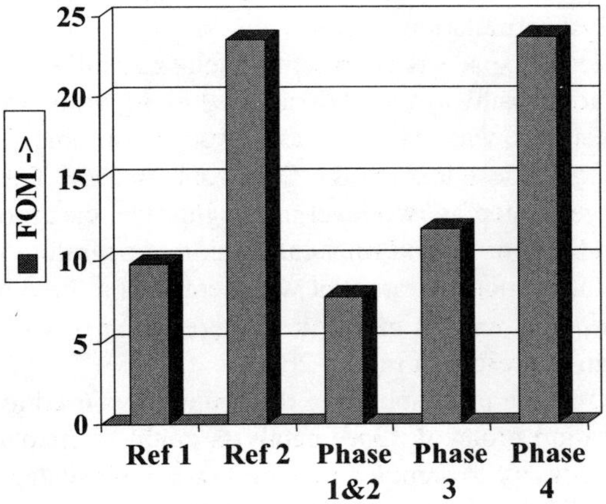

FIGURE 26.17. Combinatorial catalyst development: figure of merit (FOM). The change in the maximum FOM achieved during four successive phases of a combinatorial catalyst development study. The activity in phase 4 equals that of the best reference catalyst.

samples prepared at the combinatorial scale. This effort required the evaluation of very general catalyst preparation variables. These included the types of reagents that were used, the order in which components were added to the starting support, the overall order of the major preparation steps, and the component levels required to have moderate catalyst activity. The progress of the experimentation of this initial stage is represented in Figure 26.17.

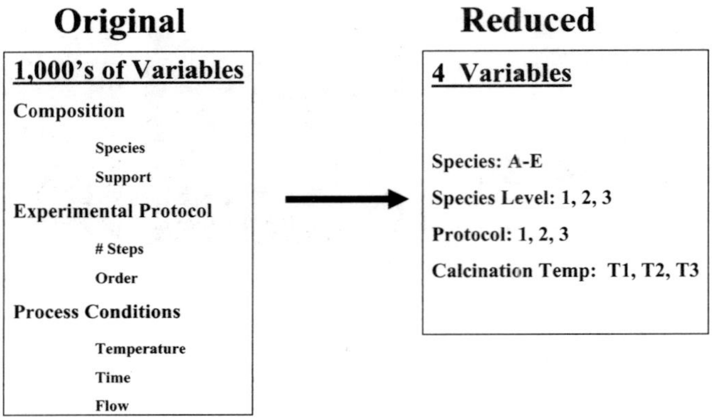

FIGURE 26.18. Combinatorial catalyst preparation key input variables. Schematic representation of the design space considered in the late stages of a combinatorial catalyst development study.

A steady improvement in catalyst performance can be seen from phase 1 through phase 4. The activity in phase 4 was equal to the existing state-of-the art catalyst (Ref 2). These results allowed the experimental space to be narrowed down to a more manageable size—several key process variables were identified such as calcination temperature and reagent addition order. Subsequent work focused on these variables along with the effect of adding modifiers to the catalyst formulation.

The narrowed design space is represented schematically in Figure 26.18. In this example, two preparation pathways were examined along with single modifier systems. The calcination temperature was also used as a process variable. The actual design was done as a general factorial design using the DOE tools in JMP. The preparation path and catalyst modifiers were treated as two-level and eight-level categorical variables, respectively. The reagent addition order and final calcination temperature were treated as blocking variables. The catalyst modifier amount was treated as a three-level variable. The net result of all these variables was an initial design preparing 6×48 samples. The overall structure of this design is present in Figure 26.19.

As described above, the initial phase of the project was used to determine the appropriate catalyst preparation protocol. Once catalysts could be made on the combinatorial scale with equivalent activity to samples prepared on a bulk scale, a number of elements were examined as modifiers to the base composition. Approximately 10 modifiers were examined in the first set. The results, shown in Figure 26.20, indicated that a number of samples had activity greater than the reference catalyst (indicated by a 95% confidence interval band on the plot). The best three samples from this plot were chosen and larger amounts of each were prepared.

The three samples, designated Combi #1, #2, and #3, were then tested in a small pilot plant used for traditional catalyst research. All three samples were found to have higher activity than the reference catalyst as shown in Figure 26.21. Combi #3 was found to have the highest activity level, while still maintaining good selectivity to the desired product. The modifier X used in this catalyst sample was then subjected to an optimization study.

		Species																							
		A			B			C			D			E			F			G			H		
Calcination	Experimental	Species Level																							
Temperature	Protocol	1	2	3	1	2	3	1	2	3	1	2	3	1	2	3	1	2	3	1	2	3	1	2	3
T1	**P1**																								
	P2																								
	P3																								
T2	**P1**																								
	P2																								
	P3																								
T3	**P1**																								
	P2																								
	P3																								

Variable	Levels
Species	8
Species Level	3
Preparation Protocol	3
Calcination Temperature	3
Total Experimental	$3 \times 3 \times 3 \times 8 = 216$
Reference Samples	72
Total Samples	288

FIGURE 26.19. Combinatorial catalyst development experimental design. An example four-variable factorial design with three three-level variables and one eight-level variable.

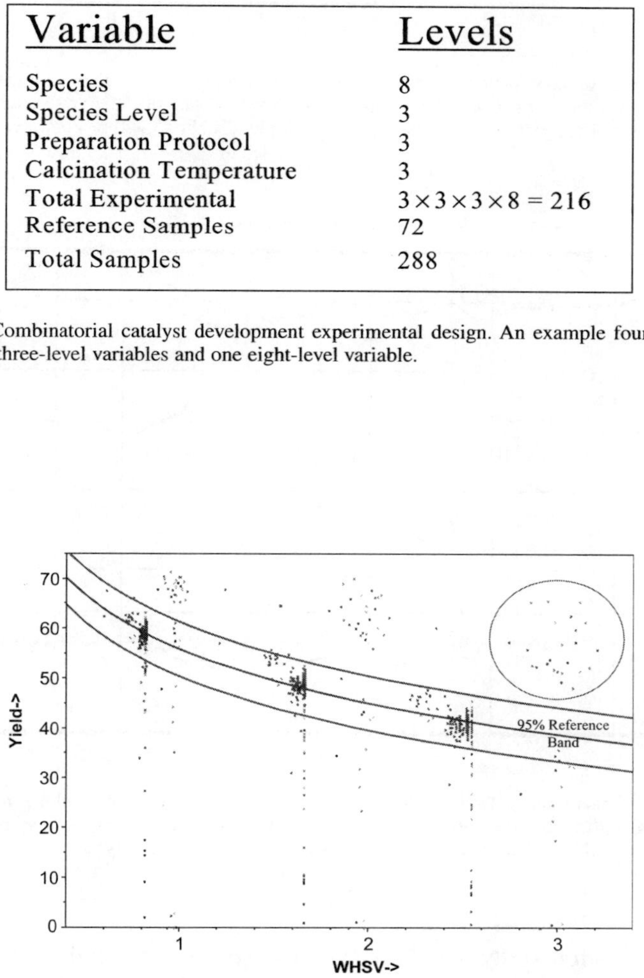

FIGURE 26.20. Combinatorial catalyst development reactor assay data. A plot of product yield versus space velocity for a combinatorial catalyst development study. A 95% confidence reference band was generated. Candidate catalysts were compared with the reference catalyst. Those lying above the reference band (circled) have higher yield than the reference catalyst and are all identified as potential catalyst leads.

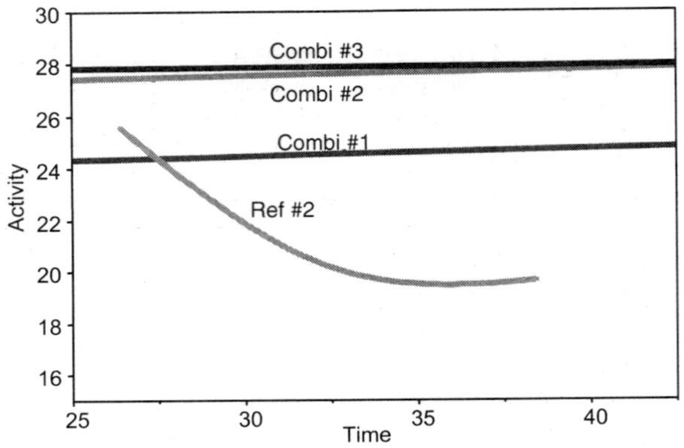

FIGURE 26.21. Scaled-up combinatorial catalyst development samples in pilot plant. Comparison of candidate catalyst leads from a combinatorial catalyst development study with a state-of-the-art reference catalyst. All three catalyst leads have better performance than the reference sample. Combi #3 will be commercialized.

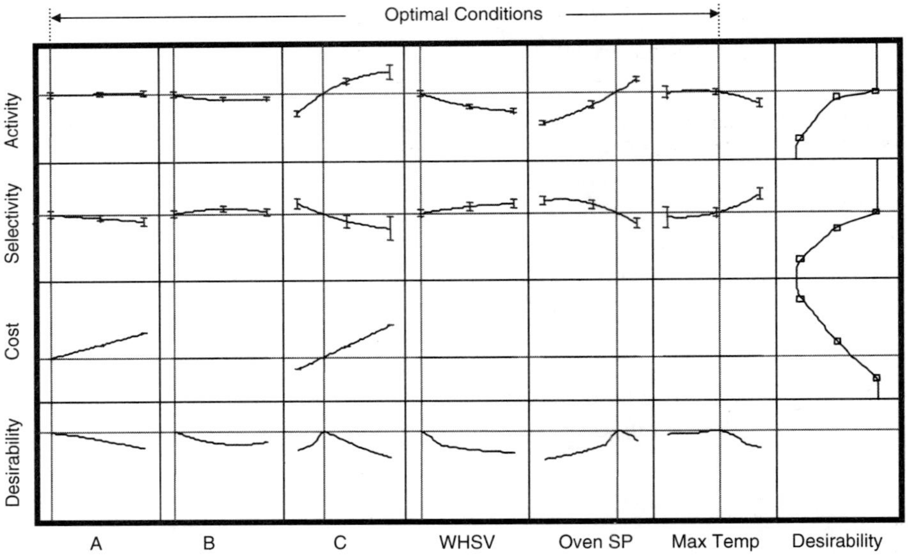

FIGURE 26.22. Combinatorial catalyst development optimization. An optimization plot generated with JMP™ (SAS Institute) software with three output variables (conversion, selectivity, and catalyst cost) plotted against six key input variables. The system was optimized to achieve maximum activity and selectivity with minimum catalyst cost.

The optimization study was based on a response surface design. The two base components were varied over a small range and modifier X was varied over a larger range and the calcination temperature was varied. Approximately 150 samples were used to develop a model in which the various components and temperatures were optimized. The optimization process is shown in Figure 26.22. This optimization was done using tools available in JMP. The catalyst composition and treatment temperature were optimized for activity, selectivity, and material costs. Including the material costs allows a rational evaluation of the trade-off that sometimes occurs between catalyst cost and performance.

The catalyst lead identified using the combinatorial system was transferred to a technology development project for further testing. Scale-up experiments were performed to prepare a commercial prototype for use in pilot plant studies in order to confirm the results observed during the combinatorial study. The commercial prototype catalyst based on Combi #3 has exhibited a significant performance enhancement over the existing zeolitic catalyst (HS-10) used for pentane and hexane isomerization. When run at equilibrium, the new catalyst, now designated PI-242, produces a product with an improvement of approximately three or four octane numbers over HS-10. PI-242 is now being offered for sale as a replacement for HS-10.

In this example, the catalyst lead generation was completed in 5 weeks through the preparation and screening of 512 samples. This is in contrast with earlier work performed by the same researcher that was carried out over 3 years using traditional methods where 271 samples were prepared, all of which failed to result in a viable lead. The success obtained using the combinational system can be attributed directly to this systematic, yet efficient, means of exploring the experimental space.

SUMMARY

During the past year, the use of the combinatorial toolbox in UOP LLC's laboratories has validated the original hypothesis that combinatorial methods would create a step change in the productivity of the laboratories. Most importantly, one of the distinctions found in the implementation of combinatorial methods in these laboratories, compared with that in the pharmaceutical industry, is the general applicability of the toolbox across the commercialization work process. In fact, there has been significant impact beyond the discovery step, which has been used to fast track the development and scale-up stages as well.

Although the use of combinatorial methods in the chemical industry is still in its infancy, tremendous benefits and opportunities for growth will be realized.

REFERENCES

1. American Chemical Society, American Institute of Chemical Engineers, Chemical Manufacturers Association, Council for Chemical Research, and Synthetic Organic Chemical Manufacturers Association. *Technology Vision 2020*: *The U.S. Chemical Industry*, December **1996**.
2. Fenniri, H. Recent advances at the interface of medicinal and combinatorial chemistry. Views on methodologies for the generation and evaluation of diversity and application to molecular recognition and catalysts. *Curr. Med. Chem.* **1996**, *3*, 343.
3. Rohrer, S. P., Birzin, E. T., Mosley, R.T., Berk, S. C., Hutchins, S. Ml., Shen, D-M., Xiong, Y., Hayes, E. C., Parmar, R. M., Foor, F., Mitra, S.W., Degrado, S. J., Shu, M., Klopp, J.M., Cai, S.J., Blake, A., Chan, W. W. S., Pasternak, A., Yang, L., Patchett, A. A., Smith, R. G., Chapman, K. T., Shaeffer, J. M. Rapid identification of subtype-selective agonists of the somatostatin receptor through combinatorial chemistry. *Science* **1998**, *282*, 737.
4. Mittach, A. Bosch, C., U.S. Patent 993,144.

5. Hanak, J. J., Gittleman, J. I., Pellicane J. P., Bozowski, S. The effect of grain size on the superconducting transition temperature of the transition metals. *Phys. Lett.* **1969**, *30A* (3), 201.
6. Hanak, J. J., Bolker, B.T.F. Calculation of composition of dilute cosputtered multicomponent films. *J. Appl. Phys.* **1973**, *44* (11), 5142.
7. Gittleman, J. I., Cohen, R. W., Hanak, J. J. Fluctuation rounding of the superconducting transition in the three dimensional regime. *Phys. Lett.* **1969**, *29A* (2), 56.
8. Hanak, J. J. The "multiple-sample concept" in materials research: Synthesis, compositional analysis and testing of entire multicomponent systems. *J. Mater. Sci.* **1970**, *5*, 964.
9. Hanak, J. J., Yocum, P. N. DC-electroluminescent flat panel display, *U.S. Gov. Rep. Announce*. **1973**, *73* (21), 128.
10. Hanak, J. J. Electroluminescence in ZnS: Mn_x: Cu_y rf-sputtered films. *J. Appl.Phys.* **1974** (Suppl. 2) (1), 809.
11. Xiang, X.-D., Sun, X., Briceno, G., Lou, Y., Wang, K.-A., Chang, H., Wallace-Freedman, W. G., Chen, S.-W., Schultz, P. G. A combinatorial approach to material discovery. *Science* **1995**, *268*, 1738.
12. Moates, F. C., Somani, M., Annamalai, J., Richardson, J. T., Luss, D., Willson, R. C. Infrared thermographic screening of combinatorial libraries of heterogeneous catalysts. *Ind. Eng. Chem. Res.* **1996**, *35*, 4801.
13. Burgess, K., Porte, A. M. Accelarated synthesis and screening of steroselective transition metal complexes. *Adv. Catal. Processes* **1997**, *2*, 69.
14. Danielson, E., Devenney, M., Giaquinta, D. M., Golden, J. H., Haushalter, R. C., McFarland, E. W., Poojary, D. M., Reaves, C. M., Weinberg, W. H., Wu, X. D. A rare-earth phosphor containing one-dimensional chairs through combinatorial methods. *Science* **1998**, *279*, 837.
15. Hideomi, K., Nobuyoki, M. Combinatorial chemistry of inorganic materials. *J.Chem (Jpn)* **1998**, *4*, 70.
16. Reddington, E., Sapeena, A., Gurau, B., Viswanthan, R., Sarongapani, S., Smotkin, E. S., Mallouk, T. E. Combinatorial electrochemistry: a highly parallel, optical screening method for discovery of better electrocatalysts. *Science* **1998**, *280*, 1735.
17. Hoffman, C., Wolf, A., Schuth, F. Parallel synthesis and testing of catalysts under nearly conventional testing conditions. *Angew. Chem. Int. Ed.* **1999**, *38* (18), 2800.
18. Klein, J., Lehmann, C. W., Schmidt, H-W., Maier, W. F. Combinatorial material libraries on the microgram scale with an example of hydrothermal synthesis. *Angew. Chem. Int. Ed.* **1998**, *38* (24), 3369.
19. Orschel, M., Klein, J., Schmidt, H-W., Maier, W. F. Detection of reaction selectivity on catalyst libraries by spatially resolved mass spectrometry. *Angew. Chem. Int. Ed.* **1999**, *38* (18), 2791
20. Akporiaye, D. E., Dahl, I. M., Karlsson, A., Wendelbo, R. Combinatorial approach to the hydrothermal synthesis of zeolites. *Angew. Chem. Int. Ed.* **1998**, *37* (5), 609–611
21. Senkan, S. High-throughput screening of solid-state catalyst libraries. *Nature* **1998**, *394*, 350.
22. Senkan, S., Kranta, K., Ozturk, S., Zengin, V., Onal, I. High-throughput testing of heterogeneous catalyst libraries using array microreactors and mass spectrometry. *Angew. Chem Int. Ed.* **1999**, *38* (18), 2794.
23. Ozturk S., Senkan S. Discovery of new fuel-lean NO reduction catalyst leads using combinatorial methodologies. *Appl. Catal. B* **2000**, *38* (3), 243–248.
24. Senkan, S., Ozturk, S. Discovery and optimization of heterogeneous catalysts by using combinatorial chemistry. *Angew. Chem Int. Ed.* **1999**, *38* (6), 791.
25. Senkan, S. Combinatorial heterogeneous catalysis—A new path in an old field. *Angew. Chem. Int. Ed.* **2001**, *40*, 312.
26. Jandeleit, B., Schaefer, D. J., Powers, T. S., Turner, H. W., Weinberg, W. H. Combinatorial materials science and catalysis. *Angew. Chem. Int. Ed.* **1999**, *38*, 2494.
27. Rodemerck, U., Ignaszewski, P., Lucas, M., Claus, P., Baerns, M. Parallel synthesis and fast screening of heterogeneous catalysis. *Top. Catal.* **2000**, *13* (3), 249.
28. Choi, K., Gardner, D., Hilbrandt, N., Bein, T. Combinatorial methods for the synthesis of aluminophosphate molecular sieves. *Angew. Chem. Int Ed.* **1999**, *38* (19), 2891.
29. Bein, T. Efficient assays for combinatorial methods for the discovery of catalysts. *Angew. Chem. Int. Ed.* **1999**, *38* (3), 323.
30. Lewis, G.J., Akporiaye, D. E., Bem, D. S., Bratu, C., Dahl, I. M., Karlsson, A., Murray, R. C., Patton, R. L., Plassen, M., Wendelbo, R. Mixed alkali templating in the Si/Al = 3 and 10 systems: A combinatorial study. *Stud. Surf. Sci. Catal.* **2001**, *135*, 597.
31. Akporiaye, D., Dahl, I., Karlsson, A., Plassen, M., Wendelbo, R., Bem, D. S., Broach, R. W., Lewis, G. J., Miller, M. A., Moscoso, J. Combinatorial chemistry—The emperor's new clothes? *J. Micropor. Mesopor. Mater.* **2001**, *48* (1–3), 367.

32. Holmgren, J. S., Bem, D. S., Bricker, M. L., Gillespie, R. D., Lewis, G. R., Akporiaye, D., Dahl, I., Karlsson, A., Plassen, M., Wendelbo, R. Application of combinatorial tools to the discovery and commercialization of microporous solids: facts and fiction. *Stud. Surf. Sci. Catal.* **2001**, *135*, 461–470.
33. Plassen, M., Akporiaye, D., Bem, D. S., Dahl, I. M., Karlsson, A., McGonegal, C., Miller, M., Lewis, G., Wendelbo, R. Automating a combinatorial hydrothermal synthesis and characterization. *NI Week 2001*, **2001**, Austin, TX.
34. Nayar, A., Liu, R., Allen, R. J., McCall, M. J., Willis, R. R., Smotkin, E. S. Laser-activated membrane introduction mass spectrometry for high-throughput evaluation of bulk heterogeneous catalysts. *Anal. Chem.* **2002**, *74* (9), 1933.
35. Sun, B., Chan, C., Ramnarayanan, R., Leventry, W. M., Mallouk, T. E., Bare, S. R., Willis, R. R. Split-pool method for synthesis of solid-state material combinatorial libraries. *J. Combi. Chem.* **2002**, *4* (6), 569.
36. Kuechl, D. E., Willis, R. R., Bem, D. S., Holmgren, J. S. Optimization of material synthesis using combinatorial methods. *Abstracts, 222nd ACS National Meeting, Chicago*, **2001**, Washington, DC: American Chemical Society.
37. Basaldella, E. I., Kikot, A., Tara, J. C. Effect of aluminum concentration on crystal size and morphology in the synthesis of a Na/Al zeolite. *Mater. Lett.* **1997**, *31*, 83–86.
38. Montgomery, D.C. *Design and Analysis of Experiments*, 4th edn, **1997**, New York: John Wiley, p. 604.
39. US Patent 6,342,185 January 29, **2002**.

Chapter 27

Combinatorial Development of Organic Clear Coatings for Plastic Substrates and Scale-Up of Combinatorial Leads

Bret J. Chisholm, Radislav A. Potyrailo, and Hariklia Reitz

INTRODUCTION

The development process for producing new organic coatings typically requires extensive experimentation. This is a consequence of the complexity of most coating systems resulting from the wide array of properties that must be imparted to the material. Even relatively subtle changes in property requirements may require screening of many coating compositions since coating components are typically highly interactive resulting in complex nonlinear structure–property relationships. Owing to these facts, the application of combinatorial methods to organic coating development appeared to be very useful [1,2].

Recently, a few publications involving the application of various aspects of the combinatorial process to coating development have appeared. Vratsanos et al. [3] developed a high-throughput screening (HTS) method for evaluating scrub resistance. Schrof et al. [4] reported on the development of an automated coating formulation process and HTS of properties such as hardness, opacity, and yellowness index. Ashley et al. [5] developed combinatorial methods for studying dewetting of thin films on chemically modified gradient energy surfaces. Krupicka et al. [6] evaluated the application of nanotesting techniques and indentation and scratch tests for HTS of coating mechanical properties. Chisholm, Potyrailo, and co-workers [7–15] have demonstrated the successful implementation of the full combinatorial process for the development of organic clear coatings for plastic substrates. Figure 27.1 displays a schematic representation of the developed "combinatorial factory." The following sections describe this combinatorial factory.

Bret J. Chisholm and Radislav A. Potyrailo • General Electric Company, Global Research Center, Schenectady, NY 12301, USA
Hariklia Reitz • General Electric Company, GE Silicones, Waterford, NY 12188, USA
Contact author: Radislav A. Potyrailo, email: Potyrailo@crd.ge.com

High-Throughput Analysis, edited by Potyrailo and Amis
Kluwer Academic Publishers, New York, 2003.

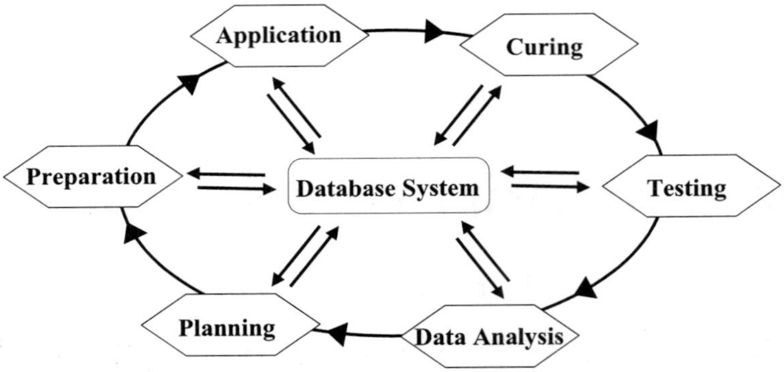

FIGURE 27.1. Schematic representation of the "combinatorial factory" developed for discovery of organic clear coatings for plastics.

PLANNING OF COMBINATORIAL RUNS

It can be understood by considering the complexity of most coating systems and the wide array of commercially available coating components that the compositional space involved in most coating development efforts can be extremely large. For example, a relatively simple class of coatings can be considered to illustrate this point. At a minimum, a clear coating composition designed to undergo curing by exposure to UV light would consist of an acrylate and/or methacrylate mixture, a photoinitiator, and a surface tension modifier to enhance leveling of the coating. Typically, a mixture of acrylates would be required to obtain an optimum balance between properties such as viscosity, mechanical properties, adhesion to the substrate, abrasion resistance, optical properties, shrinkage, film thickness, and cost. A coating comprised of three different acrylates would most likely provide the balance of properties required. A ternary mixture experimental design such as that shown in Figure 27.2 could be used to screen a three-component acrylate system. With this design, the number of coating compositions derived from the three acrylates is 66. Considering that there are at least a few hundred different acrylates and methacrylates commercially available, the number of possible coating compositions that could be screened as ternary mixtures is enormous. In addition, if the type and level of photoinitiator were varied and/or if process variables such as UV dose and/or UV intensity were included, the number of samples to be screened would be increased even further. As a result, it is essentially impossible to cover the entire compositional space for most coating systems. Thus, well-thought-out experimental designs are required to maximize the success of combinatorial factory operation. Typical experimental designs included both formulation and process conditions effects [15].

The combinatorial factory developed for organic clear coatings was designed around a 48-element array format. Thus an individual experiment with the factory consists of the production of 48 coatings. To aid in the planning of a combinatorial experiment, a software program was written that allows the user to input the experimental plan by selecting raw materials from a database and specifying the composition of each coating based on the cured film, the concentration of raw material stock solutions, and the thickness of the

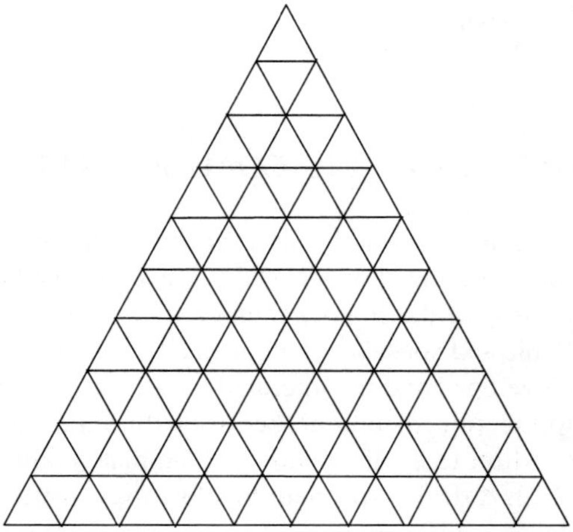

FIGURE 27.2. A ternary mixture experimental design.

coatings. With this information, the program calculates the volume elements of the stock solutions that must be transferred to prepare the array of liquid coatings (see next section). The computer file generated by the software can be read directly by the liquid-handling robot, alleviating the need to program the robot manually for the preparation of the array of liquid coatings.

PREPARATION OF LIQUID COATINGS

The preparation of an array of 48 liquid coatings is done robotically using a commercially available liquid-handling robot. The process consists of manual preparation of solutions of the individual raw materials (stock solutions) and robotic preparation of the coating solutions by transfer of aliquots of the stock solutions to elements of an array of glass vials. The volume of stock solutions and coating solutions is typically around 10 mL and 1–3 mL, respectively. The concentrations of the stock solutions are kept low enough to provide low viscosities for accurate delivery of the solutions. It has been found that the accuracy of liquid transfer decreases with increasing solution viscosity [16].

Mixing of the coating solutions is accomplished by repeated take-up and dispensing of about 60% of the coating volume into the vial using the liquid-handling robot. The quality of the mixing process was evaluated and has been described elsewhere [17]. In general, for the low-viscosity coating solutions prepared, mixing quality was quite adequate. As mentioned in the previous section, the robot uses a computer file generated by the experimental planning software to prepare the desired coating compositions. The typical time required to prepare an array of liquid coatings is about 45 min.

With the array of liquid coatings prepared, the next step in the process involves transfer of about 300 μL of each coating solution to a 48 well plate with approximately the same dimensions as the plastic substrate. This step was incorporated into the process to increase the rate of transfer of liquid coatings to the substrate since it minimizes pipette tip travel.

APPLICATION OF LIQUID COATINGS TO THE SUBSTRATE

Typical methods of application of clear coatings to plastic substrates include dip coating, flow coating, and spraying. For all of these coating methods, coating thickness is dependent on the viscosity of the coating solution. For HTS, it was important that coating thickness be controllable and consistent within an array. As a result, a process with aspects of film casting was developed and incorporated into the combinatorial factory.

The coating application technique begins with adhering a rubber template to a 9 mm × 12 mm plastic film to create a substrate template composite containing 48 wells as shown in Figure 27.3. A device was built to allow for consistent positioning of the template on the substrate. Precise positioning was required because of the robotic deposition of the liquid coating to the substrate. The substrate–template composite is placed on the deck of the liquid-handling robot and an aliquot of each liquid coating robotically dispensed into wells created by the substrate–template composite. The volume of coating solution dispensed into a well is typically 25 μL.

In order to spread the liquid coatings uniformly over the substrate within the wells, a device, referred to as an Omega Coater [18], that exerts a uniform centrifugal force to all elements of the array was designed and built. In addition, the Omega Coater allows for controlled evaporation of the solvent from the coatings. With this application

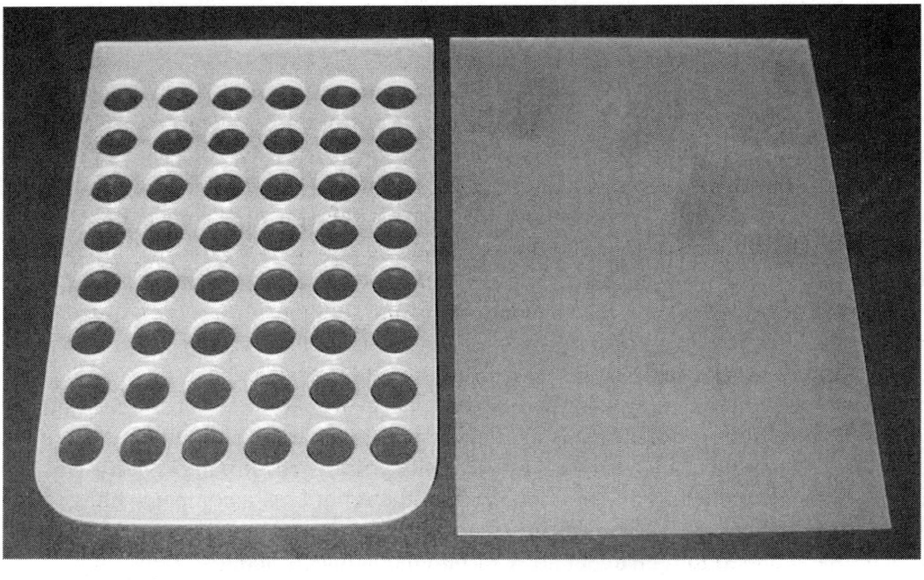

FIGURE 27.3. Elements of application of 48 coating arrays through a rubber template (left) onto a 9 mm × 12 mm plastic film (right).

technique, coating thickness is simply a function of the solids content of the coating and the aliquot size.

CURING

Arrays of liquid coatings have been cured using either heat or UV radiation. Thermal curing has been achieved within the Omega Coater. Heat is applied to the backside of the substrate, and the heating device allows for rapid heating of the array.

UV curing has been achieved by simply passing the array of liquid coatings through a conventional UV processor (Fusion UV Systems Inc.) with the template still adhering to the subtrate. After curing, the rubber template is peeled from the substrate to produce the array of solid coatings. While curing in the absence of centrifugal force results in a slight increase in coating thickness at the edge of the coating due to meniscus formation, it has been found that this imperfection does not affect the results of property testing. Figure 27.4 displays a reflected light image of an array of organic clear coatings produced using the combinatorial factory in conjunction with UV curing.

PROPERTY EVALUATION

For the class of coatings of interest, the most important properties include optical clarity, abrasion resistance, adhesion, and weatherability. HTS methods have been

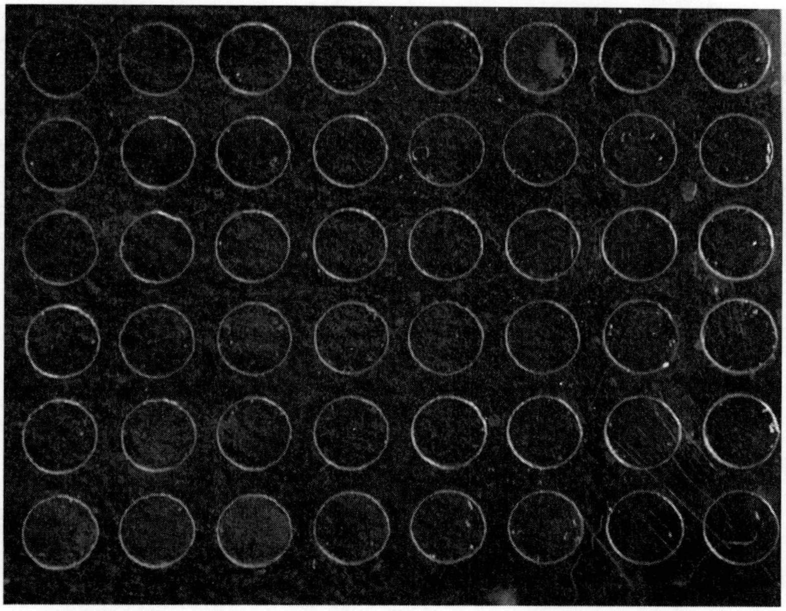

FIGURE 27.4. A general view of an array of coatings before performance testing.

developed for each of these properties. With regard to screening, of the properties, optical clarity, adhesion, and abrasion resistance were used for primary screening, while weatherability was used for secondary screening. In addition, high-throughput (HT) optimization of process parameters was performed. Details regarding the primary screening methods and HT process optimization are described in the sections to follow.

Optical Clarity

The optical clarity of clear coatings on transparent plastic substrates is typically evaluated by measuring the haze of coated materials using a hazemeter (ASTM D1003). Percent haze is the ratio of diffuse transmitted light to transmitted light as illustrated in Figure 27.5. Since this technique requires transparent substrates and is not obviously adaptable to HTS, an alternative technique was evaluated for HTS of optical clarity.

The HTS technique developed for optical clarity involved the measurement of scattered light using a fiber-optic probe and a translation stage as shown in Figure 27.6 [12]. The array is attached to a translation stage using a holding apparatus that allows for consistent positioning of coating arrays relative to the fiber optic probe which is fixed. The probe contains both the light source and detector, as shown in Figure 27.6. The translation stage is programmed to position each coating under the fiber-optic probe. Scattered light is automatically detected from each coating element in the array. The time required to obtain all 48 spectra from an array is about 3 min.

The coating arrays were screened for optical clarity before abrasion testing to reject the coatings with a high initial scatter from being further tested for abrasion resistance and after abrasion testing to determine the most abrasion-resistant coatings [12]. A video demonstrating the operation of this automated fiber-optic system for evaluation of the optical clarity of a coating array after an abrasion test is included on the CD (Video *Abrasion-modified optical clarity screening*). The developed spectroscopic setup was optimized to quantify haze levels from zero to 30% [12]. This range of haze was obtained in our traditional-scale experiments. Optimization included a combination of three

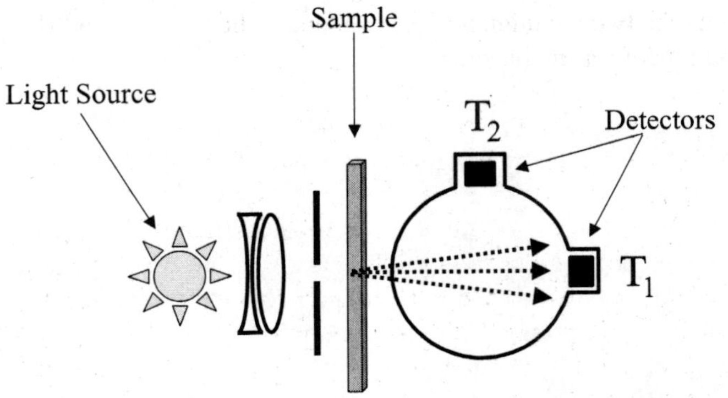

FIGURE 27.5. Schematic of a standard method for determination of transmission haze (ASTM D1003). T_1 and T_2 are detectors for detection of the transmitted and scattered portions of light, respectively.

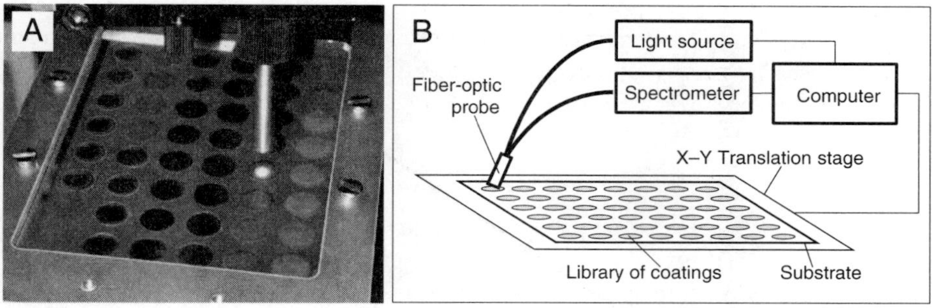

FIGURE 27.6. (A) General view and (B) schematic of the optimized HT analysis system for determination of optical clearity of coatings arrays. (Adapted with permission from ref. 12. Copyright 2002 American Chemical Society).

parameters that in concert provided the best sensitivity and precision of measurements of scattered light. The first optimization parameter was the probe angle. It was optimized with respect to the highest change in the detector response over the range of measured haze. The second optimization parameter was the distance from the probe to the coating. It was optimized with respect to the largest possible illumination spot size. This step was important to average the local heterogeneities of abraded coating surface and thus to improve the precision of measurements. Spatial or sample averaging is a common practice for measurements of heterogeneous solid materials to reduce the measurement errors. Too small a spot size (~1 mm over the longer direction of the illumination ellipse) resulted in a relatively insufficient averaging of the illuminated area and thus a relatively low measurement precision. Too large a spot size (~8 mm) resulted in large contributions from the coating edges. Thus the compromise spot size was found to be about 4–6 mm. The third optimization parameter was the spectral acquisition conditions that provided the highest signal-to-noise in the repetitive measurements under the smallest integration time (20 ms).

To validate the light scattering method, a series of coatings possessing various levels of haze were measured using the HTS method. Figure 27.7 illustrates the correlation between haze and the intensity of scattered light measured using the HTS method. The correlation between the two techniques was excellent, indicating that the HTS method can be used as a good model for the hazemeter.

Abrasion Resistance

Several different abrasion methods were investigated for the HTS of abrasion resistance. The method that was found to be most suitable was a modified version of the oscillating sand abrasion test which is described in ASTM F735. According to the ASTM method, a coated sample is attached to the bottom of a container with the coated side up, 1000 mL of a specific sand is added to the container, and the pan is oscillated back and forth at a specific frequency for a given amount of time. The major modification made to this method for the HTS of coating arrays was a change in the mode of oscillation to increase the degree of sand motion.

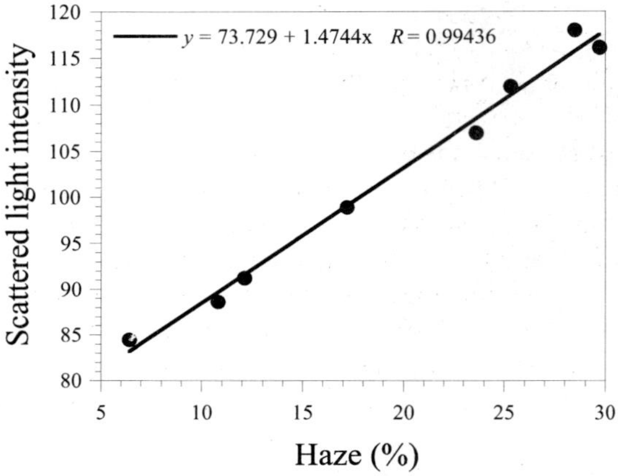

FIGURE 27.7. Correlation between haze and the intensity of scattered light measured using the HTS method. (Adapted with permission from ref. 8. Copyright 2002 Elsevier).

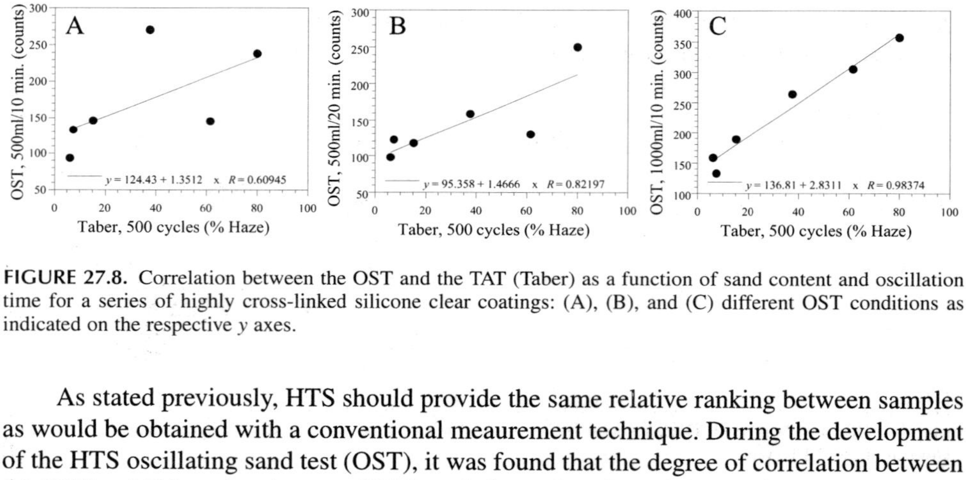

FIGURE 27.8. Correlation between the OST and the TAT (Taber) as a function of sand content and oscillation time for a series of highly cross-linked silicone clear coatings: (A), (B), and (C) different OST conditions as indicated on the respective *y* axes.

As stated previously, HTS should provide the same relative ranking between samples as would be obtained with a conventional meaurement technique. During the development of the HTS oscillating sand test (OST), it was found that the degree of correlation between the OST and Taber abrasion test (TAT) varied as a function of the conditions used for the OST. Figure 27.8 illustrates the correlation between the OST and the TAT as a function of sand content and oscillation time for a series of highly cross-linked silicone clear coatings. As can be seen from Figure 27.8, the degree of correlation between the OST and the TAT varied substantially as a function of the conditions used for the OST. In general, the correlation improved as the severity of the oscillating sand abrasion process increased (i. e., longer time and larger sand volume). To understand this observation, an investigation of the mechanism of abrasion for the two methods was undertaken.

Figure 27.9 displays the results of an experiment in which a silicone clear coating was abraded using the TAT and the surface to the coating. Images were collected as a function of the number of revolutions during the abrasion process. The method for the contrast enhancement between the coating and substrate involves the use of a fluorescent tag as detailed below in the next section. In this method, the coating appears bright while the substrate is dark. Figure 27.9 shows that, in addition to scratching and marring of the coating surface, the TAT resulted in a significant amount of delamination of the coating from the substrate. Further, the size of the delaminated areas increased as function of the number of revolutions, indicating that delaminated areas served as sites for further delamination.

A similar investigation conducted for the OST showed that the use of relatively mild abrasion conditions (i.e., 500 mL sand and 10 min oscillating time) only resulted in scratching and marring of the surface of the coating, while abrasion of the same coating using more aggressive conditions (1000 mL sand and 10 min oscillation time) promoted delamination of the coating from the substrate. These microscopic investigations explained the variations shown in Figure 27.8 regarding the correlation between the OST and TAT as a function of the conditions of the OST. In general, to obtain a good correlation between the HTS method and the TAT the conditions of the HTS method must be adjusted to promote delamination of the coating from the substrate for coatings possessing poor adhesion. Figure 27.10 shows the relative ranking of a series of coatings produced with and without a primer. In general, the silicone clear coatings used for the study do not adhere well to polycarbonate without a primer layer. The data shown in Figure 27.10 indicate that the relative ranking obtained with the TAT is preserved when the OST is used. A comparison of the relative standard deviation (RSD) of the samples for the two abrasion tests indicated a lower RSD for the OST [8]. The lower RSD for the OST allows for greater differentiation between samples compared with the TAT.

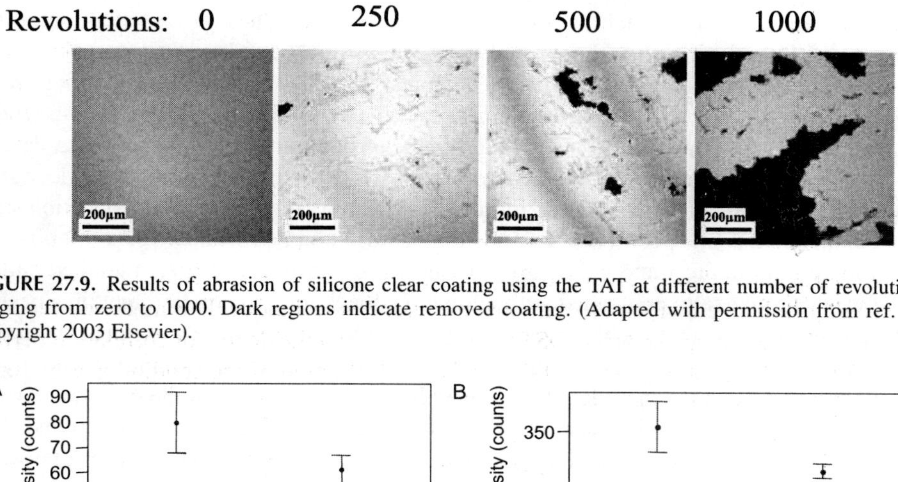

FIGURE 27.9. Results of abrasion of silicone clear coating using the TAT at different number of revolutions ranging from zero to 1000. Dark regions indicate removed coating. (Adapted with permission from ref. 10. Copyright 2003 Elsevier).

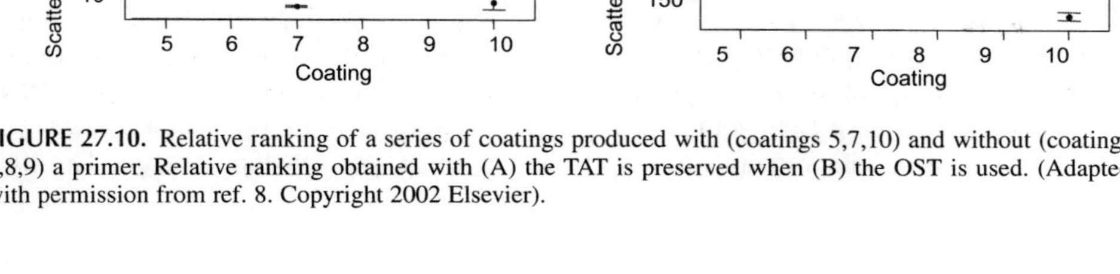

FIGURE 27.10. Relative ranking of a series of coatings produced with (coatings 5,7,10) and without (coatings 6,8,9) a primer. Relative ranking obtained with (A) the TAT is preserved when (B) the OST is used. (Adapted with permission from ref. 8. Copyright 2002 Elsevier).

Adhesion

Our HT adhesion testing methodology adapts the principles of a well-accepted crosshatch adhesion test method (ASTM D3359) but provides previously unavailable capabilities of high speed and reproducibility through a robotic automation, an expanded range of types of tested coatings through the coating tagging strategy, and an improved quantitation through high signal-to-noise automatic imaging. The crosshatch adhesion test is the industry standard test for measuring the adhesion of clear coatings to plastic substrates. The test consists of scribing a lattice pattern with 11 cuts in each direction through the coating and into the plastic substrate, applying a pressure-sensitive adhesive over the lattice, and quickly removing the tape in a single motion. The degree of adhesion is determined by comparing the area of coating removed by the tape with standards illustrated in the ASTM procedure.

There are several shortcomings of this ASTM test, which include difficulty in visually identifying the area of removed coating due to the high transparency of the coating and substrate, inability to prepare consistent lattice patterns of the same quality owing to the manual nature of lattice preparation, and the ability to discriminate between samples with marginally different adhesive properties.

The work flow for our HT testing, measurement, and data analysis methodology for the quantitative determination of adhesion loss of combinatorial arrays of coatings has been developed [15] and is illustrated schematically in Figure 27.11. The approach included fabrication of coatings arrays with spectroscopic tags, robotic application of integrity-degrading steps such as a crosshatch pattern, an adhesion loss step, automatic imaging of the resulting optical properties of coatings arrays, and a decision-making step. As an additional step for prescreening of coatings, an initial evaluation can be performed (e.g., a screen for clarity) [12]. We implemented these developments for the routine screening applications of coating libraries with a typical throughput of ~100 coatings a day.

Automatic crosshatching of the coating arrays was performed using a robotic system developed in-house and illustrated in Figure 27.12. In this system, an x–y translation stage was coupled to a z-axis positioner. The positioner was equipped with two sets of knives set at 1 mm spacing in the x and y directions as shown in Figure 27.12(b). An applied cutting force was automatically controlled with a force feedback mechanism to maintain a desired pressure required to cut through all the coatings samples and to provide uniform and reproducible cutting conditions. As a result, cutting of just the coating and not the underlying substrate was performed in under 10 min. A video of the operation of the crosshatch robot (Video *GE-built crosshatch robot*) is included on the CD. Adhesion loss was induced by periodically exposing crosshatched coating arrays to boiling water (15 min exposure) followed by rapid freezing of the array at −25°C. Adhesion loss was further induced by tape pulling the delaminated coating elements. The tape pull on the array was done with 2.5 cm wide tape that covered two rows (16 samples) at a time.

Determination of adhesion loss of coating arrays was performed by the automatic imaging of individual coating elements in the array followed by the determination of regions with removed coatings [15]. Image acquisition from the coating arrays was performed using the imaging system depicted in Figure 27.13 [15]. This system included an x–y translation stage, a light source, an imaging detector, and an associated computer to provide control of image acquisition and movement of the translation stage. The light

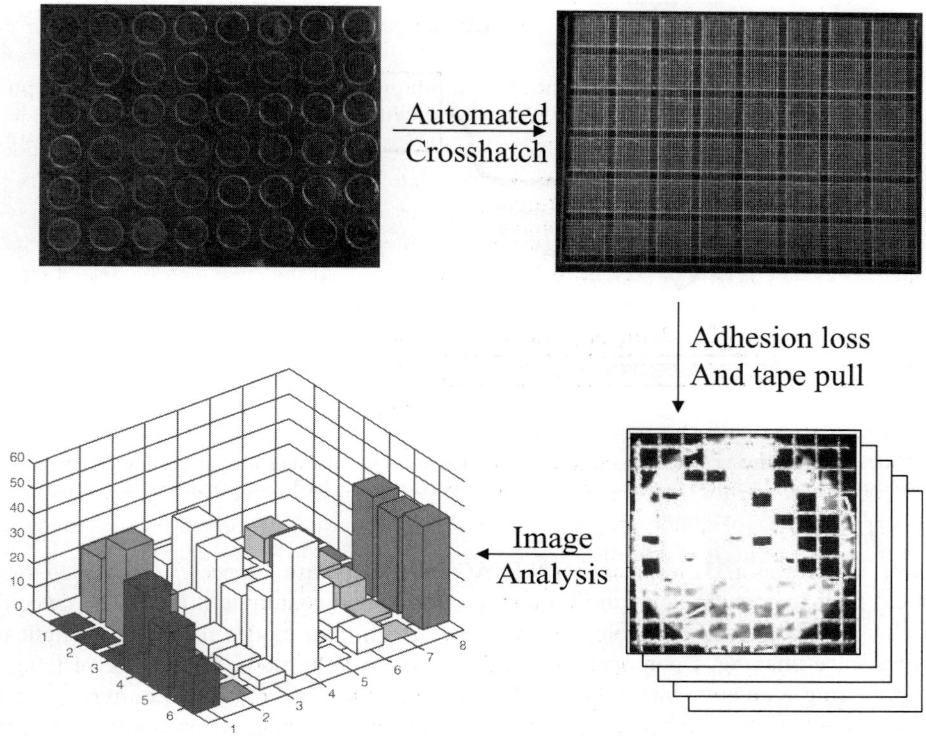

FIGURE 27.11. The work flow developed for our HT testing, measurement, and data analysis methodology for the quantitative determination of adhesion loss of combinatorial arrays of coatings.

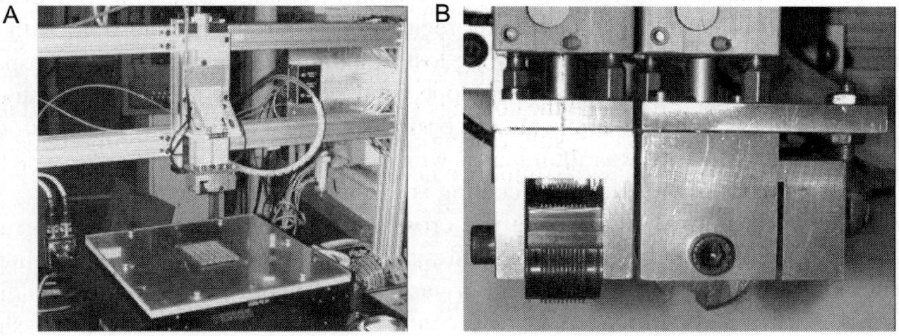

FIGURE 27.12. Robotic system built at GE Global Research for application of a crosshatch pattern onto coating arrays for adhesion testing: (A) general view of the system; (B) two perpendicular sets of 11 knives spaced 1 mm from each other. (Adapted from ref. 15 with permission. Copyright 2003 American Chemical Society).

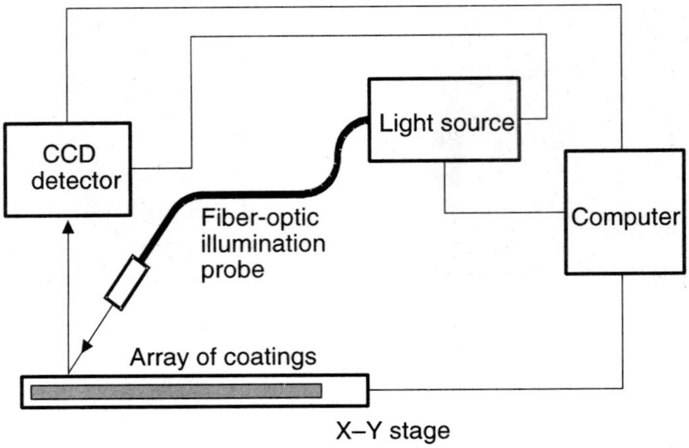

FIGURE 27.13. Automated imaging system for quantitation of adhesion loss in coating elements in a combinatorial array. (Adapted from ref. 15 with permission. Copyright 2003 American Chemical Society).

source, which was a 532 nm compact Nd:YAG laser (Nanolase, France), was coupled to an optical fiber to provide uniform illumination of a single coating at a time. The detector (ICCD camera, Andor Technologies) was operated in two modes for reflected light or fluorescence imaging. Upon operation in the reflected-light mode, a sequence of images from coating elements was collected for determination of crosshatch patterns. Upon operation in the fluorescence imaging mode, a sequence of images from coating elements was collected for determination of adhesion loss (see the video *Auto adhesion screening* included on the CD). The *x–y* stage control, image acquisition, and analysis were achieved with a computer using a program written in LabVIEW (National Instruments, Austin, TX). IMAQ Vision Builder and Advanced IMAQ Vision (National Instruments, Austin, TX) were used to develop image analysis algorithms.

One of the requirements for the determination of coating adhesion loss included quantifying removal of transparent coating regions on a transparent substrate with high measurement reproducibility. To quantify the loss of a transparent coating on a transparent substrate, a liquid coating formulation was doped with an inert dye at a low concentration. Thus measurements of coating removal were easily performed [19]. After the adhesion test was performed, the coating was illuminated with a wavelength of radiation at which the color or fluorescence of the dye in the coating is visible with an optical detector.

Another requirement was to apply a crosshatch pattern onto an entire array of coatings in a highly reproducible automated manner because manual crosshatching has inherent day-to-day and operator-to-operator variability. This need was met by building a robot to perform this operation. A typical coating array before and after a crosshatch (Figure 27.11) demonstrates an excellent quality of the crosshatch pattern across all coating elements in the array. Improvements were also required in the development of the adhesion-loss-inducing step that needed to be both rapid and correlate well with the traditional (and low-throughput) adhesion-loss method.

The final requirement was to provide improved quantitation of adhesion loss because the standard method involved manual visual determination of the removed regions only with a rough gradation of coating performance by five levels that cover coating removal from zero to 100%. In the HT system developed, this improved quantitation was achieved through the high signal-to-noise automatic imaging of the coating regions after a cross-hatch with a 10×10 cross-hatch pattern and determination of the percentage of coating the removed with accuracy to better than 1%. The high spatial resolution of the used CCD camera provided the required determination of loss of coating elements of less than 1 mm^2. Automation of quantitation was based on counting the number of removed squares or square segments. A video *Fluorescence imaging of adhesion loss* is included on the CD and demonstrates a sequence of fluorescence images of individual coatings in the 6×4 array after the adhesion test.

Curing Process Optimization

The automated analytical system developed has been also implemented for the HT optimization of processing conditions such as curing parameters in the fabrication of UV-cured coatings. Selection of optimum process conditions of combinatorial arrays of coatings is essential to correlate HTS and conventional processes and to achieve the desired physical properties of coatings. For monitoring the curing conditions of each coating in the array, a viscosity-sensitive fluorophore 4,4′-bis(2-benzoxazolyl)-stilbene was incorporated into coating formulations. This fluorescence-tagging approach permitted a gradient temperature heater and a UV curing system to be combined with the full capabilities of our HTS system, including generation and analysis of spectroscopic data. This investigation demonstrated the possibility of rapid decoupling of temperature and radiation effects in curing of UV-curable coating formulations by using multiple coatings and process conditions simultaneously.

An experimental setup for curing and automated monitoring of curing of arrays of coatings is shown in Figure 27.14 [13]. Fluorescence measurements of curing of the array

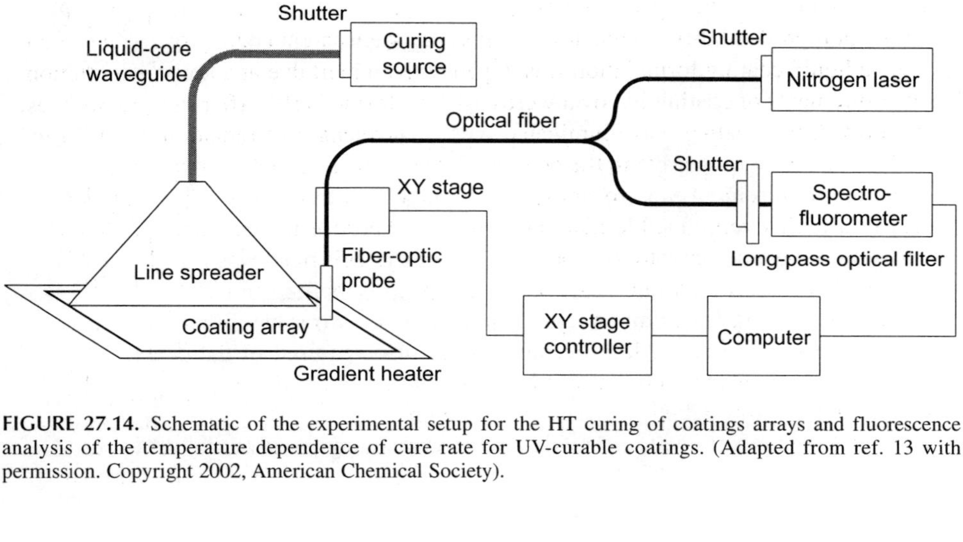

FIGURE 27.14. Schematic of the experimental setup for the HT curing of coatings arrays and fluorescence analysis of the temperature dependence of cure rate for UV-curable coatings. (Adapted from ref. 13 with permission. Copyright 2002, American Chemical Society).

of coatings were performed using a nitrogen laser, a portable spectrofluorometer, and a translation stage. The laser (Photon Technology International, London, ON), with an emission wavelength of 337.1 nm, was operated with a 20 Hz repetition rate with a maximum average power of about 50 mW, a maximum energy per pulse of about 2.8 mJ, and a pulsewidth of about 800 ps. The spectrofluorometer (Ocean Optics Inc., Dunedin, FL, Model ST2000) was equipped with a 200 μm slit, 600 grooves mm $^{-1}$ grating blazed at 400 nm and covering the spectral range from 250 to 800 nm with efficiency greater than 30%, and a linear CCD array detector. Light from the laser was focused into one of the arms of a "six-around-one" bifurcated fiber-optic reflection probe (Ocean Optics Inc., Model R400-7-UV/VIS). Emission light from the sample was collected when the common end of the fiber-optic probe was positioned near the sample at a 45° angle to the normal to the surface. The second arm of the probe was coupled to the spectrometer. Excitation light was blocked from entering the spectrometer with a long-pass optical filter. The common end of the fiber optic probe was arranged on an x–y translation stage (Parker Automation). The data acquisition and an automatic control of the x–y translation stage were achieved with a computer using a program written in LabVIEW (National Instruments, Austin, TX). The program provided an adequate control of the data-acquisition parameters and real-time communication with the translation stage. Four replicate measurements of fluorescence spectra were performed between pulses of curing radiation. Processing of collected spectra was performed using KaleidaGraph (Synergy Software, Reading, PA) and Matlab (Mathworks Inc., Natick, MA). For the evaluation of curing conditions, an intensity ratio at two wavelengths of the fluorescence emission spectra (432 and 488 nm) of the fluorophore was calculated as the ratio I_{488}/I_{432} ratio after a baseline subtraction.

Temporal evolution of the fluorescence intensity ratio I_{488}/I_{432} of BOS fluorophore in the curing process of the array of coatings at different temperatures ranging from 26 to 115°C is presented in Figure 27.15[13]. The vertical lines represent periodic 60 s

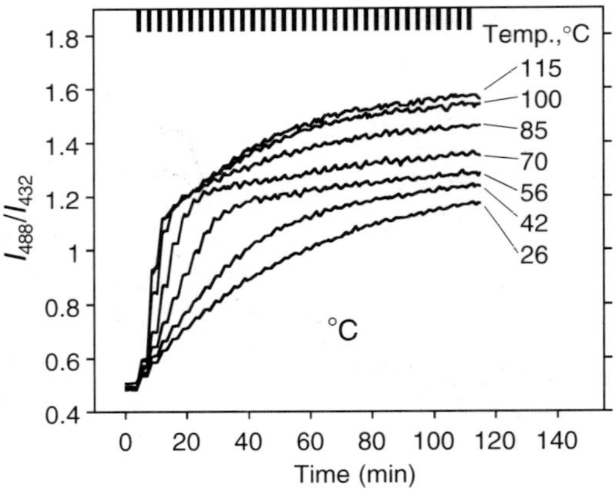

FIGURE 27.15. Temporal evolution of the fluorescence intensity ratio I_{488}/I_{432} of BOS fluorophore in the curing process of the array of coatings at different temperatures ranging from 26 to 115°C. The vertical lines represent periodic 60 s exposures of the array of coatings to UV–Vis curing radiation. (Adapted from ref. 13 with permission. Copyright 2002 American Chemical Society).

exposures of the array of coatings to UV–Vis curing radiation. The increase in the ratio I_{488}/I_{432} of the fluorescence intensities at 488 and 432 nm indicates the curing of the studied array of coatings. This figure illustrates several important findings. First, the curing rate of the coating formulations has strong temperature dependence. Secondly, at the advanced phase of the curing experiment, the formulations had different degrees of cure proportional to the temperature. Thirdly, an additional curing mechanism is present at curing temperatures above 70°C. This additional curing mechanism is evidenced by the change in the slope of the curing kinetics curve. An optimal temperature range for the radiation curing was found to be 90–100°C. This is the minimal temperature at which the maximum of curing can be performed and above which a further temperature increase does not significantly increase the cured condition. Determination of this smallest curing temperature is important from the throughput and production standpoint in order to increase the throughput in screening of different coating formulations and to minimize the energy requirements needed for adequate curing of coatings.

Examination of the initial region of the temporal evolution of the fluorescence intensity ratio I_{488}/I_{432} of BOS fluorophore in the curing process of the array of coatings at different temperatures was also performed to study the temperature-induced curing effect in the absence of curing radiation. The results of this analysis are shown in Figure 27.16. Three horizontal bars represent exposures of coatings to UV–Vis curing radiation. This figure shows that temperature alone does not contribute to curing of coatings. This is evidenced from the zero slope of the curing process before the first UV–Vis pulse is applied (0–4 min) and during the measurements of fluorescence between the curing pulses. A similar approach can be also useful for the HT analysis of "dark reactions" to evaluate any possible benefits of post-baking in radiation-cured coating formulations.

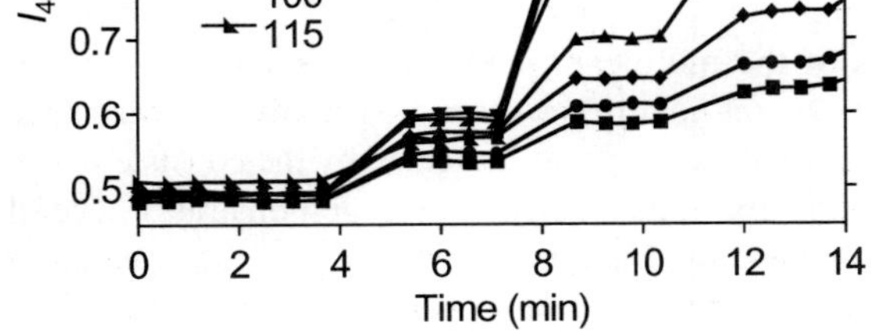

FIGURE 27.16. Initial region of the temporal evolution of the fluorescence intensity ratio I_{488}/I_{432} of BOS fluorophore in the curing process of the array of coatings at different temperatures ranging from 26 to 115°C. The three horizontal bars represent 60 s exposures to UV–Vis curing radiation. (Adapted from ref. 13 with permission. Copyright 2002 American Chemical Society).

DATA ANALYSIS AND MANAGEMENT

As mentioned earlier in this chapter, a software program was developed to facilate the coating formulation process, to generate instructions for the liquid-handling robot automatically, and store coating compositional information. In addition to these capabilities, the software program allows for easy input of the raw data obtained from the testing component of the experimental process. For each experiment, the program combines the coating compositional information with the test results to produce a report that facilitates the data analysis process.

Typically, experiments are designed such that the coatings are evaluated relative to a control or "benchmark" coating. Good performance is based on a comparison of the data obtained for the experimental coatings to the data obtained for the benchmark. Thus every array produced generally contains a control or benchmark coating. The inclusion of a control coating in every array provides a means of identifying the presence of systematic sources of error. In addition, each coating composition is generally replicated in the array to allow for the statistical significance of the results to be determined.

Figure 27.17 displays the results of abrasion resistance testing of an array of coatings in which each coating was evaluated in triplicate. The array also contained a benchmark (control) coating, as indicated with a dotted circle in Figure 27.17. Analyzing the data in this format allows for rapid identification of the "hits," which were defined as those coatings that possess equivalent or better abrasion resistance relative to the benchmark coating. In Figure 27.17, these best coatings are identified by solid circles.

For the more reliable ranking of coating performance, we implemented a method of multilevel performance screening [12,20]. Multiple levels of testing conditions when applied to combinatorial screening provide more reliable ranking of performance of combinatorial materials libraries by comparing the ranking of materials at various levels of test conditions. Further, this method provides a valuable information about the smallest required test level needed for the fastest reliable screening. The advantages of the new multilevel performance screening were initially demonstrated using model coating compositions. To produce arrays for the multilevel testing, several daughter arrays were fabricated from a single mother set of coating formulations. Each daughter array was exposed to different abrasion conditions. Each array contained four replicates of 12 different coating formulations. The libraries of coatings were abrasion tested at different abrasion conditions using the OST. Different levels of abrasion were induced by performing the test for 10, 20, and 30 min. These test conditions were previously determined to be adequate for the abrasion testing.

Typical results of the multilevel performance screening experiments are presented in Figure 27.18 for several coatings with representative performance. An initial optical clarity is very similar for all coatings, as indicated by the consistent low light scatter at zero abrasion time. Further, this figure conclusively demonstrates several advantages of the multilevel performance testing. First, it illustrates that a substrate unprotected with a coating reaches its maximum haze level upon abrasion at a smallest level (10 min) and this value stays the same under more aggressive abrasion conditions. Meanwhile, the abrasion-induced haze for coatings generally increases under more aggressive abrasion conditions. Second, the relative abrasion resistance of certain types of coatings depends on the abrasion test conditions. This fact is illustrated by the relative change in slope of plotted

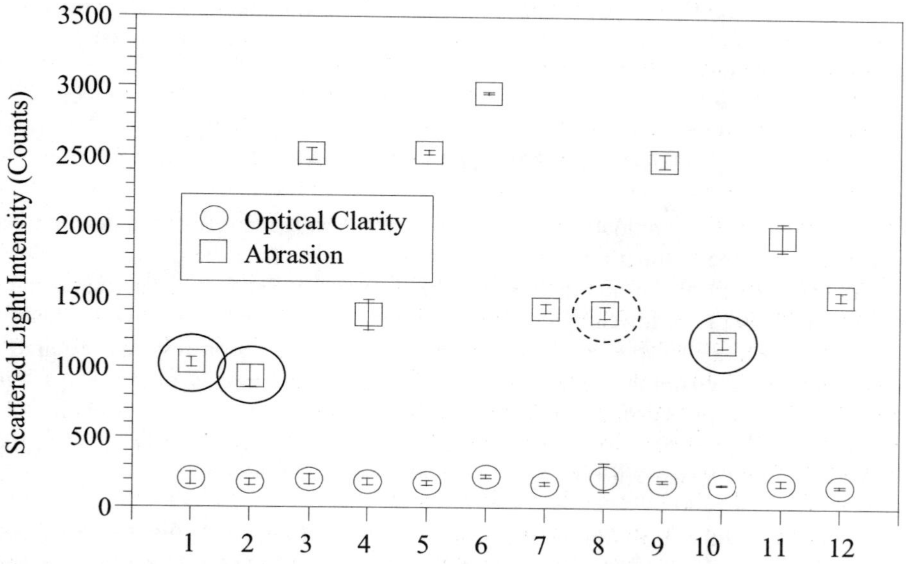

FIGURE 27.17. Results of abrasion resistance testing of an array of coatings in which each coating was evaluated in triplicate. The array also contained a benchmark coating indicated with a dotted circle. Solid circles are identified as "hits".

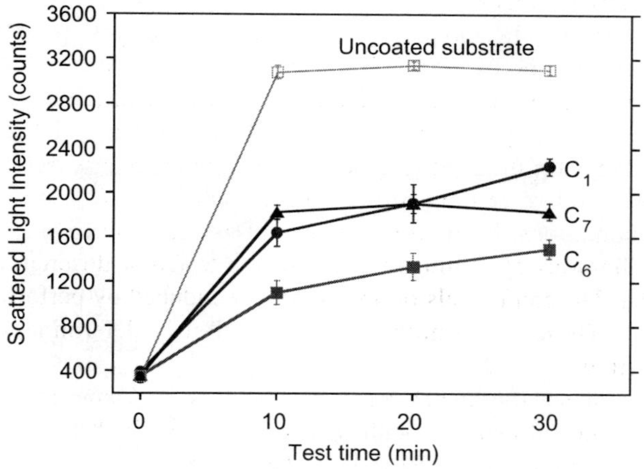

FIGURE 27.18. Analysis of abrasion resistance of coatings arrays using a multilevel abrasion testing for representative coatings C_1, C_6, C_7, and an uncoated substrate. Error bars are 1 SD ($n = 4$). (Adapted from ref. 12 with permission. Copyright 2002 American Chemical Society).

responses for different coatings. Third, under the applied test conditions, the best coating (the lead) can be defined as the one which consistently has a minimal abrasion-induced scatter for all tested conditions.

SCALE-UP OF HT LEAD COATINGS

As a result of the combinatorial studies, three leads were identified as potential lower-cost formulations with superb adhesion performance to replace the current commercial material[15]. These lead formulations were scaled up in GE Specialty Materials laboratories (Waterford, NY). Followed by conventional conditions for adhesion-loss evaluations, the coatings were also subjected to much more abusive adhesion tests as required by several automotive customers. The adhesion testing was completed in two steps. First, coated panels were placed in a humidity chamber for one heat–freeze–heat cycle (85°C, 95% humidity, 8 hr; −20°C, 0% humidity, 2 hr; 85°C, 45% humidity, 4 hr). Next, these coated panels were placed in boiling water and tested for tape pull adhesion at various intervals.

The results of the scale-up tests are summarized in Figure 27.19 [15]. The coating performance was ranked by the standard ASTM method (ASTM D3359) by five levels of coating performance from 5 (best) to zero (worst). These data demonstrate that the coatings identified using the combinatorial process showed a very great improvement in adhesion over the standard material. The laboratory-scale formulations 1 and 2 were based on two different materials discovered using the combinatorial process. The variation in these formulations was induced by changing the ratio of the formulation components.

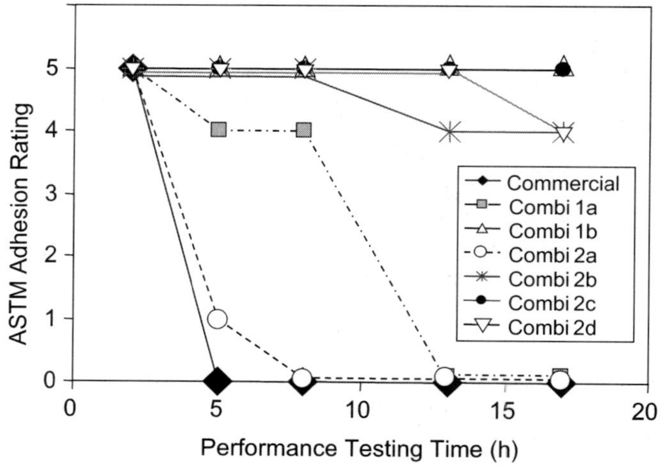

FIGURE 27.19. Results of scaling up of combinatorially developed coatings with high adhesion performance: performance test, boiling water. Adhesion varies from 5 (best) to zero (worst). Combi 1a–b and 2a–d are laboratory-scale formulations based on two different formulations discovered using the combinatorial process. (Adapted from ref. 15 with permission. Copyright 2003 American Chemical Society).

CONCLUSIONS

Coupling of combinatorial chemistry methods with industrially accepted abrasion and adhesion test methods and HT analysis provided a powerful set of tools for the accelerated discovery of new coating materials with improved abrasion resistance and adhesion. Our "combinatorial factory" for coatings development was operated by two chemists and had a maximum daily throughput of one to four arrays of 48 coatings each depending on the screening need. This approach has led to an important productivity improvement of at least 10 times over a conventional coating development process, which at best has a throughput of five to ten coatings per day with the same number of chemists. The key steps in the "combinatorial factory" process included experimental planning, coating preparation, application, and curing; testing against a performance parameter of interest; measurement of a resulting coating property; and data visualization.

Combinatorial methodologies in materials science provide important time savings in materials development in the initial discovery and optimization phase. The scalability of these materials to industrial-scale levels is the most important aspect for the acceptance of the combinatorial and HT methodologies in industry.

While the HT methodologies were reborn in 1995 with the pioneering work of Xiang et al. [21], few results have been published to date describing the scaling up of combinatorial materials leads. In this chapter, we demonstrate that the combinatorial developments progress from the small scale into traditional scale for other types of materials such as high-performance organic coatings for automotive applications.

ACKNOWLEDGMENTS

We thank our co-workers, whose names are cited in the references, for their contributions and Douglas Forman for preparing the videos. This work has been supported by the National Institute of Standards and Technology Advanced Technology Program (Grant 70NANB9H3038).

REFERENCES

1. Pilcher, G. R. *J. Coat. Technol.* **2001**, *73*, 135.
2. Wicks, D. A., Bach, H. In S. F. Thames (Ed.), *Proc. 29th International Waterborne, High-Solids, and Powder Coatings Symposium*, **2002**, Hattiesburg, MS: The University of Southern Mississippi. p. 1.
3. Vratsanos, L. A., Rusak, M., Rosar, K., Everett, T., Listemann, M. In *Proc. 2001 Athens Conference on Coatings Science and Technology*, **2001**, New Paltz, NY: Institute of Materials Science. p. 35.
4. Schrof, W., Lehmann, S., Hadeler, J., Oetter, G., Dralle-Voss, G., Beck, E., Paulus, W., Bentz, S. In *Proc. 2001 Athens Conference on Coatings Science and Techology*, **2001**, p. 283.
5. Ashley, K., Seghal, A., Amis, E. J., Raghavan, D., Karim, A. *Mater. Res. Soc. Symp. Proc.* **2002**, *700*, 151.
6. Krupicka, A., Favaro, G., Hult, A. *Polym. Prepr.* **2001**, *42* (2), 643.
7. Chisholm, B. J., Potyrailo, R. A., Cawse, J. N., Brennan, M. J., Molaison, C. A., Shaffer, R. E., Whisenhunt, D. W., Olson, D. R. In S.F. Thames (Ed.), *Proc. 29th International Waterborne, High-Solids, and Powder Coatings Symposium, New Orleans, LA*, **2002** New Paltz, NY: Institute of Materials Science pp. 125–137.
8. Chisholm, B. J., Potyrailo, R. A., Cawse, J. N., Shaffer, R. E., Brennan, M. J., Moison, C., Whisenhunt, D. W., Flanagan, W. P., Olson, D. R., Akhave, J. R., Saunders, D. L., Mehrabi, A., Licon, M. *Prog. Org. Coat.* **2002**, *45*, 313–321.

9. Chisholm, B. J., Potyrailo, R. A., Shaffer, R. E., Cawse, J. N., Brennan, M. J., Molaison, C. A. In *Proc. 2002 Athens Conference on Coatings Science and Technology*, **2002**, New Paltz, NY: Institute of Materials Science pp. 23–33.
10. Chisholm, B. J., Potyrailo, R. A., Shaffer, R. E., Cawse, J. N., Brennan, M. J., Molaison, C. A. *Prog. Org. Coat.* **2003**, *47*, 112–119.
11. Potyrailo, R. A., Chisholm, B. J., Olson, D. R., Brennan, M. J., Molaison, C. A. *Proc. SPIE—Int. Soc. Opt. Eng.* **2002**, *4578*, 145–152.
12. Potyrailo, R. A., Chisholm, B. J., Olson, D. R., Brennan, M. J., Molaison, C. A. *Anal. Chem.* **2002**, *74*, 5105–5111.
13. Potyrailo, R. A., Olson, D. R., Medford, G. F., Brennan, M. J. *Anal. Chem.* **2002**, *74*, 5676–5680.
14. Potyrailo, R. A., Morris, W. G., Chisholm, B. J., Cawse, J. N., Brennan, M. J., Molaison, C. A., Hassib, L., Flanagan, W. P., Reitz, H., Medford, G. F. Presented at Pittsburgh Conference on Analytical Chemistry and Applied Spectroscopy, New Orleans, Louisiana, March **2002**, Paper 751.
15. Potyrailo, R. A., Chisholm, B. J., Morris, W. G., Cawse, J. N., Flanagan, W. P., Hassib, L., Molaison, C. A., Ezbiansky, K., Medford, G., Reitz, H. *J. Comb. Chem.* **2003**, *5*, 472–478.
16. Brennan, M., Cawse, J. N. Unpublished data, **2000**.
17. Chisholm, B. J. Unpublished data, **2000**.
18. Saunders, D. L., Akhave, J. R., Potyrailo, R. A., Olson, D. R., Flanagan, W. P. *An Improved Combinatorial Testing Method and Apparatus for Coat Material Formulations and Methods*, World Patent Application WO0132320, **2001**.
19. Potyrailo, R. A., Olson, D. R., Brennan, M. J., Cawse, J. N., Chisholm, B. J. *Method for Determination of Structural Defects of Coatings*, US Patent 6,538,725 BZ, **2003**.
20. Potyrailo, R. A., Pickett, J. E. *Angew. Chem. Int. Ed.* **2002**, *41*, 4230–4233.
21. Xiang, X.-D., Sun, X., Briceño, G., Lou, Y., Wang, K.-A., Chang, H., Wallace-Freedman, W. G., Chen, S.-W., Schultz, P. G. *Science* **1995**, *268*, 1738–1740.

ABOUT THE EDITORS

Radislav A. Potyrailo received his Engineering degree in Optoelectronics from Kiev Polytechnic Institute, Ukraine in 1985 and Ph.D. degree in Analytical Chemistry from Indiana University (Bloomington, IN) in 1998. Dr. Potyrailo joined GE Global Research Center in 1998. He has been leading GE's effort in the design and implementation of high-throughput spectroscopic and sensor-based analytical instrumentation and lab-on-a-chip technologies. His research interests cut across the boundaries of combinatorial materials science, microfabricated analytical instrumentation, chemometrics, and functional materials. In 2001, Dr. Potyrailo has been appointed as an industrial Adjunct Assistant Professor of Chemistry at Indiana University, Bloomington, IN. Dr. Potyrailo has co-authored two monographs, has published over 30 peer-reviewed technical papers, and has 25 US Patents. Dr. Potyrailo organized two symposia on Combinatorial Materials at Pittcon (2000–01). He is the founder and chair of the First Gordon Research Conference on Combinatorial and High-Throughput Materials Science (2002) and co-organizer of Combinatorial Materials Symposia at the 2003 MRS Fall Meeting and at the 2004 ACS Spring Meeting.

Eric J. Amis received his Ph.D. in Chemistry from the University of Wisconsin-Madison and B.S. from Willamette University, Salem, Oregon. He has been Chief of the Polymers Division of the National Institute of Standards and Technology since May 1999. Before joining NIST in 1995 to lead the Polymer Blends and Processing Program, Amis spent 11 years on the faculty in physical chemistry at the University of Southern California. Prior to that he completed postdoctoral training at the University of Wisconsin and as a National Research Council Associate at NIST's predecessor, the National Bureau of Standards. From 1992 to 2002 he was the Editor of the *Journal of Polymer Science: Polymer Physics Edition.* He is a Fellow of the American Physical Society and a past Chair of the APS Division of Polymer Physics. In 2002 he was awarded the Silver Medal from the US Department of Commerce for Leadership in advancing new technical programs in the Polymers Division at NIST. Amis has chaired three Gordon Research Conferences: Macromolecular Dynamics, Polymer West, and Combinatorial and High-Throughput Materials Science (as co-founder). His research is primarily in the areas of solution rheology combined with light, neutron and X-ray scattering methods to investigate the physics of complex systems such as biomembranes, polyelectrolytes, associating polymers, gels, polymer crystallization, and dendritic polymers. He has published over 180 scientific papers. Most recently, he initiated a program to apply combinatorial and high-throughput methods to materials physics and biomaterials, which led to the establishment of the NIST Combinatorial Methods Center and a NIST initiative in metrology for tissue engineering.

Subject Index